Lecture Notes in Computer Scie

T0238498

Commenced Publication in 1973
Founding and Former Series Editors:
Gerhard Goos, Juris Hartmanis, and Jan van Leeuwen

Advanced Research in Computing and Software Science

Subline of Lectures Notes in Computer Science

Giuseppe F. Italiano   Tiziana Margaria-Steffen
Jaroslav Pokorný   Jean-Jacques Quisquater
Roger Wattenhofer (Eds.)

SOFSEM 2015: Theory and Practice of Computer Science

41st International Conference on Current Trends
in Theory and Practice of Computer Science
Pec pod Sněžkou, Czech Republic, January 24-29, 2015
Proceedings

 Springer

Volume Editors

Giuseppe F. Italiano
University of Rome Tor Vergata, Italy
E-mail: italiano@disp.uniroma2.it

Tiziana Margaria-Steffen
University of Limerick, Ireland
E-mail: tiziana.margaria@lero.ie

Jaroslav Pokorný
Charles University of Prague, Czech Republic
E-mail: pokorny@ksi.mff.cuni.cz

Jean-Jacques Quisquater
Catholic University of Louvain, Belgium
E-mail: quisquater@dice.ucl.ac.be

Roger Wattenhofer
ETH Zurich, Switzerland
E-mail: wattenhofer@ethz.ch

ISSN 0302-9743   e-ISSN 1611-3349
ISBN 978-3-662-46077-1   e-ISBN 978-3-662-46078-8
DOI 10.1007/978-3-662-46078-8
Springer Heidelberg New York Dordrecht London

Library of Congress Control Number: 2014958899

LNCS Sublibrary: SL 1 – Theoretical Computer Science and General Issues

Typesetting: Camera-ready by author, data conversion by Scientific Publishing Services, Chennai, India

Printed on acid-free paper

Springer is part of Springer Science+Business Media (www.springer.com)

Preface

This volume contains the invited and contributed papers selected for presentation at the 41st Conference on Current Trends in Theory and Practice of Computer Science (SOFSEM 2015), which was held January 24–29, 2015, in Pec pod Sněžkou, in the Czech Republic.

SOFSEM (originally SOFtware SEMinar) is devoted to leading research and fosters cooperation among researchers and professionals from academia and industry in all areas of computer science. SOFSEM started in 1974 in the former Czechoslovakia as a local conference and winter school combination. The renowned invited speakers and the growing interest of the authors from abroad gradually changed SOFSEM in the mid-1990s to an international conference with proceedings published in the Springer LNCS series. SOFSEM became a well-established and fully international conference maintaining the best of its original winter school aspects, such as a higher number of invited talks and an in-depth coverage of novel research results in selected areas of computer science. SOFSEM 2015 was organized around the following four tracks:

- Foundations of Computer Science (chaired by Roger Wattenhofer)
- Software and Web Engineering (chaired by Tiziana Margaria)
- Data, Information, and Knowledge Engineering (chaired by Jaroslav Pokorný)
- Cryptography, Security, and Verification (chaired by Jean-Jacques Quisquater)

With its four tracks, SOFSEM 2015 covered the latest advances in research, both theoretical and applied, in leading areas of computer science. The SOFSEM 2015 Program Committee consisted of 69 international experts from 23 different countries, representing the track areas with outstanding expertise.

An integral part of SOFSEM 2015 was the traditional SOFSEM Student Research Forum (chaired by Roman Špánek), organized with the aim of presenting student projects on both the theory and practice of computer science, and to give the students feedback on the originality of their results. The papers presented at the Student Research Forum were published in separate local proceedings.

In response to the call for papers, SOFSEM 2015 received 101 submissions from 31 different countries. The submissions were distributed in the conference tracks as follows: 59 in the Foundations of Computer Science, 11 in the Software and Web Engineering, 17 in the Data, Information, and Knowledge Engineering, and 14 in the Cryptography, Security, and Verification. From these, 31 submissions fell in the student category.

After a detailed reviewing process (using the EasyChair conference system for an electronic discussion), a careful selection procedure was carried out within each track. Following strict criteria of quality and originality, 42 papers were selected for presentation, namely: 26 in the Foundations of Computer Science, four in the Software and Web Engineering, eight in the Data, Information, and Knowledge Engineering, and four in the Cryptography, Security, and Verification. Based on the recommendation of the chair of the Student Research Forum, 12 student papers were chosen for the SOFSEM 2015 Student Research Forum.

As editors of these proceedings, we are grateful to everyone who contributed to the scientific program of the conference, especially the invited speakers and all the authors of contributed papers. We also thank the authors for their prompt responses to our editorial requests. SOFSEM 2015 was the result of a considerable effort by many people. We would like to express our special thanks to:

- The members of the SOFSEM 2015 Program Committee and all external reviewers for their careful reviewing of the submissions
- Roman Špánek for his preparation and handling of the Student Research Forum
- The SOFSEM Steering Committee, chaired by Július Štuller, for guidance and support throughout the preparation of the conference
- The Organizing Committee, consisting of Martin Řimnáč (Chair), Július Štuller, Pavel Tyl, Dana Kuželová and Milena Zeithamlová, for the generous support and preparation of all aspects of the conference
- Springer's LNCS series for its continued support of the SOFSEM conferences.

We are greatly indebted to the Action M Agency, in particular Milena Zeithamlová, for the local arrangements of SOFSEM 2015. We thank the Institute of Computer Science of the Academy of Sciences of the Czech Republic in Prague, for its invaluable support of all aspects of SOFSEM 2015. Finally, we are very grateful for the financial support of the Czech Society for Cybernetics and Informatics.

October 2014

Giuseppe F. Italiano
Tiziana Margaria
Jaroslav Pokorný
Jean-Jacques Quisquater
Roger Wattenhofer

Organization

SOFSEM 2015 Committees

Steering Committee

Ivana Černá	Masaryk University, Brno, Czech Republic
Brian Matthews	STFC Rutherford Appleton Laboratory, UK
Miroslaw Kutylowski	Wroclaw University of Technology, Poland
Jan van Leeuwen	Utrecht University, The Netherlands
Branislav Rovan	Comenius University, Bratislava, Slovakia
Petr Šaloun	Technical University of Ostrava, Czech Republic
Július Štuller, *Chair*	Institute of Computer Science, Academy of Sciences, Czech Republic

Program Committee

PC General Chair

Giuseppe F. Italiano	University of Rome Tor Vergata, Italy

Track Chairs

Roger Wattenhofer	ETH Zurich, Switzerland
Tiziana Margaria-Steffen	University of Limerick, Ireland
Jaroslav Pokorný	Charles University in Prague, Czech Republic
Jean-Jacques Quisquater	Catholic University of Louvain, Belgium

Student Research Forum Chair

Roman Špánek	Technical University of Liberec, Czech Republic

PC Members

Elena Andreeva	Leuven-Heverlee, Belgium
Zohra Anagnostopoulos	Lamia, Greece
Zohra Bellahsène	Montpellier, France
Petr Berka	Prague, Czech Republic
Malgorzata Biernacka	Wroclaw, Poland
Laura Bocchi	London, UK
Goetz Botterweck	Limerick, Ireland
Samia Bouzefrane	Paris, France

Markus Schordan   Livermore, USA
Cristina Seceleanu   Vasteras, Sweden
Martin Stanek   Bratislava, Slovakia
Srikanta Tirthapura   Ames, USA
Massimo Tisi   Nantes, France
A Min Tjoa   Wien, Austria
Remco Veltkamp   Utrecht, The Netherlands
Claire Vishik   Wakefield, USA
Peter Vojtáš   Prague, Czech Republic
Manuel Wimmer   Wien, Austria
Stefan Wolf   USI, Switzerland
Grigory Yaroslavtsev   Philadelphia, USA
Franco Zambonelli   Modena, Italy

Additional Reviewers

Aghiles Adjaz   Salvatore La Torre
Andris Ambainis   Anissa Lamani
Nikola Benes   Francois Le Gal
Tomas Brazdil   Lvzhou Li
Broňa Brejová   Yuanzhi Li
Witold Charatonik   Vahid Liaghat
Yijia, Chen   Kaitai Liang
Rajesh Chitnis   Peter Ljunglöf
Patrick Hagge Cording   Daniel Lokshtanov
Francesco Corman   Łukasik, Ewa
Anindya De   Ladislav Maršík
Kord Eickmeyer   Hernan Melgratti
Constantin Enea   Benjamin Mensing
Patryk Filipiak   Oscar Morales
Klaus-Tycho Förster   Ehab Morsy
Matthias Függer   Dejan Nickovic
Ariel Gabizon   Petr Novotný
Georgios Georgiadis   Thomas Nowak
Mohsen Ghaffari   Jan Obdrzalek
Alexander Golovnev   Svetlana Obraztsova
Nick Gravin   Gabriel Oksa
Dusan Guller   Hirotaka Ono
Abel Gómez   Jan Otop
Christoph Haase   Radha K.R. Pallavali
Petr Hlineny   Katarzyna Paluch
Fatiha Houacine   Debmalya Panigrahi
Bart M.P. Jansen   Panagiotis Papadakos
Artur Jeż   Dana Pardubska
Christian Kissig   Martin Perner
Kim-Manuel Klein   Oleg Prokopyev

Haridimos Kondylakis
Matthias Kowal
Kyriakos Kritikos
Julius Köpke
Randolf Schaerfig
André Schulz
Manfred Schwarz
Ayumi Shinohara
Jiri Srba
Frank Stephan
Przemysław Stpiczyński
Mária Svoreňová
Li-Yang Tan
Bangsheng Tang

Jibran Rashid
Vojtech Rehak
Saket Saurabh
A.C. Cem Say Alceste Scalas
Alceste Scalas
Yushi Uno
Søren Vind
Imrich Vrto
Kira Vyatkina
Magnus Wahlström
Kyrill Winkler
Abuzer Yakaryilmaz
Anastasios Zouzias
Damien Zufferey

Organization

SOFSEM 2015 was organized by the *Institute of Computer Science*, Academy of Sciences of the Czech Republic, Prague, and Action M Agency, Prague.

Organizing Committee

Martin Řimnáč, *Chair*	Institute of Computer Science, Prague, Czech Republic
Július Štuller	Institute of Computer Science, Prague, Czech Republic
Pavel Tyl	Technical University Liberec, Czech Republic
Dana Kuželová	Institute of Computer Science, Prague, Czech Republic
Milena Zeithamlová	Action M Agency, Prague, Czech Republic

Supported by

ČSKI – Czech Society for Cybernetics and Informatics   **ČSKI**

SSCS – Slovak Society for Computer Science

Table of Contents

Regular Papers

Foundations of Computer Science

Software and Web Engineering

Data, Information, and Knowledge Engineering

Cryptography, Security, and Verification

What is Computation: An Epistemic Approach*

Jiří Wiedermann[1] and Jan van Leeuwen[2]

[1] Institute of Computer Science of AS CR, Prague, Czech Republic
`jiri.wiedermann@cs.cas.cz`
[2] Dept. of Information and Computing Sciences, Utrecht University, The Netherlands
`J.vanLeeuwen1@uu.nl`

> *"How can one possibly analyze computation in general? The task seems daunting, if not impossible."* Y. Gurevich [14]

Abstract. Traditionally, computations are seen as processes that transform information. Definitions of computation subsequently concentrate on a description of the mechanisms that lead to such processes. The bottleneck of this approach is twofold. First, it leads to a definition of computation that is too broad and that precludes a separation of entities that, according to prevailing opinions, do perform computation from those which don't. Secondly, it also leads to a 'machine-dependent' notion of computation, complicating the identification of computational processes. We present an alternative view of computation, viz. that of a knowledge generating process. From this viewpoint, computations create knowledge within the framework of 'more or less' formalized epistemic theories. This new perception of computation allows to concentrate upon the meaning of computations – what they do for their designers or users. It also enables one to see the existing development of computers and information technologies in a completely new perspective. It permits the extrapolation of the future of computing towards knowledge generation and accumulation, and the creative exploitation thereof in all areas of life and science. The flux of our ideas on computation bring challenging new problems to the respective research, with wide connotations in the field of artificial intelligence, in cognitive sciences, and in philosophy, epistemology and methodology of science.

1 Introduction

Why do we compute? What do we compute? These two seemingly innocent questions were recently posed by Samson Abramsky in his contribution to the book commemorating the hundredth anniversary of Alan Turing's birth [1] in 2012. These questions can be made more concrete, e.g., why are we using computers? What are we computing with them? What is the meaning of computations performed with the help of computers? Here, and also in the sequel, we do not just have numerical computations ('computations with numbers') in mind, but

---

* This work was partially supported by RVO 67985807 and the GA ČR grant No. P202/10/1333.

G.F. Italiano et al. (Eds.): SOFSEM 2015, LNCS 8939, pp. 1–13, 2015.

any computations performed by whatever kind of computer. Of course, there are numerous replies possible and each of us will have an answer why he/she is making use of computations. However, what we are after is not a subjective answer pertinent to some specific use of commonly used computers. We want to have an answer that is grounded in some systematic theory, as part of a deeper understanding of the notion of computation, applicable to whatever use of whatever sort of computers. So far we do not seem to have a satisfactory answer obeying the latter conditions. This is related to fact that from the viewpoint of computer science, or computability theory for that matter, we in fact do not know quite well what computation 'is', in general. This is very unsatisfactory, since computation is the central notion in many scientific disciplines. The reasons for this unsatisfying state of affairs have accumulated during the past few decades.

Namely, computation is no longer what it used to be a one or two decades ago. Up until the end of the nineteen eighties no expert was bothered by the question what computation was. The answer was clear – computation is what is described by the generally accepted mathematical model of computation: the Turing machines, or any computational model equivalent to it [11,22].

With the advent of new computing technologies, networking, and advances in physics and biology, computation became understood as a far broader, far more common, and far more complex phenomenon than was modeled by means of Turing machines. In fact, it has become harder and harder to see these newer notions of computation through the lens of Turing machines (cf. [28]). Examples include biologically inspired models, physically inspired models and, last but not least, 'technologically enabled' models such as the Internet. One has to consider non-numerical computational models and devices, but also computations done on paper or by heart, proofs, computation with real numbers, continuous computations, geometrical constructions using compass and ruler, etc. The question is then, what is computation? What device performing computations is the 'right' one? How can computation be defined in such a way that every computational device realizes it in its specific way? Is there anything that all these computations have in common?

The scientific community, especially in informatics, physics and philosophy, has, of course, reacted to these new trends. (Un)surprisingly, instead of agreeing on a joint view of computation, the community has split into several opinion groups. For example, Frailey [12] maintains the radical position that computation is realized by whatever process. Other computer scientists, like Bajcsy [4] and Rosenbloom [16], define computation as a process which transforms information. Other researchers require additional properties, or state that computation is about symbol manipulation (e.g., Fortnow [11], Denning, [9], Conery [8] or philosopher Searle [3]). Still others, - like A.V. Aho [2], or Searle (again) [18] - require that there must be some computational model supporting the computation. Fredkin [13] has put it like this: *The thing about a computational process is that we normally think of it as a bunch of bits that are evolving over time plus an engine – the computer.* Deutsch [10] holds an even tougher view: there must also be a physical realization of a computational model. Finally, Zenil [32] has put programmability at the center of the discussion and of the definition of computation. Many opinions exist but many fall short in capturing the full notion as intuitively understood nowadays.

The previous efforts in defining computation have several things in common. First, all seem to agree that computation is a process. Secondly, all definitions tend to express computation as *'what the underlying hardware is doing'* or, in other words, *HOW the process of computation is realized.* This does not give much insight because 'what the hardware does' is performing operations on data; this forces us to see meaningless operations with data as computation. However, we are primarily interested in *WHAT the computation does*, i.e. what it does for the designers, users, observers. What a computation does, is only expressed by the design of the implementing system. Knowing how a computation does what it does is less interesting.

The intriguing question remains: what is it a computation does? Our answer [29] is simple: *computation generates knowledge.* It generates knowledge over some domain of interest for which the underlying computational system was designed or developed, or in which the system itself has evolved. Of course, the notion of knowledge itself is as hard to define as is computation. It has been debated ever since the Greek philosophers captured its many forms. For our purposes a general definition as given in Wikipedia will be good enough [31]. It stresses that knowledge is 'a familiarity with something or someone' and 'the theoretical or practical understanding of a subject'. Skills and behaviour are also considered to be knowledge.

Following this definition, knowledge essentially is an *observer-dependent* entity. Thus, the notion of computation as we defined it here is essentially observer dependent as well. This is a clear contrast to viewing computation as information processing, which we rejected above. It is far more fitting to our intuition.

The arguments to support our understanding of computation by means of examples from the history of computing will be given in Section 2. In Section 3 we concentrate on the internal structure of knowledge from the epistemic viewpoint. Section 4 presents a formal definition of computation as knowledge generation. In Section 5 we discuss the potential benefits of the epistemic view of computation. Section 6 contains conclusions.

This paper excerpts the presentation of our ideas in [29] and [30]. The readers interested in more details of the topics outlined here are kindly referred to our original works ([29,30]). In a forthcoming paper we digress on the possibility of a new theory of computation based on our philosophy [25].

2 Computation as Knowledge Generation

In Section 1 we asserted our main thesis: *computation is the process of knowledge generation.* In this section we review some examples of computational systems and how they can be seen as processes utilizing and producing knowledge.

In Table 1 (from [29]) we give an overview of a number of computational systems, together with the respective knowledge domains and types of knowledge they produce. The items in the first part of the table (*contemporary comput-ing systems*) clearly show an increasing 'growth' in knowledge generation: the further down one gets in the table, the more general and less formal the under-lying knowledge domains are and the bigger part of reality are captured. From this point of view, classical computational systems as we know them are only very primitive systems for generating knowledge. Newer systems are vastly more versatile.

The items in the second part of the table capture *natural computing systems*. These systems are not designed by people but exist in the natural world. They obviously belong to the class of systems (processes) producing knowledge ac-cording to our definition, i.e. they are computational. Note that we would not

Table 1. Computation as knowledge generation (cf. [29])

Computational system	Underlying knowledge domain	What knowledge is produced
Contemporary computing systems		
Acceptors	Formal languages	Language membership
Recognizers	Formal languages	Membership function
Translators	Functions, relations	Function value
Scientific computing	Mathematics	Solutions
Theorem provers	Logic	Proofs
Operating systems	Computer's devices and peripheries	Management of computer's own activities
Word processors and graphical editors	Graphical layout, spelling, grammar	Editing skills
Database and information systems	Relations over structured finite domains	Answers to formalized queries
Control systems	Selected domains of human activity	Monitoring, control
Search engines	Relations over unstructured potentially unbounded domains	Answers to queries in natural language
Artificial cognitive systems	Real world, science	Conjectures, explanations
Natural computing systems		
Living systems, cells	Real world	Life, behavior, intelligence
Brain, mind, social networks	Knowable world	Knowledge of the world
The Universe	Science	Living systems
Non-Turing computing systems		
Compass and ruler	Euclidean geometry	Euclidean constructions
BSS machine [6]	Theory of real numbers	Values of real functions
Oracles [22]	A set $A \subseteq \Sigma^*$	Characteristic function of A
Super-Turing computations	Formal languages in Σ_2	Language membership

have been able to include these examples under most of the classical definitions of computation, as the underlying 'computational mechanisms' are not known.

The items in the last part of the table seek the limits of our definition, resulting from attempts to falsify our thesis with the help of somewhat exotic examples which cannot be realized by Turing machines but are, nonetheless, still regarded as computations. It is clear that the examples still fit. They point to the fact that it is generally not a good idea to require that there must be a physical realization of the computational model at hand. This is in a good agreement with the practices in geometry, mathematics and even in (higher order) computability theory where processes similar to the ones we considered are routinely considered as computational processes.

3   The Structure of Knowledge

It goes without saying that once we concentrate on *WHAT* computations do instead of on *HOW* they do what they do, we lose the opportunity to investigate possible finer details of the effectuating processes as they are studied in e.g. computability theory or in computational complexity. On the other hand, our approach opens the way towards the investigation of other, so far mostly neglected aspects of computations: the insight into the character of the knowledge generating mechanisms used by computations.

In order to achieve its goals, a computation requires familiarity with the knowledge domain for which it is designed. In the sequel we will assume that this domain is given in the form of a *theory*. We do not restrict ourselves to theories in the strict formal sense of mathematics only. In this section we outline the broader analysis from [30] on the theories and structures that computations may exploit.

We will view 'theories' as an analytical tool for expressing knowledge. This may include e.g. describing, understanding, explaining and answering queries, providing solutions and predictions in areas of science or life, or the generation or control of behavior. A theory will normally consist of a collection of facts, sentences, statements, patterns of behavior, or linguistic descriptions and principles needed for deriving other statements using formal or informal inference rules. However, a theory could also have a form of a semantic network, or of a set of conditions and restrictions holding for a computation. A theory could also be a map, a scheme, and so on. Knowledge produced within the scope of a theory will have to fit the 'language' of the underlying domain. New knowledge that is generated may be kept in a *knowledge base* and become an integral part of the theory.

In general there is no a priori need for a theory to be correct or truthful. A theory, in the general sense we are using, can even be based on erroneous, unproven or non-verified beliefs and facts. Nevertheless, whatever 'knowledge' is generated within a flawed theory is formally considered to be knowledge, and thus truthful, within that theory.

A possible way of viewing the highly generalized notion of a theory that we use here, is to see it as a model of the world in which a computation is rooted

Table 2. The structure of knowledge (cf. [30])

	Mathematics, logic, & computer science	Philosophy and natural sciences	Mind and humanoid cognitive systems
Domain of discourse	Abstract entities	Ideas, empirical data	Perception, cognition
Elements of knowledge	Axioms, definitions	Facts, observations	Stimuli, multimodal concepts, episodic memories
Inference rules	Deductive system, programming languages	Rational thoughts, logics	Rules and associations formed by statistical learning
Final form of knowledge	Predicates, theorems, proofs, solutions	Statements, theorems, hypotheses, explanations, predictions, theories	Conceptualization, behavior, communication, natural language (NL), thinking, knowledge of the world formed mostly in a NL and in form of scientific theories

(cf. [26,27]). An important characteristic is that knowledge according to a theory can be generated time and again from the same base facts and principles, e.g. by computations that do so. In evolving domains, the theory corresponding to the underlying domain will have to evolve along with it.

Table 2 illustrates the structure of the theories in various knowledge domains. In the table, from left to right, the domains range from *theory-full domains* with formal theories to *theory-less domains* that admit no formal description for what they capture (cf. [23]). The examples shows the different levels of formalization, completeness and truthfulness of the theories that may underlie respective domains.

In cases where heterogeneous knowledge is used, natural language is an important mediator among theories. Semantics is of crucial importance here. Semantics is a form of knowledge and thus it is to be represented by a theory again. From this viewpoint all computations, including computations generating knowledge based on natural language understanding, bear a homogeneous structure. The knowledge framework behind the latter computations will normally be based on *cooperating theories*.

In general, theories depend not only on the knowledge at hand, but also on the context in which this knowledge is used and even on the history of past uses of this knowledge. In the case of embodied cognitive systems the context does not only refer to the grammatical context, but to the entire perceptual situation. All this leads to a complex intertwining of the respective theories. In general we do not know much about cooperating theories [26]. But here one can see the benefit of viewing computations as knowledge generating processes again: whatever deep and detailed (classical) view of the mechanisms realizing a computation can in no way contribute to the elucidation of the semantics of the computation.

The just presented view of the structure of computations contains one more principal observation. It illustrates that *computations generate knowledge from the knowledge in which computations are rooted* (see also Section 4). One might say that computation is *knowledge in action.*

4   An Epistemic Definition of Computation

Given the extensional definition given in Section 1, knowledge is an observer-dependent notion. After all, the decision of knowledge being based on a 'familiarity with something or someone' clearly is in the eye of the beholder, especially when it concerns knowledge that is not generally accepted. Therefore, computation as a process generating knowledge must be observer-relative, too ([30]).

As designers/observers of a computation, we must be aware of how a particular computation is related to the specific knowledge it generates. In response to its input a computation is not allowed to generate completely arbitrary knowledge. Each computation is required to generate knowledge over the domain for which the underlying system was designed or into which it has evolved. Similar to how intelligent behavior of an embodied robot arises from the interaction between brain, body and world, so is knowledge generated by computation in its interaction with the underlying knowledge domain.

More formally, there must be a way to *verify* the correspondence between a given computation and the domain over which the computation generates its output in the form of knowledge. For this, every computation must exploit some cognisance of the underlying knowledge domain. It is obliged to only use the facts, statements, rules and laws that describe the knowledge domain and that hold in this domain. This is what is meant when we said above that a *computation is rooted in its knowledge domain.* As illustrated in Table 2, the required attributes of computations can take different forms, depending on our knowledge of the underlying knowledge domain and on our ability to formally describe it, including the rules and laws holding in this domain.

We will now argue how the verification obligation might be expressed, i.e. that a given computational process generates knowledge that is expressed by means of the theory of the underlying domain. Of course, there must be an *explanation* (e.g., a proof) that the computational process works as expected. The explanation should express that the process generates knowledge that can be inferred from the underlying theory. The latter is also the key to a more formal definition of computation.

Before given the definition, we need one more notion. By a *'piece of knowledge'* we will denote any constant, term or expression which belongs to the theory or which can be derived using the respective rules and laws of the given theory. In [25] the collection of all items of knowledge possibly pertaining to a given computation, will be termed the *meta space* of the underlying theory. Although we will make use of the terminology used in mathematical logic below, one has to bear in mind that our notions of 'theory' and 'inference rules' are much broader than in logic and also include informal theories and informal rules of rational thinking.

Definition ([30]): *Let T be a theory, let ω be a piece of knowledge serving as the input to a computation, and let $\kappa \in T$ be a piece of knowledge from T denoting the output of a computation. Let Π be a computational process and let E be an explanation. Then we say that process Π acting on input ω generates the piece of knowledge κ if and only if the following two conditions hold:*

* – $(T, \omega) \vdash \kappa$, *i.e., κ is provable within T from ω, and*
* – E is the (causal) explanation that Π generates κ on input ω.*

We say that the 5-tuple $C = (T, \omega, \kappa, \Pi, E)$ is a computation *rooted in theory T which on input ω generates knowledge κ using the computational process Π with explanation E. The device or mechanism realizing process Π is called a* computer.

Under suitable conditions, computations may be *composed* to obtain new computations that fit the definition. Compositionality is an important property for the controlled behaviour of classes of computations (cf. [25]).

In the definition above, ω may be a set of numbers, a query in a formal or natural language, or a statement whose validity we are looking for, and so on. The computational process Π is a parameter of the computation. This captures that the same knowledge may be generated within the same theory by different computational processes. A change of computational process will most likely result in a different explanation. Whatever Π has to know about T must either be encoded in the design of Π and in ω or Π must have access to T. The condition $(T, \omega) \vdash \kappa$ implies that T is closed with respect to the inferences in T. The means that, once κ has been computed, it can be added as an explicit piece of knowledge to T, thus extending the knowledge base of T. In [30] we have argued by means of concrete examples that both conditions in the definition above are necessary.

The proposed definition of computation corresponds very well to the contemporary theory (and hopefully also the practice) of programming. The designer of a program must be aware of theory T, of the required result κ and of the fact that $(T, \omega) \vdash \kappa$. Then there is a computational model for which one has to design a computational process Π generating the required knowledge κ. One has to deliver also the evidence E, since otherwise one cannot be sure that the program does what was assumed.

Notice that the closure of T can be used e.g. to model interactive computations where after each interaction, the knowledge base is updated by the recently computed piece of knowledge. When a computation can modify the underlying theory we speak of an *evolving computation*. In this way one can model potentially infinite, interactive and evolutionary computations (cf. [28]). The formalism also enables one to define *universal computations* for some domain D, i.e. computations where the same computation process Π is used for generating the corresponding pieces of knowledge for all $\omega \in D$. (For more details see also [25].)

Finally, observe the natural way in which our approach accommodates previous efforts to define computation, by giving a common procedural platform for all kinds of computation. Using the previous notation, in the majority of the classical approaches to computation, a computation would look like this: $C = (\Pi)$. No other conditions are required from Π. In our approach we have found a different common denominator of all computations: this is the respective knowledge generation aspect.

Example. In [30] we give several examples of how our definition can be used to explain cases of 'computation' which have proved to be hard using classical definitions. Here we only discuss the example of the computing rock ([7,19]). Searle [19] describes the example as follows:

> '*Consider the example ... of a rock falling off a cliff. The rock satisfies the law $s = \frac{1}{2}gt^2$, and that fact is observer independent. But notice, we can treat the rock as a computer if we like. Suppose we want to compute the height of the cliff. We know the rule and we know the gravitational constant. All we need is a stop watch. And we can then use the rock as simple analog computer to compute the height of the cliff.*'

How does this conform to our definition of computation? First of all, for computing the height of a cliff, a rock alone is not enough. In addition to it we need both a stop watch and a person to observe the falling rock, operate the stop watch and know how to compute the distance traveled by the falling rock during the fall. Thus, the 'computer' consists of a rock *and* of a person endowed with the abilities just described. The theory behind the computation and the explanation are quite complex if all details are to be mentioned, from Newtonian physics to the visual observation ability of the observer and his capability to perform arithmetic operations, and so on. But, in principle, all these details can be delivered with sufficient plausibility. We conclude that the whole system as described indeed performs a computation according to our definition. (Actually, considering the complexity of the components of this analog computer, one could hardly call it 'a simple analog computer' as Searle does.)

It is important to realize that the conclusion above was possible only due to our insight into the entire process. An observer who has no understanding of stop watches or of the physical laws obeyed by falling bodies, can never come to such a conclusion. The example also is a clear instance of a computation that is observer-relative. In [30] several other examples are given, including e.g. the analysis of Searle's Chinese Room problem ([17]), which relies on a careful understanding of computation.

5   Discussion

There is no doubt that the classical view of computation, essentially based on the notion of Turing machines, has proved to be a very potent paradigm. It has lead to computability and computational complexity theory as we know them today.

It has learned us what can and what cannot be computed by the underlying models of computation and how efficiently this can or cannot be done. All this is carried out in the observer independent framework. The broadest framework is probably provided by Gurevich' approach using *Abstract State Machines* [5,14].

However, we seem to be reaching the limit of this approach. Namely, when it comes to solving the omnipresent problems e.g. related to artificial intelligence and especially cognition, we do not know at all what the actual potential of our computers is and what undiscovered mechanisms may be available one time to compute. For instance, we have no good clue of how to program computers in order to (learn them to) think, to be conscious, to acquire, understand and use natural languages or to create new knowledge. All these abilities are considered to be observer-relative qualities.

Our approach opens the road towards approaching the latter problems. As seen from our definition of computation, this is because it concentrates on the meaning of computation – what they do from the viewpoint of an observer or designer, and how they achieve their goals. There are many advantages of defining computation as a knowledge generating process:

(i) It gives a better way to distinguish objects which perform computation from those that do not. For example, according to our definition one can assign a computing ability to a rock (cf. [7]) only when it provably generates knowledge. In [30] we have shown that a rock can be seen as an analog computer in a scenario in which it is heated in order to 'compute' its own melting point. Under a different scenario a rock might be able to compute different knowledge – e.g., its momentum, weight, volume, and so on. More examples of our approach to unconventional computations are given in [30].

(ii) It offers a framework for modeling/discussing the question of observer dependency. Namely, an observer can be modeled by the same means as a computation. In this approach an observer is seen as a computation that 'observes' an other computation. In this way, an observer has some information about the observed computation as input (as required by our definition) and his/her task is to decide, whether it is a computation or not. This, of course, depends on the observer's own knowledge. A case analysis of the related computational scenarios reveals a number of new non-trivial insights that until now were not accessible to such a treatment. For instance, one can prove that a so-called *universal observer* whose decisions always agree with those of every other observer, for any computation, does not exist. (For details, see [30].)

(iii) It offers independence of the notion of computation from the underlying mechanisms. The definition covers not only all known instances of computation but also many hitherto unknown instances. It seems that for understanding computation one should investigate 'natural' rather than 'artificial' computations.

(iv) It supports the thinking about computation at a high level of abstraction, which is important for the design of artificial systems and for understanding other natural systems developed by some evolutionary process.

(v) It answers certain problems from cognitive science whose 'intractability' was due, as it seems now, to the use of the classical definition of computation.

For example, there is a widely discussed question in cognitive science what cognition is, if not computation [24]. As long as cognition is seen as an ability to gain, collect, produce and exploit knowledge, then it corresponds to our definition of computation in its most crystal form. Under such a view the original problem dissolves.

(vi) It puts the semantics of computations in the foreground which, classically, has so far been viewed as something secondary by which a computation can somehow be endowed *a posteriori* in the programming process. In fact, we have identified a new intermediate stage between computations (in the classical sense) and intelligence, viz. the ability to produce knowledge. Intuitively, the ability to produce knowledge is a prerequisite of intelligence. What ingredients make intelligence stronger than computation, in our sense? For a further discussion of this question, see our analysis of Searle's Chinese Room problem in [30].

(vii) It indicates that the formal framework that works for theory-full domains can be extended to theory-less domains. In doing so, natural language plays the central role: it enables dealing with knowledge domains and epistemic theories for which no formalization is available. Moreover, natural languages offer semantic means for bridging the gaps between seemingly unrelated knowledge domains, enabling one to draw analogies between such domains to be used as knowledge creation mechanisms (cf. [27]).

(viii) It offers a novel way of analysing prevailing trends in the history of information technologies when viewing them via their ability and potential to generate knowledge. It shows a steady shift towards interactivity, communication in natural languages, and to knowledge production.

(ix) Last but not least, our definition has great philosophical and methodological merit since it concentrates on the sense of computation, viz. knowledge generation, promoting computation to the key notion which is behind all progress.

6   Conclusion

In [15], David Deutsch is quoted as saying:

> ... the creation of knowledge [...] now has to be understood as one of the fundamental processes in nature; that is, [...] fundamental in the sense that one needs to understand them in order to understand the universe in a fundamental way.

In this paper we have argued that computation is the fundamental process underlying this, which explains why computation is the far-reaching process as claimed in computer science already for many years.

We believe that the time that computation was seen as an intrinsically physical process only has passed and that it is necessary to consider computation as an observer-relative process as well. This is because we are increasingly facing problems where, due to their nature, such a framework is required. This is especially the case of computations related to AGI (*artificial general intelligence*) which are all firmly rooted in theory-less, observer-dependent domains. We expect that changing the philosophy of computation towards that of viewing them as knowledge generating processes will help in the understanding and the creation of new intelligent information technologies.

References

1. Abramsky, S.: Two puzzles about computation. In: Barry Cooper, S., van Leeuwen, J. (eds.) Alan Turing: His Work and Impact, pp. 53–56. Elsevier (2013)
2. Aho, A.V.: Computation and computational thinking. Ubiquity 2011, Article No. 1 (2011)
3. Almond, P.: Machines like us, an interview by Paul Almond with John Searle. Machines Like Us (March 2009), http://machineslikeus.com/interviews/machines-us-interviews-john-searle-0
4. Bajcsy, R.: Computation and information. Comput. J. 55(7), 825 (2012)
5. Blass, A., Gurevich, Y.: Algorithms: a quest for absolute definitions. Bull. EATCS (81), 195–225 (2003)
6. Blum, L., Shub, M., Smale, S.: On a theory of computation and complexity over the real numbers: NP-completeness, recursive functions and universal machines. Bulletin of the American Mathematical Society 21(1), 1–46 (1989)
7. Chalmers, D.J.: Does a rock implement every finite-state automaton? Synthese 108, 309–333 (1996)
8. Conery, J.S.: Computation is symbol manipulation. Comput. J. 55(7), 814–816 (2012)
9. Denning, P. J.: What is computation? (opening statement). Ubiquity 2010, Article No. 1 (October 2010)
10. Deutsch, D.: What is computation (How) does nature compute? In: Zenil, H. (ed.) A Computable Universe: Understanding and Exploring Nature as Computation, pp. 551–566. World Scientific Publishing Company (2012)
11. Fortnow, L.: The enduring legacy of the Turing Machine. Comput. J. 55(7), 830–831 (2012)
12. Frailey, D.J.: Computation is process. Ubiquity 2010, Article No. 5 (November 2010)
13. Fredkin, E.: What is Computation? (How) Does Nature Compute (Transcription of a live panel discussion, with participants Calude, C.S., Chaitin, G.J., Fredkin, E., Leggett, T.J., de Ruyter, R., Toffoli, T., Wolfram, S.). In: Zenil, H. (ed.) A Computable Universe: Understanding and Exploring Nature as Computation, pp. 673–726. World Scientific Publishing Company (2012)
14. Gurevich, Y.: Foundational analyses of computation. In: Cooper, S.B., Dawar, A., Löwe, B. (eds.) CiE 2012. LNCS, vol. 7318, pp. 264–275. Springer, Heidelberg (2012)
15. Peach, F.: Interview with David Deutsch. Philosophy Now (30), (December 2000/January 2001)
16. Rosenbloom, P.S.: Computing and computation. Comput. J. 55(7), 820–824 (2012)
17. Searle, J.: Minds, Brains and Programs. Behavioral and Brain Sciences 3, 417–457 (1980)

18. Searle, J.: Is the brain a digital computer? Proceedings and Addresses of the American Philosophical Association 64, 21–37 (1990)
19. Searle, J.: The explanation of cognition. Royal Institute of Philosophy Supplement 42, 103 (1997)
20. Searle, J.: The Rediscovery of the Mind. MIT Press, Cambridge (1992)
21. Turing, A.M.: On computable numbers, with an application to the Entscheidungsproblem. Proc. London Math. Soc. Series 2 42, 230–265 (1936)
22. Turing, A.M.: Systems of logic based on ordinals. Proc. London Math. Soc. Series 2 45, 161–228 (1939)
23. Valiant, L.: Probably Approximately Correct: Nature's Algorithms for Learning and Prospering in a Complex World. Basic Books, New York (2013)
24. van Gelder, T.: What might cognition be, if not computation? The Journal of Philosophy 92(7), 345–381 (1995)
25. van Leeuwen, J., Wiedermann, J.: Knowledge, representation and the dynamics of computation. In: Dodig-Crnkovic, G., Giovagnoli, R. (eds.) Representation and Reality: Humans, Animals and Machines. Springer (to appear, 2015)
26. Wiedermann, J.: On the road to thinking machines: Insights and ideas. In: Cooper, S.B., Dawar, A., Löwe, B. (eds.) CiE 2012. LNCS, vol. 7318, pp. 733–744. Springer, Heidelberg (2012)
27. Wiedermann, J.: The creativity mechanisms in embodied agents: An explanatory model. In: 2013 IEEE Symposium Series on Computational Intelligence (SSCI), pp. 41–45. IEEE (2013)
28. Wiedermann, J., van Leeuwen, J.: How we think of computing today. In: Beckmann, A., Dimitracopoulos, C., Löwe, B. (eds.) CiE 2008. LNCS, vol. 5028, pp. 579–593. Springer, Heidelberg (2008)
29. Wiedermann, J., van Leeuwen, J.: Rethinking computation. In: Proc. 6th AISB Symp. on Computing and Philosophy: The Scandal of Computation - What is Computation?, AISB Convention 2013 (Exeter, UK), AISB, pp. 6–10 (2013)
30. Wiedermann, J., van Leeuwen, J.: Computation as knowledge generation, with application to the observer-relativity problem. In: Proc. 7th AISB Symposium on Computing and Philosophy: Is Computation Observer-Relative?, AISB Convention 2014 (Goldsmiths, University of London), AISB (2014)
31. Wikipedia (2013), http://en.wikipedia.org/wiki/Knowledge
32. Zenil, H.: What is nature-like computation? A behavioural approach and a notion of programmability. In: Philosophy & Technology (Special Issue on History and Philosophy of Computing). Springer (2013)

Progress (and Lack Thereof) for Graph Coloring Approximation Problems

Magnús M. Halldórsson

School of Computer Science, Reykjavik University, 101 Reykjavik, Iceland
mmh@ru.is

Abstract. We examine the known approximation algorithms for the classic graph coloring problem in general graphs, with the aim to extract and restate the core ideas. We also explore a recent edge-weighted generalization motivated by the modeling of interference in wireless networks. Besides examining the current state-of-the-art and the key open questions, we indicate how results for the classical coloring problem can be transferred to the approximation of general edge-weighted graphs.

1 Introduction

Graph coloring is among the most fundamental NP-hard optimization problems. It forms the basic model of conflict resolution, or scheduling with conflicts, and the allocation of scarce resources, such as access to spectrum in wireless computing.

Formally, given an input graph $G = (V, E)$ with vertex set V and edge set E, a proper (vertex) *coloring* is an assignment $\pi : V \to \{1, 2, \ldots, \}$ such that adjacent vertices receive different colors, i.e., $uv \in E$ implies that $\pi(u) \neq \pi(v)$. The objective of the *coloring problem* is to minimize the number of colors used, i.e., the largest value $\pi(v)$. The *chromatic number* $\chi(G)$ is the minimum number of colors used in a proper coloring of G. Let n denote $|V|$.

The aim of this note is to survey the state of affairs in the development of approximation algorithms for graph coloring. Namely, we focus on inexact algorithms that offer performance guarantees: the *performance (ratio)* of a coloring algorithm is largest ratio between the number of colors used by the algorithm on an instance G to the chromatic number of G. This ratio may be a function of some parameter of the graph, most commonly n, the number of vertices. We consider here general graphs, i.e., the class of all graphs, rather than (possibly) more manageable graph families.

We survey the known *polynomial-time* approximation algorithms for general graphs, with the aim to gather and restate the key ideas used in these algorithms. This is a study that extends back to early seventies, but curiously enough, the trail dries up by 1990. In the interim, research in lower bounds has brought extensive and illustrious advances, which may motivate renewed efforts on upper bounds.

We also introduce a recently proposed generalization of coloring involving edge-weighted graphs. This is motivated by scheduling in wireless networks.

G.F. Italiano et al. (Eds.): SOFSEM 2015, LNCS 8939, pp. 14–23, 2015.

2 Approximation Algorithms for Coloring General Graphs

It is useful to break the task of coloring the whole graph into *progress* steps. Informally, an operation yields *progress towards a T-coloring* if a repetition of such operations results in a valid T-coloring.

We first note that the coloring task can be reduced to the apparently easier task of finding a large *independent set*, i.e., a subset of mutually non-adjacent vertices. This set can be assigned a fresh color, and the task repeated until all the vertices have been colored. In general, such a reduction carries a log-factor overhead (as known from the approximation of the Set Cover problem). However, since we are content with performance ratios that are a *polynomial* in n when treating general graphs, the overhead factor actually reduces to a constant.

Example: Suppose we are given a $\sqrt{n}$-approximation algorithm for independent sets in χ-colorable graphs. We argue that repeated application results in a $4\sqrt{n}$-approximation for coloring. Namely, $\chi\sqrt{n}$ applications (using that many colors) reduce the number of vertices to at most $n/2$. Halving the number of vertices further takes at most another $\chi\sqrt{n/2}$ applications. Continuing, the number of colors used form a geometric sequence of $\chi\sqrt{n}(1 + \sqrt{2}^{-1} + \sqrt{2}^{-2} + \cdots) \leq \chi\sqrt{n}/(1 - 1/\sqrt{2}) < 3.5\chi\sqrt{n}$, which yields a $3.5\sqrt{n}$-approximation.

Observation 1. *Repeated application of a ρ-approximation algorithm for independent sets in χ-colorable graphs results in a $O(\rho)$-approximation algorithm for coloring, whenever $\rho = \Omega(n^\epsilon)$, for some $\epsilon > 0$.*

Johnson. One simple progress step involves identifying a smallest degree vertex v. We add v to our independent set solution and recurse on the set $\overline{N}[v]$ of vertices non-adjacent to v.

Johnson [17] analyzed several graph coloring heuristics and showed that the *minimum-degree greedy* heuristic attained a non-trivial performance ratio. The heuristic can be viewed as a repeated application of the above progress step. His observation was that for a minimum degree vertex v, the number $\overline{N}[v]$ of non-neighbors is at least $n/\chi - 1$, since all nodes that belong to the largest color class A must be non-adjacent to the other nodes in the class, and A must contain at least n/χ vertices. By repeatedly selecting a minimum degree vertex and eliminating its neighbors, we obtain an independent set whose size $S(n)$, as a function of the number of nodes n in the graph, can be given by the recurrence relation $S(n) \geq 1 + S(n/\chi)$ and $S(1) = 1$. The solution of this recurrence is $S(n) = \log_\chi n$. It follows that

Observation 2. *Selecting the minimum-degree vertices makes progress towards a $O(n/\log_\chi n)$-coloring.*

The performance ratio of the minimum-degree algorithm is therefore at most $O(n/(\chi \cdot \log_\chi n))$. This is maximized when χ is constant, for a performance ratio of $O(n/\log n)$.

Wigderson. Wigderson [22] observed that 3-colorable graphs could be colored with $O(\sqrt{n})$-colors. Namely, we can consider two cases depending on vertex degrees. If there is a vertex v of degree at least $\sqrt{n}$, then we can make progress as follows. We note that the graph $G[N(v)]$ induced by the neighbors of v is 2-colorable; thus by assigning it two fresh colors, we make progress towards a $O(\sqrt{n})$-coloring. On the other hand, if there is a vertex of degree less than $\sqrt{n}$, we can make progress towards a $O(\sqrt{n})$-coloring in the same way as argued for the min-degree heuristic.

By generalizing his argument, we can always make progress towards a $O(\chi n^{\chi-1})$-coloring by considering two cases depending on vertex degree: either there is a node of degree at most $n^{\chi-1}$, or there is a node v of higher degree in which case its sizable neighborhood is $\chi - 1$ colorable and by induction we make progress.

Observation 3. *We can always make progress towards a $O(\chi n^{\chi-1})$-coloring of χ-colorable graphs.*

If we now combine this bound with Johnson's, selecting the better of the two bounds, we obtain an algorithm with performance ratio at most $O(\min(n^{\chi-1}, n/(\chi \log_\chi n)))$, which is easily seen to be $O(n(\log \log n / \log n)^2)$.

Berger and Rompel. Even if there are no truly low-degree vertices, we may still find a *set* of independent vertices whose combined neighborhood is relatively small, allowing us to generalize the min-degree approach of making progress.

Consider the largest color class A, whose size is clearly at least n/χ. If we pick a subset S in A, then none of the other vertices $A \setminus S$ are adjacent to nodes in S, obviously. Thus, in particular, we can see that if we define $\overline{N(S)}$ to be the set of vertices non-adjacent to all nodes in S, we get that $|\overline{N(S)}| \geq n/\chi - |S|$. Since S will generally be much smaller than n/χ, finding such a set S allows us to make progress towards finding a $\Omega(|S| \cdot \log_\chi n)$-independent set, resulting in a $O(n/(|S| \log_\chi n))$-coloring.

But how to find such a set of non-trivial size? We have no certificate of what it means to be in the largest color class A. Here we are helped by abundance: A is large so it contains a lot of subsets. In particular:

Observation 4. *A random subset of size $K = \log_\chi n$ is contained in A with probability at least $1/n$.*

Thus, we can pick random subsets until we find one that satisfies the properties of belonging to A. Berger and Rompel [4] derandomized this argument to obtain a deterministic method with the same performance.

It follows that we can strengthen the min-degree approach considerably:

Observation 5. *One can always make progress towards a $O(n/(\log_\chi n)^2)$-coloring.*

If we combine this progress bound with Wigderson's, we obtain a coloring with performance ratio of $\min(\chi n^{\chi-1}, n/(\chi(\log_\chi n)^2))$. This is maximized when $\chi = \Theta(\log n / \log \log n)$, for a performance of $O(n(\log \log n / \log n)^3)$.

Note that the hardest cases for all these approximation algorithms are when $\chi(G)$ is $\Theta(\log n / \log \log n)$. Wigderson's approach then fails to deliver, leaving us with a $O(\log n / \log \log n)^2$-sized independent set.

Further Improvement. One may wonder if Observation 4 can be leveraged to get stronger properties on the subset S beyond the basic bound of n/χ on the non-neighborhood size. The key observation (from [9]) is that when the non-neighborhood size is small, nearly all of it must belong to the independent set A. But then, one can bring to bear algorithms to approximately find large independent sets. Specifically, the Ramsey method of Boppana and Halldórsson [6] achieves equivalent approximation factor for the independent set problem as Wigderson's method gave for the coloring problem.

We are thus led to the following strategy. We search for an independent set S of size $K = \log_\chi n$ that satisfies properties that hold for subsets of A. Then, one of two things must happen:

1. The Ramsey method finds a large independent set in $\overline{N(S)}$ (specifically of size $\Omega(\log^3 n)$), or
2. The non-neighborhood $\overline{N(S)}$ is larger than we previously argued.

Specifically, in the latter case the non-neighborhood must be of size $\Omega(\log_\chi n \cdot \log n / \log(\chi \log \log n / \log n))$, which in the range of interest for χ is $\Omega(\log_\chi n \cdot \log n)$.

When combined with Wigderson's bound, we obtain a performance ratio of $O(n(\log \log n)^2 / \log^3 n)$, shaving off a loglog-factor.

Additional Results. Progress on lower bounds on approximability, based on the PCP theory [2], has been extensive since the early nineties. Most of the work is on the somewhat easier independent set problem, while Feige and Kilian [7] showed how to extend some of the results to chromatic number, in particular giving $n^{1-\epsilon}$-hardness, for any $\epsilon > 0$. The strongest hardness to date for independent sets is $n/2^{(\log n)^{3/4+\gamma}}$, for any $\gamma > 0$, by Khot and Ponnuswami [20].

The most promising direction for improved approximation algorithms for independent sets and coloring has for long been *semi-definite programming* (SDP), such as the θ-function of Lovász [21] and various hierarchies and strengthenings. For many families of problems, the best possible results achievable in polynomial time are obtained by SDPs. SDPs have been useful for coloring graphs of low chromatic number, including the best approximation known for 3-coloring of $\tilde{O}(n^{3/14})$ [5]. For general graphs, however, all the results on SDP for independent sets or coloring have been negative, with stronger lower bounds than the known inapproximability bounds.

We are led to ask what may possibly help algorithms for coloring general graphs. It is curious that it is almost a quarter of a century since the appearance of the last improved approximation results ([8] in 1990). The bold conjecture in [9] that the best possible approximability is only $\Theta(n/polylog(n))$ may still be validated. That still leaves some room for improvement.

3   Coloring Edge-Weighted Graphs

We address a generalization of the classic graph coloring problem. A coloring can be viewed as a partition of the vertices into sets of ultimate sparsity: each vertex has degree *less than 1* (i.e., zero) from the other nodes in the same color class. We assume the same constraint, with the only difference that degrees are now weighted. In fact, the graph need not be symmetric, so we have an edge-weighted digraph. Thus, we now seek a partition into fewest possible vertex subsets such that each node has weighted in-degree less than 1 from other nodes in the same set.

Our motivation for this problem comes from modeling interference in wireless networks. Whereas the classical TCS approach is to model interference as a pairwise binary property (i.e., one that can be captured by a graph), a more refined view commonly used by engineering communities is that interference is a many-to-one relationship. Having a conversation in a room where another discussion is going on may work fine, but once the room becomes crowded with speakers, listening becomes progressively harder. In other words, it is the *cumulative* effect of multiple transmitters that matters when assessing whether a message can be properly decoded.

Formally, we are given a digraph $H = (V, E)$ with non-negative weights $w : E \to \mathbb{R}^+$ on the edges. A subset S of vertices is *independent* if the in-degree within S of each node is strictly less than 1, i.e., if $\sum_{u \in S} w(u, v) < 1$, for each $v \in S$. A coloring is, as before, a partition into independent sets, and we aim to use the fewest colors possible. Notice that in the case of symmetric 0/1-weights, the definition coincides with classical graph coloring.

We will first focus on coloring approximation in terms of the parameters n and χ, before examining the sparser instances that relate to the specific wireless applications.

3.1   Coloring General (Edge-Weighted) Graphs

The question of how well we can handle general edge-weighted graphs is interesting from a basic science standpoint, even if the results are too weak for most applications.

A very simple approach yields an easy $O(n/\log n)$-approximation (proposed for the wireless scheduling problem in [16], while the generic approach was perhaps first stated in [10]): Partition the graph into $n/\log n$ vertex-disjoint sets, and color each set optimally with a fresh set of colors. Since such an optimal coloring of a graph on n vertices can be obtained in time about 3^n [16], using the technique of inclusion-exclusion, the time complexity of coloring the $\log n$-sized sets is polynomial.

The only other reported result involves the case of graphs that contain an independent set of size $(1 + \epsilon)n/2$, for some $\epsilon > 0$. In this case, a semi-definite programming formulation results in an algorithm to produce an independent set of size $\Omega(\epsilon n)$ [13]. This is, however, insufficient to provide non-trivial bounds even for 2-colorable graphs.

We propose here different approaches that emulate some of the results obtained for ordinary graphs. We skip Johnson's approach and start with Berger and Rompel's. Recall that $\chi = \chi(H)$ is the chromatic number of the edge-weighted graph H.

Proposition 1. *There is an algorithm that finds a $O(n/(\log_{\chi \log n} n)^2)$-coloring of an edge weighted graph H.*

Let k be such that $k = \chi(H) \cdot \log_k^2 n$, and note that $\log k = \Theta(\log(\chi \log n))$. Let $X = \log_k^2 n$. We form a (classical) graph G on the same vertex set as H, where $uv \in E(G)$ iff $w(u,v) + w(v,u) \geq 1/X$. We then apply the algorithm of Berger and Rompel on G to obtain an independent set I in G, but retain at most X of the nodes in I. This set is then a feasible (edge-weighted) independent set in H, since all weights of edges within I are less than $1/X$.

It remains to argue a lower bound on the size of the independent set I found. Consider a color class C in H and a node v in C. Observe that v has fewer than X neighbors within C in G, thus C induces a subgraph in G of maximum degree less than X. It follows that $\chi(G) \leq X \cdot \chi(H) = k$. The Berger-Rompel algorithm produces a set I of size $\Omega(\log^2_{\chi(G)} n) = \Omega(X)$. Hence, we make progress towards a $O(n/X)$-coloring, as claimed.

We next turn to Wigderson's approach. We start with the case of 2-colorable graphs, the first case that is not computationally easy in the edge-weighted setting. With hindsight, we actually aim to handle a more general case. A graph H is said to be *t-almost k-colorable*, for parameters t and k, if there is a subset of $tn^{1/k}$ vertices whose removal leaves the graph k-colorable.

Proposition 2. *There is an algorithm that finds a $O(t\sqrt{n})$-coloring of t-almost 2-colorable edge-weighted graphs.*

Set $X = \sqrt{n}$ and form again the graph $G = G_X$ defined as before. Again, we see that an independent set in G of size at most X is independent in H. If there is a vertex of degree at most $3(t+1)n/X$ in G, we can clearly make progress towards a $O(tX)$-coloring, as desired.

On the other hand, consider a vertex v of degree at least $3(t+1)n/X$ and let N_v denote its set of neighbors. Our aim is to apply the SDP approach on the induced subgraph $H[N_v]$, and for that purpose we need to show that its independence number is high. Let C denote the set of (at most tX) vertices whose removal makes H to be 2-colorable, and let A and B denote the two color classes. Suppose v is in A, without loss of generality, and let $N_A = N_v \cap A$ be the subset of N_v from A. Note that N_A is of size at most X (since the sum of the incoming weights from nodes in A is less than 1 and each such edge weight is at least $1/X$). The set $N_v - N_A - C$ must be a subset of B, and its fraction of N_v is at least

$$\frac{|N_v - N_A - C|}{|N_v|} \geq 1 - \frac{|X| + |tX|}{|N_v|} \geq 1 - \frac{1+t}{3(t+1)} = \frac{2}{3}.$$

Thus, the induced subgraph $H[N_v]$ contains a feasible subset of at least two thirds of the nodes. We can then apply the SDP result of [13] with $\epsilon = 1/3$ to obtain an independent set I of size $\Omega(\epsilon|N_v|) = \Omega(tX)$, making progress towards a $O(n/(tX)) = O(\sqrt{n}/t)$-coloring, as desired.

We now apply the approach recursively, along the lines of Wigderson.

Proposition 3. *There is an algorithm that finds an independent set of size $\Omega(n^{1/k}/(t+k))$ in a t-almost k-colorable edge-weighted graphs. Thus, we can make progress towards a $O((t+k)n^{1-1/k})$-coloring.*

When $k = 2$, the claim holds from Prop. 2, so assume $k \geq 3$. Let $X = n^{1/k}$ and form $G = G_X$ as before. If there is a vertex of degree at most n/X in G, then by selecting it we make progress towards a $O(n/X) = O(n^{1-1/k})$ coloring. Otherwise, consider the set N_v of neighbors of an arbitrary vertex v, which is of size greater than $n/X = n^{1-1/k}$. We claim that $H[N_v]$ is $(t+1)$-almost $k-1$-colorable. Namely, N_v might contain the at most tX nodes whose removal turns H into a k-colorable graph H', and it could contain at most X nodes that belong to the same color class in H' as v, but together this amounts to only $(t+1)n^{1/k} \leq (t+1)(|N_v|^{k/(k-1)})^{1/k} = (t+1)|N_v|^{1/(k-1)}$, which satisfies $(t+1)$-almost $(k-1)$-colorability. By applying our method by induction on $H[N_v]$, we obtain an independent set I of size at least

$$c\frac{|N_v|^{1/(k-1)}}{(t+1)+(k-1)} = c\frac{|N_v|^{1/(k-1)}}{t+k} \geq c\frac{n^{1/k}}{t+k} ,$$

for some absolute constant c, thus making progress towards a $(t+k)n^{1-1/k}$-coloring, as desired.

In particular, we obtain a $O(kn^{1-1/k})$-coloring of k-colorable graphs.

Finally, we can combine the two approaches (Props. 1 and 3) to get approximation results for coloring general edge-weighted graphs that almost matches the best ratio known for ordinary graphs. The maximum of the two bounds is achieved when $\chi = \theta(\log n/\log \log n)$.

Corollary 1. *There is a $O(n(\log \log n/\log n)^3)$-approximate algorithm for coloring edge-weighted graphs.*

3.2   Better Solvable Cases

The instances that arise in wireless settings are not completely arbitrary. They have structure that distinguish them from general instances. Let us examine some of the structural properties that help in getting better solutions.

Wireless transmissions take place in physical space, and interferences generally speaking decreases with distance. One may expect there to be limits to how much a single transmission can be disturbed. Graph theoretically, it is natural to consider graphs of *low maximum in-degree* $\Delta^-(H)$. It was shown in [14] that one can color such graphs using $\lfloor 2\Delta^- + 1 \rfloor^2$ colors. Very recently, an optimal upper bound of $\lfloor 2\Delta^- + 1 \rfloor$ colors was obtained [3], which can be made constructive up to an arbitrarily small error term. A tight bound of $\lceil \Delta + 1 \rceil$ was also recently given for undirected edge-weighted graphs [1]. That resolves these questions. However, there is no clear link between maximum degree and the chromatic number, thus it does not directly address the question of efficient approximability.

Another parameter of sparsity that connects well with classical colorings is inductiveness. A graph is ρ-*inductive* if there is an ordering of the vertices so that for each vertex v, the in-degree of v from the nodes succeeding it in the order is at most ρ. Hoefer et al. [15] were among the first to treat wireless scheduling problems as edge-weighted graphs. They also introduced measures related to inductiveness in the study of the corresponding independent set problem. Earlier, Kesselheim and Vöcking [19] had actually shown that for wireless instances (with certain regimes of fixed power assignments to the transmitters), $\rho(H) = O(\log n \chi(H))$. This leaves the question of how good colorings can be obtained in terms of the inductiveness parameter ρ.

Kesselheim and Vöcking [19] gave a *distributed* algorithm that uses at most $O(\rho \log n)$ colors, resulting in a $O(\log^2 n)$-approximation. The bound on ρ was later tightened to the optimal $\rho(H) = O(\chi)$ [11], resulting in a $O(\log n)$-approximation for coloring. However, this turns out to be best possible in terms of inductiveness alone; namely, there are instances such that $\chi(H) = \Omega(\rho(H) \log n)$ [3]. Thus, better-than-logarithmic solutions will need to take additional properties of wireless instance into account.

Constant factor approximations are known for the independent set problem of edge-weighted graphs for instances derived by wireless settings. Specifically, it holds for the main two variants based on the choice of power assigned to the senders: power depends only on the intended transmission distance [12], or the power can be arbitrary [18]. This immediately implies logarithmic approximations for the corresponding coloring problems. These results utilize the fact that the transmission take place between units embedded in a *metric* space. Each node of the graph corresponds to a communication *link*, a sender-receiver pair located in the metric, while the weight of an edge represents the disturbance that one transmission has on another transmission link. In particular, in the setting where all senders use the same power, the weight of the directed edge (u, v) from link (s_u, r_u) to link (s_v, r_v) is proportional to $(d(s_v, r_v)/d(s_u, r_v))^\alpha$, where α is a fixed positive constant.

The key question is whether the metric property can be brought to bear to obtain better approximation ratios than logarithmic. The holy grail would be to give an absolute constant factor approximation. So far, all attempts have failed. Still, there are no obvious *a priori* reasons why this cannot succeed.

Additionally, it would be interesting to characterize other classes of instances that admit efficient approximations.

4   Conclusions

We have explored two graph coloring problems: The classical one and a recent edge-weighted variation motivated by wireless applications. Much remains to be done to deepen our understanding, even for the classical version.

Acknowledgments. The author thanks Christian Konrad and Tigran Tonoyan for helpful comments.

References

1. Araujo, J., Bermond, J.-C., Giroire, F., Havet, F., Mazauric, D., Modrzejewski, R.: Weighted improper colouring. Journal of Discrete Algorithms 16, 53–66 (2012)
2. Arora, S., Lund, C., Motwani, R., Sudan, M., Szegedy, M.: Proof verification and hardness of approximation problems. In: FOCS, pp. 14–23 (1992)
3. Bang-Jensen, J., Halldórsson, M.M.: A note on vertex coloring edge-weighted digraphs. Preprints on graph, hypergraphs and computing, Institute Mittag-Leffler (2014)
4. Berger, B., Rompel, J.: A better performance guarantee for approximate graph coloring. Algorithmica 5(4), 459–466 (1990)
5. Blum, A., Karger, D.: An $\tilde{O}(n^{3/14})$-coloring algorithm for 3-colorable graphs. Information Processing Letters 61(1), 49–53 (1997)
6. Boppana, R.B., Halldórsson, M.M.: Approximating maximum independent sets by excluding subgraphs. In: Gilbert, J.R., Karlsson, R. (eds.) SWAT 1990. LNCS, vol. 447, pp. 13–25. Springer, Heidelberg (1990)
7. Feige, U., Kilian, J.: Zero knowledge and the chromatic number. J. Comput. Syst. Sci. 57, 187–199 (1998)
8. Halldórsson, M.M.: A still better performance guarantee for approximate graph coloring. Technical Report 90–44, DIMACS (1990)
9. Halldórsson, M.M.: A still better performance guarantee for approximate graph coloring. Inform. Process. Lett. 45, 19–23 (1993)
10. Halldórsson, M.M.: Approximation via partitioning. Research Report IS-RR-95-0003F, JAIST (1995)
11. Halldórsson, M.M., Holzer, S., Mitra, P., Wattenhofer, R.: The power of non-uniform wireless power. In: SODA (2013)
12. Halldórsson, M.M., Mitra, P.: Wireless Capacity with Oblivious Power in General Metrics. In: SODA (2011)
13. Halldórsson, M.M., Mitra, P.: Wireless capacity with arbitrary gain matrix. Theor. Comput. Sci. 553, 57–63 (2014)
14. Halldórsson, M.M., Wattenhofer, R.: Wireless Communication Is in APX. In: Albers, S., Marchetti-Spaccamela, A., Matias, Y., Nikoletseas, S., Thomas, W. (eds.) ICALP 2009, Part I. LNCS, vol. 5555, pp. 525–536. Springer, Heidelberg (2009)
15. Hoefer, M., Kesselheim, T., Vöcking, B.: Approximation algorithms for secondary spectrum auctions. In: SPAA, pp. 177–186 (2011)
16. Hua, Q.-S., Lau, F.: Exact and approximate link scheduling algorithms under the physical interference model. In: FOMC, pp. 45–54. ACM (2008)

17. Johnson, D.S.: Worst case behavior of graph coloring algorithms. In: Proc. 5th Southeastern Conf. on Combinatorics, Graph Theory, and Computing. Congressus Numerantium X, pp. 513–527 (1974)
18. Kesselheim, T.: A Constant-Factor Approximation for Wireless Capacity Maximization with Power Control in the SINR Model. In: SODA (2011)
19. Kesselheim, T., Vöcking, B.: Distributed contention resolution in wireless networks. In: Lynch, N.A., Shvartsman, A.A. (eds.) DISC 2010. LNCS, vol. 6343, pp. 163–178. Springer, Heidelberg (2010)
20. Khot, S., Ponnuswami, A.K.: Better inapproximability results for maxClique, chromatic number and min-3Lin-deletion. In: Bugliesi, M., Preneel, B., Sassone, V., Wegener, I. (eds.) ICALP 2006. LNCS, vol. 4051, pp. 226–237. Springer, Heidelberg (2006)
21. Lovász, L.: On the Shannon capacity of a graph. IEEE Trans. Inform. Theory IT-25(1), 1–7 (1979)
22. Wigderson, A.: Improving the performance guarantee for approximate graph coloring. J. ACM 30(4), 729–735 (1983)

Recent Results in Scalable
Multi-Party Computation

Jared Saia and Mahdi Zamani

Dept. of Computer Science, University of New Mexico, Albuquerque, NM, USA 87131
{saia,zamani}@cs.unm.edu

Abstract. Secure multi-party computation (MPC) allows multiple parties to compute a known function over inputs held by each party, without
any party having to reveal its private input. Unfortunately, traditional
MPC algorithms do not scale well to large numbers of parties. In this
paper, we describe several recent MPC algorithms that are designed to
handle large networks. All of these algorithms rely on recent techniques
from the Byzantine agreement literature on forming and using *quorums*.
Informally, a quorum is a small set of parties, most of which are trustworthy. We describe the advantages and disadvantages of these scalable
algorithms, and we propose new ideas for improving practicality of current techniques. Finally, we conduct simulations to measure bandwidth
cost for several current MPC algorithms.

1   Introduction

In *secure multi-party computation (MPC)*, a set of parties, each having a secret
value (input), want to compute a common function over their inputs, without
revealing any information about their inputs other than what is revealed by the
output of the function.

In this paper, we focus on *scalable* MPC algorithms, which are designed to
be resource-efficient (*e.g.*, in terms of bandwidth, computation, and latency)
for large networks. Scalable MPC is of importance for many applications over
modern networks. For example, how can peers in BitTorrent auction off resources
without hiring an auctioneer? How can we design a decentralized Twitter that
enables provably anonymous broadcast? How can we perform data mining over
data spread over large numbers of machines?

Although much theoretical progress has been made in the MPC literature to
achieve scalability (*e.g.*, [1,9,3,22,23,25]), practical progress is slower. In particular, most known schemes suffer from either poor or unknown communication
and computation costs in practice.

Most large-scale distributed systems are composed of nodes with limited resources. This makes it of extreme importance to *balance* the protocol load across
all parties involved. Also, large networks tend to have weak admission control
mechanisms which makes them likely to contain Byzantine nodes. Thus, a key
variant of the MPC problem that we consider will be when a certain hidden
fraction of the nodes are controlled by a Byzantine adversary.

G.F. Italiano et al. (Eds.): SOFSEM 2015, LNCS 8939, pp. 24–44, 2015.
© Springer-Verlag Berlin Heidelberg 2015

1.1   Problem Statement

In the MPC problem, a set of n parties, each holding a private input, jointly evaluate a function f over their inputs while ensuring,

1. Upon termination of the protocol, all parties have learned the correct output of f; and

2. No party learns any information about other parties' inputs other than what is revealed from the output.

   We assume the identities of the n parties are common knowledge, and there is a private and authenticated communication channel between every pair of parties. We consider two communication models. In the *synchronous* model, there is an upper bound, known to all parties, on the length of time that a message can take to be sent through a channel. In the *asynchronous* model, there is no such upper bound.

   Usually, a certain fraction of the parties are controlled by a *Byzantine*[1] adversary. These parties can deviate arbitrarily from the protocol. In particular, they can send incorrect messages, stop sending any messages, share information amongst themselves, and so forth. Their goal is to thwart the protocol by either obtaining information about the private inputs, or causing the output of the function to be computed incorrectly. We say the adversary is *semi-honest* if the adversary-controlled parties are curious to learn about other parties' secret information, but they strictly follow the protocol. We say that the parties controlled by the adversary are *malicious* (or *Byzantine* or *dishonest*). The remaining parties are called *semi-honest* (or simply, *honest*).

   The adversary is either *computationally-bounded* or *computationally-unbounded*. The former is typically limited to only *probabilistic polynomial-time (PPT)* algorithms, and the latter has no computational limitations. The adversary is either assumed to be *static* or *adaptive*. A static adversary is limited to selecting the set of dishonest parties at the start of the protocol, while an adaptive adversary does not have this limitation.

1.2   Measures of Effectiveness

The following metrics are typically used to measure the effectiveness of MPC protocols.

Resource Costs. These include *communication cost* (number of messages sent and size of each message), *computation cost*, and *latency* (number of rounds of communication). We remark that load-balancing may be important for all of these resources.

Fault Tolerance. These metrics measure to what degree a protocol can tolerate adversarial attack. They include: the number of nodes that an adversary can take over (without sacrificing correctness); the type(s) of faults, *i.e.*, Byzantine, crash

---
[1] Also known as *active* or *malicious*.

faults, randomly, or adversarially distributed; the number of bits in messages that can be corrupted by an adversary; and the amount of churn that the protocol can tolerate

1.3   MPC and Byzantine Agreement

In the Byzantine setting, the MPC problem is tightly related to the problem of *Byzantine agreement (BA)*, where a group of n parties each holding an input value want to agree on a common value. In a celebrated result, Pease, Shostak, and Lamport [53] proved that perfectly-secure BA can be achieved as long as less than one third fraction of the parties is corrupted. There are several interesting connections between BA and MPC:

1. BA can be seen as MPC for the simplest type of function: a function that must return a bit equal to the input bit of at least one honest party. However, BA is simpler that MPC in that it is not necessary to maintain privacy of inputs.

2. MPC protocols strongly rely on the use of a *broadcast channel* which is typically realized using a BA protocol.[2] Most MPC results so far assume the existence of a broadcast channel. Unfortunately, this requirement is highly problematic in settings, where the number of parties is large.

3. Several recent MPC schemes [3,9,25,62] crucially build upon the notion of *quorums*[3] for achieving scalability. A quorum is a polylogarithmic set of parties, where the number of corrupted parties in each quorum is guaranteed not to exceed a certain fraction. King *et al.* [43] show how to use BA to efficiently create a collection of quorums.

Paper Organization. The rest of this paper is organized as follows. In Section 2, we review related work with a focus on MPC for many parties (*i.e.*, scalable MPC). In Section 3, we describe key open problems in scalable MPC. In Section 4, we describe algorithmic tools used in current scalable MPC algorithms. Section 5 describes recent quorum-based results for scalable MPC, and defines and analyzes new techniques for improving these results. Finally, we conclude in Section 6.

2   Related Work

Due to the large body of work, we do not attempt a comprehensive review of the MPC literature here, but rather focus on work that is relevant to scalable MPC.

---

[2] The standard definition of MPC (as given in Section 1.1) implies Byzantine agreement. Goldwasser and Lindell [36] show that a relaxed definition of MPC allows MPC without a broadcast channel (and hence without Byzantine agreement).

[3] Also known as *committees*.

The MPC problem was first described by Yao [59]. He described an algorithm for MPC with two parties in the presence of a semi-honest adversary. Goldreich *et al.* [37] propose the first MPC protocol that is secure against a Byzantine adversary. This work along with [15,33] are all based on cryptographic hardness assumptions and are often regarded as the first generic solutions to MPC. These were followed by several cryptographic improvements [12,17,39] as well as information theoretically-secure protocols [2,6,11,13] in late 1980s and 1990s. Unfortunately, these methods all have poor communication scalability. In particular, if there are n parties involved in the computation, and the function f is represented by a circuit with m gates, then these algorithms require each party to send a number of messages and perform a number of computations that is $\Omega(mn)$.

In 2000s, exciting improvements were made to the cost of MPC, when m (*i.e.*, the circuit size) is much larger than n [21,22,27]. For example, Damgard *et al.* [22] give an algorithm with computation and communication cost that is $\tilde{O}(m)$ plus a polynomial in n. Unfortunately, the additive polynomial in these algorithms is large (at least $\Omega(n^6)$) making them impractical for large n.

Depending on the model of computation, every function can be represented in terms of some *elementary operations* such as arithmetic operations (*e.g.*, addition, multiplication), Boolean operations (*e.g.*, and, or), RAM instructions (*e.g.*, get-value, set-value), etc. Informally speaking, every MPC protocol specifies how a group of elementary operations can be computed securely. The function is computed securely via composition of these secure operations. From this perspective, we classify the broad range of MPC approaches into two categories: techniques that evaluate circuits (Boolean or arithmetic), and techniques that evaluate RAM programs.

2.1   Circuit-Based Techniques

We subdivide the set of circuit-based methods into three categories based on their main approach for achieving privacy: garbled circuits, secret sharing, and fully homomorphic encryption. Although some protocols such as [9,28] may fall into more than one category, most protocols follow only one as their main approach.

Garbled Circuits. The idea of garbled circuits dates back to the two-party MPC proposed by Yao [59].[4] One party is called the *circuit generator* and the other one is called the *circuit evaluator*. For each wire in the circuit, the generator creates a mapping that maps each possible value of that wire to another value (called the *garbled value*). The generator then sends this mapping to the evaluator. The evaluator evaluates the circuit using the mapping to compute the *garbled output*. Next, the generator computes another mapping (called *translation*) that maps all possible garbled outputs to their actual values. In the final round, the generator sends the translation to the evaluator, and the evaluator sends the garbled output to the generator. Both parties can compute the actual

---

[4] The term "garbled circuits" is due to Beaver, Micali, and Rogaway [12].

output at the same time without learning anything about each other's inputs. This algorithm is only secure in the semi-honest setting.

Yao's original model has been the basis for several secure computation algorithms mostly for the two-party setting with computational hardness assumptions [37,40,44,48,50]. In a line of research, Lindell and Pinkas give the first proof of Yao's protocol [49] and present a two-party approach based on garbled circuits that uses the cut-and-choose technique to deal with malicious parties [48,50].

Secret Sharing. In secret sharing, one party (called the *dealer*) distributes a secret amongst a group of parties, each of whom is allocated a *share* of the secret. Each share reveals nothing about the secret to the party possessing it, and the secret can only be reconstructed when a sufficient number of shares are combined together.

Many MPC schemes build upon the notion of secret sharing (most notably, [6,11,13,22,25,28,39]). Informally speaking, each party secret shares its input among all parties using a secret sharing scheme such as Shamir's scheme [56]. Then, all parties perform some intermediate computation on the received shares and ensure that each part now has a share of the result of the computation. In the final stage, all parties perform a final computation on the intermediate results to find the final result. In the Byzantine setting, each stage of this approach usually requires several rounds of communication used to verify consistency of the shares distributed by each party (using a *Verifiable Secret Sharing (VSS)* scheme like [11,18]) and to perform complicated operations such as multiplication.

Ben-Or *et al.* [11] show that every functionality can be computed with information-theoretic security in the presence of a semi-honest adversary controlling less than half of the parties, and in the presence of a Byzantine adversary controlling less than a third of the parties. They propose a protocol for securely evaluating an arithmetic circuit that represents the functionality. First, the parties secret-share their inputs with each other using Shamir's scheme [56]. For the Byzantine case, an interactive VSS protocol is proposed using bivariate polynomials. The parties emulate the computation of each gate of the circuit by computing shares of the gate's output from the shares of the gate's inputs.

Given shares of the input wires, an addition gate's output is computed without any interaction simply by asking each party to add their local shares together. Unfortunately, multiplying two polynomials results in a polynomial that has a higher degree and is not completely random. Ben-Or *et al.* [11] emulate a multiplication gate computation by running interactive protocols for degree reduction and polynomial randomization.

The efficiency of [11] was later improved by others in similar and different settings [6,22,39]. Unfortunately, these protocols still do not scale well with n and incur large communication and computation costs in practice.

Dani *et al.* [24] propose an MPC protocol for evaluating arithmetic circuits in large networks. The protocol is unconditionally-secure against a Byzantine adversary corrupting less than $(1/3-\epsilon)n$ of the parties, for some positive constant ϵ. The protocol creates a set of quorums using the quorum building algorithm of [43]. For each gate in the circuit, a quorum is used to compute the output of the gate using the MPC of [11] among parties of the quorum. The protocol ensures that all parties in the quorum learn the output of gate masked with a uniformly random value which is secret-shared among all parties of the quorum. Thus, no party learns any information about the output, but the parties together have enough information to provide the input for computation of the masked output of the next gate. This procedure is repeated for every level of the circuit. At the top level, the output is computed and sent down to all parties through all-to-all communication between the quorums. Assuming a circuit of depth d with m gates, this protocol requires each party to send (and compute) $\tilde{O}(m/n + \sqrt{n})$ bits (operations) with latency $O(d + \mathsf{polylog}(n))$.

This protocol was later modified in [25] in order to support asynchronous communication incurring the same asymptotic costs but tolerating less than $(1/8 - \epsilon)n$ malicious parties. In the new model, the adversary has control over the latency of the communication channels and can arbitrarily delay messages sent over them. However, all messages sent by the parties are assumed to be eventually delivered but with indefinite delays.

The main challenge in this model is that the parties require a distributed mechanism to learn when sufficient number of inputs are received in order to start the computation over those inputs. To this end, Dani *et al.* [25] propose to count the number of *ready inputs* using a distributed data structure called τ-*counter*, where $\tau = n - t$ is the threshold on the number inputs to be received before the circuit is evaluated, and $t < n/8$.

Fully Homomorphic Encryption. A *fully homomorphic encryption (FHE)* scheme allows to perform secure computation over encrypted data without decrypting it. Gentry [30] proposed the first FHE scheme based on the hardness of lattice problems. Since then, many techniques have been proposed to improve the efficiency of FHE [10,32,58]. Unfortunately, current techniques are still very slow and can only evaluate small circuits. This restriction is primarily due to noise management techniques (such as bootstrapping [30]) used to deal with a noise term in ciphertexts that increases slightly with homomorphic addition and exponentially with homomorphic multiplication.

In particular, if the circuit has a sufficiently small multiplicative depth, then it is possible to use current FHE schemes in practice without using the expensive noise management techniques. Such a scheme is often called *somewhat homomorphic encryption (SHE)* [58], which requires significantly less amount of computation than an FHE with noise management.

Damgard *et al.* [28] propose a Byzantine-resilient MPC scheme using SHE in an offline phase to compute Beaver multiplication triples [6]. These triples are later used in the online phase to compute multiplication gates efficiently. One drawback of this scheme is that when cheating happens in the network, the protocol cannot guarantee termination. Malicious parties can take advantage of this to prevent the protocol from termination.[5]

---

[5] In general, if the majority of parties are malicious, then the termination of MPC (*i.e.*, output delivery) cannot be guaranteed.

Asharov *et al.* [1] describe a constant-round MPC scheme using a *threshold FHE (TFHE)* technique that provides Byzantine-resilience and circuit-independent communication cost. All parties first encrypt their inputs under the FHE scheme of Brakerski *et al.* [10] and send the encrypted values to all other parties. Then, each party evaluates the desired function over the encrypted inputs via homomorphism, and eventually participates in a distributed decryption protocol to decrypt the output. Although providing constant rounds of communication, this scheme does not scale well with the number of parties and the circuit size due to all-to-all communication overhead (*i.e.*, $\Omega(n^2)$) and high computation overhead of the FHE of [10] for large-depth circuits. To overcome the high computation cost, the authors propose to outsource circuit computation to a powerful party (*e.g.*, the "cloud"). While this is helpful for the semi-honest setting, it requires expensive zero-knowledge proofs to enforce honest behavior in the Byzantine setting.

Boyle *et al.* [9] describe a synchronous MPC protocol for evaluating arithmetic circuits. The protocol is computationally-secure against an adversary corrupting up to $(1/3-\epsilon)$ fraction of parties, for some fixed positive ϵ. As network size scales, it becomes infeasible to require each party communicate with all other parties. To this end, the protocol of [9] uses quorums to achieve sublinear ($\mathsf{polylog}(n)$) *communication locality* which is defined as the total number of point-to-point communication channels that each party uses in the protocol. Interestingly, the communication costs are independent of circuit size. This is achieved by evaluating the circuit over encrypted values using an FHE scheme. Unfortunately, the protocol is not fully load-balanced as it evaluates the circuit using only one quorum (called *supreme committee*) for performing general MPC. The protocol requires each party to send $\mathsf{polylog}(n)$ messages of size $\tilde{O}(n)$ bits and requires $\mathsf{polylog}(n)$ rounds.

Chandran *et al.* [14] address two limitations of the protocol of [9]: adaptive adversary and optimal resiliency (*i.e.*, $t < n/2$ malicious parties). They achieve both of these by replacing the *common reference string (CRS)* assumption of [9] with a different setup assumption called *symmetric-key infrastructure (SKI)*, where every pair of parties share a uniformly-random key that is unknown to other parties. The authors also show how to remove the SKI assumption at a cost of increasing the communication locality by $O(\sqrt{n})$. Although this protocol provides small communication locality, the bandwidth cost seems to be super-polynomial due to large non-constant message sizes.

2.2   RAM-Based Techniques

Most MPC constructions model algorithms as circuits. Unfortunately, a circuit can at best model the *worst-case* running time of an algorithm because a circuit can only be created by unrolling loops to their worst-case runtime [35]. Moreover, circuits incur at least a linear computation complexity in the total size of the input, while a sublinear overhead is crucial for achieving scalability in most large-scale applications. In addition, most algorithms have already been described in terms of instructions (programs) to a *random access memory (RAM)* machine[6] [19], not circuits. These all bring the following question to the mind: Is it possible to securely evaluate RAM programs instead of circuits? Luckily, the answer is "yes". Goldreich and Ostrovsky [38] show that by constructing a RAM with secure access, one can evaluate arbitrary RAM programs privately. Such a RAM is often called an *Oblivious RAM (ORAM)*. This is typically considered in a setting, where a group of parties (clients) want to access a data storage (RAM) held by another party (a server).

To build an ORAM, content encryption alone is not sufficient because the party holding the data (or an eavesdropper) can obtain critical information about the queries by analyzing the access patterns, even though the data is encrypted. Therefore, techniques are required to hide the access patterns to the data storage, meaning that no party is able to distinguish between any subsets of the data requests. More precisely, the following information must remain private: (1) the locations of accessed data items, (2) the order of data requests, (3) the number of requests to the same location, and (4) the type of access (*e.g.*, get-value, set-value).

Goldreich and Ostrovsky [38] propose a two-party technique for securely outsourcing data to a remote storage. The client's access pattern to the remote storage is hidden by continuously shuffling and re-encrypting data as they are accessed. The authors show that any program in the standard RAM model can be compiled into a program for an ORAM using an *ORAM simulator* with an overhead that is polylogarithmic in the size of the memory. Although asymptotically-efficient, the algorithm of [38] is not practical due to large constant factors.

Several techniques have been proposed to improve the overhead of ORAM protocols in general [52,55,57] and for secure two-party computation [31,34,47]. Damgard *et al.* [26] propose the first ORAM algorithm in the multi-party setting. Unfortunately, their algorithm requires each party to communicate and maintain information of size equivalent to all parties' inputs. Boyle *et al.* [3] describe a scalable technique for secure computation of RAM programs in large networks by performing local communications in quorums of parties. For securely evaluating a RAM program Π, their protocol incurs a total communication and computation of $\mathsf{poly}(n) + \tilde{O}(Time(\Pi))$ while requiring $\tilde{O}(|x| + Space(\Pi)/n)$ memory per party, where $Time(\Pi)$ and $Space(\Pi)$ are time and space complexity of Π respectively, and $|x|$ denotes the input size.

3   Open Problems

In this section, we describe several open problems in the domain of scalable MPC. These problems are roughly ordered from easiest to hardest. We describe some partial progress on the first two problems later in this paper.

---

[6] A RAM machine has a *lookup functionality* for accessing memory locations that takes $O(1)$ operations. Given an array A of N values and an index $x \in \{1, ..., N\}$, the lookup functionality returns $A[x]$.

Share Renewal. In the protocol of [25], each gate of the circuit is assigned a quorum Q, and the parties in Q are responsible for computing the function associated with that gate. Then, they send the result of this computation to any quorums associated with gates that need this result as input. Let Q' be one such quorum. It is necessary to securely send the output from Q to Q' without revealing any information to any individual party or to any coalition of adversarial parties. Inspired by [41], we refer to this problem as *share renewal*, because it involves generating a fresh sharing of a secret-shared value among a new set of parties.

Dani *et al.* [25] handle this problem by masking the result in Q and unmasking the result in Q'. Unfortunately, they do not provide an explicit construction of their method, and simple constructions are very expensive in terms of communication and computation costs [62]. Boyle *et al.* [9] overcome this problem by sending their encrypted inputs to only one quorum which does all of the computation using FHE. This results in large computation and communication costs for parties in that quorum. In Section 5, we give some ideas for solving this problem efficiently.

Secure Multiplication. Consider an arithmetic circuit representing the desired function to be computed. Using a linear secret sharing scheme (such as [56]), addition gates can be computed with no communication by simply adding the two input shares. On the other hand, known secret sharing schemes are not *multiplicatively homomorphic* meaning that the product of two shares is not necessarily a valid and secure share of the product of the corresponding secrets. Designing an efficient technique for secure multiplication is an important building block for secret-sharing-based MPC. We are not aware of a perfectly-secure technique for secure multiplication that requires only constant rounds of communication.

Byzantine-Resilient Threshold Decryption. Consider n parties that have jointly encrypted a message using some encryption scheme. In threshold decryption, for some parameter $x < n$, it is required that any subset of x parties can decrypt the message, while any subset of strictly less than x parties learn nothing about the encrypted message. Threshold decryption of a (fully) homomorphic encryption can be used as a primitive for constructing efficient MPC protocols.

Unfortunately, known techniques for Byzantine-resilient threshold decryption (such as [1,16]) suffer from large communication overhead, due to zero-knowledge proofs used for ensuring honest behavior. A key open problem is to reduce this communication overhead.

Las Vegas MPC. Recently, several randomized MPC algorithms have been proposed (such as [9,14,25]) with Monte Carlo guarantees. In particular, the output is correct with high probability[7]. Alternatively, one may try to design a Las Vegas MPC algorithm. For this type of algorithm, the output must be correct

---

[7] An event occurs *with high probability*, if it occurs with probability at least $1 - 1/n^c$, for any $c > 0$ and all sufficiently large n.

with probability 1, but the latency can be a random variable. It is not clear that a quorum-based approach will be effective for solving this open problem.

Oblivious Parallel RAM. While parallelism has been extensively used in various computer architectures for accelerating computations, most ORAM models of computation (see Section 2.2) are not parallel. In the *Parallel RAM (PRAM)* model of computation, several parties, running in parallel, want to access the same shared external memory. This separates the program into two parts: control flow and shared memory. The goal is to parallelize control flow via oblivious access to the share memory in order to reduce computational time and latency.

In general, any PRAM program can be converted into a RAM program to be then evaluated by a standard ORAM. Unfortunately, this transformation incurs a large computational overhead. Boyle *et al.* [4] take a first step towards addressing this problem by describing an oblivious PRAM scheme that compiles any PRAM program into an *Oblivious PRAM (OPRAM)* program. This compiler incurs polylogarithmic overhead in the number of parties and the memory size. The algorithm is based on the ORAM construction of Shi *et al.* [55] which requires cryptographic assumptions. It remains unknown if one can design a perfectly-secure oblivious PRAM with resource costs that scale well with the network size and are independent of memory size.

Large Inputs. It is also interesting to consider MPC when each party can have very large inputs. Following is a concrete problem in this domain. Let M be a sparse adjacency matrix for some graph, and let the columns of M be partitioned among the parties. This problem motivates the following questions: Can we securely and efficiently compute the shortest path between a given pair of vertices over M? We can also consider other graph-theoretic problems. Even simpler, can we securely compute the dot product of two sparse vectors with resource costs proportional only to the number of non-zero entries in the vectors? We are not aware of any algorithms for even these simple types of problems.

4   Algorithmic Tools

In this section, we describe key algorithmic tools used in scalable MPC protocols.

4.1   Verifiable Secret Sharing

An (n, t)-*secret sharing* scheme, is a protocol in which a dealer who holds a secret value shares the secret among n parties such that any set of t parties cannot gain any information about the secret, but any set of at least $t+1$ parties can reconstructs it. An (n, t)-*verifiable secret sharing (VSS)* scheme is an (n, t)-secret sharing scheme with the additional property that after the sharing phase, a dishonest dealer is either disqualified or the honest parties can reconstruct the secret, even if shares sent by dishonest parties are spurious. Katz *et al.* [42] propose a constant-round VSS protocol based on Shamir's secret sharing [56]. This result is described in Theorem 1.

Theorem 1. [42] *There exists a synchronous linear (n,t)-VSS scheme for $t <$ $n/3$ that is perfectly-secure against a static adversary. The protocol requires one broadcast and three rounds of communication.*

For practicality, one can use the cryptographic VSS of Kate *et al.* [46] called *eVSS*[8], which is based on Shamir's scheme and the hardness of the discrete logarithm problem. Since eVSS generates commitments over elliptic curve groups, it requires smaller message sizes than other DL-based VSS scheme such as [39].

Theorem 2. [46] *There exists a synchronous linear (n,t)-VSS scheme for $t <$ $n/2$ that is secure against a computationally-bounded static adversary. In worst case, the protocol requires two broadcasts and four rounds of communication.*

4.2 Secure Broadcast

In the Byzantine setting, when parties have only access to secure pairwise channels, a protocol is required to ensure secure (reliable) broadcast. Such a broadcast protocol guarantees all parties receive the same message even if the broadcaster (dealer) is dishonest and sends different messages to different parties. It is known that a BA protocol can be used to perform secure broadcasts. The BA algorithm of Braud-Santoni *et al.* [8] gives the following result.

Theorem 3. [8] *There exists an unconditionally-secure protocol for performing secure broadcasts among n parties. The protocol has $\tilde{O}(n)$ amortized communication and computation complexity, and it can tolerate up to $(1/3 - \epsilon)n$ Byzantine parties, for any positive ϵ.*

The algorithm of [8] achieves this result by relaxing the load-balancing requirements. If concerned with load-balancing, one can instead use the load-balanced BA algorithm of King *et al.* [45] with $O(\sqrt{n})$ blowup.

4.3 Quorum Building

As an intermediate result in a Byzantine agreement paper, King *et al.* [43] give a protocol that can be used to bring all parties to agreement on a collection of n quorums. This protocol is based on the almost-everywhere Byzantine agreement[9] of King *et al.* [45].

Theorem 4. [45] *Suppose there are n parties, $b < 1/4 - \epsilon$ of which are malicious, for any fixed positive ϵ. Then, there is a polylogarithmic (in n) bounded degree network and a protocol such that,*

- *With high probability, a $1 - O(1/\ln n)$ fraction of the honest parties agree on the same value (bit or string).*

---

[8] Stands for efficient VSS.

[9] King *et al.* [45] relax the requirement that all uncorrupted parties reach agreement at the end of the protocol, instead requiring that a $1 - o(1)$ fraction of uncorrupted parties reach agreement. They refer to this relaxation as almost-everywhere agreement.

- *Every honest party sends and processes only a polylogarithmic (in n) number of bits.*
- *The number of rounds required is polylogarithmic in n.*

Their main technique is to divide the parties into groups of polylogarithmic size; each party is assigned to multiple groups. In parallel, each group then uses a leader election algorithm [29] to *elect* a small number of parties from within their group to move on. This step is recursively repeated on the set of elected parties until size of the remaining parties in this set becomes polylogarithmic. At this point, the remaining parties can solve the *semi-random-string agreement problem*[10] (similarly, they can run Byzantine agreement protocol to agree on a bit). Provided the fraction of dishonest parties in the set of remaining parties is less than $1/3$ with high probability, these parties succeed in agreeing on a semi-random string. Then, these parties communicate the result value to the rest of the parties.

In general, one can use any BA algorithm to build a set of quorums as described in [43]. Theorem 5 gives a quorum building protocol using the BA algorithm of Braud-Santoni *et al.* [8].

Theorem 5. [8] *There exists an unconditionally-secure protocol that brings all good parties to agreement on n good quorums with high probability. The protocol has $\tilde{O}(n)$ amortized communication and computation complexity*[11]*, and it can tolerate up to $(1/3 - \epsilon)n$ Byzantine parties, for any positive ϵ. If at most $(\alpha - \epsilon)$ fraction of parties are Byzantine, then each quorum is guaranteed to have $T \leq \alpha N$ Byzantine parties.*

5 Quorum-Based MPC

In Table 1, we review recent MPC results that provide sublinear communication locality. All of these results rely on some quorum building technique for creating a set of quorums each with honest majority. In the rest of this section, we describe a few ideas for improving efficiency of the synchronous MPC of [24]. These techniques are proved in [62]. We also conduct experiments to show the effectiveness of our techniques when compared with the protocols of [24] and [9].

Although the protocol of [24] asymptotically scales well with n, it is inefficient in practice due to large hidden factors in its cost complexities. Moreover, the paper does not provide an explicit construction of the proposed share renewal technique (see Section 3). Unfortunately, simple constructions seem very expensive in terms of communication and computation costs since parties in the second quorum need to jointly and securely unmask each input.

---

[10] In *semi-random-string agreement problem*, we want to reach a situation where, for any positive constant ϵ, $1/2 + \epsilon$ fraction of parties are good and agree on a single string of length $O(\log n)$ with a constant fraction for random bits.

[11] Amortized communication complexity is the total number of bits exchanged divided by the number of parties.

Table 1. Recent MPC results with sublinear communication locality

Protocol	Security	Resiliency Bound	Adaptive Adversary?	Async?	Assumes Broadcast Channel?	Total Message Complexity	Total Computation Complexity	Latency	Msg Size	Load-Balanced?
[24]	Perfect	$(1/3 - \epsilon)n$	No	No	No	$\tilde{O}(m + n\sqrt{n})$	$\tilde{O}(m + n\sqrt{n})$	$O(d + \mathrm{polylog}(n))$	$O(\ell)$	Yes
[9]	Crypto	$(1/3 - \epsilon)n$	No	No	No	$\tilde{O}(n)$	$\tilde{\Omega}(n) + \tilde{\Omega}(\kappa m d^3)$†	$\tilde{O}(1)$	$O(n\ell \cdot \mathrm{polylog}(n))$	No
[25]	Perfect	$(1/8 - \epsilon)n$	No	Yes	No	$\tilde{O}(m + n\sqrt{n})$	$\tilde{O}(m + n\sqrt{n})$	$O(d + \mathrm{polylog}(n))$	$O(\ell)$	Yes
[62]	Perfect	$(1/6 - \epsilon)n$	No	No	No	$\tilde{O}(m)$ or $O(m \log^4 n)$‡	$\tilde{O}(m)$ or $O(m \log^5 n)$‡	$O(d)$ or $O(\kappa + \ell)$‡	$O(\ell)$	Yes
[3]	Perfect	$(1/3 - \epsilon)n$	No	No	Yes	$\mathrm{poly}(n) + \tilde{O}(Time(\Pi))$	$\mathrm{poly}(n) + \tilde{O}(Time(\Pi))$	$\tilde{O}(Time(\Pi))$	$O(\ell)$	Yes
[14]	Crypto	$n/2$	Yes	No	No	$O(n \log^{1+\epsilon} n)$§ or $O(n\sqrt{n} \log^{1+\epsilon} n)$	$\Omega(n \log^{1+\epsilon} n)$§ or $\Omega(n\sqrt{n} \log^{1+\epsilon} n)$	$O(\log^{\epsilon'} n)$	$\Omega(\log^{\log n} n)$§ or $\Omega(\sqrt{n}^{\log n})$	Yes

Parameters.

n: number of parties.
ℓ: size of a field element.
d: depth of the circuit.
κ: the security parameter.
ϵ, ϵ': positive constants.
$Time(\Pi)$: worst-case running time of RAM program Π.

Notes.

† Using the FHE scheme of [10].
‡ The latter is achieved using cryptographic assumptions.
§ Assuming a symmetric-key infrastructure.

In [62], we propose a simple and efficient technique for share renewal. We also implement an efficient secure multiplication protocol to reduce the bandwidth and computation costs of [24]. In the rest of this section, we sketch the main properties of this new algorithm. Full details of the algorithms are in [62].

Consider n parties in a fully-connected synchronous network with private channels. Let f be any deterministic function over n inputs in some finite field $\mathbb{F}_p$ for prime $p = \mathsf{poly}(n)$, where f is represented as an arithmetic circuit with m gates and depth d. We prove the following theorem in [62].

Theorem 6. *There exists an unconditionally-secure n-party protocol that can compute f with high probability, while ensuring,*

- *The protocol tolerates up to $(1/6 - \epsilon)n$ malicious parties.*
- *Each party sends (computes) $\tilde{O}(m/n)$ messages (operations).*
- *Each message is of size $O(\log p)$ bits.*
- *The protocol has $O(d)$ rounds of communication.*

While our techniques are perfectly-secure, one can use the efficient VSS of Theorem 2 to achieve a cryptographic variant of Theorem 6 with better efficiency.

Theorem 7. *There exists an n-party protocol secure against a PPT adversary such that the protocol can compute f with high probability, while ensuring,*

- *The protocol tolerates up to $(1/6 - \epsilon)n$ malicious parties.*
- *Each party sends $O\left(\frac{m}{n}\log^4 n\right)$ messages and computes $O\left(\frac{m}{n}(\kappa+\log p)\log^4 n\right)$ operations, where κ is the security parameter.*
- *Each message is of size $O(\kappa + \log p)$ bits.*
- *The protocol has $O(d)$ rounds of communication.*

In this section, we represent each shared value $s \in \mathbb{F}_p$ by $\langle s \rangle = (s_1, ..., s_n)$ meaning that each party P_i holds a share s_i generated by the VSS scheme of Theorem 1 during its sharing phase. Using the natural component-wise addition of representations, we define $\langle a \rangle + \langle b \rangle = \langle a + b \rangle$. For multiplication, we define $\langle a \rangle \cdot \langle b \rangle = \mathsf{Multiply}(\langle a \rangle, \langle b \rangle)$, where $\mathsf{Multiply}$ is a protocol defined later in this section.

5.1   Share Renewal

Recalling from Section 3, let Q and Q' be the quorums involved in the share renewal process. In [62], we propose a different approach than [24] by reducing the share renewal problem to the simpler problem of generating a fresh sharing of the output of Q among parties of Q'. In other words, parties in Q generate a new set of random shares that represents the same secret as the output of Q, and distribute this new sharing among parties of Q'.

The high-level idea is to first generate a random polynomial that passes through the origin, and then add it to the polynomial that corresponds to the shared secret. The result is a new polynomial that represents the same secret

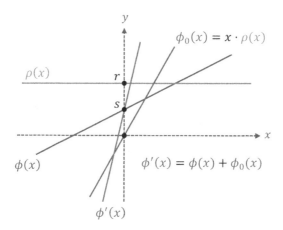

Fig. 1. Share renewal technique [62]

but has coefficients that are chosen randomly and completely independent of the coefficients of the original polynomial. Combined with the VSS scheme of Theorem 1 in a group of n parties with $t < n/3$ dishonest parties, this protocol has one round of communication and requires each party to send $O(n)$ field elements.

This idea was first proposed by Ben-Or *et al.* [11]. The solution provided in [11] requires a zero-knowledge proof, where each party is asked to prove distribution of shares over a polynomial with zero free-coefficient. Unfortunately, such a proof is either round-expensive (as in [11]) or requires a weaker adversarial model for the problem to be solved efficiently (*e.g.*, see [41]). On the other hand, by relaxing the resiliency bound by only one less dishonest party, we can generate a random polynomial that passes through the origin without requiring the zero-knowledge step.

Let $\phi(x)$ be the original polynomial. The idea is to first generate a random polynomial $\rho(x)$ of degree $\deg(\phi) - 1$, and then compute a new polynomial $\phi_0(x) = x \cdot \rho(x)$ that is of degree $\deg(\phi)$ and passes through the origin. Finally, the fresh polynomial is computed from $\phi(x) + \phi_0(x)$. The polynomial $\rho(x)$ can be simply generated by asking parties to agree on a secret-shared uniform random value (using the protocol GenRand described in [62]) over a random polynomial of degree $\deg(\phi) - 1$. Figure 1 depicts this idea for the special case of $d = 1$.

Theorem 8. [62] *Let Q and Q' be two quorums of size N, where Q holds a shared value $\langle s \rangle = (s_1, ..., s_N)$ over a polynomial ϕ of degree $d = N/3$. There exists a protocol that can generate a new shared value $\langle s' \rangle = (s'_1, ..., s'_N)$ in Q' such that $s' = s$. The protocol is secure against a computationally-unbounded Byzantine adversary corrupting less than a $1/6$ fraction of the parties in each quorum.*

5.2   Secure Multiplication

The secure multiplication protocol of [62] (denoted by Multiply) is based on a well-known technique proposed by Beaver [6]. The technique generates a shared multiplication triple $(\langle u \rangle, \langle v \rangle, \langle w \rangle)$ such that $w = u \cdot v$. The triple is then used to convert multiplications over shared values to additions.

The only difference between Multiply and Beaver's multiplication method is that Beaver generates shared random elements u and v on polynomials of degree d and multiplies them to get a polynomial of degree $2d$ for w. Then, a degree reduction algorithm is run to reduce the degree from $2d$ to d. Instead, we choose polynomials of degree $d/2$ for u and v to get a polynomial of degree d for w. In our protocol, since we require less of $1/6$ fraction of the parties be dishonest in each quorum, we can do this without revealing any information to the adversary. We note that the first step of Multiply is independent of the inputs and thus, can be performed in an offline phase to generate a sufficient number of multiplication triples.

Theorem 9. [62] *Given two secret-shared values $\langle a \rangle$ and $\langle b \rangle$, the protocol Multiply correctly generates a shared value $\langle c \rangle$ such that $c = a \cdot b$. The protocol is perfectly-secure against an adversary corrupting less than a $1/6$ fraction of the parties.*

5.3   Simulation Results

To study the effectiveness of our techniques and compare our new MPC scheme to previous work, we simulate our protocol along with the protocols of Dani *et al.* [24] and Boyle *et al.* [9]. We use these protocols to solve the *secure multi-party sorting (MPS)* problem. MPS is useful in many applications such as anonymous communication [7,51,54], privacy-preserving statistical analysis [20] (*e.g.*, top-k queries [5]), auctions [60], and location-based services [61]. It is often important for these applications to be run among *many* parties. For example, MPS is a critical component of communications algorithms that could enable the creation of large anonymous microblogging services without requiring trusted authorities (*e.g.*, an anonymous Twitter).

We run our protocol for inputs chosen from $\mathbb{Z}_p$ with a 160-bit prime p for getting about 80 bits of security. We set the parameters of our protocol in such a way that we ensure the probability of error for the quorum building algorithm of [8] is smaller than 10^{-5}. For the sorting circuit, we set $k = 2$ to get $\epsilon < 10^{-8}$ for all values of n in the experiment. Clearly, for larger values of n, the error becomes superpolynomially smaller, e.g., for $n = 2^{25}$, we get $\epsilon < 10^{-300}$. For all protocols evaluated in this section, we assume cheating (by malicious parties) happens in every round of the protocols. This is essential for evaluating various strategies used by these protocols for tolerating active attacks.

Figure 2 illustrates the simulation results obtained for various network sizes between 2^5 and 2^{30} (*i.e.*, between 32 and about 1 billion). To better compare the protocols, the vertical and horizontal axis of the plot are scaled logarithmically.

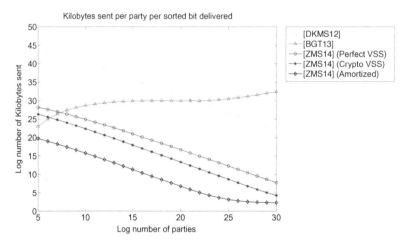

Fig. 2. Bandwidth cost of scalable MPC protocols

The x-axis presents the number of parties and the y-axis presents the number of Kilobytes sent by each party for delivering one anonymous bit.

In this figure, we report results from three different versions of our protocol. The first plot (marked with circles) belongs to our unconditionally-secure protocol (Theorem 6) that uses the perfectly-secure VSS scheme of Katz *et al.* [42]. The second plot (marked with stars) represents our computationally-secure protocol (Theorem 7) which uses the cryptographic VSS of Kate *et al.* [46]. The last plot (marked with diamonds) shows the cost of the cryptographic protocol with amortized (averaged) setup cost.

We obtain this by running the setup phase of our protocol once and then using the setup data to run the online protocol 100 times. The total number of bits sent was then divided by 100 to get the average communication cost. To achieve better results, we also generated a sufficient number of random triples in the setup phase. Then, the triples were used by our multiplication subprotocol in the online phase to multiply secret-shared values efficiently.

Our protocols significantly reduce bandwidth costs when compared to the protocols of [24] and [9]. For example, for $n = 2^{20}$ (about 1 million parties[12]), the amortized protocol requires each party to send about 64KB of data per anonymous bit delivered (about 8MB for our non-amortized version) while the protocols of [24] and [9] each send more than one Terabytes of data per party and per sorting bit delivered.

6   Conclusion

We described recent MPC algorithms that are efficient, even with many parties. In particular, we reviewed the most important results that achieve scalability

---

[12] This is less than 1% of the number of active Twitter users. An intriguing application of our protocol is an anonymous version of Twitter.

via quorums. To draw distinctions between various schemes, we described different approaches used in the literature for solving MPC. We described six open problems whose solutions would improve efficiency of scalable MPC schemes. Additionally, we described constructive techniques to improve efficiency of current quorum-based techniques. A drawback of most MPC results is the lack of empirical studies. We addressed this by implementing and benchmarking a number of recent methods as well as our new techniques.

Acknowledgments. The authors would like to acknowledge supports from NSF under grants CCF-1320994, CCR-0313160, and CAREER Award 644058. We are also grateful for valuable comments by Mahnush Movahedi from University of New Mexico, David Evans from University of Virginia, Elette Boyle from Cornell University, and Aniket Kate from Saarland University.

References

1. Asharov, G., Jain, A., López-Alt, A., Tromer, E., Vaikuntanathan, V., Wichs, D.: Multiparty computation with low communication, computation and interaction via threshold FHE. In: Pointcheval, D., Johansson, T. (eds.) EUROCRYPT 2012. LNCS, vol. 7237, pp. 483–501. Springer, Heidelberg (2012)
2. Ben-Or, M., Canetti, R., Goldreich, O.: Asynchronous secure computation. In: Proceedings of the Twenty-Fifth ACM Symposium on the Theory of Computing (STOC) (1993)
3. Boyle, E., Chung, K.-M., Pass, R.: Large-scale secure computation. Cryptology ePrint Archive, Report 2014/404 (2014)
4. Boyle, E., Chung, K.-M., Pass, R.: Oblivious parallel ram. Cryptology ePrint Archive, Report 2014/594 (2014)
5. Burkhart, M., Dimitropoulos, X.: Fast privacy-preserving top-k queries using secret sharing. In: Proceedings of 19th International Conference on Computer Communications and Networks (ICCCN), pp. 1–7 (2010)
6. Beaver, D.: Efficient multiparty protocols using circuit randomization. In: Feigenbaum, J. (ed.) CRYPTO 1991. LNCS, vol. 576, pp. 420–432. Springer, Heidelberg (1992)
7. Berman, R., Fiat, A., Ta-Shma, A.: Provable unlinkability against traffic analysis. In: Juels, A. (ed.) FC 2004. LNCS, vol. 3110, pp. 266–280. Springer, Heidelberg (2004)
8. Braud-Santoni, N., Guerraoui, R., Huc, F.: Fast Byzantine agreement. In: Proceedings of the 2013 ACM Symposium on Principles of Distributed Computing, PODC 2013, New York, NY, USA, pp. 57–64 (2013)
9. Boyle, E., Goldwasser, S., Tessaro, S.: Communication locality in secure multiparty computation. In: Sahai, A. (ed.) TCC 2013. LNCS, vol. 7785, pp. 356–376. Springer, Heidelberg (2013)
10. Brakerski, Z., Gentry, C., Vaikuntanathan, V.: Fully homomorphic encryption without bootstrapping. In: Proceedings of the 3rd Innovations in Theoretical Computer Science Conference, ITCS 2012, New York, USA, pp. 309–325 (2012)
11. Ben-Or, M., Goldwasser, S., Wigderson, A.: Completeness theorems for non-cryptographic fault-tolerant distributed computing. In: Proceedings of the Twentieth ACM Symposium on the Theory of Computing (STOC), pp. 1–10 (1988)

12. Beaver, D., Micali, S., Rogaway, P.: The round complexity of secure protocols. In: Proceedings of the Twenty-second Annual ACM Symposium on Theory of Computing, STOC 1990, New York, USA, pp. 503–513 (1990)
13. Chaum, C., Crépeau, C., Damgård, I.: Multiparty unconditionally secure protocols. In: Proceedings of the Twentieth Annual ACM Symposium on Theory of Computing (STOC), pp. 11–19 (1988)
14. Chandran, N., Chongchitmate, W., Garay, J.A., Goldwasser, S., Ostrovsky, R., Zikas, V.: Optimally resilient and adaptively secure multi-party computation with low communication locality. Cryptology ePrint Archive, Report 2014/615 (2014)
15. Chaum, D., Damgård, I., van de Graaf, J.: Multiparty computations ensuring privacy of each party's input and correctness of the result. In: Pomerance, C. (ed.) CRYPTO 1987. LNCS, vol. 293, pp. 87–119. Springer, Heidelberg (1988)
16. Cramer, R., Damgård, I., Nielsen, J.B.: Multiparty computation from threshold homomorphic encryption. In: Pfitzmann, B. (ed.) EUROCRYPT 2001. LNCS, vol. 2045, pp. 280–299. Springer, Heidelberg (2001)
17. Canetti, R., Friege, U., Goldreich, O., Naor, M.: Adaptively secure multi-party computation. Technical report, Cambridge, MA, USA (1996)
18. Chor, B., Goldwasser, S., Micali, S., Awerbuch, B.: Verifiable secret sharing and achieving simultaneity in the presence of faults. In: Proceedings of the 26th Annual Symposium on Foundations of Computer Science, SFCS 1985, pp. 383–395. IEEE Computer Society, Washington, DC (1985)
19. Cook, S.A., Reckhow, R.A.: Time-bounded random access machines. In: Proceedings of the Fourth Annual ACM Symposium on Theory of Computing, STOC 1972, New York, USA, pp. 73–80 (1972)
20. Du, W., Atallah, M.J.: Secure multi-party computation problems and their applications: A review and open problems. In: Proceedings of the 2001 Workshop on New Security Paradigms, NSPW 2001, pp. 13–22. ACM, New York (2001)
21. Damgård, I.B., Ishai, Y.: Scalable secure multiparty computation. In: Dwork, C. (ed.) CRYPTO 2006. LNCS, vol. 4117, pp. 501–520. Springer, Heidelberg (2006)
22. Damgård, I., Ishai, Y., Krøigaard, M., Nielsen, J.B., Smith, A.: Scalable multiparty computation with nearly optimal work and resilience. In: Wagner, D. (ed.) CRYPTO 2008. LNCS, vol. 5157, pp. 241–261. Springer, Heidelberg (2008)
23. Damgård, I., Ishai, Y., Krøigaard, M.: Perfectly secure multiparty computation and the computational overhead of cryptography. In: Gilbert, H. (ed.) EUROCRYPT 2010. LNCS, vol. 6110, pp. 445–465. Springer, Heidelberg (2010)
24. Dani, V., King, V., Movahedi, M., Saia, J.: Brief announcement: breaking the $o(nm)$ bit barrier, secure multiparty computation with a static adversary. In: Proceedings of the 2012 ACM Symposium on Principles of Distributed Computing, PODC 2012, pp. 227–228. ACM Press, New York (2012), Full version: http://arxiv.org/abs/1203.0289
25. Dani, V., King, V., Movahedi, M., Saia, J.: Quorums quicken queries: Efficient asynchronous secure multiparty computation. In: Chatterjee, M., Cao, J.-n., Kothapalli, K., Rajsbaum, S. (eds.) ICDCN 2014. LNCS, vol. 8314, pp. 242–256. Springer, Heidelberg (2014)
26. Damgård, I., Meldgaard, S., Nielsen, J.B.: Perfectly secure oblivious RAM without random oracles. In: Ishai, Y. (ed.) TCC 2011. LNCS, vol. 6597, pp. 144–163. Springer, Heidelberg (2011)
27. Damgård, I., Nielsen, J.B.: Scalable and unconditionally secure multiparty computation. In: Menezes, A. (ed.) CRYPTO 2007. LNCS, vol. 4622, pp. 572–590. Springer, Heidelberg (2007)

28. Damgård, I., Pastro, V., Smart, N.P., Zakarias, S.: Multiparty computation from somewhat homomorphic encryption. In: Safavi-Naini, R., Canetti, R. (eds.) CRYPTO 2012. LNCS, vol. 7417, pp. 643–662. Springer, Heidelberg (2012)
29. Feige, U.: Noncryptographic selection protocols. In: FOCS, pp. 142–153 (1999)
30. Gentry, C.: Fully homomorphic encryption using ideal lattices. In: Proceedings of the 41st Annual ACM Symposium on Theory of Computing, STOC 2009, New York, USA, pp. 169–178 (2009)
31. Gentry, C., Goldman, K.A., Halevi, S., Julta, C., Raykova, M., Wichs, D.: Optimizing ORAM and Using It Efficiently for Secure Computation. In: De Cristofaro, E., Wright, M. (eds.) PETS 2013. LNCS, vol. 7981, pp. 1–18. Springer, Heidelberg (2013)
32. Gentry, C., Halevi, S., Smart, N.P.: Fully homomorphic encryption with polylog overhead. In: Pointcheval, D., Johansson, T. (eds.) EUROCRYPT 2012. LNCS, vol. 7237, pp. 465–482. Springer, Heidelberg (2012)
33. Galil, Z., Haber, S., Yung, M.: Cryptographic computation: Secure fault tolerant protocols and the public-key model. In: Pomerance, C. (ed.) CRYPTO 1987. LNCS, vol. 293, pp. 135–155. Springer, Heidelberg (1988)
34. Gordon, S.D., Katz, J., Kolesnikov, V., Krell, F., Malkin, T., Raykova, M., Vahlis, Y.: Secure two-party computation in sublinear (amortized) time. In: Proceedings of the 2012 ACM Conference on Computer and Communications Security, CCS 2012, New York, USA, pp. 513–524 (2012)
35. Goldwasser, S., Kalai, Y.T., Popa, R.A., Vaikuntanathan, V., Zeldovich, N.: How to run turing machines on encrypted data. In: Canetti, R., Garay, J.A. (eds.) CRYPTO 2013, Part II. LNCS, vol. 8043, pp. 536–553. Springer, Heidelberg (2013)
36. Goldwasser, S., Lindell, Y.: Secure computation without agreement. In: Malkhi, D. (ed.) DISC 2002. LNCS, vol. 2508, pp. 17–32. Springer, Heidelberg (2002)
37. Goldreich, O., Micali, S., Wigderson, A.: How to play any mental game. In: Proceedings of the Nineteenth Annual ACM Symposium on Theory of Computing, STOC 1987, New York, USA, pp. 218–229 (1987)
38. Goldreich, O., Ostrovsky, R.: Software protection and simulation on oblivious RAMs. J. ACM 43(3), 431–473 (1996)
39. Gennaro, R., Rabin, M.O., Rabin, T.: Simplified VSS and fast-track multiparty computations with applications to threshold cryptography. In: Proceedings of the 17th Annual ACM Symposium on Principles of Distributed Computing, PODC 1998, pp. 101–111. ACM (1998)
40. Huang, Y., Evans, D., Katz, J., Malka, L.: Faster secure two-party computation using garbled circuits. In: Proceedings of the 20th USENIX Conference on Security, SEC 2011, pp. 35–35. USENIX Association, Berkeley (2011)
41. Herzberg, A., Jarecki, S., Krawczyk, H., Yung, M.: Proactive secret sharing or: How to cope with perpetual leakage. In: Coppersmith, D. (ed.) CRYPTO 1995. LNCS, vol. 963, pp. 339–352. Springer, Heidelberg (1995)
42. Katz, J., Koo, C.-Y., Kumaresan, R.: Improving the round complexity of VSS in point-to-point networks. In: Aceto, L., Damgård, I., Goldberg, L.A., Halldórsson, M.M., Ingólfsdóttir, A., Walukiewicz, I. (eds.) ICALP 2008, Part II. LNCS, vol. 5126, pp. 499–510. Springer, Heidelberg (2008)
43. King, V., Lonargan, S., Saia, J., Trehan, A.: Load balanced scalable byzantine agreement through quorum building, with full information. In: Aguilera, M.K., Yu, H., Vaidya, N.H., Srinivasan, V., Choudhury, R.R. (eds.) ICDCN 2011. LNCS, vol. 6522, pp. 203–214. Springer, Heidelberg (2011)

44. Kamara, S., Mohassel, P., Raykova, M.: Outsourcing multi-party computation. Cryptology ePrint Archive, Report 2011/272 (2011)
45. King, V., Saia, J., Sanwalani, V., Vee, E.: Towards secure and scalable computation in peer-to-peer networks. In: Proceedings of the 47th Annual IEEE Symposium on Foundations of Computer Science, FOCS 2006, pp. 87–98. IEEE Computer Society, Washington, DC (2006)
46. Kate, A., Zaverucha, G.M., Goldberg, I.: Constant-size commitments to polynomials and their applications. In: Abe, M. (ed.) ASIACRYPT 2010. LNCS, vol. 6477, pp. 177–194. Springer, Heidelberg (2010)
47. Lu, S., Ostrovsky, R.: Distributed oblivious RAM for secure two-party computation. In: Sahai, A. (ed.) TCC 2013. LNCS, vol. 7785, pp. 377–396. Springer, Heidelberg (2013)
48. Lindell, Y., Pinkas, B.: An efficient protocol for secure two-party computation in the presence of malicious adversaries. In: Naor, M. (ed.) EUROCRYPT 2007. LNCS, vol. 4515, pp. 52–78. Springer, Heidelberg (2007)
49. Lindell, Y., Pinkas, B.: A proof of security of yaos protocol for two-party computation. Journal of Cryptology 22(2), 161–188 (2009)
50. Lindell, Y., Pinkas, B.: Secure two-party computation via cut-and-choose oblivious transfer. In: Ishai, Y. (ed.) TCC 2011. LNCS, vol. 6597, pp. 329–346. Springer, Heidelberg (2011)
51. Movahedi, M., Saia, J., Zamani, M.: Secure anonymous broadcast. ArXiv e-prints (2014)
52. Pinkas, B., Reinman, T.: Oblivious RAM revisited. In: Rabin, T. (ed.) CRYPTO 2010. LNCS, vol. 6223, pp. 502–519. Springer, Heidelberg (2010)
53. Pease, M., Shostak, R., Lamport, L.: Reaching agreements in the presence of faults. Journal of the ACM 27(2), 228–234 (1980)
54. Rackoff, C., Simon, D.R.: Cryptographic defense against traffic analysis. In: Proceedings of the Twenty-fifth Annual ACM Symposium on Theory of Computing, STOC 1993, pp. 672–681 (1993)
55. Shi, E., Chan, T.-H.H., Stefanov, E., Li, M.: Oblivious RAM with $o((\log n)^3)$ worst-case cost. In: Lee, D.H., Wang, X. (eds.) ASIACRYPT 2011. LNCS, vol. 7073, pp. 197–214. Springer, Heidelberg (2011)
56. Shamir, A.: How to share a secret. Commun. ACM 22(11), 612–613 (1979)
57. Stefanov, E., van Dijk, M., Shi, E., Fletcher, C., Ren, L., Yu, X., Devadas, S.: Path ORAM: An extremely simple oblivious RAM protocol. In: Proceedings of the 2013 ACM SIGSAC Conference on Computer and Communications Security, CCS 2013, New York, USA, pp. 299–310 (2013)
58. van Dijk, M., Gentry, C., Halevi, S., Vaikuntanathan, V.: Fully homomorphic encryption over the integers. In: Gilbert, H. (ed.) EUROCRYPT 2010. LNCS, vol. 6110, pp. 24–43. Springer, Heidelberg (2010)
59. Yao, A.C.: Protocols for secure computations. In: Proceedings of the 23rd Annual Symposium on Foundations of Computer Science, SFCS 1982, Washington, DC, USA, pp. 160–164 (1982)
60. Zhang, B.: Generic constant-round oblivious sorting algorithm for MPC. In: Boyen, X., Chen, X. (eds.) ProvSec 2011. LNCS, vol. 6980, pp. 240–256. Springer, Heidelberg (2011)
61. Zamani, M., Movahedi, M.: Secure location sharing. In: Proceedings of the 10th ACM International Workshop on Foundations of Mobile Computing, FOMC 2014, New York, NY, USA, pp. 1–10 (2014)
62. Zamani, M., Movahedi, M., Saia, J.: Millions of millionaires: Multiparty computation in large networks. Cryptology ePrint Archive, Report 2014/149 (2014)

Online Bipartite Matching in Offline Time⋆
(Abstract)

Piotr Sankowski

University of Warsaw, Warszawa, Poland
`sank@mimuw.edu.pl`

Abstract. I will present our (with Bartłomiej Bosek, Dariusz Leniowski and Anna Zych) recent results on the problem of maintaining maximum size matchings in incremental bipartite graphs (FOCS'14). In this problem a bipartite graph G between n clients and n servers is revealed online. The clients arrive in an arbitrary order and request to be matched to a subset of servers. In our model we allow the clients to switch between servers and want to maximize the matching size between them, i.e., after a client arrives we find an augmenting path from a client to a free server. Our goals in this model are twofold. First, we want to minimize the number of times clients are reallocated between the servers. Second, we want to give fast algorithms that recompute such reallocation.

As for the number of changes, we propose a greedy algorithm that chooses an augmenting path π that minimizes the maximum number of times each server in π was used by augmenting paths so far. We show that in this algorithm each server has its client reassigned $O(\sqrt{n})$ times. This gives an $O(n^{3/2})$ bound on the total number of changes, what gives a progres towards the main open question risen by Chaudhuri *et al.* (INFOCOM'09) who asked to prove $O(n \log n)$ upper bound. Next, we argue that the same bound holds in the decremental case. Moreover, we show incremental and decremental algorithms that maintain $(1 - \varepsilon)$-approximate matching with total of $O(\varepsilon^{-1}n)$ reallocations, for any $\varepsilon > 0$.

Finally, we address the question of how to efficiently compute paths given by this greedy algorithm. We show that by introducing proper amortization we can obtain an incremental algorithm that maintains the maximum size matching in total $O(\sqrt{n}m)$ time. This matches the running time of one of the fastest static maximum matching algorithms that was given by Hopcroft and Karp (SIAM J. Comput '73). We extend our result to decremental case where we give the same total bound on the running time. Additionally, we show $O(\varepsilon^{-1}m)$ time incremental and decremental algorithms that maintain $(1 - \varepsilon)$-approximate matching for any $\varepsilon > 0$. Observe that this bound matches the running time of the fastest approximate static solution as well.

⋆ This work was partially supported by ERC StG project PAAl 259515, FET IP project MULTIPEX 317532, NCN grant 2013/11/D/ST6/03100 and N206 567940.

G.F. Italiano et al. (Eds.): SOFSEM 2015, LNCS 8939, p. 45, 2015.
© Springer-Verlag Berlin Heidelberg 2015

Quo Vadis Explicit-State Model Checking

Jiří Barnat

Faculty of Informatics Masaryk University, Brno, Czech Republic
`barnat@fi.muni.cz`

Abstract. Model checking has always been the flag ship in the fleet of automated formal verification techniques. It has been in the center of interest of formal verification research community for more than 25 years. Focusing primarily on the well-known state space explosion problem, a decent amount of techniques and methods have been discovered and applied to push further the frontier of systems verifiable with a model checker. Still, the technique as such has not yet been matured enough to become a common part of a software development process, and its penetration into the software industry is actually much slower than it was expected. In this paper we take a closer look at the so called *explicit-state model checking*, we briefly recapitulate recent research achievements in the field, and report on practical experience obtained from using our explicit state model checker DIVINE. Our goal is to help the reader understand what is the current position of explicit-state model checking in general practice and what are the strengths and weaknesses of the explicit-state approach after almost three decades of research. Finally, we suggest some research directions to pursue that could shed some light on the future of this formal verification technique.

1 Introduction

Methods for ensuring quality of various software and hardware products are inseparable part of the development process. For both software and hardware developers it is often the case that the only technique used to detect system flaws is testing, and, considering the *cost of poor quality* for a given product, it is quite often a valid and reasonable choice. However, for those cases where the consequence of a possible design error or an implementation bug is too high, the standard testing approach is insufficient, either due to the inherent principal incompleteness of error detection, or because the amount of tests to be done to decrease the probability of an undiscovered error to an acceptable level is simply too large. In such the cases formal verification methods come in place.

Model checking [28] is a formal verification procedure that takes the model of a system under verification and a single piece of system specification as inputs. For these the procedure decides whether the system meets the given specification or not. In the negative case, i.e. when there is a behaviour of the system violating the spec, a witness of such the violation, the so called counterexample, is (optionally) returned.

G.F. Italiano et al. (Eds.): SOFSEM 2015, LNCS 8939, pp. 46–57, 2015.

The strong benefit of the model checking approach is that when a verification procedure successfully proceeds, it provides user of a model checker with the confidence of satisfaction of validity of the given spec for the system at the level that is equal to the confidence given by a mathematical proof. Moreover, the decision procedure is fully automated (made by a computer) once the inputs are put to a form suitable for the model checker in use.

These obvious benefits of model checking approach naturally do not come for free. The standard work-flow of model checking requires user to provide model checker with a formal description of both the system and specification to be processed. Unfortunately, the experience with model checking shows that describing the system to be verified in a form acceptable by a model checker de facto amounts to re-formulating the relevant parts of the system in the modelling language of the model checker. While perhaps not very difficult, this is a step that is hardly automated and thus requires non-trivial human effort. Similarly, formalising specification in the context of model checking requires system engineers to express individual system properties as temporal logic formulae. Depending on the temporal logic used, we speak of the technique as of LTL (Linear Temporal Logic) model checking, CTL (Computational Tree Logic) model checking, etc. However, once the inputs to the model checker are in proper form, the decision about validity of a single system property can be fully automated.

The principle behind the automated decision procedure is to let computer fully explore all internal configurations of the system under verification. The so called *reachable state space* is a set of system configurations that the system may evolve to from a given set of initial states. With proper analysis of reachable state space the model checker may either proof the absence of erroneous reachable configuration or proof the conformance of the system's behaviour with the specification given as a temporal logic formula.

Unfortunately, real-world systems have reachable state space as large as their full analysis is beyond the capabilities of contemporary computing platforms. As a result in many cases the verification by model checking ends up with a failure due to insufficient computing resources, memory in particular. The fact that the size of the state space tends to grow exponentially with the size of system description, let us say in some programming language, is generally referred to as the *state space explosion problem*. Actually there are two fundamental reasons for the exponential grow — processing of inputs and control-flow non-determinism.

A lot of attention has been paid to the development of approaches to fight the state space explosion problem [31] in the field of automated formal verification [48]. Many techniques, such as state compaction [35], compression [37], state space reduction [29,33,47], symbolic state space representation [23], etc., were introduced to reduce the memory requirements of a single model-checking task. With the invention of application of binary decision diagrams to model checking [46] the field of model checking has got split into symbolic and explicit-state (enumerative) branch. While CTL has become the native specification logic within the symbolic branch (namely due to the SMV model checker [27]), LTL remain closely tied with the explicit-state model checking, also due to the well known explicit-state model checker SPIN [37]. Nevertheless, excursions in both directions exist, see e.g. [22,30,53].

Henceforward, we primarily focus on the explicit-state branch and LTL model checking. Due to Vardi and Wolper [52], the problem of LTL model checking reduces to the problem of checking Büchi automata emptiness, which in turns amounts to the detection of an accepting cycle in a directed graph of the reachable state space producted with a monitor of an LTL property violation. Unfortunately, accepting cycle detection algorithms as used in explicit-state model checkers, such as SPIN [37], DIVINE [3], or LTSmin [18], has to construct and store the whole graph of the product. Hence, those model checkers suffer primarily from high memory demands caused by the state space explosion problem.

This paper touches three main directions taken recently with respect to the LTL explicit-state model checking. In Section 2, we briefly recapitulate research effort spent in fighting state space explosion by means of parallel and distributed memory processing. Even though the state space explosion is a serious problem, surprisingly, it is not always the primary one that prevents model checking from being used in practice. Another quite hampering factor in the model checking scheme is the need of formal modelling. To address this problem we discuss, in Section 3, a direct application of explicit-state model checking to an LLVM bitcode. LLVM bitcode is used as an internal compiler representation of a program, and hence, it is automatically obtained from a source code with a common compiler. In Section 4 we notice that explicit-state model checking is typically used as an instance of unit-testing, and we discuss an extension towards symbolic representations that would push explicit-state model checking back to formal verification. Finally, Section 5 offers a few final remarks and hopefully gives some clues for the future of explicit-state model checking.

2   Parallel Processing and State Space Explosion

There is no doubt that the range of verification problems that can be solved with logic model checking tools has increased significantly over the last few decades. Though surprisingly, this increase is not only based on newly discovered algorithmic advances, but is strongly attributed to the increase in the speed of a single processor core and available size of the main memory [39]. Realising that the efficiency of explicit-state model checking strongly depends on the speed of computing hardware used, and supported with the fact that the speed of a single CPU core is not going to scale in the future, no option was left out than to go for parallel processing.

The main obstacle for direct extension of existing sequential LTL model checkers towards parallel architectures lied in the fact that a time-optimal parallel and scalable algorithm for Büchi emptiness problem is unlikely to exist [49]. (This is still an open problem.) As a result the pioneering work in parallel and distributed-memory LTL model checking [8] employed parallel scalable, but time-unoptimal algorithms for accepting cycle detection. While in the sequential case, algorithms for accepting cycle detection, such as Nested DFS [32] or various versions of Tarjan's algorithm for SCC decomposition [51], relies on the depth-first-search post-order, distributed-memory algorithms are built on top of reachability procedures, value propagation or topological sort [9,21].

Distributed-memory processing cannot fight the state space explosion problem alone and must be combined with other techniques. One of the most successful technique to fight the state space explosion in explicit-state model checking is Partial Order Reduction [47]. As a matter of fact, new topological sort proviso had to be developed in order to maintain efficiency of partial order reduction in the distributed-memory setting [6]. Another important algorithmic improvement relates to classification of LTL formulae [25]. For some classes of LTL formulae the parallel algorithms could be significantly improved [2]. For weak LTL formulae, the OWCTY algorithm [24] even matches the optimal sequential algorithms in terms of complexity. However, this algorithm suffers from not being an on-the-fly algorithm. Since the on-the-fly verification is an important practical aspect, a modification of this algorithm that allows for on-the-fly verification in most verification instances has been also developed [5].

All the distributed-memory algorithms has been implemented as part of parallel and distributed-memory LTL model checker DIVINE [12]. However, the focus of DIVINE on non-DFS-based algorithms and distributed-memory processing, does not require DIVINE to be used in distributed-memory only. As a matter of fact, DIVINE runs smoothly also in a shared-memory setting [4].

Since the lack of ability of parallel processing would mean a significant drawback for other explicit-state model checkers considering the contemporary computing hardware, they have also undergone a parallel processing face-lift. Namely, the SPIN model checker has been adapted to parallel processing with the so called stack-slicing [38] and piggybacking [41] techniques. Though, the most innovative extension of SPIN with respect to parallel processing was the so called swarm verification [39,40] that took the step towards the map-reduce pattern in model checking. In particular, a single verification task is cloned as many times as is the number of available processors, and for each clone the order of exploration of the state space is altered. In such a swarm of parallel tasks, the probability of early detection of an accepting cycle is significantly increased.

A completely different approach was chosen for LTSmin model checker [43]. Authors of which has successfully adapted sequential Nested DFS algorithm to parallel shared-memory processing. The idea is to run Nested DFS algorithm freely in parallel and then detect and recover from situations that could violate the soundness of computation. Even though such an approach cannot scale in general, practical measurements showed a superior results on shared-memory architectures [34,42].

Yet another parallel computing platform has become popular recently – general purpose graphical processing units (GPGPUs). Though, this platform was never meant for acceleration of memory demanding computations, the raw computing power of it is rather attractive. A series of results regarding acceleration of LTL model checking has been published recently employing non-DFS-based algorithms for accepting cycle detection [1,10,11] and accelerated state space generation [54].

3   Model Checking without Modeling

Recent formal verification research activities put a strong emphasis on direct applicability of verification methods in practice. This is witnessed, e.g., by Software Verification Competition [17] – a mainstream activity in the program analysis community. The strong drive to make formal method applications approachable by the general software development and engineering community highlights the fact that the most important factor of using formal methods in practice is their ease of use. Hence, formal methods must be applied on artefacts that software engineers and developers naturally work with, i.e. at the level of source code.

Moreover, should the model checking method spread massively and a model checkers become regular utility for software developers, it has to implement a full programming language specification, so that the programs the software developers write and run are also valid inputs for the model checker. Programming languages are rather complex in their full specification, and still engineers in pursuit of more elegant and more maintainable code balance on the boundaries of what is allowed and what is not in a particular programming language. Therefore, introducing substantial constraints on a programming language in order to enable model checking typically results in complete elimination of the model checking process from the development cycle.

A new approach to verification of C/C++ programs without the explicit need of modeling has been presented in [7]. The suggested solution effectively chains our parallel and distributed-memory model checker DIVINE with CLang compiler using the LLVM [44] infrastructure, intermediate bitcode representation (LLVM IR) in particular. Even though, LLVM IR has not precise semantics, the fact is that real-world compilers achieve an enviable level of agreement in this respect, despite numerous optimisation passes they all implement.

Using the LLVM IR as input for model checking thus not only enables model checking without the tedious process of system modelling, but it also provides a stable modelling language with reasonable well defined semantics. Within such a setup model checkers, such as DIVINE, may offer full LTL model checking of virtually unmodified C/C++ source codes.

The only limitation regarding input to the model checker is the need for completeness of the C/C++ program description. As a matter of fact, the model checker cannot verify programs that do calls to external libraries for which it has no source code available. Similarly, any calls to the kernel of operating system, such as processing of input and output are beyond the scope of this approach, unless the external environment is somehow added to the program and simulated without actual performance of Input/Output instructions. In principle, such a usage scenario resembles the well known unit testing approach.

Note that DIVINE internally provides an implementation of majority of the POSIX thread APIs (pthread.h), which in turns enables verification of unmodified multithreaded programs. In particular, DIVINE explores all possible thread interleavings systematically at the level of individual bitcode instructions. This allows DIVINE, for example, to virtually prove an absence of deadlock or assertion violation in a given multithreaded piece of code, which is a feat that is impossible to achieve with the standard testing techniques.

The main disadvantage of modelling the systems to be verified at the level of LLVM bitcode is the very fine grained nature of this language is subject to the thread of massive state space explosion. However, the low-level nature of LLVM entails a granularity of operations much finer than what is necessary for verification. τ-reduction [7] proposes a collection of heuristics that lead to a coarser granularity without losing correctness of the overall verification. The basic idea behind τ-reduction is that effects of instructions are observable from other threads only when the thread accesses main memory via load or store instructions. Model checker can benefit from this fact by pretending that most instructions are invisible and executing more of them within one step.

Among the latest extension of our LLVM bitcode model checker is the internal support for the full C++11 exception handling mechanism and C++11 threading [50].

4   Symbolic Data in Explicit-State Model Checking

For programs that do not process input data, the entire state space to be explored is derived from the program source code itself. Let us refer to these programs as to closed systems. For closed systems, model checking equals to formal verification, as it can guarantee that no system execution violates the model-checked property. As mentioned above this is particularly interesting for multithreaded programs, since for those programs regular testing approach is insufficient to detect all concurrency related issues due to the non-deterministic nature of thread scheduling.

However, for programs that read input data (the so called open systems), explicit-state model checking approach is in trouble. Note that even for a simple program that reads only a single 32-bit integer value, the enumeration of all possible input values would result into an unmanageable state space explosion. Hence, the idea of closing an open system with an environment process that would feed the program with all possible inputs is, in the case of explicit-state model checking, out of use. Though, for open systems that are executed over some concrete input data, i.e. the usual way the systems are tested; application of explicit-state model checking may have some benefits. In those cases model checking can detect inappropriate behaviour of the system after a system call failure, errors in exception handling, and/or other errors related to the control-flow non-determinism in general.

Still open systems represent a verification challenge. The way explicit-state model checking can address this problem is the so called *control explicit, data symbolic* (CEDS) approach [13]. The idea of it is to let a model checker track explicitly only the control-flow part of a system configuration while data values are stored symbolically. In other words, for each control-flow point the model checker keeps a set of possible data values that are valid for the particular control-flow point, the so called *multi-state*. Such a set-based representation of data allow for efficient handling of both sources of non-determinism present in the state space graph.

Naturally, the way the set of values are represented may differ. Explicit sets, i.e. when set members are enumerated explicitly, are very fast for small ranges but as mentioned above fail to scale. On the other hand, symbolic sets, represented, e.g., by first-order formulae in the bit-vector theory scale well to arbitrary range of input variables, but their usage make the model checker dependent on the efficiency of satisfiability modulo theory solvers.

Moreover, the detection of accepting cycles as prescribed by the automata-based approach to LTL model checking requires deciding *equality* of two multi-states during the state space traversal. In the explicated state space, this operation is trivial to be implemented efficiently with a hash table. However, when the sets of values are represented with logic formulae the decision of equivalence of multi-states is quite troublesome as the formulae lack unique canonical form. Since two equal multi-states may have different memory representations, use of efficient hashing is prohibited.

For those symbolic representations that allow at least linear ordering of multi-states, a logarithmic search would be possible, however, when using bitvector formulae this is not the case. The only obviously available option is a linear search in which the potentially new multi-state associated with a given control-flow point is compared with every other multi-state generated so far and associated to the same control-flow point. Note that the complexity of equality operation for multi-states may be very high [14].

Unfortunately, the CEDS approach is not without limitations. For example, it is unclear how to deal with dynamic allocation of memory blocks of which size is prescribed with a symbolically treated input variable. Nevertheless, for programs avoiding these allocations, the CEDS approach provides complete and efficient automatic verification procedure. As a matter of fact, we have implemented the CEDS approach by integrating DIVINE model checker and Z3 SMT solver [36] and applied our new tool successfully to verify some multithreaded programs with input [14]. We were also able to apply LTL model checking for verification of open embedded systems, simulink diagrams in particular [15].

Regarding verification of LTL properties, other symbolic approaches exist, the standard symbolic model checking [26], interpolation [45], and IC3 approach [19,20] are the most relevant. According to our experimental comparison [16] there is no clear winner among these approaches in terms of the speed of verification and applicability. In other words, extending explicit-state model checking with symbolic data representation feature makes the technique a competitive approach in the formal verification field in general.

5   Conclusions and Future Directions

The cost of deployment of formal verification (integration of a formal verification method into a development cycle) and also quite often really questionable performance are the key factors that prevent formal verification methods such as LTL model checking from being massively used in practice. While ease of use and readiness for immediate applicability have often minimal value from the academic point of view, for industry, these are the most important factors considered in many situations. As a matter of fact, a service that would include the tedious step of manual modelling is to be refused immediately by many practitioners.

Another show-stopper for academic tools are restrictions put on inputs that are processed. A tool that cannot deal with dynamic memory allocation can never be expected to be useful in practical verification of software systems. Should a formal verification tool be considered for massive use, the methods the tool builds on must be as complete as to be able to process a full-scale programming language including exception handling mechanism, dynamic memory allocation, object-oriented principles, etc.

Though, formal verification tools that are limited to some degree may still be successfully employed in many specific situations. Both practitioners and academic should learn how to find and communicate these specific setups in order to avoid a failure deployment of a model checker due to exaggerated expectations of practitioners, as well as to avoid missed opportunities due to the lack of advertisement and reporting on successful applications of model checkers in practice.

As for the explicit-state model checking approach, we have identified some of the obstacles preventing both the ease of use and efficiency of explicit-state model checkers in this paper. We showed two directions to take that counteract these problems, the connection to LLVM intermediate representation and extending the model checker with symbolic representations of data. Still there is much to do in the future.

Less theoretical, but by no means less important for verification of larger programs, is input preprocessing. The verification effort must start with pruning away those parts of the input programs that cannot influence the decision about the correctness. Especially given the low-level nature of LLVM, clever heuristics for detecting irrelevant code could lead to considerably smaller control-flow graphs. Methods such as slicing or automated abstractions will become a common part of the model checking work-flow to alleviate the burden of state space explosion as much as possible.

The technology evolution also must not be neglected. Should explicit-state model checking have some future, we predict that it must be able to fully utilise the power of future computing platforms, such as network clusters and clouds. History showed that the raw computing power cannot be underestimated, therefore we predict that new methods and techniques to allow trading of space requirements for computation time will be needed.

Finally, it is clear that there is no winning approach in the field of formal verification. An integration of techniques such as explicit-state model checking, symbolic execution and abstract interpretation is the next logical step towards the formal verification approach of the future. However, a key factor of success of such the combination is to preserve the general ability to process inputs in some form of full-scale programming language. As for approaches presented in this paper, the full combination of LLVM model checking and CEDS approach is yet to be seen.

References

1. Barnat, J., Bauch, P., Brim, L., Češka, M.: Employing Multiple CUDA Devices to Accelerate LTL Model Checking. In: 16th International Conference on Parallel and Distributed Systems (ICPADS 2010), pp. 259–266. IEEE Computer Society (2010)
2. Barnat, J., Brim, L., Černá, I.: Property driven distribution of Nested DFS. In: Proc. Workshop on Verification and Computational Logic, number DSSE-TR-2002-5 in DSSE Technical Report, pp. 1–10. University of Southampton, UK (2002)
3. Barnat, J., Brim, L., Havel, V., Havlíček, J., Kriho, J., Lenčo, M., Ročkai, P., Štill, V., Weiser, J.: DiVinE 3.0 – An Explicit-State Model Checker for Multi-threaded C & C++ Programs. In: Sharygina, N., Veith, H. (eds.) CAV 2013. LNCS, vol. 8044, pp. 863–868. Springer, Heidelberg (2013)
4. Barnat, J., Brim, L., Ročkai, P.: Scalable shared memory LTL model checking. International Journal on Software Tools for Technology Transfer (STTT) 12(2), 139–153 (2010)
5. Barnat, J., Brim, L., Ročkai, P.: A Time-Optimal On-the-Fly Parallel Algorithm for Model Checking of Weak LTL Properties. In: Breitman, K., Cavalcanti, A. (eds.) ICFEM 2009. LNCS, vol. 5885, pp. 407–425. Springer, Heidelberg (2009)
6. Barnat, J., Brim, L., Ročkai, P.: Parallel Partial Order Reduction with Topological Sort Proviso. In: Software Engineering and Formal Methods (SEFM 2010), pp. 222–231. IEEE Computer Society Press (2010)
7. Barnat, J., Brim, L., Ročkai, P.: Towards LTL Model Checking of Unmodified Thread-Based C & C++ Programs. In: Goodloe, A.E., Person, S. (eds.) NFM 2012. LNCS, vol. 7226, pp. 252–266. Springer, Heidelberg (2012)
8. Barnat, J., Brim, L., Stříbrná, J.: Distributed LTL model-checking in SPIN. In: Dwyer, M.B. (ed.) SPIN 2001. LNCS, vol. 2057, pp. 200–216. Springer, Heidelberg (2001)
9. Barnat, J., Brim, L., Černá, I.: Cluster-Based LTL Model Checking of Large Systems. In: de Boer, F.S., Bonsangue, M.M., Graf, S., de Roever, W.-P. (eds.) FMCO 2005. LNCS, vol. 4111, pp. 259–279. Springer, Heidelberg (2006)
10. Barnat, J., Brim, L., Češka, M.: DiVinE-CUDA: A Tool for GPU Accelerated LTL Model Checking. Electronic Proceedings in Theoretical Computer Science (PDMC 2009) 14, 107–111 (2009)
11. Barnat, J., Brim, L., Češka, M., Lamr, T.: CUDA accelerated LTL Model Checking. In: 15th International Conference on Parallel and Distributed Systems (ICPADS 2009), pp. 34–41. IEEE Computer Society (2009)
12. Barnat, J., Brim, L., Češka, R.P.: DiVinE: Parallel Distributed Model Checker (Tool paper). In: Parallel and Distributed Methods in Verification and High Performance Computational Systems Biology (HiBi/PDMC 2010), pp. 4–7. IEEE (2010)

13. Barnat, J., Bauch, P.: Control Explicit—Data Symbolic Model Checking: An Introduction. CoRR, abs/1303.7379 (2013)
14. Barnat, J., Bauch, P., Havel, V.: Model Checking Parallel Programs with Inputs. In: 22nd Euromicro International Conference on Parallel, Distributed, and Network-Based Processing (PDP), pp. 756–759. IEEE (2014)
15. Barnat, J., Bauch, P., Havel, V.: Temporal Verification of Simulink Diagrams. In: Proceedings of 15th IEEE International Symposium on High Assurance Systems Engineering (HASE), pp. 81–88 (2014)
16. Bauch, P., Havel, V., Barnat, J.: LTL Model Checking of LLVM Bitcode with Symbolic Data. To appear in Proceedings of MEMICS 2014. LNCS, p. 12. Springer (2014)
17. Beyer, D.: Status Report on Software Verification - (Competition Summary SV-COMP 2014). In: Ábrahám, E., Havelund, K. (eds.) TACAS 2014 (ETAPS). LNCS, vol. 8413, pp. 373–388. Springer, Heidelberg (2014)
18. Blom, S., van de Pol, J., Weber, M.: LTSMIN: Distributed and Symbolic Reachability. In: Touili, T., Cook, B., Jackson, P. (eds.) CAV 2010. LNCS, vol. 6174, pp. 354–359. Springer, Heidelberg (2010)
19. Bradley, A.R.: SAT-based model checking without unrolling. In: Jhala, R., Schmidt, D. (eds.) VMCAI 2011. LNCS, vol. 6538, pp. 70–87. Springer, Heidelberg (2011)
20. Bradley, A., Somenzi, F., Hassan, Z., Yan, Z.: An Incremental Approach to Model Checking Progress Properties. In: Proc. of FMCAD, pp. 144–153 (2011)
21. Brim, L., Barnat, J.: Platform Dependent Verification: On Engineering Verification Tools for 21st Century. In: Parallel and Distributed Methods in verifiCation (PDMC). EPTCS, vol. 72, pp. 1–12 (2011)
22. Brim, L., Yorav, K., Žídková, J.: Assumption-based distribution of CTL model checking. STTT 7(1), 61–73 (2005)
23. Burch, J.R., Clarke, E.M., McMillan, K.L., Dill, D.L., Hwang, L.J.: Symbolic model checking: 10^{20} states and beyond. Information and Computation 98(2), 142–170 (1992)
24. Černá, I., Pelánek, R.: Distributed explicit fair cycle detection (Set based approach). In: Ball, T., Rajamani, S.K. (eds.) SPIN 2003. LNCS, vol. 2648, pp. 49–73. Springer, Heidelberg (2003)
25. Černá, I., Pelánek, R.: Relating hierarchy of temporal properties to model checking. In: Rovan, B., Vojtáš, P. (eds.) MFCS 2003. LNCS, vol. 2747, pp. 318–327. Springer, Heidelberg (2003)
26. Cimatti, A., Clarke, E., Giunchiglia, E., Giunchiglia, F., Pistore, M., Roveri, M., Sebastiani, R., Tacchella, A.: NuSMV 2: An OpenSource Tool for Symbolic Model Checking. In: Brinksma, E., Larsen, K.G. (eds.) CAV 2002. LNCS, vol. 2404, pp. 241–268. Springer, Heidelberg (2002)
27. Cimatti, A., Clarke, E., Giunchiglia, F., Roveri, M.: NuSMV: a new Symbolic Model Verifier. In: Halbwachs, N., Peled, D. (eds.) CAV 1999. LNCS, vol. 1633, pp. 495–499. Springer, Heidelberg (1999)
28. Clarke, E., Grumberg, O., Peled, D.: Model Checking. MIT press (1999)
29. Clarke, E.M., Enders, R., Filkorn, T., Jha, S.: Exploiting symmetry in temporal logic model checking. Form. Methods Syst. Des. 9(1-2), 77–104 (1996)
30. Clarke, E.M., Grumberg, O., Hamaguchi, K.: Another Look at LTL Model Checking. In: Dill, D.L. (ed.) CAV 1994. LNCS, vol. 818, pp. 415–427. Springer, Heidelberg (1994)

31. Clarke, E.M., Grumberg, O., Jha, S., Lu, Y., Veith, H.: Progress on the State Explosion Problem in Model Checking. In: Wilhelm, R. (ed.) Informatics: 10 Years Back, 10 Years Ahead. LNCS, vol. 2000, pp. 176–194. Springer, Heidelberg (2001)
32. Courcoubetis, C., Vardi, M., Wolper, P., Yannakakis, M.: Memory-Efficient Algorithms for the Verification of Temporal Properties. Formal Methods in System Design 1, 275–288 (1992)
33. Emerson, E.A., Sistla, A.P.: Symmetry and model checking. Form. Methods Syst. Des. 9(1-2), 105–131 (1996)
34. Evangelista, S., Laarman, A., Petrucci, L., van de Pol, J.: Improved Multi-Core Nested Depth-First Search. In: Chakraborty, S., Mukund, M. (eds.) ATVA 2012. LNCS, vol. 7561, pp. 269–283. Springer, Heidelberg (2012)
35. Geldenhuys, J., de Villiers, P.J.A.: Runtime efficient state compaction in SPIN. In: Dams, D., Gerth, R., Leue, S., Massink, M. (eds.) SPIN 1999. LNCS, vol. 1680, pp. 12–21. Springer, Heidelberg (1999)
36. Havel, V.: Generic Platform for Explicit-Symbolic Verification. Master's thesis, Faculty of Informatics, Masaryk University, Czech Republic (2014)
37. Holzmann, G.J.: The Spin Model Checker: Primer and Reference Manual. Addison-Wesley (2004)
38. Holzmann, G.J.: A Stack-Slicing Algorithm for Multi-Core Model Checking. ENTCS 198(1), 3–16 (2008)
39. Holzmann, G.J., Joshi, R., Groce, A.: Swarm Verification. In: Automated Software Engineering (ASE 2008), pp. 1–6. IEEE (2008)
40. Holzmann, G.J., Joshi, R., Groce, A.: Swarm Verification Techniques. IEEE Transactions on Software Engineering 37(6), 845–857 (2011)
41. Holzmann, G.J.: Parallelizing the Spin Model Checker. In: Donaldson, A., Parker, D. (eds.) SPIN 2012. LNCS, vol. 7385, pp. 155–171. Springer, Heidelberg (2012)
42. Laarman, A., Langerak, R., van de Pol, J., Weber, M., Wijs, A.: Multi-core Nested Depth-First Search. In: Bultan, T., Hsiung, P.-A. (eds.) ATVA 2011. LNCS, vol. 6996, pp. 321–335. Springer, Heidelberg (2011)
43. Laarman, A., van de Pol, J., Weber, M.: Boosting Multi-Core Reachability Performance with Shared Hash Tables. In: Formal Methods in Computer-Aided Design (FMCAD 2010), pp. 247–255. IEEE (2010)
44. Lattner, C., Adve, V.: LLVM: A Compilation Framework for Lifelong Program Analysis & Transformation. In: International Symposium on Code Generation and Optimization (CGO), Palo Alto, California (2004)
45. McMillan, K.L.: Interpolation and SAT-Based Model Checking. In: Hunt Jr., W.A., Somenzi, F. (eds.) CAV 2003. LNCS, vol. 2725, pp. 1–13. Springer, Heidelberg (2003)
46. McMillan, K.L.: Symbolic model checking. Kluwer (1993)
47. Peled, D.: Ten years of partial order reduction. In: Vardi, M.Y. (ed.) CAV 1998. LNCS, vol. 1427, pp. 17–28. Springer, Heidelberg (1998)
48. Pelánek, R.: Fighting state space explosion: Review and evaluation. In: Cofer, D., Fantechi, A. (eds.) FMICS 2008. LNCS, vol. 5596, pp. 37–52. Springer, Heidelberg (2009)
49. Reif, J.H.: Depth-first search is inherrently sequential. Information Processing Letters 20(5), 229–234 (1985)
50. Ročkai, P., Barnat, J., Brim, L.: Model Checking C++ with Exceptions. In: Electronic Communications of the EASST, Proceedings of 14th International Workshop on Automated Verification of Critical Systems (to appear, 2014)

51. Tarjan, R.: Depth first search and linear graph algorithms. SIAM Journal on Computing, 146–160 (1972)
52. Vardi, M.Y., Wolper, P.: An Automata-Theoretic Approach to Automatic Program Verification. In: IEEE Symposium on Logic in Computer Science, pp. 322–331. Computer Society Press (1986)
53. Visser, W., Barringer, H.: Practical CTL* Model Checking: Should SPIN be Extended? STTT 2(4), 350–365 (2000)
54. Wijs, A., Bošnački, D.: GPUexplore: Many-Core On-the-Fly State Space Exploration Using GPUs. In: Ábrahám, E., Havelund, K. (eds.) TACAS 2014. LNCS, vol. 8413, pp. 233–247. Springer, Heidelberg (2014)

The Dos and Dont's of Crowdsourcing Software Development

Brian Fitzgerald and Klaas-Jan Stol

Lero—The Irish Software Engineering Research Centre
University of Limerick, Ireland
{bf,klaas-jan.stol}@lero.ie

1   Introduction

In 1957, the eminent computer scientist, Edsger W. Dijkstra, sought to record his profession as "Computer Programmer" on his marriage certificate. The Dutch authorities, although probably more progressive than most, refused on the grounds that there was no such profession. Ironically, just a decade later, the term "software crisis" had been coined, as delegates at a NATO Conference in Garmisch [1] reported a common set of problems, namely that software took too long to develop, cost too much to develop, and the software which was eventually delivered did not meet user expectations. Despite the advances in technology over the past 50 years, this remains problematic, as evidenced by the following quote from the US President's Council of Advisors on Science & Technology (PCAST) in 2012.

> "The problem of predictable development of software with the intended functionality that is reliable, secure and efficient remains one of the most important problems in [ICT]"

A number of initiatives have emerged over the years to address the software crisis. Outsourcing of the software development activity has been on the increase in recent years according to US[1] and European[2] reports. However, in many cases outsourcing of software development has not been successful [2,3]. The success of the open source movement which has proven surprisingly successful at developing high quality software in a cost effective manner [4] has been an inspiration for a number of specific forms of software outsourcing, including opensourcing [5], innersourcing [6] and crowdsourcing [7].

2   Open-Source-Inspired Outsourcing

The conventional wisdom of software engineering suggests that given the inherent complexity of software, it should be developed using tightly co-ordinated, centralized teams, following a rigorous development process. In recent times, the Open Source Software

---

[1] IT Outsourcing Statistics: 2012/2013.
[2] European IT Outsourcing Intelligence Report.

G.F. Italiano et al. (Eds.): SOFSEM 2015, LNCS 8939, pp. 58–64, 2015.

(OSS) phenomenon has attracted considerable attention as a seemingly agile, practice-led initiative that appears to address these three aspects of the so-called "software crisis": cost, time-scale and quality. In terms of costs, OSS products are usually freely available for public download. From the point of view of development speed, the collaborative, parallel efforts of globally-distributed co-developers has allowed many OSS products to be developed much more quickly than conventional software. In terms of quality, many OSS products are recognized for their high standards of reliability, efficiency and robustness, and the OSS phenomenon has produced several "category killers" (i.e., products that remove any incentive to develop any competing products) in their respective areas—Linux and Apache spring to mind. The OSS model also seems to harness the most scarce resource of all—talented software developers, many of whom exhibit a long-standing commitment to their chosen projects. It is further suggested that the resulting peer review model helps ensure the quality of the software produced.

This brief synopsis illustrates why the OSS topic would be of such interest to the software engineering community, and also provides a hint as to why it would have greater research appeal and interest, particularly in an outsourcing context where companies seek to take advantage of resources beyond co-located developers on a single site. As mentioned, the OSS phenomenon has inspired other forms of outsourcing, of which crowdsourcing is one. This work will focus on crowdsourcing in software development.

3   Crowdsourcing Software Development

Software engineering no longer takes place in small, isolated groups of co-located developers, all working for the same employer, but increasingly takes place in a globalized context across organizations and communities involving many people. One emerging approach to getting work done is crowdsourcing, a sourcing strategy that has emerged since the nineties [8]. Driven by Web 2.0 technologies, organizations can tap into a workforce consisting of anyone with an Internet connection. Customers, or requesters, can advertise chunks of work, or tasks, on a crowdsourcing platform, where suppliers (i.e., individual workers) select those tasks that match their interests and abilities [9].

Crowdsourcing has been adopted in a wide variety of domains, such as design and sales of T-shirts [10] and pharmaceutical research and development [11] and there are numerous crowdsourcing platforms through which customers and suppliers can find each other [12]. One of the best known crowdsourcing platforms is Amazon Mechanical Turk (AMT) [13]. On AMT, chunks of work are referred to as Human Intelligence Tasks (HIT) or micro-tasks. Typical micro-tasks are characterized as self-contained, simple, repetitive, short, requiring little time, cognitive effort and specialized skills, and crowdsourcing has worked particularly well for such tasks [14]. Examples include tagging images, and translating fragments of text. Consequently, remuneration of work is typically in the order of a few cents to a few US dollars.

In contrast to micro-tasks, software development tasks are often interdependent, complex, heterogeneous, and can require significant periods of time, cognitive effort and various types of expertise. Yet, there are cases of crowdsourcing complex tasks; for instance, InnoCentive deal with problem solving and innovation projects, which may yield payments of thousands of US dollars [10]. A number of potential benefits may arise through the use of crowdsourcing in general, and these would also be applicable in the context of software development specifically:

- *Cost reduction* through lower development costs for developers in certain regions, and also through the avoidance of the extra cost overheads typically incurred in hiring developers;
- *Faster time-to-market* through accessing a critical mass of necessary technical talent who can achieve follow-the-sun development across time zones, as well as parallel development on decomposed tasks, and who are typically willing to work at weekends, for instance;
- *Higher quality* through broad participation: the ability to get access to a broad and deep pool of development talent who self-select on the basis that they have the necessary expertise, and who then participate in contests where the highest quality 'winning' solution is chosen;
- *Creativity and open innovation*: there are many examples of "wisdom of crowds" creativity whereby the variety of expertise available ensures that more creative solutions can be explored, which often elude the fixed mindset that can exist within individual companies, a phenomenon known as 'near-field repurposing of knowledge.'

Given that the first three benefits above (cost, time and quality) directly address the three central problematic areas of the so-called 'software crisis' [15], it is not surprising that a number of authors have argued that crowdsourcing may become a common approach to software development [16,17].

We conducted a case study of a major multinational company who commissioned a crowdsourcing software development initiative using the TopCoder platform [7]. Below we present a number of lessons learned in the form of Dos and Dont's.

4   The Dos in Crowdsourcing Software Development

4.1   Do Familiarize Yourself with the Crowdsourcing Process

The software development approach in crowdsourcing can be significantly different from that which organizations use for their internal development. For example, the crowdsourcing software development process at TopCoder is a waterfall process and it is not trivial to integrate this with the agile type approach which characterizes the majority of in-house development today. It is important to become familiar with the crowdsourcing process at the outset, so that architects, developers and project managers can prepare and discuss internally what needs to be done for a smooth interaction with the crowd.

There are several new roles which emerge when crowdsourcing software development. For example at TopCoder, the interaction with crowd contestants is mediated by co-pilots who are experienced members of the crowd community and platform specialists who interact with customer companies. Also, while the concept of first and second prizes is clear, concepts such as Reliability Bonus and Digital Run points are not so obvious but have significant financial implications for the customer. The level at which prize money should be pitched for competitions, and the preparing of specifications and reviewing of competitions is also something which needs to be understood in the crowdsourcing software development process.

The warranty periods for crowdsourced work can also be problematic. For example, TopCoder operate a five day warranty period after a competition winning entry has been selected, during which the customer has to accept or reject the submission. This requires discipline at the customer end to ensure that submissions can be internally reviewed. There is also a 30-day warranty period during which problems can be reported. However, it can be difficult to operate this longer warranty period usefully as some much additional interdependent development work would have been done in the intervening 30 days and this would make it difficult to roll back the crowdsourcing element.

4.2 Do Provide Clear Documentation to the Crowd

Documentation clearly plays an important role, as this is the key channel through which crowd developers will know what to develop. The documentation that specifies the context and the requirements for the software development task at hand must be easy to understand and provide sufficient information for crowd developers to do their task. Finding the right balance is important; giving either too little or too much information will result in a deliverable that is likely to be unacceptable. Overwhelming the crowd with information is likely to scare them off, resulting in few or even no submissions. Also, the crowd tend not to have a recurrent relationship with customers. Thus the kind of tacit organizational knowledge that one can take for granted for in-house development does not occur in crowdsourcing. Consequently, far more comprehensive and explicit documentation of requirements is necessary.

4.3 Do Assign a Special Person to Answer Questions from the Crowd

Interacting with the crowd can be a very time-consuming activity. In the case study we conducted, the single point of contact had to have both technical and project management skills, and consequently such a liaison ended up being a senior resource. However, a significant amount of time of this senior person was taken up by answering technical questions on the Q&A forum through which crowd developers asked for clarification. Therefore, a better approach would be to allocate a person who would be well informed about the technical intricacies of the project but who would not have a senior role, and hence be a cheaper resource. The nature of interaction which takes place episodically, perhaps once per day, through the rather narrow Q&A forum also requires quite a lot of discipline on the part of the person charged with that responsibility.

5 The Don'ts in Crowdsourcing Software Development

5.1 Don't Stay Anonymous

A crowdsourcing customer may be concerned about potential IP "leaking," and giving away the company's "Secret Sauce." As a result, a customer may choose to disguise their participation by staying anonymous, using a pseudonym in contest descriptions. However, a significant downside of that tactic is that such contests may attract very little interest and participation from the crowd. For crowd developers, it can be particularly interesting to work for blue chip companies as doing so allows them build their resumes. It is not uncommon for developers to use their TopCoder 'rating' on their resume as evidence of their technical skills and know-how. By staying anonymous, however, a customer may be much less appealing to work for. Also, the anonymity may offer an inadequate level of protection anyway in that the specifications for a competition may effectively reveal the company's identify anyway.

5.2   Don't Reject Submissions If They Can Easily be Fixed

Once a contest is over, the customer may have five days to accept the 'winning' submission. This means that there is only limited time for a customer to fully analyze and test the deliverable before an accept or reject decision must be made. If a customer decides the deliverable is not of the expected quality, it may be rejected. However, a possible negative side effect is that crowd developers may not participate in future contests of this customer, as doing so involves a risk of spending time and not getting paid for it. If a customer is not yet ready to handle the incoming deliverable, the customer can, of course, just accept the deliverable. After accepting, there is an additional warranty period of 30 days during which identified defects can be reported and fix without additional cost. However, taking this route can pose significant overhead in receiving, checking and integrating the fixed deliverable. Therefore, a customer is probably better off to fix minor defects internally rather than using the warranty period.

5.3   Don't Underestimate the Cost

The cost of crowdsourcing software should not be underestimated. Using the TopCoder platform, for example, the cost of a single contest can be much higher than merely the prize money for the First Prize. Assuming a first place prize of 1,000 USD, the prize money for the second place is 500 USD. Add to that a Reliability Bonus of 200 USD, a Digital Run contribution of 450 USD, a specification review of 50 USD, a review board of 800 USD, and co-pilot fees of 600 USD, a single contest would cost 3,600 USD.

5.4   Don't Expect Miracles

Finally, it is important to stress that crowdsourcing software development does not represent the much sought after 'silver bullet.' Expected benefits from crowdsourcing include high-quality and innovative solutions in a faster time-scale and low cost. Indeed, given TopCoder's workforce of around 700,000 developers, one would expect a significant number of participants for each contest, and consequently, high-quality of innovative deliverables. However, our findings suggest quite a different picture. For the 53 contests held by our case study company, there were a total of 720 registrants, and a total of only 84 submissions, less than two on average. Furthermore, there were more than 500 issues reported with these submissions. The case company was also quite disappointed by the level of innovation–rather than the expected high-quality HTML5 code (HTML5 has many novel features compared to HTML4), few HTML5 features were actually used due to the portability constraints set forth by the customer company.

6 Conclusion

Crowdsourcing software development is not as straightforward as crowdsourcing micro-tasks found on platforms such as Amazon Mechanical Turk. Given the complexity of software development, we should not be surprised that the difficulties in 'common' (in-house) software development settings are exacerbated when outsourced to a crowd. Yet, little is known about crowdsourcing software development, and our suggested dos and don'ts are based on a single case study. More research is necessary–to that end, we developed a research framework that identifies the key perspectives and concerns [18].

Acknowledgments. This work was supported by Science Foundation Ireland grant 10/CE/I1855 to Lero—The Irish Software Engineering Research Centre, Enterprise Ireland grant IR/2013/0021 to ITEA2-SCALARE, and the Irish Research Council.

References

1. Naur, P., Randell, B. (eds.): Report on a conference sponsored by the NATO SCIENCE COMMITTEE (1968)
2. Nakatsu, R., Iacovou, C.: A comparative study of important risk factors involved in offshore and domestic outsourcing of software development projects: A two-panel delphi study. Information & Management 46, 57–68 (2009)
3. Tiwana, A., Keil, M.: Control in internal and outsourced software projects. Journal of Management Information Systems 26, 9–44 (2009)
4. Feller, J., Fitzgerald, B., Hissam, S., Lakhani, K.: Perspectives on Free and Open Source Software. MIT Press (2005)
5. Ågerfalk, P.J., Fitzgerald, B.: Outsourcing to an unknown workforce: Exploring opensourcing as a global sourcing strategy. MIS Quarterly 32 (2008)
6. Stol, K., Avgeriou, P., Babar, M., Lucas, Y., Fitzgerald, B.: Key factors for adopting inner source. ACM Trans. Softw. Eng. Methodol., 23 (2014)
7. Stol, K., Fitzgerald, B.: Two's company, three's a crowd: A case study of crowdsourcing software development. In: Proc. 36th Int'l Conf. Software Engineering, pp. 187–198 (2014)
8. Greengard, S.: Following the crowd. Communications of the ACM 54, 20–22 (2011)
9. Hoffmann, L.: Crowd control. Communications of the ACM 52 (2009)
10. Howe, J.: Crowdsourcing: Why the Power of the Crowd is Driving the Future of Business. Crown Business (2008)
11. Lakhani, K.R., Panetta, J.: The principles of distributed innovation. Innovations: Technology, Governance, Globalization 2 (2007)
12. Doan, A., Ramakrishnan, R., Halevy, A.: Crowdsourcing systems on the world-wide web. Communications of the ACM 54 (2011)
13. Ipeirotis, P.: Analyzing the amazon mechanical turk marketplace. XRDS 17, 16–21 (2010)
14. Kittur, A., Smus, B., Khamkar, S., Kraut, R.: Crowdforge: Crowdsourcing complex work. In: Proceedings of the ACM Symposium on User Interface Software and Technology (2011)

15. Fitzgerald, B.: Open source software: Lessons from and for software engineering. IEEE Computer 44, 25–30 (2011)
16. Begel, A., Herbsleb, J.D., Storey, M.A.: The future of collaborative software development. In: Proceedings of the ACM Symposium on Computer-Supported Collaborative Work (2012)
17. Kazman, R., Chen, H.M.: The metropolis model: A new logic for development of crowd-sourced systems. Communications of the ACM 52 (2009)
18. Stol, K., Fitzgerald, B.: Researching crowdsourcing software development: perspectives and concerns. In: Proc. 1st Int'l Workshop on Crowdsourcing in Software Engineering (2014)

Adaptively Approximate Techniques in Distributed Architectures

Barbara Catania and Giovanna Guerrini

University of Genoa, Italy
{barbara.catania,giovanna.guerrini}@unige.it

Abstract. The wealth of information generated by users interacting with the network and its applications is often under-utilized due to complications in accessing heterogeneous and dynamic data and in retrieving relevant information from sources having possibly unknown formats and structures. Processing complex requests on such information sources is, thus, costly, though not guaranteeing user satisfaction. In such environments, requests are often relaxed and query processing is forced to be adaptive and approximate, either to cope with limited processing resources (QoS-oriented techniques), possibly at the price of sacrificing result quality, or to cope with limited data knowledge and data heterogeneity (QoD-oriented techniques), with the aim of improving the quality of results. While both kinds of approximation techniques have been proposed, most adaptive solutions are QoS-oriented. Additionally, techniques which apply a QoD-oriented approximation in a QoD-oriented adaptive way (called ASAP - Approximate Search with Adaptive Processing - techniques), though demonstrated potentially useful in getting the right compromise between precise and approximate computations, have been largely neglected. In this paper, we first motivate the problem and provide a taxonomy for classifying approximate and adaptive techniques according to the dimensions pointed out above. Then, we show, through some concrete examples, the benefits of using ASAP techniques in two different contexts.

Keywords: query processing, approximate technique, adaptive technique, Quality of Data, Quality of Service.

1 Introduction

The last decade has seen the raise of new applications and novel processing environments characterized by high heterogeneity, limited data knowledge, extremely high variability and unpredictability of data characteristics and dynamic processing conditions. All such characteristics are shared by most data management applications under distributed architectures, including data integration applications, web services, data streams, P2P systems, and hosting.

Query processing in such new application contexts is characterized by two main features: *adaptivity*, in order to adapt the processing to dynamic conditions that prevent the selection of a single optimal execution strategy, and *approximation*, in order to cope with data heterogeneity, limited data knowledge

G.F. Italiano et al. (Eds.): SOFSEM 2015, LNCS 8939, pp. 65–77, 2015.

during query specification, and limited resource availability, which make precise answers impossible to compute or unsatisfactory from a user point of view.

As discussed in [9], approximate and adaptive techniques can be classified into two main groups. When they are targeted to improve the quality of result, either in terms of completeness or in terms of accuracy, we refer to the techniques as *Quality of Data (QoD)-oriented* techniques. By contrast, when they are used in order to cope with limited or constrained resource availability during processing, we refer to the techniques as *Quality of Service (QoS)-oriented* techniques. For example, in order to maximize completeness or accuracy, QoD parameters have to be taken into account for adapting or approximating query specification and processing. Often, both QoD and QoS parameters are taken into account, in order to provide the best trade-off between resource usage and data quality.

While both QoS-oriented and QoD-oriented approximate techniques have been proposed, most adaptive solutions are QoS-oriented. QoS-oriented approximation is often applied in an adaptive way, that is, when targeted at achieving a QoS goal, e.g., related to load, throughput, or memory, approximation is applied adapting to runtime conditions, possibly ensuring that certain QoD constraints are met or, less frequently, with a QoD-oriented goal (that is, minimizing the introduced inaccuracy). By contrast, very few approaches apply QoD-oriented approximation techniques in an adaptive way.

In [8,9], we claimed instead that QoD-oriented adaptive approaches for QoD-oriented approximation techniques, called *ASAP (Approximate Search with Adaptive Processing) techniques* in the following, may help in getting the right compromise between precise and approximate computations. More generally, we claimed that ASAP can be defined as a new framework under which QoD-oriented approximation techniques, which may adaptively change, at run-time, the degree of the applied approximation, can be defined. For example, this can be achieved by providing execution plans which interleave both precise and approximate evaluations in the most efficient way, dynamically taking decisions concerning when, how, and how much to approximate, with the goal of improving the quality of result with efficiency guarantees. Unfortunately, as far as we know, no general solution has been proposed so far for the problem described above. Some preliminary work has been presented in [19], where the use of adaptive techniques for combining exact (fast) and approximate (accurate) joins when performing dynamic integration has been proposed.

Our group is currently interested in investigating ASAP approaches in different scenarios, taking care of problems related to data heterogeneity and limited data knowledge in different potentially distributed architectures. In this paper, after summarizing existing approximate and adaptive approaches with respect to the type of quality they are targeted to, we consider two different instantiations of the concept of ASAP technique. The first one, that we call *ASAP in the small*, concerns the definition of a specific ASAP technique for a given application context, namely advanced architectures with a limited degree of distribution, like data stream management systems. The second one, that we call *ASAP in the large*, concerns a vision related to environments characterized by a higher degree of distribution and data heterogeneity. The paper is then concluded by some final considerations and discussion about on-going work on this subject.

2   A Taxonomy for Approximate and Adaptive Techniques

A query processing technique is said to be *adaptive* if the way in which a query is executed may change on the basis of the feedbacks obtained from the environment during evaluation. Adaptive techniques can be characterized by the following components: (i) *subject*, i.e., the elements in the processing affected by the adaptation (e.g., the query execution plan, or the assignment of load to processors); (ii) *target*, i.e., what the technique attempts at adapting to, that is, the properties monitored and the feedbacks collected during evaluation (e.g., data characteristics, arrival rates, network condition, processors load); (iii) *goal* or *aim*, i.e., the parameter(s) appearing in the objective function, that is, what the technique attempts to maximize/minimize (e.g., result data quality, time, throughput, energy consumption).

A query is said to be *relaxed* if its result is either stretched or shrunk when the results of the original query are too few or too many. Preference-based queries, such as top-k or skyline queries, can be considered relaxed queries as well: they can be thought as a shrinking with respect to the overall set of possible results, since they reduce the cardinality of the ranked result set, or as a stretching approach with respect to the set of optimal results. A processing technique is said to be *approximate* if it does not produce the exact result but an approximate one, possibly with some guarantees on the "distance" of the generated solution from the exact one.

Approximate queries or techniques can be characterized by the following components: (i) *subject*, representing the query processing task or the data to which approximation is applied (e.g., query specification, through rewriting or preferences, or processing algorithm); (ii) *target*, representing the information used for the approximation (e.g., ranking function, set of relevant attributes, similarity function, pruning condition, used summary); (iii) *goal* or *aim*, i.e., the parameter(s) appearing in the object function of the technique, that is, what it attempts to maximize/minimize (e.g., result data quality, time, throughput).

As pointed out in the introduction, depending on their aim, approximate and adaptive techniques can be classified into *Quality of Data (QoD)-oriented* techniques, when they are finalized at improving the quality of result, either in terms of completeness or in terms of accuracy, and *Quality of Service (QoS)-oriented* techniques, when they are used in order to cope with limited or constrained resource availability during query processing.

While both QoS-oriented and QoD-oriented approximate techniques have been proposed, most adaptive solutions are QoS-oriented. In the following, each group of proposals is discussed, pointing out the main considered subjects.

QoD-Oriented Approximate Techniques. They provide approximate answers in situations where precise results are not always satisfactory for the user.

High data heterogeneity and limited user knowledge about such data, indeed, may cause precise evaluation produce empty/few answers or too many answers. A solution consists in modifying traditional queries, such as selections and joins, by relaxing their definition or by approximating their evaluation, in order to improve result quality, in terms of completeness and relevance with respect to the original query. Query rewritings like those presented in [6,22,29] and preference-based queries, such as top-k and skyline [7,16,25] are examples of QoD-oriented approximation techniques which relax the original query definition with the goal of returning a more satisfactory answer set. A third group of QoD-oriented approximate techniques concerns processing algorithms for executing a traditional query (e.g., a join) by using ad hoc query processing algorithms which automatically apply the minimum amount of relaxation based on the available data, in order to return a non-empty result similar to the user request. Most QoD-oriented ApQP techniques concern the join operator [18] and face approximate match issues for strings [14] or numeric values [26]. In defining QoD-oriented techniques, QoS guarantees have to be provided, in order to cope with the available resources in the most efficient way.

QoS-Oriented Approximate Techniques. They provide approximate answers, with accuracy guarantees, to computationally expensive operations also in environments characterized by limited or unavailable resources, where a precise result can be obtained only at the price of a unacceptably high response time, communication overhead, occupied space, or it cannot be obtained at all. QoS-oriented techniques have been mainly defined for queries to be executed over either huge amount of data (as in data warehousing systems and in stream-based management systems) or complex data (like spatial data) or because corresponding to very expensive computations (as multi-way joins). Concerning the subject, four main distinct aspects have been considered: query rewriting, e.g., those presented in the data stream context [23]; data reduction, where data themselves are approximated with the aim of reducing or simplifying the dataset over which queries have to be executed [13,27], including load shedding [4,28]; processing algorithms, which modify traditional and non approximate processing techniques in order to generate an approximate result in an efficient way, with respect to the available resources [1,3].

QoS-Oriented Adaptive Techniques. In adaptive query processing, the way in which a query is executed is changed on the basis of the feedbacks obtained from the environment during evaluation. The classical *plan-first execute-next* approach to query processing is replaced either by giving away the notion of query plan at all, as in routing based approaches, where each single data item can participate to the production of the final result taking its own way (*route*) through operators composing the query, or by a *dynamic optimization* process, in which queries are on-the-fly re-optimized through a two-steps solution. Some approaches (e.g., [4,5,24,28]) introduce approximation, thus they have an impact on QoD, and some others [15,20] process approximate operators (i.e., top-k), but the aim of the adaptation is QoS.

ASAP Techniques. Some of the techniques discussed above exhibit an adaptive behavior and, at the same time, introduce some approximation. Specifically, some approaches to QoS-oriented adaptive processing of QoD-oriented approximate queries have been proposed, but target and goal of the adaptation is QoS, namely processing efficiency [15, 20]. Additionally, adaptive approaches have been proposed for some QoS-oriented approximation techniques (namely, load shedding and data summarization) [4, 5, 24, 28] but only few of them take QoD-information into account as aim [4]. Constraints on data are exploited in [5] to improve QoS (specifically, to reduce memory overhead). However, a QoD-oriented adaptation target is rarely considered, with the exception of [19].

Thus, QoD-oriented adaptive approaches for QoD-oriented approximation techniques (i.e., ASAP techniques) have been so far neglected. However, we claim that such techniques could be very relevant for various data management applications in different application contexts. In the following sections, we will present two examples of ASAP techniques showing their potential.

3   ASAP in the Small

The first example of ASAP technique we present refers to a, potentially distributed, architecture for data stream management. In this context, similarly to [19], ASAP techniques may help in defining adaptive techniques which combine exact (fast) and approximate (accurate) relaxed queries over dynamic (stream) data. In the following, we point out which data and which requests we are going to consider; we then present the targeted problem and we introduce an ASAP technique as a possible solution to the identified problem.

Data. A data stream is a continuous, unbounded, and potentially infinite sequence of data, e.g., tuples. In a data stream, each item is associated with a timestamp, either assigned by the source dataset or by the Data Stream Management System (DSMS), at arrival time. Queries over data streams can be either one-time, if they are evaluated once on a given subset of data, or continuous, if they are continuously evaluated as soon as new data arrive.

Request. According to the STREAM DSMS [2], continuous queries can be evaluated over data streams and time-varying relations. Continuous queries are evaluated, according to the relational semantics, at each time instant on the relation states and on the subsets of the data streams available at that instant. Window operators are applied on data streams in order to compute, at each time instant, a subset of the data items arrived so far in the stream.

The basic idea of a skyline-based approach to query relaxation of selection and join operations is to use a relaxing distance function d (usually, a numeric function) to quantify the distance of each tuple (pair of tuples) from the specified conditions. The relaxed version of the query provides a non-empty answer while being 'close', according to funcion d, to the original query formulated by the user [17]. Unfortunately, skyline queries for data streams are blocking operators, that require the usage of a specific window-based operator in order to compute, at each instant of time, the finite subset of data from which the best itemset (i.e., the skyline set) is computed. A skyline set is computed in terms of a dominance relation with respect to a given set of attributes, by returning those items that are not dominated by any other item.[1] An example of a relaxed operator for the data stream context, based on a skyline operator, has been proposed in [11], where the concept of relaxation skyline (r-skyline), first introduced in [17] for stored data, has been extended to deal with data streams and queries composed of selection and window-based join operations.

Example 1. Consider an application of habitat monitoring and assume that sensors have been located inside nests, returning several properties of the place around the nest, including light. Assume the user is interested in detecting the nests under a light above a certain threshold. Suppose also this monitoring should last for a long period, thus a continuous selection query is issued. Suppose the query is submitted during daytime, a given light threshold and a given humidity level is chosen, leading to the following query:

```
SELECT idNest, light, humidity
FROM SensorNest [RANGE 2 min]
WHERE light >= 50 and humidity <= 60
```

In the previous query, [RANGE 2 Minutes] is a window operator that, when applied on stream SensorNest, at each time instant returns the set of tuples arrived in the last 2 minutes which satisfy the specified conditions.

Table 1 reports a portion of data stream SensorNest. For each tuple, the arrival time τ, expressed in minutes, is shown. By assuming that the previous query is executed in a precise way, at each instant of time a non-empty result is returned, thus facing a few answer problem. On the other hand, if we interpreted the query as a r-skyline query with respect to both the selection conditions $C_1 \equiv$ SensorNest.humidity $<= 60$ and $C_2 \equiv$ SensorNest.light $>= 50$, the computation returns different itemsets. As an example, at time 4, tuples s_4 and s_3 belong to the window and we get the following distance values, just computing differences between attribute values and query constants: $d(s_3, C_1) = 3$, $d(s_3, C_2) = 5$; $d(s_4, C_1) = 5$, $d(s_4, C_2) = 10$. According to the classical notion of dominance, and assuming to prefer lower values, it follows that s_3 dominates s_4 and s_3 is returned as result at time 4.   ◊

The Targeted Problem. Precise queries in a data stream management context guarantee a very efficient execution for selection operations, since they are not blocking operations, i.e., they do not require window-based operators for

---

[1] Given a set of points, each corresponding to a list of values for the relevant attributes, a point A dominates a point B if it is better in at least one dimension and equal to or better than B in all the others, with respect to some ordering [7].

Table 1. SensorNest data stream

tuple	idNest	humidity %	light %	τ
s_1	0001	49	45	1
s_2	0002	47	49	2
s_3	0001	63	45	3
s_4	0003	65	40	4
s_5	0004	66	76	5
s_6	0005	70	70	6
s_7	0006	70	66	7
...	...	...	...	...

their computation, which, on the other hand, is mandatory for join. At the same time, they guarantee maximal accuracy by definition. However, users may not be acquainted of the actual data arriving in streaming and, as a consequence, they may issue queries that, for specific instant of times, return an empty result set, thus potentially decreasing user satisfaction. With reference to the previous example, assuming that 50 is a suitable value for daytime, during light hours, probably, a non empty answer is computed. However, at sunset, the light is getting low and few (or no) data may be returned as answer. Two scenarios may arise to increase user satisfaction: (i) this is exactly what the user wants and no modification to the query has to be specified; (ii) the user may anyway want some results to be returned, the closest to the specified conditions. In the second case, the system should modify the query in order to provide a non-empty result with accuracy guarantees. This behavior can be obtained, for example, by changing selection conditions, through user interaction, similarly to what has been done in [22] for stored data. Unfortunately, this approach is suitable neither for a data stream management context since the query is usually specified once and continuously executed over arriving data nor for a more general distributed environment, where the limited user knowledge about data often make this approach unfeasible. A different approach would be that of executing an r-skyline query, even for selection, thus always obtaining the best result and avoiding the empty/few answer problem, at the price of a costly window-based computation.

From the previous considerations, it follows that, as soon as we want to combine processing efficiency (a QoS parameter) with result accuracy (a QoD parameter), a trade-off arises: the definition of skyline-based relaxation techniques may help in solving the empty answer problem but the price to pay is the introduction of a window-based computation and therefore, in general, a decrease of performance.

The ASAP Proposal. In order to combine the benefits of both precise and relaxed queries, an ASAP approach can thus be considered. The idea is to rely on the usage of an adaptive processing approach, in order to switch from precise selection operations to skyline-based ones as soon as, based on some dynamically monitored QoD parameters, the system understands that this is needed for improving result quality. The same technique may then switch from a skyline-based computation to a precise one to reduce result size as soon as a QoD parameter indicates that a precise computation can generate a result with QoD guarantees. In the resulting approach, the target of both adaptation and approximation is thus QoD-oriented, while the aim of both adaptation and approximation is both QoD and QoS since the techniques aim at achieving the best trade-off between a QoD parameter, namely result completeness, and a QoS parameter, namely response time, by recurring to (more expensive) skyline-based computations only when needed.

The idea is to model adaptive query processing as a finite state automa in which each state corresponds to one (possibly relaxed) query to be executed [12]. Transition from one state to another is performed during processing using heuristics with the aim of maximizing accuracy, defined according to specific user or system constraints and statistics, computed over already processed data. The overall ASAP framework relies on three main components: (i) *monitor*, which collects aggregate values, related to selectivity and precision of the query in execution; (ii) *assessor*, which determines whether some QoD conditions are satisfied; (iii) *responder*, which, based on assessor predicates, determines whether the query plan should be modified in a certain instant of time.

In order to give an idea of how the ASAP processing works, we suppose that the QoD-oriented user request corresponds to a precise continuous query annotated with specific QoD constraints, over parameters to be computed in a continuous way, like: σ_{avg}, average result cardinality (selectivity constraint); π_{max}, maximal distance of the returned tuples from the specified query conditions (precision constraint); μ, weight for selectivity and precision (trade-off constraint). We can then consider the very simple ASAP automata presented in Figure 1, containing just two states: one corresponding to a precise query and one corresponding to the r-skyline query obtained by relaxing all conditions appearing in the original request. More complex state machines can of course be provided by increasing the number of the considered states, i.e., the number of relaxed queries taken into account.

The computation then proceeds as follows:

- Statistics computed by the monitor may quantify how far the current result is from the selectivity and precision constraints associated with the original request. Based on such statistics, say σ for selectivity and π for precision, an accuracy measure has been provided, which, given an annotated precise query Q and a, either precise or relaxed, query Q measures how far is Q' result with respect to Q' result. A higher accuracy for Q' implies an higher user satisfaction in obtaining Q' result.

- The assessor can then determine whether, during the computation, some QoD conditions are satisfied. The following are examples of some relevant predicates: (i) sel^+/sel^-: too many/too less results are generated by the query at hand, with respect to initially specified selectivity constraints; (ii) $relax^+/relax^-$: the distance of the returned tuples is too high or the non returned tuples are quite close to the initial query and thus can be returned.

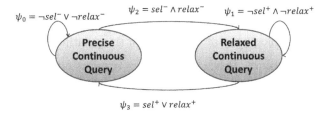

$\psi_0 = \neg sel^- \vee \neg relax^-$ $\psi_2 = sel^- \wedge relax^-$ $\psi_1 = \neg sel^+ \wedge \neg relax^+$

$\psi_3 = sel^+ \vee relax^+$

Fig. 1. An example of an ASAP automata

- Finally, the monitor component establishes whether a transition has to be performed, thus the query in execution has to be changed, in order to increase accuracy. This reasoning should rely on predicates computed by the assessor component and by the accuracy measure used to drive the process.

By varying the trade-off constraint, more emphasis can be given to either selection and precision, thus driving the process towards relaxed or precise computations.

4   ASAP in the Large

By increasing the complexity of the considered environment, moving towards highly distributed architectures, all concepts introduced in Section 3 become less clear. Indeed, while from one hand, user interactions with the network and its many applications generate a valuable amount of information, facts, and opinions with a great socio-economic potential, from the other hand, this huge wealth of information is currently being exploited much below its potential because of the difficulties in accessing data to retrieve relevant information. ASAP in the large can thus be interpreted as a step towards the realization of an entity-relationship search paradigm for uncontrolled and wide information domains, with an impact on qualitative and quantitative performance of systems for processing strongly interrelated and heterogeneous data in distributed dynamic environments. Under those new scenarios, data and requests can be characterized as follows.

Data Sources. Data from different sources are highly heterogeneous in terms of structure, semantic richness, and quality. They are often geo-referenced, time-variant, and dynamic. Information sources, which could be represented according to a graph-based data model, may contain: (i) strongly related and semantically complex but relatively static data (e.g., Linked Open Data); (ii) unstructured data, or data with a simple and defined structure; (iii) data dynamically generated by a multitude of diverse people (e.g., social networks, microblogs); (iv) highly dynamic data generated by public or private institutions linked to the territory (data streams).

Requests. Complex requests expressing relationships among the entities of user interest have to be represented, possibly relying on a graph-based query language. Such requests are often vaguely specified, since users cannot reasonably know format and structure of data encoding the relevant information. For example, the user may ask for the nearest shops selling the book which her friend Luca likes or for the biography of the author of the painting she is watching.

The Targeted Problem. Processing complex requests on heterogeneous and dynamic information sources can be costly since it first requires a request interpretation, then processing has to be performed on available sources deemed relevant (to reduce processed data volumes), and finally results should be aggregated in a consistent answer and returned to the user. Additionally, the answer may not guarantee the user satisfaction since the request could have been incorrectly interpreted, processed on inaccurate, incomplete, unreliable data, or even it could have required a processing time inadequate to the urgency of the request. As pointed out in [21], one solution to these problems relies on user intervention. However, depending on the request urgency and the specific application scenario, such intervention may not always be possible. The problem thus arises to define approaches for providing approximate answers to shared and complex information needs, even vaguely and imprecisely specified, operating on the full spectrum of relevant content, overcoming the difficulties related to heterogeneity and dynamism while requiring a limited user involvement.

The ASAP Proposal. In [10], we claimed that the ASAP paradigm can be effectively used to tackle the problem described above. In order to limit user intervention, a specific type of QoD-oriented approximation has been proposed, namely *Wearable Queries* (WQs). WQs integrate explicit requests with profile (information provided by the user as well as induced by the system, e.g., user habits) and context (spatio-temporal coordinates of the request, its motivation, and its environment, e.g., in terms of potential interaction and urgency). The computed result should minimize the distance between the returned items and the specified context and user information. To this aim, WQs computation should take into account data specificities, with a particular reference to quality, geolocalization, and freshness of data sources and specific data items, in order to select relevant data sources and provide results at the appropriate level of detail.

To enable search in a huge space of highly heterogeneous and poorly controlled sources, an adaptive pay-as-you-go approach, influenced by quality, dynamics, and specificities of the considered sources, needs to be adopted. The devised approach, which constitutes a completely new QoD-oriented adaptive approach applied to QoD-oriented approximate searches (thus, a new ASAP approach), generates and incrementally refines mappings between sources, according to the requirements induced by the submitted requests, thus avoiding the prohibitive costs of full integration. The role of the proposed QoD-adaptive approach is therefore twofold: the space of sources is incrementally adapted to the peculiarities of the submitted requests and, simultaneously, requests, specified as WQs, are processed by incrementally adapting them to the peculiarities of the space of sources and its evolution over time.

In order to maximize accuracy during the adaptive process, we explore a yet unexplored coordinate, namely metadata corresponding to synthetic representations of similar WQ executions repeated over time (called Profiled Wearable Query Patterns - PWQPs). Similar requests are very common in dynamic contexts, each time information needs are widespread among different users, because induced by an event, the interests of a community, or a place (e.g., during or after an exceptional event -environmental emergencies or flash mobbing initiatives-). The ability to take advantage of the experience gained by prior processing in new searches allows response times and interpretation errors to be limited, thus reducing the possibility of producing unsatisfactory answers.

For processing WQs, we envisage an mechanism that moves at each step, in the large space of possible approximate answers, towards the sources deemed capable of producing the best solutions with respect to profile and context of the request, quality and dynamism of the sources and knowledge gained from previous executions. The process is incremental, i.e., first it attempts to exploit PWQPs, then makes a coarse-grain selection of sources, and later it focuses approximation efforts on the description of the selected sources. The results are composed or reconciled through mappings, selected or generated on-the-fly. The envisioned solution couples this mechanism with a method for assessing the quality of each individual processing. This information, together with any explicit user feedback, is used for updates and refinements.

5   Conclusions

In this paper, after revising and classifying approximate and adaptive processing techniques with respect to the quality parameters they take into account, we introduced ASAP techniques and we showed that they can be successfully used in both specific and more general application contexts, characterized by an higher complexity of the environment and of the data sources at hand. We remark that ASAP is not a new concept, rather, it can be interpreted as a revision of existing processing approaches focusing on QoD parameters, which could be effectively and efficiently used in emerging contexts. Several issues are still open and require further investigation, especially under the "ASAP in the large" vision. In this context, we are currently investigating issues concerning data source characterization, for Linked and crowdsourced (social) data, as well as automatic acquisition of approximate geo-spatial contexts for crowdsourced (social) data.

References

1. Amato, G., Rabitti, F., Savino, P., Zezula, P.: Region Proximity in Metric Spaces and its Use for Approximate Similarity Search. ACM Trans. Inf. Syst. 21(2), 192–227 (2003)
2. Arasu, A., Babcock, B., Babu, S., Datar, M., Ito, K., Motwani, R., Nishizawa, I., Srivastava, U., Thomas, D., Varma, R., Widom, J.: STREAM: The Stanford Stream Data Manager. IEEE Data Eng. Bull. 26(1), 19–26 (2003)

3. Arya, S., Mount, D.M., Netanyahu, N.S., Silverman, R., Wu, A.Y.: An Optimal Algorithm for Approximate Nearest Neighbor Searching Fixed Dimensions. J. ACM 45(6), 891–923 (1998)

4. Babcock, B., Datar, M., Motwani, R.: Load Shedding for Aggregation Queries over Data Streams. In: ICDE, pp. 350–361 (2004)

5. Babu, S., Srivastava, U., Widom, J.: Exploiting k-Constraints to Reduce Memory Overhead in Continuous Queries over Data Streams. ACM Trans. Database Syst. 29(3), 545–580 (2004)

6. Belussi, A., Boucelma, O., Catania, B., Lassoued, Y., Podestà, P.: Towards Similarity-Based Topological Query Languages. In: Grust, T., et al. (eds.) EDBT 2006. LNCS, vol. 4254, pp. 675–686. Springer, Heidelberg (2006)

7. Börzsönyi, S., Kossmann, D., Stocker, K.: The Skyline Operator. In: ICDE, pp. 421–430 (2001)

8. Catania, B., Guerrini, G.: Towards Adaptively Approximated Search in Distributed Architectures. In: Vakali, A., Jain, L.C. (eds.) New Directions in Web Data Management 1. SCI, vol. 331, pp. 171–212. Springer, Heidelberg (2011)

9. Catania, B., Guerrini, G.: Approximate queries with adaptive processing. In: Catania, B., Jain, L.C. (eds.) Advanced Query Processing, Volume 1: Issues and Trends. ISRL, vol. 36, pp. 237–269. Springer, Heidelberg (2013)

10. Catania, B., Guerrini, G., Belussi, A., Mandreoli, F., Martoglia, R., Penzo, W.: Wearable queries: adapting common retrieval needs to data and users. In: 7th International Workshop on Ranking in Databases (co-located with VLDB 2013), DBRank 2013, Riva del Garda, Italy, August 30, p. 7. ACM (2013)

11. Catania, B., Guerrini, G., Pinto, M.T., Podestà, P.: Towards relaxed selection and join queries over data streams. In: Morzy, T., Härder, T., Wrembel, R. (eds.) ADBIS 2012. LNCS, vol. 7503, pp. 125–138. Springer, Heidelberg (2012)

12. Catania, B., Guerrini, G., Pomerano, D.: An adaptive approach for processing relaxed continuous queries (in preparation)

13. Chaudhuri, S., Das, G., Narasayya, V.R.: Optimized Stratified Sampling for Approximate Query Processing. ACM Trans. Database Syst. 32(2), 9 (2007)

14. Chaudhuri, S., Ganti, V., Kaushik, R.: A Primitive Operator for Similarity Joins in Data Cleaning. In: ICDE, p. 5 (2006)

15. Ilyas, I.F., Aref, W.G., Elmagarmid, A.K., Elmongui, H.G., Shah, R., Vitter, J.S.: Adaptive Rank-aware Query Optimization in Relational Databases. ACM Trans. Database Syst. 31(4), 1257–1304 (2006)

16. Ilyas, I.F., Beskales, G., Soliman, M.A.: A Survey of Top-k Query Processing Techniques in Relational Database Systems. ACM Comput. Surv. 40(4) (2008)

17. Koudas, N., Li, C., Tung, A.K.H., Vernica, R.: Relaxing Join and Selection Queries. In: VLDB, pp. 199–210 (2006)

18. Koudas, N., Srivastava, D.: Approximate Joins: Concepts and Techniques. In: VLDB, p. 1363 (2005)

19. Lengu, R., Missier, P., Fernandes, A.A.A., Guerrini, G., Mesiti, M.: Time-completeness Trade-offs in Record Linkage using Adaptive Query Processing. In: EDBT, pp. 851–861 (2009)

20. Marian, A., Amer-Yahia, S., Koudas, N., Srivastava, D.: Adaptive Processing of Top-k Queries in XML. In: ICDE, pp. 162–173 (2005)

21. Mass, Y., Ramanath, M., Sagiv, Y., Weikum, G.: IQ: The Case for Iterative Querying for Knowledge. In: Proc. of CIDR 2011, pp. 38–44 (2011)

22. Mishra, C., Koudas, N.: Interactive Query Refinement. In: EDBT, pp. 862–873 (2009)

23. Motwani, R., Widom, J., Arasu, A., Babcock, B., Babu, S., Datar, M., Manku, G.S., Olston, C., Rosenstein, J., Varma, R.: Query Processing, Approximation, and Resource Management in a Data Stream Management System. In: CIDR (2003)
24. Olston, C., Jiang, J., Widom, J.: Adaptive Filters for Continuous Queries over Distributed Data Streams. In: SIGMOD Conference, pp. 563–574 (2003)
25. Papadias, D., Tao, Y., Fu, G., Seeger, B.: Progressive Skyline Computation in Database Systems. ACM Trans. Database Syst. 30(1), 41–82 (2005)
26. Silva, Y.N., Aref, W.G., Ali, M.H.: The similarity join database operator. In: ICDE, pp. 892–903 (2010)
27. Spiegel, J., Polyzotis, N.: TuG Synopses for Approximate Query Answering. ACM Trans. Database Syst. 34(1) (2009)
28. Tatbul, N., Çetintemel, U., Zdonik, S.B., Cherniack, M., Stonebraker, M.: Load Shedding in a Data Stream Manager. In: VLDB, pp. 309–320 (2003)
29. Zhou, X., Gaugaz, J., Balke, W.-T., Nejdl, W.: Query Relaxation using Malleable Schemas. In: SIGMOD Conference, pp. 545–556 (2007)

Back to the Future – Should SQL
Surrender to SPARQL?

Rainer Manthey

Institute of Computer Science III, University of Bonn, Germany
manthey@cs.uni-bonn.de

Abstract. In this paper, we will take a closer look at the essential differences between two of the most prominent database query languages today, SPARQL and
SQL, and at their underlying data models, RDF resp. the relational model (RM).
There is an enormous "hype" around SPARQL/RDF at the moment claiming all
kinds of advantages of these "newcomers" over the long-established SQL/RM
setting. We discover that many of these claims are not justified, at least not as
far as data representation and querying is concerned. Our conclusion will be that
SQL/RM are well able to serve the same purpose as SPARQL/RDF if treated
fairly, and if presenting itself properly. We omit all aspects of navigation over
distributed or federated data resources, though, as SQL isn't (yet) made for this
task.

1 Introduction

The query language SPARQL [1] and RDF [2], its underlying data representation format,
have been at the forefront of interest of researchers in databases and information systems
for a number of years by now. This is not so much because this specific language and this
specific format are so revolutionary, but because both have been intimately associated
with notions like Semantic Web, Linked Open Data, Ontologies, Internet Information
Systems, Social Media and the like: Ideas and trends that are really motivating and moving not just researchers, but our entire "information society".

Consequently, getting good funding for proposals addressing these issues is comparatively easy these days (and increasingly harder for other, "old-fashioned" topics like
SQL). Even when discussing curricula for computer science students nowadays, academic teachers have to decide whether to switch from "good old" relational databases
and SQL [3] to "cool" RDF databases and SPARQL already in their introductory lectures on information systems. Making such a step at the core of academic education
would really mean for the SQL community to "surrender" to the new trend, because
you lose the fight if you lose the "youth". And fight there is, despite the increasing
number of SPARQL-to-SQL contributions, e.g. [4,5], seemingly bringing peace back,
but in reality attempting to reduce SQL to a kind of "DB assembler", hidden under the
surface, but offering SPARQL as "the" new interface to every database.

Well-established vendors of relational DBMSs have been eager to respond to this
trend not by abandoning relational technology but by "embracing" and integrating the
new concepts – probably in the hope to push them from the agenda similarly to the
successful rejection of "attacks" by object-oriented databases in the 1990s and from

G.F. Italiano et al. (Eds.): SOFSEM 2015, LNCS 8939, pp. 78–101, 2015.

XML databases during the previous decade. Not to speak about another "new kid on the block", the NoSQL movement, claiming the RDF approach as a "member of their club"! In his controversial, but influential paper of 2013, C. Mohan discusses (and partially discredits) this most recent "attack" on SQL using a title even more polemical than ours: "History Repeats Itself: Sensible and NonsenSQL Aspects of the NoSQL Hoopla" [6]. Will the vendors' strategy of "friendly enclosure" be successful this time, or has the dawn of the relational approach come at last?

"The End of an Era" is the heading of a chapter in a recent SIGMOD Record paper entitled "The relational model is dead, SQL is dead, and I don't feel so good myself" [7], summarizing opinions expressed by well-known colleagues during a panel held in Florence in 2012. The present paper aims at contributing to this "meta-scientific" discussion, too, and its title is, of course, a provocative one as well, obviously inspired by the Mohan title as well as the "Florence" title.

Being controversial in science is only possible in contributions that have not been submitted to, but invited for a conference (like the two papers mentioned above, and like the present one). Such opportunities are rare, but can be used to speak out, and to write about ideas otherwise not "sellable" in our current publication "culture". SPARQL has not been my topic in research at all, up till now, but SQL has. Writing this paper, for me went along with learning RDF and SPARQL – but sometimes learners, particularly experienced ones, ask questions and observe seemingly obvious things that specialists in a field don't ask, resp. see, anymore, or never asked or saw in the first place.

Surprisingly enough, there have been hardly any attempts up till now to compare the two languages and the two data models, not taking into account all the SPARQL-to-SQL translations (always in one direction, why not the other way round?). A recent "comprehensive comparative study" [8] to me looks neither comprehensive nor deep (enough).

The careful reader will have noticed the switch from the common, impersonal style of an author indirectly (or collectively, "we") speaking about himself to the first person style ("I") in the last section – normally, a no-go in science. Wherever personal, subjective opinion rather than objective facts is communicated in this paper, I will consciously use this style, returning to the classical indirect style in the other parts.

Last not least, why the title's prefix, "Back to the Future"? Of course, this is on the one hand a tribute to the famous movie by Robert Zemeckis of 1985 [9], Mohan's "History Repeats Itself" is also mirrored. On the other hand, the question mark at the end of the title is intended to apply to its prefix as well: Are we really returning, when promoting SPARQL as vigorously as we do now, to a dèja-vu situation, namely that phase in the early 1980s when SQL became popular and quickly dominating, leaving none of the other competing languages a chance to seriously survive? One might begin to think so. Any assumption that "back to the future" is also intended to ring the "re-inventing wheels" bell is purely speculative and can't be confirmed or even encouraged by this author – but can't be prevented either.

The main part of this paper is structured as follows: First we will restrict the scope of our comparative investigation of the two approaches to just the data representation and data retrieval context in chapter 2. Then we will discuss the data models underlying SPARQL und SQL, respectively – RDF versus the relational model (RM) – in chapter

3. In chapter 4, the focus is on the conceptualizations from which databases in each of these models are derived, formally represented using an ontology or an ER diagram. In chapter 5 we will briefly look at graph databases, a particular kind of databases SPARQL/RDF claim to serve particularly well (as opposed to relational databases). In chapter 6 at last, the two languages proper will be contrasted. We will conclude with a summary of our findings and with an answer to the question posed in the title of this paper.

2   On Data: Everywhere and Anyhow?

One of the obstacles for properly judging about SPARQL's merits, in particular in comparison with SQL, is the circumstance that SPARQL is intimately bound to two dimensions of discussion which SQL is not, or at least not obviously addressing: data manipulation in a physically distributed and federated environment (even extending to the entire internet), and data manipulation in a heterogeneous environment as far as data representation formats are concerned. Behind all of this there is the dream or vision to be able to access any kind of data residing anywhere at any time. We all have this dream, and presently "googling" for keywords is as close as we come to making this dream become real – not a too pleasant answer though. Accessing all of Google's contents using SQL queries would be much more goal-directed and informative an approach, provided we were all "SQL literate". Pattern matching and browsing enormous lists of matches hoping for a needle in the haystack is a very poor style of retrieval indeed. However, SQL never tried to go that direction, you can't "surf the web" in an SQL query, and you are bound to data stored in tables and represented in tuples.

SPARQL on the other hand can deal with URIs and URLs and has the syntactic means to (express commands that make an interpreter) navigate to a particular data resource before actually starting to evaluate the query proper. Like SQL, SPARQL is bound to data being represented in a single format – this time triples rather than tuples and datasets rather than tables. However, RDF enthusiasts claim that using this even more primitive format for representing data than the tuple format, one is able to accommodate data from a wide range of less rigorous semi-structured formats (RDF itself isn't semi-structured at all, but as rigorously structured as one could imagine). The claim of RDF to be the best (only?) approach to accommodating data of all kinds, even including unstructured data, even not requiring a schema for data contained, is probably as attractive as the world-wide reach of its access mechanism. Of course, users of the "sparqling the web" approach have to be able to "speak" a formal language, too, as opposed to Google surfers who just need to enter a search key (getting served by pattern matching only in return). So wrt to accessibility of the approach to a wider audience of users SPARQL is as "upper class" as SQL.

The interesting (but hardly ever posed) question is whether a completely new language is needed for doing so. Or would a suitable extension to SQL do (have done, if you think the train can't be stopped anymore) the same job – of course, only if coming with the necessary DBMS technology for web-wide distributed processing as well as near-universal format transformation? In order to be able to investigate (if not answer) this question fairly and seriously we will exclude the question about distribution and navigation from our discussion. SQL isn't made for this (not yet, at least, we are sure it could be), so we shouldn't compare apples and pears. Thus, we will assume that a single SPARQL backend is around, no URI and no explicit prefix of a SPARQL query will appear in this paper. We will just deal with data residing at a single site. This might look to SPARQL aficionados as losing an essential characteristic of "their" language. But I think it is inevitable to do so if (fairly and meaningfully) comparing SPARQL and SQL – and there remains enough to be discussed and investigated anyway!

As far as data formats are concerned we will compare the two approaches, the relational one as well as the RDF one, without any limitations, as both are rigorously structured formats if viewed as what they are. The question of representing semi- or unstructured "data" will be excluded as well, for space reasons and because the answer is quite obvious: Translating un- or semi-structured documents into RDF or RM databases requires a high amount of pre-processing for both models, starting from a conceptualization of the resp. application domain and extracting individual facts from the document. We don't discuss the issues arising in the context of storage strategies in this paper either. Whether column or row stores are used, whether triple stores require new DBMS technologies or just new indexing and optimization strategies in relational DBMS is a serious matter, but not essential for comparing the formats and the languages, it is orthogonal and would blow the scope of this paper.

3 On Syntax: Triples vs. Tuples

Both languages, SPARQL as well as SQL, are based on a particular formalism for representing individual data elements, the Resource Description Format (RDF) and the relational model (RM), respectively. Both, RDF and RM, are data models – even though RDF isn't called a model, but a format instead. This is the case at least if we exclude all aspects of navigation in a communication network consisting of distributed "resources", and that's what we do in this paper. Comparing the languages is impossible without comparing the models. Thus, we start by considering the foundations of RDF and RM.

We will do so, by first stating what is objectively (and hopefully undoubtedly) "true" about the two models, then we will interpret our findings and compare these interpretations. Comparing is a subjective activity for us: Without a particular interpretation of what we see, a meaningful comparison will remain on the surface. In the remainder of this paper we will thus subdivide discussions into these two categories (objective, subjective) throughout. Therefore, each of the following chapters will consist of two sections entitled "Facts", and "Views", respectively, as will this one.[1] A word of "warning" ahead: Parts of this chapter will appear like tutorials for beginners, which doesn't mean that I expect our readers to be ignorant of the basics listed. But if we want to compare two models, we better remind ourselves of these basics in order to properly compare them on the basis of what they are, rather than what we might have been told about them.

---

[1] I am well aware of the double meaning of both notions, *fact* and *view*, in the context of a database paper. I hope readers will understand which side of the coin they are looking at whenever they read these terms.

3.1 Facts

SPARQL queries are evaluated over collections of triples, i.e., tuples of length 3. Triple collections are called datasets. They may or may not contain duplicates, so "datasets" may indeed be "data-multisets". We will exclude duplicates, however, in this study for simplicity's sake. Every RDF database is a set of datasets. One of them – called the default dataset – is considered unnamed and is part of every RDF database, other named datasets may be added at the discretion of the database designer. We will not make this distinction (unnamed vs. named) in the following, but assume there is a set of uniquely named datasets in each RDF database, without loss of generality.

SQL queries, on the other hand, are evaluated over collections of tuples of arbitrary length (not just length 3). Tuple collections are called relations in RM, considering relations subsets of products of domains as done in mathematics. SQL calls tuples rows, and relations tables. Tables may contain duplicates (unlike relations), and again we will exclude duplicates from this study, so that tables are proper relations for the time being. All tables in a relational database are uniquely named. It should be clear already at this point, that RDF and RM databases are very similarly organized on the top level: relations as tuple sets vs. datasets as triple sets. This seems obvious, but is noteworthy considering the way RDF is introduced all over the place.

Tables – at least from the viewpoint of SQL[2] – require (locally unique) names for their columns, called *attributes*. The choice of attributes is left to database designers, as the choice of table names. Attributes are expected to reflect the meaning of the data items in the resp. column using terminology from the real-world domain from which the resp. data are coming. Thus, tuples structurally turn into *records*, where each component is prefixed by a *selector* (here the resp. attribute), thus making order of components irrelevant (due to the uniqueness of attributes). Attributes, like table names are considered meta-data, making up the *schema* of the resp. relational database together with the names of the resp. tables and the associated data types. Each column, and thus each attribute, is associated with a unique *data type*.

RDF datasets do not have attributes, at least not explicitly. Thus, RDF databases don't (seem to) have a schema in the sense of the relational model, apart from the names of named datasets, if any. Implicitly, however, the roles of the three columns of a triple are well-distinguished from each other, and are fixed (even their order matters). Triples are supposed to represent elementary statements about the underlying application domain syntactically reflecting the structure of basic sentences in natural language: (*Subject*, *Predicate*, *Object*). These three grammatical concepts may be considered (meta-)attributes of every RDF dataset, though not explicitly mentioned in most cases.

For comparing relational tuples and RDF triples on an equal basis, one could well treat triples as records, too, using *subject*, *predicate*, and *object* as (meta-)selectors. Thus, datasets could be properly compared with tables. And RDF databases would have a schema, too, though the (meta-)attributes of the individual datasets were all the same and did not relate to a specific domain of interpretation (but to generic grammatical con-

---

[2] Other relational languages, such as QbE and Datalog, don't use attributes, but assume columns to have a fixed order instead. We will come back to these co-existing views of rows/tuple in chapter 4.

cepts instead). Thus, it appears more appropriate to call this (implicit) schema of RDF databases a *meta-schema*. As all meta-attributes of all datasets in this meta- schema are identical in RDF, there is no need to declare them.

The *data types* underlying the columns in a table can be freely chosen in a relational database by the database designer. The choice is guided by the meaning which the chosen attribute for the resp. column has in the application domain of the resp. database. Data types consist of atomic (non-decomposable), symbolic values, or sequences of such values. The former restriction is known as the First Normal Form (1NF) requirement, the necessity to have symbolic values arises from the simple fact that tuples have to be stored in electronic devices and printed on paper or displayed on screens of such devices.

In an RDF database, the claim for atomicity and symbol-valued domains is retained for obvious reasons (recall that for the scope of this discussion we exclude all aspects of navigation over the web, thus all URIs are considered irrelevant). Free choice of domains for columns[3] is not an issue, at least "technically": As the three (meta-)column names are generic, column domains ought to reflect this genericity (i.e., application independence). They ought to be as general as possible, allowing for any symbolic value to be stored in each of the three columns, at least in principle. Thus, *string* would possibly be the best choice as data type for each column of a dataset, at least in our restricted context without URIs but just literals as values.

However, the roles to be played by the entries in each of the three columns (subject, predicate, object) imply a certain categorization of the potential entries in the columns of an RDF dataset. Subjects and objects in a triple are supposed to come from a different category than predicates. There is a category of *things* (roughly similar to the categories of entities or objects in other conceptual frameworks), and a category of *interrelationships* between things (including properties of things, which are represented by other things).[4] Meta-attributes subject and object are thing-valued, whereas predicate is relationship-valued. This is a semantic categorization, however, which has to be made concrete for each application domain for which an RDF database is designed anew – very similar to the choice of attributes in relational schemas. As the choice of data types for subject, predicate and object in RDF has to be made on the syntactic level, generically, however, data type choice is not free for designers in RDF.

A last syntactic issue remains to be mentioned: Certain attribute values in a row of an RM table can be missing. Formally, this attribute is considered to have a *null value* in the resp. row, representing a piece of data which exists in the underlying application domain but is unknown at present (at least that's how SQL interprets nulls). In certain

---

[3] Even though *column* is a notion from the relational context, we will speak of columns of a dataset (or a triple) in RDF, too, from now on, as it should be clear from the previous discussion, that datasets are special cases of tables, and triples special cases of tuples, at least structurally.

[4] We avoid speaking of *resources* in this context, rather using the fuzzy, but more neutral term *thing* instead. Resource is "RDF speak" in the W3C documents, as the origin of this approach is the representation of networks of information "resources". As we consciously exclude all aspects of distribution, navigation and data origin in this paper, and rather focus on the ability of the two approaches to represent data about real-world domains, this deviation seems appropriate.

situations, due to the design of the resp. database, there may be a lot of null values in a table, making this table rather sparse. In RDF, there is nothing like a null value ("blank nodes" are a different affair, we omit them in this context, too).

3.2 Views

Now, what does all of this mean? Is RDF any "better" than RM? Are there any databases that can be built using RDF, but not using RM, or vice versa? As the above discussion has been focusing on structural (syntactic) aspects, just touching semantics slightly here and there, we might expect a partial answer only, but no less than this we do expect.

A simplistic (?) answer to the "better" question is very easily given: Triples are tuples, and datasets are tables – thus every RDF database is a relational database, making very restricted use of the potential of the relational model, due to its restriction to the "Power of Three". Following this line of reasoning, the data representation strategy underlying RDF is no more than a very particular schema design strategy in the relational context: Restrict yourself to triple tables and follow the subject-predicate-object style when designing attributes of your "triple tables", alias datasets. RM seems to be the winner straightaway, and RDF seems to be dispensable, "emperor's new clothes". What this view would mean for SPARQL vs. SQL in the end can be guessed quite easily.[5]

Taking the "RM subsumes RDF" view, it would be quite astonishing if arbitrary (>3 columns) tables could be represented by (not as) datasets of triples, too, thus making RDF in turn able to make RM dispensable, not by subsumption, but by translation into a different format. There is a conceptually quite simple way of mapping relational databases to RDF databases, however, well-known by most researchers today: Map each table to a named dataset of its own; inside the table, map tuples to subjects, attributes (i.e., column names) to predicates, and attribute values to objects. Thus, each n-tuple in a relational database is mapped to n triples in its RDF counterpart. Doing so, translates a table with (simplified) one key and n non-key attributes, and with m rows into a dataset with $n \times m$ triples, one for each non-key attribute value.

There is a small technical problem, though: How to express entire tuples as entries in the subject column of a triple (data) set? Attributes are strings by definition, so they fit into the predicate column, attribute values are symbolic, too, so they can at least be cast into strings, and thus fit into the object column, but tuples? We at least need a symbolic identifier for each tuple to be entered into the subject column as a "surrogate" for the tuple. TIDs, tuple identifiers, used internally by most relational DBMSs for identifying rows in tables, could do the job. If table schemas happen to have single-attribute primary keys, the values of such key attributes may serve as unique symbolic representatives of the individual tuples in the subject column, too. If such key attributes exist, they don't need to be represented as predicates, their values not as objects, but as the subjects of all triples arising from the mapping of the respective table to triple set format.

TIDs are "syntactic" identifiers not related to any concept in the underlying application domain, keys on the other hand are semantic concepts arising from the application

---

[5] This isn't the entire story though, that's why I speak in conditional style here. In the next section, when semantics are considered, this view will turn out to be too narrow, but not false. For the time being, it is valid, however, as we just look at structure.

domain. But at the end, TIDs can be viewed as "artificial" semantic attributes, added during the design process for reducing potential multi-attribute keys to single-attribute ones (or adding artificial keys if none have been identified among the natural attributes). At the end, this is a minor problem: "Normalizing" each table schema in such a way that it has a single-attribute primary key leads to a dataset design which represents each n-tuple by n-1 triples[6]. Thus, each relational table can be transformed into an RDF dataset, and, thus, RDF is as expressive as RM. This, together with the first observation (RM subsumes RDF) leads to the view: Both models are equally expressive – whether the one or the other is chosen depends on additional arguments still to be discovered (or might even be just a matter of taste). Certainly the number of individual data items representing a piece of information would be much larger if going the RDF way with its "narrow tables" approach, as if being able to use arbitrarily "wide" tuples.

The existence of null values in a table doesn't provide any problem for transforming tables in RDF. If in a certain tuple the value of a certain attribute is null, this value is not represented by a triple at all. This attribute is thus treated as not being applicable to the resp. tuple, a tiny semantic difference as compared to the "exists, but currently unknown" view SQL has of null values, but not an obstacle.

There is one issue already visible here which is definitely worth closer consideration, however: In tables attributes are meta-data, in RDF datasets they are turned into data (if represented as predicates of triples, as suggested in the RM-to-RDF discussion above). This is an important, maybe the most important idea introduced by RDF as far as data representation is concerned. It is no difference between the two models, as using (application domain) attributes as values of a meta-attribute (serving as column name) is a valid choice in any relational database as well, no need to switch to RDF instead. We will come back to this issue in the next chapter. To summarize our findings up till now:

1. Both formats, RDF and RM, are in principle equally well-suited for representing any kind of data composed from atomic values.
2. RDF is strictly speaking not required as a new format, but can be purely interpreted as a schema design strategy for relational databases, characterized by turning application domain attributes or relationship roles into data rather than meta-data.
3. Missing information in a relational tuple (represented by a null value) is simply missing in the triple representation, thus leading to a more compact representation of otherwise sparse tables.

All of this is rather unspectacular so far. The only point that may cause controversy is the view that RDF up till here is dispensable as a model of its own! But we didn't treat semantics yet – let us not forget that RDF/SPARQL are the "icons" of the Semantic Web movement! And we didn't talk about graph databases either – to be discussed in chapter 5. Nevertheless, there was no need for a subjective "I" statement till now!

---

[6] If any attribute has a null value for any of its attributes, this attribute value is simply omitted in the triple representation, thus reducing the number of triples is needed. At least this appears to be a viable strategy unless we want to extend RDFs data model by a null value equivalent.

4   On Semantics: Ontologies vs. Schemas

As we saw in chapter 3, both, RDF and RM, are providing certain choices of data structures able to accommodate data syntactically, regardless of the meaning of the data. Both are generic formats, semantically neutral and uniform, even rigorous (which means that there is just a single type of structure available in both models, tables vs. datasets resp. triples vs. tuples).

For being able to use such "syntactic" models for representing data from a particular application domain, a prior semantic modelling of the resp. domain is required. *Concepts* (i.e., specific terminology, resp. vocabulary) having a particular significance and meaning in the context of the domain at hand must be identified and classified according to the categories of a certain meta-model (e.g., class, property, attribute, relationship, subclass etc.). The process of generating such a model of the specificities of a domain is often called its *conceptualization*, the meta-model used is called a *conceptual model*, and the result of the conceptualization a *conceptual schema* of the resp. domain[7]. Once the conceptualization is in place, a systematic derivation of data structures for accommodating data about the instances of the concepts at a particular moment in time takes place.

This chapter is looking more closely at such conceptualizations in general and on the way how to derive data structures in either RDF or RM from them. Both approaches to data representation, the RDF and the RM one, follow this style of design, but they do so quite differently – which possibly makes them appear more different than they essentially are.

4.1   Facts

In the RM context, there is a well-established design approach for conceptualizing an application domain before designing a relational schema, which has been in place since almost 40 years. The Entity-Relationship Model [10], or one of its many EER extensions and variants, e.g., UML class diagrams, is traditionally chosen for conceptualization. ER diagrams (graphical representations of ER schemas) are used to formally express the result of a particular conceptualization. Without going into too many details about notation variants and possible extensions, we will consider the following as the main categories of ER concepts: Entity types with properties, is_a hierarchies of sub- and super-entity types and inheritance of properties, relationship types (possibly with properties, too, and roles). Key properties for entity types as well as cardinalities for relationship types are noteworthy annotations in ER diagrams representing special constraints on instance sets (called populations) of the types.

Entities, in the philosophy of this approach, are "things" in the same spirit as discussed briefly in the previous chapter when motivating the subject-object roles in RDF

---

[7] The term "schema" is used in the database community, not in the AI-oriented W3C context. Here the result of modeling is a "model", which sounds more reasonable. Unfortunately, the term "model" has been used for the resp. meta-models in the DB context (e.g., relational model, ER model). We use "schema", as it is so specific that no misinterpretations are to be expected. "Model" has many facets in other communities and disciplines.

triples which can be played by "things" in the domain underlying the resp. database. In the ER context, however, there is a second category of things, called *values*. Values are "things with restricted rights": They have to be printable symbols (or symbol sequences considered atomic, too). Values don't have properties, they can't participate in relationships, but can only serve as properties of entities. Values can be stored in a database, entities cannot. Entities are considered (abstractions of) concrete or imagined "things" from the application domain usually not printable (such as people and events, e.g.). They can't be stored as such, but only their property values as well as information about the relationships they have to other entities represent them indirectly in a database. Properties of an entity are all "value-valued", an entity can't be a property of another entity. Only *relationships* can establish links between two or more entities.[8] Relationships resemble entities in as far as they can be characterized by properties themselves, however, relationships cannot play the role of participants in other relationships. In summary, ER is a rather "middle-of-the-road" collection of modelling techniques, some of which are a bit special (not to be found like this in other approaches), but most of them to be found in lots of other modelling approaches, too.

ER modelling is a kind of "pre-processing" step before a database schema proper is designed (these two modelling phases are called conceptual and logical design, respectively). The main idea for deriving relational DB schemas from ER diagrams is to represent types by tables and instances of the types by tuples. Thus the table corresponding to a type contains tuple representations of the population of the resp. type at the resp. moment in time. Properties are turned into attributes, i.e., each property of a type (regardless whether entity or relationship type) has its own column in the table corresponding to the resp. type. Key properties become primary key (PK) attributes in the corresponding tables. Each entity is thus represented by the tuple of all its (current) property values. Subtypes of an entity type are represented by splitting the entity table of the uppermost entity type in a hierarchy into sub-tables for the individual sub-types. Inheritance is considered by either having multiple copies of property columns for each sub-type or by linking sub- and super-type tables by means of PK columns.

Relationships are represented relationally by tuples, too, consisting of the keys of the participating entities plus the properties of the resp. relationship. The PK values of the participating entities form foreign keys (FK) in relationship tables referencing the entity tables of the related entity types. Cardinalities may lead to a reduced number of tables due to including relationship tables into entity tables in case of functional constraints on the resp. relationship type. This way of turning types into tables has been adopted by the vast majority of the RM community, they are taught to beginners in database textbooks, and almost all major relational DBMS products come along with design tools following exactly this mapping approach – however, there are alternatives around, going exactly this "classical" way if moving from entities and relationships to tuples is not necessary, but recommendable (or better: recommended in the RM field).

---

[8] In UML, there are no relationships in the ER sense, instead properties may be entity-valued. Relationship types of arity $\succ 2$ have to be "simulated" by means of entity types the instances of which are "simulating" the instances of the resp. relationship types. Such properties are called *associations* in UML.

In the RDF context, matters appear to be very different, at least as they are presented in most sources. This is mainly due to the terminology used and to the way of presenting the techniques used, going back to a different tradition in computer science, the tradition of knowledge representation in AI (in particular as followed in description logics). Here, there is no fixed (or at least traditionally used) meta-model coming with the RDF structural model, but instead any kind of conceptualization seems to be acceptable in principle, most favourably a conceptualization in form of an ontology (quite often specified in terms of the OWL [11] style of ontology specification).

"An ontology is defined as a formal, explicit specification of a shared conceptualization... It provides a common vocabulary to denote the types, properties and interrelationships of concepts in a domain." [12] Following this "definition" – which has been taken by the anonymous authors of the source of this quotation in Wikipedia straightaway from serious scientific publications – obviously, every (extended) ER diagram represents an ontology, as it consists of exactly "types, properties and interrelationships of concepts in a domain".[9]

We will not go into the same degree of detail as for ER modelling above in the context of ontologies, mainly because the general idea of an ontology and the style of conceptualization followed in ER modelling is not essentially different. In ontologies, too, objects/individuals of the application domain under consideration are characterized by means of properties assigned to the individuals. Furthermore, similar individuals (as far as the property structure is concerned) sharing additional semantically motivated aspects not expressible by properties alone, are summarized into collections with a given name, called classes.

There are usually no distinctions made between individuals resembling entities and those resembling values in ontologies. There is also no explicit mention of relationships as individuals of their own, instead properties may relate individuals arbitrarily (as in UML [13]), n-ary relationships can be (and have to, not always painless) "re-ified" as individuals. Class hierarchies with inheritance of properties (and sometimes of other class-related concepts) are to be found in nearly all ontological approaches. Whatever concrete style of specifying an ontology is used (OWL, RDF Schema, or others) – essentially these are the categories of conceptualization if coming from the AI rather than the DB tradition, and they are – again essentially – no different than the techniques of the ER approach. The entity-value as well as the entity-relationship distinction in ER are dispensable – as in UML's class diagrams, a direct derivative of ER.

What matters much more is the question how ontologies/conceptual schemas are turned into RDF representations of data about the resp. application domain, because here we have the interesting differences from the relational approach of turning the conceptual (ER) schema into a logical (RM) schema. The basic strategy for mapping ontologies to RDF is based on the observation that all information available about the "things" in an application domain can be expressed as functional associations between two "things": A property (in the ER sense) of an entity (a "thing" which is characterized by other "things") is a mapping of the characterized "thing" (the "subject") to the characterizing one (the "object"). The name of the property becomes the name of the

---

[9] That's why this observation is already stated under the heading "Facts" rather than in the "Views" section of this chapter.

(functional) mapping. If two "things" are simply related to each other, without the one "characterizing" the other, the same mapping idea can be used, as long as the association between the two is unique, at least with respect to this particular kind of association carrying some specific meaning in the application domain. In such a case, the association name plays the role of the function name. That's what the triple idea is mainly about: Representing a single functional association between two "things".

There are two essential characteristics of this alternative way of deriving data structures from a conceptualization:

1. Each property or association is represented by a data element of its own (in the RDF case by a triple), rather than by a data element collectively representing all properties and all functional associations for a given "thing" (in a single RM tuple).
2. The names of the properties resp. associations are represented as data (predicate entries in the subject-predicate-object sense) rather than meta-data (such as RM attributes).

Note, that these two design decisions are orthogonal to each other: Neither is causally depending on the other. You can choose to represent subject-object pairs by individual tuples without storing the corresponding property names as data (i.e., RDF predicate values). If doing so, all subject-object pairs with the same predicate can go into a dataset of its own, called by the name of this common predicate. This would lead to pairs rather than triples for representing each individual predicate value (object) in relation to its subject. You can also choose to represent concepts (such as properties or associations) as data without "fragmenting" the information about a common subject into as many data elements as there a concepts applicable to that subject. The result would be a tuple with one column representing the resp. subject and two columns per property/association: one for the property name (alias predicate) and one for the property value (alias object). Whether such a representation makes much sense is questionable, however.

4.2 Views

The ER (Meta-)Model has been introduced during a development phase of computer science in the late 1970s during which a rather large body of so-called semantic data models [14] emerged. Semantic models were the predecessors of the better known object-oriented data models, which became popular in the 1980s and 1990s. Most of the semantic models disappeared and have been forgotten, just ER (and its more modern derivative in UML) survived. The OO approach disappeared because it came along with the attempt to introduce new DBMS technology directly supporting OO concepts rather than retaining relational storage techniques and treating objects and classes as pre-processing concepts for relations. OO products couldn't replace RM products, just a few ideas of this period are still around in so-called object-relational extensions to RM today.

The "ontology movement" in computer science is much younger than the semantic modelling approaches, going back to the early 1990s. Influenced by much older ideas in philosophy, from where the term "ontology" has been taken, the use of this term became particular popular in the AI field. It took quite some time before you could find statements like *"The methods used in the construction of ontologies thus conceived are derived ... from earlier initiatives in database management systems."* [15]

What is the relationship between ER and OWL, ontologies and ER schemas?[10] Is RDF "bound to" ontologies like relational DBs are (seems to be) "bound to" ER diagrams? Is RDF more "powerful" because it is "supporting" ontologies rather than "just" ER diagrams? Is there a standard way of mapping an ontology to an "RDF schema" (are there any such schemas at all) like mapping an ER diagram to a particular relational schema? Can RDF accommodate ER diagrams, and can relational databases accommodate OWL ontologies? A plethora of questions all relevant to our topic!

My claim in this respect is very simple: Both "schools" of conceptual modelling are essentially achieving the same result with the same means. Most (if not all) presumed differences between ontological and semantic modelling are terminological ones arising from different traditions rather than from objective differences. Thus, RDF databases might as well be designed if starting from an ER schema, whereas relational database design may start from almost any ontology, too. Whether the ontology is formally represented by a diagram (such diagrammatical techniques exist in the ontology context, too) or the ER diagram is written out in textual form doesn't matter either, this is purely a matter of convenience! What matters is how similar concepts are treated when designing data structures that accommodate instance data after the resp. database has been "populated".

The "competition" I think we ought to concentrate on is between the strategy of mapping ER diagrams to relational schemas on the one hand, and the strategy of representing data about concepts in an ontology as RDF triples – not between ER and, say OWL (as the arguably most prominent ontological formalism) as such. The most essential question here is whether it is beneficial that properties of "things" are turned into data (as they are in case of, say, OWL-to-RDF, where they become values of meta-attribute *predicate*) or into meta-data (as they are in case of ER-to-RM, where they become attributes). The RDF approach makes properties become accessible to queries (such as: Which properties – not property values – does a particular "thing" have?). If questions like this are to be expected, the RDF design is preferable, otherwise the RM representation is much more compact. The argument that property-as-data makes it easier to represent missing property values doesn't really convince me, null values do a good job, too, at least in cases where they remain the exception rather than the rule.

The other important question is whether a strongly fragmented design of the data (as in triples, representing minimal pieces of information in a single syntactic entity) is preferable over a more compact one, where all properties of a single thing are represented together in one tuple. The major findings in this chapter are for me the following:

1. Both, RDF and RM databases, require a formal conceptualization as the basis of the design of a particular database. In RM this conceptualization is made explicit in the schema (table and column names), in RDF a major part goes into the data.

---

[10] Note that whenever I speak about ER here and in the following, I mean EER, i.e., one of the many extended ER models that have been proposed over the last 2-3 decades. At least entity type hierarchies with inheritance are part of all of these extensions.

2. Ontologies are no stronger as techniques for formalizing conceptualizations than semantic data models (as ER) from which RM database schemas are derived.
3. All steps made towards an RDF database if starting from an ontology can be made in an analogous manner if ending up with a relational schema.
4. A serious investigation of the benefits of representing properties as data (which could be done when designing RM databases as well, not just when using RDF) is still missing: Which kind of queries benefit from this and what is the price to be paid? Are null values better or worse than non-applicable predicates in RDF? Is being "schemaless" really an advantage to be aimed at?

5   On Graphs and Data – or Data and Graphs

Let us turn to graphs next, another key concept of our topic, at least from the perspective of RDF (and of SPARQL, claimed to be particularly suited for representing what is called graph databases). "RDF is a directed, labelled graph data format for representing information in the Web." [16] This sounds as if RDF is really "great" for graph data (whatever that may be) – and by implicit assumption as if RM is not. So let us look at the relationship between graphs and data first – and ask what "graph databases" could be.

5.1   Facts

Triples can be visualized as atomic graphs, i.e., as pairs of labelled nodes connected by a directed labelled edge. The source node is labelled by the subject of the triple, the destination node by its object, and the edge by the predicate. As "things" may play both, the subject as well as the object role (in different, as well as in the same triple), sets of triples can be visualized by arbitrary directed graphs, including cyclic ones. Thus, triple sets correspond to forests of connected graphs. If triple sets can be visualized by such directed graphs, the converse is true as well: Every such graph can be textually represented by a triple set. This is why RDF seems to be so well-suited for representing graph databases, whatever that means exactly.

What about relational databases and graphs? Do they have to "leave the field" to RDF in this respect, simply because they don't "have" trees, just rows? Not at all! It is well-known from basic courses on relational databases that rows can be formally modelled in two different ways: Either as tuples in the proper mathematical sense, i.e., as elements of products of symbol sets (domains), or as trees of depth 1 (we just wrote about them a few lines before in the RDF context). In the tree view, the row itself is represented by an (unlabelled) root node, each of the attribute values labels one of the leaves of the tree, and the attributes proper (i.e., the selectors in the record sense) label the edges between root and leaves, the edges being directed from root to leaves. Whereas order matters in the tuple view (where attributes don't play a role), it doesn't matter in the tree view (due to the uniqueness of attributes). Here, attributes can be understood as functions mapping rows to attribute values in a unique manner.

Thus, tuples "are" (atomic directed) graphs as much as triples "are"! The out-degree of rows in a table can be higher than 1, as opposed to that of triples, viewed as graphs, where the out-degree is fixed to 1. It is more difficult, however, to view tables, i.e., sets of rows, as (complex) graphs. This is due to the fact that attribute values (i.e., leaves of "row trees") cannot at the same time be roots of trees representing other rows. Root nodes in this "classical" style of representing rows/tuples as trees are not labelled by symbolic strings, they are unlabelled, which means they correspond to abstractions, representing the row as a whole. They are "artificial things" (non-printable) which are what variables in SQL are bound to: entire rows. Thus, there are no inter-row links, and tables are forests of depth 1 trees, each row represented by an isolated component of the overall graph.

Navigating from one row to the other via common "things" (object in the first, subject in the second statement) appears to be very simple in RDF, visually just following paths in the graph representation. In relational databases, joining tables is needed in order to achieve the same "hopping" from one "thing" to the other. Most joins are performed along foreign key-primary key (FK-PK) connections between tables: Value equality of one attribute of the one table (the foreign key attribute) and a different attribute in the other table (primary key attribute there) does the job. Existence of a join partner for the FK value is guaranteed due to the inclusion dependency being implied by the FK property, uniqueness of the matching row follows from the PK property in the "foreign" table referenced.

5.2 Views

Is it therefore true that tables are not suited for representing complex graphs (and thus relational databases are not suited to represent "graph databases"), even though rows are trees like triples? A small mental "shift of perspective" reveals that this is not, at least not necessarily true. If we "press" relational schemas a little bit by requiring each table to have a primary key, and a single-attribute PK in particular, there is room for a slightly modified tree view of tuples. Most tables satisfy this condition quite naturally anyway, and in other cases artificial surrogates – as proposed, e.g., in Codd's RM/T extension of the relational model [17] – do the job. This is the same idea as used for representing tuples by sets of triples in 3.1 above: The TIDs we mentioned up there are such symbolic surrogates for entire tuples.

If we represent a tuple as a tree, the root of which is (labelled by, and thus representing) the PK value rather than an (unnamed) abstraction of "the tuple as such", then FK references from an attribute value of one tuple to the PK value of the other are no longer using leaf nodes of isolated, unconnected trees but can go straight to the roots of other trees representing referenced tuples. Thus collections of joinable tuples (that is, joinable via FK-PK links) form general graphs, even cyclic ones are possible. Every graph representable by an RDF dataset is equally well representable by a relational graph (and vice versa)! The myth of RDF as the new answer to the challenge of graph databases has vanished! Navigation from triple to triple is as easy as navigation from tuple to tuple.

Switching to a different graphical representation of rows/tuples when looking at relational databases is really crucial here. This (mental) switch doesn't mean that the relational format has changed anyhow: It is just important to understand the particular role played by FK-PK connections between different rows as being the "real" links which are encoded into row form. This means that the PK value (a single one, no composite keys accepted!) of a row is regarded as the root of the tree "visualizing" a row. FK attributes (again always single column FKs!) are like entity-valued properties – the restriction to values as the only "things" that can play the object role in ER is essentially responsible for the (mis-)conception of rows (representing individual entities) as being unrelated (in the graphical sense) to other rows (or entities). As soon as we re-establish such links into relational thinking, RM is "back in business" as a serious format for representing collections of link data! We will see in the next chapter (on queries in SPARQL vs. queries in SQL) that a minor extension of the SQL syntax based on FK-PK links could even dramatically reduce the syntactic effort needed for queries navigating through paths in a graph.

Finally: Why am I using the proviso "whatever graph databases are" all the time? There is a big terminological misunderstanding underway at present, or better said: Even researchers are using their terms in a sloppy way (not for the first time!). *Graph database* is confused with *Graph DBMS*! A DBMS that provides specialized storage and data access technology which speeds up navigation through large collections of data which can be visualised as graphs (such as WWW structures, or links in a social network, e.g.) is one thing – a database consisting of such data is a completely different thing! A database is not a database (management) system. RDF is a data model, not a piece of DBMS technology!

You don't (necessarily) gain query evaluation speed when representing individual edges of a graph in RDF triple form rather than as RM tuples. Even the usage of a triple store doesn't help (as such) to gain performance as compared to using a relational store (organized as row store or column store or in a different manner internally). Only special evaluation mechanisms for special "graph queries" (whatever they are) will help gaining evaluation speed. Whether you need a new generation of DBMSs or whether improving the query optimizer and the ac-cess mechanisms of an existing DBMS product can be equally successful remains to be seen. But this is not our topic here, what matters: Data models and DB languages don't speed up query execution! As far as "real" graph databases are concerned: If every set of tuples and every set of triples has a directed graph counterpart (and vice versa), where does this leave the idea of RDF being better suited for representing inherently "graphy" data than RM?

Before changing topic let me again summarize what matters in the "Views" section of this chapter:

1. As a data representation format, RDF is no better for representing data that is (or can be) visualized as graphs than RM. Tuples have a natural graph visualization, too, which means that "graphy" data can be represented in tuples as in triples. None of the two formats is particularly "graph-friendly".

2. A single tuple is a more compact representation of atomic graphs which all share the same subject node, and where the predicates (labelling the edges emerging from this node) are encoded in the column headings (alias attribute) once for all rows in the table.

3. If turning away from the traditional graph view of a tuple as having an unlabelled root which represents the entire tuple towards a view where the PK value of the

resp. tuple is the root node we can even build connected complex graphs out of tables where individual tuples are graphically linked via FK references.

4. Don't mix DBMS technology (storage, access, query optimization) able to speed up graph queries over data interpreted as graphs with DBs containing "graph data". RDF and SPARQL are no "graph DBMS", but a data model and a query language, resp.

6   On Queries: Patterns vs. Literals

Let us now – finally, quite late, admittedly – turn to the query languages appearing in the title of this paper: SPARQL and SQL. Let me recall that we restricted this discussion of the core language features of SPARQL to a single-backend situation, thus we will not explicitly use any PREFIX clause in SPARQL queries. Both languages are huge if you look at the resp. grammars, SQL more so than SPARQL, but the latter rapidly catching up. Our goal is to come up with principle statements about the relationship between the two linguistic approaches, not side-tracked by "nitty-gritty" detail, but focusing on the essential characteristics, those that really matter.

6.1   Facts

On the surface, SPARQL queries have the look and feel of SQL queries, as the three main SQL keywords – SELECT, FROM and WHERE – are used, even retaining the basic semantics they have in SQL: SELECT clauses specify the structure of the output data elements, FROM specifies the input datasets (if there is a choice), and WHERE specifies which input data are to be considered in the output. Semantically, the analogy with SQL queries holds, too, as results of SPARQL queries are derived, unnamed datasets (as for SQL, where results are derived, unnamed tables)[11]. DISTINCT eliminates duplicates in SPARQL as in SQL. UNION allows for duplicate-free combination of two datasets into a common one, as in SQL. ORDER BY determines the sequence of presentation of output datasets, as in SQL. Since SPARQL 1.1, aggregate functions and GROUP BY clauses have been "inherited" from SQL, too.

However, that's how far the commonalities go! The basic mechanism of accessing triples in SPARQL is (at least syntactically) fundamentally different from that of SQL for accessing tuples. Whereas variables in SQL stand for entire tuples, and attributes (i.e., column names) are used as functions applied to such tuple variables for accessing tuple components, variables in SPARQL stand for triple components, and attributes are not used at all, mainly because they remain implicit in RDF (Subject, Predicate, Object), and are the same for all triples anyway.

Instead the basic syntactic features in the WHERE part (responsible for specifying those input triples from which the query results are to be constructed) are so-called *triple patterns*, sequences of three components separated by a space, each component being either a literal (in our restricted context without URIs and blank nodes) or a variable

---

[11] This is the "good, old" idea of an algebra in maths: The output of an expression is of the same kind as its input, thus allowing for nested expressions "consuming" the results of their sub-expressions.

(a string prefixed by a question mark, or an equivalent symbol). Thus, triple patterns are tuples potentially with variables represented without the ordinary tuple delimiters known from mathematics (i.e., without round brackets enclosing the tuple and without commas separating components).

Triple patterns are evaluated by *matching* them against variable-free triples in the input dataset, literals in the input replacing variables, each replacement generating a new candidate result. As in mathematics, different occurrences of the same variable in a query represent the same binding (different from the usage of variables in most programming languages offering "destructive assignment"). The SELECT clause determines those variables in the WHERE part the bindings to which ("matches" in the above sense) are considered relevant for the output (the remaining variables are treated as local to the WHERE part). In addition, a FILTER clause can be added after the triple pattern consisting of one or more comparisons restricting the range of variable bindings by comparing them to delimiter values using the well-known operators $>$, $>$, $=$ (or by other built-in functions). We postpone the discussion of the OPTIONAL clause in SPARQL queries till later.

In SQL, the WHERE part consists of comparisons between terms, whereas accessing tables and binding variables to tuples in these tables is achieved in the FROM part only. If a SPARQL triple pattern contains literals (i.e., constants from the perspective of logical syntax), such literals match only in case they appear identically in the input triples against which the patterns are matched. Thus, the same effect could be achieved by having nothing but variables in each triple pattern and pushing the equality condition to a subsequent FILTER part. If doing so, the "real" roles of the three parts become visible: SELECT specifies the structure of output elements in both cases, WHERE in SPARQL corresponds to FROM in SQL, and FILTER in SPARQL corresponds to WHERE in SQL. SPARQL's FROM clause is only relevant if several distinct datasets are used as input. [12]

Such linguistic details aside, the fundamental difference between both languages' basic data access mechanisms is in the role played by variables: In case of SQL they represent entire tuples, in case of SPARQL they access triple components rather than triples as a whole. The difference is well-motivated as we stated above, however, this difference in usage of variables for matching against structures is well-known in logic and in database theory: It is what distinguishes Codd's TRC (tuple relational calculus) [18] from Lacroix/Pirotte's DRC (domain relational calculus) [19] . TRC corresponds well with the tree representation of table rows, with attributes being selectors marking the edges of the trees turned into functions in TRC. DRC corresponds with the tuple view of table rows requiring a positional syntax where queries contain tuples with variables and where order of components matters. SQL is based on TRC, SPARQL is based on DRC – at least if you accept the "R" (standing for "relational") in the RDF context, but triples are tuples from the perspective of maths, so datasets are relations!

---

[12] Not such a particularly brilliant choice of keyword semantics, unless the SPARQL designers intended to vex the SQL community!

6.2 More Facts

SPARQL didn't "invent" the DRC style, however – there have been DRC languages in the relational DB field since long. Query-by-Example [20] was an early relational query language in the 1970s, which retained attention when MS Access offered a query design interface based on a drag-and-drop manipulation style which reflects QbE ideas. Much more significant, however, is Datalog [21] as a DRC language, designed in the 1970s, too, but having survived in the academic context since then: Datalog is the language of deductive database research, and – due to its close correspondence with pure PROLOG – it is the language used when discussing database problems in the area of Logic Programming.

Each triple pattern corresponds to a Datalog literal[13] without explicit predicate (i.e., relation name). A triple pattern like, e.g.,

?x friend_of ?y.

evaluated over an (implicit) dataset called ds here corresponds to the Datalog literal

ds(X,'friend_of',Y).

The dataset/relation name has to be made explicit, brackets are used for delimiting the parameter list, commas separate the list components, variables are written as (or starting with) capitals (as in Prolog), strings are quoted (as in SQL). Such patterns are evaluated over triples, represented as variable-free triple patterns, e.g.,

'John' friend_of 'Jack'.

In Datalog, a triple would be a fact, represented as a ground literal

ds('John','friend_of', 'Jack').

Matching a triple pattern against a triple in the dataset exactly corresponds to evaluation of a Datalog literal over a set of Datalog facts. This technique has been called *unification* in the context of the resolution inference technique in clausal logic since the 1950s [22], the term has been retained in Logic Programming and in Datalog.

In the "SPARQL world", matching is often called graph matching, referring to the graph representation of triples. As we saw in the previous chapter, tuples in RM (and thus in SQL and Datalog) have a graph representation, too – if we choose this "graph view", unification is a graph matching technique as well. If we look at triples in RDF in the linearized ("Turtle") form used above, it becomes obvious that graph matching is the same as unification. So again, it is all a matter of perspective and of terminology! But in essence, there is nothing new in evaluating (and writing down) triple patterns over triples as compared to evaluating (query) literals over sets of ground literals (facts) in Datalog. The main argument in favour of the Datalog terminology would be that it has been around since more than half a century in logic (which makes it "boring", as opposed to a new terminology making SPARQL innovative and "cool".

SPARQL WHERE parts consisting of more than one triple pattern are called basic graph patterns, consisting of triple patterns (each terminated by a dot), enclosed in set braces, e.g.:   ...WHERE {?x name ?n. ?x age 53. }

---

[13] Unfortunately, the term "literal" is ambiguous: In mathematical logic (and thus in Prolog and Datalog) a literal is an atomic formula (with or without negation) consisting of a predicate, i.e., a relation name, and a parameter list, where each parameter is a logical term, i.e., a constant, or a variable, or a function application. In SPARQL, a literal is a string, i.e., a constant in logical terminology.

Such a composite condition is to be read conjunctively, i.e., both triple patterns are supposed to match with a triple each, assigning the same binding to the two occurrences of variable ?x. This corresponds one-to-one to the body of a Datalog rule, where there are commas delimiting the individual literals, and comma stands for logical AND:

$$\ldots ds(X, \text{'name'}, N), ds(X, \text{'age'}, 53).$$

Dots are not needed at the end of each Datalog literal as there a list brackets terminating the component list of each such literal, but there is a separator (the conjunctive comma) explicitly between the literals. Set braces are not needed as delimiters for the entire complex condition, but there is a dot at the end of the rule body. But again: Basic group patterns are conjunctions of atomic statements, as are rule bodies in Datalog. All differences are syntactic sugar!

To conclude the story: In SPARQL, there may be FILTER clauses associated with individual triple patterns or entire basic graph patterns, together forming a *group graph pattern*, e.g.: $\ldots$ WHERE $\{$?x name ?n. ?x age ?a. FILTER a $<$50 $\}$

In Datalog, such comparison literals are "and-ed" with the database literals without any "warning" by means of an extra keyword (FILTER):

$$\ldots ds(X, \text{'name'}, N), ds(X, \text{'age'}, A), A <60.$$

But during evaluation, database literals (like the two ds-literals) are treated differently from comparison literals: Database literals generate variable bindings, whereas comparison literals "consume" such bindings (generated before by database literals in the same conjunction) and test them for truth or falsehood.

Strictly speaking, every match of an RDF literal (corresponding to a Datalog constant) in a triple pattern against a triple is an equality test rather than a binding generation. So if being consequent, only variables should be used in triple patterns and the testing part moved into a FILTER, such that the first example of a basic group pattern above would turn into a group graph pattern as follows:

$$\ldots \text{WHERE } \{ \text{?x name ?n. ?x age ?n. FILTER ?n} = 53 \}$$

In Datalog, explicit equality test rather than test by matching is possible, too:

$$\ldots ds(X, \text{'name'}, N), ds(X, \text{'age'}, A), A = 53.$$

Thus, the extra FILTER clause might even be useful as it makes the different modes of evaluation of atomic expressions explicit. However, FILTER is only useful if applied in the strict sense sketched here, where every testing/filtering step is made explicit in a FILTER clause. SQL is following such a strict strategy of "separating concerns": Variable bindings are created during evaluation of the FROM clause only. The WHERE clause in SQL is purely selective in testing the generated bindings for satisfying the specified condition. Datalog combines the two tasks (generation of bindings and testing conditions) inside a uniform conjunctive rule body without distinguishing the two parts syntactically.

Negation is a "notorical culprit" in Datalog, causing all kinds of difficulties and being responsible for concepts like safe rules, negation-as-failure and stratification. Syntactically, it is quite simple: Literals can be prefixed by a negation operator (either a symbol like #, or ?, or !, or simply the keyword **not**). Negative literals are "controlled" like comparison literals in that all variables they contain must be bound by evaluating positive database literals in the same conjunction before evaluating the negative literals.

The corresponding evaluation technique called *negation-as-failure* is a test technique, so negative literals are tests, too.

In SPARQL, there was no explicit negation initially – quite surprising! Since SPARQL 1.1, there is a NOT EXISTS construct which has to go into FILTER clauses, but uses triple patterns as parameter – the only exception from the rule that triple patterns (corresponding to positive database literals in Datalog) belong into the WHERE part proper. The syntax (using an explicit existential quantifier) has been derived from SQL (where it is used with an entire embedded query as operand, testing its answer table for non-emptiness). The semantics, however, is very close to Datalog's negation-as-failure test style. There is a tiny, but remarkable difference: If the pattern "inside" a NOT EXISTS expression contains a local variable, not bound at the time of evaluation, any match for this variable leads to the EXISTS being successful (as in SQL, where a single tuple in the embedded sub-query makes the EXISTS succeed). In Datalog, an auxiliary rule is needed as existential quantifiers are implicit at the beginning of every rule body (binding their local variables), never directly after a **not**, however.

For space reasons, we omit the (interesting) discussion of OPTIONAL clauses in SPARQL queries here. OPTIONAL is tightly related to the issue of non-applicable predicates in RDF (vs. null values in RM). They syntactically compete with IS NULL tests in SQL (Datalog doesn't "have" null values traditionally, for no obvious reasons).

6.3   Views (at Last)

The main observation I can identify in the present chapter is the observation that SPARQL is essentially much closer to Datalog than to SQL, despite the many efforts to include ever more SQL keywords into SPARQL. There have been numerous papers written on defining the semantics of SPARQL using Datalog or on using Datalog-like techniques for extending SPARQL by a rule concept (just to mention [23] as a particularly striking example of such a paper). But all of these contributions overlook (at least that is how I perceive them) the enormous closeness of SPARQL's triple patterns to Datalog literals and of SPARQL's graph matching to Datalog's unification. In this respect, SPARQL is quite far away from the basic syntactic style of SQL. On the other hand, we know that Datalog and SQL (at least as far as the common core concepts are concerned) are very close siblings, in as much as Datalog expressions can be easily and systematically mapped to (core) SQL expressions, and vice versa.

If Datalog were "sugared" by means of a whole bunch of SQL keywords (as SPARQL has been), the closeness between Datalog and SPARQL would be much more obvious. I believe myself, that introducing a more semantically motivated strategy of working with different datasets in RDF (not just because they come from different endpoints, but because they are concerned with important, distinct classes of "things" in one and the same domain) will lead to a SPARQL syntax where triple patterns are indeed prefixed directly by dataset names, not indirectly by using identifiers introduced in PREFIX clauses. Then a further step towards Datalog would have been done, revealing that triple patterns are indeed (like) Datalog literals.

A final idea at the end of this chapter: In SPARQL 1.1, the new (?) concept of a *path expression* has been introduced, allowing (among others) for a more compact style of expressing navigation along paths in a data graph. Rather than introducing variables for all the intermediate "things" through which a path runs (being object in the one, and subject in the next triple in a chain), a sequence of the predicates involved is simply formed. This technique isn't that new either, as it has been used in quite a number of OO query languages in the past. In the SQL context, however, this kind of technique is still missing. If following our proposal to exploit FK-PK links between rows in different (copies of) tables, a similar kind of path expression could be added to SQL's syntax (conservatively, quite important for standardization). Rather than having a single attribute attached to TRC variables (e.g., S.name), a sequence of such (FK) attributes in postfix notation could be allowed (e.g., S.father.mother.husband.name), abbreviating queries considerably by saving variables and – most importantly – JOIN occurrences. Furthermore, evaluation of such path expressions could be done more efficiently (by the DBMS) than via join algorithms.

7   Conclusion: A Tale of Two Languages

This paper has been an unusual one, at least in computer science. It consists almost entirely of text, even stronger, of narrative text, without any formalism, and with very few examples only. Texts like this are called essays in philology. So this was an essay about two models and two languages. Essays are a rare species in our branch of science, but I thought this genre to be appropriate regarding the contents and the intention of the paper.

Most of the paper has been rather unspectacular in style (at least that is how I perceive it). This didn't just happen to be the case, but the paper has been written like this on purpose. As most contributions on RDF/SPARQL (and Semantic Web & Co.) come along with a rather "hyped" style (at least that is how I perceive it), I intended to respond to this style in a "matter of fact" manner, trying to set a notable contrast to so many examples of "W3C hoopla" (to borrow the unserious expression from Mohan). Probably there is still quite a bit of controversial argumentation around in my "Views" parts, which causes protest by part of my audience, but that's intended, too.

I don't want to repeat all the summaries of chapters 3 to 6 here. Just so much: For me it was quite revealing to discover how close the RDF data model is to a specific schema design strategy for an RM database, and how similar the way from an ontology to an RDF database schema is to the way from an ER diagram to a relational schema. Discovering that the "graph database debate" is almost an idle fight, at least on the level of models and languages, was enlightening, too – let the SQL people enhance their query optimizers and storage architectures for "graph queries" (unfortunately not further discussed in this paper), and we will see what is left of the "graph DBMS" debate! The discovery that SPARQL finally did successfully, what Datalog couldn't achieve commercially – namely to base a relational query language an DRC rather than on TRC (as SQL did) was interesting, too. But why not position SPARQL directly in the research line established by Datalog rather than opportunistically "stealing" all kinds of SQL keywords? If SQL manages to overcome those weaknesses that motivated SPARQL designer to follow Datalog, than the many SQL keywords in SPARQL may finally turn against SPARQL, as SQL is already around all the time.

So let us finally answer the question in the title of this paper: *Should SQL surrender to SPARQL?* Remember that the motivation for this question – which wasn't meant purely rhetorical – was the observation that RDF/SPARQL topics have been dominating the research agendas of funding agencies, program committees and researchers alike for several years by now. My answer is quite clear: ***Certainly not!***

I think that there is just one notable piece of progress which SPARQL could claim in its favour, exactly the issue we omitted in this paper: Providing linguistic features for navigating over a network of distributed data resources contributing jointly for providing the answers to a given query. If SQL is extended by similar features, it would be competitive again – but this issue was just what I decided to omit from this paper.

References

1. Prud'hommeaux, E., Seaborne, A. (eds.): SPARQL Query Language for RDF, W3C Recommendation (January 15, 2008), http://www.w3.org/TR/rdf-sparql-query/
2. Klyne, G., Carroll, J. (eds.): Resource Description Framework (RDF): Concepts and Abstract Syntax, W3C Recommendation (February 10, 2004), http://www.w3.org/TR/2004/REC-rdf-concepts-20040210/
3. Date, C., Darwen, H.: A Guide to the SQL Standard, 4th edn. Addison Wesley (1997)
4. Prud'hommeaux, E., Bertails, A.: A Mapping of SPARQL to Conventional SQL, http://www.w3.org/2008/07/MappingRules/StemMapping
5. Chebotko, A., Lu, S., Fotouhi, F.: Semantics Preserving SPARQL-to-SQL Translation, DKE 68(10), 973–1000 (2009)
6. Mohan, C.: History Repeats Itself: Sensible and NonsenSQL Aspects of the NoSQL Hoopla. In: Proc. EDBT, pp. 11–16 (2013)
7. Atzeni, P., Jensen, C.S., et al.: The Relational Model is Dead, SQL is Dead, and I Don't Feel so Good Myself. ACM SIGMOD Record 42(3), 64–68 (2013)
8. Kumar, N.V., Kumar, A., Abhishek, K.: A Comprehensive Comparative Study of SPARQL and SQL. IJCSIT 2(4), 1706–(2011)
9. http://en.wikipedia.org/wiki/Back_tot_the_Future
10. Chen, P.: The Entity-Relationship Model – Toward a Unified View of Data. TODS 1(1), 9–36 (1976)
11. McGuinness, D., van Harmelen, F. (eds.): OWL Web Ontology Language Overview, W3C Recommendation (February 10, 2004), http://www.w3.org/TR/2004/REC-owl-features-20040210/
12. http://en.wikipedia.org/wiki/Ontology(informationscience)
13. The Unified Modeling Language User Guide. Addison Wesley (2005)
14. Hull, R., King, R.: Semantic Database Modelling: Survey, Applications, and Research Issues. ACM Computing Surveys 19(3), 201–260 (1987)
15. Smith, B.: Ontology. In: Floridi, L. (ed.) Blackwell Guide to the Philosophy of Computing and Information, pp. 155–166. Blackwell, Oxford (2003)
16. Harris, S., Seaborne, A. (eds.): SPARQL 1.1 Query Language, W3C Recommendation (March 21, 2013) http://www.w3.org/TR/sparql11-query/
17. Codd, E.: Extending the database relational model to capture more meaning. TODS 4(4), 397–434 (1979)
18. Codd, E.: A Relational Model for Large Shared Data Banks. CACM 13(6), 377–387 (1970)
19. Lacroix, M., Pirotte, A.: Domain-Oriented Relational Languages. In: Proc. 3rd VLDB, pp. 370–378 (1977)
20. Zloof, M.: Query by Example. In: Proc. IFIPS, pp. 431–438 (1975)

21. Ceri, S., Gottlob, G., Tanca, L.: What You Always Wanted to Know About Datalog (And Never Dared to Ask). TKDE 1(1), 146–166 (1989)
22. Robinson, J.A.: A Machine-Oriented Logic Based on the Resolution Principle. JACM 12(1), 23–41 (1965)
23. Bry, F., Furche, T., et al.: RDFLog: It's Like Datalog for RDF. In: Proc. WLP, pp. 17–26 (2008)

Balancing Energy Consumption for the Establishment of Multi-interface Networks[*]

Alessandro Aloisio[1] and Alfredo Navarra[2]

[1] Dipartimento di Ingegneria e Scienze dell'Informazione e Matematica,
Università degli Studi dell'Aquila, Italy
`alessandro.aloisio@univaq.it`
[2] Dipartimento di Matematica e Informatica,
Università degli Studi di Perugia, Italy
`alfredo.navarra@unipg.it`

Abstract. In heterogeneous networks, devices can communicate by means of multiple interfaces. By choosing which interfaces to activate (switch-on) at each device, several connections might be established. A connection is established when the devices at its endpoints share at least one active interface. Interfaces are associated with a cost defining the percentage of energy consumed to switch-on the corresponding interface.

In this paper, we consider the case where each device is limited to activate at most a fixed number p of its available interfaces in order to accomplish the required task. In particular, we consider the so-called Coverage problem. Given a network $G = (V, E)$, nodes V represent devices, edges E represent connections that can be established. The aim is to activate at most p interfaces at each node in order to establish all the connections defined by E. Parameter p implies a sort of balanced consumption among devices so that none of them suffers - in terms of consumed energy - for being exploited in the network more than others.

We provide an $\mathcal{NP}$-completeness proof for the feasibility of the problem even considering the basic case of $p = 2$ and unitary costs for all the interfaces. Then we provide optimal algorithms that solve the problem in polynomial time for different graph topologies and general costs associated to the interfaces.

1 Introduction

As technology advances and hardware costs reduce, very powerful devices are available for a wide range of applications. Moreover, heterogeneous devices may communicate to each other by means of different protocols and interfaces. The connection among heterogeneous devices might result as a fundamental mean for the communication among different local area networks that all together form

[*] Work partially supported by Research Grant 2010N5K7EB - PRIN 2010 - "ARS TechnoMedia", and "Reti di sensori e architetture distribuite di controllo e comunicazione wireless", MIUR Art 10, DM 593 - Prot MIUR DM38036.

G.F. Italiano et al. (Eds.): SOFSEM 2015, LNCS 8939, pp. 102–114, 2015.

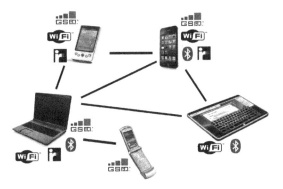

Fig. 1. Example of multi-interface network composed of heterogeneous devices connected by means of different interfaces

a wide area network, or the Internet. In this paper, we are interested in networks composed by heterogeneous devices that exploit different communication interfaces in order to establish desired connections. It is very common to find devices equipped with Bluetooth, Wi-Fi and GPRS interfaces but very few applications usually take advantage of such available heterogeneity. Selecting the best interfaces for specific connections depends on several factors. Namely, the choice might depend on the availability of a specific interface on some devices, the required communication bandwidth, the cost (in terms of energy consumption) for maintaining an active interface, the neighborhood, and so forth. Since devices are usually portable or mobile, a lot of effort must be devoted to energy consumption issues in order to prolong the network lifetime. In fact, the failure of a device due to draining batteries is something that could be delayed if suitable solutions are provided when building the connections of a desired network in accordance to specific requirements. This introduces challenging and natural optimization problems that must take care of different parameters at the same time. In general, a network of devices will be described by a graph $G = (V, E)$, where V represents the set of devices and E is the set of possible connections defined according to the distance between devices and the available interfaces that they share. Each $v \in V$ is associated with a set of available interfaces $\mathcal{W}(v)$. The set of all the possible available interfaces in the network is then determined by $\bigcup_{v \in V} \mathcal{W}(v)$; we denote the cardinality of this set by k. We say that a connection is established when the endpoints of the corresponding edge share at least one active interface. If an interface α is activated at some node u, then u consumes some energy $c(\alpha)$ for maintaining α as active, and it provides a maximum communication bandwidth $b(\alpha)$ with all its neighbors that share interface α. An example of a network instance is shown in Figure 1, where mobile phones, smartphones, tablets and laptops can communicate in a point-to point way by means of different interfaces and protocols such as IRdA, Bluetooth, Wi-Fi, GSM, Edge, UMTS and Satellite. All the possible connections can be established by means of at least one interface. Note that, some devices are not directly connected even though they share some interfaces. This can be due to many factors like for instance obstacles, distances, or used protocols. In order to provide a full instance, one should provide (if necessary) the cost for each device to activate a specific interface, and the corresponding bandwidth that can be handled. While bandwidths usually concern the interfaces, the energy spent by each device to activate a specific interface may vary substantially. In order to make easier the model, the costs might be referred to the percentage of battery consumed by each device, and hence it might be considered the same for each device with respect to a specific interface among the whole network. Nevertheless, different assumptions may lead to completely different problems that point out specific peculiarities of the composed networks.

In this paper, we are interested in the so-called *Coverage* problem, see [12]. It consists in finding the cheapest way to establish all the connections defined by an input graph G, no matter the interface used to accomplish each connection, no bandwidths requirements are provided. The problem only asks to ensure that for each edge of G, there is a common active interface at its endpoints. The objective is to minimize the overall cost of activation in the network.

Here, we add one further constraint to the original setting, in order to maintain under control the energy spent by single devices. We introduce parameter p with the meaning that no device can activate more than p interfaces. That is, instead of finding the solution that minimizes the overall cost due to the activation of the interfaces along the whole network, our aim is to minimize the overall cost subject to constraint p. The new setting is in favor of a more balanced energy consumption among nodes, hence prolonging the network lifetime.

Related Work. Multi-interface networks have been extensively studied during the last years. Many basic problems of standard network optimization are reconsidered in such a setting [3], focusing on issues related to routing [9] and network connectivity [5,10]. In [4,12], the so called *Coverage* problem has been studied. It consists in finding the cheapest way to establish all the connections defined by an input graph G, no matter the interface used to accomplish each connection. The problem only asks to ensure that for each edge of G, there is a common active interface at its endpoints. The objective is to minimize the overall cost of activation in the network. Another interesting objective function concerns the minimization of the maximum cost on a single node [8].

Connectivity issues have been addressed in [2,14]. The problem consists in finding the cheapest way to ensure the connectivity of the entire network. In other words, it aims to find at each node a subset of the available interfaces that must be activated in order to guarantee a path between every pair of nodes in G while minimizing the overall cost of the interfaces activated among the whole network. As for Coverage, another studied objective function is that of minimizing the maximum cost paid on a single node [8]. The Connectivity problem corresponds to a generalization of the well-know Minimum Spanning Tree problem.

In [14], the attention has been devoted to the so called *Cheapest path* problem. Here, the goal is to find a minimum cost subset of available interfaces that can

Table 1. Complexity of the *CMI(2)* problem. Parameters n, k and Δ are the number of nodes, the number of interfaces, and the maximum node degree of the input instance of *CMI(2)*, respectively.

Graph class	Unitary/Arbitrary Costs	Complexity of *CMI(2)*
Feasibility on graphs with $\Delta \geq 4$	unitary	$\mathcal{NP}$-complete
Complete bipartite graphs	arbitrary	solvable in $O(k^4 n)$
Complete graphs	arbitrary	solvable in $O(k^3 n)$
Rings	arbitrary	solvable in $O(k^3 n)$
Trees	arbitrary	solvable in $O(\Delta k^2 n)$
Paths	arbitrary	solvable in $O(kn)$

be activated in some nodes in order to guarantee a connecting path between two specified nodes of the network. This problem corresponds to a generalization of the well-know Shortest Path problem between two nodes in standard networks.

Another interesting study [13] investigates a basic problem like the *Maximum Matching*. As in its classical version, the problem asks for the maximum subset of connections that can be established concurrently without sharing any common node. That is, a solution is provided by a set of disjoint edges of the input graph, and hence each node appears in the solution at most once.

Bandwidth constraints have been addressed in [7], by studying flow problems on multi-interface networks. In that paper, each interface has been associated with one further parameter expressing the bandwidth that the interface can manage. The *Maximum Flow* problem as well as the *Minimum Cost Flow* problem aim to guarantee a connection between two given nodes, taking into account bandwidth constraints.

Our Results. In this paper, we are interested in the Coverage problem with the further constraint that each node of the network can activate at most p interfaces. From now on, the problem is denoted by *CMI(p)*, according to the name given in the original paper where the Coverage on multi-interface networks has been introduced [12]. In particular, we refer to what was called the *unbounded case*, where there was no bound on the number k of interfaces available over all the network. In our notation, we refer to that problem as *CMI(∞)*, since there was no restriction on the number of interfaces activated at the single nodes, that is $p = \infty$. Moreover, since this is the first study on this new variant of the problem, we start evaluating what can be computed for *CMI(2)*. The results contained in this paper are summarized in Table 1.

It turns out that already the feasibility of *CMI(2)* is $\mathcal{NP}$-complete for graphs of maximum degree $\Delta \geq 4$ (even if they are bipartite with interfaces of unitary costs). This reveals a completely different behavior with respect to the previously studied *CMI(∞)* where feasibility was trivially solved by the definition of the problem. In fact, an instance of *CMI(∞)* – as well as of *CMI(2)* – guarantees that each connection appearing in the input graph can be established. It follows that in $CMI(\infty)$ there always exists a feasible solution, by activating all the interfaces available at each node. Contrary, the constraint introduced in $CMI(2)$ by means of parameter $p = 2$ makes the feasibility problem very difficult. We then show how the problem can be optimally solved for many graph classes like complete graphs, complete bipartite graphs, rings, and trees for arbitrary costs associated to the interfaces. Interestingly, in all such graph classes but rings, $CMI(\infty)$ was difficult to be computed or even to approximate [14].

Outline. The next section provides definitions and notation in order to formally describe the $CMI(p)$ problem. Section 3 contains all the research contributions. We provide hardness results for general graphs, and optimal algorithms for various graphs classes that work in polynomial time. Finally, Section 4 contains conclusive remarks and a discussion of interesting open problems.

2   Definitions

For a graph $G = (V, E)$, we denote by V its node set of cardinality n, and by E its edge set of cardinality m. Unless otherwise stated, G is assumed to be undirected, connected, and without multiple edges or loops. For each node $v \in V$, let $\deg(v)$ be the degree of v, and Δ be the maximum degree among all the nodes. A global assignment of the interfaces to the nodes in V is given in terms of an appropriate interface assignment function W, as follows.

Definition 1. *A function $W : V \rightarrow 2^{\{1,2,\dots,k\}}$ is said to* cover *graph G if for each $\{u, v\} \in E$ we have $W(u) \cap W(v) \neq \emptyset$.*

The cost of activating an interface is assumed to be identical for all nodes and given by a cost function $c : \{1, \dots, k\} \rightarrow \mathbb{R}_+$, i.e., the cost of interface i is denoted as c_i. The considered $CMI(p)$ optimization problem is so formulated:

$CMI(p)$: Coverage in Multi-Interface Networks
Input: A graph $G = (V, E)$, an allocation of available interfaces $W : V \rightarrow 2^{\{1,\dots,k\}}$ covering graph G, an interface cost function $c : \{1, \dots, k\} \rightarrow \mathbb{R}_+$, an integer $p \geq 1$.
Solution: If possible, an allocation of active interfaces $W_A : V \rightarrow 2^{\{1,\dots,k\}}$ covering G such that for all $v \in V$, $W_A(v) \subseteq W(v)$ and $\lvert W_A(v) \rvert \leq p$; Otherwise, a negative answer.
Goal: Minimize the total cost of the active interfaces, $c(W_A) = \sum_{v \in V} \sum_{i \in W_A(v)} c_i$.

Note that we can consider two variants of the above problem: The cost function c can span over $\mathbb{R}_+$, or $c(i) = 1$, with $i = 1, \dots, k$ (*unit cost case*). In both cases we assume $k \geq 2$, since the case $k = 1$ admits an obvious unique solution (all nodes must activate their unique interface).

As mentioned above, $CMI(p)$ is a special case of the more general $CMI(\infty)$ problem (see [6,12]), where each node cannot activate more than p interfaces. Interestingly, on the one hand the basic variant with $p = 2$ results more difficult in general than $CMI(\infty)$. On the other hand, there are special graph classes that turn out to be much more affordable. For trees and complete graphs, for instance, $CMI(\infty)$ has been proved to be APX-hard and not approximable within $O(\log k)$, respectively, while $CMI(2)$ is polynomially solvable.

3   Complexity

In this section, we first study the hardness of the problem, and then we provide different algorithms designed for various graph topologies.

Theorem 1. *Finding a feasible solution for CMI(2) is $\mathcal{NP}$-complete for graphs with $\Delta \geq 4$, even for the unit cost case and bipartite graphs.*

Proof. The proof proceeds by a polynomial time reduction of the well-known *3-SAT* problem with bounded occurrences. The problem is known to be $\mathcal{NP}$-complete [11] and it can be stated as follows:

	3-SAT : 3-Satisfiability with bounded occurrences
Input:	Set S of variables, collection C of clauses over S. Each $c \in C$ satisfies $|c| \leq 3$ and, for each $s \in S$, there are at most 3 clauses in C that contain either s or $\bar{s}$.
Question:	Is there a satisfying truth assignment for C?

For our reduction we consider the unit cost interface case. Given an instance (C, S) of *3-SAT*, we transform it into an instance (G, W) of $CMI(2)$ with unit cost interfaces.

Each variable in S is represented in G by means of two interfaces i and $\bar{i}$ corresponding to the two literals generated by the variable. As shown in Figure 2, graph G can be divided into four main levels. At the first level there

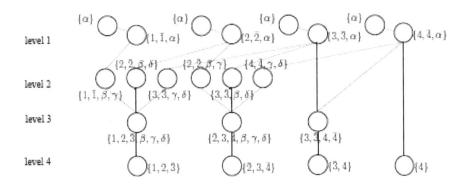

Fig. 2. A polynomial transformation from *3-SAT* to *CMI(2)*. The graph depicted in the figure is obtained from the input set of clauses $C = \{(1\lor2\lor3), (\bar{2}\lor3\lor\bar{4}), (\bar{3}\lor4), 4\}$.

are $|S|$ nodes. Each node is associated with three interfaces, two corresponding to the two literals of the represented variable, plus a new interface α common to all the nodes of this first level. At the second level of the graph, clauses composed of three literals are represented by three nodes, one for each literal. Each node is associated with the two interfaces related to the variable that generates the appearing literal plus two new interfaces among the set $\{\beta, \gamma, \delta\}$. The first node holds $\{\beta, \gamma\}$, the second $\{\beta, \delta\}$, and the third $\{\gamma, \delta\}$. At level three there is one node for each clause composed of three literals and one for each clause composed of two literals. The first type of nodes hold six interfaces, three corresponding to the literals appearing in the clause plus $\{\beta, \gamma, \delta\}$. The second type of nodes hold four interfaces related to the two variables appearing in the clause. Finally, the fourth level is composed of $|C|$ nodes, each associated with the interfaces corresponding to the literals appearing in the represented clause. For each node of the first level there is one further node holding only interface α. Moreover, each node of the first level is connected to all the nodes of the second level that share some interfaces, to the nodes of the third level corresponding to clauses composed of only two literals with which they share some interfaces, and to the nodes of the fourth level corresponding to clauses composed of only one literal with which they share some interface. The three nodes of the the second level corresponding to one clause are all connected to the node of the third level with six interfaces corresponding to the same clause. Finally, nodes of the third level are connected to those of the fourth level according to the represented clauses, i.e. one connection for each node.

Since we are considering 3-SAT with bounded occurrences of the variables, the originated graph admits a maximum degree $\Delta = 4$. In fact, nodes of level one corresponding to variables in 3-SAT can have at most three connections toward the other levels, one for each clause that contains the considered variable. For the other levels, the maximum degree of four simply follows by construction. The number of nodes in G is proportional to the number of literals plus the number of clauses appearing in the input instance of 3-SAT, while the number of interfaces is exactly $k = 2|S|+4$, hence the transformation requires polynomial time. Moreover, G is bipartite since each level is connected to at most one level above and one below. Given a satisfying truth assignment for an instance of 3-SAT, we obtain an activation function that solves the corresponding instance of $CMI(2)$, and viceversa.

($\Rightarrow$) All the nodes holding interface α activate it. At each node of level one, we activate the interface corresponding to the literals set to true by the truth assignment for each variable. Hence, all nodes at level one have now activated two interfaces. All the connections from level one are then satisfied by activate at the connected nodes the interfaces corresponding to the literals set to true by the truth assignment for each variable. At level two, one further interface is chosen among $\{\beta, \gamma, \delta\}$ according to the connections with level three. That is, consider the three nodes at level two connected to the same node of level three corresponding to one clause. Chose one of the literals l that makes true such a clause. Then, nodes at level two not holding l activate the common interface they have among $\{\beta, \gamma, \delta\}$. In doing so, the connections between nodes of levels two and three are established by activating at the nodes of level three the same interface extracted from $\{\beta, \gamma, \delta\}$ and interface l. Finally, connections between nodes of level three and four are satisfied again by activating at the nodes of level four the interfaces corresponding to the literals set to true by the truth assignment for each variable. Hence, all the connections are established and the constraint on $p = 2$ holds.

($\Leftarrow$) Given an activation function that solves the instance of *CMI(2)*, we can easily obtain a satisfying truth assignment for corresponding instance of *3-SAT*. In fact, the constraints implied by the described instances impose to activate the interface corresponding to one literal for each variable. Such an assignment must satisfy all the clauses of the *3-SAT* instance since nodes of the fourth level are associated only to the interfaces corresponding to the literals appearing in the represented clauses. Since each node of the fourth level activates at least one interface in order to establish its unique connection, then at least one literal for each clause is set to true.   □

Similarly to the approach adopted in [1], as the feasibility of *CMI(2)* has been shown to be $\mathcal{NP}$-complete, it is interesting to understand whether the optimization problem is harder than *CMI(∞)* even for specific graph topologies.

Theorem 2. *CMI(2) is solvable on complete graphs in $O(k^3 n)$ time.*

Proof. The proof is constructive, and it is divided into two main steps in order to check whether there exists a solution that makes use of one, or three interfaces over all the network. In fact, using just two or more than three different interfaces over all the network does not provide any advantage. Let $G = (V, E)$ be a complete graph. In order to construct a feasible solution using only two interfaces over all the network means to assign to each node either one or two interfaces. If there exists a solution with a node activating only one interface, it means that all the nodes must hold such an interface. It follows that the solution with all the nodes activating one same interface is cheaper. Hence, the only alternative is to activate exactly two interfaces at all the nodes.

If the solution is composed of just two interfaces over all the network, it means that all the nodes activate the same two interfaces, and hence the solution with only one interface is cheaper.

If the final optimal solution contains exactly three interfaces over all the network, then all the nodes activate two interfaces among the chosen three. In this way, all the nodes share a common active interface. Again, if a node activates only one interface then all the nodes must activate the same interface against the assumption on the optimality. The two interfaces to be activated at each node are the cheapest ones among those available. Overall, we need $\binom{k}{3}$ trials for a total complexity of $O(k^3 n)$.

Solutions with more than three interfaces over all the network are certainly not optimal as each node can activate at most $p = 2$ interfaces and every pair of nodes must share one active interface.   □

In contrast with the hardness result on bipartite graphs, the next theorem addresses the case of complete bipartite graphs.

Theorem 3. *CMI(2) is solvable on complete bipartite graphs in $O(k^4 n)$ time.*

Proof. The proof is constructive, and it is divided into five main steps in order to check whether there exists a solution that makes use of one, two, three, four interfaces over all the network, plus a final case to show that using more than four different interfaces does not provide any advantage. Let $G = (V, E)$ be a complete bipartite graph with nodes V that can be represented as two sets X and Y so that each edge in E has one endpoint in X and the other in Y, for any $u \in X$ and $v \in Y$.

Among all the solutions provided by the next phases, the algorithm will chose the cheapest one.

One interface case. In linear time it is possible to check whether all the nodes of the input graph hold one (or more) same interface, and we save the one of minimum cost, if any.

Two interfaces case. If the final optimal solution contains exactly two interfaces over all the nodes, it must be structured as follows: All the nodes in X (or in Y, resp.) activate the same two interfaces α, β, and every node in Y (or in X, resp.) activates either α or β. No other combination is possible to establish all the connections. These are at most $2\binom{k}{2}$ trials.

Three interfaces case. If the final optimal solution contains exactly three interfaces over all the nodes, it must be structured as follows. Consider three interfaces α, β, and γ. There are two possible sub-cases:

- All the nodes in both X and Y activate two interfaces (the cheapest ones if possible) among the selected triple
- in X (or in Y, resp.) all the nodes activate two interfaces in the form $\{\alpha, \beta\}$, $\{\alpha, \gamma\}$ while in Y (or X, resp.) there can be nodes activating just interface $\{\alpha\}$ or $\{\beta, \gamma\}$.

By activating the interfaces as above, each connections is clearly established. There are $\binom{k}{3}$ triples, and for each one, only the second subcase generates 6 trials, 3 given by fixing interface α, times 2 given by the choice of X and Y.

Four interfaces case. If the final optimal solution contains exactly four interfaces over all the nodes, it must be structured as follows. Consider four interfaces α, β, γ, and δ. There are two possible sub-cases:

- Nodes in X activate one couple of interfaces among $\{\alpha, \beta\}$ or $\{\gamma, \delta\}$, nodes in Y activate one couple among $\{\alpha, \gamma\}$, $\{\alpha, \delta\}$, $\{\beta, \gamma\}$, or $\{\beta, \delta\}$.
- Nodes in X activate one couple of interfaces among $\{\alpha, \beta\}$, $\{\alpha, \gamma\}$, or $\{\beta, \delta\}$, nodes in Y activate one couple among $\{\alpha, \beta\}$, $\{\alpha, \delta\}$, or $\{\beta, \gamma\}$.

By activating the interfaces as above, each connections is clearly established. There are $\binom{k}{4}$ choices for the interfaces, and for each one, the first subcase generates 6 trials, 3 given by the ways of mixing the four interfaces, times 2 given by the choice of X and Y. The second subcase gives rise to 12 trials, 6 given by the ways of mixing the four interfaces according to the description, times 2 given by the choice of X and Y.

More than four interfaces case. First we note that there cannot be a node in X activating only one interface as otherwise the solution must belong to either the one interface or the two interfaces cases.

Assume in X (or similarly in Y) there are two nodes activating overall four different interfaces by means of two different couples of interfaces $\{\alpha, \beta\}$ and $\{\gamma, \delta\}$. Then, in Y all nodes must activate one interface from the first couple and another from the second couple. Overall, only four interfaces are activated.

Let us consider a node in X that activates interfaces $\{\alpha, \beta\}$. Then, every node in Y must activate at least one interface among α and β. In order to make use of five interfaces over all the network, and by the above case, without loss of generality there must be in X a node activating $\{\alpha, \gamma\}$. It follows that every node in Y that activates β but not α must activate γ. If there are no of such nodes, all nodes in Y can activate α, and hence we are back to the one or two or three interfaces cases. In order to make use of five interfaces in Y, there must be a node in Y activating interfaces $\{\alpha, \delta\}$, that is, a new interface δ can be paired only with α. Hence, we have two nodes in Y activating two different couples of interfaces $\{\beta, \gamma\}$ and $\{\alpha, \delta\}$. By the above arguments, this implies four interfaces over all the network.

Once computed the optimal solutions for each of the above cases, the cheapest one is selected. By summing up all the required trials that the algorithm must evaluate, the outcoming time complexity is $O\left(\binom{k}{4}n\right) = O(k^4 n)$.   □

One interesting consequence of the above theorem is that *CMI(2)* can be optimally solved on star graphs where *CMI(∞)* is already *APX*-hard.

Theorem 4. *CMI(2) is solvable on trees in $O(k^2 \Delta n)$ time.*

Proof. Make the input tree rooted by choosing arbitrarily one node r as the root. Apart for the leaves, to each node holding j interfaces, we associate a list of dimension at most $j + \binom{j}{2}$. That is, j entries corresponding to the j interfaces and $\binom{j}{2}$ entries corresponding to the possible couples of interfaces that it can activate in a solution. A leaf with j interfaces is associated with a list of dimension just j since a leaf is never required to activate two interfaces. The i-th entry records the cost of activating the i-th interface belonging to the leaf. Each node x whose children are only leaves can fulfill its list with the costs of the solution for the subtree rooted at x when activating the interfaces corresponding to the considered entry of the list, if possible. In the case a considered entry - corresponding to either one or two interfaces activated at x - can not provide a feasible solution, the entry is set to ∞. This check costs $\deg(x)$ for each entry, for a total cost of $\left(j + \binom{j}{2}\right)\deg(x)$. Now, every internal node whose children have been already evaluated can be considered for updating its list. By proceeding this way until the root, the optimal solution is the cheapest one calculated at r. In total, the algorithm requires $O(k^2 \Delta n)$ steps.   □

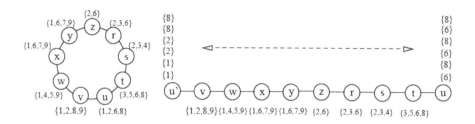

Fig. 3. All the six possible path instances arising from the shown ring cut at $\{u,v\}$

By applying the above theorem on paths, it follows that *CMI(2)* can be solved on such graphs in $O(k^2n)$ steps. It also follows that on rings, *CMI(2)* can be easily solved by exploiting the result on paths. Before providing more details on the resolution algorithm for rings, it is better to point out that another approach to solve *CMI(2)* with arbitrary costs on paths can be that used in [14] designed for the Cheapest Path problem. In fact, when considering paths as input, *CMI(∞)*, *CMI(2)* and Cheapest Paths coincide. This induces a time complexity of $O(kn)$ by means of a transformation of the input path to an instance where to apply the standard Dijkstra's algorithm for shortest paths.

Corollary 1. *CMI(2) is solvable on paths in $O(kn)$ time.*

The algorithm for paths can be used to solve *CMI(2)* on rings with a time complexity of $O(k^3n)$ by comparing the solutions found on the paths obtained by cutting the ring at one arbitrary edge, and taking care of the costs required to establish the excluded connection.

Theorem 5. *CMI(2) is solvable on rings in $O(k^3n)$ time.*

Proof. Consider one arbitrary node u whose neighbors are t and v in the input ring R. The algorithm works on the path P obtained by cutting R on $\{u,v\}$, adding a new neighbor u' to v, and defining a new covering function W^P for P. As shown in Figure 3, the new covering function W^P is equivalent to the original function W defined on R but for nodes u and u'. Let $\alpha \in W(u) \cap W(t)$ and $\beta \in W(u) \cap W(v)$ with possibly $\alpha \equiv \beta$, then for each of such occurrences that are at most $k + \binom{k}{2}$, $W^P(u) = \alpha$ and $W^P(u') = \beta$. Let W_A^P be the activation function evaluated by the algorithm for the cheapest solution arising among all the considered paths. Then, the activation function W_A for R is equivalent to W_A^P but for node u (and of course node u' that does not appear in R) to which it assigns $W_A(u) = W_A^P(u) \bigcup W_A^P(u')$. The computed solution is clearly optimal, as it equals to an exhaustive search among all possible settings. From the above description, and by Corollary 1, we obtain a time complexity of $O\left(\left(k + \binom{k}{2}\right)kn\right) = O(k^3n)$. $\qquad\square$

4 Conclusion

In the context of multi-interface networks, we have proposed a variant to the Coverage problem. Our modification to the original model concerns the further constraint for which each node can activate at most p interfaces. The new defined problem *CMI(p)* turns out to be much more difficult in general with respect to the basic *CMI(∞)*. In fact, the feasibility version of *CMI(p)* has been shown to be $\mathcal{NP}$-complete. However, we provide polynomial time optimal algorithms for *CMI(2)* with arbitrary costs with respect to many important graph topologies where the original *CMI(∞)* was proven to be hard even to be approximated. Namely, we have solved the problem on complete graphs, complete bipartite graphs, trees, rings, and paths.

As future work, it would be interesting to study what can be done when the input instance is assured to admit a solution. Moreover, since the completeness proof holds for graphs with $\Delta \geq 4$, while we solved the problem when $\Delta \leq 2$ (that is, paths and rings), it remains to show what happens for subcubic graphs, i.e. $\Delta \leq 3$. Distributed algorithms also deserve investigation.

References

1. Aloisio, A., Arbib, C., Marinelli, F.: Cutting stock with no three parts per pattern: Work-in-process and pattern minimization. Discrete Optimization 8, 315–332 (2011)
2. Athanassopoulos, S., Caragiannis, I., Kaklamanis, C., Papaioannou, E.: Energy-efficient communication in multi-interface wireless networks. Theory of Computing Systems 52, 285–296 (2013)
3. Bahl, P., Adya, A., Padhye, J., Walman, A.: Reconsidering wireless systems with multiple radios. SIGCOMM Computer Communication Review 34(5), 39–46 (2004)
4. Caporuscio, M., Charlet, D., Issarny, V., Navarra, A.: Energetic Performance of Service-oriented Multi-radio Networks: Issues and Perspectives. In: Proc. of the 6th Int'l Workshop on Software and Performance (WOSP), pp. 42–45. ACM (2007)
5. Cavalcanti, D., Gossain, H., Agrawal, D.: Connectivity in multi-radio, multi-channel heterogeneous ad hoc networks. In: Proc. of the 16th Int'l Symp. on Personal, Indoor and Mobile Radio Communications (PIMRC), pp. 1322–1326. IEEE (2005)
6. D'Angelo, G., Di Stefano, G., Navarra, A.: Multi-interface wireless networks: Complexity and algorithms. In: Ibrahiem, S.R., El Emary, M.M. (eds.) Wireless Sensor Networks: From Theory to Applications, pp. 119–155. CRC Press, Taylor & Francis Group (2013)
7. D'Angelo, G., Di Stefano, G., Navarra, A.: Flow problems in multi-interface networks. IEEE Transactions on Computers 63, 361–374 (2014)
8. D'Angelo, G., Di Stefano, G., Navarra, A.: Minimize the maximum duty in multi-interface networks. Algorithmica 63(1–2), 274–295 (2012)
9. Draves, R., Padhye, J., Zill, B.: Routing in multi-radio, multi-hop wireless mesh networks. In: Proc. of the 10th Int'l Conf. on Mobile Computing and Networking (MobiCom), pp. 114–128. ACM (2004)
10. Faragó, A., Basagni, S.: The effect of multi-radio nodes on network connectivity–a graph theoretic analysis. In: Proc. of the 19th Int'l Symp. on Personal, Indoor and Mobile Radio Communications (PIMRC), pp. 1–5. IEEE (2008)

11. Garey, M.R., Johnson, D.S.: Computers and Intractability, A Guide to the Theory of NP-Completeness. W.H. Freeman and Company, New York (1979)
12. Klasing, R., Kosowski, A., Navarra, A.: Cost Minimization in Wireless Networks with a Bounded and Unbounded Number of Interfaces. Networks 53(3), 266–275 (2009)
13. Kosowski, A., Navarra, A., Pajak, D., Pinotti, C.M.: Maximum matching in multi-interface networks. Theoretical Computer Science 507, 52–60 (2013)
14. Kosowski, A., Navarra, A., Pinotti, M.C.: Exploiting Multi-Interface Networks: Connectivity and Cheapest Paths. Wireless Networks 16(4), 1063–1073 (2010)

The Maximum k-Differential Coloring Problem

Michael A. Bekos[1], Michael Kaufmann[1], Stephen Kobourov[2], and Sankar Veeramoni[2]

[1] Wilhelm-Schickard-Institut für Informatik - Universität Tübingen, Germany
{bekos,mk}@informatik.uni-tuebingen.de
[2] Department of Computer Science - University of Arizona, Tucson AZ, USA
{kobourov,sankar}@cs.arizona.edu

Abstract. Given an n-vertex graph G and two positive integers $d, k \in \mathbb{N}$, the (d, kn)-differential coloring problem asks for a coloring of the vertices of G (if one exists) with distinct numbers from 1 to kn (treated as *colors*), such that the minimum difference between the two colors of any adjacent vertices is at least d. While it was known that the problem of determining whether a general graph is $(2, n)$-differential colorable is NP-complete, our main contribution is a complete characterization of bipartite, planar and outerplanar graphs that admit $(2, n)$-differential colorings. For practical reasons, we also consider color ranges larger than n, i.e., $k > 1$. We show that it is NP-complete to determine whether a graph admits a $(3, 2n)$-differential coloring. The same negative result holds for the $(\lfloor 2n/3 \rfloor, 2n)$-differential coloring problem, even in the case where the input graph is planar.

1 Introduction

Several methods for visualizing relational datasets use a map metaphor where objects, relations between objects and clusters are represented as cities, roads and countries, respectively. Clusters are usually represented by colored regions, whose boundaries are explicitly defined. The 4-coloring theorem states that four colors always suffice to color any map such that neighboring countries have distinct colors. However, if not all countries of the map are contiguous and the countries are not colored with unique colors, it would be impossible to distinguish whether two regions with the same color belong to the same country or to different countries. In order to avoid such ambiguity, this necessitates the use of a unique color for each country; see Figure 1.

However, it is not enough to just assign different colors to each country. Although human perception of color is good and thousands of different colors can be easily distinguished, reading a map can be difficult due to color constancy and color context effects [19]. Dillencourt et al. [6] define a good coloring as one in which the colors assigned to the countries are visually distinct while also ensuring that the colors assigned to adjacent countries are as dissimilar as possible. However, not all colors make suitable choices for coloring countries and a "good" color palette is often a gradation of certain map-like colors [4]. In more restricted scenarios, e.g., when a map is printed in gray scale, or when the countries in a given continent must use different shades of a predetermined color, the color space becomes 1-dimensional.

This 1-dimensional fragmented map coloring problem is nicely captured by the *maximum differential coloring problem* [5,15,16,23], which we slightly generalize in

G.F. Italiano et al. (Eds.): SOFSEM 2015, LNCS 8939, pp. 115–127, 2015.

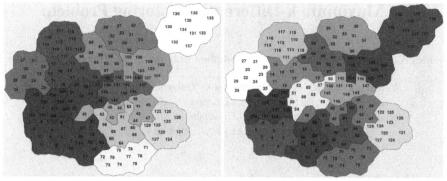

(a) Colored with random assignment of colors   (b) Colored with max. differential coloring

Fig. 1. Illustration of a map colored using the same set of colors obtained by the linear interpolation of blue and yellow. There is one country in the middle containing the vertices 40-49 which is fragmented into three small regions.

this paper: Given a map, define the *country graph* $G = (V, E)$ whose vertices represent countries, and two countries are connected by an edge if they share a non-trivial geographic boundary. Given two positive integers $d, k \in \mathbb{N}$, we say that G is (d, kn)-differential colorable if and only if there is a coloring of the n vertices of G with distinct numbers from 1 to kn (treated as *colors*), so that the *minimum color distance* between adjacent vertices of G is at least d. The *maximum k-differential coloring* problem asks for the largest value of d, called the *k-differential chromatic number* of G, so that G is (d, kn)-differential colorable. Note that the traditional *maximum differential coloring problem* corresponds to $k = 1$.

A natural reason to study the maximum k-differential coloring problem for $k > 1$ is that using more colors can help produce maps with larger differential chromatic number. Note, for example, that a star graph on n vertices has 1-differential chromatic number (or simply *differential chromatic number*) one, whereas its 2-differential chromatic number is $n + 1$. That is, by doubling the number of colors used, we can improve the quality of the resulting coloring by a factor of n. This is our main motivation for studying the maximum k-differential coloring problem for $k > 1$.

Related Work. The maximum differential coloring problem is a well-studied problem, which dates back in 1984, when Leung et al. [15] introduced it under the name "separation number" and showed its NP-completeness. It is worth mentioning though that the maximum differential coloring problem is also known as "dual bandwidth" [23] and "anti-bandwidth" [5], since it is the complement of the *bandwidth minimization problem* [17]. Due to the hardness of the problem, heuristics are often used for coloring general graphs, e.g., LP-formulations [8], memetic algorithms [1] and spectral based methods [13]. The differential chromatic number is known only for special graph classes, such as Hamming graphs [7], meshes [20], hypercubes [20,21], complete binary trees [22], complete m-ary trees for odd values of m [5], other special types of trees [22], and complements of interval graphs, threshold graphs and arborescent comparability graphs [14]. Upper bounds on the differential chromatic number are given by Leung et al. [15] for connected graphs and by Miller and Pritikin [16] for bipartite graphs. For a more detailed bibliographic overview refer to [2]. Note that in addition to map-coloring, the maximum differential coloring problem is motivated by the *radio frequency assignment problem*, where n transmitters have to be assigned n frequencies, so that interfering transmitters have frequencies as far apart as possible [12].

Our Contribution. In Section 2, we present preliminary properties and bounds on the k-differential chromatic number. One of them guarantees that any graph is $(1, n)$-differential colorable; an arbitrary assignment of distinct colors to the vertices of the input graph guarantees a minimum color distance of one (see Lemma 1). So, the next reasonable question to ask is whether a given graph is $(2, n)$-differential colorable. Unfortunately, this is already an NP-complete problem (for general graphs), since a graph is $(2, n)$-differential colorable if and only if its complement has a Hamiltonian path [15]. This motivates the study of the $(2, n)$-differential coloring problem for special classes of graphs. In Section 3, we present a complete characterization of bipartite, outer-planar and planar graphs that admit $(2, n)$-differential colorings.

In Section 4, we double the number of available colors. As any graph is $(2, 2n)$-differential colorable (due to Lemma 1; Section 2), we study the $(3, 2n)$-differential coloring problem and we prove that it is NP-complete for general graphs (Theorem 4; Section 4). We also show that testing whether a given graph is $(k + 1, kn)$-differential colorable is NP-complete (Theorem 5; Section 4). On the other hand, all planar graphs are $(\lfloor n/3 \rfloor + 1, 2n)$-differential colorable (see Lemma 3; Section 2) and testing whether a given planar graph is $(\lfloor 2n/3 \rfloor, 2n)$-differential colorable is shown to be NP-complete (Theorem 6; Section 4). In Section 5, we provide a simple ILP-formulation for the maximum k-differential coloring problem and experimentally compare the optimal results obtained by the ILP formulation for $k = 1$ and $k = 2$ with GMap, which is a heuristic based on spectral methods developed by Hu et al. [10]. We conclude in Section 6 with open problems and future work.

2   Preliminaries

The maximum k-differential coloring problem can be easily reduced to the ordinary differential coloring problem as follows: If G is an n-vertex graph that is input to the maximum k-differential coloring problem, create a disconnected graph G' that contains all vertices and edges of G plus $(k - 1) \cdot n$ isolated vertices. Clearly, the k-differential chromatic number of G is equal to the 1-differential chromatic number of G'. A drawback of this approach, however, is that few results are known for the ordinary differential coloring problem, when the input is a disconnected graph. In the following, we present some immediate upper and lower bounds on the k-differential chromatic number for connected graphs.

Lemma 1. *The k-differential chromatic number of a connected graph is at least k.*

Proof. Let G be a connected graph on n vertices. It suffices to prove that G is (k, kn)-differential colorable. Indeed, an arbitrary assignment of distinct colors from the set $\{k, 2k, \ldots, kn\}$ to the vertices of G guarantees a minimum color distance of k.   □

Lemma 2. *The k-differential chromatic number of a connected graph $G = (V, E)$ on n vertices is at most $\lfloor \frac{n}{2} \rfloor + (k-1)n$.*

Proof. The proof is a straightforward generalization of the proof of Yixun and Jin-jiang [23] for the ordinary maximum differential coloring problem. One of the vertices of G has to be assigned with a color in the interval $[\lceil \frac{n}{2} \rceil, \lceil \frac{n}{2} \rceil + (k-1)n]$, as the size of this interval is $(k-1)n + 1$ and there can be only $(k-1)n$ unassigned colors. Since G is connected, that vertex must have at least one neighbor which (regardless of its color) would make the difference along that edge at most $kn - \lceil \frac{n}{2} \rceil = \lfloor \frac{n}{2} \rfloor + (k-1)n$. □

Lemma 3. *The k-differential chromatic number of a connected m-colorable graph $G = (V, E)$ on n vertices is at least $\lfloor \frac{(k-1)n}{m-1} \rfloor + 1$.*

Proof. Let $C_i \subseteq V$ be the set of vertices of G with color i and c_i be the number of vertices with color i, $i = 1, \ldots, m$. We can show that G is $(\lfloor \frac{(k-1)n}{m-1} \rfloor + 1, kn)$-differential colorable by coloring the vertices of C_i with colors from the following set: $[(\sum_{j=1}^{i-1} c_j) + 1 + (i-1)\lfloor \frac{(k-1)n}{m-1} \rfloor, (\sum_{j=1}^{i} c_j) + (i-1)\lfloor \frac{(k-1)n}{m-1} \rfloor]$ □

3 The (2,n)-Differential Coloring Problem

In this section, we provide a complete characterization of (i) bipartite graphs, (ii) ou-terplanar graphs and (iii) planar graphs that admit $(2, n)$-differential coloring. Central to our approach is a result of Leung et al. [15] who showed that a graph G has $(2, n)$-differential coloring if and only if the complement G^c of G is Hamiltonian. As a consequence, if the complement of G is disconnected, then G has no $(2, n)$-differential coloring.

In order to simplify our notation scheme, we introduce the notion of *ordered differential coloring* (or simply *ordered coloring*) of a graph, which is defined as follows. Given a graph $G = (V, E)$ and a sequence $S_1 \to S_2 \to \ldots \to S_k$ of k disjoint subsets of V, such that $\cup_{i=1}^{k} S_i = V$, an *ordered coloring* of G implied by the sequence $S_1 \to S_2 \to \ldots \to S_k$ is one in which the vertices of S_i are assigned colors from $(\sum_{j=1}^{i-1} |S_j|) + 1$ to $\sum_{j=1}^{i} |S_j|$, $i = 1, 2, \ldots, k$.

Theorem 1. *A bipartite graph admits a $(2, n)$-differential coloring if and only if it is not a complete bipartite graph.*

Proof. Let $G = (V, E)$ be an n-vertex bipartite graph, with $V = V_1 \cup V_2$, $V_1 \cap V_2 = \emptyset$ and $E \subseteq V_1 \times V_2$. If G is a complete bipartite graph, then its complement is disconnected. Therefore, G does not admit a $(2, n)$-differential coloring. Now, assume that G is not complete bipartite. Then, there exist at least two vertices, say $u \in V_1$ and $v \in V_2$, that are not adjacent, i.e., $(u, v) \notin E$. Consider the ordered coloring of G implied by the sequence $V_1 \setminus \{u\} \to \{u\} \to \{v\} \to V_2 \setminus \{v\}$. As u and v are not adjacent, it follows that the color difference between any two vertices of G is at least two. Hence, G admits a $(2, n)$-differential coloring. □

Lemma 4. *An outerplanar graph with $n \geq 6$ vertices, that does not contain $K_{1,n-1}$ as a subgraph, admits a 3-coloring, in which each color set contains at least 2 vertices.*

Proof. Let $G = (V, E)$ be an outerplanar graph with $n \geq 6$ vertices, that does not contain $K_{1,n-1}$ as a subgraph. As G is outerplanar, it admits a 3-coloring [18]. Let $C_i \subseteq V$ be the set of vertices of G with color i and c_i be the number of vertices with color i, that is $c_i = |C_i|$, for $i = 1, 2, 3$. W.l.o.g. let $c_1 \leq c_2 \leq c_3$. We further assume that each color set contains at least one vertex, that is $c_i \geq 1$, $i = 1, 2, 3$. If there is no set with less than 2 vertices, then the lemma clearly holds. Otherwise, we distinguish three cases:

Case 1: $c_1 = c_2 = 1$ and $c_3 \geq 4$. W.l.o.g. assume that $C_1 = \{a\}$ and $C_2 = \{b\}$. As G is outerplanar, vertices a and b can have at most 2 common neighbors. On the other hand, since G has at least 6 vertices, there exists at least one vertex, say $c \in C_3$, which is not a common neighbor of a and b. W.l.o.g. assume that $(b, c) \notin E$. Then, vertex c can be colored with color 2. Therefore, we derive a new 3-coloring of G for which we have that $c_1 = 1$, $c_2 = 2$ and $c_3 \geq 3$.

Case 2: $c_1 = 1$, $c_2 = 2$ and $c_3 \geq 3$: W.l.o.g. assume that $C_1 = \{a\}$ and $C_2 = \{b, b'\}$. First, consider the case where there exists at least one vertex, say $c \in C_3$, which is not a neighbor of vertex a. In this case, vertex c can be colored with color 1 and a new 3-coloring of G is derived with $c_1 = c_2 = 2$ and $c_3 \geq 3$, as desired. Now consider the more interesting case, where vertex a is a neighbor of all vertices of C_3. As G does not contain $K_{1,n-1}$ as a subgraph, either vertex b or vertex b' is not a neighbor of vertex a. W.l.o.g. let that vertex be b, that is $(a, b) \notin E$. As G is outerplanar, vertices a and b' can have at most 2 common neighbors. Since G has at least 6 vertices and vertex a is a neighbor of all vertices of C_3, there exist at least one vertex, say $c \in C_3$, which is not adjacent to vertex b', that is $(b', c) \notin E$. Therefore, we can color vertex c with color 2 and vertex b with color 1 and derive a new 3-coloring of G for which we have that $c_1 = c_2 = 2$ and $c_3 \geq 2$, as desired.

Case 3: $c_1 = 1$, $c_2 \geq 3$ and $c_3 \geq 3$: W.l.o.g. assume that $C_1 = \{a\}$. Then, there exists at least one vertex, say $c \in C_2 \cup C_3$, which is not a neighbor of vertex a. In this case, vertex c can be colored with color 1 and a new 3-coloring of G is derived with $c_1 = c_2 = 2$ and $c_3 \geq 3$, as desired. □

Lemma 5. *Let $G = (V, E)$ be an outerplanar graph and let V' and V'' be two disjoint subsets of V, such that $|V'| \geq 2$ and $|V''| \geq 3$. Then, there exist two vertices $u \in V'$ and $v \in V''$, such that $(u, v) \notin E$.*

Proof. The proof follows from the fact that an outerplanar graph is $K_{2,3}$ free. □

Theorem 2. *An outerplanar graph with $n \geq 8$ vertices has $(2, n)$-differential coloring if and only if it does not contain $K_{1,n-1}$ as subgraph.*

Proof. Let $G = (V, E)$ be an outerplanar graph with $n \geq 8$ vertices. If G contains $K_{1,n-1}$ as subgraph, then the complement G^c of G is disconnected. Therefore, G does not admit a $(2, n)$-differential coloring. Now, assume that G does not contain $K_{1,n-1}$ as subgraph. By Lemma 4, it follows that G admits a 3-coloring, in which each color set contains at least two vertices. Let $C_i \subseteq V$ be the set of vertices with color i and $c_i = |C_i|$, for $i = 1, 2, 3$, such that $2 \leq c_1 \leq c_2 \leq c_3$. We distinguish the following cases:

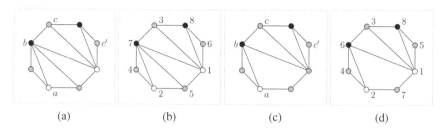

Fig. 2. (a) An outerplanar graph colored with 3 colors, white, black and grey (Case 1 of Thm. 2), and, (b) its $(2, n)$-differential coloring. (c) Another outerplanar graph also colored with 3 colors, white, black and grey (Case 2 of Thm. 2), and, (d) its $(2, n)$-differential coloring.

Case 1: $c_1 = 2$, $c_2 = 2$, $c_3 \geq 4$. Since $|C_1| = 2$ and $|C_3| \geq 4$, by Lemma 5 it follows that there exist two vertices $a \in C_1$ and $c \in C_3$, such that $(a, c) \notin E$. Similarly, since $|C_2| = 2$ and $|C_3 \setminus \{c\}| \geq 3$, by Lemma 5 it follows that there exist two vertices $b \in C_2$ and $c' \in C_3$, such that $c \neq c'$ and $(b, c') \notin E$; see Figure 2a-2b.

Case 2: $c_1 \geq 2$, $c_2 \geq 3$, $c_3 \geq 3$. Since $|C_1| = 2$ and $|C_3| \geq 3$, by Lemma 5 it follows that there exist two vertices $a \in C_1$ and $c \in C_3$, such that $(a, c) \notin E$. Similarly, since $|C_2| \geq 3$ and $|C_3 \setminus \{c\}| \geq 2$, by Lemma 5 it follows that there exist two vertices $b \in C_2$ and $c' \in C_3$, such that $c \neq c'$ and $(b, c') \notin E$; see Figure 2c-2d.

For both cases, consider the ordered coloring implied by the sequence $C_1 \setminus \{a\} \to \{a\} \to \{c\} \to C_3 \setminus \{c, c'\} \to \{c'\} \to \{b\} \to C_2 \setminus \{b\}$. As $(a, c) \notin E$ and $(b, c') \notin E$, it follows that the color difference between any two vertices of G is at least two. Hence, G admits a $(2, n)$-differential coloring. □

The next theorem gives a complete characterization of planar graphs that admit $(2, n)$-differential colorings. Due to space constraints, the detailed proof (which is similar to the one of Theorem 2) is given in the full version [3].

Theorem 3. *A planar graph with $n \geq 36$ vertices has a $(2, n)$-differential coloring if and only if it does not contain as subgraphs $K_{1,1,n-3}$, $K_{1,n-1}$ and $K_{2,n-2}$.*

Sketch of Proof. It can be shown that a planar graph G with $n \geq 36$ vertices, that does not contain as subgraphs $K_{1,1,n-3}$, $K_{1,n-1}$ and $K_{2,n-2}$, admits a 4-coloring, in which two color sets contain at least 2 vertices and the remaining two at least 5 vertices [3]. This together with a property similar to the one presented in Lemma 5 for outerplanar graphs implies that the complement of G is Hamiltonian and hence G has a $(2, n)$-differential coloring [3]. □

4 NP-completeness Results

In this section, we prove that the $(3, 2n)$-differential coloring problem is NP-complete. Recall that all graphs are $(2, 2n)$-differential colorable due to Lemma 1.

Theorem 4. *Given a graph $G = (V, E)$ on n vertices, it is NP-complete to determine whether G has a $(3, 2n)$-differential coloring.*

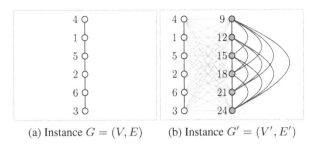

(a) Instance $G = (V, E)$ (b) Instance $G' = (V', E')$

Fig. 3. (a) An instance of the $(3, n)$-differential coloring problem for $n = 6$; (b) An instance of the $(3, 2n')$-differential coloring problem constructed based on graph G

Proof. The problem is clearly in NP. In order to prove that the problem is NP-hard, we employ a reduction from the $(3, n)$-differential coloring problem, which is known to be NP-complete [15]. More precisely, let $G = (V, E)$ be an instance of the $(3, n)$-differential coloring problem, i.e., graph G is an n-vertex graph with vertex set $V = \{v_1, v_2, \ldots, v_n\}$. We will construct a new graph G' with $n' = 2n$ vertices, so that G' is $(3, 2n')$-differential colorable if and only if G is $(3, n)$-differential colorable; see Figure 3.

Graph $G' = (V', E')$ is constructed by attaching n new vertices to G that form a clique; see the gray colored vertices of Figure 3b. That is, $V' = V \cup U$, where $U = \{u_1, u_2, \ldots, u_n\}$ and $(u, u') \in E'$ for any pair of vertices u and $u' \in U$. In addition, for each pair of vertices $v \in V$ and $u \in U$ there is an edge connecting them in G', that is $(v, u) \in E'$. In other words, (i) the subgraph, say G_U, of G' induced by U is complete and (ii) the bipartite graph, say $G_{U \times V}$, with bipartition V and U is also complete.

First, suppose that G has a $(3, n)$-differential coloring and let $l : V \rightarrow \{1, \ldots, n\}$ be the respective coloring. We compute a coloring $l' : V' \rightarrow \{1, \ldots, 4n\}$ of G' as follows: (i) $l'(v) = l(v)$, for all $v \in V' \cap V$ and (ii) $l'(u_i) = n + 3i, i = 1, 2, \ldots, n$. Clearly, l' is a $(3, 2n')$-differential coloring of G'.

Now, suppose that G' is $(3, 2n')$-differential colorable and let $l' : V' \rightarrow \{1, \ldots, 2n'\}$ be the respective coloring (recall that $n' = 2n$). We next show how to compute the $(3, n)$-differential coloring for G. W.l.o.g., let $V = \{v_1, \ldots v_n\}$ contain the vertices of G, such that $l'(v_1) < \ldots < l'(v_n)$, and $U = \{u_1, \ldots u_n\}$ contains the newly added vertices of G', such that $l'(u_1) < \ldots < l'(u_n)$. Since G_U is complete, it follows that the color difference between any two vertices of U is at least three. Similarly, since $G_{U \times V}$ is complete bipartite, the color difference between any two vertices of U and V is also at least three. We claim that l' can be converted to an equivalent $(3, 2n')$-differential coloring for G', in which all vertices of V are colored with numbers from 1 to n, and all vertices of U with numbers from $n + 3$ to $4n$.

Let U' be a maximal set of vertices $\{u_1, \ldots, u_j\} \subseteq U$ so that there is no vertex $v \in V$ with $l'(u_1) < l'(v) < l'(u_j)$. If $U' = U$ and $l'(v) < l'(u_1), \forall v \in V$, then our claim trivially holds. If $U' = U$ and $l'(v) > l'(u_j), \forall v \in V$, then we can safely recolor all the vertices in V' in the reverse order, resulting in a coloring that complies with our claim. Now consider the case where $U' \subsetneq U$. Then, there is a vertex $v_k \in V$ s.t.

$l'(v_k) - l'(u_j) \geq 3$. Similarly, we define $V' = \{v_k, \ldots, v_l \in V\}$ to be a maximal set of vertices of V, so that $l'(v_k) < \ldots < l'(v_l)$ and there is no vertex $u \in U$ with $l'(v_k) < l'(u) < l'(v_l)$. Then, we can safely recolor all vertices of $U' \cup V'$, such that: (i) the relative order of the colors of U' and V' remains unchanged, (ii) the color distance between v_l and u_1 is at least three, and (iii) the colors of U' are strictly greater than the ones of V'. Note that the color difference between u_j and u_{j+1} and between v_{k-1} and v_k is at least three after recoloring, i.e., $l'(u_{j+1}) - l'(u_j) \geq 3$ and $l'(v_k) - l'(v_{k-1}) \geq 3$. If we repeat this procedure until $U' = U$, then the resulting coloring complies with our claim. Thus, we obtain a $(3, n)$-differential coloring l for G by assigning $l(v) = l'(v), \forall v \in V$.   □

Theorem 5. *Given a graph $G = (V, E)$ on n vertices, it is NP-complete to determine whether G has a $(k + 1, kn)$-differential coloring.*

Sketch of Proof. Based on an instance $G = (V, E)$ of the $(k+1, n)$-differential coloring problem, which is known to be NP-complete [15], construct a new graph $G' = (V', E')$ with $n' = kn$ vertices, by attaching $n(k - 1)$ new vertices to G, as in the proof of Theorem 4. Then, using a similar argument as above, we can show that G has a $(k + 1, n)$-differential coloring if and only if G' has a $(k + 1, kn')$-differential coloring.

The NP-completeness of 2-differential coloring in Theorem 4 was about general graphs. Next, we consider the complexity of the problem for planar graphs. Note that from Lemma 2 and Lemma 3, it follows that the 2-differential chromatic number of a planar graph on n-vertices is between $\lfloor \frac{n}{3} \rfloor + 1$ and $\lfloor \frac{3n}{2} \rfloor$ (a planar graph is 4-colorable). The next theorem shows that testing whether a planar graph is $(\lfloor 2n/3 \rfloor, 2n)$-differential colorable is NP-complete. Since this problem can be reduced to the general 2-differential chromatic number problem, it is NP-complete to determine the 2-differential chromatic number even for planar graphs.

Theorem 6. *Given an n-vertex planar graph $G = (V, E)$, it is NP-complete to determine if G has a $(\lfloor 2n/3 \rfloor, 2n)$-differential coloring.*

Proof. The problem is clearly in NP. To prove that the problem is NP-hard, we employ a reduction from the well-known 3-coloring problem, which is NP-complete for planar graphs [11]. Let $G = (V, E)$ be an instance of the 3-coloring problem, i.e., G is an n-vertex planar graph. We will construct a new planar graph G' with $n' = 3n$ vertices, so that G' is $(\lfloor 2n'/3 \rfloor, 2n')$-differential colorable if and only if G is 3-colorable.

Graph $G' = (V', E')$ is constructed by attaching a path $v \rightarrow v_1 \rightarrow v_2$ to each vertex $v \in V$ of G; see Figure 4a-4b. Hence, we can assume that $V' = V \cup V_1 \cup V_2$, where V is the vertex set of G, V_1 contains the first vertex of each 2-vertex path and V_2 the second vertices. Clearly, G' is a planar graph on $n' = 3n$ vertices. Since G is a subgraph of G', G is 3-colorable if G' is 3-colorable. On the other hand, if G is 3-colorable, then G' is also 3-colorable: for each vertex $v \in V$, simply color its neighbors v_1 and v_2 with two distinct colors different from the color of v. Next, we show that G' is 3-colorable if and only if G' has a $(\lfloor 2n'/3 \rfloor, 2n')$-differential coloring.

First assume that G' has a $(\lfloor 2n'/3 \rfloor, 2n')$-differential coloring and let $l : V' \rightarrow \{1, \ldots, 2n'\}$ be the respective coloring. Let $u \in V'$ be a vertex of G'. We assign a color

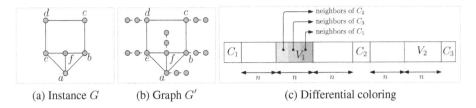

(a) Instance G (b) Graph G' (c) Differential coloring

Fig. 4. (a) An instance of the 3-coloring problem; (b) An instance of the $(\lfloor 2n'/3 \rfloor, 2n')$-differential coloring problem constructed based on graph G; (c) The $(\lfloor 2n'/3 \rfloor, 2n')$-differential coloring of G', in the case where G is 3-colorable

$c(u)$ to u as follows: $c(u) = i$, if $2(i-1)n + 1 \leq l(u) \leq 2in$, $i = 1, 2, 3$. Since l is a $(\lfloor 2n'/3 \rfloor, 2n')$-differential coloring, no two vertices with the same color are adjacent. Hence, coloring c is a 3-coloring for G'.

Now, consider the case where G' is 3-colorable. Let $C_i \subseteq V$ be the set of vertices of the input graph G with color i, $i = 1, 2, 3$. Clearly, $C_1 \cup C_2 \cup C_3 = V$. We compute a coloring l of the vertices of graph G' as follows (see Figure 4c):

- Vertices in C_1 are assigned colors from 1 to $|C_1|$.
- Vertices in C_2 are assigned colors from $3n + |C_1| + 1$ to $3n + |C_1| + |C_2|$.
- Vertices in C_3 are assigned colors from $5n + |C_1| + |C_2| + 1$ to $5n + |C_1| + |C_2| + |C_3|$.
- For a vertex $v_1 \in V_1$ that is a neighbor of a vertex $v \in C_1$, $l(v_1) = l(v) + 2n$.
- For a vertex $v_1 \in V_1$ that is a neighbor of a vertex $v \in C_2$, $l(v_1) = l(v) - 2n$.
- For a vertex $v_1 \in V_1$ that is a neighbor of a vertex $v \in C_3$, $l(v_1) = l(v) - 4n$.
- For a vertex $v_2 \in V_2$ that is a neighbor of a vertex x $v_1 \in V_1$, $l(v_2) = l(v_1) + 3n + |C_2|$.

From the above, it follows that the color difference between (i) any two vertices in G, (ii) a vertex $v_1 \in V_1$ and its neighbor $v \in V$, and (iii) a vertex $v_1 \in V_1$ and its neighbor $v_2 \in V_2$, is at least $2n = \lfloor \frac{2n'}{3} \rfloor$. Thus, G' is $(\lfloor 2n'/3 \rfloor, 2n')$-differential colorable.   □

5   An ILP for the Maximum k-Differential Coloring Problem

In this section, we describe an integer linear program (ILP) formulation for the maximum k-differential coloring problem. Recall that an input graph G to the maximum k-differential coloring problem can be easily converted to an input to the maximum 1-differential coloring by creating a disconnected graph G' that contains all vertices and edges of G plus $(k-1) \cdot n$ isolated vertices. In order to formulate the maximum 1-differential coloring problem as an integer linear program, we introduce for every vertex $v_i \in V$ of the input graph G a variable x_i, which represents the color assigned to vertex v_i. The 1-differential chromatic number of G is represented by a variable OPT, which is maximized in the objective function. The exact formulation is given below. The first two constraints ensure that all vertices are assigned colors from 1 to n. The third constraint guarantees that no two vertices are assigned the same color, and the forth constraint maximizes the 1-differential chromatic number of the graph. The first three constraints also guarantee that the variables are assigned integer values.

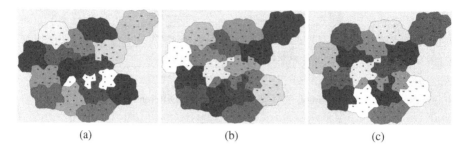

| (a) | (b) | (c) |

Fig. 5. A map with 16 countries colored by: (a) GMap [10], (b) ILP-n, (c) ILP-2n

$$
\begin{aligned}
\textbf{maximize} \quad & OPT \\
\textbf{subject to} \quad & x_i & \leq n & \quad \forall v_i \in V \\
& x_i & \geq 1 & \quad \forall v_i \in V \\
& |x_i - x_j| \geq 1 & & \quad \forall (v_i, v_j) \in V^2 \\
& |x_i - x_j| \geq OPT & & \quad \forall (v_i, v_j) \in E
\end{aligned}
$$

Note that a constraint that uses the absolute value is of the form $|X| \geq Z$ and therefore can be replaced by two new constraints: (i) $X + M \cdot b \geq Z$ and (ii) $-X + M \cdot (1-b) \geq Z$, where b is a binary variable and M is the maximum value that can be assigned to the sum of the variables, $Z + X$. That is, $M = 2n$. If b is equal to zero, then the two constraints are $X \geq Z$ and $-X + M \geq Z$, with the second constraint always true. On the other hand, if b is equal to one, then the two constraints are $X + M \geq Z$ and $-X \geq Z$, with the first constraint always true.

Next, we study two variants of the ILP formulation described above: ILP-n and ILP-2n, which correspond to $k = 1$ and $k = 2$, and compare them with GMap, which is a heuristic based on spectral methods developed by Hu et al. [10].

Our experiment's setup is as follow. We generate a collection of $1,200$ synthetic maps and analyze the performance of ILP-n and ILP-2n, on an Intel Core i5 1.7GHz processor with 8GB RAM, using the CPLEX solver. For each map a country graph $G_c = (V_c, E_c)$ with n countries is generated using the following procedure. (1) We generate $10n$ vertices and place an edge between pairs of vertices (i,j) such that $\lfloor \frac{i}{10} \rfloor = \lfloor \frac{j}{10} \rfloor$, with probability 0.5, thus resulting in a graph G with approximately n clusters. (2) More edges are added between all pairs of vertices with probability p, where p takes the values $1/2, 1/4 \ldots 2^{-10}$. (3) Ten random graphs are generated for different values of p. (4) Graph G is used as an input to a map generating algorithm (available as the `Graphviz` [9] function `gvmap`), to obtain a map M with country graph G_c. A sample map generated by the aforementioned procedure is shown in Figure 5.

Note that the value of p determines the "fragmentation" of the map M, i.e., the number of regions in each country, and hence, also affects the number of edges in the country graph. When p is equal to $1/2$, the amount of extra edges is enough to make almost all regions adjacent and therefore the country graph is a nearly complete graph, whereas for p equal to 2^{-10}, the country graph is nearly a tree. To determine a suitable range for the number of vertices in the country graph, we evaluated real world datasets, such as those available at `gmap.cs.arizona.edu`. Even for large graphs with over $1,000$ vertices, the country graphs tend to be small, with less than 16 countries.

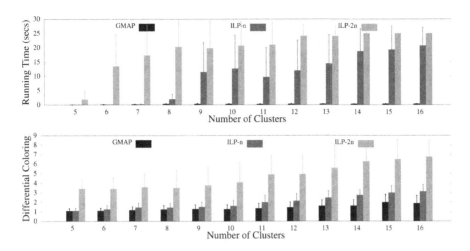

Fig. 6. Running-time results and differential coloring performance for all algorithms

Figure 6 summarizes the experimental results. Since n is ranging from 5 to 16, the running times of both ILP-n and ILP-2n are reasonable, although still much higher than GMap. The color assignments produced by ILP-n and GMap are comparable, while the color assignment of ILP-2n results in the best minimum color distance. Note that in the presence of twice as many colors as the graph's vertices, it is easier to obtain higher color difference between adjacent vertices. However, this comes at the cost of assigning pairs of colors that are more similar to each other for non-adjacent vertices, as it is also the case in our motivating example from the Introduction where G is a star.

6   Conclusion and Future Work

In this paper, we gave complete characterizations of bipartite, outerplanar and planar graphs that admit $(2, n)$-differential colorings (which directly lead to polynomial-time recognition algorithms). We also generalized the problem for more colors than the number of vertices in the graph and showed that it is NP-complete to determine whether a graph admits a $(3, 2n)$-differential coloring. Even for planar graphs, the problem of determining whether a graph is $(\lfloor 2n/3 \rfloor, 2n)$-differential colorable remains NP-hard.

Several related problems are still open: (i) Is it possible to characterize which bipartite, outerplanar or planar graphs are $(3, n)$-differential colorable? (ii) Extend the characterizations for those planar graphs that admit $(2, n)$-differential colorings to 1-planar graphs. (iii) Extend the results above to (d, kn)-differential coloring problems with larger $k > 2$. (iv) As all planar graphs are $(\lfloor \frac{n}{3} \rfloor + 1, 2n)$-differential colorable, is it possible to characterize which planar graphs are $(\lfloor \frac{n}{3} \rfloor + 2, 2n)$-differential colorable? (v) Since it is NP-complete to determine the 1-differential chromatic number of a planar graph [2], a natural question to ask is whether it is possible to compute in polynomial time the corresponding chromatic number of an outerplanar graph.

Acknowledgement. The work of M.A. Bekos is implemented within the framework of the Action "Supporting Postdoctoral Researchers" of the Operational Program "Education and Lifelong Learning" (Action's Beneficiary: General Secretariat for Research and Technology), and is co-financed by the European Social Fund and the Greek State.

References

1. Bansal, R., Srivastava, K.: Memetic algorithm for the antibandwidth maximization problem. Journal of Heuristics 17, 39–60 (2011)
2. Bekos, M., Kaufmann, M., Kobourov, S., Veeramoni, S.: A note on maximum differential coloring of planar graphs. Journal of Discrete Algorithms (2014)
3. Bekos, M., Kobourov, S., Kaufmann, M., Veeramoni, S.: The maximum k-differential coloring problem. Arxiv report (2014), http://arxiv.org/abs/1409.8133
4. Brewer, C.: ColorBrewer - Color Advice for Maps, http://www.colorbrewer.org
5. Calamoneri, T., Massini, A., Török, L., Vrt'o, I.: Antibandwidth of complete k-ary trees. Electronic Notes in Discrete Mathematics 24, 259–266 (2006)
6. Dillencourt, M.B., Eppstein, D., Goodrich, M.T.: Choosing colors for geometric graphs via color space embeddings. In: Kaufmann, M., Wagner, D. (eds.) GD 2006. LNCS, vol. 4372, pp. 294–305. Springer, Heidelberg (2007)
7. Dobrev, S., Královic, R., Pardubská, D., Török, L., Vrt'o, I.: Antibandwidth and cyclic antibandwidth of hamming graphs. Electronic Notes in Discrete Mathematics 34, 295–300 (2009)
8. Duarte, A., Martí, R., Resende, M., Silva, R.: Grasp with path relinking heuristics for the antibandwidth problem. Networks 58(3), 171–189 (2011)
9. Ellson, J., Gansner, E., Koutsofios, E., North, S., Woodhull, G.: Graphviz and dynagraph static and dynamic graph drawing tools. In: Jünger, M., Mutzel, P. (eds.) Graph Drawing Software, pp. 127–148. Springer, Heidelberg (2004)
10. Gansner, E., Hu, Y., Kobourov, S.: Gmap: Visualizing graphs and clusters as maps. In: 2010 IEEE Pacific Visualization Symposium (PacificVis), pp. 201–208 (2010)
11. Garey, M.R., Johnson, D.S.: Computers and Intractability: A Guide to the Theory of NP-Completeness. W. H. Freeman & Co., New York (1979)
12. Hale, W.: Frequency assignment: Theory and applications. Proceedings of the IEEE 68(12), 1497–1514 (1980)
13. Hu, Y., Kobourov, S., Veeramoni, S.: On maximum differential graph coloring. In: Brandes, U., Cornelsen, S. (eds.) GD 2010. LNCS, vol. 6502, pp. 274–286. Springer, Heidelberg (2011)
14. Isaak, G.: Powers of hamiltonian paths in interval graphs. Journal of Graph Theory 27, 31–38 (1998)
15. Leung, J.Y.-T., Vornberger, O., Witthoff, J.D.: On some variants of the bandwidth minimization problem. SIAM Journal on Computing 13(3), 650–667 (1984)
16. Miller, Z., Pritikin, D.: On the separation number of a graph. Networks 19(6), 651–666 (1989)
17. Papadimitriou, C.: The NP-Completeness of the bandwidth minimization problem. Computing 16, 263–270 (1975)
18. Proskurowski, A., Syso, M.: Efficient vertex- and edge-coloring of outerplanar graphs. SIAM Journal on Algebraic Discrete Methods 7(1), 131–136 (1986)
19. Purves, D., Lotto, R.B.: Why we see what we do: An empirical theory of vision. Sinauer Associates (2003)

20. Raspaud, A., Schröder, H., Sýkora, O., Török, L., Vrt'o, I.: Antibandwidth and cyclic antibandwidth of meshes and hypercubes. Discrete Math. 309(11), 3541–3552 (2009)
21. Wang, X., Wu, X., Dumitrescu, S.: On explicit formulas for bandwidth and antibandwidth of hypercubes. Discrete Applied Mathematics 157(8), 1947–1952 (2009)
22. Weili, Y., Xiaoxu, L., Ju, Z.: Dual bandwidth of some special trees. Journal of Zhengzhou University (Natural Science) 35(3), 16–19 (2003)
23. Yixun, L., Jinjiang, Y.: The dual bandwidth problem for graphs. Journal of Zhengzhou University (Natural Science) 35(1), 1–5 (2003)

Exact Algorithms for 2-Clustering with Size Constraints in the Euclidean Plane*

Alberto Bertoni, Massimiliano Goldwurm, and Jianyi Lin

Dipartimento di Informatica, Università degli Studi di Milano,
Via Comelico 39/41, 20135 Milano, Italy
`jianyi.lin@unimi.it`

Abstract. We study the problem of determining an optimal bipartition $\{A, B\}$ of a set X of n points in $\mathbb{R}^2$ that minimizes the sum of the sample variances of A and B, under the size constraints $|A| = k$ and $|B| = n - k$. We present two algorithms for such a problem. The first one computes the solution in $O(n\sqrt[3]{k}\log^2 n)$ time by using known results on convex-hulls and k-sets. The second algorithm, for an input $X \subset \mathbb{R}^2$ of size n, solves the problem for all $k = 1, 2, \ldots, \lfloor n/2 \rfloor$ and works in $O(n^2 \log n)$ time.

Keywords: algorithms for clustering, cluster size constraints, data analysis, Euclidean distance, machine learning.

1 Introduction

The general Clustering Problem consists in finding an optimal partition of a set X of n points in m clusters, i.e. a partition of X in m subsets that minimizes the sum of the dispersion of points around the centroid in each subset. This is a fundamental problem in many research areas like data mining, image analysis, pattern recognition and bioinformatics [18]. Clustering is a classical method in unsupervised machine learning, frequently used in statistical data analysis [5,10].

A computational analysis of the problem depends on a variety of parameters: the dimension d of the point space (usually $\mathbb{R}^d$), the distance or semi-distance used to measure the dispersion of points, the number m of clusters (which may be arbitrary, as part of the instance, or fixed in advance), the size of the clusters and possibly others constraints [2,20,21]. In most cases the problem is difficult. For instance, assuming the squared Euclidean semi-distance, when the dimension d is arbitrary the general Clustering Problem is NP-hard even if the number m of clusters is fixed to 2 [1,6]. The same occurs if m is arbitrary and the dimension is $d = 2$ [14]. The problem is solvable in polynomial time when fixing both m and d [11]. Moreover there exists a well-known, usually fast, heuristic for finding an approximate solution, called k-Means [13], which however requires exponential time in the worst case [19].

* This research has been supported by project PRIN #H41J12000190001 "Automata and formal languages: mathematical and applicative aspects".

G.F. Italiano et al. (Eds.): SOFSEM 2015, LNCS 8939, pp. 128–139, 2015.

Thus, a natural goal of research in this context is to study particular cases with suitable hypothesis on the input that allow us to design polynomial time algorithms. Here, we consider the Clustering Problem in $\mathbb{R}^2$, assuming squared Euclidean semi-distance, when the number of clusters is $m = 2$ and their size is given by the instance. We call it *Size Constrained 2-Clustering* in $\mathbb{R}^2$ (2-SCC-2 for short).

More precisely, an instance of this problem is given by a set $X \subset \mathbb{R}^2$ of n points in general position and an integer k such that $1 \leq k \leq n/2$, while the solution is a bipartition $\{A, B\}$ of X such that $|A| = k$, that minimizes the total weight $W(A) + W(B)$, where $W(A)$ (respectively, $W(B)$) is the sum of the squares of the ℓ_2-distances of all points $a \in A$ (resp. $b \in B$) from the centroid of A (resp. B). A more formal description is given in Section 3. Recall that the unconstrained version of the same problem, with an arbitrary number of clusters, is NP-hard [14].

The relevance of the 2-clustering problems is due to the wide spread of hierarchical clustering techniques, that repeatedly apply the 2-clustering as the key step. The 2-clustering problem with cluster size constraints has been already studied in [12,4], where it is shown that in dimension 1 the problem is solvable in polynomial time for every norm ℓ_p with integer $p \geq 1$, while there is some evidence that the same result does not hold for non-integer p. It is also known that for arbitrary dimension d the same problem is NP-hard even assuming equal sizes of the two clusters.

In this work we show two results. First, we describe an algorithm that solves 2-SCC-2 in $O(n\sqrt[3]{k}\log^2 n)$ time. This is obtained by using known results of computational geometry concerning in particular dynamic data structures for convex hulls [17,16] and the enumeration of k-sets in $\mathbb{R}^2$ [9,7]. Then, we present an algorithm for the full-version of the problem, i.e. a procedure yielding a solution for all $k = 1, 2, \ldots, \lfloor n/2 \rfloor$, which works in $O(n^2 \log n)$ time. Both algorithms are based on a separation result on the clusters of optimal solutions for 2-SCC-2, presented in Section 3 and proved in [4], which intuitively extends to the bidimensional case the so-called String Property of the optimal clusterings on the real line [15]. The results we present here are obtained by assuming the Euclidean norm and this hypothesis is crucial our proofs. We observe that the Euclidean norm is an essential hypothesis in our proofs and the results we present here do not seem to hold under different assumptions. For instance, in the case of Manhattan distance (ℓ_1 norm), the separation result on the plane yields a $O(n^3 \log n)$ time algorithm for the full-version of the problem [3].

2   Preliminary Notions

In this section we fix our notation and recall some known results of computational geometry [8,17].

For any point $a \in \mathbb{R}^2$, we denote by a_x and a_y the abscissa and the ordinate of a, respectively. We denote by $\|a\|$ the usual Euclidean norm of point a, i.e. $\|a\| = (a_x^2 + a_y^2)^{1/2}$. We also fix a total order on points in $\mathbb{R}^2$: for every $a, b \in \mathbb{R}^2$, we set $a <_o b$ if either $a_y < b_y$ or $a_y = b_y \wedge a_x < b_x$. Clearly, every point a defines a vector of length $\|a\|$ oriented from the origin to a.

Given two points $a, b \in \mathbb{R}^2$, the oriented line segment from a to b is called *oriented edge* and is identified by the pair (a, b). By a little abuse of language we also denote by (a, b) the *straight line* through the two points oriented from a to b. We define the (positive) *phase* of (a, b) as the angle between the oriented edges $(a, (a_x + 1, a_y))$ and (a, b), measured counter-clockwise. We denote it by $phase(a, b)$. Clearly, we have $0 \leq phase(a, b) < 2\pi$. Note that two oriented edges have the same phase if and only if they are parallel and have the same orientation.

We also define the *slope* of (a, b) as the remainder of the division $phase(a, b)/\pi$ and we denote it by $slope(a, b)$. Observe that (a, b) and (b, a) have the same slope and, more generally, two oriented edges have the same slope if and only if they are parallel (with either equal or opposite orientation).

Moreover, for every pair of oriented edges (a, b), (c, d), we say that (a, b) is *on the right* of (c, d) and write (a, b)Right(c, d) if the position of the straight line (c, d) is obtained by a counter-clockwise rotation of the straight line (a, b) (around their intersection point) smaller than π. On the contrary, if such a rotation is greater than π we say that (a, b) is *on the left* of (c, d) and write (a, b)Left(c, d). Note that relations *Right* and *Left* correspond respectively to the positive and negative sign of the cross product $(a, b) \times (c, d)$, which is determined by the well-known right-hand rule.

Clearly, once the coordinates of points are known, one can compute in constant time both phase and slope of a point, as well as establish the validity of relations *Right* and *Left* between two oriented edges.

Now, let us consider a finite set $X \subset \mathbb{R}^2$: we say that X is in *general position* if it does not contain 3 collinear points. Moreover, if X consists of n points, for any integer $0 \leq k \leq n$ a k-*set* of X is a subset $A \subseteq X$ of cardinality $|A| = k$ such that $A = X \cap H$ for a suitable half-space $H \subset \mathbb{R}^2$. This means A is separable from its complement $\bar{A}$ by a straight-line. Determining the maximum number of k-sets of a family of n points in $\mathbb{R}^2$ is a central problem in combinatorial geometry, first posed in [9].

Recall that the intersection of an arbitrary collection of convex sets is convex. Then for any set $A \subseteq \mathbb{R}^d$, the *convex hull* or *convex closure* of A, denoted by $Conv(A)$, is defined as the smallest convex subset of $\mathbb{R}^d$ containing A, i.e.

$$Conv(A) = \bigcap \{Y \subseteq \mathbb{R}^d : A \subseteq Y, Y \text{ is convex}\}$$

It is well-known that the convex closure of a finite set A of points in $\mathbb{R}^d$ is a polytope [17] determined by the intersection of finitely many half-spaces; in particular in $\mathbb{R}^2$, $Conv(A)$ is a convex polygon. It is possible to identify a polygon by giving its vertices, and hence the determination of the convex closure $Conv(A)$ of a given set $A \subset \mathbb{R}^2$ consists in finding the vertices of the associated polygon. We recall that the convex hull of a set X of n points in $\mathbb{R}^2$ can be computed in time $O(n \log n)$ [17].

3   Problem Definition and First Properties

To give a formal definition of the problem, we recall that a *cluster* of a finite set $X \subset \mathbb{R}^2$ is a non-empty subset $A \subset X$, while the pair $\{A, \bar{A}\}$ is a *2-clustering* of X, where $\bar{A} = X \smallsetminus A$ is the complement of A. Assuming the Euclidean norm $\|\cdot\|$, the *centroid* of A is the value $C_A \in \mathbb{R}^2$ defined by

$$C_A = \operatorname*{argmin}_{\mu \in \mathbb{R}^2} \sum_{a \in A} \|a - \mu\|^2.$$

It turns out that C_A is the mean value of points in A: $C_A = \frac{\sum_{a \in A} a}{|A|}$. Moreover, we denote by $W(A)$ the *weight* of A, that is

$$W(A) = \sum_{a \in A} \|a - C_A\|^2$$

Note that $\frac{W(A)}{|A|-1}$ is the traditional sample variance of A, once we interpret the elements of A as sample points picked up from a random variable in $\mathbb{R}^2$ (rather than in $\mathbb{R}$). Hence, it represents a natural measure of the dispersion of points in A around their mean value.

Then, the 2-SCC-2 problem is defined as follows:

> Given a set $X \subset \mathbb{R}^2$ of cardinality n (in general position) and an integer k, $1 \le k \le n/2$, find a 2-clustering $\{A, \bar{A}\}$ of X, with $|A| = k$, that minimizes the weight $W(A, \bar{A}) = W(A) + W(\bar{A})$.

By the observation above, since the size of the clusters is constrained, this is equivalent to looking for the 2-clustering (A, B) that minimizes the sum of the sample variances of A and B.

The weight of a 2-clustering in the plane can be computed using the following proposition, the proof of which is here omitted for sake of brevity.

Proposition 1. *Let $\{A, B\}$ be a 2-clustering of a set $X \subset \mathbb{R}^2$ of n points such that $|A| = k$ for some $k \in \{1, 2, \dots, \lfloor n/2 \rfloor\}$. Then*

$$W(A, B) = \sum_{p \in X} \|p\|^2 - \frac{n}{k(n-k)} \|S_A\|^2$$

where $S_A = \sum_{a \in A} a$.

As a consequence, solving the 2-SCC-2 problem for an instance (X, k) is equivalent to determining a subset $A \subseteq X$ of size k that maximizes the value $\|S_A\|$.

Another property we use in the present work is the following separation result between the clusters of optimal solutions of the 2-SCC-2 problem, proved in [4].

Proposition 2 (Separation Result). *Let $\{A, B\}$ be an optimal solution of the 2-SCC-2 problem for an instance $X \subset \mathbb{R}^2$ with constraint $|A| = k$. Then, there exists a constant $c \in \mathbb{R}$ such that, for every $p \in X$,*

$$p \in A \quad \Rightarrow \quad 2p_x(C_{Bx} - C_{Ax}) + 2p_y(C_{By} - C_{Ay}) < c + \|C_B\|^2 - \|C_A\|^2,$$

$$p \in B \quad \Rightarrow \quad 2p_x(C_{Bx} - C_{Ax}) + 2p_y(C_{By} - C_{Ay}) > c + \|C_B\|^2 - \|C_A\|^2.$$

As a consequence, both clusters of any optimal solution $\{A, B\}$ of the 2-SCC-2 problem are separated by the straight line of equation

$$2x(C_{Bx} - C_{Ax}) + 2y(C_{By} - C_{Ay}) = c + \|C_B\|^2 - \|C_A\|^2$$

for some constant $c \in \mathbb{R}$. This implies that A is a k-set and B is a $(n - k)$-set.

This result can be interpreted as a natural extension of the well-known String Property, stating that all optimal clusterings on the real line consist of contiguous subsets of the input set (see for instance [15]).

Moreover, Propositions 1 and 2 imply that the 2-SCC-2 problem for an instance (X, k) can be solved by computing a k-set $A \subseteq X$ with maximum $\|S_A\|$.

4   Algorithm for Constrained 2-Clustering in $\mathbb{R}^2$

In this section we present an efficient technique, based on Proposition 2, to solve the 2-SCC-2 problem as defined in Section 3. Here the input is given by a set $X \subset \mathbb{R}^2$ of $n \geq 4$ points in general position and an integer k such that $1 < k \leq \lfloor n/2 \rfloor$. Our purpose is to show that the algorithm works in $O(n\sqrt[3]{k}\log^2 n)$ time.

We first introduce some preliminary notions. Given two disjoint polygons it is easy to see that there exist 4 straight lines that are tangent to both polygons: they are called *bitangents*. Two bitangents keep one polygon on one side and the other polygon on the other side, while the other two bitangents keep both polygons on the same side. Bitangents as well as straight lines can be oriented.

Given $X \subset \mathbb{R}^2$ and two points $a, b \in X$ we define

$$X^r(a, b) = \{p \in X \mid (a, p)\text{Right}(a, b)\},$$
$$X^l(a, b) = \{p \in X \mid (a, p)\text{Left}(a, b)\}.$$

In other words, $X^r(a, b)$ is the set of points in X on the right-hand side of (a, b), while $X^l(a, b)$ is the set of points in X on the left-hand side of (a, b).

Definition 1. *Given a set $X \subset \mathbb{R}^2$ of cardinality $n \geq 4$ in general position, we say that the oriented edge (a, b) with $a, b \in X$, $a \neq b$, is a $(k - 1)$-set edge if $|X^r(a, b)| = k - 1$ (and hence $|X^l(a, b)| = n - k - 1$).*

Setting $A = X^r(a, b) \cup \{a\}$ and $\bar{A} = X^l(a, b) \cup \{b\}$, it is clear that A is a k-set of X and the straight line (a, b) is the unique bitangent between $\text{Conv}(A)$ and $\text{Conv}(\bar{A})$ that keeps $A \setminus \{a\}$ on its right. Indeed, as illustrated in Figure 1a, two of the other three bitangents between $\text{Conv}(A)$ and $\text{Conv}(\bar{A})$ do not separate A from $\bar{A}$, and the remaining one does not keep $A \setminus \{a\}$ on its right-hand side. This proves the following proposition for any set $X \subset \mathbb{R}^2$ of $n \geq 4$ points and any $k = 2, \ldots, \lfloor n/2 \rfloor$.

Proposition 3. *There exists a bijection between the $(k - 1)$-set edges of X and the k-sets of X: each $(k - 1)$-set edge (a, b) can be associated with the k-set $\{a\} \cup X^r(a, b)$, while any k-set A corresponds to the unique oriented edge (a, b), bitangent to $\text{Conv}(A)$ and $\text{Conv}(\bar{A})$, such that $A \setminus \{a\} = X^r(a, b)$.*

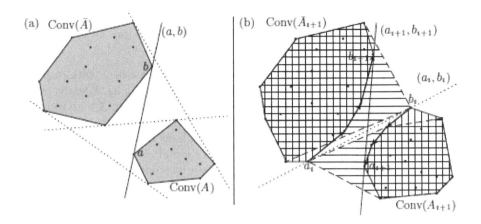

Fig. 1. (a) There are 4 bitangents between $\mathrm{Conv}(A)$ and $\mathrm{Conv}(\bar{A})$, but only (a, b) separates A and $\bar{A}$ keeping $A \smallsetminus \{a\}$ on its right-hand side. (b) Given the k-set A_i (polygon with horizontal lines) associated to the $(k-1)$-set edge (a_i, b_i) we can compute the subsequent k-set A_{i+1} (polygon with vertical lines) associated to (a_{i+1}, b_{i+1}) by removing a_i and inserting b_i in the convex hull. Segments with arrow represent oriented edges scanned by procedure NextBitangent on the perimeter of the convex hull.

Let us denote by $A(a, b)$ the k-set corresponding to the $(k-1)$-set edge (a, b).

Since two $(k-1)$-set edges of X cannot have the same phase, we can consider the sequence

$$E_{k-1} = \{(a_1, b_1), ..., (a_h, b_h)\}$$

of all $(k-1)$-set edges of X ordered according with increasing phase.

The following proposition yields the key property for computing all k-sets $A(a_i, b_i)$ for $i = 1, 2, \ldots, h$. The proof is here omitted but its validity should be evident from Figure 1b.

Proposition 4. *For every* $i = 1, 2, ..., h$, *we have*

$$A(a_{1+\langle i \rangle_h}, b_{1+\langle i \rangle_h}) = A(a_i, b_i) \cup \{b_i\} \smallsetminus \{a_i\}$$

where $\langle i \rangle_h$ *is the remainder of the division* i/h.

As a consequence, setting

$$S_i = \sum_{p \in A(a_i, b_i)} p \tag{1}$$

we have $S_{1+\langle i \rangle_h} = S_i - a_i + b_i$.

Thus, in order to solve the 2-SCC-2 problem for an instance (X, k), we can simply determine the first oriented edge $(a_1, b_1) \in E_{k-1}$ and design a procedure for computing $(a_{1+\langle i \rangle_h}, b_{1+\langle i \rangle_h})$ from (a_i, b_i). This allows us to compute all pairs $(a_i, b_i) \in E_{k-1}$ and hence all values $\|S_i\|$'s one after the other, taking the largest one. The overall computation is described by Algorithm 2.

Let us start by showing the computation of (a_1, b_1).

Proposition 5. *Given a set $X \subset \mathbb{R}^2$ of $n \geq 4$ points in general position, for any $k = 2, \ldots, \lfloor n/2 \rfloor$ the $(k-1)$-set edge of smallest phase can be obtained in $O(n \log n)$ time.*

Proof. First, it is easy to determine the k-th smallest point a in X with respect to the total order $<_o$, together with set $A = \{q \in X \mid q \leq_o a\}$. Clearly, A is a k-set of X. Similarly, we can determine the $(k+1)$th element b in X with respect to $<_o$, i.e. the smallest point in the set $\bar{A} = X \smallsetminus A$. To determine the $(k-1)$-set edge of A, the algorithm first computes the convex hulls $\mathrm{Conv}(A)$ and $\mathrm{Conv}(\bar{A})$. Then, starting from a and b, it moves two points u and v counter-clockwise on the perimeter of $\mathrm{Conv}(A)$ and $\mathrm{Conv}(\bar{A})$, respectively, stopping at the first edge on $\mathrm{Conv}(A)$ (respectively, on $\mathrm{Conv}(\bar{A})$) that is on the right (respectively, on the left) of the oriented edge (u, v).

The procedure is formally described by Algorithm 1 given below, where for every point p on the perimeter of a convex hull, $\mathrm{Succ}(p)$ is the counter-clockwise successor of p on the same perimeter.

Algorithm 1. FirstSetEdge(a,b)

1: $u := a; u' := \mathrm{Succ}(u)$
2: $v := b; v' := \mathrm{Succ}(v)$
3: **while** $((u, u')\mathrm{Left}(u, v) \ \vee \ (v, v')\mathrm{Right}(u, v))$ **do**
4:     **if** $(u, u')\mathrm{Left}(u, v)$ **then**
5:         **if** $(u, u')\mathrm{Left}(u, v')$ **then**
6:             $u := u'; u' :=\mathrm{Succ}(u)$
7:         **else**
8:             $v := v'; v' :=\mathrm{Succ}(v)$
9:     **else**
10:         $v := v'; v' :=\mathrm{Succ}(v)$
11: **return** (u, v)

At each loop iteration the procedure checks whether the current edges (u, u') and (v, v') on the two perimeters verify the exit condition

$$(u, u')\mathrm{Right}(u, v) \ \wedge \ (v, v')\mathrm{Left}(u, v),$$

which guarantees $A \smallsetminus \{u\} = X^r(u, v)$ and $\bar{A} \smallsetminus \{v\} = X^l(u, v)$. Hence, in the affirmative case, (u, v) is the required $(k-1)$-set edge.

The most expensive operation in the overall procedure is the computation of the convex hulls $\mathrm{Conv}(A)$ and $\mathrm{Conv}(\bar{A})$, which can be done in $O(n \log n)$ time [17]. Also note that the while loop at lines 3–10 requires $O(n)$ steps.   □

Once the initial $(k-1)$-set edge is determined our general procedure computes all the subsequent $(k-1)$-set edges in the order defined by E_{k-1}. For each $(a_i, b_i) \in E_{k-1}$, the procedure computes the squared norm of S_i (defined in Equation 1), maintaining in q the largest value. The details are given in Algorithm 2, where procedure NextBitangent, called at lines 10 and 16, yields the successive $(k-1)$-set edge in the required order.

Algorithm 2. Solving 2-SCC-2

Input: a set $X \subset \mathbb{R}^2$ of $n \geq 4$ points in general position; an integer $1 < k \leq \lfloor n/2 \rfloor$.
Output: the solution $\pi = \{A, B\}$ of the 2-SCC-2 problem on instance X with constraint $|A| = k$.

1: Compute the k-th smallest point a in X with respect to $<_o$
2: $A := \{p \in X : p \leq_o a\}$; $\bar{A} := X \smallsetminus A$
3: Compute the smallest point b in $\bar{A}$ with respect to $<_o$
4: $\mathcal{A} := \text{Conv}(A)$
5: $\bar{\mathcal{A}} := \text{Conv}(\bar{A})$
6: $(a_1, b_1) := \text{FirstSetEdge}(a, b)$
7: $S := \sum_{x \in A} x$
8: $q := \|S\|^2$
9: $(x, y) := (a_1, b_1)$
10: $(r, s) := \text{NextBitangent}(a_1, b_1)$
11: **while** $(r, s) \neq (a_1, b_1)$ **do**
12:     $S := S - r + s$
13:     **if** $q < \|S\|^2$ **then**
14:         $q := \|S\|^2$
15:         $(x, y) := (r, s)$
16:     $(r, s) := \text{NextBitangent}(r, s)$
17: $\pi := \{X^r(x, y) \cup \{x\}, X^l(x, y) \cup \{y\}\}$
18: **return** π

Procedure NextBitangent is defined by Algorithm 3, which uses function $Succ$ as in Algorithm 1. Such a procedure first computes the convex hulls $\mathcal{A}$ and $\bar{\mathcal{A}}$ of the new k-set A and of its complement by two insert and delete operations. Then, in the main loop, the procedure determines the $(k-1)$-set edge of the new k-set by following a path counter-clockwise on the perimeter of the two convex hulls. As in Algorithm 1, the exit condition

$$(u, u')\text{Right}(u, v) \ \wedge \ (v, v')\text{Left}(u, v)$$

guarantees that (u, v) is the required $(k-1)$-set edge.

The correctness proof of Algorithm 2, which is here omitted for lack of space, is based on the fact that distinct k-sets must have $(k-1)$-set edges with different phases.

The analysis of time complexity of Algorithm 2 requires the following result on the number of k-sets of X, given in [7].

Theorem 1. *For any set X of n points in $\mathbb{R}^2$ and any $k \in \mathbb{N}$, $1 \leq k \leq \lfloor \frac{n}{2} \rfloor$, the number of k-sets of X is less than $6.48n\sqrt[3]{k}$.*

Proposition 6. *The time complexity required by Algorithm 2 on an input of $n \geq 4$ points with $1 < k \leq \lfloor \frac{n}{2} \rfloor$ is $O(n\sqrt[3]{k} \cdot \log^2 n)$.*

Proof. First recall that, by Proposition 5, the first part of the procedure from line 1 to line 8, can be executed in time $O(n \log n)$. The remaining time required

---

Algorithm 3. NextBitangent(r, s)

Input: a $(k-1)$-set edge (r, s) of X computed in Algorithm 2.
Output: the subsequent $(k-1)$-set edge of X in phase order.
1: $\mathcal{A} :=$ Insert(Delete$(\mathcal{A}, r)$,s)
2: $\bar{\mathcal{A}} :=$ Insert(Delete$(\bar{\mathcal{A}}, s)$,r)
3: $u := s;\ u' := \text{Succ}(u)$
4: $v := r;\ v' := \text{Succ}(v)$
5: **repeat**
6:    **if** (u, u')Left(u, v) **then**
7:        **if** (u, u')Left(u, v') **then**
8:            $u := u';\ u' :=$Succ(u)
9:        **else**
10:            $v := v';\ v' :=$Succ(v)
11:    **else**
12:        $v := v';\ v' :=$Succ(v)
13: **until** (u, u')Right$(u, v)\ \wedge\ (v, v')$Left(u, v)
14: **return** (u, v)

---

by the algorithm is dominated by calls to procedure NextBitangent. The computation maintains, as permanent structure, the convex hulls $\mathcal{A}$ and $\bar{\mathcal{A}}$, which are updated at lines 1–2 of each call of NextBitangent in $O(\log^2 n)$ time by using the data structure introduced in [16]. Since there is just one call to NextBitangent for each $(k-1)$-set edge, by Proposition 3 and Theorem 1 the total cost of all updates of $\mathcal{A}$ and $\bar{\mathcal{A}}$ is $O(n\sqrt[3]{k}\log^2 n)$.

The time cost of the other operations of NextBitangent is due to the repeat-until loop of Algorithm 3. Here, the key observation is that each edge (u, u') inside the repeat-until loop is a $(n-k)$-set edge, because the phase of the opposite edge (u', u) is included between the phases of two consecutive $(k-1)$-set edges. These edges (u, u') scan counter-clockwise the boundary of $\mathcal{A}$ and each of them is considered just by one call of NextBitangent. The same occurs for the k-set edges (v, v') scanning counter-clockwise the boundary of $\bar{\mathcal{A}}$. Therefore, the time required by the main loop in all calls to NextBitangent is at most proportional to $|E_k| + |E_{n-k}| = 2|E_k|$, which is again $O(n\sqrt[3]{k})$ by Theorem 1. Thus, the time cost of all calls to NextBitangent turns out to be $O(n\sqrt[3]{k}\log^2 n)$.   □

5   Algorithm for the Full Problem

In this section we present an algorithm that, for an input $X \subset \mathbb{R}^2$ of n points in general position, computes an optimal 2-clustering $\{A_k, \bar{A}_k\}$ of X such that $|A_k| = k$, for each $k = 1, 2, \ldots, \lfloor n/2 \rfloor$. We prove that the algorithm works in time $O(n^2 \log n)$.

The result is based on Propositions 1 and 2 and on the following relationship between oriented edges and 2-clusterings of X including a k-set. Given two distinct points $a, b \in X$, we associate the oriented edge (a, b) with the 2-clustering $\{A, \bar{A}\}$ of X where either A or $\bar{A}$ equals the set $\mathcal{R}(a, b)$ defined by

$$\mathcal{R}(a, b) = \begin{cases} X^r(a, b) \cup \{b\} & \text{if } a <_o b \\ X^l(a, b) \cup \{b\} & \text{otherwise.} \end{cases}$$

Note that here $\{A, \bar{A}\}$ is an unordered pair of sets. Moreover, A and $\bar{A}$ are, respectively, a k-set and a $(n - k)$-set for some $k \in \{1, 2, \ldots, n-1\}$. Also observe that the 2-clusterings associated with (a, b) and (b, a) are always different.

Proposition 7. *Given a set $X \subset \mathbb{R}^2$ of n points in general position, let $\{A, \bar{A}\}$ be a 2-clustering of X where A is a k-set for some $k \in \{1, 2, \ldots, n - 1\}$. Then $\{A, \bar{A}\}$ is the 2-clustering of X associated with an oriented edge (a, b) with $a, b \in X$.*

Proof. Since A is a k-set we can consider a bitangent (u, v) that separates A and $\bar{A}$ with $u \in \bar{A}$ and $v \in A$. Assume that $u <_o v$: if $A = X^r(u, v) \cup \{v\}$ then (u, v) is the required oriented edge because $A = \mathcal{R}(u, v)$; otherwise, A equals $X^l(u, v) \cup \{v\}$ and the same 2-clustering $\{A, \bar{A}\}$ is associated with (v, u) because $\bar{A} = \mathcal{R}(v, u)$. A symmetric reasoning holds in case $v <_o u$.   □

Note that in the previous proof we can choose the bitangent (u, v) separating A and $\bar{A}$ in two different ways. This proves that every k-sets of X is associated with two oriented edges. Hence, such a correspondence is quite different from the bijection of Proposition 3 introduced in the previous section.

By the proposition above, one can design an algorithm that scans all k-sets by considering in some order all oriented edge outgoing from each point. In order to compute efficiently the weights of the clusters we introduce a special order among the oriented edges $E = \{(a, b) \mid a, b \in X, a \neq b\}$: for every $(u, v), (w, z) \in E$, we set $(u, v) <_e (w, z)$ if either $u <_o w$ or $u = w$ and $slope(u, v) < slope(w, z)$.

Proposition 8. *For any instance set $X \subset \mathbb{R}^2$ of n points, the 2-SCC-2 problem for all $k = 1, 2, \ldots, \lfloor n/2 \rfloor$ can be solved in $O(n^2 \log n)$ time.*

Proof. Consider procedure Full 2-SCC-2 defined by Algorithm 4. It computes points $S_{\mathcal{R}(a,b)}$ (as defined in Proposition 1) for all oriented edges in $(a, b) \in E$, taken according with the total order $<_e$. For each $k = 1, 2, \ldots, \lfloor n/2 \rfloor$, the procedure maintains the optimal value $q[k] = \|S_A\|^2 / k(n - k)$, where A is a k-set of X and keeps in $e[k]$ the associated oriented edge. By Propositions 2 and 7, this guarantees that all possible 2-clusterings of X are considered.

The computation first considers all points in X in the order $<_o$ and, for each $a \in X$, it determines $R = S_{A(a)}$, where $A(a) = \{u \in X \mid u <_o a\}$. Then, it computes $S_{\mathcal{R}(a,b)}$ for every edge (a, b) such that $b \in X \smallsetminus \{a\}$ in the order $<_e$: for any pair of consecutive edges $(a, b), (a, c)$, the value $S_{\mathcal{R}(a,c)}$ is obtained from $S_{\mathcal{R}(a,b)}$ by adding or subtracting c according whether $a <_o c$ or $c <_o a$ (see instructions 10 and 14, respectively). Note that such a computation only requires constant time.

The time complexity of the algorithm is dominated by the operation of sorting the oriented edges with the same starting point a. This can be done in $O(n \log n)$ time for each $a \in X$. Note that the other operations, in the inner for-loop, require at most constant time and hence they are executed $O(n^2)$ many times. Therefore, the overall time of the algorithm is $O(n^2 \log n)$. $\qquad\square$

Algorithm 4. Full 2-SCC-2

Input: a set $X \subset \mathbb{R}^2$ of n points in general position
Output: the sequence $(e[1], \ldots, e[\lfloor n/2 \rfloor])$ of oriented edges, where each $e[k]$ is associated with the solution $\{A_k, \bar{A}_k\}$ of 2-SCC-2 for X such that $|A_k| = k$
1: **for** $k = 1, 2, \ldots, \lfloor n/2 \rfloor$ **do**
2:     $q[k] := 0$
3: $T := \sum_{p \in X} p$
4: Sort X according with $<_o$ and let $(a_1, a_2, \ldots, a_n)$ be the ordered sequence
5: **for** $i = 1, 2, \ldots, n$ **do**
6:     $R := \sum_{j < i} a_j$
7:     $g := i - 1$
8:     Sort the set $X \setminus \{a_i\}$ according with $slope(a_i, \cdot)$ and let $(b_1, b_2, \ldots, b_{n-1})$ be the ordered sequence
9:     **for** $j = 1, 2, \ldots, n - 1$ **do**
10:         **if** $a_i <_o b_j$ **then** $\begin{cases} R := R + b_j \\ g := g + 1 \end{cases}$
11:         **if** $g \le n - g$ **then** $\begin{cases} m := g \\ S := R \end{cases}$
12:                 **else** $\begin{cases} m := n - g \\ S := T - R \end{cases}$
13:         **if** $q[m] < \frac{\|S\|^2}{g(n-g)}$ **then** $\begin{cases} q[m] := \frac{\|S\|^2}{g(n-g)} \\ e[m] := (a_i, b_j) \end{cases}$
14:         **if** $b_j <_o a_i$ **then** $\begin{cases} R := R - b_j \\ g := g - 1 \end{cases}$
15: **return** $(e[1], \ldots, e[\lfloor n/2 \rfloor])$

References

1. Aloise, D., Deshpande, A., Hansen, P., Popat, P.: NP-hardness of Euclidean sum-of-squares clustering. Machine Learning 75, 245–249 (2009)
2. Basu, S., Davidson, I., Wagstaff, K.: Constrained Clustering: Advances in Algorithms, Theory, and Applications. Chapman and Hall/CRC (2008)
3. Bertoni, A., Goldwurm, M., Lin, J., Pini, L.: Size-constrained 2-Clustering in the Plane with Manhattan Distance. In: Proc. 15th Italian Conference on Theoretical Computer Science. CEUR Workshop Proceedings, vol. 1231, pp. 33–44. CEUR-WS.org (2014) ISSN 1613-0073
4. Bertoni, A., Goldwurm, M., Lin, J., Saccà, F.: Size Constrained Distance Clustering: Separation Properties and Some Complexity Results. Fundamenta Informaticae 115(1), 125–139 (2012)

5. Bishop, C.: Pattern Recognition and Machine Learning. Springer (2006)
6. Dasgupta, S.: The hardness of k-means clustering. Technical Report CS2007-0890, Department of Computer Science and Engineering, University of California, San Diego (2007)
7. Dey, T.: Improved Bounds for Planar k-Sets and Related Problems. Discrete & Computational Geometry 19(3), 373–382 (1998)
8. Edelsbrunner, H.: Algorithms in Combinatorial Geometry. EATCS monographs on theoretical computer science. Springer (1987)
9. Erdős, P., Lovász, L., Simmons, A., Straus, E.G.: Dissection graphs of planar point sets. In: A Survey of Combinatorial Theory (Proc. Internat. Sympos., Colorado State Univ., Fort Collins, Colo., 1971), pp. 139–149. North-Holland, Amsterdam (1973)
10. Hastie, T., Tibshirani, R., Friedman, J.: The Elements of Statistical Learning: Data Mining, Inference, and Prediction, 2nd edn. Springer (2009)
11. Inaba, M., Katoh, N., Imai, H.: Applications of weighted voronoi diagrams and randomization to variance-based k-clustering (extended abstract). In: Proceedings of the Tenth Annual Symposium on Computational Geometry, SCG 1994, USA, pp. 332–339 (1994)
12. Lin. J.: Exact algorithms for size constrained clustering. PhD Thesis, Dottorato di ricerca in Matematica, Statistica e Scienze computationali, Università degli Studi di Milano. Ledizioni Publishing (2013)
13. MacQueen, J.B.: Some method for the classification and analysis of multivariate observations. In: Proceedings of the 5th Berkeley Symposium on Mathematical Structures, pp. 281–297 (1967)
14. Mahajan, M., Nimbhorkar, P., Varadarajan, K.: The planar k-means problem is NP-hard. Theoretical Computer Science 442, 13–21 (2012)
15. Novick, B.: Norm statistics and the complexity of clustering problems. Discrete Applied Mathematics 157, 1831–1839 (2009)
16. Overmars, M.H., van Leeuwen, J.: Maintenance of configurations in the plane. J. Comput. Syst. Sci. 23(2), 166–204 (1981)
17. Preparata, F., Shamos, M.: Computational geometry: an introduction. Texts and monographs in computer science. Springer (1985)
18. Theodoridis, S., Koutroumbas, K.: Pattern Recognition. Academic Press, Elsevier (2009)
19. Vattani, A.: K-means requires exponentially many iterations even in the plane. In: Proceedings of the 25th Symposium on Computational Geometry (SoCG) (2009)
20. Wagstaff, K., Cardie, C.: Clustering with instance-level constraints. In: Proc. of the 17th Intl. Conf. on Machine Learning, pp. 1103–1110 (2000)
21. Zhu, S., Wang, D., Li, T.: Data clustering with size constraints. Knowledge-Based Systems 23(8), 883–889 (2010)

Local Routing in Convex Subdivisions[*]

Prosenjit Bose[1], Stephane Durocher[2], Debajyoti Mondal[2], Maxime Peabody[1],
Matthew Skala[1], and Mohammad Abdul Wahid[3]

[1] Carleton University, Ottawa, Canada
jit@scs.carleton.ca, maximepeabody@cmail.carleton.ca
[2] University of Manitoba, Winnipeg, Canada
{durocher,jyoti,mskala}@cs.umanitoba.ca
[3] Bits in Glass, Calgary, Canada
wahidrahman@gmail.com

Abstract. In various wireless networking settings, node locations determine a network's topology, allowing the network to be modelled by a geometric graph drawn in the plane. Without any additional information, local geometric routing algorithms can guarantee delivery to the target node only in restricted classes of geometric graphs, such as triangulations. In order to guarantee delivery on more general classes of geometric graphs (e.g., convex subdivisions or planar subdivisions), previous local geometric routing algorithms required $\Theta(\log n)$ state bits to be stored and passed with the message. We present the first local geometric routing algorithm using only one state bit to guarantee delivery on convex subdivisions and the first local geometric memoryless routing algorithm that guarantees delivery on edge-augmented monotone subdivisions (including all convex subdivisions) when the algorithm has knowledge of the incoming port (the preceding node on the route).

1 Introduction

1.1 Local Geometric Routing

A *local routing algorithm* determines a sequence of forwarding decisions that defines a path in a network from a source node to a given target node, where each internal node along the path selects one of its neighbours to extend the path as a function of its local network neighbourhood and limited information about the target node. Additional information available to each node on the path may include the identity of its neighbour that forwarded the message (the incoming port on which the message arrived) as well as a small number of state bits passed with the message (which may be modified locally before forwarding). In various wireless networking settings, the locations of nodes and physical proximity between nodes determine the pairs of nodes that can communicate;

---

[*] This research was supported in part by the Natural Sciences and Engineering Research Council of Canada (NSERC). A preliminary version of the algorithm presented in Section 2 appeared in Mohammad Abdul Wahid's M.Sc. thesis [19].

G.F. Italiano et al. (Eds.): SOFSEM 2015, LNCS 8939, pp. 140–151, 2015.
© Springer-Verlag Berlin Heidelberg 2015

that is, the network is determined geometrically. The network's geometric properties can provide navigational cues, enabling a local routing algorithm to use this additional geometric information to guide a message towards its destination. Each node may know its location, allowing every node on the path to make a forwarding decision as a function of the relative locations of its neighbours, the target node, and itself. We refer to such algorithms as *local geometric routing algorithms*. This paper examines the problem of defining local geometric routing algorithms that guarantee delivery from any source node to any target node on specific classes of geometric graphs.

1.2   Model and Definitions

We represent a network by an undirected graph G drawn in the plane, where each vertex is represented by a point and each edge is represented by a (straight) line segment connecting the vertices at its endpoints. Let $V(G)$ denote the set of vertices (points) of G and let $E(G)$ denote its set of edges (line segments), where $n = |V(G)|$ and $m = |E(G)|$. To simplify the discussion, we assume that vertices are in general position. By that we mean that no three points are collinear and no two points have the same x-coordinate or the same y-coordinate.

We require G to be connected for a route to exist between any pair of nodes. The drawing need not be planar, although some of our discussion relates to planar subdivisions. A drawing of a graph G in the plane is a *planar subdivision* (also *planar drawing*, *plane graph*, or *planar straight-line graph*) if each edge in $E(G)$ is drawn as a line segment and any two edges intersect only at their common endpoint. A planar subdivision partitions the plane into *faces*. When each internal face is a convex polygon and the boundary of the outer face is the convex hull, the drawing is a *convex subdivision*. When each internal face is a triangle, the subdivision is a *triangulation*. When each internal face is an x-monotone polygon (but not necessarily convex) and the boundary of the outer face is also an x-monotone polygon, the drawing is a *monotone subdivision*. Recall that a polygon is x-monotone if the intersection of its interior with any vertical line gives a connected region (i.e., a line segment). Every convex subdivision is also a monotone subdivision.

When G contains a spanning subgraph that is a convex subdivision, (respectively, a monotone subdivision), then we say G is an *edge-augmented convex subdivision* (respectively, an *edge-augmented monotone subdivision*); in this case, G corresponds to a convex subdivision to which zero or more edges have been added joining pairs of vertices in the underlying convex subdivision, possibly creating edge crossings. Edge-augmented convex subdivisions are not planar in general. Any routing algorithm that guarantees delivery on edge-augmented convex subdivisions also guarantees delivery on convex subdivisions.

Using notation similar to that previously defined [3,10,11], a local geometric routing algorithm can be expressed as a *routing function* $f : V(G) \times V(G) \times \mathscr{P}(V(G)) \to V(G)$, where $\mathscr{P}()$ denotes the power set, with arguments $f(u, t, N(u))$ such that $u \in V(G)$ is the vertex for which a forwarding decision is being made (i.e., the node presently holding the message), $t \in V(G)$ is the target vertex, and $N(u) \subseteq V(G)$ is the set of neighbours of u in G. Upon receiving a message destined for a node t, a node u forwards the message to its neighbour $w = f(u, t, N(u))$.

If u knows which of its neighbours forwarded the message, then we say the routing algorithm is *predecessor-aware* and represent the corresponding routing function as $f(u, v, t, N(u))$, where $v \in V(G)$ denotes the neighbour of u that last forwarded the message. Otherwise, we say the routing algorithm is *predecessor-oblivious*. Furthermore, if c state bits are passed with the message then we say the routing algorithm is *c-bit* local and the routing function becomes $f(u, t, N(u), e)$ (or $f(u, v, t, N(u), e)$ if predecessor-aware), where $e \in \{0, 1\}^c$. We focus on the case $c = 1$. If no bits are passed with the message then we say the routing algorithm is *stateless*. Note that no state information is stored at a node after it has forwarded a message; that is, the network is *memoryless*. When a message is forwarded, its destination t and the c state bits are passed with it. All other information is available locally at node u. Randomized solutions exists (e.g., [10]); in this work we restrict attention to deterministic routing algorithms.

1.3   Related Work

When applying a local geometric algorithm that is stateless and predecessor-oblivious, every time a node u receives a message destined for a given target node t, u always forwards the message to its same neighbour. Consequently, stateless predecessor-oblivious routing algorithms that guarantee delivery are limited to restricted classes of geometric graphs. These include greedy routing [12] and compass routing [15], both of which succeed on any Delaunay triangulation but fail on more general triangulations [6], as well as greedy-compass routing [2], which succeeds on any triangulation. In a triangulation each node knows the complete set of edges bounding every face on which it is adjacent. Beyond triangulations are convex subdivisions, where faces remain convex, but a node only knows two edges bounding every face on which it is adjacent. Every stateless and predecessor-oblivious local geometric routing algorithm fails on some convex subdivision [2]. Conseqently, local routing algorithms require additional reference beacons, or the ability to store learned route information in state bits, to support successful navigation on convex subdivisions or, more generally, on planar subdivisions.

Face routing [15] succeeds on any planar subdivision, but requires both predecessor-awareness and $\Theta(\log n)$ state bits (assuming vertex coordinates can be stored using $\Theta(\log n)$ bits per vertex). Variants of face routing succeed on unit disc graphs [7] and some quasi unit disc graphs [16]. Some local geometric algorithms define a route (or a graph traversal) on planar and near-planar subdivisions by performing a depth-first traversal of a locally defined spanning tree [1,4,9,17]; all such algorithms known require $\Theta(\log n)$ state bits. For graphs drawn in three-dimensional space, every stateless and predecessor-aware local geometric routing algorithm fails on some unit ball graph [11].

If $\Theta(\log n)$ state bits are available, then geometric information is not necessary to support local routing: by storing an index into a polynomial-length universal traversal sequence, predecessor-oblivious routing is possible on any graph, not restricted to belonging to any particular class of drawings [8]; this requires each node to be able to reconstruct the traversal sequence. Without geometric information, a stateless routing algorithm requires knowledge of a large neighbourhood around each node to guarantee delivery. Specifically, a predecessor-aware stateless routing algorithm requires each node to have knowledge of the induced subgraph within graph distance $n/3$ of itself; for predecessor-oblivious algorithms the distance increases to $n/2$ [3]. There exists a small set of graphs such that every stateless routing algorithm whose knowledge is limited to a smaller neighbourhood around each node fails on one or more of these graphs.

In addition to knowing the target node t, knowledge of the source node s also determines whether local routing is possible. Applying the right-hand rule along the edges of the sequence of faces that intersect the line segment from s to t gives a stateless predecessor-aware local geometric routing that succeeds on convex subdivisions (requiring knowledge of s) [17]; this is essentially face routing applied to a specific class of graphs that does not require backtracking. To succeed on planar subdivisions that are not convex, face routing is occasionally forced to backtrack [7,15], requiring $\Theta(\log n)$ state bits. Knowledge of s is significant even when geometric information is not available. For example, given s, a stateless predecessor-aware local routing algorithm only requires knowledge of the induced subgraph within graph distance $n/4$ of each node to guarantee delivery in any graph, instead of distance $n/3$ without knowledge of s [3].

Although similar to local routing, *online routing* [5] differs by the fact that each node u along the route has complete information about the subgraph explored prior to arriving at u. Storing such information in a message requires $\Theta(n \log n)$ state bits in general.

Guaranteeing delivery on geometric graphs beyond triangulations requires state information or predecessor-awareness. In this paper we seek to bridge the gap between stateless predecessor-oblivious local routing algorithms, which cannot guarantee delivery even on convex subdivisions, and $\Theta(\log n)$-bit local routing algorithms. Specifically, we examine whether navigation is possible when a local routing algorithm is provided a single state bit and whether it is possible when enabled with predecessor awareness. In each case we seek to define a routing algorithm and to identify broad classes of geometric graphs on which the algorithm guarantees delivery. For surveys on local geometric routing, see Morin [17], Guan [14], Urrutia [18], and Frey et al. [13].

1.4   Overview of Results

No stateless predecessor-oblivious local geometric routing algorithm can guarantee delivery on convex subdivisions [2]; to succeed on convex subdivisions and, therefore, on more general classes of graphs, such as planar subdivisions, a local geometric routing algorithm must have the ability to store state information or be provided with predecessor awareness. To the authors' knowledge, prior to this work no predecessor-oblivious c-bit local geometric routing algorithm was known to guarantee delivery on convex subdivisions for any $c \in o(\log n)$.

Table 1. Graphs on which local routing is possible. New contributions appear in bold

bits	predecessor oblivious	predecessor aware
0	triangulations [2] impossible on convex subdiv. [2]	**(edge-aug.) convex subdiv.,** **(edge-aug.) monotone subdiv.**
1	**convex subdiv.**	beyond edge-aug. monotone subdiv.: unknown
$O(1)$	beyond convex subdiv.: unknown	
$O(\log n)$	all graphs [8]	planar [15], unit disc [7], all graphs [8]

Similarly, no predecessor-aware stateless local geometric routing algorithm was known to guarantee delivery on convex subdivisions. This paper presents the first predecessor-aware stateless local geometric routing algorithm and the first 1-bit predecessor-oblivious local geometric routing algorithm to guarantee delivery on any non-trivial class of geometric graphs beyond triangulations. See Table 1. In Section 2 we present a predecessor-oblivious local geometric routing algorithm that uses one state bit ($c = 1$) and guarantees delivery on any convex subdivision. In Section 3, we present a stateless predecessor-aware local geometric routing algorithm that guarantees delivery on any edge-augmented monotone subdivision. We conclude with a discussion in Section 4.

2   Using One State Bit

We describe a predecessor-oblivious one-bit geometric local routing algorithm, called OneBit, that guarantees delivery on any convex subdivision. Let u denote the node holding the message, i.e., the node making a forwarding decision. As in compass routing [15] and greedy-compass routing [2], we refer to the *clockwise* (respectively, *counterclockwise*) *neighbour* of u relative to t, denoted $cw(u)$ (respectively, $ccw(u)$) defined as the node $v \in N(u)$ that forms the smallest clockwise (counterclockwise) angle $\angle tuv$. Let H_s denote the closed half-plane containing u whose boundary is the vertical line ℓ_t through t. See Figure 1A. Algorithm OneBit never forwards the message to a node outside H_s, enabling all nodes along the route to identify H_s consistently relative to ℓ_t, regardless of whether the source node s is left or right of ℓ_t.

Algorithm OneBit uses one state bit, denoted c, to determine whether node u should forward the message from u to $cw(u)$ or to $ccw(u)$. The state bit c can be initialized arbitrarily at the source node s, e.g., $c \leftarrow 0$. Node s does not need to know it is the source; the algorithm can initialize c arbitrarily if it is not assigned a value in $\{0, 1\}$. See Algorithm 1 and Figure 1B. The resulting route corresponds to a sequence of clockwise forwarding decisions (when $c = 0$), which we call an *clockwise chain*, followed by a sequence of counterclockwise forwarding decisions (when $c = 1$), which we call a *counterclockwise chain*, followed by another clockwise chain (when $c = 0$ again), and so on, until the message

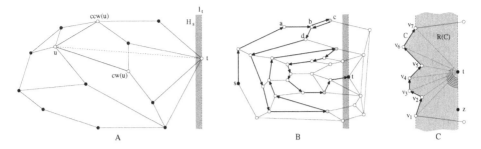

Fig. 1. (A) The clockwise and counterclockwise neighbours of u are defined relative to the line segment from u to t. The state bit c remains unchanged since both $ccw(u)$ and $cw(u)$ are in H_s; node u forwards the message to $ccw(u)$ if $c = 0$ and to $cw(u)$ if $c = 1$. **(B)** The bold arrows denote the sequence of local forwarding decisions made by Algorithm OneBit from s to t on this convex subdivision. Nodes at which the state bit toggles are shaded grey. Some edges can be traversed once in each direction on a route; e.g., see the subsequence $a \rightarrow b \rightarrow c \rightarrow b \rightarrow d$. **(C)** The increasing sequence of angles and the region $R(C)$ determined by a clockwise chain C.

reaches the target node t. Note that a clockwise chain proceeds in a counterclockwise direction relative to t, and vice versa. The algorithm toggles the state bit whenever continuing the chain would send the message outside H_s. As we show, each chain in the resulting sequence is contained within a region bounded by the preceding chain, giving a convergence towards t. We refer to the first and last vertices on a chain according to the chronological order of the sequence of forwarding decisions as its *head* and *tail*, respectively, where the tail of the ith chain is the head of the $(i + 1)$st chain.

Lemma 1. *For every node u in a convex subdivision, if $u \neq t$, then $cw(u) \in H_s$ or $ccw(u) \in H_s$.*

Proof. The lemma follows from the fact that the half-plane H_s is closed and is bounded by the vertical line ℓ_t through t and that every face is convex. □

Lemma 2. *Every clockwise (respectively, counterclockwise) chain C terminates, either at t or at the head of an oppositely oriented chain.*

Proof. Without loss of generality, suppose C is a clockwise chain corresponding to the sequence of vertices $v_1, \ldots, v_k$. By construction, the nodes $v_1, \ldots, v_k$ are all contained in H_s and corresponds to a sequence of increasing angles $\angle v_1 t z < \cdots < \angle v_k t z$, where z is any point that lies below t on ℓ_t. See Figure 1C. As defined in Algorithm 1, the chain C terminates when the tail node u has no clockwise neighbour. By Lemma 1, u must have a counterclockwise neighbour v, which defines the head of the subsequent counterclockwise chain. □

Lemma 2 implies that the forwarding sequence cannot continue indefinitely (i.e., it cannot cycle) without a change of state. Consequently, every chain C has a head and a tail. Given a chain C, let $R(C)$ denote the region bounded by C,

---

Algorithm 1. OneBit(u, c, t)

Preconditions: u is the node holding the message, $N(u)$ is its set of neighbours,
$\quad$ $c \in \{0, 1\}$ is the state bit passed with the message, t is the target node, and $cw(u)$
$\quad$ and $ccw(u)$ denote the clockwise and counterclockwise neighbours of u, respectively.
Postconditions: Forward the message from u to w with state bit c', where $w \in N(u)$.

1: $c' \leftarrow c$
2: **if** $[c' = 0$ **and** $ccw(u) \notin H_s]$ **or** $[c' = 1$ **and** $cw(u) \notin H_s]$ **then**
3: $\quad$ $c' \leftarrow$ **not** c' (The current chain cannot be continued in H_s: change states.)
4: **end if**
5: **if** $t \in N(u)$ **then** (The target node t is adjacent to u.)
6: $\quad$ $w \leftarrow t$ (Forward the message to the target node t.)
7: **else if** $c' = 0$ **then** (State 0)
8: $\quad$ $w \leftarrow ccw(u)$ (Forward the message along a counterclockwise chain.)
9: **else** (State 1)
10: $\quad$ $w \leftarrow cw(u)$ (Forward the message along a clockwise chain.)
11: **end if**

---

the respective vertical rays emanating away from its head and tail, and ℓ_t. See
Figure 1C. We say a chain is *complete* if it originated and terminated as a result
of toggling the state bit. Consequently, all chains are complete, except the first
(whose head is the source node s) and the last (whose tail is the target node t).

Lemma 3. *If C_i and C_j are any two chains in a route such that C_i is complete
and C_i precedes C_j, then $R(C_j) \subseteq R(C_i)$*

Proof. The result follows by induction on the sequence of chains between C_i
and C_j. Consider the case when C_i and C_j are consecutive chains. Without
loss of generality, suppose C_i is a clockwise chain. Let u denote any node in C_i
other than the tail. Node u forwards the message to its neighbour $v = cw(u)$.
Therefore, $ccw(v)$ exists. That is, either $ccw(v) = u$ or $ccw(v) = u'$ such that
$\angle tvu' < \angle tvu$. See Figure 2A. That is, no two chains can cross. They can,
however, share a common sequence of adjacent vertices. $\qquad\square$

Lemma 4. *If C_i and C_j are any two oppositely oriented chains, then $C_j \neq C_i$.*

Proof. We prove the lemma by contradiction. By definition of $R(C)$ and
Lemma 2, no point of C lies in the interior of $R(C)$ and, consequently, for any
chains C_i and C_j, $R(C_i) \neq R(C_j)$ if and only if $C_i \neq C_j$. Suppose $C_i = C_j$,
where C_i and C_j are two oppositely oriented chains. Therefore, for every edge
$\{u, v\}$ in both C_i and C_j, $u = cw(v)$ and $v = ccw(u)$ (or vice versa). Consequently, no internal vertex on the chains can have an edge into the interior of
$R(C_i) = R(C_j)$. See Figure 2B. Since every face is convex, t must have a neighbour in H_s. Therefore, some node on C_i must have a neighbour that is t or that
lies in the interior of $R(C_i)$, deriving a contradiction. $\qquad\square$

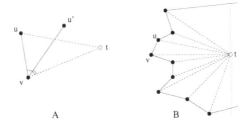

Fig. 2. Illustrations in support of Lemmas 3 (**A**) and 4 (**B**)

Theorem 1. *Given any convex subdivision G and any vertices $\{s,t\} \subseteq V(G)$, Algorithm OneBit is a predecessor-oblivious geometric local routing algorithm that uses one state bit to determine a sequence of forwarding decisions from s to t in G.*

Proof. Each chain is finite by Lemma 2. In particular, the first chain terminates. Each pair of subsequent chains, C_i and C_j, is complete, except for the last chain which terminates at t. Lemmas 3 and 4 imply that $R(C_j)$ is a proper subset of $R(C_i)$. Consequently, the sequence of chains converges towards ℓ_t. Since the vertices of each chain are vertices of G, of which there are a finite number, the result follows. □

3   Using Predecessor Awareness

We describe a predecessor-aware stateless geometric local routing algorithm, called PredAware(u,v,t), that guarantees delivery on any edge-augmented monotone subdivision.

Let G be an edge-augmented monotone subdivision. We define a partial order $\mathcal{P}$ over the vertex set $V(G)$ as follows. For each $u \in V(G)$, let ℓ_u denote the vertical line through u, let z_u' denote a point on ℓ_u above u, and let H_u^- and H_u^+ denote the respective left and right half-planes bounded by ℓ_u. Let the *ith left neighbour* of u, denoted $\text{left}_u(i)$, be the node in $v \in N(u) \cap H_u^-$ that forms the ith smallest convex angle $\angle vuz_u'$. The *parent* of u is its first left neighbour, $\text{left}_u(1)$. Similarly, let the *ith right neighbour* of u, denoted $\text{right}_u(i)$, be the node $v \in N(u) \cap H_u^+$ that forms the ith smallest convex angle $\angle vuz_u'$. See Figure 3A. If u has no left neighbours, then u is a *root*. If u has no right neighbours, then u is a *leaf*. For all nodes u and v, $u = \text{left}_v(i)$ for some i if and only if $v = \text{right}_u(j)$ for some j; in particular, this inverse relationship exists if and only if u and v are neighbours and u lies to the left of ℓ_v. The left neighbour relation, $\prec$ (or equivalently, the right neighbour relation) assigns an orientation to each edge in $E(G)$ such that $u \prec v$ if $\{u,v\} \in E(G)$ and $u_x < v_x$, where a_x denotes the x-coordinate of point a. That is, each (previously undirected) edge $\{u,v\} \in E(G)$, where $u_x < v_x$, is assigned the orientation (u,v). Since x-coordinates belong to a total order, the corresponding directed graph is acyclic, which defines the partial order $\mathcal{P}$ on the vertex set $V(G)$. See Figure 3C.

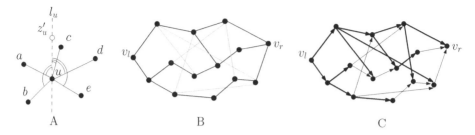

Fig. 3. (A) Nodes a and b are the respective 1st and 2nd left neighbours of u. Node a is the parent of u. Nodes c, d, and e are the respective 1st, 2nd, and 3rd right neighbours of u. **(B)** An edge-augmented monotone subdivision G, where the underlying monotone subdivision is shown in bold. **(C)** The corresponding directed acyclic graph on G with parent edges defining a spanning tree in bold.

Lemma 5. $\mathcal{P}$ *defines a single-source (single root) single-sink (single leaf) directed acyclic graph over G.*

Proof. Let M be the monotone subdivision underlying G. Recall that the boundary of the exterior face of any monotone subdivision is monotone. Hence M has a leftmost and a rightmost node, which we denote by v_l and v_r, respectively. Since these nodes remain incident to the outer face even after edge augmentation, they are also the leftmost and rightmost vertices in G. Observe that for each vertex $v \notin \{v_l, v_r\}$ of G, there is a monotone path from v_l to v_r that passes through v. Hence every vertex $v \notin \{v_l, v_r\}$ has a left and a right neighbour.

The existence of a directed cycle in G would imply some edge oriented from right to left. By definition, all edges are oriented from left to right. Therefore, the resulting edge orientations on G determine a directed acyclic graph with a unique source v_l and a unique sink v_r. □

The term "source" as used in Lemma 5 refers to a vertex with in-degree zero and non-zero out-degree. Throughout the rest of the paper, a source node refers to the initial node s from which a message is routed to a target node t.

Algorithm PredAware traverses every edge of G using the partial order $\mathcal{P}$ defined on G. The orientation of each edge and, consequently, the partial order $\mathcal{P}$, can be determined locally by any node u that knows its coordinates and those of its set of neighbours $N(u)$. The route corresponds to a depth-first traversal of a spanning tree of G, resulting in a complete traversal of the graph's vertices (implying guaranteed delivery). Specifically, for each node $u \in V(G)$ that is not the leftmost node (which is the tree root) the edge $\{u, \text{left}_u(1)\} \in E(G)$ (i.e., the edge from u to its parent) corresponds to a tree edge.

Lemma 6. *The set of parent edges defines a spanning tree on G.*

Proof. By Lemma 5, the graph G is a directed acyclic graph with a single source. Each vertex other than the root has a single parent edge, which is an in-edge. The result follows. □

Let u denote the node holding the message, i.e., the node making a forwarding decision. Let $v \in N(u)$ denote the neighbour of u that last forwarded the message to u. At the start of the route (when $u = s$ initially), suppose $v = \varnothing$. The tree traversal algorithm is described in Algorithm 2.

Algorithm 2. PredAware(u, v, t)

Preconditions: u is the node holding the message, $N(u)$ is its set of neighbours, $v \in N(u)$ is u's neighbour that last forwarded the message, t is the target node, left$_u(i)$ and right$_u(i)$ denote the ith left and right neighbours of u, respectively.

Postconditions: Forward the message from u to w, where $w \in N(u)$.

1: **if** $t \in N(u)$ **then** (The target node t is adjacent to u.)
2:     $w \leftarrow t$ (Forward the message to the target node t.)
3: **else if** $v = \varnothing$ **then** (There is no predecessor: initiate the route.)
4:     **if** right$_u(1) \neq \varnothing$ **then** (u has a right neighbour.)
5:         $w \leftarrow$ right$_u(1)$ (Forward the message to u's first right neighbour.)
6:     **else** (u has no right neighbour.)
7:         $w \leftarrow$ left$_u(1)$ (Forward the message to u's parent.)
8:     **end if**
9: **else if** $v =$ left$_u(1)$ **then** (u's parent passed the message into a new subtree of u.)
10:     **if** right$_u(1) \neq \varnothing$ **then** (u has a right neighbour.)
11:         $w \leftarrow$ right$_u(1)$ (Forward the message to u's first right neighbour.)
12:     **else** (u has no right neighbour.)
13:         $w \leftarrow v$ (Return the message to u's parent.)
14:     **end if**
15: **else if** $v =$ left$_u(i)$ for some $i \geq 2$ **then** (This edge is not in the spanning tree; return the message.)
16:     $w \leftarrow v$ (Return the message to the sender v.)
17: **else** (Traversal of u's ith subtree is complete. Traverse u's $(i + 1)$st subtree.)
18:     **if** right$_u(i + 1) \neq \varnothing$ **then** (u has an $(i + 1)$st right neighbour.)
19:         $w \leftarrow$ right$_u(i + 1)$ (Forward the message to u's $(i + 1)$st right neighbour.)
20:     **else if** left$_u(1) \neq \varnothing$ **then** (u has no $(i + 1)$st right neighbour but has a parent.)
21:         $w \leftarrow$ left$_u(1)$ (Forward the message to u's parent.)
22:     **else** (u has neither an $(i + 1)$st right neighbour nor a parent: u is the root.)
23:         $w \leftarrow$ right$_u(1)$ (Forward the message to u's first right neighbour.)
24:     **end if**
25: **end if**

Theorem 2. *Given any edge-augmented monotone subdivision G and any vertices $\{s, t\} \subseteq V(G)$, Algorithm PredAware is a predecessor-aware stateless geometric local routing algorithm that determines a sequence of forwarding decisions from s to t in G. Furthermore, Algorithm PredAware performs a traversal of G.*

Proof. Upon receiving the message from its parent, each node u sequentially forwards the message to each of its right neighbours in clockwise order (see lines 9–14 and 17–24 in Algorithm 2). Upon receiving the message from its ith right neighbour, u forwards the message to its $(i + 1)$st right neighbour. If u has no

$(i+1)$st neighbour, then u returns the message to its parent. If u receives the message from a left neighbour other than its parent, then u returns the message to the sender; this indicates that the message was sent along a non-tree edge, and the message is returned immediately. Therefore, the route is extended only when a node u receives the message from its parent, by forwarding the message to each of u's right neighbours. A node's set of right neighbours includes all of its children in the spanning tree on the set of parent edges. Algorithm 2 generates a depth-first recursive traversal of the set of parent edges, which, by Lemma 6, corresponds to a spanning tree of G. The resulting sequence of forwarding decisions is a preorder (depth-first) traversal of the spanning tree.   □

Although local algorithms exist for various classes of geometric graphs that construct a spanning tree on which a depth-first tree traversal determines a graph traversal sequence (e.g., [1,17]), these all require $\Theta(\log n)$ state bits. Algorithm PredAware is stateless. Its ability to guarantee delivery on a monotone subdivision is due to predecessor awareness.

4   Discussion and Directions for Future Research

The algorithms presented in this paper reduce the gap between the classes of geometric graphs on which guaranteed delivery is possible without state bits and those on which it is possible with $O(\log n)$ state bits. Several questions remain to be answered to close this gap. See Table 1.

If nodes have distinct labels, then identifying t in the message requires $\Omega(\log n)$ bits. That data is *static* and is not modified by the routing algorithm. The goal of this research is to minimize the state bits *modified dynamically* by the algorithm. With sufficient state bits, a routing algorithm can record the complete partial graph that has been explored (e.g., $O(n \log n)$ state bits). Braverman's local routing algorithm [8] guarantees delivery on any graph using $\Theta(\log n)$ state bits, regardless of geometry, and without requiring predecessor awareness. In many cases, $\Theta(\log n)$ bits is an allowable cost.

We seek to identify and characterize classes of geometric graphs on which delivery can be guaranteed using few states. In this paper we showed that guaranteed delivery is possible on convex subdivisions using only one state bit and without predecessor awareness, and on edge-augmented monotone subdivisions using only predecessor awareness and no state bits. Routing in planar subdivisions (and other classes of geometric graphs) allows a local routing algorithm to capitalize on the inherent *geometry* to guarantee delivery using fewer states than are necessary on arbitrary graphs. This leads to some natural open questions. Is geometric local routing on planar subdivisions possible using c state bits, where $c \in o(\log n)$ or $c \in O(1)$? On what classes of geometric graphs can a geometric local routing algorithm guarantee delivery using $O(1)$ state bits? On what classes of geometric graphs can a stateless geometric local routing algorithm guarantee delivery using predecessor awareness? With both predecessor awareness and $O(1)$ state bits, can a local routing algorithm guarantee delivery on more general classes of graphs than if it were predecessor-aware and stateless?

Finally, measuring and bounding a local routing algorithm's dilation (worst-case ratio of actual route length to shortest path length) is of interest. Can $O(1)$ dilation be guaranteed on convex subdivisions with one state bit?

References

1. de Berg, M., van Kreveld, M., van Oostrum, R., Overmars, M.: Simple traversal of a subdivision without extra storage. Int. J. Geog. Inf. Sci. 11(4), 359–373 (1997)
2. Bose, P., Brodnik, A., Carlsson, S., Demaine, E.D., Fleischer, R., López-Ortiz, A., Morin, P., Munro, I.: Online routing in convex subdivisions. Int. J. Comp. Geom. & Appl. 12(4), 283–295 (2002)
3. Bose, P., Carmi, P., Durocher, S.: Bounding the locality of distributed routing algorithms. Dist. Comp. 26(1), 39–58 (2013)
4. Bose, P., Morin, P.: An improved algorithm for subdivision traversal without extra storage. Int. J. Comp. Geom. & Appl. 12(4), 297–308 (2002)
5. Bose, P., Morin, P.: Competitive online routing in geometric graphs. Theor. Comp. Sci. 324, 273–288 (2004)
6. Bose, P., Morin, P.: Online routing in triangulations. SIAM J. Comp. 33(4), 937–951 (2004)
7. Bose, P., Morin, P., Stojmenović, I., Urrutia, J.: Routing with guaranteed delivery in ad hoc wireless networks. Wireless Net. 7(6), 609–616 (2001)
8. Braverman, M.: On ad hoc routing with guaranteed delivery. In: Proc. ACM PODC, vol. 27, p. 418 (2008)
9. Chavez, E., Dobrev, S., Kranakis, E., Opatrny, J., Stacho, L., Urrutia, J.: Traversal of a quasi-planar subdivision without using mark bits. J. Interconn. Net. 5(4), 395–408 (2004)
10. Chen, D., Devroye, L., Dujmović, V., Morin, P.: Memoryless routing in convex subdivisions: Random walks are optimal. Comp. Geom.: Theory & Appl. 45(4), 178–185 (2012)
11. Durocher, S., Kirkpatrick, D., Narayanan, L.: On routing with guaranteed delivery in three-dimensional ad hoc wireless networks. Wireless Net. 16, 227–235 (2010)
12. Finn, G.G.: Routing and addressing problems in large metropolitan-scale internetworks. Technical Report ISI/RR-87-180, Information Sciences Institute (1987)
13. Frey, H., Ruehrup, S., Stojmenovic, I.: Routing in wireless sensor networks. In: Misra, S., Woungag, I., Misra, S. (eds.) Guide to Wireless Ad Hoc Networks, ch.4, pp. 81–111. Springer (2009)
14. Guan, X.: Face routing in wireless ad-hoc networks. PhD thesis, Univ. Toronto (2009)
15. Kranakis, E., Singh, H., Urrutia, J.: Compass routing on geometric networks. In: Proc. CCCG, vol. 11, pp. 51–54 (1999)
16. Kuhn, F., Wattenhofer, R., Zollinger, A.: Ad-hoc networks beyond unit disk graphs. In: Proc. ACM DIALM-POMC, pp. 69–78 (2003)
17. Morin, P.: Online routing in geometric graphs. PhD thesis, Carleton Univ. (2001)
18. Urrutia, J.: Handbook wireless net. & mob. comp. In: Stojmenovic, I. (ed.) Routing with Guaranteed Delivery in Geometric and Wireless Networks, pp. 393–406. John Wiley & Sons, Inc. (2002)
19. Wahid, M.A.: Local geometric routing algorithms for edge-augmented planar graphs. Master's thesis, Univ. Manitoba (2013)

Nondeterministic Modal Interfaces*

Ferenc Bujtor[1], Sascha Fendrich[2], Gerald Lüttgen[2], and Walter Vogler[1]

[1] Institut für Informatik, University of Augsburg, Germany
{walter.vogler,ferenc.bujtor}@informatik.uni-augsburg.de
[2] Software Technologies Research Group, University of Bamberg, Germany
{gerald.luettgen,sascha.fendrich}@swt-bamberg.de

Abstract. Interface theories are employed in the component-based design of concurrent systems. They often emerge as combinations of Interface Automata (IA) and Modal Transition Systems (MTS), e.g., Nyman et al.'s IOMTS, Bauer et al.'s MIO, Raclet et al.'s MI or our MIA. In this paper, we generalise MI to *nondeterministic* interfaces, for which we resolve the longstanding conflict between unspecified inputs being allowed in IA but forbidden in MTS. With this solution we achieve, in contrast to related work, an *associative* parallel composition, a *compositional* preorder, a conjunction on interfaces with *dissimilar alphabets* supporting perspective-based specifications, and a quotienting operator for decomposing *nondeterministic* specifications in a single theory.

1 Introduction

Interface theories [2,7,8,15,16,18] support the component-based design of concurrent systems and offer a semantic framework for, e.g., software contracts [1] and web services [4]. Several such theories are based on de Alfaro and Henzinger's *Interface Automata* (IA) [10], whose distinguishing feature is a parallel composition on labelled transition systems with inputs and outputs, where receiving an unexpected input is regarded as an error, i.e., a communication mismatch. All states are pruned from which entering an error state cannot be prevented by the environment, rather than leaving the parallel composition fully undefined as in [2].

Various researchers have combined IA with Larsen's *Modal Transition Systems* (MTS) [14], which features may- and must-transitions to express allowed and required behaviour, resp. In a refinement of an interface, all required behaviour must be preserved and no disallowed behaviour may be added. Whereas in IA outputs are optional, they may now be enforced in theories combining IA and MTS, such as Nyman et al.'s IOMTS [15], Bauer et al.'s MIO [2], Raclet et al.'s *Modal Interfaces* (MI) [18] and our *Modal Interface Automata* (MIA) [16,17]. In this paper we extend MI to nondeterministic systems, yielding the most general approach to date and permitting new applications, e.g., for dealing with races in networks. We built upon our prior work in [17], from which we adopt disjunctive must-transitions that are needed for operationally defining conjunction, which is another key operator in interface theories and supports perspective-based specification.

---

* Research support was provided by the DFG (German Research Foundation) under grants LU 1748/3-1 and VO 615/12-1.

G.F. Italiano et al. (Eds.): SOFSEM 2015, LNCS 8939, pp. 152–163, 2015.

Combining IA and MTS is, however, problematic since unspecified inputs are forbidden in MTS, but allowed in IA with arbitrary behaviour afterwards. In IOMTS [15], the MTS-view was adopted and, as a consequence, compositionality of refinement wrt. the parallel operator $\|$ was lost. In [17] we followed the IA-view but found that resolving the conflict is essential for a more flexible conjunction. In our new MIA, we can optionally express the IA-view for state p and input i by an i-may-transition from p to a *special, universal state e* that can be refined in any way; we will need this option when defining $\|$. There is a similar idea in MI [18], but an ordinary state is used there with the consequence that $\|$ is not associative. In contrast to the somewhat related demonic completion as used, e.g., in [11], we do not enforce input-enabledness. With the new feature, our interface theory allows for a proper distinction between may- and must-transitions for inputs, unlike [16,17]. This enables us to define conjunction also on interfaces with dissimilar alphabets via alphabet extension.

As in MI, our MIA is equipped with a multicast parallel composition, where one output can synchronise with several inputs. We also develop a quotienting operator as a kind of inverse of parallel composition $\|$. For a specification P and a given component D, quotienting constructs the most general component Q such that $Q \| D$ refines P. Quotienting is a very practical operator because it can be used for decomposing concurrent specifications stepwise, specifying contracts [3], and reusing components. In contrast to [18], our quotienting permits *nondeterministic* specifications and complements $\|$ rather than a simpler parallel product without pruning.

In summary, our new interface theory MIA generalises and improves upon existing theories combining IA and MTS: parallel composition is commutative and associative (cf. Section 3), quotienting also works for nondeterministic specifications (cf. Section 4), conjunction properly reflects perspective-based specification (cf. Sections 5 and 6), and refinement (cf. Section 2) is compositional and permits alphabet extension (cf. Section 6). A technical report of this paper [5] contains all proofs, more explanations and examples; it also introduces a disjunction and an action scoping operator.

2   Modal Interface Automata: The Setting

In this section we define MIA and its supported operations. Essentially, MIAs are state machines with disjoint input and output alphabets and two transition relations, *may* and *must*, as in MTS [14]. May-transitions describe permitted behaviour, while must-transitions describe required behaviour. Unlike previous versions of MIA [16,17] and other similar theories, we introduce the *universal state e* as an extra constituent.

Definition 1 (Modal Interface Automata). *A* Modal Interface Automaton *(MIA) is a tuple* $(P, I, O, \longrightarrow, \dashrightarrow, p_0, e)$, *where*

- *P is the set of states containing the* initial state p_0 *and the* universal state e,
- *I and O are disjoint sets, the* alphabets of *input* and *output actions, not containing the special internal action* τ, *and* $A =_{df} I \cup O$ *is called the* alphabet,
- $\longrightarrow \subseteq P \times (A \cup \{\tau\}) \times (\mathscr{P}_{fin}(P) \setminus \emptyset)$ *is the* disjunctive must-transition *relation, with* $\mathscr{P}_{fin}(P)$ *being the set of finite subsets of P*,
- $\dashrightarrow \subseteq P \times (A \cup \{\tau\}) \times P$ *is the* may-transition *relation.*

We require (a) for all $\alpha \in A \cup \{\tau\}$ *that* $p \xrightarrow{\alpha} P'$ *implies* $\forall p' \in P'.\ p \dashrightarrow{\scriptstyle\alpha} p'$ *(syntactic consistency) and that (b) e appears in transitions only as the target of input may-transitions.*

Cond. (a) states that whatever is required should be allowed; this syntactic consistency is a natural and standard condition (cf. [14]). Cond. (b) matches the idea for e explained in the introduction. We use this state in the context of parallel composition to represent communication errors. Note that our disjunctive must-transitions have a single label, in contrast to Disjunctive MTS [13].

In the sequel, we identify a MIA $(P, I, O, \longrightarrow, \dashrightarrow, p_0, e)$ with its state set P and, if needed, use index P when referring to one of its components, e.g., we write I_P for I. Similarly, we write, e.g., I_1 instead of I_{P_1} for MIA P_1. In addition, we let i, o, a, ω and α stand for representatives of the alphabets $I, O, A, O \cup \{\tau\}$ and $A \cup \{\tau\}$, resp.; we write $A = I/O$ when highlighting inputs I and outputs O in an alphabet A, and we define $\hat{a} =_{\mathrm{df}} a$ and $\hat{\tau} =_{\mathrm{df}} \varepsilon$ (the empty word). Furthermore, outputs and internal actions are called *local* actions since they are controlled locally by P. For convenience, we let $p \xrightarrow{a} p'$, $p \not\xrightarrow{\beta}$ and $p \not\dashrightarrow{\scriptstyle\beta}$ denote $p \xrightarrow{a} \{p'\}$, $\nexists p'.\ p \xrightarrow{a} p'$ and $\nexists p'.\ p \dashrightarrow{\scriptstyle a} p'$, resp. In figures, we often refer to an action a as a? if $a \in I$, and as a! if $a \in O$. Must-transitions (may-transitions) are drawn using solid, possibly splitting arrows (dashed arrows); any depicted must-transition also implicitly represents the underlying may-transition(s).

We now define *weak* must- and may-transition relations that abstract from transitions labelled by τ. The following definition is equivalent to the one in [17].

Definition 2 (Weak Transition Relations). *We define* weak *must-transition and* weak *may-transition relations,* $\Longrightarrow$ *and* $=\!\Rightarrow$ *resp., as the smallest relations satisfying the conditions* $P' \overset{\varepsilon}{\Longrightarrow} P'$ *for finite* $P' \subseteq P$, $p =\!\overset{\varepsilon}{\Rightarrow} p$ *as well as:*

(a) $P' \overset{\hat{\alpha}}{\Longrightarrow} P''$, $p'' \in P''$ *and* $p'' \xrightarrow{\tau} P'''$ *implies* $P' \overset{\hat{\alpha}}{\Longrightarrow} (P'' \setminus \{p''\}) \cup P'''$,

(b) $P' \overset{\varepsilon}{\Longrightarrow} P'' = \{p_1, \ldots, p_n\}$ *and* $\forall j.\ p_j \xrightarrow{a} P_j$ *implies* $P' \overset{a}{\Longrightarrow} \bigcup_{j=1}^{n} P_j$,

(c) $p =\!\overset{\varepsilon}{\Rightarrow} p'' \dashrightarrow{\scriptstyle\tau} p'$ *implies* $p =\!\overset{\varepsilon}{\Rightarrow} p'$,

(d) $p =\!\overset{\varepsilon}{\Rightarrow} p'' \dashrightarrow{\scriptstyle\alpha} p''' =\!\overset{\varepsilon}{\Rightarrow} p'$ *implies* $p =\!\overset{\alpha}{\Rightarrow} p'$.

For $\{p'\} \overset{\hat{\alpha}}{\Longrightarrow} P''$ we often write $p' \overset{\hat{\alpha}}{\Longrightarrow} P''$. Mostly for inputs a, we also use relation compositions $\xrightarrow{a}\overset{\varepsilon}{\Longrightarrow}$ and $\dashrightarrow{\scriptstyle a}=\!\overset{\varepsilon}{\Rightarrow}$ resp., i.e., where leading τs are disallowed. Observe that $p \xrightarrow{a}\overset{\varepsilon}{\Longrightarrow} P'$ implies $p \overset{a}{\Longrightarrow} P'$, and $p \dashrightarrow{\scriptstyle a}=\!\overset{\varepsilon}{\Rightarrow} p'$ implies $p =\!\overset{a}{\Rightarrow} p'$.

Now we define our refinement relation. It is a weak alternating simulation conceptually similar to the observational modal refinement found, e.g., in [12]. A notable aspect, originating from IA [10], is that inputs must be matched immediately, i.e., only trailing τs are allowed. Intuitively, this is due to parallel composition requiring that a signal sent from one system must be received immediately; otherwise, it is considered an error (a communication mismatch). Since one wishes not to introduce new errors during refinement, a refined system must immediately provide all specified inputs.

We treat the universal state e as completely underspecified, i.e., we decree that any state refines it. This is only possible since e is not an ordinary state. We define our refinement preorder for MIAs with common input and output alphabets; we relax this in Section 6.

Definition 3 (MIA Refinement). *Let P, Q be MIAs with common input/output alphabets. A relation $\mathscr{R} \subseteq P \times Q$ is a MIA-refinement relation if for all $(p, q) \in \mathscr{R}$ with $q \neq e_Q$:*

(i) $p \neq e_P$,

(ii) $q \xrightarrow{i} Q'$ *implies* $\exists P'. p \xrightarrow{i} \overset{\varepsilon}{\Longrightarrow} P'$ *and* $\forall p' \in P' \exists q' \in Q'. (p', q') \in \mathscr{R}$,

(iii) $q \xrightarrow{\omega} Q'$ *implies* $\exists P'. p \overset{\hat{\omega}}{\Longrightarrow} P'$ *and* $\forall p' \in P' \exists q' \in Q'. (p', q') \in \mathscr{R}$,

(iv) $p \dashrightarrow{}^{i} p'$ *implies* $\exists q'. q \dashrightarrow{}^{i} = \overset{\varepsilon}{\Longrightarrow} q'$ *and* $(p', q') \in \mathscr{R}$,

(v) $p \dashrightarrow{}^{\omega} p'$ *implies* $\exists q'. q = \overset{\hat{\omega}}{\Longrightarrow} q'$ *and* $(p', q') \in \mathscr{R}$.

We write $p \sqsubseteq q$ and say that p MIA-refines q if there exists a MIA-refinement relation $\mathscr{R}$ such that $(p, q) \in \mathscr{R}$, and we let $p \sqsupseteq\sqsubseteq q$ stand for $p \sqsubseteq q$ and $q \sqsubseteq p$. Furthermore, we extend these notations to MIAs, write $P \sqsubseteq Q$ if $p_0 \sqsubseteq q_0$, and use $\sqsupseteq\sqsubseteq$ analogously.

MIA refinement $\sqsubseteq$ is a preorder and the largest MIA-refinement relation. The preorder property is quite subtle to prove due to the weak transition relations.

3   Parallel Composition

IA [9,10] is equipped with an interleaving parallel operator, where an action occurring as an input in one interface is synchronised with the same action occurring as an output in some other interface; the synchronised action is hidden, i.e., labelled by τ. Since our work builds upon MI [18] we instead consider here a parallel composition, where the synchronisation of an interface's output action involves all concurrently running interfaces that have the action as input. We define a parallel operator $\parallel$ on MIA in two stages. First, a standard product $\otimes$ between two MIAs is introduced.

Definition 4 (Parallel Product). *MIAs P_1, P_2 are composable if $O_1 \cap O_2 = \emptyset$. For such MIAs we define the product $P_1 \otimes P_2 = ((P_1 \times P_2) \dot{\cup} \{e_{12}\}, I, O, \longrightarrow, \dashrightarrow, (p_{01}, p_{02}), e_{12})$, where $I =_{df} (I_1 \cup I_2) \setminus (O_1 \cup O_2)$ and $O =_{df} O_1 \cup O_2$ and where $\longrightarrow$ and $\dashrightarrow$ are the smallest relations satisfying the following conditions:*

(PMust1) $(p_1, p_2) \xrightarrow{\alpha} P'_1 \times \{p_2\}$ *if* $p_1 \xrightarrow{\alpha} P'_1$ *and* $\alpha \notin A_2$

(PMust2) $(p_1, p_2) \xrightarrow{\alpha} \{p_1\} \times P'_2$ *if* $p_2 \xrightarrow{\alpha} P'_2$ *and* $\alpha \notin A_1$

(PMust3) $(p_1, p_2) \xrightarrow{a} P'_1 \times P'_2$ *if* $p_1 \xrightarrow{a} P'_1$ *and* $p_2 \xrightarrow{a} P'_2$ *for some a*

(PMay1) $(p_1, p_2) \dashrightarrow{}^{\alpha} (p'_1, p_2)$ *if* $p_1 \dashrightarrow{}^{\alpha} p'_1$ *and* $\alpha \notin A_2$

(PMay2) $(p_1, p_2) \dashrightarrow{}^{\alpha} (p_1, p'_2)$ *if* $p_2 \dashrightarrow{}^{\alpha} p'_2$ *and* $\alpha \notin A_1$

(PMay3) $(p_1, p_2) \dashrightarrow{}^{a} (p'_1, p'_2)$ *if* $p_1 \dashrightarrow{}^{a} p'_1$ *and* $p_2 \dashrightarrow{}^{a} p'_2$ *for some a.*

From the parallel product, parallel composition is obtained by pruning, i.e., one removes errors and states leading up to errors via local actions, so called *illegal* states. This cuts all input transitions leading to an illegal state.

In [6] we have shown that de Alfaro and Henzinger have defined pruning in an inappropriate way in [9]. We remedied this by cutting not only an i-transition from some state p to an illegal state, but also all other i-transitions from p. Now, in [6,9], p can be refined by a state with an i-transition and arbitrary behaviour afterward; we express this by introducing an i-may-transition to the universal state.

$$P: \quad p_0 \xrightarrow{a?} \xrightarrow{b!} \qquad\qquad Q: \quad q_0 \rightleftharpoons b? \qquad\qquad R: \quad r_0 \circlearrowright j?$$

$$\begin{array}{ccc} j? & j? & j? \\ \circlearrowright & \circlearrowright & \circlearrowright \\ (p_0 \parallel q_0) \parallel r_0 \xdashrightarrow{a?} \mathrm{tt} \parallel r_0 \rightleftharpoons a?, b! & & p_0 \parallel (q_0 \parallel r_0) \xdashrightarrow{a?} \mathrm{tt} \rightleftharpoons a?, b!, j? \end{array}$$

Fig. 1. Differences of our state e to tt in [18], where $A_P = \{a\}/\{b\}$, $A_Q = \{b\}/\emptyset$ and $A_R = \{j\}/\emptyset$

Definition 5 (Parallel Composition). *Given a parallel product $P_1 \otimes P_2$, a state (p_1, p_2) is a* new error *if there is some $a \in A_1 \cap A_2$ such that (a) $a \in O_1$, $p_1 \xrightarrow{a}$ and $p_2 \not\xrightarrow{a}$, or (b) $a \in O_2$, $p_2 \xrightarrow{a}$ and $p_1 \not\xrightarrow{a}$. It is an* inherited error *if one of its components is a universal state, i.e., if it is of the form (e_1, p_2) or (p_1, e_2).*

We define the set $E \subseteq P_1 \times P_2$ of illegal *states as the least set such that $(p_1, p_2) \in E$ if (i) (p_1, p_2) is a new or inherited error or (ii) $(p_1, p_2) \xdashrightarrow{\omega} (p'_1, p'_2)$ and $(p'_1, p'_2) \in E$.*

Should the initial state be an illegal state, i.e., $(p_{01}, p_{02}) \in E$, then e_{12} becomes the initial – and thus the only reachable – state of the parallel composition $P_1 \parallel P_2$.

Otherwise, $P_1 \parallel P_2$ is obtained from $P_1 \otimes P_2$ by pruning illegal states as follows. If there is a state $(p_1, p_2) \notin E$ with $(p_1, p_2) \xdashrightarrow{i} (p'_1, p'_2) \in E$ for some $i \in I$, then all must- and may-transitions labelled i and starting at (p_1, p_2) are removed, and a single transition $(p_1, p_2) \xdashrightarrow{i} e_{12}$ is added. Furthermore, all states in E, all unreachable states (except for e_{12}), and all their incoming and outgoing transitions are removed. If $(p_1, p_2) \in P_1 \parallel P_2$, we write $p_1 \parallel p_2$ and call p_1 and p_2 compatible.

In [18], Raclet et al. use a similar approach to pruning: they introduce a state we denote as tt, which has only input may-transitions as incoming transitions. Furthermore, it has a may-loop for every action of the parallel composition so that it can be refined by any state, much like our universal state (cf. Def. 3(i)). To see the difference, condsider the MIAs P, Q, R in Figure 1, where we construct $(P \parallel Q) \parallel R$ according to [18]. Since tt is an ordinary state, it is combined with r_0 inheriting the j-must-loop. In our approach, the combination with r_0 is an inherited error, and e does not have any must-transitions.

More importantly, there is the severe problem that parallel composition in [18] is not associative. Consider again the systems P, Q and R in Fig. 1; their parallel compositions shown are not equivalent according to $\sqsupseteq\sqsubseteq$ (and the equivalence in [18]). Note that our example does not rely on the multicast aspect of our parallel composition; it works just as well for IA parallel composition.

Theorem 6 (Associativity of Parallel Composition). *Parallel composition is associative in the sense that, for MIAs P, Q and R, if $(P \parallel Q) \parallel R$ is defined, then $P \parallel (Q \parallel R)$ is defined as well and they are isomorphic, and vice versa.*

Theorem 7 (Compositionality of Parallel Composition). *Let P_1, P_2 and Q_1 be MIAs and $P_1 \sqsubseteq Q_1$. Assume that Q_1 and P_2 are composable, then (a) P_1 and P_2 are composable, and (b) $P_1 \parallel P_2 \sqsubseteq Q_1 \parallel P_2$, and $P_1 \parallel P_2$ is compatible if $Q_1 \parallel P_2$ is.*

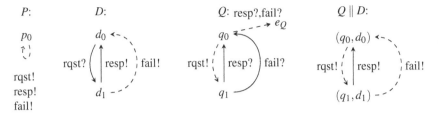

Fig. 2. $Q = P//D$ with $q_0 = p_0//d_0$ and $q_1 = p_0//d_1$, where the alphabets are $A_P = \emptyset/\{\text{rqst, resp, fail}\}$, $A_D = \{\text{rqst}\}/\{\text{resp, fail}\}$, $A_Q = \{\text{resp, fail}\}/\{\text{rqst}\}$ and $A_{Q\|D} = \emptyset/\{\text{rqst, resp, fail}\}$

4 Quotienting

The quotient operation is a kind of inverse or adjoined operation to parallel composition. It equips the theory with a means for component reuse and incremental, component-based specification. To describe the participants in a quotient operation we use the letters P for the specification, D for the divisor (the already implemented component) and Q for the quotient or its refinements. Given MIAs P and D, the quotient is the coarsest MIA Q such that $Q \| D \sqsubseteq P$ holds; we call this inequality the *defining inequality of the quotient*. We write $P//D$ for the quotient if it exists.

We demonstrate quotienting with the simple client-server application of Figure 2. The server takes the role of the already given component D. It can receive a request and answers with a response. Additionally, the server may implement a failure as answer. When composed in parallel, client Q and server D are supposed to form a *closed* system, i.e., all shared actions are outputs. Thus, the parallel composition of client and server must refine the overall specification P. A specification for the client is then obtained as the quotient $Q = P//D$. Figure 2 gives a preview of this Q according to our construction below. Client Q may implement the sending of a request, and if so, it must be receptive for a response and a failure. If one of the latter two transitions were of may-modality, this would cause a communication mismatch in the parallel composition with D. The may-transitions resp? and fail! from q_0 to e_Q only exist to make Q as coarse as possible; they disappear in the parallel composition with D. Now, it is easy to check that the defining inequality $Q \| D \sqsubseteq P$ is satisfied. The example also shows that, in general, we do not have equality of $(P//D) \| D$ and P.

We define the quotient for a restricted set of MIAs, namely where the specification P has no τs and where the divisor D is may-deterministic without τs. We call D *may-deterministic* if $d \overset{\alpha}{\dashrightarrow} d'$ and $d \overset{\alpha}{\dashrightarrow} d''$ implies $d' = d''$. Due to syntactic consistency, a may-deterministic MIA has no disjunctive must-transitions, i.e., the target sets of must-transitions are singletons. In addition, we exclude the pathological case where P has some state p and input i with $p \overset{i}{\dashrightarrow} e_P$ and $\exists p' \neq e_P. p \overset{i}{\dashrightarrow} p'$. Recall that transitions $p \overset{i}{\dashrightarrow} e_P$ are meant to express the following situation: (a) input i is not specified at p, but at the same time (b) p shall be refinable as in IA [10] by a state with an i-transition and arbitrary subsequent behaviour. Despite these restrictions, our quotient significantly generalises that of MI [18], which considered deterministic specifications and divisors only. In the following, we call MIAs P, D satisfying our restrictions a *quotient pair*.

4.1   Definition and Main Result

Like most other operators we define the quotient in two stages, where $\mathrm{may}_P(p,\alpha)$ stands for $\{p' \in P \mid p \overset{\alpha}{\dashrightarrow}_P p'\}$.

Definition 8 (Pseudo-quotient). *Let* $(P,I_P,O_P,\longrightarrow_P,\dashrightarrow_P,p_0,e_P)$, $(D,I_D,O_D,\longrightarrow_D,$ $\dashrightarrow_D,d_0,e_D)$ *be a quotient pair with* $A_D \subseteq A_P$ *and* $O_D \subseteq O_P$, *and* $I =_{df} I_P \cup O_D$ *and* $O =_{df} O_P \setminus O_D$. *The* pseudo-quotient *of* P *over* D *is defined as the MIA* $(\{(e_P,e_D)\},I,O,$ $\emptyset,\emptyset,(e_P,e_D),(e_P,e_D))$ *if* $p_0 = e_P$. *Otherwise,* $P \oslash D =_{df} (P \times D, I, O, \longrightarrow, \dashrightarrow, (p_0,d_0),$ $(e_P,e_D))$, *where the transition relations are defined by:*

(QMust1)	$(p,d) \overset{a}{\longrightarrow} P' \times \{d\}$	if	$p \overset{a}{\longrightarrow}_P P'$ and $a \notin A_D$
(QMust2)	$(p,d) \overset{a}{\longrightarrow} P' \times \{d'\}$	if	$p \overset{a}{\longrightarrow}_P P'$ and $d \overset{a}{\longrightarrow}_D d'$
(QMust3)	$(p,d) \overset{a}{\longrightarrow} P' \times \{d'\}$	if	$P' =_{df} \mathrm{may}_P(p,a) \neq \emptyset$, $e_P \notin P'$, $d \overset{a}{\dashrightarrow}_D d'$ and $a \in O_D$
(QMay1)	$(p,d) \overset{a}{\dashrightarrow} (p',d)$	if	$p \overset{a}{\dashrightarrow}_P p' \neq e_P$ and $a \notin A_D$
(QMay2)	$(p,d) \overset{a}{\dashrightarrow} (p',d')$	if	$p \overset{a}{\dashrightarrow}_P p' \neq e_P$ and $d \overset{a}{\longrightarrow}_D d'$
(QMay3)	$(p,d) \overset{a}{\dashrightarrow} (p',d')$	if	$p \overset{a}{\dashrightarrow}_P p'$, $e_P \notin \mathrm{may}_P(p,a)$, $d \overset{a}{\dashrightarrow}_D d'$ and $a \notin O_P \cap I_D$
(QMay4)	$(p,d) \overset{a}{\dashrightarrow} (e_P,e_D)$	if	$e_P \in \mathrm{may}_P(p,a)$ (note: $a \in I_P \subseteq I$)
(QMay5)	$(p,d) \overset{a}{\dashrightarrow} (e_P,e_D)$	if	$p \neq e_P$, $d \overset{a}{\not\longrightarrow}_D$ and $a \in A_D \setminus (O_P \cap I_D)$

Regarding the definition of the alphabets we follow [8] and [18]; there is, however, a choice regarding the input alphabet, which we discuss in Sec. 6. The intuition behind a state (p,d) in $P \oslash D$ is that (p,d) composed in parallel with d refines state p, and that (p,d) should be coarsest wrt. MIA refinement satisfying this condition. With this in mind, we now justify some of the above rules intuitively.

Rule (QMust1) is necessary due to the following consideration. If P has an a-must-transition where a is unknown to D, this can only originate from an a-must-transition in the quotient Q that we wish to construct; in order to be most permissive, each $p' \in P'$ must have a match in $Q \parallel D$. The corresponding consideration is true for Rule (QMay1), which also establishes syntactic consistency for Rule (QMust1).

Rule (QMust3) ensures that (p,d) and d are compatible in case of an output of d. An application of this rule can be seen in Fig. 2 for action fail? at $q_1 = p_0//d_1$. Syntactic consistency results from Rules (QMay2) and (QMay3); note that $a \in O_D$ implies $a \notin I_D$.

Rule (QMay5) makes $P \oslash D$ as coarse as possible. The input a-may-transitions introduced here just disappear in $(P \oslash D) \parallel D$, since a is blocked by D. This can be seen in Fig. 2 for actions resp? and fail? at $q_0 = p_0//d_0$ and in $Q \parallel D$ at (q_0,d_0).

$P \oslash D$ is indeed a MIA. We have already argued for syntactic consistency. All rules ensure $p \neq e_P$; hence, $e_{P \oslash D}$ has no outgoing transitions. Incoming transitions of $e_{P \oslash D}$ can only arise from Rules (QMay4) or (QMay5), which are only applicable for $a \in I$.

Up to now, we have only defined the pseudo-quotient. Considering a candidate pair (p,d), for some combinations of modalities and assignments of actions to input or output, it is impossible that p is refined by a state resulting from a parallel composition with d. We call such states *impossible states* and remove them from the pseudo-quotient states. For example, for $p \xrightarrow{a}$ and $d \dashrightarrow{a}$ such that $d \not\xrightarrow{a}$, no parallel composition with d refines p. While may-transitions can be refined by removing them and disjunctive transitions can be refined to subsets of their targets to prevent the reachability of impossible states, all states having a must-transition to only impossible states must also be removed.

Definition 9 (Quotient). *Let $P \oslash D$ be the pseudo-quotient of P over D. The set $G \subseteq P \times D$ of impossible states is defined as the least set refining the following rules:*

(G1)	$p \xrightarrow{a}_P p$ and $d \not\xrightarrow{a}_D d$ and $a \in A_D$	*implies*	$(p,d) \in G$
(G2)	$p \neq e_P$ and $p \not\dashrightarrow_P p$ and $d \dashrightarrow{a}_D d$ and $a \in O_D$	*implies*	$(p,d) \in G$
(G3)	$p \neq e_P$ and $d = e_D$	*implies*	$(p,d) \in G$
(G4)	$(p,d) \xrightarrow{a}_{P \oslash D} R'$ and $R' \subseteq G$	*implies*	$(p,d) \in G$

The quotient $P/\!/D$ *is obtained from $P \oslash D$ by deleting all states $(p,q) \in G$. This also removes any may- or must-transition exiting and any may-transition entering a deleted state. Deleted states are also removed from targets of disjunctive must-transitions. If $(p,d) \in P/\!/D$, we write $p/\!/d$. If $(p_0,d_0) \notin P/\!/D$, the quotient P over D is not defined.*

Rule (G1) is obvious since (p,d) cannot ensure that $p \xrightarrow{a}_P p$ is matched if d has no a-must-transition, as an a-may-transition or even a forbidden action at d can in no case compose to a refinement of a must-transition at p. Rule (G2) captures the situation where d has an output a that is forbidden at p. Offering an a-must-input in the quotient would lead to a transition in the parallel composition with d, while not offering it would lead to an error; both would not refine p. Rule (G3) captures the division by e_D: state e_D in parallel with any state is universal and does not refine $p \neq e_P$. Finally, Rule (G4) propagates back all impossibilities that cannot be avoided by refining.

Note that $P/\!/D$ is a MIA. Quotienting yields the coarsest MIA satisfying the defining inequality; proving this statement involves showing that the definedness of $\|$ and $/\!/$ is mutually preserved across refinement. Operator $/\!/$ is also monotonous at the left.

Theorem 10 ($/\!/$ is a Quotient Operator wrt. $\|$). *Let P, D be a quotient pair and Q be a MIA such that $A_D \subseteq A_P$, $O_D \subseteq O_P$, $O_Q = O_P \setminus O_D$ and $I_Q = I_P \cup O_D$. Then, $Q \sqsubseteq P/\!/D$ iff $Q \| D \sqsubseteq P$.*

Theorem 11 (Monotonicity of $/\!/$ wrt. $\sqsubseteq$). *Let P_1, P_2, D be MIAs with $P_1 \sqsubseteq P_2$. If $P_1/\!/D$ is defined and P_2, D are a quotient pair, then $P_2/\!/D$ is defined and $P_1/\!/D \sqsubseteq P_2/\!/D$.*

4.2 Discussion

For $Q \| D \sqsubseteq P$ to hold, $Q \| D$ and P must have the same input alphabet and the same output alphabet. Thus, we must have $O_Q = O_P \setminus O_D$ and $I_Q \supseteq I_P \setminus I_D$. Concerning the input actions in D, quotient Q can listen to them but does not have to. Hence,

$I_Q \subseteq I_P \setminus I_D \cup A_D = I_P \cup O_D$. The more inputs Q has, the easier it is to supply the behaviour ensuring $Q \parallel D \sqsubseteq P$. Thus, we have chosen the input alphabet $I_P \cup O_D$ for our quotient $P//D$, just as is done in [8] and [18]. When comparing some Q to $P//D$ in Thm. 10, Q necessarily has the same input and output alphabets as $P//D$, by Def. 3.

Quotient operators for interface theories have already been discussed by Raclet et al. [18] and Chilton et al. [7]. Our quotient $Q = P//D$ is most similar to [18], where D is assumed to be may-deterministic, P and D have no internal transitions, and $I_Q = I_P \cup O_D$. However, also P must be may-deterministic there, whereas we additionally allow nondeterminism and disjunctive must-transitions in P.

In addition, we have corrected some technical shortcomings of MI [18]. Its quotient operation ignores compatibility so that quotienting is an adjoint to the parallel product but *not* to parallel composition. This has been recognised in a technical report [3], which unfortunately employs a changed setting without a universal state.

5   Conjunction

Besides parallel composition and quotienting, conjunction is one of the most important operators of interface theories. It allows one to specify different perspectives of a system separately, from which an overall specification can be determined. More formally, the conjunction should be the coarsest specification that refines the given perspective specifications, i.e., it should characterise the greatest lower bound of the refinement preorder. In the sequel, we define conjunction on MIAs with common alphabets, as we did for MIA refinement. Similar to parallel composition, we first present a conjunctive product and, in a second step, remove state pairs with contradictory specifications.

Definition 12 (Conjunctive Product). *Consider two MIAs* $(P, I, O, \longrightarrow_P, \dashrightarrow_P, p_0, e_P)$ *and* $(Q, I, O, \longrightarrow_Q, \dashrightarrow_Q, q_0, e_Q)$ *with common alphabets. The* conjunctive product *is defined as* $P \& Q =_{df} (P \times Q, I, O, \longrightarrow, \dashrightarrow, (p_0, q_0), (e_P, e_Q))$, *satisfying the following rules plus the symmetric rules of (OMust1), (IMust1), (EMust1), (May1), (EMay1):*

(OMust1)	$(p,q) \xrightarrow{\omega} \{(p',q') \mid p' \in P', q \overset{\hat{\omega}}{\Longrightarrow}_Q q'\}$	if $p \xrightarrow{\omega}_P P'$ and $q \overset{\hat{\omega}}{\Longrightarrow}_Q$	
(IMust1)	$(p,q) \xrightarrow{i} \{(p',q') \mid p' \in P', q \dashrightarrow = \overset{\varepsilon}{\Longrightarrow}_Q q'\}$	if $p \xrightarrow{i}_P P'$ and $q \dashrightarrow = \overset{\varepsilon}{\Longrightarrow}_Q$	
(EMust1)	$(p,e_Q) \xrightarrow{\alpha} P' \times \{e_Q\}$ if $p \xrightarrow{\alpha}_P P'$		
(May1)	$(p,q) \dashrightarrow (p',q)$	if $p \overset{\tau}{\Longrightarrow}_P p'$	
(OMay)	$(p,q) \overset{\omega}{\dashrightarrow} (p',q')$	if $p \overset{\omega}{\Longrightarrow}_P p'$ and $q \overset{\omega}{\Longrightarrow}_Q q'$	
(IMay)	$(p,q) \overset{i}{\dashrightarrow} (p',q')$	if $p \dashrightarrow = \overset{\varepsilon}{\Longrightarrow}_P p'$ and $q \dashrightarrow = \overset{\varepsilon}{\Longrightarrow}_Q q'$	
(EMay1)	$(p,e_Q) \overset{\alpha}{\dashrightarrow} (p',e_Q)$	if $p \overset{\alpha}{\dashrightarrow}_P p'$	

Note that this definition is similar to the one in [17], except for the treatment of inputs and the universal state. The conjunctive product is inherently different from the parallel product. Single transitions are defined through weak transitions, e.g., as in Rules (OMust1), (IMust1) and (May1), and τ-transitions synchronise by Rule (OMay). Furthermore, as given by Rules (EMust1) and (EMay1), a universal state is a neutral element for the conjunctive product, whereas it is absorbing for the parallel product.

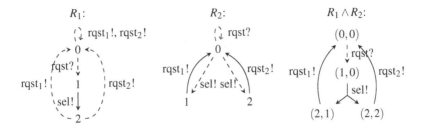

Fig. 3. Conjunction on MIAs may lead to disjunctive transitions

Definition 13 (Conjunction). *Given a conjunctive product P&Q, the set $F \subseteq P \times Q$ of inconsistent states is defined as the least set satisfying for all $p \neq e_P$ and $q \neq e_Q$:*

(F1) $(p \xrightarrow{o}_P$ and $q \overset{o}{\not\Rightarrow}_Q)$ or $(p \overset{o}{\not\Rightarrow}_P$ and $q \xrightarrow{o}_Q)$ implies $(p,q) \in F$

(F2) $(p \xrightarrow{i}_P$ and $q \overset{i}{\not\rightarrow}_Q)$ or $(p \overset{i}{\not\rightarrow}_P$ and $q \xrightarrow{i}_Q)$ implies $(p,q) \in F$

(F3) $(p,q) \xrightarrow{\alpha} R'$ and $R' \subseteq F$ implies $(p,q) \in F$

The conjunction $P \wedge Q$ is obtained by deleting all states $(p,q) \in F$ from P&Q. This also removes any may- or must-transition exiting and any may-transition entering a deleted state; in addition, deleted states are removed from targets of disjunctive must-transitions. We write $p \wedge q$ for (p,q) of $P \wedge Q$; all such states are defined and consistent by construction. If $(p_0,q_0) \in F$, then the conjunction of P and Q does not exist.

An example of conjunction is given in Fig. 3. MIAs R_1 and R_2 can be understood as requirements for a server front-end that routes between a client and at least one of two back-ends. MIA R_1 specifies that, after getting a client's request (rqst?), a back-end selection (sel!) must be performed, after which the request can be forwarded to one of the two back-ends (rqst$_1$!, rqst$_2$!). MIA R_2 specifies that, with the selection, it is decided to which one of the back-ends the request will be forwarded (rqst$_1$!, rqst$_2$!).

In $R_1 \wedge R_2$, the selection process (sel!) is given by a *disjunctive* must-transition. Such a requirement cannot be specified in a deterministic theory, such as MI [18] which our theory extends. Although one might approximate the disjunctive sel! by individual selection actions sel$_1$! and sel$_2$! for each back-end, the conjunction would either have both actions as may-transitions and thus allow one to omit both, or would have both actions as must-transitions, disallowing a server application with only one back-end.

Theorem 14 ($\wedge$ is And). *Let P and Q be MIAs with common alphabets. Then, (i) ($\exists R.$ $R \sqsubseteq P$ and $R \sqsubseteq Q$) iff $P \wedge Q$ defined. Further, in case $P \wedge Q$ is defined and for any R: (ii) $R \sqsubseteq P$ and $R \sqsubseteq Q$ iff $R \sqsubseteq P \wedge Q$.*

Clearly, conjunction is commutative. Further, as a consequence of the above theorem, (i) it is also associative and (ii) MIA refinement is compositional wrt. conjunction.

6 Alphabet Extension

So far, MIA refinement is only defined on MIAs with the same alphabets. This is insufficient for supporting perspective-based specification, where an overall specification is conjunctively composed of smaller specifications, each addressing one 'perspective' (e.g., a single system requirement) and referring only to actions that are relevant to that perspective. Hence, it is useful to extend conjunction and thus MIA refinement to dissimilar alphabets in such a way that we can add new inputs and outputs in a refinement step. For this purpose we introduce alphabet extension as an operation on MIAs, similar to [17] and also to *weak extension* in [18]. More precisely, we add may-loops for all new actions to each state, except the universal state.

Definition 15 (Alphabet Extension & Refinement). *Given a MIA* $(P, I, O, \longrightarrow, \dashrightarrow, p_0, e)$ *and disjoint action sets* I' *and* O' *satisfying* $I' \cap A = \emptyset = O' \cap A$, *where* $A =_{df} I \cup O$, *the* alphabet extension *of* P *by* I' *and* O' *is given by* $[P]_{I', O'} =_{df} (P, I \cup I', O \cup O', \longrightarrow, \dashrightarrow', p_0, e)$ *for* $\dashrightarrow' =_{df} \dashrightarrow \cup \{(p, a, p) \mid p \in P \setminus \{e\}, a \in I' \cup O'\}$. *We often write* $[p]_{I', O'}$ *for* p *as state of* $[P]_{I', O'}$, *or conveniently* $[p]$ *in case* I', O' *are understood from the context.*

For MIAs P *and* Q *with* $p \in P$, $q \in Q$, $I_P \supseteq I_Q$ *and* $O_P \supseteq O_Q$, *we define* $p \sqsubseteq' q$ *if* $p \sqsubseteq [q]_{I_P \setminus I_Q, O_P \setminus O_Q}$. *Since* $\sqsubseteq'$ *extends* $\sqsubseteq$ *to MIAs with different alphabets, we write* $\sqsubseteq$ *for* $\sqsubseteq'$ *and abbreviate* $[q]_{I_P \setminus I_Q, O_P \setminus O_Q}$ *by* $[q]_P$; *the same notations are used for* P *and* Q.

Compositionality of parallel composition as in Thm. 7 is preserved by the extended refinement relation as long as alphabet extension does not yield new communications.

Theorem 16 (Compositionality of Parallel Composition). *Let* P_1, P_2, Q *be MIAs such that* Q *and* P_2 *are composable and* $P_1 \sqsubseteq Q$. *Assume further that, for* $I' =_{df} I_1 \setminus I_Q$ *and* $O' =_{df} O_1 \setminus O_Q$, *we have* $(I' \cup O') \cap A_2 = \emptyset$. *Then: (a)* P_1 *and* P_2 *are composable, and (b) if* Q *and* P_2 *are compatible, then so are* P_1 *and* P_2 *and* $P_1 \parallel P_2 \sqsubseteq Q \parallel P_2$.

Our conjunction operator may be lifted to conjuncts with dissimilar alphabets by defining $P \wedge' Q =_{df} [P]_Q \wedge [Q]_P$; the lifted operator $\wedge'$ satisfies the analogue of Thm. 14.

7 Conclusions and Future Work

We presented an extension of Raclet et al.'s modal interface theory [18] to *nondeterministic* systems. To do so we resolved, for the first time properly, the conflict between unspecified inputs being allowed in interface theories derived from de Alfaro and Henzinger's Interface Automata [10] but forbidden in Modal Transition Systems [14]. To this end, we introduced a special universal state, which enabled us to achieve compositionality (in contrast to [15]) as well as associativity (in contrast to [18]) for parallel composition; this also allowed for a more practical support of perspective-based specification when compared to [16,17]. In addition, we defined a quotienting operator that permits the decomposition of *nondeterministic* specifications and takes *pruning* in parallel composition into account (in contrast to [18]).

Regarding future work, we wish to explore the choice of alphabets for quotienting and relax the determinism requirement on divisors. We also intend to implement our theory in MICA (see http://www.irisa.fr/s4/tools/mica/) or the MIO Workbench [2].

Acknowledgments. We thank the reviewers for their comments and suggestions.

References

1. Bauer, S.S., David, A., Hennicker, R., Guldstrand Larsen, K., Legay, A., Nyman, U., Wą-sowski, A.: Moving from specifications to contracts in component-based design. In: de Lara, J., Zisman, A. (eds.) FASE 2012. LNCS, vol. 7212, pp. 43–58. Springer, Heidelberg (2012)
2. Bauer, S.S., Mayer, P., Schroeder, A., Hennicker, R.: On weak modal compatibility, refinement, and the MIO Workbench. In: Esparza, J., Majumdar, R. (eds.) TACAS 2010. LNCS, vol. 6015, pp. 175–189. Springer, Heidelberg (2010)
3. Benveniste, A., Caillaud, B., Nickovic, D., Passerone, R., Raclet, J.B., Reinkemeier, P., Sangiovanni-Vincentelli, A., Damm, W., Henzinger, T.A., Larsen, K.G.: Contracts for system design. Tech. Rep. 8147, INRIA (November 2012)
4. Beyer, D., Chakrabarti, A., Henzinger, T.A., Seshia, S.A.: An application of web-service interfaces. In: ICWS, pp. 831–838. IEEE (2007)
5. Bujtor, F., Fendrich, S., Lüttgen, G., Vogler, W.: Nondeterministic modal interfaces. Tech. Rep. 2014-06, Institut für Informatik, Universität Augsburg (2014)
6. Bujtor, F., Vogler, W.: Error-pruning in interface automata. In: Geffert, V., Preneel, B., Rovan, B., Štuller, J., Tjoa, A.M. (eds.) SOFSEM 2014. LNCS, vol. 8327, pp. 162–173. Springer, Heidelberg (2014)
7. Chen, T., Chilton, C., Jonsson, B., Kwiatkowska, M.Z.: A compositional specification theory for component behaviours. In: Seidl, H. (ed.) ESOP 2012. LNCS, vol. 7211, pp. 148–168. Springer, Heidelberg (2012)
8. Chilton, C.: An Algebraic Theory of Componentised Interaction. Ph.D. thesis, Oxford (2013)
9. de Alfaro, L., Henzinger, T.A.: Interface automata. In: FSE, pp. 109–120. ACM (2001)
10. de Alfaro, L., Henzinger, T.A.: Interface-based design. In: Engineering Theories of Software-Intensive Systems. NATO Science Series, vol. 195. Springer (2005)
11. De Nicola, R., Segala, R.: A process algebraic view of input/output automata. Theor. Comput. Sci. 138(2), 391–423 (1995)
12. Hüttel, H., Larsen, K.G.: The use of static constructs in a modal process logic. In: Meyer, A.R., Taitslin, M.A. (eds.) Logic at Botik 1989. LNCS, vol. 363, pp. 163–180. Springer, Heidelberg (1989)
13. Larsen, K., Xinxin, L.: Equation solving using modal transition systems. In: LICS, pp. 108–117. IEEE (1990)
14. Larsen, K.G.: Modal specifications. In: Sifakis, J. (ed.) CAV 1989. LNCS, vol. 407, pp. 232–246. Springer, Heidelberg (1990)
15. Larsen, K.G., Nyman, U., Wąsowski, A.: Modal I/O automata for interface and product line theories. In: De Nicola, R. (ed.) ESOP 2007. LNCS, vol. 4421, pp. 64–79. Springer, Heidelberg (2007)
16. Lüttgen, G., Vogler, W.: Modal interface automata. LMCS 9(3) (2013)
17. Lüttgen, G., Vogler, W.: Richer interface automata with optimistic and pessimistic compatibility. ECEASST 66 (2013), an extended version has been submitted to Acta Informatica
18. Raclet, J.B., Badouel, E., Benveniste, A., Caillaud, B., Legay, A., Passerone, R.: A modal interface theory for component-based design. Fund. Inform. 108(1-2), 119–149 (2011)

Group Search on the Line

Marek Chrobak[1,*], Leszek Gąsieniec[2], Thomas Gorry[2], and Russell Martin[2]

[1] Dept of Computer Science and Engineering, University of California,
Riverside, CA, USA
[2] Dept of Computer Science, University of Liverpool, Liverpool, United Kingdom
Russell.Martin@liverpool.ac.uk

Abstract. In this paper we consider the *group search problem*, or *evacuation problem*, in which k mobile entities ($\mathcal{ME}$s) located on the line perform search for a specific destination. The $\mathcal{ME}$s are initially placed at the same origin on the line L and the target is located at an unknown distance d, either to the left or to the right from the origin. All $\mathcal{ME}$s must *simultaneously* occupy the destination, and the goal is to minimize the time necessary for this to happen. The problem with $k = 1$ is known as the *cow-path* problem, and the time required for this problem is known to be $9d - o(d)$ in the worst case (when the cow moves at unit speed); it is also known that this is the case for $k \geq 1$ unit-speed $\mathcal{ME}$s. In this paper we present a clear argument for this claim by showing a rather counter-intuitive result. Namely, independent of the number of $\mathcal{ME}$s, group search cannot be performed faster than in time $9d - o(d)$. We also examine the case of $k = 2$ $\mathcal{ME}$s with different speeds, showing a surprising result that the bound of $9d$ can be achieved when one $\mathcal{ME}$ has unit speed, and the other $\mathcal{ME}$ moves with speed at least $1/3$.

Keywords: evacuation, group search, mobile entity.

1 Introduction

Search problems are well-studied within the fields of operations research, computing, and mathematics. Indeed, nearly sixty years ago Bellman [6] asked a question that can be stated as follows: "A hiker is lost in a forest whose dimensions are known to her. What is the best path for her to follow to escape the forest?"

In general, search problems deal with a searcher looking for a hidden object (or "target"), with a goal of minimizing the time required to find it. Many versions of this problem can be considered, including variations in the environment (e.g., a geometric setting vs. a graph), whether the target is fixed or mobile, or if the target is a point in space or a boundary of a region or other curve, the use of a deterministic or randomized search strategy, and whether or not the searcher(s) have access to additional tools to aid the search (such as markers to drop in the environment) [3,4,5,7,8,11,12,15,16].

---

* Research partially supported by National Science Foundation grants CCF-1217314 and OISE-1157129.

G.F. Italiano et al. (Eds.): SOFSEM 2015, LNCS 8939, pp. 164–176, 2015.

Search also naturally leads into the *rendezvous problem*, where two or more searchers seek to meet in an environment, and that problem lends itself to additional considerations of the inherent abilities of the searchers themselves, such as whether they have the same speed or different speeds, their ability to communicate and see their environment (typically over a limited distance), and if the searchers are able to follow the same or different search strategy, e.g. do the searchers have unique identifiers so they can adopt their own search method, or are they indistinguishable and therefore must use the same (randomized or deterministic) strategy? [1,9,10]. The book by Alpern and Gal [2] is a good survey of known results for both the search and rendezvous problems.

The focus of this paper is on the *group search problem* or *evacuation problem*, where k mobile entities, all starting from the origin on the line, must find and *simultaneously* gather at the target located at an unknown distance d from the origin. The inspiration for the name comes from consideration of an evacuation procedure of a building (one, say, that is on fire). We note that some of the mobile entities might find the target and move away from it, only to return later. Strictly speaking, the "simultaneous gathering" condition can be dispensed with by noting that the evacuation procedure has finished only when all mobile entities have reached and "exited" at the target.

In the case of the line that we consider, the most relevant previous results are in relation to the *cow-path problem*, a search problem that was introduced by Baeza-Yates, et al. in 1988 [3] and has since been considered in the same form and in different variations in [4,5,11,13,14,16,17]. The cow-path problem involves a single cow, Eloise[1], who is standing at a crossroads (defined as the origin) with w paths leading off into unknown territory. Traveling with unit speed, the goal of Eloise is to locate a target destination (say, a tasty patch of grass) that is at distance d from the origin in as small a time as possible. Eloise faces three difficulties: (1) she does not know the value of d, (2) she does not know which of the w paths leads to the goal, and (3) her eyesight is not very good, so she will not know she has found the goal until she is standing in it.

Baeza-Yates, et al. [3,4] studied the cow-path problem, and proposed a deterministic algorithm they called *Linear Spiral Search* (detailed later) as a solution. In the case that $w = 2$ (two paths), this algorithm will find the goal in time at most $9d$, and they showed that this is optimal up to lower order terms. In the same work, the authors considered the case of $w > 2$ paths, showing an optimal (up to lower order terms) result of $\left(1 + 2\frac{w^w}{(w-1)^{w-1}}\right) d$ time bound to find the target using a deterministic search strategy.

Let us move away from cows, and into the world of *mobile entities* ($\mathcal{ME}$s) in what follows, opening up our entities to (possibly) have more computational power, memory, and/or communication ability than the average cow. We will use the phrase *mobile agent* (or more simply *agent*) interchangeably with *mobile entity* in what follows. We will also use the words *target* or *destination* to denote the goal of the search.

---

[1] From the book *Eloise and the Old Blue Truck*, by Kennon Graham, illustrated by Florence Sarah Winship, a childhood favorite of one of the authors.

In [5] Baeza-Yates and Schott examined other variations of the cow-path problem in these stronger settings. They note the straightforward fact that if d is known by the $\mathcal{ME}$ then, in the worst case, it must travel for $3d$ units of time (for $w = 2$ paths). They also considered cases involving two or more $\mathcal{ME}$s having uniform speed. If the $\mathcal{ME}$s are able to communicate arbitrarily far away, then a total distance of at least $2d$ must be traveled to find the destination, and $4d$ if both $\mathcal{ME}$s must reach the destination. Baeza-Yates and Schott showed the total distance traveled when no communication is present, *and* both $\mathcal{ME}$s must reach the goal is also $9d$, the same time it would take a single $\mathcal{ME}$.

The previous results all applied to deterministic search algorithms. Kao, et al. [14] examined the first randomized algorithm for the cow-path problem and, for the case of $w = 2$ paths, obtained an optimal randomized $4.59112d$ bound for the search time. Those authors also give a bound for $w > 2$ paths, which they conjecture to be an optimal randomized strategy.

The cow-path problem, with either one or two $\mathcal{ME}$s, cannot be solved in time smaller than $9d$ (up to lower order terms), where d is the distance from the origin to the destination, and the $\mathcal{ME}$s have unit speed. This result is proved, and re-proved in various fashions, in [4,5,11,13,14,16,17]. However, [5] seems to claim that if the number of $\mathcal{ME}$s is greater than two, then the evacuation procedure can be performed in a smaller time. We would dispute this claim and in the first part of this paper will give a proof showing that $9d$ is also optimal (up to lower order terms) when the number of $\mathcal{ME}$s is at least 2. In doing so, we present an alternative way of proving the lower bound of $9d$ than the papers previously mentioned have done.

In the second part of this paper, we would initiate the study of the evacuation problem where mobile entities have different maximum speeds. We show the somewhat surprising result (to the authors at least) that when there are two (or more) mobile entities, one with unit speed and the others having maximum speed at least $1/3$, then the evacuation problem can still be performed in time at most $9d$. The authors believe that this is the first result regarding the evacuation problem with mobile entities having different maximum speeds, and hope to inspire further work in this direction. Indeed, the authors know of no prior work in the field of search, rendezvous, or evacuation that considers mobile entities with differing maximum speeds.

1.1   Our Results

We consider k $\mathcal{ME}$s on the line, all starting at the origin. We work in the restricted setting where communication between $\mathcal{ME}$s is only possible when they are in contact (i.e. occupy the same location), but we consider that any communication occurs instantaneously.

We examine the *evacuation problem* where all k $\mathcal{ME}$s must simultaneously occupy a target located at an unknown distance d from the origin. The aim is to achieve this goal in as small a time as possible. We assume here that d is a positive integer, but most all of our results can be generalized for rational or real values of d, provided that d is not too small.

In Section 2 we consider $k(\geq 2)$ $\mathcal{ME}$s having a uniform speed that (by rescaling time) we will set to 1. We briefly recall the Linear Spiral Search method described in [3,4] in which a single $\mathcal{ME}$ can find a target in time at most $9d$, and also recall a coordinated method for two $\mathcal{ME}$s to solve evacuation in time at most $9d$.

We give a new proof of a lower bound of $9d - o(d)$ for the evacuation time of two or more $\mathcal{ME}$s having unit speed (Theorem 3).

In Section 3 we will look at the case when $k = 2$ and the $\mathcal{ME}$s have different maximum speeds. We will normalize the speeds of the $\mathcal{ME}$s by setting the speed of the fastest $\mathcal{ME}$ to 1 and then setting the speed of the slower $\mathcal{ME}$ proportionately. We demonstrate that, provided the speed of the slower $\mathcal{ME}$ is at least $1/3$, then the $9d$ evacuation time bound can still be achieved (Theorem 4).

In our considerations we will use time-space diagrams to support our reasoning and the proofs. A time-space is a $2d$-plane with the horizontal axis representing location on the line L and the vertical axis refers to the time t. We are only interested in the half-plane where the values of time are positive. In this context, the trajectory adopted by a $\mathcal{ME}$ can be described as a function of time t to give a location on L.

2   Multiple $\mathcal{ME}$s with Uniform Speeds

As mentioned earlier, there has been much previous work done in this problem before. However, the goal of this section is to provide a clear and complete explanation to the claim of the $9d - o(d)$ worst case in this setting for multiple searchers with uniform speeds. We will use d to denote the destination as well as the (unknown) distance to that destination. This should not cause confusion as the meaning should be clear from the context. We recall our assumption that $d \geq 1$.

For completeness we first recall what is known in the case of one or two searchers.

2.1   A Basic Strategy for a Single Mobile Agent

As a brief reminder, a search strategy in the case of a single $\mathcal{ME}$ for two paths was outlined in [4], referred to as Linear Spiral Search by the authors in that paper. This search strategy is given as Algorithm 1, where for simplicity we consider the two paths to be a line in this case.

This deterministic search strategy for a single $\mathcal{ME}$ yields the search time of $9d$, which is optimal up to lower order terms [4, Theorem 2.1].

2.2   Evacuating Two Mobile Entities on the Line

For the evacuation problem with two $\mathcal{ME}$s on the line (or two paths), there are at least two strategies that will yield a $9d$ upper bound for the problem.

Algorithm 1. A "doubling strategy" for a single mobile agent

$r \leftarrow 1$;
$dir \leftarrow$ left   /* $dir \in \{$left, right$\}$ */ ;
while *(Destination not found)* **do**
      Walk distance r in direction dir and return to the origin /* (Stop at
      destination if found.) */ ;
      Reverse dir ;
      $r \leftarrow 2 \cdot r$;
endw

One strategy is that each entity ignores the existence of the other entity and simply executes their own version of Algorithm 1, independently of the other (in fact, each $\mathcal{ME}$ can independently begin by going left or right at the start of their own procedure). This will clearly give an evacuation time of at most $9d$, since an entity that finds the destination simply waits for the other.

A second strategy coordinates the use of the searchers to find the target. One such coordinated strategy was first described and analyzed in [5]: Two searchers start exploring in opposite directions with a speed of $1/3$, then the searcher who finds the target uses their maximum speed to catch up to and inform the other agent, and then both travel to the target at maximum speed.

This procedure also guarantees the $9d$ upper bound for completing evacuation of the pair of searchers. Furthermore, this coordinated strategy gives the total distance of $12d$ traveled by the pair, as opposed to a worst-case total distance of $18d$ by the use of Algorithm 1 in an uncoordinated fashion. (See [5] for details.)

2.3 A Lower Bound

In [5], the authors make the statement "if we have more [than two] robots, we can have two robots searching and coming back to certain points, while other robots can carry messages between the searchers until the goal is found. In this case, the goal can be reached in a smaller time."

It is unclear if the authors of [5] are claiming that the evacuation problem can be solved in time smaller than $9d - o(d)$ using more than two searchers, and no proof of any such claim is offered. In any case, we would dispute such a claim, and here we want to give a new proof that $9d$ is a lower bound (up to lower order terms) on the evacuation problem (for any number of agents). We remind the reader that the lower bound of $9d - o(d)$ was proven to be optimal for a single agent in [4]. We want to investigate the lower bound for two or more agents with maximal speed 1.

So here we assume that there are at least two agents performing the group search problem. To facilitate our proof, we first define some notation. We suppose that the agents performing the evacuation procedure are following some fixed (but unknown) algorithmic procedure, which may or may not be coordinated. The only restrictions we impose are the ones mentioned earlier, that agents can only communicate when they occupy the same point, that this communication is instantaneous, and that the maximum speed is 1.

Definition 1. *For* $t > 0$, *we define* $\alpha(t) \in \left[0, \frac{\pi}{2}\right]$ *as the angle, measured in radians, as follows:*

$$\alpha(t) = \sup_{t' \geq t} \left\{ \arctan\left(\frac{x}{t'}\right) : (x, t') \in E \right\},$$

where E is the set of all pairs (x, t') such that some agent is at distance x from the origin at time t'.

In other words, $\alpha(t)$ defines a symmetric cone (centered around the origin) of size $2\,\alpha(t)$ in the time/space diagram that contains all terrain that is ever explored from time t to time ∞ during the evacuation procedure, if we assume that the evacuation target does not actually exist, so that the agents will be exploring the x-axis forever.

We note the following facts without proof.

Fact 1. *For any sequence $0 < t_1 < t_2 < \ldots$, we have $\alpha(t_1) \geq \alpha(t_2) \geq \ldots$. In other words, for any increasing sequence of numbers $\{t_i\}_{i=1}^{\infty}$, the sequence $\{\alpha(t_i)\}_{i=1}^{\infty}$ is a non-increasing sequence.*

Fact 2. $\lim_{i \to \infty} \alpha(t_i) \overset{\text{def}}{=} \alpha$ *exists, and is independent of the particular increasing series of numbers $\{t_i\}_{i=1}^{\infty}$ chosen.*

The previous fact follows as the non-increasing sequence $\{\alpha(t_i)\}_{i=1}^{\infty}$ is bounded below (by 0), and, hence by the monotone convergence theorem, has a limit. (We can alternatively express Facts 1 and 2 in terms of the tangents of the angles.)

Theorem 3. *Suppose that there are two agents performing the evacuation. If $\tan(\alpha) \neq \frac{1}{3}$, where α is the limit defined in Fact 2, then there exists $\delta > 0$ and arbitrarily large values of d such that the evacuation procedure takes time at least $(9 + \delta)d$.*

Proof. For the sake of this proof, we may suppose that there is an "adversary" who decides where to place the evacuation target, provided that this point has not yet been visited by any agent in the evacuation procedure.

Suppose $\varepsilon > 0$. (We shall say more about how we select ε later.) Given this ε, let us pick t_0 and $t_1 > t_0$ large enough so that:

(a) $|\alpha(t_0) - \alpha| < \varepsilon$,
(b) the position at time t_1, $z(t_1)$, of an agent Z satisfies $\left| \arctan\left(\frac{|z(t_1)|}{t_1}\right) - \alpha \right| < \varepsilon$
   (Remark: We assume, without loss of generality, that the value of $\alpha(t)$ in an interval around t_1 is defined by an agent located to the right of the origin. Otherwise, we may consider a similar argument to the one that follows where $\alpha(t_1)$ is defined by an agent to the left of the origin. Hence, under our assumption, the agent could be at the point labeled Z in Fig. 2.3.), and
(c) the line from Z extending backwards in time at a 45° angle to the time-axis that intersects the cone defined by the angles $\pm(\alpha \pm \varepsilon)$ does so after the time t_0. (See Fig. 2.3.)

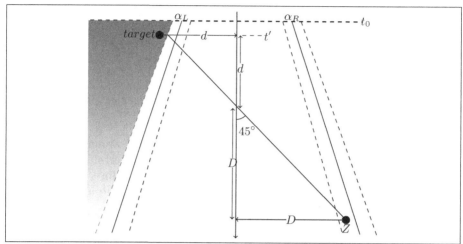

Fig. 1. Time/space diagram of configuration used to establish the lower bound

The purpose of the last condition is that the shaded region shown in the figure has *not* been visited by any agent (as has the corresponding region on the right-hand side of the figure, but we have not shaded that region).

Given this choice of t_0 and t_1, the evacuation point is placed slightly inside of the shaded region, as shown in Fig. 2.3.

With this configuration as labeled, we can make the following conclusions:

The earliest time that the evacuation point, at distance d from the origin, can be found is at time t', where t' satisfies $\tan(\alpha+\varepsilon+\zeta') = \frac{d}{t'}$, for a small $\zeta' > 0$ (to guarantee the target is in the unexplored region). Note that we can also choose ζ' so that $\zeta' < \varepsilon$.

This means that the earliest time that agent Z could learn about the existence of the target is at the time $t' + d + D$, where D satisfies $\left|\frac{D}{t'+d+D} - \tan(\alpha)\right| < \varepsilon$. This is because Z is at the specified location in the diagram at time $t' + d + D$, and obtaining the information at an earlier time would violate the "speed of light" in this timeframe (which is the maximum speed of 1, indicated by the line that makes a 45° angle with the time axis).

Finally, this means that the earliest that agent Z could arrive at the evacuation point is at time $t' + 2 \cdot (d + D)$, since Z would require time $d + D$ to travel to the evacuation point from its current location at full speed.

The remaining part of this argument is some calculations in order to attempt to lower bound the sum $t' + 2 \cdot (d + D)$. First we have that

$$\tan(\alpha + \varepsilon + \zeta') = \frac{d}{t'} \quad \text{as already noted, and} \tag{1}$$

$$\tan(\alpha \pm \zeta'') = \frac{D}{D + d + t'} \quad \text{for a small } 0 < \zeta'' < \varepsilon \text{ (with sign} \tag{2}$$

$$\text{depending upon the exact location of } Z).$$

Using Taylor's Theorem (see your favorite beginning calculus book), we note that we can write

$$\tan(\alpha + \varepsilon + \zeta') = \tan\alpha + (\sec^2\alpha)(\varepsilon + \zeta') + g(\delta')(\varepsilon + \zeta')^2 \quad \text{and} \tag{3}$$

$$\tan(\alpha \pm \zeta'') = \tan\alpha \pm (\sec^2\alpha)(\zeta'') \pm g(\delta'')(\zeta'')^2 \tag{4}$$

where $g(z) = 2\sec^2 z \tan z$, $0 < \delta' < \varepsilon + \zeta'$, and $0 < \delta'' < \zeta''$.

The signs in (4) depend upon the position of Z. For the location of Z as given in Fig. 2.3, we have that (4) is actually (using the appropriate signs):

$$\tan(\alpha - \zeta'') = \tan\alpha - (\sec^2\alpha)(\zeta'') - g(\delta'')(\zeta'')^2. \tag{5}$$

(The case for Z located on the other side of the angle labeled α_R is similar to the analysis we give below, and is left to the reader.)

Then, for a given ε, we can find (small) constants C_1 and C_2 (that depend upon ε) such that

$$\tan(\alpha + \varepsilon + \zeta') \leq \tan\alpha + (\sec^2\alpha)(\varepsilon + \zeta') + C_1(\varepsilon + \zeta')^2 \tag{6}$$

$$\text{and} \quad \tan\alpha - (\sec^2\alpha)(\zeta'') - C_2(\zeta'')^2 \leq \tan(\alpha - \zeta'') \tag{7}$$

for all $0 < \zeta' < \varepsilon$ and $0 < \zeta'' < \varepsilon$. We note that $C_1 \to 0$ as $\varepsilon + \zeta' \to 0$ and, similarly, $C_2 \to 0$ as $\zeta'' \to 0$.

In what follows, in order to simplify the notation somewhat, we will let $x = \tan\alpha$ and recall, of course, that $\sec^2\alpha = 1 + \tan^2\alpha = 1 + x^2$.

Therefore, from (1) and (2), and (6) and (7) we can write:

$$t' \geq \frac{d}{x + (1 + x^2)(\varepsilon + \zeta') + C_1(\varepsilon + \zeta')^2}, \quad \text{and} \tag{8}$$

$$\frac{D}{D + d + t'} \geq x - (1 + x^2)(\zeta'') - C_2(\zeta'')^2, \quad \text{from which we get} \tag{9}$$

$$D \geq \frac{(d + t')\left(x - (1 + x^2)(\zeta'') - C_2(\zeta'')^2\right)}{1 - x + (1 + x^2)(\zeta'') + C_2(\zeta'')^2.} \tag{10}$$

(Again, in the other case to consider for the location of the agent Z, one can obtain similar inequalities to use as lower bounds.)

The earliest time that agent Z can complete the evacuation procedure is $t' + 2 \cdot (d + D)$ which, using (8) and (10) is lower bounded by the function, after some simplification,

$$d \cdot h(x, \varepsilon, \zeta', \zeta'', C_1, C_2) \overset{\text{def}}{=} d \left\{ \frac{1}{x + (1 + x^2)(\varepsilon + \zeta') + C_1(\varepsilon + \zeta')^2} + 2 \right.$$

$$\left. + \left(\frac{2\left(x - (1 + x^2)(\zeta'') - C_2(\zeta'')^2\right)}{1 - x + (1 + x^2)(\zeta'') + C_2(\zeta'')^2} \right) \cdot \left(1 + \frac{1}{x + (1 + x^2)(\varepsilon + \zeta') + C_1(\varepsilon + \zeta')^2} \right) \right\}.$$

Recall that if $\varepsilon + \zeta' \to 0$, then $C_1 \to 0$, and if $\zeta'' \to 0$, then $C_2 \to 0$.

So let us consider the function $f(x) = \frac{1}{x} + 2 + \left(\frac{2x}{1-x}\right) \cdot \left(1 + \frac{1}{x}\right)$. We claim that $f(x) \geq h(x, \varepsilon, \zeta', \zeta'', C_1, C_2)$ for any $x \in (0, 1)$, and for all small enough ε

(and hence ζ', ζ'', C_1, and C_2), and that $h(x, \varepsilon, \zeta', \zeta'', C_1, C_2)$ increases to $f(x)$ as $\{\varepsilon, \zeta', \zeta'', C_1, C_2\}$ all approach 0.

Elementary calculus tells us that $f(x)$ is minimized, under the restriction that $0 < x < 1$, when $x = \frac{1}{3}$. In this case, we have $f\left(\frac{1}{3}\right) = 9$, and $f(x) > 9$ for any other value of $x \in (0, 1) - \left\{\frac{1}{3}\right\}$.

We therefore claim that for any other value of $x \in (0, 1) - \left\{\frac{1}{3}\right\}$, since $f(x) > 9$, and since h increases with decreasing values of $\{\varepsilon, \zeta', \zeta'', C_1, C_2\}$, we can find (suitably small) ε (and corresponding C_1 and C_2 for all $0 < \zeta', \zeta'' < \varepsilon$) such that $h(x, \varepsilon, \zeta', \zeta'', C_1, C_2) > 9$ as well. So if $\tan(\alpha) \neq \frac{1}{3}$, we can find a $\varepsilon > 0$, and a corresponding $\delta > 0$ so that $h(x, \varepsilon, \zeta', \zeta'', C_1, C_2) \geq 9 + \delta$.

Finally, for this δ defined above (by choosing an appropriate ε), since $\alpha(t)$ decreases to α, we can find an infinite sequence of pairs of times $\{(t_{2i}, t_{2i+1})\}_{i=0}^{\infty}$ so that (t_{2i}, t_{2i+1}) would all satisfy conditions (a), (b), and (c) above (with t_{2i} in place of t_0 and t_{2i+1} in place of t_1).

This means that there is an infinite sequence of distances d_i such that the evacuation time (having a target at distance d_i) will be at least $(9 + \delta)d_i$.   □

Intuitively, Theorem 3 tells us that the leftmost and rightmost boundaries of the region explored by the agents must (in the limit) grow an average of $1/3$ unit distance per unit of time in order to successfully accomplish evacuation in time (at most) $9d$.

For more than two agents, we may consider the leftmost and rightmost agent at any time. The region that has been explored by a set of agents will still consist of a single connected segment in the line. Hence, we can conclude the following result just by considering the leftmost and rightmost agent at any moment in time, and repeating the proof of Theorem 3.

Corollary 1. *For two or more agents, if* $\tan(\alpha) \neq \frac{1}{3}$, *then there exists* $\delta > 0$ *and arbitrarily large* d *such that evacuation takes time at least* $(9 + \delta)d$.

3 Agents Having Different Maximum Speeds

Now we consider two cases involving mobile entities having different maximum speeds. As before, by rescaling, we assume the maximum speed is 1. We call a mobile entity with maximum speed 1 a "fast $\mathcal{ME}$". A mobile entity with speed s, where $0 < s < 1$ shall be called a "slow $\mathcal{ME}$". We use the notation $\mathcal{FME}$ to refer to the "fast" mobile entity. Similarly, we will use $\mathcal{SME}$ to refer to the "slow" mobile entity.

Section 3.1 deals with the special case of one fast $\mathcal{ME}$ and one slow $\mathcal{ME}$. In the case that $s \geq \frac{1}{3}$, we show that evacuation can still be accomplished in time $9d$, a fact that these authors found surprising when we first discovered it.

Section 3.2 deals with the case of two fast $\mathcal{ME}$s and one slow $\mathcal{ME}$. Even in this case, if the slow $\mathcal{ME}$ is not too slow (in particular, if $s \geq \frac{1}{5}$), then evacuation can still be performed in time $9d$.

3.1 One Fast, One Slow

For one $\mathcal{FME}$ and one $\mathcal{SME}$ we will show that, provided $s \geq \frac{1}{3}$, the $9d$ evacuation time bound still holds using a coordinated strategy for the two mobile entities. A picture that hints at the strategies of the two $\mathcal{ME}$s can be seen in Fig. 2, but we give some brief discussion of each strategy in what follows.

The $\mathcal{FME}$'s Strategy. The $\mathcal{FME}$ searches for the evacuation point as if the $\mathcal{SME}$ is not there, using the doubling strategy described in Algorithm 1 (always traveling at its maximum speed). The $\mathcal{FME}$ follows this strategy until the evacuation target is located. Having found the target, it immediately seeks to make contact with the $\mathcal{SME}$ (still moving at its maximum speed of 1). Both $\mathcal{ME}$s then walk together to the evacuation target with the full speed s of the $\mathcal{SME}$. Fig. 2 shows the exploration path the $\mathcal{FME}$ takes to find the target as the solid black line, which is simply the doubling strategy from before.

One point to keep in mind is that the $\mathcal{FME}$ knows the strategy of the $\mathcal{SME}$, so the $\mathcal{FME}$ knows in which direction to travel in order to find and inform the $\mathcal{SME}$ once it locates the target.

The $\mathcal{SME}$'s Strategy. The slow mobile entity is obviously unable to mimic the path of the $\mathcal{FME}$ due to its reduced maximum speed. Somewhat counterintuitively, even if $s > \frac{1}{3}$, the $\mathcal{SME}$ is instructed to use speed $\frac{1}{3}$ and follow the dashed path outlined in Fig. 2. It follows such a path until it is informed by the $\mathcal{FME}$ of the location of the evacuation target, and then proceeds at maximum speed, i.e. s, to reach that destination.

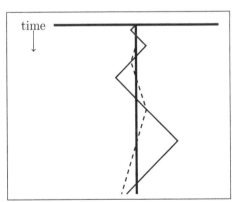

Fig. 2. An optimal strategy for two different speeds, where the slower agent has a speed at least 1/3 the speed of the fast agent

The $\mathcal{SME}$ is following its own "doubling strategy", but this takes more time to execute than it does for the $\mathcal{FME}$. In particular, initially the $\mathcal{SME}$ stays at the *origin* for 4 units of time, and then begins its own movements. After that, it uses a "doubling strategy" to move to distances $1, 2, 4, 8, 16, \ldots$ from the origin (on opposite sides of the origin, i.e. moving to distance 1 to the left, taking three units of time, returning to the origin, then to distance 2 to the right and returning, then to distance 4 to the left, etc.) Recall that the $\mathcal{SME}$ is moving at speed $1/3$, and, hence takes time $2 \cdot 3 \cdot 2^k$ to execute one portion of its "doubling" move, i.e. moving to distance 2^k and returning to the origin.

Observe that the $\mathcal{SME}$ and $\mathcal{FME}$ will meet at certain pre-defined times and locations during their trajectories. All of the meeting points, aside from the first one at the origin while the $\mathcal{SME}$ is not moving, occur at locations that were originally extreme points (i.e. turning points) of the trajectory of the $\mathcal{FME}$. For example, the two agents will meet at distances $1, 2, 4, \ldots$ from the origin (again, on opposite sides of the origin).

Under this strategy, the $\mathcal{SME}$ will never discover the evacuation point before the $\mathcal{FME}$ does so, and therefore must simply keep walking in this way until the $\mathcal{FME}$ comes to inform it of the location of the evacuation target and take it there with the maximum speed of the $\mathcal{SME}$.

We also remark that by following the particular outlined strategy, the $\mathcal{SME}$ is at the origin at the same moment that the $\mathcal{FME}$ is at one of the turning points of its movements.

Still $9d$ Evacuation, When $s \geq \frac{1}{3}$.

Theorem 4. *The coordinated strategy outlined above for the $\mathcal{SME}$ and $\mathcal{FME}$ gives a 9d upper bound for time of the evacuation problem, as long as $s \geq \frac{1}{3}$.*

Proof. We can think of the evacuation procedure as a three-step process where (1) the $\mathcal{FME}$ locates the evacuation target, (2) the $\mathcal{FME}$ informs the $\mathcal{SME}$ of that location, and (3) the two agents proceed (back) to the target.

We assume that $d \geq 2$. (The $9d$ bound for $d = 1$ is easy to verify.) Note that the $\mathcal{FME}$ will locate the target between successive "peaks" on the same side of the origin (or it will find the destination just at a "peak"), so we define k to be the integer such that $2^{k-2} < d \leq 2^k$. In particular, we can write $d = 2^{k-2} + \varepsilon$ for some $0 < \varepsilon \leq 3 \cdot 2^{k-2}$.

The "discovery phase" to locate the target will take time (at most)

$$2 \cdot 1 + 2 \cdot 2 + \cdots + 2 \cdot 2^{k-1} + d \;=\; 2 \sum_{i=0}^{k-1} 2^i + d \;=\; 2 \cdot (2^k - 1) + d.$$

At the time when the $\mathcal{FME}$ locates the evacuation target, the distance between the $\mathcal{FME}$ and $\mathcal{SME}$ is $\frac{4}{3}\varepsilon$. Why? The two agents crossed paths at the meeting point that is 2^{k-2} away from the origin, and since that meeting the $\mathcal{FME}$ has moved distance ε and the $\mathcal{SME}$ has moved a distance of $\frac{\varepsilon}{3}$ (as it is moving at speed $\frac{1}{3}$). After the $\mathcal{FME}$ locates the target, it immediately reverses direction to inform the $\mathcal{SME}$. At that time, the two agents are $\frac{4}{3}\varepsilon$ apart and the distance between the pair will decrease at a rate of $\frac{2}{3}$ (the relative speeds between the agents). Therefore, the time for the $\mathcal{FME}$ to inform the $\mathcal{SME}$ is $\frac{4}{3}\varepsilon \div \frac{2}{3} = 2\varepsilon$. Note that this also means when the $\mathcal{FME}$ informs the $\mathcal{SME}$, they are at distance 2ε from the evacuation target.

Finally, the two agents return to the target to complete the evacuation procedure. Thus, assuming the $\frac{1}{3}$ worst-case speed of the $\mathcal{SME}$, this final "exit portion" will take time $2\varepsilon \div \frac{1}{3} = 6\varepsilon$.

Therefore, the entire evacuation procedure (in the worst-case, with a $\frac{1}{3}$ speed for the $\mathcal{SME}$) will take time at most

$$
\begin{aligned}
2(2^k - 1) + d + 2\varepsilon + 6\varepsilon &= 2(4 \times 2^{k-2} - 1) + d + 8\varepsilon \\
&= 2\left(4(2^{k-2} + \varepsilon) - 4\varepsilon - 1\right) + d + 8\varepsilon \\
&= 8d - 8\varepsilon - 2 + d + 8\varepsilon \\
&= 9d - 2. \qquad \Box
\end{aligned}
$$

We conjecture that when $s < 1/3$, then the evacuation time for the pair is strictly larger than $9d$, i.e. there exists a constant $\delta > 0$ such that the evacuation time is at least $(9 + \delta)d - o(d)$.

3.2 Two (or More) Fast Agents, Many Slow Agents

We finish with a remark about evacuating two (or more) fast agents, together with one or more slow $\mathcal{ME}$s.

With (at least) two $\mathcal{FME}$s, this pair can perform the coordinated evacuation procedure mentioned in Section 2.2. Once a fast $\mathcal{ME}$ discovers the evacuation target and proceeds to inform the other $\mathcal{FME}$, any slow agents that have remained at the origin can be informed as the $\mathcal{FME}$ passes through the origin. It takes an $\mathcal{FME}$ time $4d$ to find the target and return to the origin, and another $5d$ time to catch up to the other $\mathcal{FME}$, inform it, and return to the target. Hence, as long as the slow $\mathcal{ME}$s have a speed of at least $1/5$, they will arrive at the evacuation point at the exact same time as the fast pair, hence, the collection of all $\mathcal{ME}$s can still finish the evacuation in time $9d$.

4 Conclusion

As stated in the introduction, our main goal in this paper was to initiate study of the evacuation problem using mobile entities having different maximum speeds.

We demonstrated some cases where the original optimal $9d$ bound for homogeneous mobile agents is still obtainable in this new setting, provided the maximum speed of the slow agent(s) is not too small. Further work is necessary to investigate these problems, and the related, more general, search and rendezvous problems utilizing entities with different maximum speeds.

Acknowledgments. The authors would like to thank Shantanu Das, Mordechai Shalom, and Shmuel Zaks for valuable conversations about these problems during the course of a visit to Technion.

References

1. Alpern, S., Baston, V., Essegaier, S.: Rendezvous search on a graph. J. Applied Probability 36(1), 223–231 (1999)
2. Alpern, S., Gal, S.: The Theory of Search Games and Rendezvous. Kluwer Academic Publishing, Dordrecht (2003)
3. Baeza-Yates, R.A., Culberson, J.C., Rawlins, G.J.E.: Searching with uncertainty. In: Karlsson, R., Lingas, A. (eds.) SWAT 1988. LNCS, vol. 318, pp. 176–189. Springer, Heidelberg (1988)
4. Baeza-Yates, R.A., Culberson, J.C., Rawlins, G.J.E.: Searching in the plane. Information and Computation 106(2), 234–252 (1993)
5. Baeza-Yates, R.A., Schott, R.: Parallel searching in the plane. Computational Geometric Theory and Applications 5(3), 143–154 (1995)
6. Bellman, R.: Minimization problem. Bull. AMS 62(3), 270 (1956)
7. Bender, M.A., Fernández, A., Ron, D., Sahai, A., Vadhan, S.P.: The power of a pebble: Exploring and mapping directed graphs. In: STOC 1998, pp. 269–278 (1998)
8. Bose, P., De Carufel, J.-L., Durocher, S.: Revisiting the problem of searching on a line. In: Bodlaender, H.L., Italiano, G.F. (eds.) ESA 2013. LNCS, vol. 8125, pp. 205–216. Springer, Heidelberg (2013)
9. Collins, A., Czyzowicz, J., Gąsieniec, L., Labourel, A.: Tell me where I am so I can meet you sooner. In: Abramsky, S., Gavoille, C., Kirchner, C., Meyer auf der Heide, F., Spirakis, P.G. (eds.) ICALP 2010. LNCS, vol. 6199, pp. 502–514. Springer, Heidelberg (2010)
10. Dieudonné, Y., Pelc, A.: Anonymous meeting in networks. In: SODA 2013, pp. 737–747 (2013)
11. Ghosh, S.K., Klein, R.: Online algorithms for searching and exploration in the plane. Computer Science Review 4(4), 189–201 (2010)
12. Hammar, M., Nilsson, B.J., Schuierer, S.: Parallel searching on m rays. Comput. Geom. 18(3), 125–139 (2001)
13. Jeż, A., Łopuzański, J.: On the two-dimensional cow search problem. Information Processing Letters 131(11), 543–547 (2009)
14. Kao, M.Y., Reif, J.H., Tate, S.R.: Searching in an unknown environment: An optimal randomized algorithm for the cow-path problem. Information and Computation 109(1), 63–79 (1996)
15. Koutsoupias, E., Papadimitriou, C.H., Yannakakis, M.: Searching a fixed graph. In: Meyer auf der Heide, F., Monien, B. (eds.) ICALP 1996. LNCS, vol. 1099, pp. 280–289. Springer, Heidelberg (1996)
16. Li, H., Chong, K.P.: Search on lines and graphs. In: Proc. 48th IEEE Conference on Decision and Control, 2009 held Jointly with the 2009 28th Chinese Control Conference (CDC/CCC 2009), vol. 109(11), pp. 5780–5785 (2009)
17. Temple, T., Frazzoli, E.: Whittle-indexability of the cow path problem. In: American Control Conference (ACC), pp. 4152–4158 (2010)

Online Makespan Scheduling with Sublinear Advice⋆

Jérôme Dohrau

Department of Computer Science, ETH Zurich, Switzerland
dohrauj@ethz.ch

Abstract. We study online makespan scheduling with a fixed number of parallel machines. Jobs arrive in an online fashion in consecutive time steps, and must be scheduled both immediately and definitely. In contrast to the number of machines, the number of jobs is not known in advance. This paper focuses on the *advice complexity* of the problem. Basically, we ask how much additional information may help us to obtain solutions of high quality. Our main result is the construction of a $(1 + \varepsilon)$-competitive online algorithm with advice that reads a constant number of advice bits, for any $\varepsilon > 0$; here, "constant" means with respect to the input size, but our bound does depend on the number of machines and ε. This result is particularly interesting since it shows some very significant threshold behavior; it is known that, to be a little better, namely optimal, a linear number of advice bits is necessary. We also show that the *advice* can be derived from the input in polynomial time (with respect to the input size).

Keywords: online algorithms, advice complexity, makespan scheduling.

1 Introduction

Many problems that are met in practice are intrinsically online which means that parts of the output need to be computed while only a prefix of the definite input is known. A prominent example is the class of scheduling problems where *jobs* need to be assigned to a number of resources (which we call *machines*). Every now and then a new job appears and an algorithm must assign it to a machine without knowing what comes next. Furthermore, each job has a processing time, and every machine has a load which is the sum of all processing times of jobs that are assigned to this machine. A solution is computed with a particular objective in mind, which usually means to either minimize some cost function or maximize some gain function. In the case of the scheduling problem considered in the work at hand, our goal is to minimize the load of the machine with the largest one, i.e., we want the machine with the highest load to finish as early as possible; we call this value the *makespan*, and accordingly speak of *online*

⋆ This work was partially supported by the SNF grant 200021-141089. A special case of the results presented in this work is a part of the author's Bachelor's thesis and was published as a technical report [8].

G.F. Italiano et al. (Eds.): SOFSEM 2015, LNCS 8939, pp. 177–188, 2015.
© Springer-Verlag Berlin Heidelberg 2015

makespan scheduling. To simplify the environment we work in, we assume that the jobs arrive in consecutive *time steps*, job after job. Before we continue, let us give a formal definition of an online minimization problem.

Definition 1 (Online Minimization Problem). *An* online minimization problem *consists of a set $\mathcal{I}$ of inputs and a cost function. Every input $I \in \mathcal{I}$ is a sequence $I = \langle x_1, \ldots, x_n \rangle$ of requests. Furthermore, a set of feasible outputs (or solutions) is associated with every I; every output is a sequence $O = \langle y_1, \ldots, y_n \rangle$ of answers. The cost function assigns a positive real value $\text{cost}(I, O)$ to every input I and any feasible output O. For every input I, we call any feasible output O for I that has smallest possible cost (i. e., that minimizes the cost function) an* optimal solution *for I.*

We call algorithms that need to work without knowing the whole input in advance *online algorithms* [1]. For makespan scheduling, the requests equal jobs which are identified with their processing times. The problem is parameterized by the number of machines m, and we assume that m is known in advance. For each request, an online algorithm answers with the index of the machine that the job is assigned to (i. e., scheduled on); this assignment must not be changed in any subsequent time step. Let us give a formal definition.

Definition 2 (Online Makespan Scheduling). *The* online makespan scheduling problem, *short MS, is defined as follows. Given are an integer $m \geq 2$ and a sequence of n positive rational numbers $I = \langle x_1, \ldots, x_n \rangle$, for some $n \in \mathbb{N}^+$, where $x_i \in \mathbb{Q}^+$ is the processing time of the i-th job on any of the m available identical machines. A feasible solution is a sequence $O = \langle y_1, \ldots, y_n \rangle$ such that $y_i \in \{1, \ldots, m\}$, for all $i \in \{1, \ldots, m\}$. Alternatively, we can represent every feasible solution as an m-tuple $(M_1, \ldots, M_m)$ where $M_j = \{i \in \{1, \ldots, n\} \mid y_i = j\}$. For every $j \in \{1, \ldots, m\}$, we say that $\sum_{i \in M_j} x_i$ is the* load *of the j-th machine. The goal is to minimize the* makespan

$$\text{cost}(I, O) := \max_{j \in \{1, \ldots, m\}} \sum_{i \in M_j} x_i.$$

of the solution O on the input I.

For the sake of easy notation, we write $\text{cost}(O)$ instead of $\text{cost}(I, O)$ as I is always clear from the context. Moreover, we sometimes refer to the optimal makespan by $T^* := \text{cost}(O^*)$, where O^* is an optimal solution for the given instance. In this paper, we consider a special kind of online algorithms that were introduced by Böckenhauer et al. [4] and Hromkovič et al. [12]. These algorithms are equipped with an additional resource in the form of a special tape, which contains binary information about the input which is written onto it by a hypothetical *oracle* that sees the whole input in advance and has unbounded computing power. Our main object of interest is the length of the information that the oracle communicates this way, and what advantage the online algorithm gains from that knowledge. Formally, these *online algorithms with advice* are defined as follows.

Definition 3 (Online Algorithm with Advice). *Consider an input I of an online minimization problem. An* online algorithm ALG *with advice computes the output sequence* $\text{ALG}^\phi(I) = \langle y_1, \ldots, y_n \rangle$ *such that y_i is computed from* $\phi, x_1, \ldots, x_i$, *where ϕ is the content of the advice tape, i.e., an infinite binary sequence. An online algorithm* ALG *is c-competitive with advice complexity $b(n)$ if there exist non-negative constants c and β such that, for every n and for any input sequence I of length at most n, there exists some advice string ϕ such that*

$$\text{cost}\Big(\text{ALG}^\phi(I)\Big) \leq c \cdot \text{cost}(\text{OPT}(I)) + \beta,$$

where OPT *is an optimal offline algorithm, and at most the first $b(n)$ bits of ϕ have been accessed during the computation of the solution $\text{ALG}^\phi(I)$. If the above inequality holds with $\beta = 0$, we call* ALG *strictly c-competitive with advice complexity $b(n)$. Moreover,* ALG *is called* optimal *if it is strictly 1-competitive.*

While online algorithms are classically investigated in the framework of *competitive analysis* [1, 20], studying the advice complexity asks another question instead of simply what is possible when computing online. The idea is to get a better understanding of what makes a certain problem hard, i.e., what is hidden in the input that must be extracted to compute a satisfiable output. Hromkovič et al. [12] proposed to use this value as a measurement for the information content of an online problem. As observed before [3], there are a few extremely successful approaches to give a mathematical definition of the information content of finite objects such as Kolmogorov complexity [7,13] or Shannon's entropy [19]. However, these are hardly helpful to measure the information content of a concrete instance with respect to the given problem.

The dependence of the achievable output quality and the number of advice bits used comes in many different flavors. For instance, for the simple online knapsack problem it was shown that one bit of additional information allows to compute a 2-competitive solution while purely deterministic algorithms are arbitrarily bad; surprisingly, any further increase of the number of advice bits does not help until a number is supplied that logarithmically grows in the input size [6]. Then, if a logarithmic number of advice bits is available, there is an online algorithm that is $(1 + \varepsilon)$-competitive, for any $\varepsilon > 0$. In this paper, we demonstrate a phase transition that is even more drastic.

Other problems studied within this (or a similar) model include online bin packing [5,18], the k-server problem [3,9,10,17], and the online set cover problem [16]. There were also some efforts made to reduce different online problems to each other using some kind of *advice preserving reduction* [2,5,9]. This allows to compare two problems to each other in terms of their advice complexity, and to propagate hardness results.

Scheduling problems have been thoroughly studied. The advice complexity of a special kind of scheduling was investigated by Böckenhauer et al. [4]. Some of the results were later generalized and improved by Komm and Královič [14,15]. For the variant we investigate in this paper, a simple greedy approach achieves a competitive ratio of $2 - 1/m$ [1]. Recently, Renault et al. [18] studied online scheduling with advice for different objectives including makespan minimization. The authors showed the existence of a $(1 + \varepsilon)$-competitive algorithm that uses $\mathcal{O}(1/\varepsilon \log_2(1/\varepsilon))$ advice bits per request, for any $\varepsilon > 0$. They also gave a complementing lower bound showing that at least $(1 - 2m/n) \log_2(m)$ advice bits per time step are necessary to obtain an optimal solution. However, the model used in their work, which was introduced by Emek et al. [9], does not allow for studying sublinear advice, i.e., less than a constant amount of advice per time step. In this model, the advice is not read from a tape, but it is supplied with every time step where the number of advice bits is fixed and the same in every step.

In the following section, we present an online algorithm with advice that achieves, for any $\varepsilon > 0$, a competitive ratio of $1 + \varepsilon$ while reading

$$\left\lceil \frac{\log\left(\frac{3m}{\varepsilon}\right)}{\log\left(1 + \frac{\varepsilon}{3m}\right)} \right\rceil \lceil \log_2(m) \rceil + m \left\lceil \log_2\left(1 + \frac{3m}{\varepsilon}\right) \right\rceil$$

advice bits.

2   Sublinear Advice

As already mentioned, different problems behave in quite different ways in terms of their advice complexity. However, it seems to be rather common that a linear amount of advice is necessary to be optimal (of course, there are exceptions, e.g., the ski rental problem where 1 bit of advice allows for a trivial optimal solution [14]). It has been shown by Renault et al. [18] that also for the makespan scheduling problem a linear amount of advice is necessary. An intriguing fact is, however, that we pay this large amount of additional information for an arbitrarily small fraction of performance: if we do not demand to be optimal, but only want to be extremely (arbitrarily) close to an optimal solution, the sufficient amount of information surprisingly drops to a constant (for a fixed number of machines). This very significant threshold behavior renders the result presented in this section particularly interesting.

2.1   Important Jobs

Before we start with the description of an algorithm achieving this goal, we briefly introduce some notation that we will use during our analysis. We start with a simple observation.

Observation 1. *An optimal solution cannot do better than perfectly balance the processing times of all jobs. Therefore, the makespan of an optimal solution is at least the average load over all machines, i.e.,*

$$T^* \geq \frac{1}{m} \sum_{i=1}^{n} x_i.$$

The analysis of the algorithm presented in this section relies on cleverly partitioning the jobs; we first describe a rather naive approach.

Definition 4. *Let $\alpha > 0$. The k-th job is called α-large if $x_k \geq \alpha T^*$. Conversely, the k-th job is called α-small if $x_k < \alpha T^*$.*

If there are only two machines, any deterministic online algorithm without advice has a competitive ratio that is at least $3/2$ [1]. It can be shown that, in the case of two machines, the knowledge of the existence and the index of a $2/3$-large job allows an online algorithm to improve its competitive ratio to $4/3$ [8]. Since an online algorithm does not know T^* in advance, it is unable to tell whether a given job is α-large or not until the last job arrives. Encoding an index requires roughly $\log_2(n)$ bits of advice and is therefore not feasible to achieve a constant advice complexity. However, by Observation 1, an online algorithm can use the prefix of the input that was presented so far to estimate a bound on the makespan of an optimal solution. This way, the algorithm can recognize potentially large jobs. The following definition formalizes this notion.

Definition 5. *Let $\alpha > 0$. The k-th job is (α, m)-important if*

$$x_k \geq \frac{\alpha}{m} \sum_{i=1}^{k-1} x_i.$$

In fact, the concept of important jobs is a generalization of large jobs in the following sense.

Observation 2. *Every α-large job is (α, m)-important.*

Proof. Assume the k-th job is α-large. We have

$$x_k \geq \alpha T^* \geq \frac{\alpha}{m} \sum_{i=1}^{n} x_i \geq \frac{\alpha}{m} \sum_{i=1}^{k-1} x_i,$$

where the second inequality follows from Observation 1. Thus, the k-th job is (α, m)-important. $\qquad\square$

2.2 The Algorithm

For any given $\varepsilon > 0$, we aim at constructing an online algorithm A_ε with advice that is strictly $(1 + \varepsilon)$-competitive and that has an advice complexity that is constant with respect to the length of the input sequence. We do this by showing how A_ε schedules the jobs of an arbitrary but fixed instance $I = \langle x_1, \ldots, x_n \rangle$. Let $\delta := \varepsilon/3$. We partition the set of jobs as follows: Let L be the set of all (δ, m)-important jobs in I, i.e.,

$$L := \left\{ k \in \{1, \ldots, n\} \,\middle|\, x_k \geq \frac{\delta}{m} \sum_{i=1}^{k-1} x_i \right\},$$

and let
$$S := \{1, \ldots, n\} \setminus L$$

be the set of all other jobs. Note that, as implied by Observation 2, the indices of all δ-large jobs are contained in L. Consequently, all jobs corresponding to the indices in S are δ-small. Since an online algorithm can recognize whether an arriving job is (δ, m)-important or not, it can use two different strategies to schedule the jobs with indices in L and the jobs with indices in S. Intuitively speaking, we will exploit that if the algorithm A_ε misplaces an (δ, m)-important job, the relative error significantly decreases as more and more (δ, m)-important jobs arrive. Below, following this idea, we prove that it is sufficient to know how to schedule the last few jobs that are (δ, m)-important and let the oracle encode the indices of these jobs. As for the jobs that are not (δ, m)-important, the algorithm A_ε reads some information about the distribution of their processing times among the machines in an optimal solution and then tries to approximate this distribution. In this case, the relative error can be restrained because the processing times of these jobs are all small with respect to the makespan of an optimal solution.

For later analysis, consider any optimal solution $O^* = \langle y_1^*, \ldots, y_n^* \rangle$ for the instance I and, for all $j \in \{1, \ldots, m\}$, let l_j^* and s_j^* denote the sum of the processing times of all jobs with their index in L or S, respectively, that are scheduled on the j-th machine, i.e., we define

$$l_j^* := \sum_{i \in M_j^* \cap L} x_i \quad \text{and} \quad s_j^* := \sum_{i \in M_j^* \cap S} x_i$$

where $M_j^* := \{i \in \{1, \ldots, n\} \mid y_i^* = j\}$. Note that the optimal makespan is given by $T^* = \max_{j \in \{1, \ldots, m\}} l_j^* + s_j^*$.

Although it has not yet been described how the algorithm A_ε schedules the jobs, let $O := A_\varepsilon(I) = \langle y_1, \ldots, y_n \rangle$ be the solution produced by A_ε on the instance I. For $j \in \{1, \ldots, m\}$, we define

$$l_j := \sum_{i \in M_j \cap L} x_i$$

where $M_j := \{i \in \{1, \ldots, n\} \mid y_i = j\}$. Let k_i, for $i \in \{1, \ldots, |L|\}$, denote the index of the i-th (δ, m)-important job. The following lemma shows that although there might be an arbitrary number of (δ, m)-important jobs only the last few of them really affect the quality of the solution.

Lemma 1. *Let $t := \lceil \log(m/\delta)/\log(1 + \delta/m) \rceil$. If $|L| > t$, then we have $r \leq \delta T^*$, where $r := \sum_{i=1}^{|L|-t} x_{k_i}$ denotes the load induced by the first $|L| - t$ jobs that are (δ, m)-important.*

Proof. Assume there are at least $t + 1$ jobs that are (δ, m)-important. First, we notice that, for every $c \in \{2, \ldots, |L|\}$, we have

$$\sum_{i=1}^{k_c} x_i = x_{k_c} + \sum_{i=1}^{k_c-1} x_i \geq \left(1 + \frac{\delta}{m}\right) \sum_{i=1}^{k_c-1} x_i \geq \left(1 + \frac{\delta}{m}\right) \sum_{i=1}^{k_c-1} x_i \tag{1}$$

by the definition of important jobs. Combining Observation 1 together with multiple applications of (1) yields

$$T^* \geq \frac{1}{m} \sum_{i=1}^{n} x_i \geq \frac{1}{m} \sum_{i=1}^{k_{|L|}} x_i \geq \frac{1}{m} \left(1 + \frac{\delta}{m}\right)^t \sum_{i=1}^{k_{|L|}-t} x_i.$$

We can now use these insights to bound the load induced by the first $|L| - t$ jobs that are (δ, m)-important from above by

$$r = \sum_{i=1}^{|L|-t} x_{k_i} \leq \sum_{i=1}^{k_{|L|}-t} x_i \leq \frac{mT^*}{\left(1 + \frac{\delta}{m}\right)^t} \leq \delta T^*,$$

which concludes the proof.   □

The goal is to schedule the last $t' := \min\{t, |L|\}$ jobs that are (δ, m)-important according to the optimal solution O^*. For this purpose, the oracle encodes a sequence of t' indices that indicate how the last t' jobs that are (δ, m)-important are to be scheduled. Clearly, these indices can be encoded using at most $t\lceil \log_2(m) \rceil$ bits of advice. Note that since $|L|$ is not known to the algorithm, it does not know when the first one of these jobs arrives. To address this problem, as elaborated in the following, the t' indices of the sequence are ordered in such a way that the algorithm does not need to know when the $(|L| - t' + 1)$-th job that is (δ, m)-important arrives. Essentially, the oracle takes the "pattern" that describes how the last t' jobs that are (δ, m)-important are to be scheduled and shifts it so that the algorithm can simply loop over these indices and naturally aligns with the pattern as soon as the $(|L| - t' + 1)$-th job that is (δ, m)-important arrives. Let $R_t(\cdot)$ denote the remainder of the Euclidean division by t. For $i \in \{1, \ldots, |L|\}$, the algorithm A_ε schedules the i-th (δ, m)-important job according to the $R'_t(i)$-th index encoded by the oracle, where $R'_t(i) := R_t(i-1) + 1$. Note that A_ε reads t' different indices for the last t' jobs that are (δ, m)-important. As for the advice string, for $i \in \{|L| - t' + 1, \ldots, |L|\}$, the $R'_t(i)$-th index encodes y_i^*. Consequently, the algorithm A_ε and the optimal solution O^* schedule the last t' jobs that are (δ, m)-important identically.

The following lemma shows that this is a good approximation to the optimal solution.

Lemma 2. *For $j \in \{1, \ldots, m\}$, we have $l_j \leq l_j^* + \delta T^*$.*

Proof. Consider an arbitrary machine with index $j \in \{1, \ldots, m\}$. If $|L| \leq t$, the algorithm A_ε and the solution O^* schedule all (δ, m)-important jobs identically and we have $l_j = l_j^*$. Otherwise, by our previous observations, l_j and l_j^* differ by at most r and, following Lemma 1, we get $l_j \leq l_j^* + r \leq l_j^* + \delta T^*$.   □

Next, we describe how the algorithm deals with the jobs that are not (δ, m)-important. For $j \in \{1, \ldots, m\}$, we define $p_j := s_j^*/s$, where $s := \sum_{j'=1}^{m} s_{j'}^*$. The ratios $p_1, \ldots, p_m$ indicate how the processing times of the jobs in S are distributed among the machines by the optimal solution O^*. Furthermore, let $c := m/\delta$ and $q_j := \lceil p_j c \rceil / c$, for all $j \in \{1, \ldots, m\}$. Note that since $q_j \in \{0/c, 1/c, \ldots, c/c\}$ the oracle can encode each q_j using $\lceil \log_2(c+1) \rceil$ bits. For all $i \in \{0, \ldots, n\}$ and all $j \in \{1, \ldots, m\}$, let $s_j^{(i)}$ denote the load induced by the jobs that (i) are scheduled on the j-th machine by the algorithm A_ε, (ii) are not (δ, m)-important and (iii) arrive at the latest in the i-th time step; formally, we define

$$s_j^{(i)} := \sum_{i' \in \{1,\ldots,i\} \cap M_j \cap S} x_{i'}.$$

For a shorter notation, let

$$s^{(i)} := \sum_{j=1}^{m} s_j^{(i)}.$$

Note that $s^{(i-1)} \leq s^{(i)}$, for all $i \in \{1, \ldots, n\}$, and $s^{(n)} = s$. If in time step i a job arrives that is not (δ, m)-important, then A_ε schedules this job on a machine with index $j \in \{1, \ldots, m\}$ such that $s_j^{(i-1)}$ does not exceed $q_j s^{(i-1)}$. In the following, we prove that in every time step there is actually such an index with the desired property. Moreover, we show that the processing times of the jobs that are not (δ, m)-important are distributed among the machines approximately equally by the algorithm A_ε and the optimal solution O^*.

Lemma 3. *For every $i \in \{0, \ldots, n\}$, there exists an index $j \in \{1, \ldots, m\}$ such that $s_j^{(i)} \leq q_j s^{(i)}$.*

Proof. Towards contradiction, we assume the contrary, i.e., $s_j^{(i)} > q_j s^{(i)}$, for all $j \in \{1, \ldots, m\}$. Under this assumption, we have

$$s^{(i)} = \sum_{j=1}^{m} s_j^{(i)} > \sum_{j=1}^{m} q_j s^{(i)} \geq s^{(i)} \sum_{j=1}^{m} p_j = s^{(i)}$$

which clearly cannot be true.   □

Lemma 4. *For all $j \in \{1, \ldots, m\}$, we have $s_j^{(n)} \leq s_j^* + 2\delta T^*$.*

Proof. Consider an arbitrary $j \in \{1, \ldots, m\}$. If no jobs are scheduled on the j-th machine or all jobs that are scheduled on the j-th machine are (δ, m)-important, then $s_j^{(n)} = 0$ and the claim follows trivially. Otherwise, let k be the index of the last job that is not (δ, m)-important and that is scheduled on the j-th machine. We have

$$s_j^{(k-1)} \leq q_j s^{(k-1)} \leq (p_j + 1/c)s = s_j^* + s/c \leq s_j^* + \delta T^*$$

where we used that $s \leq \sum_{i=1}^{n} x_i$ together with Observation 1 and $c := m/\delta$ in the last inequality. Since the k-th job is not (δ, m)-important and therefore δ-small, we get

$$s_j^{(n)} = s_j^{(k-1)} + x_k \leq s_j^* + 2\delta T^*$$

as claimed.   □

Algorithm 1. The online algorithm A_ε with advice for MS.

$k \leftarrow 0$
for $j \leftarrow 1$ **to** m **do**
   $s_j \leftarrow 0$
for $i \leftarrow 1$ **to** n **do**
   if i-th job is (δ, m)-important **then**
      $k \leftarrow k + 1$
      $k \leftarrow R_t(k-1) + 1$ where $t = \lceil \log(3m/\varepsilon)/\log(1 + \varepsilon/3m) \rceil$
      output the k-th index encoded by the oracle
   else
      $j \leftarrow$ index $j \in \{1, \ldots, m\}$ such that $s_j \le q_j \sum_{j'=1}^m s_{j'}$
      $s_j \leftarrow s_j + x_i$
      output the index j

To conclude the analysis of the algorithm A_ε, we now combine the various results elaborated above to determine the competitive ratio as well as the advice complexity of A_ε.

Theorem 1. *The algorithm A_ε for MS is strictly $(1 + \varepsilon)$-competitive and reads $b(n) \in \mathcal{O}(1)$ bits of advice. Note that $b(n)$ depends on the number of machines m and ε.*

Proof. We can use Lemma 2 and Lemma 4 to bound the makespan of the solution O produced by A_ε from above by

$$
\begin{aligned}
\text{cost}(O) &= \max_{j \in \{1, \ldots, m\}} \sum_{i \in M_j} x_i \\
&= \max_{j \in \{1, \ldots, m\}} s_j^{(n)} + l_j \\
&\le \max_{j \in \{1, \ldots, m\}} s_j^* + l_j^* + 3\delta T^* \\
&= (1 + \varepsilon)T^* = (1 + \varepsilon)\text{cost}(O^*).
\end{aligned}
$$

Thus, A_ε is strictly $(1 + \varepsilon)$-competitive.

Next, we address the advice complexity of A_ε. As described above, the algorithm reads at most $t \lceil \log_2(m) \rceil$ bits to schedule the last t' jobs that are (δ, m)-important according to the optimal solution O^*, and another $m \lceil \log_2(c+1) \rceil$ bits to read the approximative ratios $q_1, \ldots, q_m$. Thus, if we express the advice complexity $b(n)$ of the algorithm in terms of ε and m, we get

$$
b(n) \le \left\lceil \frac{\log\left(\frac{3m}{\varepsilon}\right)}{\log\left(1 + \frac{\varepsilon}{3m}\right)} \right\rceil \lceil \log_2(m) \rceil + m \left\lceil \log_2\left(1 + \frac{3m}{\varepsilon}\right) \right\rceil \in \mathcal{O}(1),
$$

which finishes the proof. Note that we assume the number m of machines to be fixed. $\qquad\square$

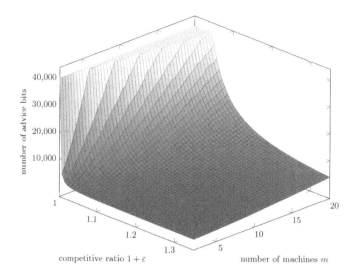

Fig. 1. The number of advice bits required depending on ε and m

For the sake of clarity, the pseudocode of the complete algorithm A_ε is shown in Algorithm 1. Moreover, the number of advice bits read by A_ε plotted as a function of the competitive ratio $1 + \varepsilon$ and the number of machines m is shown in Figure 1.

2.3   A Polynomial-Time Oracle

In our model, we make the quite strong assumption that the oracle has unbounded computing power. Indeed, a straightforward approach of generating the advice is to compute an optimal solution (which is NP-hard), to identify the (δ, m)-important jobs, and to create the advice as described in the preceding section. However, for the problem studied in this paper, we can even construct an efficient oracle that also allows to compute a solution that is arbitrarily close to an optimal one.

Since there is a polynomial-time approximation scheme (PTAS) for the makespan scheduling problem [11], the oracle can compute an approximate solution O' with makespan $T' := \text{cost}(O') \leq (1 + \varepsilon')T^*$ in polynomial time, for any $\varepsilon' > 0$. It then encodes the advice string as described in the previous subsection but with respect to the solution O' (rather than with respect to an optimal solution) and some parameter $\varepsilon'' > 0$. This can also be done in polynomial time. If this advice string is given to the algorithm $A_{\varepsilon''}$, the makespan of the algorithm's output can be bounded by $T \leq (1 + \varepsilon'')T'$. For appropriate choices of ε' and ε'', for instance $\varepsilon' := \varepsilon'' := \varepsilon/3$, this yields a solution with makespan

$$T \leq (1 + \varepsilon')(1 + \varepsilon'')T^* \leq (1 + \varepsilon)T^*$$

if ε is small enough.

Clearly, an oracle with infinite computing power can compute an optimal solution and needs to communicate less advice bits (because $\varepsilon > \varepsilon''$). However, the number of advice bits encoded with this approach is still constant in the number of jobs and only dependent on the number of machines and ε.

Since the advice used in total is constant with respect to the input length, we can give an even cruder approach. The oracle simply computes an upper bound u on the number of the advice bits, then generates all binary strings of this length, and simulates the online algorithm on each of them. Since the algorithm itself clearly runs in polynomial time (with respect to n), and there are 2^u strings to try, which again is constant for fixed ε and m, the advice string ϕ^* that leads to the best approximation is found in polynomial time. This string is not necessarily the same as the string ϕ that is produced by the approach described in Subsection 2.2. However, since ϕ is a string (or a prefix of a string) that was considered by the oracle, the solution produced by A_ε when given ϕ^* as advice is at least as good as if it were given ϕ.

Note that this last observation is independent of the problem and can be applied to every online problem whose advice complexity has the same asymptotic behavior.

3   Conclusion

In this paper, we studied the advice complexity of online makespan scheduling. We showed that, with respect to the number of jobs that arrive online, constant advice suffices to produce a near-optimal output. This is particularly interesting as it is known that linear advice [18] is necessary to obtain an optimal result. Intuitively speaking, this is due to the fact that there is only a constant number of jobs that significantly affect the quality of the solution whereas the optimality of the solution might depend on the placement of every single job. Moreover, the oracle that produces the advice works in polynomial time. It remains open whether similar results can be obtained for machines that have different speeds or objectives for scheduling problems other than makespan minimization such as maximizing the minimum load (*machine cover*) or, for $1 < p \leq \infty$, minimizing the ℓ_p norm [18].

Acknowledgments. The author thanks Hans-Joachim Böckenhauer, Juraj Hromkovič, and Dennis Komm for making this work possible, their support, and all the enlightening discussions.

References

1. Borodin, A., El-Yaniv, R.: Online Computation and Competitive Analysis. Cambridge University Press (1998)
2. Böckenhauer, H.-J., Hromkovič, J., Komm, D., Krug, S., Smula, J., Sprock, A.: The string guessing problem as a method to prove lower bounds on the advice complexity. In: Du, D.-Z., Zhang, G. (eds.) COCOON 2013. LNCS, vol. 7936, pp. 493–505. Springer, Heidelberg (2013)

3. Böckenhauer, H.-J., Komm, D., Královič, R., Královič, R.: On the advice complexity of the k-server problem. In: Aceto, L., Henzinger, M., Sgall, J. (eds.) ICALP 2011, Part I. LNCS, vol. 6755, pp. 207–218. Springer, Heidelberg (2011)
4. Böckenhauer, H.-J., Komm, D., Královič, R., Královič, R., Mömke, T.: On the advice complexity of online problems. In: Dong, Y., Du, D.-Z., Ibarra, O. (eds.) ISAAC 2009. LNCS, vol. 5878, pp. 331–340. Springer, Heidelberg (2009)
5. Boyar, J., Kamali, S., Larsen, K.S., López-Ortiz, A.: Online bin packing with advice. In: Proc. of STACS 2014, pp. 174–186 (2014)
6. Böckenhauer, H.-J., Komm, D., Královič, R., Rossmanith, P.: The online knapsack problem: advice and randomization. Theoretical Computer Science 527, 61–72 (2014)
7. Chaitin, G.J.: On the length of programs for computing finite binary sequences. Journal of the ACM 13(4), 547–569 (1966)
8. Dohrau, J.: Online makespan scheduling with sublinear advice. Technical Report, ETH Zurich (2013)
9. Emek, Y., Fraigniaud, P., Korman, A., Rosén, A.: Online computation with advice. In: Albers, S., Marchetti-Spaccamela, A., Matias, Y., Nikoletseas, S., Thomas, W. (eds.) ICALP 2009, Part I. LNCS, vol. 5555, pp. 427–438. Springer, Heidelberg (2009)
10. Gupta, S., Kamali, S., López-Ortiz, A.: On advice complexity of the k-server problem under sparse metrics. In: Moscibroda, T., Rescigno, A.A. (eds.) SIROCCO 2013. LNCS, vol. 8179, pp. 55–67. Springer, Heidelberg (2013)
11. Hochbaum, D.S., Shmoys, D.B.: Using dual approximation algoritms for scheduling problems: practical and theoretical results. Journal of ACM 34, 144–162 (1987)
12. Hromkovič, J., Královič, R., Královič, R.: Information complexity of online problems. In: Hliněný, P., Kučera, A. (eds.) MFCS 2010. LNCS, vol. 6281, pp. 24–36. Springer, Heidelberg (2010)
13. Kolmogorov, A.N.: Three approaches to the definition of the concept "quantity of information". Problemy Peredachi Informatsii 1, 3–11 (1965)
14. Komm, D.: Advice and randomization in online computation. PhD Thesis, ETH Zurich (2012)
15. Komm, D., Královič, R.: Advice complexity and barely random algorithms. Theoretical Informatics and Applications (RAIRO) 45(2), 249–267 (2011)
16. Komm, D., Královič, R., Mömke, T.: On the advice complexity of the set cover problem. In: Hirsch, E.A., Karhumäki, J., Lepistö, A., Prilutskii, M. (eds.) CSR 2012. LNCS, vol. 7353, pp. 241–252. Springer, Heidelberg (2012)
17. Renault, M.P., Rosén, A.: On Online Algorithms with Advice for the k-Server Problem. In: Solis-Oba, R., Persiano, G. (eds.) WAOA 2011. LNCS, vol. 7164, pp. 198–210. Springer, Heidelberg (2012)
18. Renault, M.P., Rosén, A., van Stee, R.: Online algorithms with advice for bin packing and scheduling problems. CoRR abs/1311.7589 (2013)
19. Shannon, C.E.: A mathematical theory of communication. Mobile Computing and Communications Review 5(1), 3–55 (2001)
20. Sleator, D.D., Tarjan, R.E.: Amortized efficiency of list update and paging rules. Communications of the ACM 28(2), 202–208 (1985)

Deterministic Rendezvous in Restricted Graphs[*]

Ashley Farrugia[1], Leszek Gąsieniec[1], Łukasz Kuszner[2], and Eduardo Pacheco[3]

[1] Department of Computer Science, University of Liverpool, UK
`l.a.gasieniec@liverpool.ac.uk`
[2] Department of Algorithms & System Modeling,
Gdańsk University of Technology, Poland
[3] School of Computer Science, Carleton University, Canada

Abstract. In this paper we consider the problem of synchronous rendezvous in which two anonymous mobile entities (robots) A and B are expected to meet at the same time and point in a graph $G = (V, E)$. Most of the work devoted to rendezvous in graphs assumes that robots have access to the same sets of nodes and edges, where the topology of connections may be initially known or unknown. In our work we assume the movement of robots is restricted by the topological properties of the graph space coupled with the intrinsic characteristics of robots preventing them from visiting certain edges in E.

We consider three rendezvous models reflecting on restricted maneuverability of robots A and B. In *Edge Monotonic Model* each robot $X \in \{A, B\}$ has weight w_X and each edge in E has a weight restriction. Consequently, a robot X is only allowed to traverse edges with weight restrictions greater that w_X. In the remaining two models graph G is unweighted and the restrictions refer to more arbitrary subsets of traversable nodes and edges. In particular, in *Node Inclusive Model* the set of nodes V_X available to robot X, for $X \in \{A, B\}$ satisfies the condition $V_A \subseteq V_B$ or vice versa, and in *Blind Rendezvous Model* the relation between V_A and V_B is arbitrary. In each model we design and analyze efficient rendezvous algorithms. We conclude with a short discussion on the asynchronous case and related open problems.

1 Introduction

In this paper we consider rendezvous problem, a challenge in which two or more mobile entities, called later *robots* have the goal to meet at the same point and time in provided space. This space can be either a network of discrete nodes between which robots can move along existing connections, or a geometric environment in which the movement of robots is restricted by the topological properties of the space. As indicated in [19] *symmetry* plays a key role in determining the feasibility and efficiency of solutions in the rendezvous problem. It is quite often that *anonymous* (indistinguishable) players find themselves in a situation where

---

[*] Research partially supported by the Polish National Science Center grant DEC-2011/02/A/ST6/00201 and by Network Sciences and Technologies initiative at University of Liverpool.

G.F. Italiano et al. (Eds.): SOFSEM 2015, LNCS 8939, pp. 189–200, 2015.

the tools and advice given to each robot are identical and rendezvous may not be feasible [5]. In this context, determining even small pieces of information that would help to distinguish between participating robots often prove to be vital in achieving rendezvous.

Rendezvous problems have been studied in a number of different settings. A vast literature includes several exhaustive surveys on the topic and other searching games, see, e.g., [4,5,22]. The work on rendezvous includes both deterministic algorithms surveyed recently in [22] as well as randomized approaches including already classical work in [2,3,8,9]. Another group of algorithms focus on geometric setting including earlier work on the line [9,10] and the plane [6,7] as well as more recent work on fat (with non-zero radius) robots [1,14]. Another interesting group of rendezvous algorithms is designed for infinite (Euclidean) spaces for both synchronized and asynchronous solutions [12, 13, 16]. An important group of rendezvous algorithms have been considered for graph based environments, see, e.g., [15, 18, 21]. However, all previous work is devoted to the case when both robots have access to the same part of the network. An interesting version of rendezvous in which robots face different costs associated with traversed edges was considered recently in [17] where the authors consider scenarios with and without communication between participating robots.

Our work refers to the extreme communicationless case of [17] in which the costs imposed on edges are either unit or infinite. We also make reference to *blind rendezvous* considered recently in the context of cognitive radio networks [11,20].

1.1   Model of Computation

We consider rendezvous of *anonymous* (indistinguishable also with respect to the control mechanism) robots in networks modeled by graphs. The network $G = (V, E)$ where the two robots are expected to rendezvous is a simple connected graph in which two nodes $s_A, s_B \in V$ are selected as the starting points for robots A and B respectively. Moreover, for each $X \in \{A, B\}$ we define its *reachability graph* also referred to as *the map* $G_X = (V_X, E_X)$, a subgraph of G in which V_X and E_X are respectively the sets of nodes and edges accessible from s_X. Moreover, agent X is only able to see its own map. Let $k_X = |V_X|$ be *the size* of map G_X and assume w.l.o.g. that $k_A \leq k_B$. While the robots are anonymous, we use extra assumptions with respect to the network nodes (and in some cases edges too). In particular, we assume that all nodes of the input network graph $G = (V = \{v_1, v_2, \ldots, v_n\}, E)$ are ordered, s.t., $v_i < v_{i+1}$ for all $i = 1, 2, \ldots n - 1$ and this order is consistent with the order of nodes in G_X, for $X \in \{A, B\}$. In particular, if $V_X = \{v_1^{(X)}, v_2^{(X)}, \ldots, v_{k_X}^{(X)}\}$, $v_a^{(X)} = v_i$, and $v_b^{(X)} = v_j$, where $v_i, v_j \in V$ and $i \leq j$, we also get $v_a^{(X)} \leq v_b^{(X)}$. Finally, let $T(V_X)$ be a rooted tree that spans all nodes in V_X in which the starting point s_X is placed in the root of $T(V_X)$ and the order on children is consistent with the order of nodes in V_X.

The actions of the two robots are synchronized. I.e., the two robots A and B have access to the global clock ticking in discrete *time steps* $0, 1, 2, \ldots$. Our algorithms start with the global clock set to time 0. During a single time step each robot assesses the node in which it resides in (including check for co-location/rendezvous with the other robot). Then it decides whether to stay at the same node or to move to one of its neighbors via an available (edge) connection. During the traversal between two connected nodes the "eyes" of the robot are closed. Consequently, since the robots cannot meet on edges rendezvous has to take place at some node. The running time of all algorithms is bounded, i.e., the robots stop within the time given to the respective rendezvous algorithms.

We consider three **models of computation** with restrictions on maps given to robots A and B.

1. **Edge Monotonic Model.** This model is motivated by the case in which each robot $X \in \{A, B\}$ has weight w_X and each edge in E has weight restriction. This setting imposes an order on edges in $E = \{e_1, e_2, \ldots e_m\}$, in which for any $1 \leq i < j \leq m$ edge e_j tolerates weights non-smaller than e_i. Let i_X be the smallest integer, s.t., e_{i_X} tolerates weight w_X. One can conclude that robot X is only allowed to traverse edges with index $\geq i_X$. Consequently in this model if rendezvous is possible $E_A \subseteq E_B$ and $V_A \subseteq V_B$ (i.e., G_A is a subgraph of G_B), see section 2.1.

2. **Node Inclusive Model.** In this model we only assume that $V_A \subseteq V_B$, i.e., the relationship between edges spanning nodes in E_A and E_B remains unspecified.

3. **Blind Rendezvous Model.** In this model we only assume that $V_A \cap V_B \neq \emptyset$. Also here the relationship between E_A and E_B is unspecified.

1.2 Our Results

In this paper we study synchronized rendezvous in three different restriction models. In section 2.1 we present optimal $O(k_A + k_B)$−time rendezvous algorithm **RV1** in the Edge Monotonic Model. In section 2.2 we present rendezvous algorithm **RV2** that meets two robots in the Node Inclusive Model in almost linear time $O((k_A + k_B) \log(k_A + k_B))$. In the Blind Rendezvous Model, see section 2.3 we show that rendezvous is not feasible. We introduce explicit labels to make rendezvous feasible and present two rendezvous algorithms **RV3** and **RV4** whose superposition allows robots to meet in time $\min\{O((k_A + k_B)^3 \log\log n, O((k_A + k_B)^2 \log n)\}$. We conclude with the final comment and a short discussion on asynchronous models in section 3.

2 Rendezvous Algorithms

In this section we design and analyze several rendezvous algorithms in the considered restriction models.

2.1 Rendezvous in Edge Monotonic Model

Recall that in this model, we adopt the order of edges in $E = \{e_1, e_2, \ldots, e_m\}$ where $e_i < e_{i+1}$. For any $l \in \{1, \ldots, m\}$, we define a sequence of subgraphs

$G(l) = (V(l), E(l))$, where $E(l) = \{e_l, e_{l+1}, ..., e_m\}$ and $V(l)$ is the set of nodes in V induced by the edges of $E(l)$, and $E(l+1) \subset E(l)$. In this model each robot X is associated with the threshold index $i_X \in \{1, ..., m\}$ determining the set of edges $E(i_X)$ traversable by X. In other words, robot X can walk only along edges from $E(i_X)$. We also define a sequence of connected components $G_X(l) = \{V_X(l), E_X(l)\}$, for $l \in \{i_X, ..., m\}$, where $V_X(l)$ is the set of nodes reachable from s_X via edges in $E(i_X)$, and $E_X(l) \subseteq E(l)$ is the maximal set of edges spanning nodes in $V_X(l)$. So in this case $V_X = V_X(i_X)$, $E_X = E_X(i_X)$, and $k_X = |V_X(i_X)|$. The following Lemma holds.

Lemma 1. *In Edge Monotonic Model either* $(V_A \subseteq V_B)$ *or* $(V_B \subseteq V_A)$, *or* $V_A \cap V_B = \emptyset$.

Proof. The lemma (statement) would be false if all of the terms $(V_A \subseteq V_B)$, $(V_B \subseteq V_A)$, and $V_A \cap V_B = \emptyset$ were false too. Assume w.l.o.g. that $V_A \cap V_B \neq \emptyset$, where $V_A = V_A(i_A)$ and $V_B = V_B(i_B)$, and $i_A \geq i_B$. Since $i_B \leq i_A$ (edges traversable by A are also traversable by B) and $V_A \cap V_B \neq \emptyset$ (the reachability graphs G_A and G_B coincide) all edges and points in $G_A(i_A)$ are also available to robot B, meaning $V_A \subseteq V_B$.

We define the concept of a *sleeve of graphs* with respect to X denoted by $SL(X)$.

Definition 1. *The sleeve of graphs* $SL(X)$ *with respect to robot* X *is the maximal sequence of connected components* $G_X(i_X)$, $G_X(i_X + 1), ..., G_X(l^*)$, *in which* $|V_X(l+1)| > |V_X(l)|/2$, *for all* $i_X \leq l^* < m$. *A subsequence* $G_X(i_X + j)$, $G_X(i_X + j + 1), ..., G_X(l^*)$, *for any* $j \in \{0, 1, ..., l^* - i_X\}$, *is called a tail of* $SL(X)$ *and the smallest (in the adopted order) node* $v^* \in V_X(l^*)$ *is called the target in* $SL(X)$.

In what follows we present a pseudo-code of the proposed rendezvous algorithm in the monotonic model. If at any time step the two robots A and B meet, the rendezvous is achieved and the two robots *halt*.

Algorithm RV1($X \in \{A, B\}$)
Step 1 Walk from s_X to the target node v^* in $SL(X)$
Step 2 Wait in v^* until conclusion of time step $2k_X$;
Step 3 Walk along the Euler tour of $T(V_X)$ and *Halt*.

Theorem 1. *If rendezvous is feasible, Algorithm* **RV1** *admits meeting in optimal time* $O(k_A + k_B)$.

Proof. Recall that $k_A \leq k_B$. According to Lemma 1 if rendezvous is feasible, i.e., $V_A \cap V_B \neq \emptyset$ we conclude that $V_A \subseteq V_B$. We consider two complementary cases:

Case 1 $[2k_A > k_B]$ Since $2k_A > k_B$ according to Definition 1 sleeve $SL(A)$ is a tail of $SL(B)$ and the two sleeves share the same target v^*. The robots A and B are initially placed in their own sleeves at distance at most $k_B < 2k_A$ from the joint target v^*. This admits rendezvous in **Step 1** in time at most k_B.

Case 2 $[2k_A \leq k_B]$ In this case, robot A halts at the latest at time step $4k_A$ on the conclusion of **Step 3**, i.e., after $2k_A$ time steps devoted to **Step 1** and **Step 2**, followed by additional $2k_A - 2$ time steps devoted to the Euler tour traversal in $T(V_A)$) in **Step 3**. Note, however, that robot B enters **Step 3** in time step $2k_B + 1 > 4k_A$, when robot A is already immobilized. Since during **Step 3** robot B visits all nodes in V_B (that include also all nodes in V_A) rendezvous must occur. □

2.2 Rendezvous in Node Inclusion Model

Recall that in this model we assume that all nodes are ordered and $k_A \leq k_B$, where $V_A \subseteq V_B$. In this model we have no order on edges and in turn the concept of sleeve of graphs cannot be applied here. Instead, one can focus on a different mechanism that will allow to distinguish between two robots and with this in mind we focus on the values of k_A and k_B. Note that if $k_A = k_B$ due to the inclusion assumption we also have $V_A = V_B$. In this case, since orders of nodes in V_A and V_B are consistent the robots can meet at the smallest (in order) node v^* in V_A and V_B that must coincide. Otherwise, the values of k_A and k_B differ and each robot X, for $X \in \{A, B\}$ can adopt k_X as its unique identifier. Furthermore, apart from unique identities there needs to be a synchronization mechanism (sizes of k_A and k_B can be dramatically different) that will allow robots to coordinate their individual moves. The rendezvous mechanism for any robot X is based on synchronized awaiting of the first stage that is long enough to accommodate actions reflecting the size k_X. In particular, we identify the power of two j_X, s.t., $2^{j_X-1} \leq k_X < 2^{j_X}$ that provide a constant estimation and the upper bound on the size of k_X. The algorithm applied to robot X operates in stages $j = 1, 2, 3, ..., j_X$, where during stages 1 through $j_X - 1$ the robot remains immobilized and in the last stage j_X it actively participates (visiting all nodes in V_X) in the rendezvous process. Note that if $j_A < j_B$ (and $V_A \subset V_B$) in stage j_B, when robot A is already immobilized, B by visiting all nodes in V_B (that is a superset of V_A) must conclude rendezvous. In the complementary case, i.e., when the estimates j_A and j_B are the same we use binary expansions $b_A[0, \ldots, j_A]$ and $b_B[0, \ldots, j_B]$ (where positions j_A, j_B are the most significant) of k_A and k_B respectively to differentiate between the robots.

Lemma 2. *If $j_A = j_B$ and $k_A < k_B$ there exists $i \in \{0, 1, \ldots, j_A = j_B\}$, s.t., $b_A[i] = 0$ and $b_B[i] = 1$.*

Proof. If for all $i \in \{0, 1, \ldots, j_A = j_B\}$, $(b_A[i] = 0) => (b_B[i] = 0)$ would imply $k_A \geq k_B$.

A pseudo-code of the rendezvous algorithm **RV2** in the inclusion model follows. If at any time step the two robots A and B meet, the rendezvous is achieved and the two robots *halt*.

---

1. **Algorithm RV2**$(X \in \{A, B\})$
2. **Step 1** Compute j_X and $b_X[0, \ldots, j_B]$.
3. **Step 2 for** $j = 1, 2, \ldots, j_X$ **do**
4.   **if** $(j = j_X)$ {*active stage*}
5.   **use** 2^{j_X} **time steps** to **walk** to and **wait** in v^*. {smallest node}
6.   (i) **for** $i = 0, 1, \ldots, j_X$ **do**
7.   **if** $(b_X[i] = 1)$
8.   (a) **use** $2 \cdot 2^{j_X}$ time steps to visit Euler tour in $T(V_X)$
9.   and **return** to v^*
10.   **else** (b) **wait** $2 \cdot 2^{j_X}$ time steps in v^*
11.   **else** (ii) **wait** $2^j \cdot (2j + 1)$ time steps where you are.

We prove the following theorem.

Theorem 2. *If rendezvous is feasible Algorithm* **RV2** *admits meeting in time* $O((k_A + k_B) \log(k_A + k_B))$.

Proof. The rendezvous algorithm runs in j_X stages controlled by the loop **for** in line 3. There are two cases. In the first case, where $j_A < j_B$, when robot B is in the active stage robot A is already immobilized, and B meets A during traversal of the Euler tour in $T(V_B)$, see line 8 of the code. Otherwise, when $j_A = j_B$ we have two subcases. In the first subcase when $k_A = k_B$ the robots meet in the shared smallest node v^*, see line 5. In the second subcase, where $k_A < k_B$, according to Lemma 2 there exists i, s.t., $b_A[i] = 0$ and $b_B[i] = 1$ when robot B visits the Euler tour in $T(V_B)$ and robot A is immobilized. Thus this subcase admits rendezvous too.

With respect to the time complexity we first observe that the execution time of algorithm **RV2** is bounded and it depends on the parameter j_X. The time complexity of each stage $j = 1, ..., j_X$ is bounded by $3 \cdot 2^j$ resulting in the total complexity $\sum_{j=1}^{j_X}(2^j \cdot (2j + 1)) \le \sum_{j=0}^{j_X}(2^j \cdot (2j_X + 1))$. This is equivalent to $(2j_X + 1)\sum_{j=1}^{j_X}(2^j) = (2j_X + 1) \cdot (2^{j_X+1} - 1) = O(k_X \cdot \log k_X)$, since $2^{j_X} - 1 \le k_X < 2^{j_X}$. This admits the time complexity $O((k_A + k_B) \log(k_A + k_B))$. □

2.3   Blind Rendezvous Model

In this section we consider rendezvous where the relationship between the maps of robots is more arbitrary. We first show that without any additional information, even if $V_A \cap V_B \ne \emptyset$, rendezvous cannot be reached.

Theorem 3. *Blind rendezvous is not feasible.*

Proof. Assume that for any $X \in \{A, B\}$ we have $V_X = \{v_1^{(X)}, v_2^{(X)}\}$ and $E_X = \{(v_1^{(X)}, v_2^{(X)})\}$, where node $v_2^{(A)}$ coincides with $v_1^{(B)}$ and where for each robot X the starting node s_X coincides with v_1^X on its own map. It is enough to observe that without any additional information the symmetry tie cannot be broken. And indeed, since the robots are anonymous (indistinguishable) whenever robot A visits $v_2^{(A)}$ robot B visits $v_2^{(B)}$, i.e., the two robots never visit the shared node simultaneously. □

One can adopt a natural assumption that the nodes in V_X apart from being ordered they also have *explicit labels*. In consequence, if a node $v_a^{(A)} \in V_A$ coincides with $v_b^{(B)} \in V_B$ they both possess the same explicit label. We assume that the labels are drawn from the set of integers $\{1, 2, \ldots, n\}$, and we use notation $b_i^{(X)}$ (or $b_i^{(X)}[0.. \log n]$) to denote the binary expansion of the explicit label of $v_i^{(X)} \in V_X$.

We also assume that n is known to both robots. Otherwise no rendezvous algorithm would stop and report infeasibility of rendezvous when $V_A \cap V_B = \emptyset$, as robots are not aware of sizes of each others maps.

Before we present two rendezvous algorithms we show that the symmetry tie problem, see Theorem 3, can be overcome if the explicit labels are available. W.l.o.g. we also assume that the order of labels is consistent with the order imposed on nodes on each map. If this is not the case a new (consistent) order for nodes in V_A and V_B can be computed on the basis of explicit labels (we only care about nodes in $V_A \cap V_B$). The following result has been shown in [11]. Our proof, however, is much simpler and based on binary representation of explicit labels.

Lemma 3. *Assume that the map of any robot $X \in \{A, B\}$ is an ordered pair of nodes $(v_1^{(X)}, v_2^{(X)})$ connected by a symmetric edge, where nodes $v_2^{(A)}$ and $v_1^{(B)}$ physically coincide and nodes $v_1^{(A)}$ and $v_2^{(B)}$ don't. In such network one can break the symmetry tie to reach rendezvous in time $O(\log \log n)$.*

Proof. We first observe that according to the imposed order $b_1^{(A)} < b_2^{(A)} = b_1^{(B)} < b_2^{(B)}$. The case where $s_A = v_2^{(A)}$ and $s_B = v_1^{(B)}$ is trivial and another case where $s_A = v_1^{(A)}$ and $s_B = v_2^{(B)}$ can be easily resolved by an algorithm that alternates between the two nodes (e.g., in every other time step). Let $1 \leq r_A \leq \log n$ be the largest integer position, s.t., $b_1^{(A)}[r_A] \neq b_2^{(A)}[r_A]$. Since $b_1^{(A)} < b_2^{(A)}$ one can conclude that $b_1^{(A)}[r_A] = 0$ and $b_2^{(A)}[r_A] = 1$. Similarly let $1 \leq r_B \leq \log n$ be the largest integer position, s.t., $b_1^{(B)}[r_B] \neq b_2^{(B)}[r_B]$. Since $b_1^{(B)} < b_2^{(B)}$ one can also conclude that $b_1^{(B)}[r_B] = 0$ and $b_2^{(B)}[r_B] = 1$. We observe that since $b_2^{(A)} = b_1^{(B)}$ one can conclude that $r_A \neq r_B$ as the respective positions cannot contain 0 and 1 at the same time. Moreover binary expansions br_A and br_B of r_A and r_B respectively are limited to $\log \log n + 1$ bits.

We consider a symmetry breaking algorithm in which in time step i each robot $X \in \{A, B\}$ moves to the other node only if $i = 2 \cdot l$ (i is even) or if $i = 2 \cdot l - 1$

(i is odd) and $br_X[l] = 1$, for $l = 1, \ldots, \log \log n + 1$. Note that since $r_A \neq r_B$ for some $1 \leq l \leq \log \log n + 1$ we must have $br_A[l] \neq br_B[l]$ and if until now the rendezvous is not reached (all previous moves were symmetric and in the last odd time step, when the symmetry was broken robots occupy different nodes) in the next even step the rendezvous is accomplished.

Corollary 1. *Note that the lemma above applies to pairs of nodes at distance 1. In a more general case, where the distance between nodes in the pair is $d \geq 1$, the symmetry breaking rendezvous takes time $O(d \log \log n)$.*

In the remaining part of this section we present two rendezvous algorithms followed by their superposition. The first algorithm **RV3** has the time complexity $O((k_A + k_B)^3 \log \log n)$ and its idea is based on the blind rendezvous algorithm from [11] where the problem was studied in complete graphs. The second algorithm **RV4** has the time complexity $O((k_A + k_B)^2 \log n)$ making it superior to **RV3** when $k_A + k_B > \frac{\log n}{\log \log n} = \tau$, where τ is the threshold value. This rendezvous algorithm resembles algorithm **RV2** however here the symmetry tie is broken with the help of explicit labels.

Blind Rendezvous in Time $O((k_A + k_B)^3 \log \log n)$. Similarly to its predecessor **RV2** also the first blind rendezvous algorithms **RV3** operates in stages accommodating geometrically increasing estimates on sizes of the input maps. This is needed as the size of the map of one robot is not known to the other. The robot starts using active stages only when the current estimate is large enough to accommodate its map. The rendezvous process terminates in time $O((k_A + k_B)^3 \log \log n)$ if the maps of both agents are smaller than the threshold value τ. Otherwise, algorithm **RV3** is followed by execution of algorithm **RV4**. If at any time step the two robots A and B meet, the rendezvous is achieved and the two robots *halt*.

```
1. Algorithm RV3(X ∈ {A, B})
2. Step 1 Compute jX and the threshold τ = log n / log log n.
3. Step 2 for j = 1, 2, . . . , ⌈log τ⌉ do
4.          if (j ≥ jX) {active stage}
5.              (i) for all pairs (a, b) ∈ {1, . . . , 2^j} × {1, . . . , 2^j}
6.                  ordered lexicographically do
7.                  if (either of v_a^(X), v_b^(X) exists)
8.                      (a) run blind rendezvous in pair (v_a^(X), v_b^(X)) or
9.                          wait appropriate O(2^j log log n) time steps
10.                         in the only existing node;
11.                     else (b) wait appropriate O(2^j log log n) time steps
12.                         where you currently are;
13.             else (ii) wait suitable O(2^{3j} · log log n) time steps where you are.
```

Theorem 4. *If $k_A + k_B < \tau = \frac{\log n}{\log \log n}$ and rendezvous is feasible, algorithm* **RV3** *admits rendezvous in time $O((k_A + k_B)^3 \log \log n)$.*

Proof. The rendezvous algorithm runs in $\lceil \log \tau \rceil$ stages controlled by the loop **for** in line 3. Robot X starts executing active stages as soon as the stages can accommodate the size of X's map. If the size of the map is too big, robot X awaits execution of the second rendezvous algorithms **RV4**, see line 9. During an active round all pairs (a, b) from the Cartesian product $\{1, \ldots, 2^j\} \times \{1, \ldots, 2^j\}$ are drawn in the lexicographic order. Only certain pairs are valid, i.e., when either of $v_a^{(X)}$ and $v_b^{(X)}$ exists. In each valid pair if only one node exists robot X remains in this node for the duration of the symmetry breaking procedure. Otherwise, if both nodes exist the breaking symmetry procedure is executed with the distance between the two nodes bounded by 2^j.

If rendezvous is feasible we must have nodes $v_a^{(A)} \in V_A$ and $v_b^{(B)} \in V_B$ that coincide by sharing the same label. If the pair $(v_a^{(X)}, v_b^{(X)})$ exists in both maps thanks to the symmetry breaking procedure eventually robot A will visit $v_a^{(A)}$ at the same time when entity B visits $v_b^{(B)}$ and the rendezvous is reached. If only one element of the pair $(v_a^{(X)}, v_b^{(X)})$ exists, i.e., either $v_a^{(A)}$ for A or $v_b^{(B)}$ for B the respective robot is asked to wait in the existing node of the pair resulting in rendezvous too. Otherwise the robots await the next pair from the Cartesian Product without movement for the period corresponding to execution of the symmetry breaking procedure. Thus the actions performed by robots A and B remain fully synchronized.

With respect to the time complexity we first observe that the execution time of algorithm **RV2** is bounded and it depends on the parameter j_X. The time complexity of each stage $j = 1, ..., j_X$ is bounded by $3 \cdot 2^j$ resulting in the total complexity $\sum_{j=1}^{j_X}(2^j \cdot (2j + 1)) \le \sum_{j=0}^{j_X}(2^j \cdot (2j_X + 1))$. This is equivalent to $(2j_X + 1) \sum_{j=1}^{j_X}(2^j) = (2j_X + 1) \cdot (2^{j_X+1} - 1) = O(k_X \cdot \log k_X)$, since $2^{j_X-1} \le k_X < 2^{j_X}$. This admits the time complexity $O((k_A + k_B) \log(k_A + k_B))$. ☐

Blind Rendezvous in Time $O((k_A + k_B)^2 \log n)$. We start with the proof of the following fact.

Lemma 4. *One can impose a periodic order $\pi(X)$ on nodes of a spanning tree $T(V_X)$, s.t., the walking distance (the number of edges to be visited) between two consecutive nodes in order $\pi(X)$ is at most 3.*

Proof. We say that the nodes located at an even distance from the root s_X are on an even level and all the remaining nodes are on an odd level. The ordering of nodes π is created according to the following principle. Starting from the root s_X we visit all nodes in $T(V_X)$ using depth-first search algorithm. The root gets label 0. When we arrive (from the parent) to an even level the currently visited node gets the next available label. In other words at even levels we use *pre-order numbering principle*. And when we arrive (from the last child) to an odd level the currently visited node gets the next available label. I.e., at odd levels we follow *post-order numbering principle*

We need to show that the labeling (ordering) procedure proposed above generates at least one new label in three consecutive steps. And indeed, if we follow the route determined by the depth-first search algorithm and we visit for the first time a node v at an even level (when the new label is generated): (case 1) if the first child of v has a child w then w (which is at distance 2 from v) gets the new label; (case 2) if the first child of v is a leaf this child (which is at distance 1 from v) gets the new label; (case 3) if the node v is a leaf but not the last child of its parent the next label goes to the (next) sibling of v (which is at distance 2); and (case 4) if v is the last child the next label goes to its parent (which is at distance 1).

Similarly, if v is visited for the last time on an odd level it gets a new label. Now (case 5) if v is the last child and its parent w is not the last child the next sibling of the parent (which is at distance 3 from v) gets the new label; (case 6) if v is the last child and its parent w is also the list child then the parent of w (at distance 2 from v) gets the new label; (case 7) and if v is the last child and its parent is the root, the periodic order is established (and the next label is at distance 1). In the remaining cases when v is not the last child (case 8) if its next sibling (at distance 2) is a leaf it gets the new label; and (case 9) if the next sibling of v has children the next label go to the first child (at distance 3 from v) of this sibling. $\qquad\square$

---

1. **Algorithm RV4**($X \in \{A, B\}$)
2. **Step 1** Determine j_X, the threshold $\tau = \frac{\log n}{\log \log n}$, and the label $b_i^{(X)}$ of s_X;
3. **Step 2 for** $j = \lceil \log \tau \rceil, 2, \ldots, \log n$ **do**
4.   **if** ($j \geq j_X$) {*active stage*}
5.   (**walk** to and **wait** in s_X) in 2^j time steps;
6.   **for** $l = 0, 1, \ldots, \log n$ **do** {test all bits}
7.   **if** ($b_i^{(X)}[l] = 1$) {walk all the time}
8.   **for** $2^{2j} \times 3$ time steps **do**
9.   **walk** to the next node in order $\pi(X)$;
10.   **else repeat** 2^j times {walk and wait for another}
11.   (**walk** to the next node in order $\pi(X)$
12.   and **wait** there) in $2^j \times 3$ time steps;
13.   **else wait** appropriate $O(2^{2j} \cdot \log n)$ time steps where you are.

The last rendezvous algorithm **RV4** operates on the following principle. At the start of each active stage robot X returns (if moved before) to the starting point s_X. If the two starting points in V_A and in V_B coincide rendezvous is accomplished. Otherwise the algorithm controls further movement of robots, s.t., during long enough ($\geq 2^j \times 3$ time steps) interval of an active stage j one of the robots, say w.l.o.g. A, visits all nodes in V_A in the periodic order $\pi(A)$ with frequency of one visit per three time steps. While the other robot B visits consecutive nodes with frequency of $2^j \times 3$ time steps. So when eventually robot B resides in the node that belongs to $V_A \cap V_B$ there is enough time for robot A to arrive in this node before B moves away. If at any time step the two robots A and B meet, the rendezvous is achieved and the two robots *halt*.

Theorem 5. *If $k_A + k_B \geq \tau = \frac{\log n}{\log \log n}$ and rendezvous is feasible, algorithm* **RV4** *admits rendezvous in time $O((k_A + k_B)^2 \log n)$.*

Proof. Lets consider the first stage that is active for both robots A and B, i.e., when $j = j_B$. Note that line 13 of the pseudo-code accommodates for the waiting time needed for two robots to stay synchronized prior to this stage. In this active stage loop for in line 6 compares consecutive bits of labels $b_i^{(A)}$ adopted by A and $b_{i'}^{(B)}$ adopted by B. There must be at least one position l on which the two labels differ. In consequence, there is a spell of $2^{2j} \times 3$ time steps during which one of the robots, say w.l.o.g. A with the bit $b_i^{(A)}[l] = 1$, visits periodically all nodes in V_A with frequency of 3 time steps per node. During the same times spell the other robot B with the bit $b_{i'}^{(B)}[l] = 0$ waits long ($\geq 2^j \times 3$ time steps) periods of time in every node of V_B. So when eventually robot B visits the node that belongs to $V_A \cap V_B$ the other robot A has enough time to arrive in this node before B moves on.

The time complexity of this first active stage is $O(2^{2j_B} \cdot \log n) = O(k_B^2 \log n)$. Since the duration of stages grows exponentially we conclude that the total time complexity is also $O(k_B^2 \log n) = O((k_A + k_B)^2 \log n)$. □

Corollary 2. *In the Blind Rendezvous Model two robots can rendezvous in time* $\min\{O((k_A + k_B)^3 \log \log n, O((k_A + k_B)^2 \log n)\}$.

Proof. The result follows directly from the superposition of **RV3** and **RV4**.

3   Conclusion

In this paper we studied deterministic synchronized rendezvous of two robots in the network environment with restrictions imposed on network edges. The restrictions prevent robots from visiting certain parts of the network. We considered three restriction models and we provided four efficient solutions in Section 2. One of the open problems is to establish the exact complexity of rendezvous in considered models and to answer whether the use of randomisation helps. One can also consider models in which maps are not known to the robots. Another interesting question refers to better understanding (including time complexity) of gathering more than two robots. In this setting while robots could meet in pairs, one mutually accessible location for gathering may not be available. It would be also good to understand the case when robots are asked to meet asynchronously. Initial studies indicate that in Edge Monotonic Model there exist rendezvous algorithms that allows robots to meet after adopting trajectories of length polynomial in $k_A + k_B$. In Node Inclusive Model the lengths of respective trajectories become exponential. Finally in Blind Rendezvous Model rendezvous is not feasible even if explicit labels are provided.

References

1. Agathangelou, C., Georgiou, C., Mavronicolas, M.: A distributed algorithm for gathering many fat mobile robots in the plane. In: Proc. PODC 2013, pp. 250–259 (2013)
2. Alpern, S.: The rendezvous search problem. SIAM J. Control and Optimization (33), 673–683 (1995)

3. Alpern, S.: Rendezvous search on labeled networks. Naval Reaserch Logistics (49), 256–274 (2002)

4. Alpern, S., Fokkink, R.: Gąsieniec, L., Lindelauf, R., Subrahmanian, V.S.: Search Theory, A Game Theoretic Perspective. Springer (2013)

5. Alpern, S., Gal, S.: The Theory of Search Games and Rendezvous. Kluwer Academic Publishers (2002)

6. Anderson, E., Fekete, S.: Asymmetric rendezvous on the plane. In: Proc. Symp. on Computational Geometry, pp. 365–373 (1998)

7. Anderson, E., Fekete, S.: Two-dimensional rendezvous search. Operations Research (49), 107–118 (2001)

8. Anderson, E., Weber, R.: The rendezvous problem on discrete locations. Journal of Applied Probability (28), 839–851 (1990)

9. Baston, V., Gal, S.: Rendezvous on the line when the players' initial distance is given by an unknown probability distribution. SIAM J. Control and Optimization (36), 1880–1889 (1998)

10. Baston, V., Gal, S.: Rendezvous search when marks are left at the starting points. Naval Reaserch Logistics (48), 722–731 (2001)

11. Chen, S., Russell, A., Samanta, A., Sundaram, R.: Deterministic Blind Rendezvous in Cognitive Radio Networks. In: Proc. ICDCS (2014)

12. Collins, A., Czyzowicz, J., Gąsieniec, L., Labourel, A.: Tell Me Where I Am So I Can Meet You Sooner. In: Abramsky, S., Gavoille, C., Kirchner, C., Meyer auf der Heide, F., Spirakis, P.G. (eds.) ICALP 2010. LNCS, vol. 6199, pp. 502–514. Springer, Heidelberg (2010)

13. Collins, A., Czyzowicz, J., Gąsieniec, L., Kosowski, A., Martin, R.: Synchronous Rendezvous for Location-Aware Agents. In: Peleg, D. (ed.) DISC 2011. LNCS, vol. 6950, pp. 447–459. Springer, Heidelberg (2011)

14. Czyzowicz, J., Gąsieniec, L., Pelc, A.: Gathering few fat mobile robots in the plane. Theoretical Computer Science 410(6-7), 481–499 (2009)

15. Czyzowicz, J., Kosowski, A., Pelc, A.: How to meet when you forget: Log-space rendezvous in arbitrary graphs. Distributed Computing (25), 165–178 (2012)

16. Czyzowicz, J., Labourel, A., Pelc, A.: How to meet asynchronously (almost) everywhere. ACM Transactions on Algorithms (8), article 37 (2012)

17. Dereniowski, D., Klasing, R., Kosowski, A., Kuszner, Ł.: Rendezvous of heterogeneous mobile agents in edge-weighted networks. In: Halldórsson, M.M. (ed.) SIROCCO 2014. LNCS, vol. 8576, pp. 311–326. Springer, Heidelberg (2014)

18. Kowalski, D., Malinowski, A.: How to meet in anonymous network. Theoretical Computer Science (399), 141–156 (2008)

19. An, H.-C., Krizanc, D., Rajsbaum, S.: Mobile Agent Rendezvous: A Survey. In: Flocchini, P., Gąsieniec, L. (eds.) SIROCCO 2006. LNCS, vol. 4056, pp. 1–9. Springer, Heidelberg (2006)

20. Lin, Z., Liu, H., Chu, X., Leung, Y.-W.: Jump-stay based channel-hopping algorithm with guaranteed rendezvous for cognitive radio networks. In: Proc. INFOCOM, pp. 2444–2452 (2011)

21. Miller, A., Pelc, A.: Time Versus Cost Tradeoffs for Deterministic Rendezvous in Networks. In: Proc. PODC, pp. 282–290 (2014)

22. Pelc, A.: Deterministic rendezvous in networks: A comprehensive survey. Networks (59), 331–347 (2012)

23. Schelling, T.: The Strategy of Conflict. Harvard Univ. Press, Cambridge (1960)

Fastest, Average and Quantile Schedule

Armin Fügenschuh[1], Konstanty Junosza-Szaniawski[2], Torsten Klug[3],
Sławomir Kwasiborski[2], and Thomas Schlechte[3]

[1] Helmut Schmidt University / University of the Federal Armed Forces Hamburg,
Holstenhofweg 85, 22043 Hamburg, Germany
http://am.hsu-hh.de
[2] Politechnika Warszawska, Matematyki
ul. Koszykowa 75, 00-662 Warszawa
http://mini.pw.edu.pl
[3] Department of Optimization, Zuse Institute Berlin,
Takustraße 7, 14195 Berlin, Germany
http://www.zib.de/

Abstract. We consider problems concerning the scheduling of a set of trains on a single track. For every pair of trains there is a minimum headway, which every train must wait before it enters the track after another train. The speed of each train is also given. Hence for every schedule - a sequence of trains - we may compute the time that is at least needed for all trains to travel along the track in the given order. We give the solution to three problems: the fastest schedule, the average schedule, and the problem of quantile schedules. The last problem is a question about the smallest upper bound on the time of a given fraction of all possible schedules. We show how these problems are related to the travelling salesman problem. We prove NP-completeness of the fastest schedule problem, NP-hardness of quantile of schedules problem, and polynomiality of the average schedule problem. We also describe some algorithms for all three problems. In the solution of the quantile problem we give an algorithm, based on a reverse search method, generating with polynomial delay all Eulerian multigraphs with the given degree sequence and a bound on the number of such multigraphs. A better bound is left as an open question.

Keywords: schedule, generating permutations with repetitions, Eulerian multigraphs.

1 Introduction

In the theory of combinatorial algorithms typically the following problems are considered: find any feasible solution, find one feasible solution, find an optimal solution, enumerate all solutions (with minimum weight), count all solutions (with a given weight). We ask a natural follow-up question: what is a quantile of the given fraction of feasible solutions, i.e., what is the minimum number a such that the weights of all feasible solutions of a given fraction are not exceeding

G.F. Italiano et al. (Eds.): SOFSEM 2015, LNCS 8939, pp. 201–216, 2015.
© Springer-Verlag Berlin Heidelberg 2015

a. For example for the travelling salesman problem the question about the 0.8-quantile is: what is the smallest number *a* such that 80% of all travelling salesman tours for a set of given cities have a weight of at most *a*.

Problems considered in this paper originate from railway track allocation in a real world application [6].

Let us consider the strategic routing of freight trains in a highly utilized network. Given a railway network that is utilized by a set of passenger trains and a model day that is partitioned into a few time slices. Each time slice represents a special traffic situation of the day and comprised several hours, for instances the morning and afternoon peak of a working day or the night with a rather small amount of passenger traffic. We classify the trains into different train types, which describe the characteristic properties of the trains,e.g., running times, headway times or special requirements for the track. The preset passenger traffic for each slice is simply described by the number of trains of a specific train type. In particular there is no information of the actual schedule of the trains. We only know that this trains used the track within the time slice. On the demand side the freight trains are defined by an origin destination pair, the departure time and its train type. The task is to find a route for each freight train that does not exceed a given distance and running time limit and minimize the expected delays. Since we are only interested in a strategic routing with a rough approximated timing, the minimal expected delays should ensure the existence of a feasible timetable or at least increase the possibility for one.

At this point the topic of the paper pops up. We need for each track an estimation of the expected delays. This could be done by estimating the possible schedules for each slice, track and train set. In our case we have a fixed set of passenger train and a huge amount of possible freight train sets. Let us denote by a configuration a vector that contains a number of trains for each freight train type. Now we are looking for an assignment of configurations to tracks and slices so that a feasible routing for the requested demand exists. Since we can assume that the cost of a schedule increases with the number of trains, it is possible to assign a configuration with much more trains in it than in the end are routed over the specific track. We could interpret an assigned configuration as capacity. Therefore a solving procedure can start with a subset of the possible configurations and can generate on demand new smaller configurations with better cost and tighter capacity to improve the overall solution. In particular, it is usefull within a column generation approach to solve a mixed integer program formulation of the problem. Finally we come to the following problem that had to be solved as sub-problem several times.

We consider a single railway track (from station A to B) and a set of trains, with their speeds and minimal headway between every pair of trains. For any given sequence of trains we can compute the least time that is needed for all trains of the sequence to arrive at B.

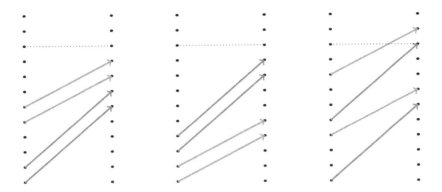

Fig. 1. Example for schedules of two slow and two fast trains and their times

Example 1. *Let us consider the following trivial example. We are given two train types with running time 3 and 5 and the headways are given in the matrix* $\begin{pmatrix} 1 & 1 \\ 3 & 1 \end{pmatrix}$. *Figure 1 shows three potential orderings of 4 trains (2 of each type).*

A natural question is which sequence gives the minimal time. Under some natural conditions with respect to the speeds of the trains and the minimal headways we prove that a sequence of trains ordered with non-increasing speed is fastest. However in the general case the problem of the fastest schedule is NP-complete. For the general case we give an algorithm for finding the fastest schedule, based on dynamic programming. Moreover we give an explicit formula for the average time taken over all possible schedules. This problem is equivalent to the problem of determining the average weight of all Hamilton cycles.

The last question and most interesting from both a practical and theoretical point of view is the question about the quantile of the schedule, i.e., what is the minimum number a such that the weights of all schedules of a given fraction are not exceeding a. For example, if a 0.8-quantile is equal to a, then 80% of all schedules can be realized in time not exceeding a. To solve the problem we take advantage of the fact that the speeds of trains and the minimal headways depend only on the type of trains. In addition, there is only a small number of types compared to the number of trains. The simplest way is to generate all sequences of trains or sequences of types of trains, and to compute the quantile directly. We solve the problem in a more sophisticated way. First we define an equivalence relation on the set of schedules such that any two schedules in the relation have the same time for finishing. Then we generate all equivalence classes and compute the time for finishing for every class. The equivalence classes of our relation directly correspond to Eulerian multigraphs with a given degree sequence. To generate the multigraphs we use the reverse search method introduced by Avis and Fukuda [2]. We prove that there are $O(n^{k^2-1})$ Eulerian multigraphs on k vertices with n edges. This bound is not tight. Any better bound on this number would give a better complexity bound of our algorithm since we generate Eulerian multigraphs with polynomial delay.

2   Preliminaries and Problem Formulation

For $n \in \mathbb{N}$ we denote $\{1 \ldots n\}$ by $[n]$ and by $[n]^*$ the set of all finite sequences of elements of the set $[n]$. Let $X = \{x_1, x_2, \ldots, x_n\}$ be the set of trains. Let the function $t \colon X \to [k]$ assign every train its type. We assume that the set X and the function t are fixed in the following. Let $l \colon [k] \to \mathbb{N}$ be a function defined by $l(i) = |t^{-1}(i)|$. Thus $l_i = l(i)$ denotes the number of trains of the i-th type. Let $r \colon [k] \to \mathbb{R}_+$ assign the running time to every train type and let $m \colon [k] \times [k] \to \mathbb{R}_+$ be the function determining the minimal headway between the trains of certain types.

Let $S(X)$ denote a set of all permutations of the set X. We will call the permutations of X schedules. The minimal running time of the schedule $y \in S(X)$ is computed as

$$RT(y) = \sum_{i=1}^{n-1} m(t(y_i), t(y_{i+1})) + r(t(y_n)).$$

We define the following three problems:

Problem 1. *Fastest-schedule*
   Input: (X, t, r, m, k).
   Output: YES if and only if there exists a schedule $s \in S(X)$ such that $RT(s) \le k$.

Problem 2. *Average-schedule*
   Input: (X, t, r, m).
   Output: $\bar{\tau} \in \mathbb{R}$ - average running time of a schedule, i.e.,

$$\bar{\tau} = \frac{\sum_{y \in S(X)} RT(y)}{|S(X)|}.$$

Problem 3. *α-quantile schedule*
   Input: (X, t, r, m, α^1).
   Output: τ - time needed to realize $\alpha|S(X)|$ schedules, i.e.,

$$\tau = \min\{rt : \frac{|\{s \in S(X) : RT(s) \le rt\}|}{|S(X)|} \ge \alpha\}.$$

---

[1] with α - a given fraction of the schedules that have to be realizable (1 means that all schedules must be realizable, 0.5 means that half of the potential schedules must be realizable)

3   Fastest Schedule Problem

Theorem 2. *Fastest-schedule is NP-complete regarding to number of train types.*

We can solve the fastest schedule problem by interpreting it as a slight modification of the TSP. The modification will state that a city from the TSP setting must be visited a given number of times. It is allowed to visit the same city multiple times in a row but there is a non zero "distance" assigned to such operation. As cities we will denote members of set $[k] \cup 0$. Distances are given by following function:

$$d(x, y) = \begin{cases} m(x, y), & x \neq 0 \wedge y \neq 0, \\ r(x), & x \neq 0 \wedge y = 0, \\ 0, & x = 0 \wedge y \neq 0. \end{cases}$$

The number of times that the city denoted by $x \in [k]$ must be visited is given by the function $l(x)$ as defined in the preliminaries. The city with number 0 must be visited exactly once. Notice that every tour in this graph corresponds to a number of schedules (the tour determines the order of train types so actual trains of the same type can be permuted). The tour length is equal to the running time of the corresponding schedules.

Such modification of the TSP can be solved using a modification of the classic dynamic programming algorithm given by Bellman [3]. Let $\mu \colon \mathbb{N} \times [k] \to \mathbb{N}$ be a function. By $\mu(t, (x_1 \ldots x_k))$ we will denote the minimal length of tour from 0 to t passing through each city $i \in [k]$ exactly x_i times (starting and ending visits are not counted). We can define μ recursively as:

$$\mu(t, (x_1, \ldots, x_k)) = \min_{i \in [k] \wedge x_i \neq 0} \{\mu(i, (x_1, \ldots x_i - 1, \ldots, x_k)) + w(i, t)\}.$$

The iterative procedure can be initiated for all i by:

$$\mu(i, (0, \ldots, 0)) = w(0, i),$$

from which we can obtain the respective next values using the recursive formula. Value of $\mu(0, (l(1), \ldots, l(k)))$ gives a solution to the Fastest-schedule problem. The computational complexity of this algorithm can be calculated by estimating the number of different parameter sets of function μ that must be calculated and the time for computing a single value. The first parameter can be picked in k ways and the second parameter is the number of solutions of the inequality: $x_1 + \ldots + x_k \leq n$. The number of solutions of this inequality can be estimated by n^k. Each value of the function μ can be computed in linear time. From above we conclude that the algorithm runs in time $O(n^{k+1})$.

The most natural candidate for an optimal solution is a schedule with trains that are ordered non-decreasing by their running time. This simple solution seems to work in real world scenario, but it can be shown that it is not correct in the general case. A question rises what conditions have to be fulfilled for this simple solution to be correct.

Theorem 3. *Let* (X, t, r, m) *be an instance of the fastest-schedule problem. If we assume that:*

$$\forall_{x_1, x_2, x_3, x_4 \in X} \; r(t(x_1)) \leq r(t(x_2)) \wedge r(t(x_1)) \leq r(t(x_3)) \wedge r(t(x_1)) \leq r(t(x_4))$$
$$\Rightarrow m(t(x_1), t(x_2)) + m(t(x_3), t(x_4)) \leq m(t(x_3), t(x_1))$$
$$+ m(t(x_1), t(x_4))$$

$$(1)$$

and

$$\forall_{x_1, x_2, x_3, x_4, x_5 \in X} \; r(t(x_1)) \leq r(t(x_2)) \leq r(t(x_3)) \wedge r(t(x_2)) \leq r(t(x_4)) \wedge r(t(x_2))$$
$$\leq r(t(x_5))$$
$$\Rightarrow m(t(x_1), t(x_2)) + m(t(x_2), t(x_3)) + m(t(x_4), t(x_5))$$
$$\leq m(t(x_1), t(x_3)) + m(t(x_4), t(x_2)) + m(t(x_2), t(x_5)) \quad (2)$$

then the schedule consisting of trains ordered not-decreasing by running time, is the solution to fastest schedule problem.

4 Average Schedule Problem

Besides the question for an optimal solution, let it be minimum or maximum, finding the running time of an *average* schedule could be of interest. This problem can be solved in polynomial time. First we reduce the problem of the average schedule to the problem of the average Hamilton cycle length in a complete graph. Let $X = \{1, \ldots, n\}$ be a set of trains. For each pair of trains $i, j \in V$ with $i \neq j$, $m(t(i), t(j))$ determines minimal headway between i and j. Let $V = X \cup \{0\}$ be the vertex set of the graph, $A = [n] \times [n] \setminus \{(i, i) : i \in [n]\}$ be the arc set and let weight function be given by:

$$w(i, j) = \begin{cases} m(t(i), t(j)), & i \neq 0 \wedge j \neq 0 \\ r(t(i)), & i \neq 0 \wedge j = 0 \\ 0, & i = 0 \wedge j \neq 0. \end{cases}$$

Observe that the tour in this graph corresponds to exactly one schedule and that the tour length is equal to this schedule running time. Hence the average time of all schedules is equal to the average Hamilton cycle length.

Theorem 4. *Let* $G = (V, E, w)$ *be a weighted undirected complete graph. Then the average tour length for a Hamiltonian cycle in* G *is* $\frac{2}{n-1} \sum_{e \in E} w_e$.

Proof. Since G is a complete graph, each edge is contained in exactly $(n-2)!$ Hamiltonian cycles. There are in total $\frac{n!}{2n}$ Hamiltonian cycles in a complete graph. Therefore the average weight of all Hamiltonian cycles is

$$\frac{2n}{n!} \cdot \sum_{e \in E} (n-2)! \cdot w_e = \frac{2}{n-1} \sum_{e \in E} w_e. \quad (3)$$

$\square$

Corollary 5. *Let* $D = (V, A, w)$ *be a weighted directed complete graph. Then the average tour length for a Hamiltonian cycle in* D *is* $\frac{1}{n-1}\sum_{a\in A} w_a$.

Proof. There are in total $\frac{n!}{n}$ Hamiltonian cycles in a complete directed graph. □

Corollary 6. *Let* $G = (V, E, w)$ *be a weighted undirected complete graph. Then the average length of a Hamiltonian path in* G *is* $\frac{2}{n}\sum_{e\in E} w_e$.

Corollary 7. *Let* $D = (V, A, w)$ *be a weighted directed complete graph. Then the average length of a Hamiltonian path in* D *is* $\frac{1}{n}\sum_{(i,j)\in A} w(i,j)$.

Proof. Let $V' := V \cup \{0\}$ as above and $A' := A \cup \{(0,i) : i \in V\} \cup \{(i,0) : i \in V\}$. Let $w'_a := w_a$ for $(i,j) \in A$, $w'_a := 0$ for $a \in A' \backslash A$. Again, there is a bijection between Hamiltonian cycles in $D' := (V', A', w')$ and Hamiltonian paths in D, from which the formula follows. □

We can conclude that average running time of a schedule for given (X, t, r, m) equals to $\frac{1}{n}\sum_{(i,j)\in A} w(i,j)$, where A and $w(i,j)$ are defined as above.

5   Schedules Quantiles

The problem "quantile schedule" is at least as hard as the fastest schedule. For $\alpha = \frac{1}{n!}$, where n is the number of trains, a solution of α-quantile is also a solution of the problem Fastest-schedule.

We need some more notations. For $b, k \in \mathbb{N}$, $s = (s_1, \dots, s_b) \in [k]^*, p, q \in [k]$, by $\Delta(s, p, q)$ we will denote $\{i \in [b-1] : p = t(s_i), q = t(s_{i+1})\}$ which is a set of all the indices on which a train type changes from p to q in the sequence s. By $\delta(s, p, q)$ we will denote $|\Delta(s, p, q)|$.

Let $\sim \subset S(X) \times S(X)$ be a relation defined on permutations of the set of the trains.

We say that:

$$y \sim z :\Leftrightarrow \forall_{p,q\in[k]} \; \delta(y, p, q) = \delta(z, p, q).$$

Lemma 8. *For any* $y, z \in S(X)$ *if* $y \sim z$ *then* $t(y_n) = t(z_n)$ *and* $t(y_1) = t(z_1)$.

Proof. Notice that the sum $\sum_{q\in[k]} \delta(y, t(y_n), q)$ is equal to the number of occurrences of a train of the type $t(y_n)$ in the schedule y on positions from 1 to $n-1$. Hence

$$\sum_{q\in[k]} \delta(y, t(y_n), q) = l_{t(y_n)} - 1.$$

By the definition of the relation $\sim$ we get:

$$\sum_{q\in[k]} \delta(z, t(y_n), q) = \sum_{q\in[k]} \delta(y, t(y_n), q) = l_{t(y_n)} - 1.$$

So the trains of the type $t(y_n)$ occur $l_{t(y_n)} - 1$ times on positions from 1 to $n-1$ in the sequence z, but trains of the type $t(y_n)$ occur $l_{t(y_n)}$ times (on positions from 1 to n) in the sequence z, hence $t(z_n) = t(y_n)$. The proof of $t(y_1) = t(z_1)$ is analogue. □

Theorem 9. *For any $y, z \in S(X)$ if $y \sim z$, then $RT(y) = RT(z)$.*

Proof. We can observe that

$$RT(y) = \left(\sum_{i=1}^{n-1} m(t(y_i), t(y_{i+1})) \right) + r(t(y_n))$$

$$= \left(\sum_{p,q \in [k]} m(p,q)\delta(y,p,q) \right) + r(t(y_n))$$

From Lemma 8 and $y \sim z$, $t(y_n) = t(z_n)$, we obtain $r(t(y_n)) = r(t(z_n))$. From the definition of $\sim$ we have that

$$\forall_{p,q \in [k]}\ \delta(y,p,q) = \delta(z,p,q)$$

from above:

$$RT(y) = \left(\sum_{p,q \in [k]} m(p,q)\delta(y,p,q) \right) + r(t(y_n))$$

$$= \left(\sum_{p,q \in [k]} m(p,q)\delta(z,p,q) \right) + r(t(z_n)) = RT(z).$$

□

Theorem 10. *There exist functions m, r such that if $y, z \in S(X)$ and $y \nsim z$, then $RT(y) \neq RT(z)$.*

Let us denote the equivalence classes of the relation $\sim$ by $[s]_\sim$. By $\delta([s]_\sim, p, q)$ we will denote the value of $\delta(y, p, q)$ for any $y \in [s]_\sim$. This notation is well-defined since from the definition of relation $\sim$ for any $y \in [s]_\sim$ it holds that $\forall p, q \in [k]\ \delta(s, p, q) = \delta(y, p, q)$.

By a block of the trains of the type i we denote a sequence of consecutive trains of type i such that a train directly before and after the block are of any type not equal to i. Given $s \in S(X)$, by $b_s(i)$ we denote number of blocks of the trains of the type i.

We define a function $R \colon S(X) \to [k]^*$ as follows: for $s \in S(X)$, $R(s)$ is a sequence obtained from s by replacing every block of trains of type i by single appearance of i. Notice that $R(s)$ is a sequence of length $\sum_{i \in [k]} b_s(i)$. Moreover notice that:

$$\delta(R(s), p, q) = \begin{cases} \delta(s, p, q), & p \neq q, \\ 0, & p = q. \end{cases}$$

It is easy to observe that if $R(y) = R(z)$ then $y \sim z$. Let $R([s]_\sim) = \{R(y) : y \in [s]_\sim\}$ and $R^{-1}(R(s)) = \{y \in S(X) : R(y) = R(s)\}$.

Lemma 11. *For any $s \in S(X)$*

$$|R^{-1}(R(s))| = \prod_{i=1}^{k} l(i)! \binom{l(i) - 1}{b_s(i) - 1}.$$

From above lemma directly follows:

Corollary 12. *For any $s \in S(X)$*

$$|[s]_\sim| = |R([s]_\sim)| \cdot \prod_{i=1}^{k} l(i)! \binom{l(i) - 1}{b_s(i) - 1}.$$

Hence to count the number of schedules in $[s]_\sim$, it is enough to count the number of sequences in $R([s]_\sim)$.

Let $G_{[s]_\sim} = (V_{[s]_\sim}, \mu_{[s]_\sim})$ be a directed multigraph where $\mu_{[s]_\sim} : V_{[s]_\sim}^2 \to \mathbb{N}$ is function assigning to vertices p, q the number of arcs from p to q. The multigraph is constructed as follows: $V_{[s]_\sim} = [k] \cup \{0\}$, for all $p, q \in [k]$, $\mu_{[s]_\sim}(p, q) = \delta([s]_\sim, p, q)$, moreover $\mu_{[s]_\sim}(t(s_n), 0) = 1$ and $\mu_{[s]_\sim}(0, t(s_1)) = 1$. By Eulerian cycle in $G_{[s]_\sim}$ we mean a vertex sequence in $G_{[s]_\sim}$ containing every pair $(p, q) \in V_{[s]_\sim}^2$ as consecutive pair pq exactly $\mu_{[s]_\sim}(p, q)$ times. By $\widehat{G}_{[s]_\sim} = (V_{[s]_\sim}, \widehat{\mu}_{[s]_\sim})$ we denote the multigraph obtained from $G_{[s]_\sim}$ by deleting all loops (for each vertex $v \in V_{[s]_\sim}$ $\widehat{\mu}_{[s]_\sim}(v, v) = 0$).

Let $G = (V, \mu)$ be a multigraph, by $deg_G^-(i) = \sum_{v \in V} \mu(v, i)$ we denote the indegree of vertex i in graph G, and by $deg_G^+(i) = \sum_{v \in V} \mu(i, v)$ we denote out degree of vertex i in graph G. It can be noted that for all $i \in [k]$ it holds that $deg_G^-(i) = deg_G^+(i)$. Moreover it can by shown that $deg_G^-(i) = b_i$ for $i \in [k]$. The following observation is the key to our algorithm.

Remark 13. *Every sequence $r \in R([s]_\sim)$ corresponds to one Euler vertex sequence in $\widehat{G}_{[s]_\sim}$.*

Remark 14. *Graph $\widehat{G}_{[s]_\sim}$ is connected for any $s \in S(X)$.*

For a multigraph $G = (V, \mu)$ we define the Kirchhoff matrix $K(G)$ as follows :

$$K(G)_{ij} = \begin{cases} deg_G^-(i), & \text{if } i = j, \\ -\mu_G(i, j), & \text{if } i \neq j. \end{cases}$$

For $i \in [n]$ we denote by K_i the matrix obtained from K by deleting the i-th row and the i-th column. By $\det(K)$ we denote determinant of matrix K. By $ec(G)$ we denote number of Euler cycles in G.

Theorem 15. (de Bruijn, van Aardenne-Ehrenfest, Smith, Tutte [1]) *Given a multigraph $G = (V, \mu)$ then the number of Eulerian cycles $ec(G)$ is given by:*

$$ec(G) = \frac{t_1(G) \prod_{v \in V} \left(deg_G^-(v) - 1 \right)!}{\prod_{i,j \in [k], i \neq j} (\mu(i,j))!},$$

where $t_v(G)$ denotes number of trees rooted at vertex v.

Theorem 16. (Tutte Matrix Tree Theorem [5]) *Given a multigraph $G = (V, \mu)$ with Kirchhoff matrix $K(G)$, then the number of trees rooted at vertex v is equal to $\det(K_v(G))$.*

Theorem 17. *For any $s \in S(X)$ it holds that*

$$|[s]_\sim| = \frac{\det \left(K_1(\widehat{G}_{[s]_\sim}) \right) \prod_{i=1}^{k} \left(deq_{\widehat{G}_{[s]_\sim}}^-(i) - 1 \right)!}{\prod_{i,j \in [k], i \neq j} (\delta(s,i,j))!} \cdot \prod_{i=1}^{k} l(i)! \binom{l(i) - 1}{deg_{\widehat{G}_{[s]_\sim}}^-(i) - 1}.$$

Proof. Follows directly from the Lemma 11 and the Theorems 15 and 16.   □

6   Algorithm

Instead of enumerating all equivalence classes of relation $\sim$, we can enumerate all connected Eulerian multigraphs with given vertex degree sequence. To generate only connected graphs with desired properties, we use the reverse search method described in the next part of the paper. Every graph identifies one $\sim$ equivalence class therefore corresponding schedules have equal running times. The algorithm generates all graphs. Then sorts them by running time of corresponding schedules. Then it finds a first equivalence class such that number of schedules in this class and in proceeding classes is at least α fraction of all schedules and returns its running time. The following algorithm solves the running time problem:

Algorithm 1. RunningTime(X, t, r, m, α)

1: Generate all graphs for X into S using reverse search.
2: Order the schedules in S by running time in ascending order
3: $allSchedulesNumber = \sum_{s \in S} |[s]_\sim|$
4: $\tau = 0$
5: $currentNumberOfSchedules = 0$
6: **while** $currentNumberOfSchedules < \alpha \cdot allSchedulesNumber$ **do**
7:    $s = S.Pop$
8:    $currentNumberOfSchedules+ = |[s]_\sim|$
9:    $\tau = RT(s)$
10: **end while**
11: **return** τ

Functions RT and l are defined in terms of t, r, m as in former part of the paper.

Theorem 18. *The Running Time algorithm returns a valid result.*

Proof. The validity of result follows from Theorem 17.   □

Theorem 19. *There are at most $O(n^{k^2-1})$ connected multigraphs with given vertex degree sequence.*

From Theorem 19 we know that there are at most $O(n^{k^2-1})$ connected multigraphs with given vertex degree sequence. All operations conducted on single graphs take polynomial time Therefore the complexity of the whole algorithm is at most $O(n^{k^2-1})$.

7   Algorithm for Generating Connected Graphs

To generate connected graphs efficiently, we can use a method of Avis and Fukuda called *Reverse Search* [2]. The main idea of this technique is to define a graph on the set of objects to generate and perform a search on its spanning tree generating one object by visiting each vertex. To be precise: a triple $(\Gamma, \widehat{S}, f)$, where $\Gamma = (\mathcal{V}, \mathcal{E})$, $\widehat{S} \in \mathcal{V}$, f is a mapping $\mathcal{V} \setminus \{\widehat{S}\} \to \mathcal{V}$, is called *local search* if $\{v, f(v)\} \in \mathcal{E}$ for each $v \in \mathcal{V} \setminus \{\widehat{S}\}$.

Local search $(\Gamma, \widehat{S}, f)$ is called *finite local search* if for each $v \in \mathcal{V} \setminus \{\widehat{S}\}$ there exists a positive integer i such that $f^i(v) = \widehat{S}$.

Let (Γ, S, f) be a finite local search with trace T. As "abstract reverse search" we call a routine of traversing T and outputting all its vertices. The traversal can be implemented in any way. In this paper we will conduct the traversal by breadth first search starting from the sink and traversing all edges in a way opposite to their direction.

By $N_\mu(v) = \{u \in V : \mu(v, u) > 0\}$ we denote the neighbourhood of vertex v. By $C(V, \mu)$ we denote the number of connected components of graph $G = (V, \mu)$. For $\mu : V \times V \to \mathbb{N}$ such that $\mu(u, v) > 0$ we define $\mu - (u, v) = \mu'$ by

$$\mu'(p, q) = \begin{cases} \mu(u, v) - 1, & \text{for } (p, q) = (u, v), \\ \mu(p, q), & \text{for } (p, q) \neq (u, v). \end{cases}$$

For a graph $G = (V, \mu)$ a traversal from u to v we call a "bridge traversal" if and only if $C(V, \mu) < C(V, \mu - (u, v))$. By non-bridge neighbours of v we denote the set $NN_\mu(v) = \{u \in N_\mu(v) : (v, u) \text{ is not a bridge traversal}\}$.

For a $G = (V, \mu) \in \mathcal{G}_l$ by $\overrightarrow{G}$ we will denote a minimal Euler cycle for graph G - a cycle generated by the following algorithm:

Assuming that we use Tarjan's [7] algorithm for finding bridges then the time complexity of above algorithm is $O(|E|^2)$ where $|E| = \sum_{v \in V} deg_G^+(v)$ is the number of edges in the graph G. The algorithm is a realization of Fleury's algorithm [4] for finding Euler cycle. The correctness of the algorithm follows directly from the correctness of Fleury's algorithm.

Let $w, x, y, z \in V$ such that $\mu(w, x), \mu(x, y), \mu(y, z) > 0$. By $t(G, (w, x, y, z)) = (V, \mu_t)$ we will denote a multigraph obtained from G by following modification

Algorithm 2. MinimalEulerCycle(V, μ)

1: $\mu_F = \mu$, $v = 0$, $u = 0$
2: $mec = $ "empty sequence" {minimal Euler cycle}
3: **repeat**
4:    $mec \mathrel{+}= u$ {append to the end of the sequence}
5:    **if** $NN_{\mu_F}(v) \neq \emptyset$ **then** $u = \min\left(NN_{\mu_F}(v)\right)$
6:    **else** $u = \min\left(N_{\mu_F}(v)\right)$ **end if**
7:    $\mu_F = \mu_F - (v, u)$
8:    $v = u,$
9: **until** $v = 0$
10: **return** mec

of μ:

$$\mu_t(p, q) = \begin{cases} \mu(p, q) - 1, & \text{for } (p, q) \in \{(w, x), (x, y), (y, z)\}, \\ \mu(p, q) + 1, & \text{for } (p, q) \in \{(w, y), (y, x), (y, z)\}, \\ \mu(p, q), & \text{otherwise.} \end{cases}$$

It can be noted that the transformation t preserves the vertex degree sequence and the connectivity of the graph. It should be noted that we did not assume that vertexes w, x, y, z are not equal, so it is possible that two, three, or all are equal. It can also be noted that for every $G \in \mathcal{G}_l$ and w, x, y, z there exist w', x', y', z', such that if $G' = t(G, (w, x-, y, z))$, then $t(G', (w', x', y', z')) = G$.

Let $\hat{s} \in \{0, \ldots, k\}^n$ and $\hat{s} = (s_0, \ldots, s_n)$ be a sequence where $s_0 \leq \ldots \leq s_n$. By $(s)_i$ we denote the i-th element of s, by $(s)_{\leq i} = (s_1, \ldots, s_i)$ we denote the sequence containing the first i elements of s. By $(s)_{\geq i} = (s_i, \ldots, s_n)$ we denote the sequence containing elements of s starting from the i-th element.
Let $P(G) = \max\{i \in \{0, \ldots, n\} : (\overrightarrow{G})_{\leq i} = (\hat{s})_{\leq i}\}$.
Let $Pv(G) = \min\{(\overrightarrow{G})_{P(G)+1}, \ldots, (\overrightarrow{G})_n\}$.
Let $PP(G) = \min\{i > P(G) : (\overrightarrow{G})_i = Pv(G)\}$.

Lemma 20. Let $G = (V, \mu) \in \mathcal{G}_l$. It holds that: $PP(G) > P(G) + 1$.

Corollary 21. Let $G = (V, \mu) \in \mathcal{G}_l$. It holds that: $PP(G) \geq 2$.

Let $f : \mathcal{G}_l \setminus \{G_{[\hat{s}]\sim}\} \to \mathcal{G}_l$ be a function declared as follows: let $h = ((\overrightarrow{G})_{PP(G)-2}, (\overrightarrow{G})_{PP(G)-1}, (\overrightarrow{G})_{PP(G)}, (\overrightarrow{G})_{PP(G)+1})$ then $f(G)$ equals to $t(G, h)$. Notice that the function f is well defined because of Corollary 21.

Remark 22. It can be noted that function f preserves the first $PP(G) - 2$ elements of $\overrightarrow{G}$, i.e., $\overrightarrow{G}_{\leq PP(G)-2} = \overrightarrow{f(G)}_{\leq PP(G)-2}$.

By $f^{-1}(G) = \{H \in \mathcal{G}_l : f(H) = G\}$ we will denote the inverse function of f.

Lemma 23. Let $G \in \mathcal{G}_l$ then $P(G) \leq P(f(G))$.

Lemma 24. Let $G = (V, \mu) \in \mathcal{G}_l$. If $P(G) = P(f(G))$ then $PP(f(G)) < PP(G)$.

Theorem 25. *For each $G \in \mathcal{G}_l$ there exists $i \in \mathbb{N}$ such that $f^i(G) = G_{[\hat{s}]_\sim}$.*

Let $\Gamma = (\mathcal{G}_l, \mathcal{E})$ be a graph. $\{G_1, G_2\} \in \mathcal{E}$ if and only if G_1 can be obtained from G_2 by applying the transformation t to G_1 or vice-versa. Let $\hat{S} = G_{[\hat{s}]_\sim}$, it is clear that for every $G \in \mathcal{G}_l \setminus \{\hat{S}\}$ it holds that $\{G, f(G)\} \in \mathcal{E}$.
From above and from Theorem 25 it follows that $(\Gamma, \hat{S}, f)$ is a finite local search and a reverse search method can be applied to generate all graphs in $\mathcal{G}_l$.

Lemma 26. *The time complexity of $f(G)$ is $O(n^2)$.*

Lemma 27. *The time complexity of $f^{-1}(G)$ is $O(n^6)$.*

Theorem 28. *Traversing Γ by the reverse search method outputs elements with maximal headway of $O(n^6)$.*

Theorem 29. *The computation complexity of the RunningTime algorithm is at most $O(n^{k^2-1})$.*

References

1. Van Aardenne-Ehrenfest, T., de Bruijn, N.G.: Circuits and trees in oriented linear graphs. Stevin, S.: Wis-en Natuurkundig Tijdschrift 28, 203 (1951)
2. Avis, D., Fukuda, K.: Reverse search for enumeration. Discrete Applied Mathematics 65(1), 21–46 (1996)
3. Bellman, R.: Dynamic programming treatment of the travelling salesman problem. Journal of the ACM (JACM) 9(1), 61–63 (1962)
4. Lucas, É.: Récréations mathématiques, vol. 1. Gauthier-Villars (1882)
5. Kirchhoff, G.R.: Über die Auflösung der Gleichungen, auf welche man bei der Untersuchung der linearen Verteilung galvanischer Ströme geführt wird. Ann. Phys. Chem. 72, 497–508 (1847)
6. Schlechte, T., Borndörfer, R., Erol, B., Graffagnino, T., Swarat, E.: Micro–macro transformation of railway networks. Journal of Rail Transport Planning & Management 1(1), 38–48 (2011)
7. Tarjan, R.E.: A note on finding the bridges of a graph. Information Processing Letters 2(6), 160–161 (1974)

Appendix

Proof (Theorem 2). First we will show that Fastest-schedule is NP-hard. Let (C, d, B) be a travelling salesman problem (TSP) instance where C is the set of cities, $d : C^2 \to \mathbb{N}$ is the distance function and B is the maximal searched tour length. For $c \in C$ by $i(c)$ we will denote index of c in an arbitrarily chosen ordering of C, and by c_i we will denote i-th city in the ordering. We define Fastest-schedule instance as follows: $X = C$ for all $c \in X, t(c) = i(c), m(i, j) = d(c_i, c_j), r = 0, k = B$. Every schedule of a given instance contains every train from X exactly once. Therefore it contains each train of every type exactly once. From above every schedule corresponds to a travelling salesman tour (a sequence of types of trains induce travelling salesman tour (TS-tour)). The running time of a schedule of this instance is equal to the length of the corresponding travelling salesman tour. Therefore the fastest schedule corresponds to the shortest TS-tour, so the answer to a given TSP problem is YES, if and only if the answer to the constructed Fastest-schedule problem is YES.

Moreover we will show that the Fastest-schedule problem is in NP. Given an instance of the Fastest-schedule problem (X, t, r, m, k) and a certificate C which is a sequence of trains, we can determine the running time of C by applying the running time formula in polynomial time. The answer to a given problem is YES, if and only if the running time of given certificate is less or equal to k. From the above it follows that the Fastest-schedule problem is NP-complete. □

Proof (Theorem 3). Without loss of generality we can assume that $t(i) \leq t(j) \Leftrightarrow r(t(i)) \leq r(t(j))$ - we relabel train types so trains of types with lower indices are faster. Let us assume that s is the solution to the Fastest-schedule problem and that s does not consist of trains ordered by train type. By s_i we will denote the i-th element of s. By s' we will denote the schedule consisting of trains ordered by train types thus ordered not-decreasing by running time. Let i be the fastest train such that $t(s_i) \neq t(s'_i)$. Let j be the index of the last appearance of a train of type $t(s_i) - 1$. If such train does not exist, i.e., s_i is the fastest train, let $j = 0$. Let us move train s_i to position $j + 1$ in s. The conditions from the theorem guarantee that such operation will not increase running time of the schedule - the condition (1) guarantees that moving the fastest train to the beginning will not increase the time, and the condition (2) guarantees that moving trains to positions grater than 1 will not increase running time. After applying this operation repeatedly, we will obtain s', with a running time that is not larger than the running time of s. □

Proof (Theorem 6). Let 0 be a vertex not in V. Let $V' := V \cup \{0\}, E' := E \cup \{\{0, i\} : i \in V\}$, and $w'_e := w_e$ for $\{i, j\} \in E$ and $w'_e := 0$ for $e \in E' \backslash E$. Then each Hamiltonian cycle in $G' := (V', E', w')$ corresponds to exactly one Hamiltonian path in G and vice versa. Therefore the average weight of all Hamiltonian paths in G equals the average weight of all Hamiltonian cycles in G', which is □

$$\frac{2}{(n + 1) - 1} \sum_{e \in E'} w_e = \frac{2}{n} \sum_{e \in E} w_e. \tag{4}$$

Proof (Theorem 10). Let us define r such that $\forall_{i \in S(X)}\ r(i) = 0$. Let H be the sequence of all pairs of train types. Let $m(H_i)$ denote the value of m for the i-th pair from the sequence H. Let $H_i = (a, b)$. By $l(H_i)$ we will denote $\max\{l(t(a)), l(t(b))\}$. We define m as follows: $m(H_1) = 1$ and $m(H_i) = 2^{\lceil \log_2 \sum_{j=1}^{i-1} l(H_j) \rceil + 1}$. The function m is constructed in such way that when we analyse RT expressed in a binary number system, we can observe that particular places correspond to a specific pair from H. For a given i values of RT at indices $2^{\lceil \log_2 \sum_{j=1}^{i-1} l(H_i) \rceil + 1} - 2^{\lceil \log_2 \sum_{j=1}^{i} l(H_j) \rceil + 1}$ depend only on the value of pair H_i. So if $y, z \in S(X)$ and $y \not\sim z$, there exist $p, q \in [k]$ such that $\delta(y, p, q) \neq \delta(z, p, q)$. Let us assume that pair (p, q) has index i in H. Binary notations of $RT(y)$ and $RT(z)$ differ on one of the following indices: $2^{\lceil \log_2 \sum_{j=1}^{i-1} l(H_i) \rceil + 1} - 2^{\lceil \log_2 \sum_{j=1}^{i} l(H_j) \rceil + 1}$. □

Proof (Lemma 11). In order to calculate the total number of schedules in $R^{-1}(R(s))$ we have to arrange the trains of every type into certain number of blocks. Type i has to be arranged into $b_s(i)$ blocks. This can be done in $l(i)!\binom{l(i)-1}{b_s(i)-1}$ ways - first we choose one of $l(i)!$ permutations and then divide it into the desired number of blocks. To obtain the final number of the schedules we have to take the product over all train types. □

Proof (Theorem 19). Given a number of trains n, a number of types k and a vertex degree sequence l, then every multigraph G with vertex sequence l can be represented as Kirchhoff matrix $K(G)$. All values in matrix $K(G)$ sum up to n, i.e., $\sum_{i,j \in [k]} K(G)_{i,j} = n$. From the former we know that the number of unique Kirchhoff matrices is at most the number of solutions of the following equation: $x_1 + \ldots + x_{k^2} = n$ (every variable in the equation corresponds to one value in $K(G)$). This equation has at most $\binom{n+k^2-1}{k^2-1} = O(n^{k^2-1})$ solutions. □

Proof (Lemma 20). From the definition of $PP(G)$ we know that $PP(G) > P(G)$, so let us suppose on the contrary that $PP(G) = P(G) + 1$. This means that $(\overrightarrow{G})_{P(G)+1} = \min\{(\overrightarrow{G})_{P(G)+1}, \ldots, (\overrightarrow{G})_n\} = \hat{s}_{P(G)+1}$, which contradicts the definition of $P(G)$. □

Proof (Lemma 23). Since from Lemma 20 we know that $PP(G) \geq P(G) + 2$, it follows from Remark 22 that $\overrightarrow{G}_{\leq P(G)} = \overrightarrow{f(G)}_{\leq P(G)}$ and thus $P(G) \leq P(f(G))$. □

Proof (Lemma 24). From Remark 22 we know that the first $PP(G) - 2$ steps of the *MinimalEulerCycle* algorithm will be in $f(G)$ the same as in G. Let $h = (w, x, y, z)$ be the sequence selected in the definition of f. After $PP(G) - 2$ steps of the *MinimalEulerCycle* algorithm in $f(G)$ we are visiting vertex w. By μ_F^G we will denote function μ_F maintained by algorithm applied to G while visiting w and by $\mu_F^{f(G)}$ when applied to $f(G)$. There are two cases:
(1) the traversal from w to x is not a bridge traversal in (V, μ_F^G), then the traversal from w to y is not a bridge in $(V, \mu_F^{f(G)})$.

(2) the traversal from w to x is a bridge traversal in (V, μ_F^G), which means that $NN_{\mu_F^G}(w) = \emptyset$. From above it can be shown that $NN_{\mu_F^{f(G)}}(w) = \emptyset$.

By definition of f, y is vertex with smallest value in $(\overrightarrow{G})_{\geq P(G)+1}$. From the definition of Fleury's algorithm we also know that $NN_{\mu_F^G}(w)$ is a subset of the set of vertices occurring in $(\overrightarrow{G})_{\geq P(G)+1}$. From above and the case analysis in the former part of the proof we know that the algorithm will traverse from w to y. From the definition of f we know that $(\overrightarrow{G})_{PP(G)} = y$. From the fact that algorithm traverses from w to y instead of x we know that $(\overrightarrow{f(G)})_{PP(G)-1} = y$. From the definition of f, it follows that $Pv(G) = y$. Because $P(G) = P(f(G))$, we know that also $Pv(G) = Pv(f(G))$, so $Pv(f(G)) = y$. From above it imminently follows that $PP(f(G)) = PP(G) - 1 < PP(G)$. □

Proof (Thorem 25). From Lemma 23 we know that $P(G) \leq P(f(G))$. Let us assume that $P(G) = P(f(G))$, then from Lemma 24 we know that $PP(f(G)) < PP(G)$. Because $PP(G) > P(G)$, there exists a finite j such that $P(G) < P(f^j(G))$. From the former observation it is clear that there exists a finite i such that $P(f^i(G)) = n$. If $P(f^i(G)) = n$, then $f^i(G) = G_{[\hat{s}]_\sim}$. □

Proof (Lemma 26). The evaluation of function $f(G)$ requires to compute $\overrightarrow{G}$ and functions P, Pv, PP. Former can be done in $O(n^2)$ and latter in $O(n)$, which gives final complexity of $O(n^2)$. □

Proof (Lemma 27). To compute $f^{-1}(G)$, we can enumerate the entire graph H such that there exist $w, x, y, z \in V$ such that $t(H, (w, x, y, z)) = G$. There are $O(n^4)$ sequences $w, x, y, z \in V$ thus we have to enumerate at most $O(n^4)$. For each graph H we have to check if $f(H) = G$ which can be done in a time proportional to $O(n^2)$. From the above we get that the time complexity of f^{-1} is $O(n^6)$. □

Proof (Theorem 28). The *trace* of local search $(\Gamma, \widehat{S}, f)$ is a directed sub-graph $T = (V, \mathcal{E}(f))$ where $\mathcal{E}(f) = \{(v, f(v)) : v \in V \setminus \{\widehat{S}\}\}$. T is simply the directed spanning tree of Γ, rooted in $\widehat{S}$, defined by f.

When performing a traversal of *trace* of Γ between outputting consecutive elements, we have to calculate f^{-1}. Aside from computing f^{-1}, the reverse search routine has to push computed vertices to a queue which can be done in $O(n)$. From the above it follows that dominating operation during reverse search is computing f^{-1} which from Lemma 27 can be done in $O(n^6)$. □

Machine Characterizations for Parameterized Complexity Classes Beyond Para-NP

Ronald de Haan* and Stefan Szeider*

Institute of Information Systems, Vienna University of Technology, Vienna, Austria

Abstract. Due to the remarkable power of modern SAT solvers, one can efficiently solve NP-complete problems in many practical settings by encoding them into SAT. However, many important problems in various areas of computer science lie beyond NP, and thus we cannot hope for polynomial-time encodings into SAT. Recent research proposed the use of fixed-parameter tractable (fpt) reductions to provide efficient SAT encodings for these harder problems. The parameterized complexity classes $\exists^k\forall^*$ and $\forall^k\exists^*$ provide strong theoretical evidence that certain parameterized problems are not fpt-reducible to SAT. Originally, these complexity classes were defined via weighted satisfiability problems for quantified Boolean formulas, extending the general idea for the canonical problems for the Weft Hierarchy.

In this paper, we provide alternative characterizations of $\exists^k\forall^*$ and $\forall^k\exists^*$ in terms of first-order logic model checking problems and problems involving alternating Turing machines with appropriate time bounds and bounds on the number of alternations. We also identify parameterized Halting Problems for alternating Turing machines that are complete for these classes.

The alternative characterizations provide evidence for the robustness of the new complexity classes and extend the toolbox for establishing membership results. As an illustration, we consider various parameterizations of the 3-coloring extension problem.

1 Introduction

The recent success of modern SAT solvers in many practical settings has placed them at the heart of an important approach to solving NP-complete problems, where problem instances are encoded to SAT and subsequently solved using a SAT solver [3,12,17,22]. However, many important computational problems lie above the first level of the Polynomial Hierarchy (PH), and thus this approach does not work to solve these problems, as polynomial-time reductions to SAT are not possible for these problems, unless the PH collapses.

Problem instances occurring in practical settings are not random, and often contain some kind of structure, which can be exploited by parameterized algorithms. Recently, the structure in problems instances was used to break the complexity barriers between the first and second level of the PH, by means of fpt-reductions [9,21]. Such fpt-reducibility results adopt a new perspective on what amounts to positive results in

---

* Supported by the European Research Council (ERC), project 239962 (COMPLEX REASON), and the Austrian Science Fund (FWF), project P26200 (Parameterized Compilation).

G.F. Italiano et al. (Eds.): SOFSEM 2015, LNCS 8939, pp. 217–229, 2015.

parameterized complexity. This new perspective (i.e., aiming at fpt-reducibility to SAT rather than fpt-solvability) greatly extends the power of positive results, as parameters can be less restrictive, and problems can be solved efficiently on larger classes of instances.

In order to provide suitable negative results, a new parameterized complexity classes $\exists^k\forall^*$ has been introduced [13,14], which lies at the basis of a hardness theory that provides such negative evidence. The class $\exists^k\forall^*$ is located above para-co-NP and below para-Σ_2^P (see Figure 1), and is based on weighted variants of quantified Boolean satisfiability problems. Several problems from various domains have already been shown hard or complete for the class $\exists^k\forall^*$ or its dual $\forall^k\exists^*$, including problems in Knowledge Representation [14], Boolean Optimization [13], and Computational Social Choice [8]. The role that $\exists^k\forall^*$ and $\forall^k\exists^*$ play in the analysis of parameterizations of problems complete for the second level of the PH, is analogous to the role that the Weft-hierarchy plays in the analysis of parameterizations of NP-complete problems. The parameterized complexity classes para-NP and para-co-NP, on the one hand, and the classes $\exists^k\forall^*$ and $\forall^k\exists^*$, on the other hand, constitute a borderline between problems that are fpt-reducible to SAT (or UNSAT) and problems that are not, similarly to the way in which the classes W[1] and FPT provide a borderline between problems that are fixed-parameter tractable and problems that are not. Neither W[1] nor $\exists^k\forall^*$ and $\forall^k\exists^*$ have a direct counterpart in classical complexity theory, and these classes thus provide a tighter complexity analysis than parameterized complexity classes that are derived from classical classes [10].

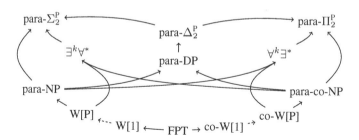

Fig. 1. Parameterized complexity classes up to the second level of the polynomial hierarchy. Arrows indicate inclusion relations. (For a definition of the classes para-DP and para-Δ_2^P, we refer to other resources [2,10,20])

1.1 New Contributions

We provide new characterizations of the parameterized complexity class $\exists^k\forall^*$ in terms of first-order model checking problems and in terms of alternating Turing machines, with appropriate time bounds. Consequently, dual characterizations hold for the parameterized complexity class $\forall^k\exists^*$. More specifically, we show the following results.

1. $\exists^k \forall^*$ is precisely the class of all parameterized problems that can be expressed in terms of checking whether a given first-order formula $\exists x_1, \ldots, x_k . \forall y_1, \ldots, y_n . \psi$ is true in a given relational structure, taking the number of existential variables as the parameter. (Theorem 1)

2. $\exists^k \forall^*$ is precisely the class of all parameterized problems that can be decided by a 2-alternating Turing machine that runs in fixed-parameter tractable time, starts in an existential state, and uses a number of nondeterministic existential steps that is bounded by a function of the parameter. (Theorem 2)

3. The Halting Problem for 2-alternating Turing machines that start in an existential state, parameterized by the number of nondeterministic existential steps, is complete for $\exists^k \forall^*$. (Theorem 3)

Theorem 1 provides an easy and convenient way for establishing membership results; we use it also in the proofs of Theorems 2 and 3 and give an example application to a combinatorial problem in Section 5.

Theorem 2 establishes the robustness of the class $\exists^k \forall^*$, in analogy to the characterization of the first two classes of the Weft-hierarchy in terms of Turing machines [4,5].

Theorem 3 provides an analogue to the Cook-Levin Theorem for the complexity class $\exists^k \forall^*$, which supports our assumption that $\exists^k \forall^* \neq$ para-co-NP, in analogy to the argumentation that the W[1]-completeness of the Halting Problem for nondeterministic Turing machines, parameterized by the number of steps, supports the assumption W[1] $\neq$ FPT (see [6] and cf. the discussion in [5]). Interestingly, our version of the Halting Problem remains $\exists^k \forall^*$-complete, independently of whether the Turing machine uses a single tape, or an arbitrary number of tapes, in contrast to versions of the Halting Problem that characterize classes of the Weft-hierarchy, where a single tape captures W[1], and an arbitrary number of tapes captures W[2] [4,5].

We would like to remark that the membership in W[1] or W[2] for some parameterized problems remained open for a long time, and was finally established by means of machine characterizations [4]. We expect that our machine characterizations for $\exists^k \forall^*$ can be of similar use.

In Section 5 we exemplify our new complexity toolbox by applying it to parameterizations of a graph coloring problem, shown to be Π_2^P-complete by Ajtai, Fagin, and Stockmeyer [1].

We provide proof sketches for the results presented in this paper. For full detailed proofs we refer to a technical report [15].

2 Preliminaries

Propositional and First-Order Logic. A *literal* is a propositional variable x or a negated variable $\neg x$. We use the standard notion of *(truth) assignments* $\alpha : \mathrm{Var}(\varphi) \to \{0,1\}$ for Boolean formulas and *truth* of a formula under such an assignment.

A *(relational) vocabulary* τ is a finite set of relation symbols. Each relation symbol R has an *arity* arity$(R) \geq 1$. A *structure* $\mathcal{A}$ of vocabulary τ, or *τ-structure* (or simply *structure*), consists of a set A called the *domain* and an interpretation $R^{\mathcal{A}} \subseteq A^{\mathrm{arity}(R)}$ for each relation symbol $R \in \tau$. We use the usual definition of truth of a first-order logic sentence φ over the vocubulary τ in a τ-structure $\mathcal{A}$. We let $\mathcal{A} \models \varphi$ denote that the sentence φ is true in structure $\mathcal{A}$. If φ is a first-order formula with free variables $\text{Free}(\varphi)$, and $\mu : \text{Free}(\varphi) \to A$ is an assignment, we use the notation $\mathcal{A}, \mu \models \varphi$ to denote that φ is true in structure $\mathcal{A}$ under the assignment μ.

The Polynomial Hierarchy. There are many natural decision problems that are not contained in the classical complexity classes P or NP. The *Polynomial Hierarchy* [18,20,23,24] contains a hierarchy of increasing complexity classes Σ_i^P, for all $i \geq 0$. We give a characterization of these classes based on the satisfiability problem of various classes of quantified Boolean formulas. A *quantified Boolean formula* is a formula of the form $Q_1 X_1 Q_2 X_2 \dots Q_m X_m \psi$, where each Q_i is either $\forall$ or $\exists$, the X_i are disjoint sets of propositional variables, and ψ is a Boolean formula over the variables in $\bigcup_{i=1}^m X_i$. The quantifier-free part of such formulas is called the *matrix* of the formula. Truth of such formulas is defined in the usual way. Let $\gamma = \{x_1 \mapsto d_1, \dots, x_n \mapsto d_n\}$ be a function that maps some variables of a formula φ to truth values. We let $\varphi[\gamma]$ denote the application of such a substitution γ to the formula φ. For each $i \geq 1$, we let QSAT_i be the problem to decide whether a given quantified Boolean formula $\varphi = \exists X_1 \forall X_2 \exists X_3 \dots Q_i X_i \psi$ is true, where Q_i is a universal quantifier if i is even and an existential quantifier if i is odd.

Input formulas to the problem QSAT_i are called Σ_i^P-formulas. For each nonnegative integer $i \leq 0$, the complexity class Σ_i^P can be characterized as the closure of the problem QSAT_i under polynomial-time reductions [23,24]. The Σ_i^P-hardness of QSAT_i holds already when the matrix of the input formula is restricted to 3CNF for odd i, and restricted to 3DNF for even i. The class Σ_0^P coincides with P, and the class Σ_1^P coincides with NP. For each $i \geq 1$, the class Π_i^P is defined as co-Σ_i^P.

Parameterized Complexity. We briefly introduce some core notions from parameterized complexity theory. For an in-depth treatment we refer to other sources [6,7,11,19]. A *parameterized problem* L is a subset of $\Sigma^* \times \mathbb{N}$ for some finite alphabet Σ. For an instance $(I, k) \in \Sigma^* \times \mathbb{N}$, we call I the *main part* and k the *parameter*. A parameterized problem L is *fixed-parameter tractable* if there exists a computable function f and a constant c such that there exists an algorithm that decides whether $(I, k) \in L$ in time $O(f(k)\|I\|^c)$, where $\|I\|$ denotes the size of I. Let $L \subseteq \Sigma^* \times \mathbb{N}$ and $L' \subseteq (\Sigma')^* \times \mathbb{N}$ be two parameterized problems. An *fpt-reduction* from L to L' is a mapping $R : \Sigma^* \times \mathbb{N} \to (\Sigma')^* \times \mathbb{N}$ from instances of L to instances of L' such that there exist some computable function $g : \mathbb{N} \to \mathbb{N}$ such that for all $(I, k) \in \Sigma^* \times \mathbb{N}$: (i) (I, k) is a yes-instance of L if and only if $(I', k') = R(I, k)$ is a yes-instance of L', (ii) $k' \leq g(k)$, and (iii) R is computable in fpt-time. We write $L \leq_{\text{fpt}} L'$ if there is an fpt-reduction from L to L'. Similarly, we call reductions that satisfy properties (i) and (ii) but that are computable in time $O(\|I\|^{f(k)})$, for some fixed computable function f, *xp-reductions*.

Parameterized complexity theory also offers complexity classes for problems that lie higher in the polynomial hierarchy. Let C be a classical complexity class, e.g., NP. The parameterized complexity class para-C is then defined as the class of all parameterized problems $L \subseteq \Sigma^* \times \mathbb{N}$, for some finite alphabet Σ, for which there exist an alphabet Π, a computable function $f : \mathbb{N} \to \Pi^*$, and a problem $P \subseteq \Sigma^* \times \Pi^*$ such that $P \in C$ and for all instances $(x, k) \in \Sigma^* \times \mathbb{N}$ of L we have that $(x, k) \in L$ if and only if $(x, f(k)) \in P$. Intuitively, the class para-C consists of all problems that are in C after a precomputation that only involves the parameter [10]. In particular, the class para-NP consists of all parameterized problems that can be fpt-reduced to the trivial parameterized variant of the propositional satisfiability problem, i.e., the problem SAT where the parameter value is a fixed constant for all instances.

The basic complexity classes $\exists^k \forall^*$ and $\forall^k \exists^*$ are defined in terms of the following weighted variant of QSAT$_2$ [13,14].

$\exists^k \forall^*$-WSAT
Instance: A quantified Boolean formula $\varphi = \exists X. \forall Y. \psi$ and an integer k.
Parameter: k.
Question: Does there exist a truth assignment α to X with weight k such that $\forall Y. \psi[\alpha]$ is true?

The class $\exists^k \forall^*$ is defined to be the closure of $\exists^k \forall^*$-WSAT under fpt-reductions. Moreover, its dual class $\forall^k \exists^*$ is defined by $\forall^k \exists^* = \text{co-}\exists^k \forall^*$.

We will also consider the variant $\exists^{\leq k} \forall^*$-WSAT of $\exists^k \forall^*$-WSAT, where the problem is to decide whether there exists a truth assignment α to X with weight at most k such that $\forall Y. \psi[\alpha]$ is true. This problem is also $\exists^k \forall^*$-complete. A proof of this can be found in the technical report [15].

Alternating Turing machines. We use the same notation as Flum and Grohe [11, Appendix A.1]. Let $m \geq 1$ be a positive integer. An *alternating Turing machine (ATM)* with m tapes is a 6-tuple $\mathbb{M} = (S_\exists, S_\forall, \Sigma, \Delta, s_0, F)$, where: $S_\exists$ and $S_\forall$ are disjoint sets; $S = S_\exists \cup S_\forall$ is the finite set of *states*; Σ is the alphabet; $s_0 \in S$ is the *initial state*; $F \subseteq S$ is the set of *accepting states*; and $\Delta \subseteq S \times (\Sigma \cup \{\$, \square\})^m \times S \times (\Sigma \cup \{\$\})^m \times \{\mathbf{L}, \mathbf{R}, \mathbf{S}\}^m$ is the *transition relation*. The elements of Δ are the *transitions*). The symbols $\$, \square \notin \Sigma$ are special symbols. "$\$$" marks the left end of any tape. It cannot be overwritten and only allows $\mathbf{R}$-transitions.[1] "$\square$" is the *blank symbol*. Intuitively, the tapes of our machine are bounded to the left and unbounded to the right. The leftmost cell, the 0-th cell, of each tape carries a "$\$$", and initially, all other tape cells carry the blank symbol. The input is written on the first tape, starting with the first cell, the cell immediately to the right of the "$\$$". A *configuration* is a tuple $C = (s, x_1, p_1, \ldots, x_m, p_m)$, where $s \in S$, $x_i \in \Sigma^*$, and $0 \leq p_i \leq |x_i| + 1$ for each $1 \leq i \leq k$. Intuitively, $\$x_i \square \square \ldots$ is the sequence of symbols in the cells of tape i, and the head of tape i scans the p_i-th cell. The *initial configuration* for an input $x \in \Sigma^*$ is $C_0(x) = (s_0, x, 1, \epsilon, 1, \ldots, \epsilon, 1)$, where ϵ denotes the empty word. A *computation step* of $\mathbb{M}$ is a pair (C, C') of configurations such that the transformation from C to C' obeys the transition relation. We omit the formal details. We write $C \to C'$ to denote that (C, C') is a computation step of $\mathbb{M}$. If $C \to C'$, we call C' a *successor configuration* of C. A *halting configuration* is a configuration that has no successor configuration. A halting configuration is *accepting* if its state is in F. A step $C \to C'$

---

[1] To formally achieve that "$\$$" marks the left end of the tapes, whenever $(s, (a_1, \ldots, a_m), s', (a'_1, \ldots, a'_m), (d_1, \ldots, d_m)) \in \Delta$, then for all $1 \leq i \leq m$ we have that $a_i = \$$ if and only if $a'_i = \$$ and that $a_i = \$$ implies $d_i = \mathbf{R}$.

is *nondeterministic* if there is a configuration $C'' \neq C'$ such that $C \to C''$, and is *existential* if C is an existential configuration. A state $s \in S$ is called *deterministic* if for any $a_1, \ldots, a_m \in \Sigma \cup \{\$, \square\}$, there is at most one $(s, (a_1, \ldots, a_m), s', (a_1', \ldots, a_m'), (d_1, \ldots, d_m)) \in \Delta$. Similarly, we call a non-halting configuration *deterministic* if its state is deterministic, and *nondeterministic* otherwise. A configuration is called *existential* if it is not a halting configuration and its state is in $S_\exists$, and *universal* if it is not a halting configuration and its state is in $S_\forall$. Intuitively, in an existential configuration, there must be at least one possible run that leads to acceptance, whereas in a universal configuration, all possible runs must lead to acceptance. Formally, a *run* of an ATM $\mathbb{M}$ is a directed tree where each node is labeled with a configuration of $\mathbb{M}$ such that: (1) The root is labeled with an initial configuration. (2) If a vertex is labeled with an existential configuration C, then the vertex has precisely one child that is labeled with a successor configuration of C. (3) If a vertex is labeled with a universal configuration C, then for every successor configuration C' of C the vertex has a child that is labeled with C'. We often identify nodes of the tree with the configurations with which they are labeled. The run is *finite* if the tree is finite, and *infinite* otherwise. The *length* of the run is the height of the tree. The run is *accepting* if it is finite and every leaf is labeled with an accepting configuration. If the root of a run ρ is labeled with $C_0(x)$, then ρ is a run *with input x*. Any path from the root of a run ρ to a leaf is called a *computation path*. The *language (or problem) accepted by* $\mathbb{M}$ is the set $Q_\mathbb{M}$ of all $x \in \Sigma^*$ such that there is an accepting run of $\mathbb{M}$ with initial configuration $C_0(x)$. $\mathbb{M}$ *runs in time* $t : \mathbb{N} \to \mathbb{N}$ if for every $x \in \Sigma^*$ the length of every run of $\mathbb{M}$ with input x is at most $t(|x|)$. A *step* $C \to C'$ is an *alternation* if either C is existential and C' is universal, or vice versa. A run ρ of $\mathbb{M}$ is *ℓ-alternating*, for an $\ell \in \mathbb{N}$, if on every path in the tree associated with ρ, there are less than ℓ alternations between existential and universal configurations. The machine $\mathbb{M}$ is *ℓ-alternating* if every run of $\mathbb{M}$ is ℓ-alternating.

3   A First-Order Model Checking Characterization

In this section, we characterize the class $\exists^k \forall^*$ in terms of first-order model checking. Consider the following parameterized problem.

$\exists^k \forall^*$-MC
Instance: A first-order logic sentence $\varphi = \exists x_1, \ldots, x_k. \forall y_1, \ldots, y_n. \psi$ over a vocabulary τ, where ψ is quantifier-free, and a finite τ-structure $\mathcal{A}$.
Parameter: The number k of existentially quantified variables of φ.
Question: Does $\mathcal{A} \models \varphi$?

We show that this problem is complete for the class $\exists^k \forall^*$. This result does not imply that $\exists^k \forall^* \subseteq A[2]$ (cf. [11]), because the parameter of the problem $\exists^k \forall^*$-MC is only the number of existential variables, not the size of the entire first-order formula.

Theorem 1. $\exists^k \forall^*$-MC *is* $\exists^k \forall^*$-*complete.*

Proof. We show $\exists^k \forall^*$-membership by giving an fpt-reduction to $\exists^k \forall^*$-WSAT. Let $(\varphi, \mathcal{A})$ be an instance of $\exists^k \forall^*$-MC, where $\varphi = \exists x_1, \ldots, x_k. \forall y_1, \ldots, y_n. \psi$ is a first-order logic sentence over vocabulary τ, and $\mathcal{A}$ is a τ-structure with domain A. We may assume without loss of generality that ψ contains only connectives $\wedge$ and $\neg$.

We construct an instance (φ', k) of $\exists^k \forall^*$-WSAT, where φ is of the form $\exists X'. \forall Y'. \psi'$. We define $X' = \{ x'_{i,a} : 1 \leq i \leq k, a \in A \}$, and $Y' = \{ y'_{j,a} : 1 \leq j \leq n, a \in A \}$. Intuitively, the variable $x'_{i,a}$ denotes whether the variable x_i is assigned to value a, and similarly, the variable $y'_{j,a}$ denotes whether y_j is assigned to value a. In order to define ψ', we will use the auxiliary function μ on subformulas of ψ, defined by letting $\mu(\chi_1 \wedge \chi_2) = \mu(\chi_1) \wedge \mu(\chi_2)$, $\mu(\neg \chi_1) = \neg \mu(\chi_1)$, and $\mu(\chi) = \bigvee_{1 \leq u \leq u}(\psi_{z_1, a_1^i} \wedge \cdots \wedge \psi_{z_m, a_m^i})$ if $\chi = R(z_1, \ldots, z_m)$ and $R^{\mathcal{A}} = \{(a_1^1, \ldots, a_m^1), \ldots, (a_1^u, \ldots, a_m^u)\}$, where for each $z \in X \cup Y$ and each $a \in A$ we let $\psi_{z,a} = x'_{i,a}$ if $z = x_i$, and we let $\psi_{z,a} = y'_{j,a}$ if $z = y_j$. Now, we define ψ' by letting $\psi' = \psi'_{\text{unique-}X'} \wedge (\psi'_{\text{unique-}Y'} \to \mu(\psi))$, where $\psi'_{\text{unique-}X'} = \bigwedge_{1 \leq i \leq k}(\bigvee_{a \in A} x'_{i,a} \wedge \bigwedge_{a, a' \in A, a \neq a'}(\neg x'_{i,a} \vee \neg x'_{i,a'}))$, and $\psi'_{\text{unique-}Y'} = \bigwedge_{1 \leq j \leq n}(\bigvee_{a \in A} y'_{j,a} \wedge \bigwedge_{a, a' \in A, a \neq a'}(\neg y'_{j,a} \vee \neg y'_{j,a'}))$. Intuitively, the formula $\psi'_{\text{unique-}X'}$ represents whether the variables $x'_{i,a}$ encode a unique assignment for each variable x_i. Similarly, the formula $\psi'_{\text{unique-}Y'}$ represents whether the variables $y'_{i,a}$ encode a unique assignment for each variable y_i. We claim that $(\mathcal{A}, \varphi) \in \exists^k \forall^*$-MC if and only if $(\varphi', k) \in \exists^k \forall^*$-WSAT.

Hardness can be shown by means of an fpt-reduction from $\exists^k \forall^*$-WSAT. A detailed proof of both membership and hardness can be found in the technical report [15]. □

4 Alternating Turing Machine Characterizations

Next, we characterize $\exists^k \forall^*$ in terms of ATMs. In particular, we consider parameterized problems related to the halting problem for a particular class of ATMs, and show that these problems are $\exists^k \forall^*$-complete. Moreover, we show that $\exists^k \forall^*$ is exactly the class of parameterized decision problems that can be decided by a certain class of ATMs.

We consider the following restrictions on ATMs. An $\exists \forall$-*Turing machine* (or simply $\exists \forall$-machine) is a 2-alternating ATM $(S_\exists, S_\forall, \Sigma, \Delta, s_0, F)$, where $s_0 \in S_\exists$. Let $\ell, t \geq 1$ be positive integers. We $\mathfrak{s}$ay that an $\exists \forall$-machine $\mathbb{M}$ *halts (on the empty string) with existential cost ℓ and universal cost t* if: (1) there is an accepting run of $\mathbb{M}$ with input ϵ, and (2) each computation path of $\mathbb{M}$ contains at most ℓ existential configurations and at most t universal configurations.

Let P be a parameterized problem. An $\exists^k \forall^*$-*machine for P* is a $\exists \forall$-machine $\mathbb{M}$ such that there exists a computable function f and a polynomial p such that: (1) $\mathbb{M}$ decides P in time $f(k) \cdot p(|x|)$; and (2) for all instances (x, k) of P and each computation path R of $\mathbb{M}$ with input (x, k), at most $f(k) \cdot \log |x|$ of the existential configurations of R are nondeterministic. We say that a parameterized problem P *is decided by some $\exists^k \forall^*$-machine* if there exists a $\exists^k \forall^*$-machine for P.

Let $m \in \mathbb{N}$ be a positive integer. We consider the following parameterized problem.

$\exists^k \forall^*$-TM-HALT^m.
Instance: An $\exists\forall$-machine $\mathbb{M}$ with m tapes, and positive integers $k, t \geq 1$.
Parameter: k.
Question: Does $\mathbb{M}$ halt on the empty string with existential cost k and universal cost t?

In addition, we consider the parameterized problem $\exists^k \forall^*$-TM-HALT$^* = \bigcup_{m \in \mathbb{N}} \exists^k \forall^*$-TM-HALT^m, i.e., the variant of the above problem where the number of tapes is given as part of the input, rather than being a fixed constant.

In the remainder of this section, we show that the class $\exists^k \forall^*$ is characterized by alternating Turing machines in the way specified by the following two theorems.

Theorem 2. *$\exists^k \forall^*$ is exactly the class of parameterized decision problems that are decided by some $\exists^k \forall^*$-machine.*

Theorem 3. *The problem $\exists^k \forall^*$-TM-HALT* is $\exists^k \forall^*$-complete, and so is the problem $\exists^k \forall^*$-TM-HALT^m for each $m \in \mathbb{N}$.*

Proof (Theorems 2 and 3). In order to show these results, concretely, we will prove the following claims:

1. $\exists^k \forall^*$-TM-HALT$^* \leq_{\text{fpt}} \exists^k \forall^*$-MC.
2. For any parameterized problem P that is decided by some $\exists^k \forall^*$-machine with m tapes, it holds that $P \leq_{\text{fpt}} \exists^k \forall^*$-TM-HALT^{m+1}.
3. There is an $\exists^k \forall^*$-machine with a single tape that decides $\exists^{\leq k} \forall^*$-WSAT.
4. Let A and B be parameterized problem. If there is an $\exists^k \forall^*$-machine for B with m tapes, and if $A \leq_{\text{fpt}} B$, then there is an $\exists^k \forall^*$-machine for A with m tapes.
5. $\exists^k \forall^*$-TM-HALT$^2 \leq_{\text{fpt}} \exists^k \forall^*$-TM-HALT^1.

These claims imply the desired results in the following way.

By Claims 1 and 2, by Theorem 1, and by transitivity of fpt-reductions, we have that any parameterized problem P that is decided by an $\exists^k \forall^*$-machine is fpt-reducible to $\exists^k \forall^*$-WSAT, and thus is in $\exists^k \forall^*$. Conversely, let P be any parameterized problem in $\exists^k \forall^*$. Then, by $\exists^k \forall^*$-hardness of $\exists^{\leq k} \forall^*$-WSAT, we know that $P \leq_{\text{fpt}} \exists^{\leq k} \forall^*$-WSAT. By Claims 3 and 4, we know that P is decided by some $\exists^k \forall^*$-machine with a single tape. From this we conclude that $\exists^k \forall^*$ is exactly the class of parameterized problems P decided by some $\exists^k \forall^*$-machine.

Together, Claims 2 and 3 imply that $\exists^{\leq k} \forall^*$-WSAT $\leq_{\text{fpt}} \exists^k \forall^*$-TM-HALT^2. Clearly, for all $m \geq 2$, $\exists^k \forall^*$-TM-HALT$^2 \leq_{\text{fpt}} \exists^k \forall^*$-TM-HALT^m. This gives us $\exists^k \forall^*$-hardness of $\exists^k \forall^*$-TM-HALT^m, for all $m \geq 2$. $\exists^k \forall^*$-hardness of $\exists^k \forall^*$-TM-HALT^1 follows from Claim 5, which states that there is an fpt-reduction from $\exists^k \forall^*$-TM-HALT^2 to $\exists^k \forall^*$-TM-HALT^1. This also implies that $\exists^k \forall^*$-TM-HALT* is $\exists^k \forall^*$-hard. Then, by Claim 1, and since $\exists^k \forall^*$-MC is in $\exists^k \forall^*$ by Theorem 1, we obtain $\exists^k \forall^*$-completeness of $\exists^k \forall^*$-TM-HALT* and $\exists^k \forall^*$-TM-HALT^m, for each $m \geq 1$.

For Claims 1–3, we describe the main idea and intuition behind the proof. A full detailed proof of these claims can be found in the technical report [15].

Proof of Claim 1 (sketch). Given an $\exists\forall$-machine $\mathbb{M}$ with m tapes and positive integers $k, t \geq 1$, we construct a structure $\mathcal{A}$ and a first-order sentence $\exists x_1, \ldots, x_{k'}.$ $\forall y_1, \ldots, y_u.\psi$ such that $\mathcal{A} \models \varphi$ if and only if $\mathbb{M}$ halts on the empty string with existential cost k and universal cost t. In order to do so, firstly, we transform $\mathbb{M}$ in such a way that each computation path contains exactly k existential configurations and exactly t universal configurations (rather than at most k existential configurations and at most t universal configurations) by adding a "clock" to it, i.e., by indexing the existential and universal states with time steps i and allowing $\mathbb{M}$ to be "idle" at each time step.

Then, we use the existential variables x_i (and the structure $\mathcal{A}$) to guess the first k many (existential) configurations and transitions of $\mathbb{M}$, and we use universal variables (and the structure $\mathcal{A}$) to represent the subsequent t many (universal) configurations and transitions. The position of the tape heads and the tape contents for the first k many configurations can be represented by formulas whose size depends only on k. This is not entirely straightforward, but can be done by adapting a technique used by Flum and Grohe [11, Theorem 7.28] to our setting. In order to represent the position of the tape heads and the tape contents for the universal configurations, we can use additional universally quantified variables, since the number of universal variables is not bounded by the parameter. Finally, it is straightforward to encode into ψ the condition that the computation path of $\mathbb{M}$ that is represented by the variables x_i and y_i must be an accepting run.

Proof of Claim 2 (sketch). Let P be a parameterized problem, and let $\mathbb{M}$ be an $\exists^k\forall^*$-machine that decides it, i.e., there exists a computable function f and a polynomial p such that for any instance (x, k) of P we have that any computation path of $\mathbb{M}$ with input (x, k) has length at most $f(k) \cdot p(|x|)$ and contains at most $f(k) \cdot \log |x|$ nondeterministic existential configurations. Let (x, k) be an instance of P. We construct an $\exists\forall$-machine $\mathbb{M}^{(x,k)}$ and positive integers $k', t \geq 1$ such that $\mathbb{M}^{(x,k)}$ accepts the empty string with existential cost k' and universal cost t if and only if $\mathbb{M}$ accepts (x, k).

In order to do so, we add symbols σ to the alphabet that represent sequences of u many nondeterministic transitions of $\mathbb{M}$, for $u \leq \lceil \log |x| \rceil$. The machine $\mathbb{M}^{(x,k)}$ firstly guesses $f(k)$ of such symbols σ. This can be done using $k' = f(k)$ many existential steps. Then, $\mathbb{M}^{(x,k)}$ simulates using (deterministic) universal steps the existential steps of $\mathbb{M}$ on input (x, k), where it simulates the nondeterministic existential steps of $\mathbb{M}$ by "reading off" the transitions of the guessed symbols σ. Finally, $\mathbb{M}^{(x,k)}$ simulates the (nondeterministic) universal steps of $\mathbb{M}$. The entire simulation of $\mathbb{M}$ on input (x, k) requires at most $t = f(k) \cdot p(|x|)$ universal steps.

Proof of Claim 3 (sketch). We describe the working of an $\exists^k\forall^*$-machine $\mathbb{M}$ for $\exists^{\leq k}\forall^*$-WSAT. Let (φ, k) be an instance of $\exists^{\leq k}\forall^*$-WSAT, where $\varphi = \exists X.\forall Y.\psi$, and $X = \{x_1, \ldots, x_n\}$. Firstly, $\mathbb{M}$ determines the size of X, and nondeterministically guesses k many bitstrings of length $\lceil \log |X| \rceil$, which it appends to the tape contents. This can be done using fpt-many existential steps, of which at most $k \cdot \lceil \log |X| \rceil$ many are nondeterministic. These bitstrings represent an assignment $\alpha : X \rightarrow \{0, 1\}$ of weight at most k in the following way: α sets exactly those x_i to true for which the tape contains a bitstring that is the binary representation of index i. Then, $\mathbb{M}$ uses polynomially many nondeterministic universal steps to guess an assignment $\beta : Y \rightarrow \{0, 1\}$.

Finally, it applies the assignment $\alpha \cup \beta$ to the formula ψ and simplifies the resulting formula, using polynomially many deterministic steps. The machine $\mathbb{M}$ accepts if and only if $\psi[\alpha \cup \beta]$ evaluates to true.

Proof of Claim 4 (sketch). Let R be the fpt-reduction from A to B, and let M be an algorithm that decides B and that can be implemented by an $\exists^k \forall^*$-machine with m tapes. Then, the composition of R and M is an algorithm that decides A. It is straightforward to verify that the composition of R and M can be implemented by an $\exists^k \forall^*$-machine with m tapes.

Proof of Claim 5 (sketch). The claim follows by the following statement, which is known from the literature [16, Thm 8.9 and Thm 8.10]. Let $m \geq 1$ be a (fixed) positive integer. For each ATM $\mathbb{M}$ with m tapes, there exists an ATM $\mathbb{M}'$ with 1 tape such that: (1) $\mathbb{M}$ and $\mathbb{M}'$ are equivalent, i.e., they accept the same language; (2) $\mathbb{M}'$ simulates n many steps of $\mathbb{M}$ using $O(n^2)$ many steps; and (3) $\mathbb{M}'$ simulates existential steps of $\mathbb{M}$ using existential steps, and simulates universal steps of $\mathbb{M}$ using universal steps. □

5   Showcase Application to a Combinatorial Problem

In this section, we will exemplify our new complexity toolbox by applying it to various parameterizations of a well-known Π_2^P-complete problem, as considered by Ajtai, Fagin, and Stockmeyer [1].

Let $G = (V, E)$ be a graph. We will denote those vertices v that have degree 1 by *leaves*. We call a (partial) function $c : V \rightarrow \{1, 2, 3\}$ a 3-*coloring (of G)*. Moreover, we say that a 3-coloring c is *proper* if c assigns a color to every vertex $v \in V$, and if for each edge $e = \{v_1, v_2\} \in E$ holds that $c(v_1) \neq c(v_2)$. Now consider the following Π_2^P-complete decision problem.

3-Col-Ext
Instance: a graph $G = (V, E)$ with n many leaves, and an integer m.
Question: can any 3-coloring that assigns a color to exactly m leaves of G (and to no other vertices) be extended to a proper 3-coloring of G?

We consider several parameterizations **p** for this problem, denoted 3-Col-Ext(**p**).

p	**parameter** (k)
degree	the degree of G, i.e., $k = \deg(G)$
#leaves	the number of leaves of G, i.e., $k = n$
#col.leaves	the number of leaves that are pre-colored, i.e., $k = m$
#uncol.leaves	the number of leaves that are not pre-colored, i.e., $k = n - m$

For most of these parameterizations, the existing parameterized complexity toolbox suffices to determine whether or not an fpt-reduction to SAT exists. The following results witness this (proofs of these results can be found in the technical report [15]). For parameterized problems that are in para-NP, an fpt-reduction to SAT exists, whereas this is not the case for problems that are hard for para-Π_2^P (unless the PH collapses).

Proposition 1. *The problems* 3-COL-EXT(degree) *and* 3-COL-EXT(#uncol.leaves) *are* para-Π_2^P-*complete. The problem* 3-COL-EXT(#leaves) *is* para-NP-*complete.*

For the remaining parameterization of the problem 3-COL-EXT the classes para-NP and para-Π_2^P seem to be of little help. On the one hand, 3-COL-EXT(#col.leaves) is unlikely to be hard for the class para-Π_2^P, for the following reason. It is straightforward to construct an xp-reduction from 3-COL-EXT(#col.leaves) to SAT. However, problems that are hard for para-Π_2^P do not allow xp-reductions to SAT, unless the PH collapses [14]. Therefore, 3-COL-EXT(#col.leaves) is not para-Π_2^P-hard, unless the PH collapses. On the other hand, at first sight it is unclear how one can come up with a more efficient reduction from 3-COL-EXT(#col.leaves) to SAT than the obvious xp-reduction. To back up this conjecture of the non-existence of an fpt-reduction to SAT for the problem 3-COL-EXT(#col.leaves), we will use the class $\exists^k\forall^*$.

In order to give evidence that the problem 3-COL-EXT(#col.leaves) does not allow an fpt-reduction to SAT, we can show that it is hard for the class $\forall^k\exists^*$. In addition, we can illustrate the use of the characterization of the parameterized complexity class $\exists^k\forall^*$ in terms of first-order model checking (Theorem 1), by using the problem $\forall^k\exists^*$-MC (which is the complement of the problem $\exists^k\forall^*$-MC) to show $\forall^k\exists^*$-membership, characterizing the complexity of 3-COL-EXT(#col.leaves) as $\forall^k\exists^*$-complete.

Theorem 4. 3-COL-EXT(#col.leaves) *is* $\forall^k\exists^*$-*complete.*

Proof. To show membership, we give an fpt-reduction from 3-COL-EXT(#col.leaves) to $\forall^k\exists^*$-MC. Let (G,m) be an instance of 3-COL-EXT(#col.leaves), where V' denotes the set of leaves of G, and where $k = m$ is the number of edges that can be pre-colored. Moreover, let $V' = \{v_1,\dots,v_n\}$ and let $V = V' \cup \{v_{n+1},\dots,v_u\}$. We construct an instance $(\mathcal{A},\varphi)$ of $\forall^k\exists^*$-MC. We define the domain $A = \{a_{v,i} : v \in V', 1 \le i \le 3\} \cup \{1,2,3\}$. Next, we define $C^{\mathcal{A}} = \{1,2,3\}$, $S^{\mathcal{A}} = \{(a_{v,i}, a_{v,i'}) : v \in V', 1 \le i, i' \le 3\}$, and $F^{\mathcal{A}} = \{(j,j') : 1 \le j, j' \le 3, j \ne j'\}$. Then, we can define the formula φ, by letting $\varphi = \forall x_1,\dots,x_k.\exists y_1,\dots,y_u.(\psi_1 \to (\psi_2 \wedge \psi_3 \wedge \psi_4))$, where $\psi_1 = \bigwedge_{1 \le j < j' \le k} \neg S(x_i, x_{i'})$, and $\psi_2 = \bigwedge_{1 \le j \le u} C(y_j)$, and $\psi_3 = \bigwedge_{v_j \in V', 1 \le i \le 3}((\bigvee_{1 \le \ell \le k}(x_\ell = a_{v_j,i})) \to (y_j = i))$, and $\psi_4 = \bigwedge_{\{v_j, v_{j'}\} \in E} F(y_j, y_{j'})$. It is straightforward to verify that $(G,m) \in$ 3-COL-EXT if and only if $\mathcal{A} \models \varphi$.

Intuitively, the assignments to the variables x_i correspond to the pre-colorings of the vertices in V'. This is done by means of elements $a_{v,i}$, which represent the coloring of vertex v with color i. The subformula ψ_1 is used to disregard any assignments where variables x_i are not assigned to the intended elements. Moreover, the assignments to the variables y_i correspond to a proper 3-coloring extending the pre-coloring. The subformula ψ_2 ensures that the variables y_i are assigned to a color in $\{1,2,3\}$, the subformula ψ_3 ensures that this coloring extends the pre-coloring encoded by the assignment to the variables x_i, and the subformula ψ_4 ensures that this coloring is proper.

Hardness can be shown by means of an fpt-reduction from $\forall^k\exists^*$-WSAT. A proof of hardness can be found in the technical report [15].   □

6   Conclusion

The classes $\exists^k\forall^*$ and $\forall^k\exists^*$ are parameterized complexity classes between the first and the second level of the PH, that can be used to give evidence that certain parameterized problems do not allow an fpt-reduction to SAT. By definition, $\exists^k\forall^*$ and $\forall^k\exists^*$ are characterized in terms of weighted variants of the quantified Boolean satisfiability problem. We provided characterizations of these classes in terms of a first-order logic model checking problem, and in terms of alternating Turing machines with appropriate time bounds and bounds on the number of alternations. Moreover, we showed how one of these alternative characterizations can be used to show membership in the class $\exists^k\forall^*$, by means of an example problem that is related to extending partial graph 3-colorings to complete, proper 3-colorings. Our alternative characterizations establish the robustness of the classes and provide new ways of showing membership.

Further research includes applying the additional characterizations we provided to show membership in $\exists^k\forall^*$ and $\forall^k\exists^*$ for further parameterized problems. In addition, it would be interesting to obtain similar characterizations for the classes $\exists^*\forall^k$-W[t], which are parameterized complexity classes that are defined analogously to $\exists^k\forall^*$, and that can be used to get similar intractability results [14,15].

References

1. Ajtai, M., Fagin, R., Stockmeyer, L.J.: The closure of monadic NP. J. of Computer and System Sciences 60(3), 660–716 (2000)
2. Arora, S., Barak, B.: Computational Complexity – A Modern Approach. Cambridge University Press (2009)
3. Biere, A., Heule, M., van Maaren, H., Walsh, T. (eds.): Handbook of Satisfiability. Frontiers in Artificial Intelligence and Applications, vol. 185. IOS Press (2009)
4. Cesati, M.: The Turing way to parameterized complexity. J. of Computer and System Sciences 67, 654–685 (2003)
5. Chen, Y., Flum, J.: A parameterized halting problem. In: Bodlaender, H.L., Downey, R., Fomin, F.V., Marx, D. (eds.) Fellows Festschrift 2012. LNCS, vol. 7370, pp. 364–397. Springer, Heidelberg (2012)
6. Downey, R.G., Fellows, M.R.: Parameterized Complexity. Monographs in Computer Science. Springer, New York (1999)
7. Downey, R.G., Fellows, M.R.: Fundamentals of Parameterized Complexity. Texts in Computer Science. Springer (2013)
8. Endriss, U., de Haan, R., Szeider, S.: Parameterized complexity results for agenda safety in judgment aggregation. In: Proceedings of the 5th International Workshop on Computational Social Choice (COMSOC 2014). Carnegie Mellon University (2014)
9. Fichte, J.K., Szeider, S.: Backdoors to normality for disjunctive logic programs. In: Proceedings of the Twenty-Seventh AAAI Conference on Artificial Intelligence (AAAI 2013), pp. 320–327. AAAI Press (2013)
10. Flum, J., Grohe, M.: Describing parameterized complexity classes. Information and Computation 187(2), 291–319 (2003)
11. Flum, J., Grohe, M.: Parameterized Complexity Theory. An EATCS Series, vol. XIV. Springer, Berlin (2006)
12. Gomes, C.P., Kautz, H., Sabharwal, A., Selman, G.: Satisfiability solvers. In Handbook of Knowledge Representation. Foundations of Artificial Intelligence, vol. 3, pp. 89–134. Elsevier (2008)
13. de Haan, R., Szeider, S.: Fixed-parameter tractable reductions to SAT. In: Sinz, C., Egly, U. (eds.) SAT 2014. LNCS, vol. 8561, pp. 85–102. Springer, Heidelberg (2014)

14. de Haan, R., Szeider, S.: The parameterized complexity of reasoning problems beyond NP. In: Baral, C., De Giacomo, G., Eiter, T. (eds.) Principles of Knowledge Representation and Reasoning: Proceedings of the Fourteenth International Conference, KR 2014, Vienna, Austria, July 20-24. AAAI Press (2014)

15. de Haan, R., Szeider, S.: The parameterized complexity of reasoning problems beyond NP. Technical Report 1312.1672v3, arXiv.org (2014)

16. Hopcroft, J.E., Motwani, R., Ullman, J.D.: Introduction to Automata Theory, Languages, and Computation, 2nd edn. Addison-Wesley Series in Computer Science. Addison-Wesley-Longman (2001)

17. Malik, S., Zhang, L.: Boolean satisfiability from theoretical hardness to practical success. Communications of the ACM 52(8), 76–82 (2009)

18. Meyer, A.R., Stockmeyer, L.J.: The equivalence problem for regular expressions with squaring requires exponential space. In: SWAT, pp. 125–129. IEEE Computer Soc. (1972)

19. Niedermeier, R.: Invitation to Fixed-Parameter Algorithms. Oxford Lecture Series in Mathematics and its Applications. Oxford University Press, Oxford (2006)

20. Papadimitriou, C.H.: Computational Complexity. Addison-Wesley (1994)

21. Pfandler, A., Rümmele, S., Szeider, S.: Backdoors to abduction. In: Rossi, F. (ed.) Proceedings of the 23rd International Joint Conference on Artificial Intelligence, IJCAI 2013. AAAI Press/IJCAI (2013)

22. Sakallah, K.A., Marques-Silva, J.: Anatomy and empirical evaluation of modern SAT solvers. Bulletin of the European Association for Theoretical Computer Science 103, 96–121 (2011)

23. Stockmeyer, L.J.: The polynomial-time hierarchy. Theoretical Computer Science 3(1), 1–22 (1976)

24. Wrathall, C.: Complete sets and the polynomial-time hierarchy. Theoretical Computer Science 3(1), 23–33 (1976)

Maximally Permissive Controlled System Synthesis for Modal Logic*

Allan C. van Hulst, Michel A. Reniers, and Wan J. Fokkink

Eindhoven University of Technology, The Netherlands
`ahulst@tue.nl`

Abstract. We propose a new method for controlled system synthesis on non-deterministic automata, which includes the synthesis for deadlock-freeness, as well as invariant and reachability expressions. Our technique restricts the behavior of a Kripke-structure with labeled transitions, representing the uncontrolled system, such that it adheres to a given requirement specification in an expressive modal logic, while all non-invalidating behavior is retained. This induces maximal permissiveness in the context of supervisory control. Research presented in this paper allows a system model to be constrained according to a broad set of liveness, safety and fairness specifications of desired behavior, and embraces most concepts from Ramadge-Wonham supervisory control, including controllability and marker-state reachability. The synthesis construction is formally verified using the Coq proof assistant.

1 Introduction

This paper concerns the controlled system synthesis on non-deterministic automata for requirements in modal logic. The controlled systems perspective treats the system under control — the *plant* — and a system component which restricts the plant behavior — the *controller* — as a single integrated entity. This means that we take a model of all possible plant behavior, and construct a new model which is constrained according to a logical specification of desired behavior — the *requirements*. The *synthesis*, of such a restricted behavioral model incorporates a number of concepts from supervisory control theory [6], which affirm the generated model as being a proper controlled system, in relation to the original plant specification. Events are strictly partitioned into being either controllable or uncontrollable, such that synthesis only disallows events of the first type. In addition, synthesis preserves all behavior which does not invalidate the requirements, thereby inducing maximal permissiveness [6]. The requirement specification formalism extends Hennessy-Milner Logic [10] with invariant, reachability, and deadlock-freeness expressions, and is also able to express the supervisory control concept of marker-state reachability [13].

The intended contribution of this paper is two-fold. First, it presents a technique for controlled system synthesis in a non-deterministic context. Second, it

---

* Supported by the EU FP7 Programme under grant agreement No. 295261 (MEALS).

G.F. Italiano et al. (Eds.): SOFSEM 2015, LNCS 8939, pp. 230–241, 2015.

defines synthesis for a modal logic which is able to capture a broad set of requirements. Regarding the first contribution, it should be noted that supervisory control synthesis is often approached using a deterministic model of both plant and controller. Notably, the classic Ramadge-Wonham supervisory control theory [13] is a well-researched example of this setup. The resulting controller restricts the behavior of the deterministic plant model, thereby ensuring that it operates according to the requirements via event-based synchronization. A controlled system can not be constructed in this way for a non-deterministic model, as illustrated by example in Figure 1. Assume that we wish to restrict all technically possible behavior of an indicator light of a printer (Figure 1a) such that after a single *refill* event, the indicator light turns *green* immediately. In the solution shown in Figure 1b, the self-loop at the right-most state is disallowed, as indicated using dashed lines, while all other behavior is preserved. Note that it is not possible to construct this maximally-permissive solution using event-based synchronization, as shown in [4]. However, an outcome as shown in Figure 1b can be obtained by applying synthesis for the property □ [*refill*] *green*, using the method described in this paper. As this example shows, the strict separation between plant and controller is not possible for non-deterministic models, and therefore we interpret the controlled system as a singular entity.

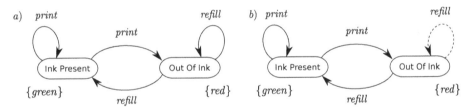

Fig. 1. Example of control synthesis in a non-deterministic context. A model for all possible behavior of an ink presence indicator light of a printer is restricted in such a way that after every *refill*, the state labeled with *green* is reached directly. Synthesis, as defined in this paper, of the property □ [*refill*] *green* upon the model in Figure 1a, results in a synthesis outcome as in Figure 1b., where disallowed behavior is indicated using dashed lines.

This requirement formalism applied in this paper, which extends Hennessy-Milner Logic with invariant and reachability operators, and also includes a test for deadlock-freeness, is able to express a broad set of liveness, safety, and fairness properties. For instance, an important liveness concept in supervisory control theory involves marker-state reachability, which is informally expressed as the requirement that it is always possible to reach a state which is said to be *marked*. This requirement is modeled as □ ◊ *marked*, using the requirement specification logic, in conjunction with assigning *marked* as a separate property to the designated states in the Kripke-model.

Safety-related requirements, which model the absence of faulty behavior, include deadlock-avoidance, expressed as □ *dlf* (i.e., invariantly, deadlock-free) and safety requirements of a more general nature. For instance, one might require that some type of communicating system is always able to perform a *receive* step, directly after every *send* step. Such a property is expressed as $\Box$ [*send*] <*receive*>*true*, using the requirement specification logic applied in this paper. In addition, we argue that this logic is able to model a limited class of fairness properties. One might require from a system which uses a shared resource that in every state, the system has access to the resource (the state has the *access* property), or it can do a *lock* step to claim the resource, after which access is achieved immediately. We synthesize the property $\Box$ (*access* $\vee$ <*lock*>*access*) in order to constrain the plant behavior in this way.

The remainder of this paper is set up as follows. We consider a number of related works on control synthesis in Section 2. Preliminary definitions in Section 3 introduce formal notions up to a clear statement of the synthesis problem. Section 4 concerns the formal definition of the synthesis construction while Section 5 lists a number of important theorems indicating correctness of the synthesis approach, with detailed proofs being available in textual form [19], as well as in computer-verified form [16].

2   Related Work

Earlier work by the same authors concerning synthesis for modal logic includes a recursive synthesis method for Hennessy-Milner Logic [17], and a synthesis method for a subset of the logic considered in this paper, with additional restrictions on combinations of modal operators [18]. This paper vastly improves previous efforts by allowing unrestricted synthesis for invariant formulas and including the new operator $\Diamond$ for reachability. It also takes into account deadlock-freeness, and uncontrollable events, thereby achieving controllability.

We analyze related work alongside the three intended improvements in this paper: 1) Allowance of non-determinism in plant specifications, 2) Expressiveness of the requirement specification formalism, and, 3) Adhering to some form of maximal permissiveness.

Ramadge-Wonham supervisory control [13] defines a broadly-embraced methodology for controller synthesis on deterministic plant models for requirements specified using automata. It defines a number of key elements in the relationship between plant and controlled system, such as controllability, marker-state reachability, deadlock-freeness and maximal permissiveness. Despite the fact that a strictly separated controller offers advantages from a developmental or implementational point of view, we argue that increased abstraction and flexibility justifies research into control synthesis for non-deterministic models. In addition, we emphasize that the automata-based description of desired behavior in the Ramadge-Wonham framework [13] does not allow the specification of requirements of existential nature. For instance, in this framework it is not possible to specify that a step labeled with a particular event *must* exist, hence the choice of modal logic as our requirement formalism.

Work by Pnueli and Rosner [12] concerns a treatment of synthesis for reactive systems, based upon a finite transducer model of the plant, and a temporal specification of desired behavior. This synthesis construction is developed further for deterministic automata in [12], but the treatment remains non-maximal. This research is extended in [2], which connects reactive synthesis to Ramadge-Wonham supervisory control using a parity-game based approach. The methodology described in [2] transforms the synthesis control problem for μ-calculus formulas in such a way that the set of satisfying models of a μ-calculus formula coincides with the set of controllers which enforce the controlled behavior. Although non-determinism is allowed in plant-specifications in [2], the treatment via loop-automata does not allow straightforward modeling of all (infinite) behaviors. Also, maximal permissiveness is not specified as a criterion for control synthesis in [2]. Interesting follow-up research is found in [3], for non-deterministic controllers over non-deterministic processes. However, the specification of desired behavior is limited to alternating automata [3], which do not allow complete coverage of invariant expressions over all modalities, or an equivalent thereof. Reactive synthesis is further applied to hierarchical [1] and recursive [11] component-based specifications. These works, which both are based upon a deterministic setting, provide a quite interesting setup from a developmental perspective, due to their focus on the re-usability of components.

Research in [20] relates Ramadge-Wonham supervisory control to an equivalent model-checking problem, resulting in important observations regarding the mutual exchangeability and complexity analysis of both problems. Despite the fact that research in [20] is limited to a deterministic setting, and synthesis results are not guaranteed to be maximally permissive, it does incorporate a quite expressible set of μ-calculus requirements. Other research based upon a dual approach between control synthesis and model checking studies the incremental effects of transition removal on the validity of μ-calculus formulas [7,14].

Research by D'Ippolito and others [8], [9] is based upon the framework of the world machine model for the synthesis of liveness properties, stated in fluent temporal logic. A distinction is made between controlled and monitored behavior, and between system goals and environment assumptions [8]. A controller is then derived from a winning strategy in a two-player game between original and required behavior, as expressed in terms of the notion of generalized reactivity, as introduced in [8]. Research in [8] also emphasizes the fact that pruning-based synthesis is not adequate for control of non-deterministic models, and it defines synthesis of liveness goals under a maximality criterion, referred to as best-effort controller. However, this maximality requirement is trace-based and is therefore not able to signify inclusion of all possible infinite behaviors. In addition, some results in [8] are based upon the assumption of a deterministic plant specification.

3   Definitions

We assume a set $\mathcal{E}$ of events and a set $\mathcal{P}$ of state-based properties. In addition, we assume a strict partition of $\mathcal{E}$ into controllable events $\mathcal{C}$ and uncontrollable events $\mathcal{U}$, such that $\mathcal{C} \cup \mathcal{U} = \mathcal{E}$ and $\mathcal{C} \cap \mathcal{U} = \emptyset$. State-based properties are used to capture state-based information, and are assigned to states using a labeling function.

Example properties are shown in Figure 1, as *red* and *green*. Figure 1 also shows examples of the events *print* and *refill*, which are assumed to be controllable in this example. Events are used to capture system dynamics, and represent actions occurring when the system transitions between states. Controllable events may be used to model actuator actions in the plant, while an uncontrollable event may represent, for instance, a sensor reading. Basic properties and events are used to model plant behavior in the form of a Kripke-structure [5] with labeled transitions, to be abbreviated as Kripke-LTS, as formalized in Definition 1. Note that we assume finiteness of the given transition relation.

Definition 1. *We define a Kripke-LTS as a four-tuple $(X, L, \longrightarrow, x)$ for state-space X, labeling function $L : X \mapsto 2^{\mathcal{P}}$, finite transition relation $\longrightarrow \subseteq X \times \mathcal{E} \times X$, and initial state $x \in X$. The universe of all Kripke-LTSs is denoted by $\mathcal{K}$.*

As usual, we will use the notation $x \xrightarrow{e} x'$ to denote that $(x, e, x') \in \longrightarrow$. The reflexive-transitive closure $\longrightarrow^*$ of a transition relation $\longrightarrow$ is defined in the following way: For all $x \in X$ it holds that $(x, x) \in \longrightarrow^*$ and if there exist $e \in \mathcal{E}$ and $y, x' \in X$ such that $x \xrightarrow{e} y$ and $y \longrightarrow^* x'$ then $(x, x') \in \longrightarrow^*$.

Two different behavioral preorders are applied in this paper. The first is the simulation preorder, which is reiterated in Definition 2. Simulation is used to signify inclusion of behavior, while synthesis may alter the transition structure due to, for instance, unfolding. Simulation as applied in this paper is a straightforward adaptation of the definition of simulation in [15].

Definition 2. *For $k' = (X', L', \longrightarrow', x')$ and $k = (X, L, \longrightarrow, x)$ we say that k' and k are related via simulation (notation: $k' \preceq k$) if there exists a relation $R \subseteq X' \times X$ such that $(x', x) \in R$ and for all $(y', y) \in R$ the following holds:*

1. *We have $L'(y') = L(y)$; and*
2. *If $y' \xrightarrow{e}' z'$ then there exists a step $y \xrightarrow{e} z$ such that $(z', z) \in R$.*

Partial bisimulation [4] is an extension of simulation such that the subset of uncontrollable events is bisimulated. For plant specification $k \in \mathcal{K}$ and synthesis result $s \in \mathcal{K}$ we require that s is related to k via a partial bisimulation. This signifies the fact that synthesis did not disallow any uncontrollable event, which implies controllability in the context of supervisory control. Research in [4] details the nature of this partial bisimulation preorder.

Definition 3. *If $k' = (X', L', \longrightarrow', x')$ and $k = (X, L, \longrightarrow, x)$, then k' and k are related via a partial bisimulation (notation: $k' \precsim k$) if there exists a relation $R \subseteq X' \times X$ such that $(x', x) \in R$ and for all $(y', y) \in R$ the following holds:*

1. *We have $L'(y') = L(y)$;*
2. *If $y' \xrightarrow{e}' z'$ then there exists a step $y \xrightarrow{e} z$ such that $(z', z) \in R$; and*
3. *If $y \xrightarrow{e} z$ for $e \in \mathcal{U}$ then there exists a step $y' \xrightarrow{e}' z'$ such that $(z', z) \in R$.*

Requirements are specified using a modal logic $\mathcal{F}$ given in Definition 5, which is built upon the set of state-based formulas $\mathcal{B}$ in Definition 4.

Definition 4. *The set of state-based formulas* $\mathcal{B}$ *is defined by the grammar:*

$$\mathcal{B} ::= true \mid false \mid \mathcal{P} \mid \neg\mathcal{B} \mid \mathcal{B} \wedge \mathcal{B} \mid \mathcal{B} \vee \mathcal{B}$$

As indicated in Definition 4, state-based formulas are constructed from a straightforward Boolean algebra which includes the basic expressions *true* and *false*, as well as a state-based property test for $p \in \mathcal{P}$. Formulas in $\mathcal{B}$ are then combined using the standard Boolean operators $\neg$, $\wedge$ and $\vee$.

Definition 5. *The requirement specification logic* $\mathcal{F}$ *is defined by the grammar:*

$$\mathcal{F} ::= \mathcal{B} \mid \mathcal{F} \wedge \mathcal{F} \mid \mathcal{B} \vee \mathcal{F} \mid [\mathcal{E}]\mathcal{F} \mid <\mathcal{E}>\mathcal{F} \mid \Box\mathcal{F} \mid \Diamond\mathcal{B} \mid dlf$$

We briefly consider the elements of the requirement logic $\mathcal{F}$. Basic expressions in Definition 4 function as the basic building blocks in the modal logic $\mathcal{F}$. Conjunction is included, having its usual semantics, while disjunctive formulas are restricted to those having a state-based formula in the left-hand disjunct. This restriction guarantees correct synthesis solutions, since it enables a local state-based test for retaining the appropriate transitions. The formula $[e]f$ can be used to test whether f holds after every e-step, while the formula $<e>f$ is used to assess whether there exists an e-step after which f holds. These two operators thereby follow their standard semantics from Hennessy-Milner Logic [10]. An invariant formula $\Box f$ tests whether f holds in every reachable state, while a reachability expression $\Diamond b$ may be used to check whether there exists a path such that the state-based formula b holds at some state on this path. Note that the sub-formula b of a reachability expression $\Diamond b$ is restricted to a state-based formula $b \in \mathcal{B}$. This is used to acquire unique synthesis solutions, due to the fact that for unrestricted reachability expressions, only an indefinite unfolding coincides with a maximal solution. The deadlock-free test *dlf* tests whether there exists an outgoing step of a particular state. Combined with the invariant operator, the formula $\Box dlf$ can be used to specify that the entire synthesized system should be deadlock-free. Deadlock-freeness is not defined as a state-based expression here since it requires information about (the existence of) outgoing transitions, which may have been removed during synthesis. Validity of formulas in $\mathcal{B}$ and $\mathcal{F}$, with respect to a Kripke-LTS $k \in \mathcal{K}$, is as shown in Definition 6.

Definition 6. *For* $k = (X, L, \longrightarrow, x) \in \mathcal{K}$ *and* $f \in \mathcal{F}$ *we define if* k *satisfies* f *(notation:* $k \vDash f$*) as follows:*

$$\frac{}{k \vDash true} \qquad \frac{p \in L(x)}{(X, L, \longrightarrow, x) \vDash p} \qquad \frac{k \nvDash b}{k \vDash \neg b} \qquad \frac{k \vDash f \quad k \vDash g}{k \vDash f \wedge g} \qquad \frac{k \vDash f}{k \vDash f \vee g} \qquad \frac{k \vDash g}{k \vDash f \vee g}$$

$$\frac{\forall x \xrightarrow{e} x' \quad (X, L, \longrightarrow, x') \vDash f}{(X, L, \longrightarrow, x) \vDash [e]f} \qquad \frac{x \xrightarrow{e} x' \quad (X, L, \longrightarrow, x') \vDash f}{(X, L, \longrightarrow, x) \vDash <e>f}$$

$$\frac{\forall x \longrightarrow^* x' \quad (X, L, \longrightarrow, x') \vDash f}{(X, L, \longrightarrow, x) \vDash \Box f} \qquad \frac{x \longrightarrow^* x' \quad (X, L, \longrightarrow, x') \vDash b}{(X, L, \longrightarrow, x) \vDash \Diamond b} \qquad \frac{x \xrightarrow{e} x'}{(X, L, \longrightarrow, x) \vDash dlf}$$

We may now formulate the synthesis problem in terms of the previous definitions in Definition 7. Research in this paper focuses on resolving this problem.

Definition 7. *Given $k \in \mathcal{K}$ and $f \in \mathcal{F}$, find $s \in \mathcal{K}$ in a finite method such that
the following holds: 1) $s \models f$, 2) $s \preceq k$, 3) $s \precsim k$, 4) For all $k' \preceq k$ and $k' \models f$
holds $k' \preceq s$; or determine that such an s does not exist.*

These four properties are interpreted in the context of supervisory control
synthesis as follows. Property 1 (*validity*) states that the synthesis result satisfies the synthesized formula. Property 2 (*simulation*) asserts that the synthesis
result is a restriction of the original behavior, while property 3 (*controllability*)
ensures that no accessible uncontrollable behavior is disallowed during synthesis. Controllability is achieved if the synthesis result is related to the original
plant-model via a partial bisimulation, which adds bisimulation of all uncontrollable events to the second property. Note that the third property implies the
second property, as can be observed in Definitions 2 and 3. However, both these
properties are of importance since the former is related to the synthesis result
being a restriction of original behavior, while the latter signifies achievement of
controllability. Property 4 (*maximality*) states that synthesis removes the least
possible behavior, and thereby induces maximal permissiveness. That is, the behavior of every alternative synthesis option is included in the behavior of the
synthesis result.

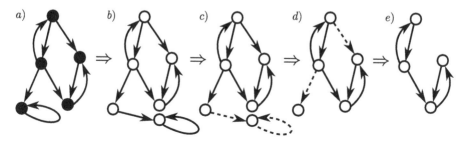

Fig. 2. Overview of the synthesis process. Steps in the original transition relation
(Figure 2a) of type $x \xrightarrow{e} x'$ are combined with reductions of the synthesized requirement (Figure 2b), resulting in transitions of type $(x, f) \xrightarrow{e}_0 (x', f')$, and possibly
inducing unfoldings. Transition are then removed (Figure 2c-2d) based upon a local
synthesizability test for formulas assigned to target states, until synthesizability holds
in every reachable state (Figure 2e).

4   Synthesis

The purpose of this section is to illustrate the formal definition of the synthesis
construction. Synthesis as defined in this paper involves three major steps, after
which a modified Kripke-LTS is obtained. If synthesis is successful, the resulting
structure satisfies all synthesis requirements, as stated in Definition 7. The first
stage of synthesis transforms the original transition relation $\longrightarrow \subseteq X \times \mathcal{E} \times X$,

for state-space X, into a new transition relation $\longrightarrow_0 \subseteq (X \times \mathcal{F}) \times \mathcal{E} \times (X \times \mathcal{F})$ over the state-formula product space. This allows us to indicate precisely which modal (sub-)formula needs to hold at each point in the new transition relation. The second step removes transitions based upon an assertion of *synthesizability* of formulas assigned to the target states of transitions. This second step is repeated until no more transitions are removed. The third and final synthesis step tests whether synthesis has been successful by evaluating whether the *synthesizability* predicate holds for every remaining state. An overview of the synthesis process is shown in Figure 2.

A formal derivation of the starting point in the synthesis process $\longrightarrow_0$ is shown in Definition 9. This definition relies upon sub-formulas under conjunction and invariant operators, as formalized in Definition 8.

Definition 8. *We say that $f \in \mathcal{F}$ is a sub-formula of $g \in \mathcal{F}$ (notation $f \in$ sub (g)) if this can be derived by the following rules:*

$$\frac{}{f \in sub\,(f)} \qquad \frac{f \in sub\,(g)}{f \in sub\,(g \wedge h)} \qquad \frac{f \in sub\,(h)}{f \in sub\,(g \wedge h)} \qquad \frac{f \in sub\,(g)}{f \in sub\,(\Box g)}$$

Definition 9. *For state-space X and original transition relation $\longrightarrow \subseteq X \times \mathcal{E} \times X$, we define the starting point of synthesis $\longrightarrow_0 \subseteq (X \times \mathcal{F}) \times \mathcal{E} \times (X \times \mathcal{F})$ as follows:*

$$\frac{x \xrightarrow{e} x'}{(x,b) \xrightarrow{e}_0 (x, true)} \qquad \frac{(x,f) \xrightarrow{e}_0 (x',f') \quad (x,g) \xrightarrow{e}_0 (x',g') \quad g' \in sub\,(f')}{(x, f \wedge g) \xrightarrow{e}_0 (x', f')}$$

$$\frac{(x,f) \xrightarrow{e}_0 (x',f') \quad (x,g) \xrightarrow{e}_0 (x',g') \quad g' \notin sub\,(f')}{(x, f \wedge g) \xrightarrow{e}_0 (x', f' \wedge g')} \qquad \frac{x \xrightarrow{e} x' \quad x \vDash b}{(x, b \vee f) \xrightarrow{e}_0 (x', true)}$$

$$\frac{(x,f) \xrightarrow{e}_0 (x',f')}{(x, b \vee f) \xrightarrow{e}_0 (x',f')} \qquad \frac{x \xrightarrow{e} x'}{(x, [e]\,f) \xrightarrow{e}_0 (x', f)} \qquad \frac{x \xrightarrow{e} x' \quad e \neq e'}{(x, [e']\,f) \xrightarrow{e}_0 (x', true)}$$

$$\frac{x \xrightarrow{e} x'}{(x, <e>f) \xrightarrow{e}_0 (x', f)} \qquad \frac{x \xrightarrow{e} x'}{(x, <e'>f) \xrightarrow{e}_0 (x', true)} \qquad \frac{(x,f) \xrightarrow{e}_0 (x',f') \quad f' \in sub\,(\Box f)}{(x, \Box f) \xrightarrow{e}_0 (x', \Box f)}$$

$$\frac{(x,f) \xrightarrow{e}_0 (x',f') \quad f' \notin sub\,(\Box f)}{(x, \Box f) \xrightarrow{e}_0 (x', \Box f \wedge f')} \qquad \frac{x \xrightarrow{e} x'}{(x, \Diamond b) \xrightarrow{e}_0 (x', true)}$$

$$\frac{x \xrightarrow{e} x'}{(x, \Diamond b) \xrightarrow{e}_0 (x', \Diamond b)} \qquad \frac{x \xrightarrow{e} x'}{(x, dlf) \xrightarrow{e}_0 (x', true)}$$

The intuitive interpretation of a derivation rule for $(x, f) \xrightarrow{e}_0 (x', f')$ in Definition 9 is an assignment of the formula f' to the state x', if f' is required for the validity of f in x, after event e. This is particularly recognizable in the derivation rules for $[e]\,f$. The derivation rules for conjunction ensure the validity of reductions for both operands. However, in order to achieve a terminating synthesis procedure, reductions of conjunctive formulas are prevented from expanding infinitely often using the sub-formula relation. This applies also to reductions of invariant formulas. The prevention of indefinite formula expansion under conjunction is essential, and only required for, finiteness of the formula-reductions for invariant expressions. Other reduction strategies typical for the synthesis approach in Definition 9 include a limitation of outgoing transitions based upon the state-based validity of left-hand disjuncts. In addition, reductions towards *true* are included for $<e>f$ and $\Diamond p$, in order to achieve maximal permissiveness, since the synthesis for a single witness does not affect other outgoing transitions, which should be left in place.

The starting point of synthesis $\longrightarrow_0$ is subjected to transition removal via a synthesizability test for formulas assigned to the target states of transitions. In generalized form, we define a formula $f \in \mathcal{F}$ to be synthesizable in the state-formula pair (x, g) if this can be derived by the rules in Definition 11. For an appropriate definition of synthesizability, it is necessary to extend the notion of sub-formulas in such a way that a state-based evaluation can be incorporated, in order to handle disjunctive formulas correctly. This leads to the sub-formula notion called *part*, which is shown in Definition 10.

Definition 10. *We say that a formula $f \in \mathcal{F}$ is a part of a formula $g \in \mathcal{F}$ in the context of a state based evaluation for $(X, L, \longrightarrow, x)$ if $f \equiv g$, or a derivation can be obtained by the following rules:*

$$\frac{f \in part\,(x, g)}{f \in part\,(x, g \wedge h)} \qquad \frac{f \in part\,(x, h)}{f \in part\,(x, g \wedge h)} \qquad \frac{x \not\models b \quad f \in part\,(x, g)}{f \in part\,(x, b \vee g)} \qquad \frac{f \in part\,(x, g)}{f \in part\,(x, \Box g)}$$

Partial formulas as shown in Definition 10 are used in the definition of synthesizability as shown in Definition 11. In particular, this is used in the definition of synthesizability for formulas of type $<e>f$. In addition, partial formulas play a major role in the correctness proofs of the synthesis method.

Definition 11. *With regard to an intermediate relation $\longrightarrow_n \subseteq (X \times \mathcal{F}) \times \mathcal{E} \times (X \times \mathcal{F})$ in the synthesis procedure, we say that a formula $f \in \mathcal{F}$ is synthesizable in the state-formula pair (x, g) (notation: $(x, g) \uparrow f$) if this can be derived as follows:*

$$\frac{x \models b}{(x, g) \uparrow b} \qquad \frac{(x, g) \uparrow f_1 \quad (x, g) \uparrow f_2}{(x, g) \uparrow f_1 \wedge f_2} \qquad \frac{x \models b}{(x, g) \uparrow b \vee f} \qquad \frac{(x, g) \uparrow f}{(x, g) \uparrow b \vee f}$$

$$\frac{}{(x, g) \uparrow [e]\,f} \qquad \frac{(x', g') \uparrow f \quad (x, g) \xrightarrow{e}_n (x', g') \quad f \in part\,(x', g')}{(x, g) \uparrow <e>f}$$

$$\frac{(x, g) \uparrow f}{(x, g) \uparrow \Box f} \qquad \frac{(x, g) \longrightarrow^*_n (x', g') \quad x' \models b}{(x, g) \uparrow \Diamond b} \qquad \frac{(x, g) \xrightarrow{e}_n (x', g')}{(x, g) \uparrow dlf}$$

It is important to note here that the *synthesizability* test serves as a partial assessment. The synthesizability predicate for f holds in the state-formula pair (x, g) if it is possible to modify outgoing transitions of (x, g) in such a way that

f becomes satisfied in (x, g). However, synthesizability is not straightforwardly definable for a number of formulas. For instance, it can not be directly assessed whether it is possible to satisfy an invariant formula. Therefore, the synthesizability test in Definition 11 is designed to operate in conjunction with the process of repeated transition removal, as shown in Figure 2. This is reflected, for instance, in the definition of synthesizability for an invariant formula $\Box f$, which only relies upon f being synthesizable. However, since synthesizability needs to hold at every reachable state for synthesis to be successful, such a definition of synthesizability for invariant formulas is appropriate due to its role in the entire synthesis process. A synthesis example for the invariant formula $\Box p \wedge [a] q$ is shown in Figure 3.

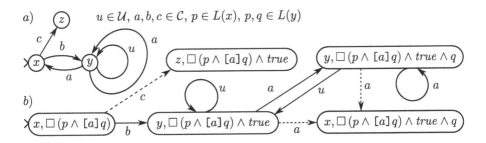

Fig. 3. Synthesis for the formula $\Box p \wedge [a] q$ upon the model in Figure 3a, resulting in the restricted behavioral model shown in Figure 3b. Note the unfolding for $[a] q$, the restricted formula-expansion for invariant formulas, and transition disabling, indicated by dashed lines, due to the state-based formula q not being synthesizable in x, and p not being synthesizable in z

Using the definitions stated before, we are now ready to define the main synthesis construction. That is, how transitions are removed from the synthesis starting point $\longrightarrow_0$, and how are the subsequent intermediate transition relations $\longrightarrow_1, \longrightarrow_2, \ldots$ constructed. In addition, more clarity is required with regard to reaching a stable point during synthesis, and verifying whether the synthesis construction has been completed successfully.

Definition 12. *For $k = (X, L, \longrightarrow, x) \in \mathcal{K}$ and $f \in \mathcal{F}$, we define the n-th iteration in the synthesis construction as follows:*

$$\frac{(x, f) \xrightarrow{e}_n (x', f') \quad e \in \mathcal{U}}{(x, f) \xrightarrow{e}_{n+1} (x', f')} \qquad \frac{(x, f) \xrightarrow{e}_n (x', f') \quad (x, f) \uparrow f}{(x, f) \xrightarrow{e}_{n+1} (x', f')}$$

The corresponding system model $S^n_{k,f}$ is defined as stated below, using the labeling function L_{proj}, such that $L_{proj}(y, g) = L(y)$, for all $y \in X$ and $g \in \mathcal{F}$.

$$S^n_{k,f} = (X \times \mathcal{F}, L_{proj}, \longrightarrow_n, (x, f))$$

One last definition remains, namely *completeness* of the synthesis construction. The formula reductions induced by Definition 9 are finite, which implies a terminating construction of the transition relation $\longrightarrow_0$. Since $\longrightarrow_0$ consists of finitely many transitions, only finitely many steps may be removed. This means that at some point, no more transitions are removed, and a stable point will be reached. If at this point, synthesizability holds at every reachable state, synthesis is successful. Otherwise, it is not. It is natural that a formal notion representing the first situation serves as a premise for a number of correctness results. This notion is formalized as *completeness* in Definition 13.

Definition 13. *For $k = (X, L, \longrightarrow, x) \in \mathcal{K}$, $f \in \mathcal{F}$ and $n \in \mathbb{N}$, we say that $S_{k,f}^n$ is* complete *if the following holds:*

$$\text{For all } (x, f) \longrightarrow_n^* (x', f') \text{ it holds that } (x', f') \uparrow f'.$$

5   Correctness

We show that synthesis as defined in this paper results in a controlled system adhering to the conditions in Definition 7. Detailed proofs are available in [19], while computer-verified proofs are available as well [16]. The synthesis method is finite (Theorem 1), and the result satisfies the synthesized requirement (Theorem 2). In addition, we show that the synthesis result is related to the original model via partial bisimulation (Theorem 3), which implies simulation. As a final result, we prove maximal permissiveness (Theorem 4). Assessing whether synthesis has been successful is done by checking whether synthesizability holds at every reachable state in the fixed point obtained as a result of Theorem 1.

Theorem 1. *For $k = (X, L, \longrightarrow, x) \in \mathcal{K}$, having finite $\longrightarrow$, and $f \in \mathcal{F}$, there exists an $n \in \mathbb{N}$ such that $S_{k,f}^n = S_{k,f}^m$ for all $m > n$.*

Theorem 2. *If $S_{k,f}^n$ is complete then $S_{k,f}^n \vDash f$.*

Theorem 3. *If $S_{k,f}^n$ is complete then $S_{k,f}^n \precsim k$.*

Theorem 4. *If $k' \preceq k$ and $k' \vDash f$ then $k' \preceq S_{k,f}^n$.*

6   Conclusions

This paper presents a novel approach to controlled system synthesis for modal logic on non-deterministic plant models. The behavior of a Kripke-structure with labeled transitions is adapted such that it satisfies the synthesized requirement. The relationship between the synthesis result and the original plant specification adheres to important notions from Ramadge-Wonham supervisory control: controllability and maximal permissiveness. The requirement specification logic also allows expressibility of deadlock-freeness and marker-state reachability. The synthesis approach, via a reduction on modal expressions combined with an iteratively applied synthesizability test for formulas assigned to target states of transitions results in an effective synthesis procedure. Our next research efforts will focus on determining the effectiveness of this procedure as well as its applicability in case studies.

References

1. Aminof, B., Mogavero, F., Murano, A.: Synthesis of hierarchical systems. In: Arbab, F., Ölveczky, P.C. (eds.) FACS 2011. LNCS, vol. 7253, pp. 42–60. Springer, Heidelberg (2012)
2. Arnold, A., Vincent, I., Walukiewicz, I.: Games for synthesis of controllers with partial observation. Theoretical Computer Science 1(303), 7–34 (2003)
3. Arnold, A., Walukiewicz, I.: Nondeterministic controllers of nondeterministic processes. In: Logic and Automata, pp. 29–52. Amsterdam University Press (2008)
4. Baeten, J., van Beek, B., van Hulst, A., Markovski, J.: A process algebra for supervisory coordination. In: Process Algebra and Coordination. EPTCS, pp. 36–55 (2011)
5. Bull, R., Segerberg, K.: Basic modal logic. In: Handbook of Philosophical Logic, pp. 1–88. Springer (1994)
6. Cassandras, C., Lafortune, S.: Introduction to Discrete Event Systems. Springer (1999)
7. Cleaveland, R., Steffen, B.: A linear-time model checking algorithm for the alternation-free modal mu-calculus. Formal Methods in System Design 2, 121–147 (1993)
8. D'Ippolito, N., Braberman, V., Piterman, N., Uchitel, S.: Synthesis of live behaviour models. In: Foundations of Software Engineering, pp. 77–86. ACM Press (2010)
9. D'Ippolito, N., Braberman, V., Piterman, N., Uchitel, S.: Synthesizing nonanomalous event-based controllers for liveness goals. ACM Transactions on Software Engineering Methodology 22(1), 1–36 (2013)
10. Hennessy, M., Milner, R.: Algebraic laws for nondeterminism and concurrency. Journal of the ACM 32(1), 137–161 (1985)
11. Lustig, Y., Vardi, M.: Synthesis from recursive-components libraries. In: Games, Automata, Logics and Formal Verification. EPTCS, pp. 1–16 (2011)
12. Pnueli, A., Rosner, R.: On the synthesis of a reactive module. In: Principles of Programming Languages, pp. 179–190. ACM Press (1989)
13. Ramadge, P., Wonham, W.: Supervisory control of a class of discrete event processes. SIAM Journal on Control and Optimization 25(1), 206–230 (1987)
14. Sokolsky, O., Smolka, S.: Incremental model checking in the modal mu-calculus. In: Dill, D.L. (ed.) CAV 1994. LNCS, vol. 818, pp. 351–363. Springer, Heidelberg (1994)
15. van Glabbeek, R.: The linear time-branching time spectrum II. In: Best, E. (ed.) CONCUR 1993. LNCS, vol. 715, pp. 66–81. Springer, Heidelberg (1993)
16. van Hulst, A.: Coq v8.3 proofs (2014), http://seweb.se.wtb.tue.nl/~ahulst/sofsem/
17. van Hulst, A., Reniers, M., Fokkink, W.: Maximal synthesis for Hennessy-Milner logic. In: Application of Concurrency to System Design, pp. 1–10. IEEE (2013)
18. van Hulst, A., Reniers, M., Fokkink, W.: Maximal synthesis for Hennessy-Milner logic with the box-modality. In: Workshop on Discrete Event Systems, pp. 278–285. IEEE (2014)
19. van Hulst, A., Reniers, M., Fokkink, W.: Maximally permissive controlled system synthesis for modal logic (2014), preprint at http://arxiv.org/abs/1408.3317/
20. Ziller, R., Schneider, K.: Combining supervisory synthesis and model checking. ACM Transactions on Embedded Computing Systems 4(2), 331–362 (2005)

Approximation Hardness of the Cross-Species Conserved Active Modules Detection Problem

Thomas Hume*, Hayssam Soueidan*, Macha Nikolski, and Guillaume Blin

University Bordeaux, CNRS / LaBRI, F-33405 Talence, France
University Bordeaux, CBiB, F-33076 Bordeaux, France
University Bordeaux, INSERM U1035, F-33076 Bordeaux, France
{guillaume.blin,thomas.hume,macha.nikolski}@labri.fr,
hayssam.soueidan@gmail.com

Abstract. Biological network comparison is an essential but algorithmically challenging approach for the analysis of underlying data. A typical example is looking for certain subgraphs in a given network, such as subgraphs that maximize some function of their nodes' weights. However, the corresponding MAXIMUM-WEIGHT CONNECTED SUBGRAPH (MWCS) problem is known to be hard to approximate. In this contribution, we consider the problem of the simultaneous discovery of maximum weight subgraphs in two networks, whose nodes are matched by a mapping: the MAXIMUM-WEIGHT CROSS-CONNECTED SUBGRAPHS (MWCCS) problem. We provide inapproximability results for this problem. These results indicate that the complexity of the problem is conditioned both by the nature of the mapping function and by the topologies of the two networks. In particular, we show that the problem is inapproximable even when the mapping is an injective function and the input graphs are two binary trees. We also prove that it remains hard to approximate when the mapping is a bijective function and the input graphs are a graph and a binary tree. We further analyze a variant of the MWCS problem where the networks' nodes are assigned both a weight and a contribution value, that we call MAXIMUM-WEIGHT RATIO-BOUNDED CONNECTED SUBGRAPH (MWRBCS). We provide a polynomial time algorithm for bounded-degree trees and an efficient dynamic programming solution for cycles. These algorithms allow us to derive a polynomial solution for MWCCS applicable when (i) MWRBCS is polynomially solvable for one of the graphs and (ii) the set of connected induced subgraphs of the other graph is polynomially enumerable.

1 Introduction

Networks of interacting units are a core concept in modern biology that enables understanding of biological processes at the system level. In their most basic form biological networks are graphs where vertices represent biological entities such as genes or proteins and edges represent interactions between these entities.

---

* These authors have contributed equally to the paper.

G.F. Italiano et al. (Eds.): SOFSEM 2015, LNCS 8939, pp. 242–253, 2015.
© Springer-Verlag Berlin Heidelberg 2015

Increasingly advanced experimental methods are used to provide evidence of existing interactions and nowadays comprehensive resources provide access to this knowledge (see for example [5] and [9]).

One of the key concepts to understand biological processes is that of *modules* within biological networks. Modules are considered to be sets of entities (genes, proteins, etc.) that function in a coordinated fashion or physically interact (for a review see [8]). The problem of finding gene modules within a biological network was first solved using simulated annealing by Ideker et al. [6].

A possible formulation for the problem of finding modules within a network is to look for connected sub-networks that maximize weights on the nodes. These weights typically represent some measure of biological activity, for example the expression level of genes. Finding the optimal (with respect to sum of weights) module in a biological network has been formally defined as the MAXIMUM (NODE-)WEIGHT CONNECTED SUBGRAPH problem (MWCS) [3].

Roughly, MWCS consists in the computation of the connected induced subgraph G' of a node-weighted graph G, such that the cumulative sum of its node's weights is maximal.

However, it is NP-hard to approximate the optimum of the MWCS problem within a constant factor $0 < \epsilon < 1$ [1]. Despite this complexity, there exist efficient exact solutions to this problem, using either reductions to the PRIZE-COLLECTING STEINER TREE problem [3], or using branch-and-cut mixed integer programming with node separation [1].

One limitation of the existing formulation is that it only considers one network at a time. Indeed, several studies have demonstrated the added value of identifying biological processes that are conserved across different conditions or even different species [12,7] as modules identified in single condition lack robustness [11]. We previously proposed a formulation for the identification of modules that are conserved across species. In our formulation, the two species are represented by two different networks with weighted nodes and we are provided with a mapping between the nodes of these networks. This mapping represents the similarity between genes or proteins across species, for example modeling orthology.

We formalized the identification of conserved modules as the MAXIMUM-WEIGHT CROSS-CONNECTED SUBGRAPHS (MWCCS) problem [4] which consists in the computation of two modules (connected subgraphs, one in each network), such that (*i*) the cumulative sum of their node weights is maximal and (*ii*) the proportion of *conserved* nodes within the solution is greater than a fixed threshold α. We consider a node in one of the modules to be conserved if it is mapped to a node in the other module. We have proposed an efficient mixed-integer programming solution for this problem and provided a fast implementation[1].

In this paper, we investigate the algorithmic complexity of the MWCCS problems. In the case of $\alpha = 0$, the MWCCS problem is as hard as the MWCS problem since it amounts to solving two independent MWCS instances. Here, we (*i*) establish the hardness of the problem when $\alpha = 1$, corresponding to a complete

---

[1] http://software.cwi.nl/xheinz

conservation requirement where all nodes in a module must admit a mapped counterpart in the other module; and (*ii*) provide polynomial exact algorithms for certain sub-cases and arbitrary α. This paper is organized as follows. We recall basic definitions and problem formulation in Section 2. In Section 3, we provide inapproximability results for this problem when $\alpha = 1$. These results indicate that the complexity of the problem is conditioned both by the nature of the mapping function and by the topologies of the two networks.

In particular, we show that the problem is inapproximable even when the mapping is an injective function and the input graphs are two binary trees. We also prove that it remains hard to approximate when the mapping is a bijective function and the input graphs are a graph and a binary tree. In Section 4, we study a variant of the MWCS problem where the networks' nodes are assigned both a weight and a contribution value, that we call MAXIMUM-WEIGHT RATIO-BOUNDED CONNECTED SUBGRAPH (MWRBCS). We provide a polynomial time algorithm for bounded-degree trees and an efficient dynamic programming solution for cycles. These algorithms allow us to derive a polynomial solution for MWCCS applicable when (*i*) MWRBCS is polynomially solvable for one of the graphs and (*ii*) the set of subgraphs of the other graph is polynomially enumerable.

2   Preliminaries

Let us first recall the basic needed material related to graphs. A graph $G = (V, E)$ consists of a set of vertices V and a set of edges (unordered pairs of vertices) E. We say that G is node-weighted if a function $w\colon V \to \mathbb{R}$ is provided. Given a graph $G = (V, E)$, its subgraph $G' = (V', E')$ is said to be *induced* if G' has exactly the edges that appear in G over the vertex set $V' \in V$, that is $E' = \{(x, y) \in E \mid x, y \in V'\}$. We denote the graph *induced* by the node set V' in G by $G[V']$.

The MWCS and MWCCS problems are formally defined as follows.

MAXIMUM (NODE-)WEIGHT CONNECTED SUBGRAPH problem (MWCS): Given a node-weighted graph $G = (V, E)$, w its node-weighting function, find a subset $V^* \subseteq V$, such that the induced graph $G[V^*]$ is connected, and $\sum_{v \in V^*} w(v)$ is maximum. Roughly, MWCS consists in the discovery of the connected subgraph of maximal weight, in a node weighted (possibly negatively) graph.

MAXIMUM-WEIGHT CROSS-CONNECTED SUBGRAPHS (MWCCS): Given two node-weighted graphs $G_1 = (V_1, E_1)$ and $G_2 = (V_2, E_2)$, w_1 and w_2 their respective node-weighting functions, a symmetric relation $M(V_1, V_2)$, and an interconnection ratio $\alpha \in [0, 1]$, MWCCS asks to find two subsets of nodes $V_1^* \subseteq V_1$ and $V_2^* \subseteq V_2$ such that:

1. the induced graphs $G_1[V_1^*]$ and $G_2[V_2^*]$ are connected, and
2. an α-fraction of the solution is M-related:
   $|U^*| \geq \alpha \times |V_1^* \cup V_2^*|$ where $U^* = \{u \in V_1^*, v \in V_2^* \mid M(u, v)\}$, and
3. $\sum_{u \in V_1^*} w_1(u) + \sum_{v \in V_2^*} w_2(v)$ is maximal.

3   Inapproximability of MWCCS

We prove the inapproximability of two specific cases of the MWCCS problem. First, we prove that if the mapping between G_1 and G_2 is an injective function and G_1 is a comb tree while G_2 is a binary tree, MWCCS is APX-hard and can not be approximated within factor 1.0014. Then, we prove that if the mapping is a bijective function, the problem is as hard to approximate as when considering a tree and a graph. These results shade light on the role of the mapping with respect to the difficulty of the problem.

Both proofs consist in an L-reduction from theAPX-hard MAX-3SAT(B) problem [10]: Given a collection $C_q = \{c_1, \ldots c_q\}$ of q clauses where each clause consists of a set of three literals over a finite set of n boolean variables $V_n = \{x_1, \ldots x_n\}$ and every literal occurs in at most B clauses, is there a truth assignment of V_n satisfying the largest number of clauses of C_q?

Proposition 1. *The* MWCCS *problem for a comb tree and a binary tree is APX-hard and not approximable within factor* 1.0014 *even when the mapping M is an injective function and a complete conservation (i.e. alpha* $= 1$*) is required.*

We first describe how we build an instance of MWCCS corresponding to an instance of MAX-3SAT(B). Given any instance (C_q, V_n) of MAX-3SAT(B), we build a comb tree $G_1 = (V_1, E_1)$ with weight function w_1, a binary tree $G_2 = (V_2, E_2)$ with weight function w_2 and a mapping M as follows.

The comb graph G_1 is defined as follows. The vertex set is $V_1 = \{r, l_i, c_j, dl_i, dc_j \mid 1 \le i \le n, 1 \le j \le q\}$. The edge set is given by the following equation.

$$E_1 = \{(c_j, dc_j), (l_i, dl_i) \mid 1 \le i \le n, 1 \le j \le q\} \cup$$
$$\{(dc_q, r), (r, dl_1)\} \cup$$
$$\{(dc_j, dc_{j+1}), (dl_i, dl_{i+1}) \mid 1 \le i < n, 1 \le j < q\}.$$

The weight function w_1 is defined as follows: for all $1 \le i \le n$ and $1 \le j \le q$, $w_1(l_i) = B$, $w_1(c_j) = 1$ and $w_1(r) = w_1(dc_j) = w_1(dl_i) = 0$.

Roughly, in G_1 there is a node for each clause (denoted by c_j) and for each literal (denoted by l_i) that represent the leaves of the comb. The spine of the comb contains dummy nodes for each clause (denoted by dc_j) and for each literal (denoted by dl_i) separated by a central node (denoted by r).

The binary tree $G_2 = (V_2, E_2)$ with weight function w_2 is defined as follows. The vertex set is $V_2 = \{r, x_i, \overline{x_i}, c_j^k, dx_i, d\overline{x_i}, dc_j^i, dc_j^{\overline{i}} \mid 1 \le i \le n, 1 \le j \le q, 1 \le k \le 3\}$. The edge set E_2 is given by the following equation.

$$E_2 = \{(r, dx_n)\} \cup$$
$$\{(c_j^{k'}, dc_j^k) \mid x_k, \text{is the } k'\text{-th literal of clause } c_j\} \cup$$
$$\{(c_j^{k'}, dc_j^{\overline{k}}) \mid \overline{x_k}, \text{is the } k'\text{-th literal of clause } c_j\} \cup$$
$$\{(dx_i, d\overline{x_{i+1}}) \mid 1 \le i < n\} \cup$$
$$\{(dx_i, d\overline{x_i}), (dx_i, x_{n-i+1}), (d\overline{x_i}, \overline{x_{n-i+1}}), (x_i, dc_1^i), (\overline{x_i}, dc_1^{\overline{i}}) \mid 1 \le i \le n\} \cup$$
$$\{(dc_j^i, dc_{j+1}^i), (dc_j^{\overline{i}}, dc_{j+1}^{\overline{i}}) \mid 1 \le i \le n, 1 \le j < q\}$$

The weight function w_2 is defined as follows: for all $1 \leq i \leq n$, $1 \leq j \leq q$ and $1 \leq k \leq 3$, $w_2(x_i) = w_2(\overline{x_i}) = -B$ and $w_2(r) = w_2(c_j^k) = w_2(dx_i) = w_2(d\overline{x_i}) = w_2(dc_j^i) = w_2(dc_j^{\overline{i}}) = 0$

Roughly, in G_2 there is a node for each literal of each clause (denoted by c_j^k) and for each value of each literal (denoted by x_i and $\overline{x_i}$). Dummy nodes for literals have been duplicated (one for each value of the literal - that is dx_i and $d\overline{x_i}$). Dummy nodes for clauses have also been duplicated (one for each value of all literals - dc_j^i and $dc_j^{\overline{i}}$). The structure is not as easy to informally describe as for G_1 but the reader may refer to an illustration provided in Figure 1.

Finally, the mapping M is an injective function from V_1 to V_2 defined as follows.

$$M(r) = r$$
$$M(l_i) = \{x_i, \overline{x_i}\}, \text{for all } 1 \leq i \leq n$$
$$M(c_j) = \{c_j^k | 1 \leq k \leq 3\}, \text{for all } 1 \leq j \leq q$$
$$M(dl_i) = \{dx_i, d\overline{x_i}\}, \text{for all } 1 \leq i \leq n$$
$$M(dc_j) = \{dc_j^i, dc_j^{\overline{i}}\}, \text{for all } 1 \leq i \leq n \text{ and } 1 \leq j \leq q$$

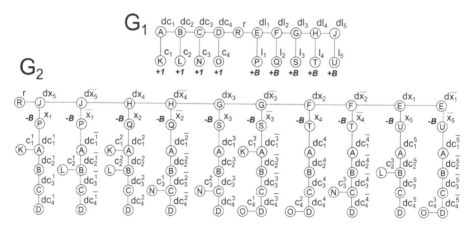

Fig. 1. Illustration of the construction of G_1, G_2, and M, given $C_q = \{(x_1 \lor x_2 \lor \neg x_3), (\neg x_1 \lor x_2 \lor x_5), (\neg x_2 \lor x_3 \lor \neg x_4), (\neg x_3 \lor x_4 \lor \neg x_5)\}$. For readability, the mapping M is not drawn but represented as labels located on the nodes: any pair of nodes (one in G_1 and one in G_2) of similar inner label are mapped in M.

Let us prove that this construction is indeed an L-reduction from MAX-3SAT(B). More precisely, we will prove the following property.

Lemma 1. *There exists an assignment of V_n satisfying at least m clauses of C_q if and only if there exists a solution to* MWCCS *of weight at least m.*

Proof. $\boxed{\Rightarrow}$ Given an assignment $\mathcal{A}$ of V_n satisfying m clauses of C_q, we construct a solution to MWCCS of weight m as follows.

$$
\begin{aligned}
\text{Let } V_1^* =\; & V_1 \setminus \{c_j \mid c_j \text{ is not satisfied by the assignment}\} \text{ and} \\
V_2^* =\; & \{r\} \cup \\
& \{c_j^k \mid c_j \text{ is satisfied by its } k\text{-th literal}\} \cup \\
& \{x_i, dc_j^i \mid x_i = 1, 1 \le j \le q\} \cup \\
& \{\overline{x_i}, dc_j^{\overline{i}} \mid x_i = 0, 1 \le j \le q\} \cup \\
& \{dx_i, d\overline{x_i} \mid 1 \le i \le n\}.
\end{aligned}
$$

By construction, $G_1[V_1^*]$ is connected since all the vertices of the spine of the comb have been kept. Moreover, $G_1[V_1^*]$ contributes $B \times n + m$ to the overall weight of the solution, that is B for each of the l_i and $+1$ for each satisfied clause. By construction, all the sub-trees rooted at x_i (resp. $\overline{x_i}$) are kept in $G_2[V_2^*]$ if $x_i = 1$ (resp. $x_i = 0$) in $\mathcal{A}$. Moreover, all the dummy nodes for literals (dx_i and $d\overline{x_i}$) and the root r have been kept. Thus, $G_2[V_2^*]$ is also connected. Furthermore, $G_2[V_2^*]$ contributes to $-B \times n$ to the overall weight of the solution since exactly one of each variable node (x_i and $\overline{x_i}$) has been kept. One can easily check that any node of V_1^* has a mapping counterpart in V_2^*. The overall solution is valid and of total weight m.

$\boxed{\Leftarrow}$ Given any solution $\{V_1^*, V_2^*\}$ to MWCCS of weight m, we construct a solution to the MAX-3SAT(B) problem satisfying at least m clauses as follows.

First, note that we can assume that any such solution to MWCCS is *canonical*, meaning that V_2^* does not contain both vertices x_i and $\overline{x_i}$ for all $1 \le i \le n$. Indeed, by contradiction, suppose there exists a solution such that $\{x_i, \overline{x_i}\} \subseteq V_2^*$ for a given $1 \le i \le n$. Then, $\{x_i, \overline{x_i}\}$ in G_2 induce a negative weight of $-2B$. This negative contribution can at most be compensated by the weight of the corresponding literal node in G_1 ($w_1(l_i) = B$) and at most B clause nodes in G_1 ($B \ge \sum w_1(c_j)$ where $x_i \in c_j$ or $\overline{x_i} \in c_j$) since every literal occurs in at most B clauses in C_q. Therefore, such local configuration does not provide any positive contribution to the solution and can be transformed into a better solution by removing one of the sub-trees rooted in $\{x_i, \overline{x_i}\}$. We will consider hereafter that m is the weight of the resulting canonical solution. We further assume that $m > 1$ since otherwise we can build a trivial assignment $\mathcal{A} = \{c_1^1 = 1\}$ of V_n that is satisfying at least one clause of C_q.

Let $\mathcal{A}$ be an assignment of V_n such that for all $1 \le i \le n$ if $x_i \in V_2^*$ then $x_i = 1$ and $x_i = 0$ otherwise. Note that, since our solution is canonical, each literal has been assigned a single boolean value in $\mathcal{A}$. Let us now prove that this assignment satisfies at least m clauses of C_q.

First, note that since our solution is canonical and we require any node of V_1^* to have a mapping counterpart in V_2^*, this implies that if $l_i \in V_1^*$ then its contribution (that is $w_1(l_i) = B$) is cancelled by the negative contribution of either x_i or $\overline{x_i}$ in V_2^* (that is $w_2(x_i) = w_2(\overline{x_i}) = -B$). Therefore, the weight m

of the solution can only be realized by m clause nodes of G_1, say $\mathcal{C}_1 \subseteq V_1^*$ – since $w_1(c_j) = 1$ for all $1 \leq j \leq q$.

As already stated, to be part of the solution any node in V_1^* has a mapping counterpart in V_2^*. Thus, for each node in $\mathcal{C}_1$, there should be a node of $\mathcal{C}_2 \subseteq \{c_j^k \mid 1 \leq j \leq q, 1 \leq k \leq 3\}$ in V_2^*. More precisely, by construction, any node c_j in V_1 has exactly three mapping counterparts in V_2 (that is $\{c_j^k \mid 1 \leq k \leq 3\}$) and for each $c_j \in \mathcal{C}_1$ at least one of these mapping counterparts has to belong to $\mathcal{C}_2$.

Finally, since both $G_1[V_1^*]$ and $G_2[V_2^*]$ have to be connected, each node in $\mathcal{C}_2$, say c_j^k, should be connected by a path to a node x_i or $\overline{x_i}$, say x_i, for some $1 \leq i \leq n$, in $G_2[V_2^*]$. By construction, this is the case if x_i is the k-th literal of the clause c_j for some $1 \leq k \leq 3$. Thus, $\mathcal{A}$ is an assignment that satisfies any clause c_j such that the clause node c_j belongs to V_1^*. As already stated $|\mathcal{C}_1| = m$. □

The above reduction linearly preserves the approximation since the weights of optimal solutions of the problems correspond and there exists an assignment of V_n satisfying at least m clauses of C_q if and only if there exists a solution to MWCCS of weight at least m. Hence, given an approximation to MWCCS, one can derive an algorithm for MAX-3SAT(B) with the same approximation ratio. Since MAX-3SAT(B), $B \geq 3$, is APX-hard [10] and MAX-3SAT(B) for $B = 6$ is not approximable within factor 1.0014 [2], so is MWCCS, which proves Proposition 1.

Let us now prove a similar result for MWCCS problem when the mapping is a bijective function.

Proposition 2. *The MWCCS problem for a graph and a tree is APX-hard and not approximable within factor* 1.0014 *even when the mapping is a bijective function and a complete conservation (i.e. alpha* $= 1$*) is required.*

Proof. Given any instance (C_q, V_n) of MAX-3SAT(B), we build a graph $G_1 = (V_1, E_1)$ with weight function w_1, a tree $G_2 = (V_2, E_2)$ with weight function w_2 and a mapping M as follows. The graph G_1 has the vertex set $V_1 = \{r, l_i, x_i, \overline{x_i}, c_j, c_j^k \mid 1 \leq i \leq n, 1 \leq j \leq q, 1 \leq k \leq 3\}$ and the edge set defined by the following equation.

$$E_1 = \{(l_i, x_i), (l_i, \overline{x_i}), (r, x_i), (r, \overline{x_i}) \mid 1 \leq i \leq n\} \cup$$
$$\{(c_j, c_j^k), (r, c_j^k) \mid 1 \leq k \leq 3, 1 \leq j \leq q\}.$$

The weight function w_1 is defined as follows: for all $1 \leq k \leq 3$, $1 \leq i \leq n$ and $1 \leq j \leq q$, $w_1(l_i) = B$, $w_1(c_j) = 1$ and $w_1(r) = w_1(c_j^k) = w_1(x_i) = w_1(\overline{x_i}) = 0$.

Roughly, in G_1 there is a node for each clause (denoted by c_j), for each of the three literals of each clause (denoted by c_j^k), for each literal (denoted by l_i) and for each valuation of each literal (denoted by x_i, $\overline{x_i}$). Clause nodes and literal nodes are separated by a central node r.

The tree G_2 is defined as follows. The vertex set is $V_2 = V_1$, the edge set is given by the following equation:

$$E_2 = \{(l_i, r), (c_j, r), (x_i, r), (\overline{x_i}, r) \mid 1 \le i \le n, 1 \le j \le q\} \cup$$
$$\{(c_j^k, x_i) \mid x_i \text{ is the } k\text{-th literal of clause } c_j\} \cup$$
$$\{(c_j^k, \overline{x_i}) \mid \overline{x_i} \text{ is the } k\text{-th literal of clause } c_j\}.$$

The weight function w_2 is defined as follows: for all $1 \le k \le 3$, $1 \le i \le n$ and $1 \le j \le q$, $w_2(x_i) = w_2(\overline{x_i}) = -B$, $w_2(r) = w_2(c_j^k) = w_2(l_i) = w_2(c_j) = 0$.

Roughly, in G_2 all the nodes except the ones in $\{c_j^k \mid 1 \le j \le q, 1 \le k \le 3\}$ form a star centered in node r. The nodes representing the literal of the clause (that is c_j^k) are connected to their corresponding variable nodes (that is x_i or $\overline{x_i}$).

Finally, the mapping M is a bijective function from V_1 to V_2 defined as the identity (that is each node in V_1 is mapped to the node of similar label in V_2).

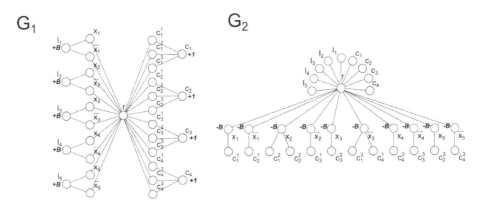

Fig. 2. Illustration of the construction of G_1, G_2, and M, given $C_q = \{(x_1 \vee x_2 \vee \neg x_3), (\neg x_1 \vee x_2 \vee x_5), (\neg x_2 \vee x_3 \vee \neg x_4), (\neg x_3 \vee x_4 \vee \neg x_5)\}$. For readability, the mapping M is not drawn but deduced from the labels of the nodes; any pair of nodes (one in G_1 and one in G_2) of similar label are mapped in M.

Let us prove that this construction is indeed an L-reduction from MAX-3SAT(B). More precisely, we will prove the following property.

Lemma 2. *There exists an assignment of V_n satisfying at least m clauses of C_q if and only if there exists a solution (not necessarily optimal) to MWCCS of weight at least m.*

Proof. $\boxed{\Rightarrow}$ Given an assignment $\mathcal{A}$ of V_n satisfying m clauses of C_q, we construct a solution to MWCCS of weight m as follows.

Let $V_1^* = V_2^* = \{c_j \mid c_j \text{ is satisfied by } \mathcal{A}\} \cup \{c_j^k \mid c_j^k \text{ is satisfying } c_j \text{ by } \mathcal{A}\} \cup \{x_i \mid x_i = 1\} \cup \{\overline{x_i} \mid x_i = 0\} \cup \{r, l_i \mid 1 \le i \le n\}$.

By construction, $G_1[V_1^*]$ and $G_2[V_2^*]$ are connected. Moreover, $G_1[V_1^*]$ contributes $B \times n + m$ to the overall weight of the solution, that is B for each of the l_i and $+1$ for each satisfied clause, while $G_2[V_2^*]$ contributes $-B \times n$ to the overall weight of the solution since exactly one of each variable node (*i.e.*, x_i and $\overline{x_i}$) has been kept. The overall solution is valid and of total weight m.

$\boxed{\Leftarrow}$ Given any solution $V^* \subseteq V_1$ to MWCCS of weight m, we construct a solution to the MAX-3SAT(B) problem satisfying at least m clauses as follows.

First, note that, as in the previous construction, we can assume that any such solution to MWCCS is *canonical* meaning that V^* does not contain both vertices x_i and $\overline{x_i}$ for any $1 \le i \le n$.

Let $\mathcal{A}$ be an assignment of V_n such that for all $1 \le i \le n$, if $x_i \in V^*$ then $x_i = 1$ and $x_i = 0$ otherwise. Note that, since our solution is canonical, each literal has been assigned a single boolean value in $\mathcal{A}$. Let us now prove that this assignment satisfies at least m clauses of C_q.

First, note that since our solution is canonical, as in the previous construction, the weight m of the solution can only be induced by m clause nodes of G_1, say $\mathcal{C}_1 \subseteq V^*$.

Since both $G_1[V^*]$ and $G_2[V^*]$ have to be connected, any solution with $m > 1$ will include node r in V^*. Thus, for each node $c_j \in \mathcal{C}_1$ there should be a node of $\{c_j^k \mid 1 \le k \le 3\}$ in $G_1[V^*]$ to connect c_j to r. In $G_2[V^*]$, in order for nodes r and c_j^k to be connected, the corresponding literal node (that is x_i or $\overline{x_i}$), say x_i – has to be kept in V^*. By construction, this is the case if x_i is the k-th literal of clause c_j. Thus, $\mathcal{A}$ is an assignment that satisfies any clause c_j such that the clause node c_j belongs to V^*. As already stated $|\mathcal{C}_1| = m$.   □

The above reduction linearly preserves the approximation and proves Proposition 2.

4   A General Algorithm for Some Polynomial Cases of MWCCS

In this section, we consider the general version of the problem where α is given in the input rather than being fixed, but where the mapping is restricted to a partial function (any element of V_1 has at most one image in V_2) and G_1 to a polynomially enumerable graph (i.e. the number of connected induced subgraphs is polynomially bounded). We will consider each subgraph of G_1 as part of a candidate solution and will try to find the best subgraph in G_2, that is a subgraph that maximizes the total weight of the candidate solution and such that at least an α-fraction of the nodes of G_1 and G_2 in the solution are M-related. The optimal solution will be the maximum among the candidate ones.

We suppose that there is a polynomial number of connected induced subgraphs of G_1. For every subgraph $G_1' = (V_1', E_1')$ of G_1, we define the corresponding G_2 *contribution function* $c \colon V_2 \to \mathbb{N}$ to be $c(v) = |\{v, u \mid M(u, v), u \in V_1'\}|$. Informally, the contribution function provides for each node of V_2 the number of inverse images plus one if at least one exists, given that G_1' is supposed to be the candidate solution.

Given $G_2 = (V_2, E_2)$, its weight-function w_2 and its contribution function c, the problem now corresponds to the discovery of the connected subgraph of maximum weight such that: $\sum_{v \in V_2'} c(v) - \alpha \times |V_1'| \geq \alpha \times |V_2'|$, where $\alpha \times |V_1'|$ is constant.

We call this problem the MAXIMUM-WEIGHT RATIO-BOUNDED CONNECTED SUBGRAPH (MWRBCS) problem and in the general case is defined formally as follow: Given a node-weighted graph $G = (V, E)$, its node-weighting function $w \colon V \to \mathbb{R}$, its contribution function $c \colon V \to \mathbb{N}$ such that $\sum_{v \in V} c(v) \leq k|V|$, $k \in \mathbb{N}$, and a ratio $\alpha \in [0, 1]$ find a subset $V^* \subseteq V$ such that:

1. the induced graph $G[V^*]$ is connected, and
2. the ratio of the sum of contributions plus some constant over the number of nodes in the solution is greater than or equal to α, that is:
   $$\sum_{v \in V^*} c(v) + C \geq \alpha \times |V^*|, \text{ and}$$
3. $\sum_{v \in V^*} w(v)$ is maximum.

Proposition 3. MWRBCS *is as difficult as* MWCS.

Proof. Indeed, when $\forall v \in V, c(v) = 1$ the MWCS and MWRBCS problems are equivalent. Thus, MWRBCS is hard to approximate for general graphs.   □

Let us show now that it is in PTIME for bounded-degree trees.

Proposition 4. MWRBCS *is solvable in* $O(n^{d+2})$ *time for d-ary trees.*

Proof. Let us consider the MWRBCS problem for a d-ary tree. We define a dynamic programming strategy with a $O(n^{d+2})$ time complexity. This leads to a polynomial algorithm for d-ary trees. The basic idea is to define a 3-dimensional table T of size $|V| \times \sum_{v \in V} c(v) \times |V|$ that stores the maximum weight of a subtree rooted in v of size s and of total contribution tc.

Formally, $\forall v \in V, 0 \leq tc \leq \sum_{v \in V} c(v), 0 \leq s \leq |V|$, let us note $v_{\langle i \rangle}$ the i-th child of v, $1 \leq i \leq d$, we have:

$$T[v][0][0] = 0$$

$$T[v][tc][s] = \max_{tc_1, \ldots, tc_d, \, s_1, \ldots, s_d} \left(w(v) + \sum_{1 \leq i \leq d} T[v_{\langle i \rangle}][tc_i][s_i] \right)$$

$$\text{s.t.} \quad tc = c(v) + \sum_{1 \leq i \leq d} tc_i$$

$$s = 1 + \sum_{1 \leq i \leq d} s_i$$

The optimal subtree can be reconstructed from the table by finding the entry with the maximal weight and where the contribution ratio is not violated, and backtracking from that entry on the selected tc_i's and s_i's from the max function. Each entry of the table can be computed in $O(n^{d-1})$ (that is, an integer partition of $|V|$ into d parts) time, and since $\sum_{v \in V} c(v) \in O(n)$, there are $O(n^3)$ of them, which leads to the overall complexity.   □

As paths and cycles are trees of degree 1, using the preceding result leads to an $O(n^3)$ algorithm for these cases. However, one can achieve a better complexity.

Proposition 5. MWRBCS *is solvable in* $O(n^2)$ *time for paths and cycles.*

Proof. Let us first consider the MWRBCS problem for paths. Leveraging the linearity of the graph structure, we define a dynamic programming strategy with an $O(n^2)$ time complexity.

The idea is to define two 2-dimensional tables T_w and T_{tc} with n^2 entries each and that store respectively, for each pair of indices, the maximum weight and the total contribution, of the corresponding graph. Let us consider a given orientation in the path with the node at the starting end as the reference node, of index 0. Every candidate solution (a subpath) in the path can then be defined as a pair of positions, the first element being the starting position as an index number, the second element being the size of the candidate solution. The main idea being that increasing the indices one by one enables us to update the weights and total contributions incrementally.

Formally, let us denote the k-th node of the graph in the predefined orientation by n_k, we have for all $0 \leq i \leq j \leq n$:

$$T_w[i][i] = 0$$
$$T_w[i][j] = w(n_{i+j-1}) + T_w[i][j-1]$$

$$T_{tc}[i][i] = 0$$
$$T_{tc}[i][j] = c(n_{i+j-1}) + T_{tc}[i][j-1]$$

The optimal subpath is defined by the indices of the entry with the maximal weight and where the contribution ratio is not violated (*i.e.*, for any (i, j) s.t. $T_{tc}[i][j] \geq \alpha \cdot j$). Each $O(n^2)$ entry of the tables can be computed in constant time, leading to the overall complexity. For cycles, the trick consists in taking any linearization of the cycle and merging two copies of the corresponding linearization as the input path. This ensures that we will consider any candidate solution (*i.e.*, simple subpath of the cycle). The time complexity is preserved. □

5   Conclusion

In this contribution we provide the first deep complexity analysis of the MWCCS problem and show several interesting results. There still remain numerous pertinent questions to be answered. First of all, generalizing the problem to more than two graphs is of interest; even if the hardness results will hold, what practical solutions can be derived? We also would like to study the complexity effect of the relaxation of the connectivity constraints. Finally, it would be relevant to further analyse the links that can be set up between MWRBCS and variants of MWCS such as the budget constraint one.

References

1. Álvarez-Miranda, E., Ljubić, I., Mutzel, P.: The maximum weight connected subgraph problem. In: Facets of Combinatorial Optimization, pp. 245–270. Springer (2013)

2. Berman, P., Karpinski, M.: On some tighter inapproximability results (extended abstract). In: Wiedermann, J., Van Emde Boas, P., Nielsen, M. (eds.) ICALP 1999. LNCS, vol. 1644, pp. 200–209. Springer, Heidelberg (1999)
3. Dittrich, M.T., Klau, G.W., Rosenwald, A., Dandekar, T., Müller, T.: Identifying functional modules in protein–protein interaction networks: an integrated exact approach. Bioinformatics 24(13), i223–i231 (2008)
4. El-Kebir, M., Soueidan,H., Hume, T., Beisser, D., Dittrich, M., Müller, T., Blin, G., Heringa, J., Nikolski, M., Wessels, L.F.A., Klau, G.W.: Conserved cross-species network modules elucidate Th17 T cell differentiation in human and mouse (under review, 2014)
5. Franceschini, A., Szklarczyk, D., Frankild, S., Kuhn, M., Simonovic, M., Roth, A., Lin, J., Minguez, P., Bork, P., von Mering, C., Jensen, L.J.: String v9.1: protein-protein interaction networks, with increased coverage and integration. Nucleic Acid Research 41(D1), 808–815 (2013)
6. Ideker, T., Ozier, O., Schwikowski, B., Siegel, A.F.: Discovering regulatory and signalling circuits in molecular interaction networks. Bioinformatics 18(suppl. 1), S233–S240 (2002)
7. Lu, Y., Rosenfeld, R., Nau, G.J., Bar-Joseph, Z.: Cross species expression analysis of innate immune response. Journal of Computational Biology 17(3), 253–268 (2010)
8. Mitra, K., Carvunis, A.-R., Ramesh, S.K., Ideker, T.: Integrative approaches for finding modular structure in biological networks. Nature Reviews Genetics 14(10), 719–732 (2013)
9. Orchard, S., Kerrien, S., Abbani, S., Aranda, B., Bhate, J., Bidwell, S., Bridge, A., Briganti, L., Brinkman, F.S.L., Cesareni, G., Chatr-Aryamontri, A., Chautard, E., Chen, C., Dumousseau, M., Goll, J., Hancock, R.E.W., Hannick, L.I., Jurisica, I., Khadake, J., Lynn, D.J., Mahadevan, U., Perfetto, L., Raghunath, A., Ricard-Blum, S., Roechert, B., Salwinski, L., Stumpflen, V., Tyers, M., Uetz, P., Xenarios, I., Hermjakob, H.: Protein interaction data curation: the international molecular exchange (imex) consortium. Nature Methods 9(4), 345–350 (2012)
10. Papadimitriou, C.H., Yannakakis, M.: Optimization, approximation, and complexity classes. Journal of Computer and System Sciences 43(3), 425–440 (1991)
11. Staiger, C., Cadot, S., Kooter, R., Dittrich, M., Müller, T., Klau, G.W., Wessels, L.F.A.: A critical evaluation of network and pathway-based classifiers for outcome prediction in breast cancer. PLoS One 7(4), e34796 (2012)
12. Waltman, P., Kacmarczyk, T., Bate, A.R., Kearns, D.B., Reiss, D.J., Eichenberger, P., Bonneau, R.: Multi-species integrative biclustering. Genome Biology 11(9), R96 (2010)

Finding Highly Connected Subgraphs*

Falk Hüffner, Christian Komusiewicz, and Manuel Sorge

Institut für Softwaretechnik und Theoretische Informatik, TU Berlin
{falk.hueffner,christian.komusiewicz,manuel.sorge}@tu-berlin.de

Abstract. A popular way of formalizing clusters in networks are *highly connected* subgraphs, that is, subgraphs of k vertices that have edge connectivity larger than $k/2$ (equivalently, minimum degree larger than $k/2$). We examine the computational complexity of finding highly connected subgraphs. We first observe that this problem is NP-hard. Thus, we explore possible parameterizations, such as the solution size, number of vertices in the input, the size of a vertex cover in the input, and the number of edges outgoing from the solution (edge isolation), and expose their influence on the complexity of this problem. For some parameters, we find strong intractability results; among the parameters yielding tractability, the edge isolation seems to provide the best trade-off between running time bounds and a small parameter value in relevant instances.

1 Introduction

A popular method of analyzing complex networks is to identify *clusters* or *communities*, that is, subgraphs that have many interactions within themselves and fewer with the rest of the graph (e. g. [18, 19]). Hartuv and Shamir [9] proposed a prominent clustering algorithm producing *highly connected* clusters, formalized as follows: the *edge connectivity* $\lambda(G)$ of a graph G is the minimum number of edges whose deletion results in a disconnected graph, and a graph G with n vertices is called *highly connected* if $\lambda(G) > n/2$. An equivalent characterization is that a graph is highly connected if each vertex has degree at least $\lfloor n/2 \rfloor + 1$ [3]. Moreover, highly connected graphs have diameter at most two [9].

We study the following problem:

HIGHLY CONNECTED SUBGRAPH
Input: An undirected graph $G = (V, E)$ and a nonnegative integer k.
Question: Is there a vertex set S such that $|S| = k$ and $G[S]$ is highly connected?

In addition to the natural application in analyzing complex networks [19], HIGHLY CONNECTED SUBGRAPH also occurs (with vertex weights) as a subproblem in a column generation algorithm for partitioning graphs into highly connected components [11].

---

* FH and MS gratefully acknowledge support by Deutsche Forschungsgemeinschaft, projects ALEPH (HU 2139/1) and DAPA (NI 369/12) respectively. Due to space constraints, proofs for results marked by ⋆ are deferred to a full version of this extended abstract.

G.F. Italiano et al. (Eds.): SOFSEM 2015, LNCS 8939, pp. 254–265, 2015.

Since HIGHLY CONNECTED SUBGRAPH is NP-hard (Theorem 1), we explore the "parameter ecology" [7] of this problem. We are looking for fixed-parameter algorithms, that is, we try to find problem parameters p that allow for a running time bounded by $f(p) \cdot |G|^{O(1)}$. The hope is that the function f grows not too fast (although it has to be superpolynomial unless P = NP), and that the parameter value p can be expected to be relatively small in interesting instances. Clearly, there is a trade-off between these goals. Similarly to NP-hardness, fixed-parameter tractability can be refuted by giving suitable reductions from hard problems of the classes W[1] or W[2]. For details, refer to the literature [5].

Results. We list the results going from the hardest parameters to the easiest, corresponding roughly to going from small expected parameter values to large ones. Let n be the number of vertices in G. For the parameter $\ell := n - k$ (the number of vertices to delete to obtain a highly connected subgraph), we obtain a strong hardness result: there is a trivial $n^{O(\ell)}$ time algorithm, but it is unlikely that $n^{o(\ell)}$ time can be achieved (Theorem 1). For the size of the solution k, a fixed-parameter algorithm is unlikely, even if we additionally consider the degeneracy of G as a parameter (Theorem 2). If we take the minimum size τ of a vertex cover for G as parameter, we obtain a fixed-parameter algorithm: the problem can be solved in $(2\tau)^\tau \cdot n^{O(1)}$ time (Theorem 3). Considering the number of vertices n, we can clearly solve the problem in $2^n \cdot n^{O(1)}$ time. We show that it is unlikely that this can be improved to $2^{o(n)} \cdot n^{O(1)}$ time (Theorem 4). If the parameter is the number γ of edges between $G[S]$ and the remaining vertices, then the problem can be solved in time $O(4^\gamma n^2)$ (Theorem 5). Finally, if we consider the number α of edges to delete to obtain a highly connected subgraph (plus singleton vertices), we obtain a $O(2^{4 \cdot \alpha^{0.75}} + \alpha^2 nm)$-time algorithm (Theorem 8). This running time is subexponential in α.

Related work. The algorithm by Hartuv and Shamir [9] partitions a graph heuristically into highly connected components; another algorithm tries to explicitly minimize the number of edges that need to be deleted for this [11]. Highly connected graphs can be seen as *clique relaxation* [18], that is, a graph class that has many properties similar to cliques, without being as restrictive. Highly connected graphs are very similar to *0.5-quasi-complete graphs* [17], that is, graphs where every vertex has degree at least $(n-1)/2$. These graphs are also referred to as *(degree-based) 0.5-quasi-cliques* [15]. Recently, also the task of finding subgraphs with high *vertex* connectivity has been examined [20].

Preliminaries. Using standard graph notation, we consider only simple undirected graphs $G = (V, E)$ with $n := |V|$ and $m := |E|$; we call n the *order* of G. We use $N(v)$ to denote the set of *neighbors* of a vertex v. For a vertex set $S \subseteq V$, we denote $G[S] := (S, \{\{u, v\} \in E \mid u, v \in S\})$ the *subgraph of G induced by S*. We use $G - S$ as shorthand for $G[V \setminus S]$. A cut (A, B) in a graph $G = (V, E)$ is a vertex bipartition, that is, $A \cap B = \emptyset$ and $A \cup B = V$. The *cut edges* are the edges between vertices in A and B; the *size* of a cut is the number of its cut edges.

2   Vertex Deletion

For finding large cliques in a graph, one successful approach is to use fixed-parameter algorithms for the parameter "number of vertices in the graph that are not in the clique" [13, 14]. We show by a reduction from HITTING SET that such fixed-parameter algorithms are unlikely for HIGHLY CONNECTED SUBGRAPH.

Theorem 1 ($\star$). HIGHLY CONNECTED SUBGRAPH *is NP-hard and W[2]-hard parameterized by* $\ell := n-k$. *Moreover, an* $n^{o(\ell)}$*-time algorithm implies FPT=W[1].*

3   Solution Size and Degeneracy

A graph has degeneracy d if every subgraph contains at least one vertex that has degree at most d. In many graphs from real-world applications, the degeneracy of a graph is very small compared to the network size [6]. For yes-instances, the degeneracy of the input graph has to be at least $\lfloor k/2 \rfloor + 1$. Therefore, HIGHLY CONNECTED SUBGRAPH is polynomial-time solvable if the input graph has constant degeneracy: trying all subgraphs with $k \leq 2d + 2$ vertices decides the problem in $n^{2d} \cdot n^{O(1)}$ time. This can be improved to the following running time.

Proposition 1 ($\star$). HIGHLY CONNECTED SUBGRAPH *can be solved in* $2^d \cdot n^{d+O(1)}$ *time where d is the degeneracy of G.*

Unfortunately, if we regard the degeneracy as a parameter instead of a constant, we obtain hardness using a reduction from CLIQUE, even if additionally the solution size k is a parameter.

Theorem 2 ($\star$). HIGHLY CONNECTED SUBGRAPH *parameterized by the combined parameter (d, k), where d is the degeneracy of G, is W[1]-hard.*

4   Vertex Cover Size

We next consider the parameter τ, the minimum size of a vertex cover of G. This parameter is interesting because it can be smaller than the number of vertices of G. We show that HIGHLY CONNECTED SUBGRAPH is solvable in $(2\tau)^\tau \cdot n^{O(1)}$ time. The algorithm first computes a vertex cover C, then determines via branching the intersection of C and the desired solution S, and then adds suitable vertices of the independent set $V \setminus C$. We identify suitable vertices in the independent set by solving an instance of SET MULTICOVER.

SET MULTICOVER
Input: A universe U with covering demands $d : U \rightarrow \mathbb{N}$, a family $\mathcal{F}$ of subsets of the universe with multiplicity values $m : \mathcal{F} \rightarrow \mathbb{N}$, and $p \in \mathbb{N}$.
Question: Is there a multiset of at most p subsets from $\mathcal{F}$ that contains each $F \in \mathcal{F}$ at most $m(F)$ times, and covers each $u \in U$ with at least $d(u)$ subsets?

Lemma 1. *A given instance of* HIGHLY CONNECTED SUBGRAPH *with a vertex cover of size τ can be solved using the answers to at most 2^τ instances of* SET MULTICOVER, *each with $|U| \leq \tau$ and $2 \max_{u \in U} d(u) \leq \tau$. Furthermore, all these instances can be computed in $O(2^\tau(n + \tau n))$ time.*

Proof. Fix some highly connected subgraph $G[S]$ of order k if it exists. First compute a minimum vertex cover C for G in $O(1.274^\tau + \tau n)$ time [4]. Enumerate all 2^τ possibilities for $C' \subseteq C$. Clearly, in one branch $C' = C \cap S$. In each branch, delete the vertices from $C \setminus C'$. Then remove vertices from the independent set $V \setminus C'$ that have $k/2$ or less neighbors in C', since they cannot be part of S. Let V' be the thus reduced vertex set. It remains to find $k' := k - |C'|$ vertices in $V' \setminus C'$ such that each vertex v in C' has more than $d(v) := k/2 - |N(v) \cap C'|$ neighbors among these k' vertices. This is an instance of SET MULTICOVER: In our case, the universe is C', the covering demands are d as defined above, the family is $\mathcal{F} = \{N(v) \cap C' \mid v \in V' \setminus C'\}$, the multiplicity of $X \in \mathcal{F}$ is the number of vertices in $V' \setminus C'$ having neighborhood X in C', and $p = k'$. If the solution to SET MULTICOVER has less than k' sets, then we can add arbitrary further vertices from $V' \setminus C'$ to make the vertex subset large enough (if $|V' \setminus C'| < k'$, then we can safely reject this branch for $C' \subseteq C$). $\qquad\square$

SET MULTICOVER with multiplicity constraints can be solved in $O((b+1)^{|U|}|\mathcal{F}|)$ time [10], where $b := \max_{u \in U} d(u)$. Note that $|C'| > k/2$, since the vertices outside of C' form an independent set and we cannot choose $k/2$ or more of them. Thus, $b < |C'| \leq \tau$. The size of $\mathcal{F}$ is at most n. Together with the enumeration of the instances, we obtain the following.

Theorem 3. HIGHLY CONNECTED SUBGRAPH *can be solved in $O((2\tau)^\tau \cdot \tau n)$ time.*

5   Number of Vertices

A trivial algorithm for HIGHLY CONNECTED SUBGRAPH is to enumerate all vertex subsets S of size k and to check for each subset whether it is highly connected. This algorithm has running time $O(2^n \cdot m)$. We now show by a reduction from CLIQUE that a running time improvement to $2^{o(n)} \cdot n^{O(1)}$ is unlikely. The idea of the reduction is to add to a CLIQUE instance some new graph that is so large, that, in the resulting instance of HIGHLY CONNECTED SUBGRAPH, every highly connected graph of size k must contain this new graph. The remaining vertices must form a clique in order to have sufficiently high degree. The following lemma shows that we can efficiently construct the graph which we need to add to the CLIQUE instance.

Lemma 2 $(\star)$. *For any two integers $a, b \in \mathbb{N}$ such that a is even, $b - 3 \geq 8$ is a power of two, and $a - 2 \geq 2b$, there is a graph $G = (X \cup W, E)$ on the disjoint vertex sets X and W, such that*
  i) $G[X]$ is connected,
  ii) $|X| = a - 2, |W| = a - b + 1,$

iii) $N_G(X) \setminus X = W$,

iv) *each vertex in X has degree a, and each vertex in W has degree $a - b$.*
Moreover, G can be constructed in time polynomial in a.

We call the graph G described in the above lemma an (a, b)-*equalizer* and the vertices in W are its *ports*.

Lemma 3. *There is a polynomial-time many-one reduction from* CLIQUE *to* HIGHLY CONNECTED SUBGRAPH *that is parameter-linear with respect to the number of vertices.*

Proof. Let (G, p) represent an instance of CLIQUE. Without loss of generality, assume that $p - 3 \geq 8$ and $p - 3$ is a power of two. Otherwise, repeatedly add a universal vertex and increase p by one, until $p - 3 \geq 8$, and $p - 3$ is a power of two. Note that this at most doubles p. Furthermore, assume that $n - 1 \geq p$; otherwise, solve the instance in polynomial time.

Denote $|V(G)| = n$. We construct the instance (G', k) of HIGHLY CONNECTED SUBGRAPH where $k = 4n - 1$. Note that the minimum degree in a highly connected graph with k vertices is $2n$. Graph G' is constructed as follows. First, copy G into G'. Then add a vertex-disjoint $(2n, p)$-equalizer. By Lemma 2, a $(2n, p)$-equalizer exists and is computable in polynomial time, because, by choice of (G, p), $2n$ is even, $p - 3 \geq 8$ is a power of two, and $2n - 2 \geq 2p$. Denote the ports of the equalizer by W and its remaining vertices by X. Add an edge between each port and each vertex in $V(G)$; this finishes the construction. The graph G' has less than $5n$ vertices, since the $(2n, p)$-equalizer has less than $4n$ vertices. It remains to show equivalence of the instances, that is,

(G, p) is a yes-instance $\Leftrightarrow$ $(G', k = 4n - 1)$ is a yes-instance.

"$\Rightarrow$": Let $G[S]$ be a clique of order p in G. Then, $G'[S \cup X \cup W]$ is highly connected: Each vertex in S is adjacent to $p - 1$ vertices in S and to $2n - p + 1$ vertices in W. Hence, each vertex in S has $2n$ neighbors in $S \cup X \cup W$, as required. Each port has $2n - p$ neighbors in $X \cup W$ and p neighbors in S. Finally, each vertex in X has $2n$ neighbors in $X \cup W$.

"$\Leftarrow$:" Let $G'[S]$ be a highly connected graph of order k in G'. There are at most n vertices in $V(G) \cap S$, thus there is at least one vertex in $S \cap X$. Since $G'[X]$ is connected and each vertex in X has degree exactly $2n$ (the minimum degree in $G'[S]$), we have $X \subseteq S$. Furthermore, since $\{v \mid X \cap N(v) \neq \emptyset\} \setminus X = W$, also $W \subseteq S$, leaving $4n - 1 - |X| - |W| = p$ vertices in $S \cap V(G)$. Since $N_{G'}(V(G)) \setminus V(G) = W$ and $|W| = 2n - p + 1$, each vertex in $S \cap V(G)$ has at least $p - 1$ neighbors in $S \cap V(G)$. Thus $G[S \cap V(G)]$ is a clique. $\square$

Using Lemma 3, we can connect the running time with respect to parameter n with the Exponential Time Hypothesis (ETH) [16].

Theorem 4. *If the Exponential Time Hypothesis (ETH) is true, then* HIGHLY CONNECTED SUBGRAPH *does not admit a $2^{o(n)} \cdot n^{O(1)}$-time algorithm.*

6   Edge Isolation

We now present a single-exponential FPT algorithm for the number γ of edges between the desired highly connected subgraph $G[S]$ and the remaining graph. In this case, S is called "γ-isolated". More formally, if $G = (V, E)$ is a graph, we call a set $S \subseteq V$ γ-isolated if $(S, V \setminus S)$ is a cut of size at most γ. To our knowledge, Ito et al. [13] were the first to consider a formal notion of isolation in the context of dense subgraph identification. There is the following difference between the isolation definitions: we count the total size of the cut $(S, V \setminus S)$, whereas previous definitions count the size of $(S, V \setminus S)$ divided by the size of S [12, 13] or the minimum of the number of outgoing edges per vertex [14]. Our isolation definition leads to the following problem.

ISOLATED HIGHLY CONNECTED SUBGRAPH
Input: An undirected graph $G = (V, E)$, nonnegative integers k and γ.
Question: Is there a k-vertex γ-isolated highly connected subgraph contained in G?

The notion of isolation is not only motivated from an algorithmic point of view but also from the application. Ideally, communities in a network have fewer connections to the rest of the network [18]. Thus, putting an additional constraint on the number of outgoing edges may yield better communities than merely demanding high edge connectivity.

In the following, it will be useful to consider an augmented version of ISO-LATED HIGHLY CONNECTED SUBGRAPH: we place integer labels on the vertices which imply that these vertices are harder to isolate. We thus additionally equip each instance of ISOLATED HIGHLY CONNECTED SUBGRAPH with a labeling $f \colon V \to \mathbb{N}$ and we call $V' \subseteq V$ γ-isolated under f if there are at most $\gamma - \sum_{v \in V'} f(v)$ edges between V' and $V \setminus V'$ in G. Without loss of generality, assume $k \geq 2$ in the following.

The algorithm first performs three reduction rules. The first simple rule removes connected components that are too small.

Rule 1. *Remove all connected components with less than k vertices from G.*

The next rule finds connected components that are either trivial solutions or cannot contain any solution since proper subgraphs violate the isolation condition.

Rule 2. *If there is a connected component $C = (V', E')$ of G that has minimum cut size at least $\gamma + 1$, then accept if C is highly connected, $|V'| = k$, and V' is γ-isolated under f. Otherwise remove C from G.*

Proof (Correctness of Rule 2). The rule is clearly correct if it accepts. If the rule removes C, then C has a minimum cut of size at least $\gamma + 1$. Thus, for every induced subgraph $C[S]$ of C that does not contain all of its vertices, set S is not γ-isolated. Hence, no subgraph of C is a solution and we can safely remove C. $\qquad\square$

Rule 3. *If G has a connected component C with a minimum cut (A, B) of size at most $k/2$, then do the following. For each $v \in A$ redefine $f(v) := f(v) + |N(v) \cap B|$ and for each $v \in B$ redefine $f(v) := f(v) + |N(v) \cap A|$. Then, delete all edges between A and B.*

Proof (Correctness of Rule 3). Any k-vertex subgraph of C with nonempty intersection with both sides of (A, B) is not highly connected as it has a minimum cut of size at most $k/2$. Hence, any highly connected induced subgraph $C[S]$ of C is either contained in $C[A]$ or in $C[B]$. If S is γ-isolated under f in G, then it is also γ-isolated under the modified f in the modified graph (and vice-versa) by the way we have redefined f.   □

Exhaustive application of these rules yields a relation between γ and $k/2$.

Lemma 4. *If Rules 1 to 3 are not applicable, then $\gamma > k/2$.*

Proof. Assume the contrary. Each connected component has a minimum cut cutting at least one edge because Rule 1 is not applicable and $k \geq 2$. Further, each connected component has a cut of size at most γ because Rule 2 is not applicable. By assumption $\gamma \leq k/2$ and, hence, each connected component has a cut of size at most $k/2$ which contradicts the inapplicability of Rule 3.   □

As shown by the following lemma, the reduction rules can be applied efficiently.

Lemma 5 (⋆). *Rules 1 to 3 can be exhaustively applied in $O((kn + \gamma)nm)$ time.*

Using the above, we can now present the branching algorithm.

Theorem 5. *There is an $O(4^\gamma n^2 + (kn + \gamma)nm)$-time algorithm for* ISOLATED HIGHLY CONNECTED SUBGRAPH.

Proof. We first reduce the instance with respect to Rules 1 to 3. By Lemma 5 this can be done in $O((kn + \gamma)nm)$ time. Next, we guess one vertex v that is in the solution S (by branching into n cases according to the n vertices). We start with $S' := \{v\}$ and try to extend S' to a solution. More precisely, we choose a vertex v' from the neighborhood of S' (that is, from $\bigcup_{u \in S'} N(u) \setminus S'$), and branch into two cases: add v' to S', or exclude v', that is, delete v' and increase $f(u)$ by one for all $u \in N(v')$. In the first case, we increase $|S'|$ by one. In the second case, we increase $\sum_{u \in S'} f(u)$ by at least one. Branching is performed until $|S'| = k$ or $\sum_{u \in S'} f(u)$ exceeds γ or the neighborhood of S' is empty. When $|S'|$ reaches k, we check whether S' is highly connected and γ-isolated under f, and if this is the case, we have found a solution. Otherwise, when $\sum_{u \in S'} f(u)$ exceeds γ or no branching is possible because the neighborhood of S' is empty, we abort the branch; in this case, clearly no superset of S' can be a solution. The height of the search tree is bounded by $k + \gamma$, and each branch can be executed in $O(n)$ time, yielding a running time bound of $O(n \cdot 2^{k+\gamma} \cdot n)$.

We now distinguish two cases: $k \leq \gamma$ and $k > \gamma$. In the first case $2^{k+\gamma} \leq 4^\gamma$, as required. If $k > \gamma$, there is at least one vertex in S that has no neighbors outside of S. Thus, instead of $S' := \{v\}$, we can start with $S' := \{v\} \cup N(v)$. Since v has more than $k/2$ neighbors in S, we have $|S'| > k/2 + 1$, and thus there are less than $k/2$ branches of adding a vertex. By Lemma 4, $2^{k/2+\gamma} \leq 4^\gamma$.   □

We now present a further way of analyzing the presented data reduction rules by giving a *Turing kernelization* [1] for ISOLATED HIGHLY CONNECTED SUB-GRAPH parameterized by γ. Informally, a Turing kernelization is a reduction of the input instance of a parameterized problem to many instances of the same problem which are small measured in the parameter. Then, the solution to the original input instance can be computed by solving the small problem instances separately.

To motivate the Turing kernelization result we first observe that ISOLATED HIGHLY CONNECTED SUBGRAPH does not admit a problem kernel, that is, a Turing kernelization which produces only one small problem instance. The disjoint union of a set of graphs has an isolated highly connected subgraph if and only if at least one of the graphs has one. Hence, ISOLATED HIGHLY CON-NECTED SUBGRAPH has a trivial OR-composition which implies the following [2].

Proposition 1. ISOLATED HIGHLY CONNECTED SUBGRAPH *does not admit a polynomial-size problem kernel with respect to γ unless $NP \subseteq coNP/poly$.*

Before describing the Turing kernelization, we give a formal definition.

Definition 1. *Let L be a parameterized problem and let $g : \mathbb{N} \to \mathbb{N}$ be a computable function. A* Turing kernelization *for L is an algorithm that, for each instance (x, k), decides whether $(x, k) \in L$ in polynomial time using an oracle for $\{(x', k') \mid |x'| + k' \le g(k) \land (x', k') \in L\}$. The sequence of queries posed to the oracle is called* Turing kernel. *We call $g(k)$ the* size *of the Turing kernel.*

We now describe the algorithm in detail. The first step is to reduce to the augmented version of ISOLATED HIGHLY CONNECTED SUBGRAPH in which we introduce the vertex labeling f. Then, apply Rules 2 and 3 exhaustively. Afterwards, apply the following reduction rule which removes high-degree vertices.

Rule 4 ($\star$). *Let (G, k, γ) be an instance of* ISOLATED HIGHLY CONNECTED SUBGRAPH *that is reduced with respect to Rules 2 and 3. If G contains a vertex v of degree at least $3\gamma - f(v)$, then remove v from G, and for each $u \in N(v)$ increase $f(u)$ by one.*

Now we construct n instances of ISOLATED HIGHLY CONNECTED SUBGRAPH that have $O(\gamma^3)$ vertices each. The original instance is a yes-instance if and only if one of these instances is a yes-instance. The idea is to exploit the fact that highly connected graphs have diameter two [9]. Thus, to find highly connected graphs, it is sufficient to explore the two-neighborhood of each vertex. More precisely, the instances are constructed as follows.

For each vertex $v \in V$, construct the graph $G_v := G[N_2[v]]$ where $N_2[v]$ is the set of all vertices that have distance at most two from v (including v). When solving the ISOLATED HIGHLY CONNECTED SUBGRAPH instances we need to determine whether a subgraph is γ-isolated. Thus, the graph G_v has to contain information on the original vertex degrees. Note that for each $u \in V$, $f(u)$ denotes the number of edges deleted during the data reduction that are incident with u. Moreover, for each vertex u in G_v, let $g(u)$ denote the number of neighbors of u in G in $V \setminus N_2[v]$. To obtain instances of ISOLATED HIGHLY

CONNECTED SUBGRAPH one may not use vertex labelings. Thus, for each u of G_v add $g(u) + f(u)$ new vertices and make them adjacent to u. This completes the construction of G_v. In this way we obtain n instances (G_v, k, γ) of ISOLATED HIGHLY CONNECTED SUBGRAPH. The following lemma shows that it is sufficient to solve these instances in order to determine whether the original ISOLATED HIGHLY CONNECTED SUBGRAPH instance $(G = (V, E), k, \gamma)$ is a yes-instance.

Lemma 6 ($\star$). *Let $(G = (V, E), k, \gamma)$ be an instance of* ISOLATED HIGHLY CONNECTED SUBGRAPH *and, for each $v \in V$, let (G_v, k, γ) denote the instance as constructed above. Then, (G, k, γ) is a yes-instance if and only if there is a $v \in V$ such that (G_v, k, γ) is a yes-instance.*

We now show that the instances have bounded size.

Lemma 7 ($\star$). *Let (G_v, k, γ) be an instance of* ISOLATED HIGHLY CONNECTED SUBGRAPH *constructed from G as described above. Then G_v has less than $(3\gamma)^3$ vertices and less than $3\gamma^4$ edges.*

Combining Lemmas 6 and 7 leads to the following.

Theorem 6. ISOLATED HIGHLY CONNECTED SUBGRAPH *admits a Turing kernel of size $O(\gamma^4)$ which has less than $(3\gamma)^3$ vertices.*

7   Edge Deletion

We now show that there is a subexponential fixed-parameter algorithm for HIGHLY CONNECTED SUBGRAPH with respect to α, the number of edges we are allowed to delete in order to obtain a highly connected graph of order k. The algorithm is a search tree algorithm which branches on whether or not a given vertex is part of the highly connected graph. Repeated application of two reduction rules (similar to Rules 2 and 3 above) ensures that the branches are effective in reducing the remaining search space. To give a precise presentation of the branching step and the reduction rules, we define the problem with an additional *seed S*, a set of vertices which have to be in the highly connected graph.

> SEEDED HIGHLY CONNECTED EDGE DELETION
> Input: An undirected graph $G = (V, E)$, a vertex set $S \subseteq V$, and non-negative integers k and α.
> Question: Is there a set $E' \subseteq E$ of at most α edges such that $G - E'$ consists only of degree-zero vertices and a $(k + |S|)$-vertex highly connected subgraph containing S?

For $S = \emptyset$ we obtain the plain edge deletion problem. The reduction rules are as follows.

Rule 5 (⋆). *If there is a connected component $C = (V', E')$ of G that has minimum cut size at least $\alpha + 1$, then accept if C is highly connected, $S \subseteq V'$, $|V' \setminus S| = k$, and the remaining connected components of G contain at most α edges. Otherwise reject.*

Rule 6 (⋆). *If there is a connected component of G that has a minimum cut of size at most $(k + |S|)/2$, then delete all cut edges and reduce α by their number.*

Similarly to the edge isolation parameter, after using the reduction rules k, $|S|$, and α are related.

Lemma 8 (⋆). *If Rules 5 and 6 are not applicable, then $\alpha > (k + |S|)/2$.*

As the ones presented in Section 6, both rules can be applied efficiently.

Lemma 9 (⋆). *Rules 5 and 6 are exhaustively applicable in $O(\alpha^2 nm)$ time.*

Exhaustively applying the reduction rules lets us bound the number of the remaining vertices linearly in α. This will be useful in the branching algorithm below.

Theorem 7 (⋆). SEEDED HIGHLY CONNECTED EDGE DELETION *admits a problem kernel with at most $2\alpha + 4\alpha/k$ vertices and $\binom{2\alpha}{2} + \alpha$ edges computable in $O(\alpha^2 nm)$ time.*

In the subexponential branching algorithm, we use the following simple branching rule. It simply takes a vertex and branches on whether or not it should be added to the seed S for the desired highly connected graph.

Branching Rule 1. *If $\alpha + k \geq 0$, then choose an arbitrary vertex $v \in V \setminus S$ and branch into the cases of adding v to S or removing v from G. That is, create the instances $I_1 = (G, S \cup \{v\}, k - 1, \alpha)$ and $I_2 = (G - v, S, k, \alpha - \deg_G(v))$. Accept if I_1 or I_2 is accepted.*

It is clear that Branching Rule 1 is correct. We now describe the complete algorithm and bound its running time.

Theorem 8. *There is an $O(2^{4 \cdot \alpha^{0.75}} + \alpha^2 nm)$-time algorithm for* HIGHLY CONNECTED EDGE DELETION.

Proof. We first apply the kernelization from Theorem 7, which entails applying Rules 5 and 6 exhaustively. Then, if $k \leq 2\sqrt{\alpha}$ we check whether $S \cup V'$ induces a highly connected subgraph, for every vertex subset $V' \subseteq V \setminus S$ of size k. We accept or reject accordingly. If $k > 2\sqrt{\alpha}$, then we apply Branching Rule 1 and recurse on the two created instances.

From the correctness of the rules it is clear that this algorithm finds a solution if there is one. Let us analyze its running time. Note that in each recursive call, except the first one, the input instance has $O(\alpha)$ vertices according to Theorem 7. Thus applying Rules 5 and 6 in a recursive call amounts to $O(\alpha^5)$ time except in the first one where it is $O(\alpha^2 nm)$ time, by Lemma 9. Next, in each recursive call we may have to check whether $S \cup V'$ is highly connected for all k-vertex subsets V'. This is done only after Rules 5 and 6 have been exhaustively applied and only if $k \leq 2\sqrt{\alpha}$. Thus, the graph G is of order at most $2\alpha + 4\alpha/k \leq 4\alpha$ (note that $k \geq 2$ without loss of generality). Hence, testing the subgraphs amounts to $O((4\alpha)^{2\sqrt{\alpha}+2})$ time. In total, the time spent per search tree node is $O((4\alpha)^{\max\{5,2\sqrt{\alpha}+2\}})$.

Now let us bound the number of leaves C of the search tree. Note that the total number of search tree nodes is within a constant factor of C. For an instance $I = (G, S, k, \alpha)$ of HIGHLY CONNECTED SUBGRAPH, consider the value $\mu(I) = k+\alpha$ in the root of the search tree, after applying Rules 5 and 6. Then, $\mu(I) \leq 3\alpha$ by Lemma 8. Let $C(\mu(I))$ denote an upper bound on the number of leaves that a search tree with a root with value $\mu(I)$ can have. Whenever we apply Branching Rule 1, μ is reduced by a certain amount. More precisely, $C(\mu(I))$ fulfills $C(0) = 1$, and $C(\mu(I)) \leq C(\mu(I_1)) + C(\mu(I_2))$. Hence, $C(\mu(I))$ is monotone. Further, since Rule 6 is not applicable, $\deg_G(v) \geq (|S| + k)/2 \geq k/2 \geq \sqrt{\alpha}$ in the application of Branching Rule 1. This implies $C(\mu(I)) \leq C(\mu(I)-1)+C(\mu(I)-\sqrt{\alpha})$. Hence $C(\mu(I))$ is at most the number of paths in $\mathbb{R}^2$ from the origin to some point (x, y) that take only steps $(1,0)$ or $(0, \sqrt{\alpha})$, where $x + y = \mu(I)$. Scaling the y-axis by a factor of $1/\sqrt{\alpha}$, computing $C(\mu(I))$ reduces to the problem of counting such paths from the origin to some (x, y') taking only steps $(1,0)$ or $(0, 1)$ such that $x + \sqrt{\alpha}y' = \mu(I)$. We now bound the number of these paths.

The number of $(0, 1)$ steps is at most $3\sqrt{\alpha}$. If the path contains i $(0, 1)$-steps, then the total number of steps in the path is $i + 3\alpha - \sqrt{\alpha}i$. Hence, there are $\binom{i+3\alpha-\sqrt{\alpha}i}{i}$ paths with exactly i steps $(0, 1)$. This implies $C(\mu(I)) \leq \sum_{i=0}^{3\sqrt{\alpha}} \binom{i+3\alpha-\sqrt{\alpha}i}{i}$. To bound this number we use the fact that $\binom{a+b}{a} \leq 2^{2\sqrt{ab}}$ [8, Lemma 9]. Hence $C(\mu(I)) \leq \sum_{i=0}^{3\sqrt{\alpha}} 2^{2\sqrt{i\cdot(3\alpha-\sqrt{\alpha}i)}}$. Consider the derivative $f(i)$ of $2\sqrt{i \cdot (3\alpha - \sqrt{\alpha}i)}$ with respect to i. We have

$$f(i) = \frac{\sqrt{\alpha}(3\sqrt{\alpha} - 2i)}{\sqrt{\sqrt{\alpha}i(3\sqrt{\alpha} - i)}}.$$

Inspecting $f(i)$ shows that $\sqrt{i \cdot (3\alpha - \sqrt{\alpha}i)}$ is maximized over $0 \leq i \leq 3\sqrt{\alpha}$ if $i = 3\sqrt{\alpha}/2$. This gives $C(\mu(I)) \leq 3\sqrt{\alpha} \cdot 2^{3\sqrt{\alpha\sqrt{\alpha}}}$. Finally,

$$3\sqrt{\alpha} \cdot 2^{3\sqrt{\alpha\sqrt{\alpha}}} \cdot (4\alpha)^{\max\{5,2\sqrt{\alpha}+2\}} \in O(2^{4\alpha^{0.75}}),$$

giving the overall running time bound of $O(2^{4\alpha^{0.75}} + \alpha^2 nm)$. □

Although the presented algorithm is a subexponential-time algorithm with relatively small constants in the exponential functions, it is unclear whether it can be useful in practice. This is because the parameter α is likely to be large in real-world instances. With further substantial running time improvements, however, one might obtain practical algorithms. For example, an algorithm with running time $O(2^{\alpha^{0.5}} \cdot nm)$ should perform well on many real-word instances.

References

1. Binkele-Raible, D., Fernau, H., Fomin, F.V., Lokshtanov, D., Saurabh, S., Villanger, Y.: Kernel(s) for problems with no kernel: On out-trees with many leaves. ACM Trans. Algorithms 8(4), 38 (2012)
2. Bodlaender, H.L., Downey, R.G., Fellows, M.R., Hermelin, D.: On problems without polynomial kernels. J. Comput. Syst. Sci. 75(8), 423–434 (2009)
3. Chartrand, G.: A graph-theoretic approach to a communications problem. SIAM J. Appl. Math. 14(4), 778–781 (1966)
4. Chen, J., Kanj, I.A., Xia, G.: Improved upper bounds for vertex cover. Theor. Comput. Sci. 411, 40–42 (2010)
5. Downey, R.G., Fellows, M.R.: Fundamentals of Parameterized Complexity. Texts in Computer Science. Springer (2013)
6. Eppstein, D., Löffler, M., Strash, D.: Listing all maximal cliques in sparse graphs in near-optimal time. In: Cheong, O., Chwa, K.-Y., Park, K. (eds.) ISAAC 2010, Part I. LNCS, vol. 6506, pp. 403–414. Springer, Heidelberg (2010)
7. Fellows, M.R., Jansen, B.M.P., Rosamond, F.A.: Towards fully multivariate algorithmics: Parameter ecology and the deconstruction of computational complexity. Eur. J. Combinatorics 34(3), 541–566 (2013)
8. Fomin, F.V., Kratsch, S., Pilipczuk, M., Pilipczuk, M., Villanger, Y.: Tight bounds for parameterized complexity of cluster editing with a small number of clusters. J. Comput. Syst. Sci. 80(7), 1430–1447 (2014)
9. Hartuv, E., Shamir, R.: A clustering algorithm based on graph connectivity. Inf. Process. Lett. 76(4–6), 175–181
10. Hua, Q.-S., Wang, Y., Yu, D., Lau, F.C.M.: Dynamic programming based algorithms for set multicover and multiset multicover problems. Theor. Comput. Sci. 411(26-28), 2467–2474 (2010)
11. Hüffner, F., Komusiewicz, C., Liebtrau, A., Niedermeier, R.: Partitioning biological networks into highly connected clusters with maximum edge coverage. IEEE/ACM Trans. Comput. 11(3), 455–467 (2014)
12. Ito, H., Iwama, K.: Enumeration of isolated cliques and pseudo-cliques. ACM Trans. Algorithms 5(4), Article 40 (2009)
13. Ito, H., Iwama, K., Osumi, T.: Linear-time enumeration of isolated cliques. In: Brodal, G.S., Leonardi, S. (eds.) ESA 2005. LNCS, vol. 3669, pp. 119–130. Springer, Heidelberg (2005)
14. Komusiewicz, C., Hüffner, F., Moser, H., Niedermeier, R.: Isolation concepts for efficiently enumerating dense subgraphs. Theor. Comput. Sci 410(38-40), 3640–3654 (2009)
15. Liu, G., Wong, L.: Effective pruning techniques for mining quasi-cliques. In: Daelemans, W., Goethals, B., Morik, K. (eds.) ECML PKDD 2008, Part II. LNCS (LNAI), vol. 5212, pp. 33–49. Springer, Heidelberg (2008)
16. Lokshtanov, D., Marx, D., Saurabh, S.: Lower bounds based on the Exponential Time Hypothesis. Bulletin of the EATCS 105, 41–71 (2011)
17. Matsuda, H., Ishihara, T., Hashimoto, A.: Classifying molecular sequences using a linkage graph with their pairwise similarities. Theor. Comput. Sci. 210(2), 305–325 (1999)
18. Pattillo, J., Youssef, N., Butenko, S.: On clique relaxation models in network analysis. Eur. J. Operational Research 226(1), 9–18 (2013)
19. Sharan, R., Ulitsky, I., Shamir, R.: Network-based prediction of protein function. Mol. Syst. Biol. 3, 88 (2007)
20. Veremyev, A., Prokopyev, O.A., Boginski, V., Pasiliao, E.L.: Finding maximum subgraphs with relatively large vertex connectivity. Eur. J. Operational Research 239(2), 349–362 (2014)

Fixing Improper Colorings of Graphs

Konstanty Junosza-Szaniawski[1], Mathieu Liedloff[2],
and Paweł Rząiewski[1]

[1] Warsaw University of Technology, Faculty of Mathematics and Information Science,
Koszykowa 75, 00-662 Warszawa, Poland
{k.szaniawski,p.rzazewski}@mini.pw.edu.pl
[2] Université d'Orléans, INSA Centre Val de Loire,
LIFO, 45067 Orléans, France
mathieu.liedloff@univ-orleans.fr

Abstract. In this paper we consider a variation of a recoloring problem, called the r-Color-Fixing. Let us have some non-proper r-coloring φ of a graph G. We investigate the problem of finding a proper r-coloring of G, which is "the most similar" to φ, i.e. the number k of vertices that have to be recolored is minimum possible. We observe that the problem is NP-complete for any $r \geq 3$, but is Fixed Parameter Tractable (FPT), when parametrized by the number of allowed transformations k. We provide an $\mathcal{O}^*(2^n)$ algorithm for the problem (for any fixed r) and a linear algorithm for graphs with bounded treewidth. Finally, we investigate the *fixing number* of a graph G. It is the maximum possible distance (in the number of transformations) between some non-proper coloring of G and a proper one.

1 Introduction

Many problems in real-life applications have a dynamic nature. When the constraints change, the previously found solution may no longer be optimal or even feasible. Therefore often there is needed to recompute the solution (preferably using the old one). This variant is called a *reoptimization* and has been studied for many combinatorial problems, e.g. TSP (see Ausiello *et al.* [1]), Shortest Common Superstring (see Bilò *et al.* [2]) or Minimum Steiner Tree (see Zych and Bilò [17]). We also refer the reader to the paper of Shachnai *et al.* [16], where the authors describe a general model for combinatorial reoptimization.

Another family of problems, in which we deal with transforming one solution to another, is *reconfiguration*. Here we are given two feasible solutions and want to transform one into another by a series of simple transformations in such a way that every intermediate solution is feasible. When we consider a reconfiguration version of the graph coloring problem, we want to transform one proper coloring into another one in such a way that at every step we can recolor just one vertex and the coloring obtained after this change is still proper.

A special attention has been paid to determining if a given graph G is r-*mixing*, i.e. if for any two proper r-colorings of G you can transform one into

G.F. Italiano et al. (Eds.): SOFSEM 2015, LNCS 8939, pp. 266–276, 2015.
© Springer-Verlag Berlin Heidelberg 2015

another (maintaining a proper r-coloring at each step). Cereceda *et al.* [8–10] characterize graphs, which are 3-mixing and they provide a polynomial algorithm for recognizing them. Determining if a graph is r-mixing for any $r \geq 4$ is PSPACE-complete [7]. There are also some results showing that a graph G is r-mixing, where r is some function of G. For example, Jerrum [14] showed that every graph G is $(\Delta(G) + 2)$-mixing. This bound has been recently refined by Bonamy and Bousquet [6], who proved that every graph is $(\chi_g(G) + 1)$-mixing, where $\chi_g(G)$ denotes the Grundy number of G, i.e. the highest possible number of colors used by a greedy coloring of G. Clearly $\chi_g(G) \leq \Delta(G) + 1$.

Another direction of research in r-mixing graphs is the maximum number of transformations necessary to obtain one r-coloring from another one (i.e. the distance between those colorings). Bonamy and Bousquet [6] show that if $r \geq \text{tw}(G)+2$ (where $\text{tw}(G)$ denotes the *treewidth* of G), then any two r-colorings of G are in distance of at most $2(n^2 + n)$, while for $r \geq \chi_g(G) + 1$, any two r-colorings are in distance of at most $4 \cdot \chi_g(G) \cdot n$.

A slightly different problem has been considered by Felsner *et al.* [13]. They also transformed one r-coloring to another one using some local changes, but did not require the initial coloring to be proper (the final one still has to be proper). Also, a vertex could be recolored to color x if it did not have any neighbor colored with x (strictly speaking, any out-neighbor, as the authors were considering directed graphs). They showed that if G is a 2-orientation (i.e. every out-degree is equal to 2) of some maximal bipartite planar graph (i.e. a plane quadrangulation), then every proper 3-coloring of G could be reached in $\mathcal{O}(n^2)$ steps from any initial (even non-proper) 3-coloring of G. Similar results hold for 4-colorings and 3-orientations of maximal planar graphs (i.e. triangulations).

In this paper we consider a slightly different problem. We start with some (possibly non-proper) r-coloring and ask for the minimum number of transformations needed to obtain a proper r-coloring (any proper r-coloring, not the specific one). We are allowed to change colors of vertices arbitrarily, provided that we recolor just one vertex in each step. We mainly focus on the computational aspects of determining if, starting with some given r-coloring of G, we can reach a proper r-coloring in at most k steps.

The paper is organized as follows. In Section 3 we show that our problem is NP-complete for any $r \geq 3$ and polynomial otherwise (here k is a part of the input). In Section 4 we provide an $\mathcal{O}^*(2^n)$ algorithm for the problem[1]. In the next two Sections we focus on the parametrized complexity (we refer the reader to the book by Downey and Fellows [12] for an introduction to the parametrized complexity theory). Namely, we show that our problem is in FPT, when parametrized by k (Section 5) and provide a linear algorithm solving the problem for graphs with bounded treewidth (Section 6). In Section 7 we investigate the *fixing number* of G, i.e. the maximum (over all initial colorings φ) distance from φ to a proper coloring of G.

---

[1] In $\mathcal{O}^*$ notation we suppress factors, which are polynomial in the input size.

2   Preliminaries

For a natural number r, by $[r]$ we denote the set $\{1, 2, .., r\}$. By r-*coloring* of a graph G we mean any assignment of natural numbers from $[r]$ (called *colors*) to vertices of G. A coloring is *proper* if no adjacent vertices get the same color. Note that there may be some colors that are not assigned to any vertex.

For two r-colorings φ, φ' let $\varphi \ominus \varphi'$ denote the set $\{v \in V : \varphi(v) \neq \varphi'(v)\}$. We also define the *distance* $\mathrm{dist}(\varphi, \varphi')$ between two r-colorings φ, φ'. It is equal to their Hamming distance, i.e. $|\varphi \ominus \varphi'|$.

Let r be some fixed integer. The problem we consider in this paper is formally defined as follows.

Problem: r-**Color-Fixing** (r-Fix)
Instance: A graph G, integer k, an r-coloring φ of $V(G)$.
Question: Does there exist a proper r-coloring φ' of G such that $\mathrm{dist}(\varphi, \varphi') \leq k$?

Such a coloring φ' is called a *witness* of $\mathcal{I}$, for $\mathcal{I} = (G, k, \varphi)$ being a YES-instance of r-Fix. Obviously, if $r < \chi(G)$ (by $\chi(G)$ we denote the *chromatic number* of G, i.e. the smallest number of colors needed to color G properly), then the answer is always No. By *recoloring* of a vertex v we mean an operation of changing the color assigned to v, obtaining another coloring φ', such that $\varphi \ominus \varphi' = \{v\}$.

In the optimization version of the problem we ask for the minimum number k of recolorings needed to transform φ into a proper coloring of G. Let $\overline{\chi}_\varphi^r(G)$ denote this minimum possible value of k. If $r < \chi(G)$, we define $\overline{\chi}_\varphi^r(G) := \infty$ for every r-coloring φ of G.

3   Complexity

First, we shall prove that if k is the part of the input, r-Fix problem is NP-complete for any $r \geq 3$. Clearly, the problem is in NP for all r. Also, a graph G with n vertices is r-colorable if and only if G can be recolored from any fixed coloring within at most n steps. Thus we obtain the following.

Observation 1. r-Fix *problem is NP-complete for any* $r \geq 3$ *(when the number k of allowed recolorings is a part of the input).*

For $r = 1$, the problem clearly reduces to determining if the graph has no edges. For $r = 2$ the problem is also polynomial time solvable. If G is not bipartite, the answer is No.

Observation 2. *Let G be a bipartite graph with bipartition classes X and Y and let φ be a 2-coloring of G. Then we have[2]*

$$\overline{\chi}_\varphi^2(G) = \sum_{\substack{C:\ connected \\ component\ of\ G}} \min\{|\left(X \ominus \varphi^{-1}(1)\right) \cap V(C)|, |\left(X \ominus \varphi^{-1}(2)\right) \cap V(C)|\}.$$

---

[2] By $\varphi^{-1}(i)$ we denote the set of vertices colored with the color i.

Proof. Let C be a connected component of G and let X', Y' denote its classes of bipartition. By φ' we denote the restriction of φ to C. To obtain a proper coloring of C, we either have to recolor the vertices from $X' \setminus \varphi'^{-1}(1)$ to color 1 and vertices from $Y' \setminus \varphi^{-1}(2)$ to color 2, or the other way around. Therefore the minimum number of recoloring operations needed to obtain a proper coloring of C is equal to $\min\{|X' \ominus \varphi'^{-1}(1)|, |X' \ominus \varphi'^{-1}(2)|\}$. Clearly $X' \ominus \varphi'^{-1}(1) = (X \cap V(C)) \ominus (\varphi^{-1}(1) \cap V(C)) = (X \ominus \varphi^{-1}(1)) \cap V(C)$ (and symmetrically for $\varphi^{-1}(2)$). We repeat this for every connected component C of G. $\square$

4   Exact Algorithm via Inclusion-Exclusion

In this section we deal with the optimization version of the problem. Note that the brute force algorithm works in time $\mathcal{O}^*(\sum_{k=0}^{n} \binom{n}{k}(r-1)^k) = \mathcal{O}^*(r^n)$. We shall obtain a better algorithm by reducing the instance of our problem to an instance of the so-called MAX WEIGHTED PARTITION problem and then solve it, using the algorithm by Björklund, Husfeldt and Koivisto [3]. A *partition* of the set N is a family of sets $S_1, \ldots, S_r$ such that $\bigcup_{i=1}^{r} S_i = N$ and $S_i \cap S_j = \emptyset$ for every $i \neq j$. Notice that we do not require for the sets S_i to be non-empty.

Problem: MAX WEIGHTED PARTITION
Instance: A set N, integer d and functions $f_1, f_2, \ldots, f_d : 2^N \rightarrow [-M, M]$ for some integer M.
Question: What is the maximum w, for which there exists a partition $S_1, S_2, \ldots, S_d$, such that $\sum_{i=1}^{d} f_i(S_i) = w$?

Let G be a graph and let φ be its r-coloring. We shall construct a corresponding instance $\mathcal{J} = (N, d, f_1, \ldots, f_d)$ of MAX WEIGHTED PARTITION problem. Set $N = V(G)$ and $d = r$. We define functions $f_1, f_2, \ldots, f_d$ as:
$$f_i(S) = \begin{cases} -|S \setminus \varphi^{-1}(i)| & \text{if } S \text{ is independent,} \\ -r \cdot n & \text{otherwise.} \end{cases}$$
In this way every partition of $V(G)$ into r independent set, corresponding to the proper r-coloring φ', has the total weight $(-\sum_{i=1}^{r} |\varphi'^{-1}(i) \setminus \varphi^{-1}(i)|)$. It is also easy to notice that any partition into independent sets has greater weight than any partition having at least one non-independent set.

The only thing left is to prove that the weight is maximized for a partition corresponding to a coloring φ', such that $\text{dist}(\varphi', \varphi)$ is minimum. To see this, notice that $\text{dist}(\varphi', \varphi) = |\{v \in V : \varphi'(v) \neq \varphi(v)\}| = |\bigcup_{i=1}^{r}\{v \in V : \varphi'(v) = i \wedge \varphi(v) \neq i\}| = \sum_{i=1}^{r} |\{v \in V : \varphi'(v) = i \wedge \varphi(v) \neq i\}| = \sum_{i=1}^{r} |\varphi'^{-1}(i) \setminus \varphi^{-1}(i)|$. Moreover, if a weight of a found partition is at most $-r \cdot n$, it contains at least one non-independent set, which means that $r < \chi(G)$ and therefore $\overline{\chi}_\varphi^r(G) = \infty$.

Now we can use the algorithm by Björklund *et al.* to find the optimal solution for $\mathcal{J}$.

Theorem 3 (Björklund, Husfeldt, Koivisto [3]). MAX WEIGHTED PARTITION *problem can be solved in time* $\mathcal{O}^*(2^n d^2 M)$ *(where n is the cardinality of the ground set), using exponential space.*

Since $d = r$ is a constant and $M = n \cdot r$, we obtain the following corollary.

Corollary 4. *For any constant r, the optimization version of the r-FIX problem can be solved in time $\mathcal{O}^*(2^n)$ (where n is the number of vertices in the input graph), using exponential space.*

If we have only polynomial space available, we get the following bounds.

Theorem 5 (Björklund, Husfeldt, Koivisto [3]). MAX WEIGHTED PARTITION *problem can be solved in time $\mathcal{O}^*(3^n d^2 M)$ (where n is the cardinality of the ground set), using polynomial space.*

Corollary 6. *For any constant r, the optimization version of the r-FIX problem can be solved in time $\mathcal{O}^*(3^n)$ (where n is the number of vertices in the input graph), using polynomial space.*

5   Parametrized Complexity

Clearly for k being a fixed integer, the problem can be easily solved in time $\mathcal{O}^*(\binom{n}{k}r^k) = \mathcal{O}^*(n^k r^k)$. To do it, we have to consider every k-element subset of vertices and check whether recoloring the chosen vertices (in r^k ways, since we are interested in recoloring *at most* k vertices and thus some colors may remain unchanged) allows us to obtain a proper coloring. Therefore our problem is in XP (when parametrized by k). In the remainder of this section we show that the problem is in FPT, i.e. we can solve it in time $f(k) \cdot n^{\mathcal{O}(1)}$ (i.e. the degree of a polynomial function of n does not depend on k).

For an improper coloring φ of G, let G^φ denote a *conflict graph* of G under the coloring φ, i.e. a graph induced by the set of edges $\{uv \in E(G) : \varphi(u) = \varphi(v)\}$. Note that the conflict graph can be found in polynomial time. Consider the following algorithm.

Algorithm 1: $\text{Fix}(r, \mathcal{I} = (G, k, \varphi))$

---

1 **if** φ *is a proper coloring of* G **then return** YES
2 **if** $k = 0$ **then return** NO
3 $xy \leftarrow$ any edge from G^φ
4 **foreach** $col \in [r] \setminus \{\varphi(x)\}$ **do**
5      $\varphi_1 \leftarrow \varphi$ with vertex x recolored to col
6   **if** $Fix(r, (G, k-1, \varphi_1)) = $ YES **then return** YES
7 **foreach** $col \in [r] \setminus \{\varphi(y)\}$ **do**
8      $\varphi_1 \leftarrow \varphi$ with vertex y recolored to col
9   **if** $Fix(r, (G, k-1, \varphi_1)) = $ YES **then return** YES
10 **return** NO

Lemma 7. *Let φ be a non-proper r-coloring of G. Then $\mathcal{I} = (G, k, \varphi)$ is a* YES-*instance of r-*FIX *if and only if for any edge $xy \in E(G^\varphi)$ there exists an r-coloring (possibly non-proper) φ_1 of G such that:*

1. $\varphi \ominus \varphi_1 = \{x\}$ *or* $\varphi \ominus \varphi_1 = \{y\}$,
2. $\mathcal{I}' = (G, k - 1, \varphi_1)$ *is a* YES-*instance of r-*FIX.

Proof. First assume that $\mathcal{I} = (G, k, \varphi)$ is a YES-instance of r-FIX and let φ' be its witness. Consider an edge xy from G^φ. By the definition of G^φ, we have $\varphi(x) = \varphi(y)$. Since φ' is proper, clearly $\varphi'(x) \neq \varphi'(y)$. Then at least one of the vertices x, y has changed its color. Without loss of generality assume that $\varphi'(x) \neq \varphi(x)$. Let φ_1 be a coloring defined as follows.

$$\varphi_1(u) = \begin{cases} \varphi(u) & \text{if } u \neq x \\ \varphi'(u) & \text{if } u = x. \end{cases}$$

It is clear that it satisfies the conditions given in lemma.

Now consider φ being an r-coloring of G and let xy be an edge from G^φ. Without loss of generality let φ_1 to be some r-coloring of G such that $\varphi \ominus \varphi_1 = \{x\}$ and the instance $\mathcal{I}' = (G, k - 1, \varphi_1)$ is a YES-instance of r-FIX.

Let φ_1' be a witness of $\mathcal{I}'$. Notice that $\text{dist}(\varphi_1', \varphi) \leq \text{dist}(\varphi_1', \varphi_1) + \text{dist}(\varphi_1, \varphi) \leq k$ and therefore $\mathcal{I} = (G, k, \varphi)$ is a YES-instance of r-FIX with witness φ_1'. □

Lemma 8. *The algorithm* Fix *solves r-*FIX *problem for any r.*

Proof. Let $\mathcal{I} = (G, k, \varphi)$ be an instance of r-FIX. If φ is a proper labeling of G, then the algorithm returns YES in line 1. Suppose then that φ is not proper. If $k = 0$, the algorithm returns NO in line 2.

Assume that $k > 0$ and the algorithm works properly for all instances with parameter smaller than k. Suppose that $\mathcal{I} = (G, k, \varphi)$ is a YES-instance of r-FIX. Let xy be an edge chosen in line 3. Then, by Lemma 7, there exist an r-coloring φ_1 of G such that $\varphi \ominus \varphi_1 = \{x\}$ or $\varphi \ominus \varphi_1 = \{y\}$ and $\mathcal{I}' = (G, k - 1, \varphi_1)$ is a YES-instance of REC. Without loss of generality assume that $\varphi \ominus \varphi_1 = \{x\}$. Let us consider an iteration of the loop in lines 4–6 color $\varphi_1(x)$. By the inductive assumption, the recursive call in line 6 returns YES and therefore the whole algorithm returns YES.

If I is a NO-instance of REC, then by the inductive assumption and Lemma 7, for every *col* the recursive calls in lines 6 and 9 return NO. Therefore the algorithm returns NO in line 10. □

Let $T(n, k)$ be the computational complexity of the algorithm *Fix*. We can write the following recursive formula:

$$T(n, k) \leq n^{O(1)} + (r - 1) \cdot T(n, k - 1) + (r - 1) \cdot T(n, k - 1).$$

Solving it, we obtain the following.

Lemma 9. *The algorithm* Fix *solves r-*FIX *problem in time $T(n, k) \leq (2(r - 1))^k \cdot n^{\mathcal{O}(1)}$.*

Corollary 10. *For any fixed r, the r-*Fix *problem is in FPT, when parametrized by k.*

Corollary 10 yields that the r-Fix problem admits a kernel. However, its size is exponential.

Open Problem 1. *Design a polynomial kernel for the r-*Fix *problem, parametrized by k or prove that it does not exist (under some reasonable complexity assumptions).*

6   Algorithm for Graphs with Bounded Treewidth

In this section we consider the optimization version of r-Fix problem for graphs with bounded treewidth. For more information about tree decompositions and treewidth, the reader is referred to Diestel's book [11]. Here we just quickly present basic definitions that we shall use.

Let $G = (V, E)$ be a graph with n vertices. A *tree decomposition* of G is a pair $(\{X_i \colon i \in I\}, T = (I, F)\})$, where T is a tree whose every node has associated a subset X_i of vertices of G with the following properties:

1. $\bigcup_{i \in I} X_i = V$,
2. for every $vw \in E$ there exists $i \in I$ such that $\{v, w\} \subseteq X_i$,
3. for every $v \in V$, the set $\{i \in I \colon v \in X_i\}$ induces a subtree in T.

The *width* of a tree decomposition $(\{X_i \colon i \in I\}, T)$ is equal to $\max_{i \in I} |X_i| - 1$, while the *treewidth* tw(G) of a graph G is the minimum width of a tree decomposition of G.

Let T be a rooted tree. This gives us a notion of „children" of nodes of T. A decomposition $(\{X_i \colon i \in I\}, T = (I, F)\})$ of $G = (V, E)$ is *nice* if every node $i \in I$ belongs to one of the following types:

1. Leaf: node i is a leaf of T and $|X_i| = 1$,
2. Introduce: node i has exactly one child j and there is a vertex $v \in V$ such that $X_i = X_j \cup \{v\}$,
3. Forget: node i has exactly one child j and there is a vertex $v \in V$ such that $X_j = X_i \cup \{v\}$,
4. Join: node i has exactly two children j_1 and j_2, and $X_i = X_{j_1} = X_{j_2}$.

Every graph G with n vertices admits a nice tree decomposition with $O(n)$ nodes and width equal to tw(G). Moreover, it can be found in linear time if tw(G) is bounded (for the details see Bodlaender [4] and Kloks [15]).

In this section we prove the following theorem.

Theorem 11. *For any fixed r, the optimization version of r-*Fix *problem can be solved in time $\mathcal{O}(n \cdot r^{t+2})$, where n is the number of vertices of the input graph and t is its treewidth.*

We shall use a standard technique of dynamic programming on a tree-decomposition. See the survey by Bodlaender and Koster [5] for more examples. Consider a graph G and its r-coloring φ. Let $(\{X_i : i \in I\}, T = (I, F))$ be a nice tree decomposition of G with width equal to $\mathrm{tw}(G)$. Let i_0 be the root of T. Moreover, let G_i denote a subgraph of G induced by the set $\bigcup_j X_j$ where j belongs to the subtree of T rooted at i.

For every node i of T we introduce a table K_i, indexed with all possible proper r-colorings of X_i. Let $f : X_i \to [r]$. If f is a proper r-coloring of $G[X_i]$, then $K_i[f]$ is the minimum number of recolorings needed to obtain from $\varphi|_{G_i}$ a proper r-coloring φ' of G_i, such that $\varphi'|_{X_i} = f$. Clearly, $\overline{\chi}_\varphi^r(G) = \min_f \{K_{i_0}[f]\}$.

We shall show how to construct tables K_i for every type of node. We traverse T in a post-order fashion, so when we consider a node i, we have already considered all its children.

Leaf Node. Let i be a leaf node and $X_i = \{v\}$. It is clear that the value of $K_i[f]$ is equal to 0 if $f(v) = \varphi(v)$ or 1 otherwise. Thus the table K_i can be computed in $\mathcal{O}(1)$ time (as r is fixed).

Introduce Node. Let i be an introduce node and j be its child node with $X_i = X_j \cup \{v\}$. Observe that from the properties of tree decompositions follows that $v \notin V(G_j)$ (property 3). Moreover, v is not adjacent to any vertex from $V(G_j) \setminus X_j$ (property 2 and 3). It is not hard to observe that $K_i[f]$ is defined as follows:

$$K_i[f] = \begin{cases} K_j[f|_{X_j}] & \text{if } f(v) = \varphi(v) \\ K_j[f|_{X_j}] + 1 & \text{otherwise.} \end{cases}$$

Observe that K_i can be computed in time $\mathcal{O}(r^{t+1})$.

Forget Node. Let i be a forget node and j be its child with $X_i \cup \{v\} = X_j$. Clearly $G_i = G_j$. Then $K_i[f]$ is the minimum of $K_j[f']$, where $f'|_{X_i} = f$. Note that there are at most r such colorings f' for each f. Therefore the table K_i can be computed in time $\mathcal{O}(r^{t+2})$.

Join Node. Let i be a join node and let j_1 and j_2 be its children. Recall that $X_i = X_{j_1} = X_{j_2}$. From the properties of tree decompositions it follows that $V(G_{j_1}) \cap V(G_{j_2}) = X_i$ (property 3) and no vertex from $V(G_{j_1}) \setminus X_i$ is adjacent to a vertex from $V(G_{j_2}) \setminus X_i$. Therefore we can recolor G_i by recoloring G_{j_1} and G_{j_2} separately and glueing obtain colorings on X_i. Thus $K_i[f] = K_{j_1}[f] + K_{j_2}[f] - (f \ominus \varphi|_{X_i})$. Clearly we can compute K_i in time $\mathcal{O}((t+1)r^{t+1})$.

Observe that for each i, the table K_i has at most r^{t+1} elements. Therefore the space complexity of the algorithm is bounded by $\mathcal{O}(n \cdot r^{t+1})$. Since we can compute each table in time $\mathcal{O}(r^{t+2})$, the total time complexity of the algorithm is $\mathcal{O}(n \cdot r^{t+2})$, which finishes the proof of Theorem 11.

7   Fixing Number

Recall that for a graph G and its r-coloring φ, by $\overline{\chi}_\varphi^r(G)$ we denote the minimum number vertices that have to be recolored to obtain some proper r-coloring of G.

An *r-fixing number* of a graph G (denoted by $\overline{\chi}^r(G)$) is a maximum value of $\overline{\chi}_\varphi^r(G)$ over all colorings $\varphi\colon V(G) \to \{1,..,r\}$.

Definition 1. *By $\overline{\chi}(G)$ we denote the* fixing number *of a graph G, defined as a maximum value of $\overline{\chi}^r(G)$ over all $r \geq \chi(G)$.*

Lemma 12. *Let φ be some r-coloring of G and φ' be an $(r+1)$-coloring of G such that $\varphi^{-1}(i) = \varphi'^{-1}(i)$ for $i \in \{1,\ldots,r-1\}$. Then $\overline{\chi}_{\varphi'}^{r+1}(G) \leq \overline{\chi}_\varphi^r(G)$.*

Proof. Recoloring the vertices in the same way as with φ makes φ' proper. □

Lemma 13. *For all graphs G and $r \geq \chi(G)$ holds $\overline{\chi}^{r+1}(G) \leq \overline{\chi}^r(G)$.*

Proof. Let φ' be an $(r+1)$-coloring of G such that $\overline{\chi}_{\varphi'}^{r+1}(G) = \overline{\chi}^{r+1}(G)$. Let φ be an r-coloring of G obtained from φ' by identifying colors r and $r+1$. By Lemma 12 we obtain the following. $\overline{\chi}^{r+1}(G) = \overline{\chi}_{\varphi'}^{r+1}(G) \leq \overline{\chi}_\varphi^r(G) \leq \overline{\chi}^r(G)$. □

Corollary 14. *For all graphs G holds $\overline{\chi}(G) = \overline{\chi}^{\chi(G)}(G)$.*

Let G be a bipartite graph with bipartition classes X, Y and let φ be its 2-coloring. Recall from Observation 2 that

$$\overline{\chi}_\varphi^r(G) = \sum_{\substack{C:\text{ connected} \\ \text{component of } G}} \min\{|\,(X \ominus \varphi^{-1}(1)) \cap V(C)|, |\,(X \ominus \varphi^{-1}(2)) \cap V(C)|\}.$$

Note that if $|\,(X \ominus \varphi^{-1}(1)) \cap V(C)| \geq |C|/2$, then $|\,(X \ominus \varphi^{-1}(2)) \cap V(C)| \leq |C|/2$ for any connected component C of G. Thus we can easily obtain the following corollary.

Corollary 15. *$\overline{\chi}(G) \leq \lfloor n/2 \rfloor$ for every bipartite graph G on n vertices.*

This result can be generalized for non-bipartite graphs.

Theorem 16. *For all G holds $\overline{\chi}(G) \leq \left\lfloor n \cdot \frac{\chi(G)-1}{\chi(G)} \right\rfloor$.*

Proof. For a graph $G = (V, E)$ set $r = \chi(G)$ and let φ an r-coloring of G such that $\overline{\chi}_\varphi^r(G) = \overline{\chi}(G)$. Consider some proper r-coloring φ' of G. Let $A_i = \varphi^{-1}(i)$ and $A_i' = \varphi'^{-1}(i)$ for all $i \in [r]$. By $B_{i,j}$ we denote $A_i \cap A_j'$. Clearly $\bigcup_{i,j} B_{i,j} = V$. Note that for any permutation σ of $[r]$, we can obtain a proper r-coloring of G from φ by recoloring all vertices but $C_\sigma = \bigcup_{i \in [r]} B_{i,\sigma(i)}$ (this proper coloring will be equivalent to φ' up to the permutation of colors).

Suppose that $|C_\sigma| < \frac{n}{r}$ for all σ. On one hand we have $|\bigcup_\sigma C_\sigma| \leq \sum_\sigma |C_\sigma| < r! \cdot \frac{n}{r}$. On the other hand, we have:

$$\left|\bigcup_\sigma C_\sigma\right| = \left|\bigcup_\sigma \bigcup_i B_{i,\sigma(i)}\right| = \left|\bigcup_i \bigcup_\sigma B_{i,\sigma(i)}\right| = \left|\bigcup_i \bigcup_j \bigcup_{\substack{\sigma \text{ s.t.} \\ \sigma(i)=j}} B_{i,j}\right|$$

$$= (r-1)! \left|\bigcup_i \bigcup_j B_{i,j}\right| = (r-1)!n.$$

This is a contradiction, so there exists σ with $|C_\sigma| \geq \frac{n}{r}$ and therefore we can obtain a proper r-coloring of G by recoloring at most $n \cdot \frac{r-1}{r}$ vertices. Since $\overline{\chi}(G)$ is an integer, we obtain our claim. $\square$

To see that this bound is attainable, consider a graph $G(m, r)$ on $n = m \cdot r$ vertices, consisting of m disjoint copies of K_r. Clearly $\chi(G(m, r)) = r$. Let φ be an r-coloring of $G(m, r)$ such that $\varphi(v) = 1$ for every vertex v. Clearly we have to recolor every vertex but one from every copy of K_r, which gives us $\overline{\chi}(G(m, r)) \geq \overline{\chi}_\varphi^r(G(m, r)) = m(r - 1) = n\frac{r-1}{r}$.

However, there are graphs G for which the value of $\overline{\chi}(G)$ is significantly smaller. For example, consider an odd cycle C_n for $n \geq 9$. Clearly $\chi(C_n) = 3$. Let φ be any coloring of C_n with $r \geq 3$ colors. Arbitrarily choose vertex v and remove it from C_n, obtaining a path P_{n-1}. Since $\chi(P_{n-1}) = 2$, then by Theorem 16 we can obtain a proper coloring of P_{n-1} by recoloring at most $\lfloor (n - 1)/2 \rfloor$ vertices. Then we can restore vertex v and, if necessary, recolor it to an available color (there is always at least one). In this way we performed at most $1 + \lfloor (n - 1)/2 \rfloor$ recoloring operations, which is roughly $\frac{n}{2}$ compared to $\frac{2n}{3}$ given by Theorem 16.

Acknowledgement. The authors are sincerely grateful to Dieter Kratsch for valuable discussion on the topic.

References

1. Ausiello, G., Escoffier, B., Monnot, J., Paschos, V.T.: Reoptimization of minimum and maximum traveling salesmans tours. J. of Discrete Algorithms 7, 453–463 (2009)
2. Bilò, D., Böckenhauer, H.-J., Komm, D., Královic, R., Mömke, T., Seibert, S., Zych, A.: Reoptimization of the Shortest Common Superstring Problem. Algorithmica 61, 227–251 (2011)
3. Björklund, A., Husfeldt, T., Koivisto, M.: Set partitioning via inclusion-exclusion. SIAM Journal on Computing 39, 546–563 (2009)
4. Bodlaender, H.: A linear time algorithm for finding tree-decompositions of small treewidth. SIAM Journal on Computing 25, 1305–1317 (1996)
5. Bodlaender H., Koster A.: Combinatorial Optimization on Graphs of Bounded Treewidth. The Computer Journal (2007)
6. Bonamy, M., Bousquet, N.: Recoloring bounded treewidth graphs. Electronic Notes in Discrete Mathematics 44, 257–262 (2013)
7. Bonsma, P., Cereceda, L.: Finding Paths Between Graph Colourings: PSPACE-Completeness and Superpolynomial Distances. In: Kučera, L., Kučera, A. (eds.) MFCS 2007. LNCS, vol. 4708, pp. 738–749. Springer, Heidelberg (2007)
8. Cereceda, L., van den Heuvel, J., Johnson, M.: Mixing 3-Colourings in Bipartite Graphs. In: Brandstädt, A., Kratsch, D., Müller, H. (eds.) WG 2007. LNCS, vol. 4769, pp. 166–177. Springer, Heidelberg (2007)
9. Cereceda, L., Heuvel, J., van den Johnson, M.: Connectedness of the graph of vertex colourings. Discrete Mathematics 308, 166–177 (2008)
10. Cereceda, L., van den Heuvel, J., Johnson, M.: Finding paths between 3-colorings. Journal of Graph Theory 67, 69–82 (2011)

276 K. Junosza-Szaniawski, M. Liedloff, and P. Rzą—

11. Diestel, R.: Graph Theory, 3rd edn. Graduate Texts in Mathematics. Springer (2005)
12. Downey, R.G., Fellows, M.R.: Parameterized Complexity. Monographs in Computer Science. Springer (1999)
13. Felsner, S., Huemer, C., Saumell, M.: Recoloring directed graphs. In: Proc. of XIII Encuentros de Geometría Computacional, pp. 91–97 (2009)
14. Jerrum, M.: A very simple algorithm for estimating the number of k-colorings of a low-degree graph. Random Structures & Algorithms 7, 157–165 (1995)
15. Kloks, T. (ed.): Treewidth. LNCS, vol. 842. Springer, Heidelberg (1994)
16. Shachnai, H., Tamir, G., Tamir, T.: A Theory and Algorithms for Combinatorial Reoptimization. In: Fernández-Baca, D. (ed.) LATIN 2012. LNCS, vol. 7256, pp. 618–630. Springer, Heidelberg (2012)
17. Zych, A., Bilò, D.: New Reoptimization Techniques applied to Steiner Tree Problem. Electronic Notes in Discrete Mathematics 37(2–1), 387–392

Efficient Online Strategies for Renting Servers in the Cloud

Shahin Kamali and Alejandro López-Ortiz

University of Waterloo, Canada
{s3kamali,alopez-o}@uwaterloo.ca

Abstract. When scheduling jobs for systems in the cloud, we often deal with jobs that arrive and depart in an online manner. Each job should be assigned to a server upon arrival. Jobs are annotated with sizes which define the amount of resources that they need. Servers have uniform capacity and, at all times, the total size of jobs assigned to a server should not exceed this capacity. This setting is closely related to the classic bin packing problem. The difference is that, in bin packing, the objective is to minimize the total number of used servers. In the cloud systems, however, the cost for each server is proportional to the length of the time interval it is rented for, and the goal is to minimize the cost involved in renting all used servers. Recently, certain bin packing strategies were considered for renting servers in the cloud [Li et al. SPAA'14]. There, it is proved that all Any-Fit strategies have a competitive ratio of at least μ, where μ is the max/min interval length ratio of jobs. It is also proved that First Fit has a competitive ratio of $2\mu + 13$, while Best Fit is not competitive at all. We observe that the lower bound of μ extends to all online algorithms. We also prove that, surprisingly, Next Fit algorithm has a competitive ratio of at most $2\mu + 1$. We also show that a variant of Next Fit achieves a competitive ratio of $K \times \max\{1, \mu/(K-1)\} + 1$, where K is a parameter of the algorithm. In particular, if the value of μ is known, the algorithm has a competitive ratio of $\mu + 2$; this improves upon the existing upper bound of $\mu + 8$. Finally, we introduce a simple algorithm called Move To Front (MTF) which has a competitive ratio of at most $6\mu + 8$. We experimentally study the average-case performance of different algorithms and observe that the typical behaviour of MTF is better than other algorithms.

1 Introduction

Bin packing is a classic problem in the context of online computation. The input is a sequence of *items* of different *sizes* which appear in a sequential, online manner. The goal is to *place* these items into a minimum number of *bins* of uniform capacity so that the total size of items in each bin is no more than the uniform capacity of bins. It is often assumed that bins have size 1 and items have positive sizes no more than 1. The problem is online in the sense that, upon receiving an item, an algorithm should place it into a bin without any knowledge about the (size of) incoming items. A simple online strategy is Next Fit (NF) in which there is a single *open* bin at each time. If an incoming item fits in the open bin, the algorithm places it there; otherwise, it *closes* the open bin and opens a new bin for the item. First Fit (FF) is an online algorithm that does not close bins. It places an incoming item in the first bin that has enough space; if such

G.F. Italiano et al. (Eds.): SOFSEM 2015, LNCS 8939, pp. 277–288, 2015.

a bin does not exist, it opens a new bin. Best Fit (BF) works similarly to First Fit except that it maintains bins in the decreasing order of their *level*. The level of a bin is the total size of items placed in the bin. Note that First Fit and Best Fit avoid opening new bins unless they have to. Algorithms with this property are called Any Fit algorithms.

In many cloud systems, a set of *jobs* appear in an online manner. These jobs should be hosted by servers of fixed capacities. Each job has a *load* which defines the amount of resources that it needs. Depending on the application, the load of a job might represent its memory requirement, GPU resource usage or bandwidth usage. In the cloud gaming systems, for example, different instances of computer games are created in an online fashion and run in cloud servers; players interact with the servers via thin clients [8,10]. Here, an instance of a game is a job which, depending on the game and the number of users involved in it, has a load. In the case of computer games, the load of a job is mainly defined through the amount of GPU resources that it demands [10].

With the above description, any online bin packing algorithm can be used to assign jobs to servers. A job of load x can be treated as an item of size x which is assigned to a server (bin) of certain capacity. In this paper, we interchangeably use terms 'job' and 'item' as well as 'server' and 'bin'. There are, however, distinctions between assigning jobs to servers and the bin packing problem. First, jobs depart the system after they *complete*. When a job arrives, it is not clear when it completes and an algorithm should place it without any knowledge about its departure time. Second, in the bin packing problem, the objective is to minimize the number of used bins. Bins can be regarded as servers that one can *buy*, and the goal is to minimize the cost by buying a small number of servers. In the cloud, however, the servers are *rented* from cloud service providers. For example, gaming companies such as OnLive [3] and GaiKai [2] offer cloud gaming services which execute in public clouds like Amazon EC2 [1]. A rented server is charged by its usage, often in an hourly or monthly basis. So, in order to minimize the cost, we need to minimize the total time that servers are rented. In doing so, an algorithm *releases* a server when all its assigned jobs complete.

Definition 1. *In the* server renting problem *, a series of jobs (items) appear in an online manner. Each job has a load (size) that defines the amount of resources that it needs. Upon its arrival, a job should be assigned to a server (bin). Servers have a uniform capacity, and the total load of jobs assigned to a server at any time should not exceed this capacity. Besides its arrival time, each job has a departure time that indicates when it completes and leaves the system. The length of the interval between the arrival and departure time of a job is called the 'length' of the job. Upon its arrival,, the length of a job is unknown to online algorithms. To assign a job to a server, an online algorithm might open (rent) a new server or place it to any of the previously opened servers. When all jobs assigned to a server depart, that server is released. The goal is to minimize the total usage time of servers . More precisely, assuming that an algorithm opens m bins $B_1, \ldots, B_m$, the total cost of the algorithm is $\sum_{i=1}^{m} t_i$, where t_i is the length of the time interval between when B_i is opened and when it is released. Without loss of generality, we assume the capacity of servers to be 1 and jobs have size at most 1. Also, we assume the length of jobs is at least Δ and at most $\mu\Delta$ where $\mu \geq 1$.*

When studying the server renting problem, we are interested algorithms with good worst-case and average-case performance. For measuring the worst-case performance, we compare online algorithms with an optimal offline algorithm OPT that knows the entire sequence (all arrival times, lengths and sizes) in advance. An algorithm is said to be c-competitive (more precisely, *asymptotic* c-competitive) if the cost of serving any input sequence never exceeds c times the cost of OPT within an additive constant.

1.1 Previous Work and Contribution

The Bin Packing problem has been widely studied over the past few decades. It is known that Next Fit is 2-competitive while Best Fit and First Fit are both 1.7-competitive [9]. Generally, any Any Fit algorithm that avoids placing items in the bin with the lowest level is 1.7-competitive (these algorithms are called Almost Any Fit). The Harmonic family of algorithms is another class of bin packing algorithms which are based on placing items of similar sizes together in the same bins. These algorithm generally have better competitive ratios than Any Fit algorithms. The best member of this family is Harmonic++ with a competitive ratio of 1.5888 [11]. However, because of their poor average-case performance, algorithms of the Harmonic family are rarely used in practice. It is known that no online algorithm can be better than 1.54037-competitive [4].

Coffman et al. [6] studied a dynamic version of the bin packing problem in which items arrive and depart. It is proved that the competitive ratio of First Fit is between 2.75 and 2.89 while no online algorithm can do better than 2.5 [6]. A discrete version of the problem, where each item has size $1/k$ for some integer k, is studied in [5]. Note that in these results, the objective is to minimize the number of opened bins.

The online server renting problem as defined above was recently introduced by Li et al. [10]. Some terms and notations in this paper are also borrowed from this paper. The authors prove that no Any Fit algorithm can be better than μ competitive. Recall that μ is the ratio between the largest length and the smallest length in the sequence. The competitive ratio of First Fit is proved to be at most at most $2\mu + 13$-competitive. On the other hand, it is proved that Best Fit does not have a bounded competitive ratio. This result is somewhat surprising as Best Fit is usually the superior algorithm for most applications of the bin packing problem. In [10], a slight modification of the First Fit algorithm is introduced which achieves a competitive ratio of at most $\frac{8}{7}\mu + 55/7$ in the general case and a competitive ratio of at most $\mu + 8$ when the value of μ is known.

In this paper, we further study the server renting problem. We first observe that the lower bound of μ holds for the competitive ratio of *any* online algorithm. In the standard bin packing, there is no harm in keeping bins open; hence, Any Fit algorithms have advantage over algorithms that close bins (e.g., Next Fit). This is not necessarily the case for the server renting problem because when a bin gets closed, as no new item is placed there, the bin gets released earlier; this might improve the cost of the algorithm. Assume items sizes are at no larger than $1/k$ for some $k \geq 1$. We show that the competitive ratio of the Next Fit algorithm is at most $\frac{k}{k-1}\mu + 1$ when $k \geq 2$ and at most $2\mu + 1$ when $k < 2$. In particular, when $k = 1$ (when there is no restriction on item sizes), the competitive ratio is at most $2\mu + 1$. Note that this is much better than $2\mu + 13$ of First Fit. We also introduce a variant of Next Fit which achieves a ratio of at most $K \times \max\{1, \frac{\mu}{K-1}\} + 1$, where K is a parameter of the algorithm. In particular, if the value of μ is known, we get an algorithm with competitive ratio of at most $\mu + 2$ which is better than $\mu + 8$ of a similar algorithm presented in [10].

Although Next Fit has a good competitive ratio, unfortunately, it has a poor average-case performance. Our experiments indicate that, for sequences generated uniformly at random, Best Fit outperforms both Next Fit and First Fit. Recall that Best Fit does not have a bounded competitive ratio. We introduce a simple algorithm, called Move To Front (MTF), which performs better than Best Fit on average. Moreover, in contrast to Best Fit, it has a bounded competitive ratio of at most $6\mu + 8$.

2 Preliminaries

In this section, we present some basic results about the server renting problem. First, using a similar argument as [10], we prove a general lower bound for competitive ratio of any online algorithm.

Theorem 1. *The competitive ratio of any online algorithm for the server renting problem is at least* $\frac{\mu}{1+\epsilon(\mu-1)}$ *where* ϵ *is a lower bound for the size of items.*

Proof. Recall that the lengths of all items are at least Δ and at most $\mu\Delta$. Consider a sequence which is defined through phases. Each phase starts with $\frac{1}{\epsilon^2}$ items of size ϵ. To place these items, any algorithm has to open at least $1/\epsilon$ bins. At time Δ, $\frac{1}{\epsilon^2} - \frac{1}{\epsilon}$ items depart in an adversarial manner so that there is a single of item of size ϵ in $1/\epsilon$ bins (some bins might get released at this time). The remaining items stay for a period of length $\mu\Delta$ and the online algorithm keeps a single bin for each of them. At time $\mu\Delta$, all items depart and the phase ends. The cost of the online algorithm for each phase is at least $\mu\Delta/\epsilon$ since it keeps $1/\epsilon$ bins for a period of $\mu\Delta$. OPT places items which have length $\mu\Delta$ together in a single bin and incurs a cost of $\mu\Delta$ for them. Other $\frac{1}{\epsilon^2} - \frac{1}{\epsilon}$ items are placed tightly together in $1/\epsilon - 1$ bins for a period of length Δ. The cost of OPT for these items will be $\Delta/\epsilon - \Delta$. In total, the cost of OPT will be $\mu\Delta + \Delta/\epsilon - \Delta$ and the competitive ratio of the algorithm will be $\frac{\mu\Delta/\epsilon}{\mu\Delta+\Delta/\epsilon-\Delta} = \frac{\mu}{1+\epsilon(\mu-1)}$. $\square$

Next, we introduce two lower bounds for the cost of OPT. We call an item x *active* at time t if t is in the interval between the arrival and the departure time of x. Let the *span* of an input sequence σ be the total length of intervals at which at least one item is active. Clearly, the cost of any algorithm for serving σ is at least equal to the span of σ. Define the *resource utilization* of an item as the product of its size and its length. The cost of any algorithm for σ is at least equal to the total resource utilization of items in σ, denoted by $util(\sigma)$. So, we have the following lower bounds for the cost of OPT.

Proposition 1. *For any input sequence* σ, *the cost of an optimal offline algorithm* OPT *is at least equal to* $span(\sigma)$ *and* $util(\sigma)$, *namely, the span of* σ *and also the total resource utilization of items in* σ.

When we allow arbitrarily small items, Theorem 1 indicates that all algorithms have a competitive ratio of at least μ. This suggests that when item sizes are larger than a fixed value, better competitive ratios can be achieved. Consider a sequence σ in which all item sizes are larger than $1/k$ for some positive value k. The cost of any algorithm is at most equal to the total length of all items denoted by $L(\sigma)$, which happens when no two items share a bin. On the other hand, the total resource utilization of items, and consequently cost of OPT, is at least $L(\sigma)/k$. So, we get the following.

Proposition 2. *[10] When items sizes are at least $1/k$ (k is a positive value), the competitive ratio of any online algorithm for the server renting problem is at most k.*

3 Next Fit Algorithm

In this section, we analyze the Next Fit algorithm for the server renting problem. Recall that for the bin packing problem Next Fit keeps one bin open at any given time. If an incoming item does not fit in the open bin, it closes the bin and opens a new bin. For the server renting problem, we distinguish between *closing* and *releasing* a bin. When an item does not fit in the open bin, the algorithm closes the bin and does not refer to it. Such a bin remains in the system (i.e., is being rented) until all items which are placed there depart and it becomes released.

Example 1. Consider a sequence $\langle a = (0.3, 1, 5), b = (0.4, 2, 6), c = (0.4, 3, 5), \ldots \rangle$. The first element of each tuple indicates the size of an item, while the second and the third elements respectively indicate the arrival and the departure times of the item. At time 1, item a arrives and is placed in the single open bin. At time 2, item b arrives and is placed in the same bin (the level of the bin will be 0.7). At time 3, item c arrives and does not fit in the open bin; hence, the current open bin is closed and a new bin is opened for c. The closed bin remains in the system (and a rental cost is paid for it) until time 6 where item b departs and the bin gets released.

Theorem 2. *Assume all items have size $1/k$ or smaller, where k is a positive value no smaller than 1. When $k \geq 2$, the competitive ratio of Next Fit for the server renting problem is at most $\frac{\mu}{1-1/k} + 1$. When $k < 2$, the ratio is at most $2\mu + 1$.*

Proof. Assume Next Fit opens m bins $B_1, \ldots, B_m$ for serving an arbitrary sequence σ. Let st_i denote the length of the time interval during which server B_i is rented. We refer to this period as the *stretch* of B_i. Let $\mathrm{NF}(\sigma)$ denote the cost of Next Fit for serving σ. We have $\mathrm{NF}(\sigma) = \sum_1^m st_i$. The stretch of B_i can be partitioned into two period. The first one is the interval between when B_i is opened and when it is closed. The second period is the interval between when B_i is closed and when it is released. Let st_i^1 and st_i^2 respectively denote the lengths of the first and the second periods of B_i ($st_i^1 + st_i^2 = st_i$). If a bin is released before being closed, the second period will be empty, i.e., $st_i^2 = 0$. Let $p \leq m$ denote the number of other bins, i.e., those which are closed before being released. We call these bins *critical* bins. Note that when a bin is closed, it takes a time of length at most $\mu\Delta$ before it gets released, i.e., the second period of each critical bin has a length of at most $\mu\Delta$ (see Figure 1). So, the total rental time for the second periods of all bins is no more than $p \times \mu\Delta$. On the other hand, the total rental time of the first periods of all bins is no more than the span of input sequence. This is because the first period of a bin starts when that of previous bin is finished, i.e., no two bins are in their first stretch period at the same time. So we have

$$\mathrm{NF}(\sigma) = \sum_{i=1}^{m} st_i = \sum_{i=1}^{m} st_i^1 + \sum_{i=1}^{p} st_i^2 \leq span(\sigma) + p \times \mu\Delta \qquad (1)$$

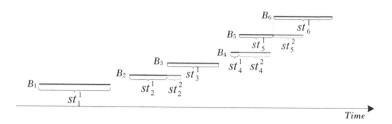

Fig. 1. The stretch of bins in a packing of the Next Fit algorithm. In this example, bins B_2, B_4, and B_5 are critical bins. The second periods of bins are highlighted in red.

Assume $k \geq 2$. At the time of being closed, all critical bins have a level of at least $1 - 1/k$; otherwise, the item that caused opening of a new bin could fit in such a bin. This implies that the number of critical bins is no more than $\frac{w(\sigma)}{1-1/k}$ where $w(\sigma)$ is the total size of items in σ. Let $util(\sigma)$ denote the total resource utilization of items in σ. Since the length of each item is at least Δ, we have $util(\sigma) \geq w(\sigma) \times \Delta$, and by Proposition 1, $w(\sigma) \leq \text{OPT}(\sigma)/\Delta$. Consequently, $p \leq \frac{\text{OPT}(\sigma)}{(1-1/k)\Delta}$. Also, by Proposition 1, $span(\sigma) \leq \text{OPT}(\sigma)$. Plugging these into Equation 1, we get the following inequality which completes the proof.

$$\text{NF}(\sigma) \leq \text{OPT}(\sigma) + \frac{\text{OPT}(\sigma)}{1 - 1/k} \times \mu$$

Next, assume $k < 2$. We define the *amortized level* of a critical bin B as the the size of the item that closes B plus the total sizes of items in B at the time Next Fit closes it. By definition of NF, the amortized level of all critical bins is more than 1. At the same time, the size of each item is added at most twice in the total amortized cost (once as a part of a critical bin and once as the time that closes a critical bin). Hence, the total sum of the amortized levels of all critical bins is at most twice the total size of sequence. This implies that the number of critical bins is no more than twice the total size of items in σ, i.e., $p \leq 2w(\sigma) \leq 2\,\text{OPT}(\sigma)/\Delta$. Applying this into Equation 1, we get $\text{NF}(\sigma) \leq \text{OPT}(\sigma) + 2\,\text{OPT}(\sigma) \times \mu$ which completes the proof. □

3.1   Improving the Competitive Ratio: Modified Next Fit Algorithm

In this section, we modify the Next Fit algorithm to improve its competitive ratio. Intuitively speaking, the competitive ratio improves for sequences formed by small items. On the other hand, as Proposition 2 implies, when all items are relatively large, the competitive ratio is independent of μ. This suggest that the competitive ratio can be improved when large and small items are treated separately. A similar approach is used in [10] to improve the competitive ratio of First Fit. Consider the following algorithm.

Modified Next Fit with parameter K: The algorithm applies the Next Fit strategy to place items. In doing so, it treats items with size smaller than $1/K$ separately from those with size larger than or equal to $1/K$.

Theorem 3. *The competitive ratio of Modified Next Fit with parameter K is at most $K \times max\{1, \mu/(K-1)\} + 1$.*

Proof. Consider a sequence σ and let σ^s and σ^l denote the subsequences of σ respectively formed by items smaller and larger or equal to K. Recall that the resource utilization of an item is the product of its length and its size, and the total resource utilization of all items in a sequence is a lower bound for the cost of OPT. As Proposition 2 suggests, the number of opened bin by Modified Next Fit for items in σ^l is no more than $k \times util(\sigma^l)$, where $util(\sigma^l)$ is the total utilization of items in σ^l.

For placing σ^s, as the proof of Theorem 2 suggests, the algorithm incurs a cost of at most $\mu\Delta \times \frac{\omega(\sigma^s)}{1-1/K} + span(\sigma^s)$, where $\omega(\sigma^s)$ is the total size of item in σ^s. This will be no more than $\mu\Delta \times \frac{util(\sigma^s)}{\Delta(1-1/K)} + span(\sigma^s)$, where $util(\sigma^s)$ is the total resource utilization of σ^s (this is because the length of all items is at least Δ). In total, the cost of the algorithm will be at most $K \times util(\sigma^l) + \mu \times \frac{util(\sigma^s)}{1-1/K} + span(\sigma^s)$. This is no more than $K \times util(\sigma) \times max\{1, \frac{\mu}{K-1}\} + span(\sigma)$ where $util(\sigma)$ is the total utilization of items in σ. Since $util(\sigma)$ and $span(\sigma)$ are lower bounds for the cost of OPT, we can conclude the cost of Modified Next Fit is at most $K \times max\{1, \frac{\mu}{K-1}\} + 1$.   □

When the value of μ is known to the algorithm, we can define K to be $\mu + 1$. In this case, the competitive ratio of Modified First Fit will be at most $\mu + 2$.

Proposition 3. *When the value of μ is known, there exists an online algorithm which achieves a competitive ratio of at most $\mu + 2$.*

4   Toward Practical Algorithms: Move to Front Algorithm

In the previous sections, we showed that Next Fit and a variant of that have promising competitive ratios. Providing these types of worst-case guarantees is important in the theoretical analysis of the problem. Nevertheless, in practice, we are interested in algorithms that also have good average-case performance. For example, in the case of the classic bin packing problem, Best Fit and First Fit are preferred over other algorithms in most applications. This is because they have acceptable worst-case performance (although not as good as the Harmonic family) and superior average-case performance. We examined different algorithms to evaluate their average-case performance for the server renting problem. Our experiments are presented in Section 5 and show that, on average, Best Fit has an evident advantage over First Fit and Next Fit. These results are in contrast with the competitive results and indicate that the worst-case behaviour and average-case behaviour of these algorithms are quite different.

In this section, We introduce a new algorithm, called Move-To-Front (MTF), for the server renting problem. We prove that, unlike Best Fit, MTF has a bounded competitive ratio. Our experiments indicate that MTF has a better average-case performance compared to other algorithms. MTF is a simple Any Fit algorithm and runs as fast as BF and FF. Hence, we believe that MTF is a practical and efficient algorithm for the problem.

Move To Front: The algorithm maintains a list of open bins. It places an item x in the first bin in the list which has enough space. If no bin has enough space, a new bin is opened. After placing x into a bin, that bin is moved to the front of the list.

Theorem 4. *Move To Front has a competitive ratio of at most* $6\mu + 8$.

Proof. Consider the packing of MTF for a sequence σ. We assume σ is *continuous* in the sense that at any given time there is at least one active item, and any algorithm maintains at least one open bin. For sequences which are not continuous, at some point, all bins of MTF and OPT are released. In this case, we can divide the sequence into continuous subsequences and apply the same argument for each of them.

We divide the span of the sequence into *periods* of length $(\mu + 1)\Delta$ (except for the last period which might be shorter). We also divide bins of MTF into critical and non-critical bins. A bin B is critical if in the interval between when B is opened and when it is released, some other bin B' is opened by MTF. Note that no two non-critical bins can be opened at the same time. Hence, the total cost of MTF for all non-critical bins is at most $span(\sigma)$ which is no more than $\text{OPT}(\sigma)$.

For each critical bin B, we define *head, tail,* and *body* of B as follows. If B is opened and released in the same period, its head is its stretch (the interval between its opening and releasing) while its body and tail are empty intervals. Otherwise, the head of B is the interval between when B is opened and the end of the period in which it is opened. The tail of B is the interval between the start of the period in which it is released and when it is released. The body of B is the interval between the head and the tail of B (see Figure 2). Let $head(B)$, $body(B)$, and $tail(B)$ indicate the lengths of head, body, and tail of B, respectively. For the cost that MTF incurs for B we have:

$$stretch(B) = head(B) + body(B) + tail(B) \le 2(\mu + 1)\Delta + body(B)$$

Assume there are m critical bins opened by MTF. The algorithm incurs a cost of at most $2 \times (\mu + 1)$ for the head and the tail of each bin. We will have:

$$\text{MTF}(\sigma) \le 2m(\mu + 1)\Delta + \sum_{b=1}^{m} body(B_b) + \text{OPT}(\sigma)$$

The last term is added for non-critical bins. Assume there are $q+1$ periods; note that the last period cannot include the body of any bin. For each period P_i among other periods $(1 \le i \le q)$, let $\alpha(P_i)$ denote the number of critical bins which have their body in P_i, i.e., critical bins which are open at the beginning of the period and remain open till the end. MTF incurs a cost of $\alpha(P_i) \times (\mu + 1)\Delta$ for bodies of critical bins in P_i. So:

$$\text{MTF}(\sigma) \le 2m(\mu + 1)\Delta + (\mu + 1)\Delta \sum_{i=1}^{q} \alpha(P_i) + \text{OPT}(\sigma) \qquad (2)$$

Next, we consider the cost of OPT for packing σ. We prove the following claims:

Claim 1: For the number of critical bins opened by MTF we have $m \le 2\,\text{OPT}(\sigma)/\Delta$.
Claim 2: For each period P, if $\alpha(P) = 1$, OPT incurs a cost of $(\mu + 1)\Delta$.
Claim 3: For each period P, if $\alpha(P) \ge 2$, OPT incurs a cost of at least $(\alpha(P) - 1)\Delta/2$.

Claim 1 implies that the first term in Equation 2 is upper bounded by $4(\mu+1)\,\text{OPT}(\sigma)$. Claims 2 implies that in the specified periods, MTF and OPT incur the same costs. Claim

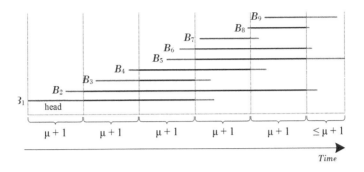

Fig. 2. The span of an input sequence is divided into periods. The red and blue intervals respectively indicate heads and tails of the bins while black intervals are bodies of the bins.

3 implies that $\alpha(P) \leq 2\,\text{OPT}(P)/\Delta + 1$ where $\text{OPT}(P)$ is the cost inured by OPT in period P. Consequently, the second term in Equation 2 is upper bounded by

$$(\mu+1)\Delta\sum_{i=1}^{q}(2\,\text{OPT}(P_i)/\Delta+1) = 2(\mu+1)\sum_{i=1}^{q}\text{OPT}(P_i) + q(\mu+1)\Delta$$
$$< 2(\mu+1)\,\text{OPT}(\sigma) + span(\sigma)$$

The last inequality holds because we have divided the stretch of σ into $q+1$ periods of length $(\mu+1)\Delta$ (except the last period which might be shorter). Note that $span(\sigma) \leq \text{OPT}(\sigma)$. Adding all terms in Equation 2, we get:

$$\text{MTF}(\sigma) \leq 4(\mu+1)\,\text{OPT}(\sigma) + 2(\mu+1)\,\text{OPT}(\sigma) + 2\,\text{OPT}(\sigma)$$
$$= (6\mu+8)\,\text{OPT}(\sigma)$$

To complete the proof, it remains to show the above claims hold.

For Claim 1, define the amortized weight of a critical bin B as the total size of items in B plus the size of the item that opens a new bin, while B is still open (by definition of critical bins, such item exists). The amortized weight of every critical bin is more than 1, and the total amortized cost of all critical bins is more than m. Each item is counted at most twice in the total amortized cost (once as the item that opens a new bin and once as the member of a critical bin). Assume $w(\sigma)$ is the total size of items; we will have $m < 2w(\sigma)$. Note that $w(\sigma) \leq util(\sigma)/\Delta \leq \text{OPT}(\sigma)/\Delta$. Hence, we get $m \leq 2\,\text{OPT}(\sigma)/\Delta$.

For Claim 2, note that OPT maintains at least one open bin any given each time; otherwise, the sequence is not continuous.

For Claim 3, let t denote the start time of P and let B^* denote the set of the $\alpha(P)$ bins that have their body stretched along P. Consider the time interval $[t+\Delta, t+(\mu+1)\Delta)$. In this interval, any of the bins in B^* receive at least one new item; otherwise, the algorithm would have released the bin (recall that the length of any item is at most $\mu\Delta$). For each bin B in B^*, except the last bin in the list maintained by the algorithm right before time $t + \Delta$, let t_B indicate the time that the bin B' receives an item for the first time (in the interval $[t + \Delta, t + (\mu + 1)\Delta))$. Here, B' is the bin that is placed right after B in the mentioned ordering. Define the *critical set* of B as the set of items in B at time t_B plus the item that was placed in B'. Note that the total size of items in the critical set of each bin is more than 1. Hence, the critical items of each bin have a resource utilization of more than Δ in the interval $[t, t + (\mu + 1)\Delta)$. Since each item belongs to the critical sets of at most two bins, the total resource utilization of critical items is at least $(\alpha(P) - 1)\Delta/2$ in the interval $[t, t + 1 + \mu)$. Note that the resource utilization is a lower bound for the cost of OPT in the same interval. □

The above proof can be extended to any algorithm that maintains a list of bins and places an incoming item in the first bin which has enough space. Such an algorithm might update the list after *placing* items. In particular, the above analysis holds for the First Fit algorithm. For the Best Fit algorithm, it fails because the order of bins changes when items *depart*. Recall that Best Fit is not competitive.

5   Average-Case Analysis: An Experimental Study[1]

In this section, we compare the average-case performance of different algorithms for the server renting problem on randomly-generated sequences. We discretize the problem by assuming that the capacity of bins is an integer E and items have integer sizes in the range $[1, E]$. Moreover, we assume items arrive in discrete time-steps in the range $[1, T - \mu]$ and their length is in the interval $[1, \mu]$. Here, T is a measure of *sparsness* and defines the rate at which the items arrive. We examine different values of μ and T for sequences of fixed length. Table 1 gives a summary of the datasets that we generated for our experiments. In all cases, sizes and lengths of items are randomly and independently taken from the indicated ranges (assuming a uniform distribution). For each setting, we run different algorithms on 10^3 randomly generated sequences. For each sequence, we compute the resource utilization of the sequence as a lower bound for the cost of OPT. We use the ratio between the cost of an algorithm and the resource utilization as a measure of performance.

Table 1. A summary of the experimental settings

Parameter	Description	Value	Note
n	length of sequences	10^5	Number of items to be packed
μ	maximum length of items	1,2,5,10,100	Lengths are picked from the range $[1, \mu]$
T	span of sequence	$10^3, 10^4, 10^5$	Arrival times are picked from the range $[1, T - \mu]$
E	bin capacity	10^3	Sizes are picked from the range $[1, E]$

The algorithms that we considered in the experiments are Next Fit, Modified Next Fit, First Fit, Modified First Fit, Harmonic, Best Fit, and Move To Front. We define the

---

[1] We thank David S. Johnson for various useful suggestions about the experimental validation of the various strategies.

parameters of Modified Next Fit and Modified First Fit to be respectively $E/(\mu + 1)$ and $E/(\mu + 7)$. These values ensure that these algorithms achieve their best possible competitive ratio (see Section 3.1 and [10]). Note that the value of μ is not known to an online algorithm and the mentioned algorithms are semi-online in this sense. We also consider the Harmonic algorithm which classify items by their sizes (using harmonic intervals) into K classes and applies the Next Fit strategy for placing items of each class; for our experiments we assume $K = 10$. A straightforward analysis shows that the competitive ratio of the Harmonic algorithm is as good as Modified Next Fit (with the same parameter K). However, similar to the bin packing problem, for the server renting problem, Harmonic seems to have a poor average-case performance.

Figure 3 shows the average-case performance ratio for different algorithms. In most cases, Move To Front is the best algorithm. Intuitively, there are two factors which define the quality of a packing. One is how well items are *aligned* to each other. A packing is well-aligned if items that arrive at the same time are placed together; this ensures that, on expectation, all items of a bin depart also at (almost) the same time. Thus, there is a more chance of saving cost through releasing bins. The second factor in defining the quality of an algorithm is how well the items are packed together. Clearly, if items are tightly packed, there is a save in cost by opening a smaller number of bins.

Next Fit results in well-aligned packings; however, it does not packs items as tightly as Any Fit algorithms do. On the other side, Best Fit results in tight packings which are not necessarily well-aligned. Move To Front provides a compromise; the packing of MTF is well-aligned because items placed in the most recent bin are expected to have (almost) same arrival times. At the same time, as an Any Fit algorithm, MTF places items almost tightly and does not open a large number of bins. For smaller values of μ, it is more important to achieve well-aligned packings. This is because, when items have roughly same lengths, there is more benefit in placing them together according to their

| | $T = 1,000$ | | | | | $T = 10,000$ | | | | | $T = 100,000$ | | | | |
	$\mu = 1$	$\mu = 2$	$\mu = 5$	$\mu = 10$	$\mu = 100$	$\mu = 1$	$\mu = 2$	$\mu = 5$	$\mu = 10$	$\mu = 100$	$\mu = 1$	$\mu = 2$	$\mu = 5$	$\mu = 10$	$\mu = 100$
Next Fit	**1.4011**	1.4618	1.4970	1.5113	1.5255	1.7186	1.6564	1.5810	1.5473	1.5263	1.9539	1.9291	1.8721	1.8014	1.5496
Modified Next Fit	1.4392	1.4820	1.4780	1.5023	1.5253	1.8069	1.8061	1.7223	1.6383	1.5366	1.9762	1.9738	1.9313	1.8601	1.5631
First Fit	1.5647	1.4561	1.3448	1.2929	1.2255	1.7485	1.6635	1.5288	1.4295	1.2624	1.9544	1.9292	1.8705	1.7950	1.4132
Modified First Fit	1.6666	1.5335	1.3952	1.3233	1.2287	1.8362	1.7622	1.6330	1.5148	1.2768	1.9726	1.9534	1.9042	1.8352	1.4293
Harmonic	1.7598	1.6937	1.5848	1.5143	1.4235	1.9555	1.9270	1.8842	1.8195	1.5170	1.9946	1.9903	1.9797	1.9726	1.8294
Best Fit	1.6659	1.5027	1.3509	1.2637	**1.1151**	1.7401	1.6585	1.5342	1.4359	**1.2107**	1.9540	1.9287	1.8696	1.7935	1.4164
Move To Front	1.4113	**1.3921**	**1.3094**	**1.2560**	1.1612	**1.7134**	**1.6323**	**1.5036**	**1.4110**	1.2251	**1.9536**	**1.9283**	**1.8689**	**1.7913**	**1.4005**

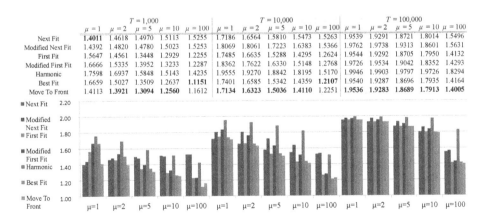

Fig. 3. Average-case performance ratio of major server renting problem algorithms, assuming a uniform distribution for the length and size of items. The bold numbers indicate the best algorithm for different values of T and μ. In all cases MTF is the best or second to the best algorithm. To make comparison easier, the numbers are plotted into a bar diagram.

arrival time so that they depart at (almost) the same time (which results in releasing their bins). This is particularly more evident for sequences with small span ($T = 1,000$). In these sequences, many items appear at the same time and almost all algorithms result in relatively good packing (regarding the number of opened bins). As a results, for smaller values of μ and T, Next Fit performs better relative to other algorithms. In particular, when $\mu = 1$ and $T = 1,000$, it slightly outperforms MTF. For larger values of μ, it is more important to avoid opening new bins; this is because each bin remains open for a relativity long period of time and one should avoid opening new bins if possible. Hence, when μ is large ($\mu = 100$), Best Fit is slightly better than MTF.

6   Concluding Remarks

We showed that the Next Fit algorithm provides promising worst case guarantees for the server renting problem. We expect that the same holds for other bounded-space algorithms, e.g., Harmonic or BBF_2 of [7]. Unfortunately, these algorithms do not have good average-case performance. We introduced Move To Front algorithm as a simple and fast Any Fit algorithm which can be regarded as an alternative to Best Fit and First Fit. Our experiments indicate that MTF outperforms other algorithms on average. The closest counterpart of MTF (regarding the average case performance) is Best Fit which does not have a bounded competitive ratio. In contrast to Best Fit, we proved that MTF has a competitive ratio of at most $6\mu + 8$. We believe this upper bound is not tight and the competitive ratio of MTF can be improved; we leave this as a future work. Another promising direction for future work is to provide theoretical upper bounds for the average-case performance of MTF.

References

1. Amazon EC2, http://aws.amazon.com/ec2/ (accessed: August 14, 2014)
2. Gaikai, http://www.gaikai.com/ (accessed: August 14, 2014)
3. OnLive, http://www.onlive.com/ (accessed: August 14, 2014)
4. Balogh, J., Békési, J., Galambos, G.: New lower bounds for certain classes of bin packing algorithms. Theoret. Comput. Sci. 440–441, 1–13 (2012)
5. Chan, W.T., Lam, T.W., Wong, P.W.H.: Dynamic bin packing of unit fractions items. Theoret. Comput. Sci. 409(3), 521–529 (2008)
6. Coffman, E.G., Garey, M.R., Johnson, D.S.: Dynamic bin packing. SIAM J. Comput. 12, 227–258 (1983)
7. Csirik, J., Johnson, D.S.: Bounded space on-line bin packing: Best is better than First. In: Proc. 2nd Symp. on Discrete Algorithms (SODA), pp. 309–319 (1991)
8. Huang, C.Y., Hsu, C.H., Chang, Y.C., Chen, K.T.: Gaminganywhere: An open cloud gaming system. In: Proc. the 4th ACM Multimedia Systems Conference, pp. 36–47 (2013)
9. Johnson, D.S., Demers, A., Ullman, J.D., Garey, M.R., Graham, R.L.: Worst-case performance bounds for simple one-dimensional packing algorithms. SIAM J. Comput. 3, 256–278 (1974)
10. Li, Y., Tang, X., Cai, W.: On dynamic bin packing for resource allocation in the cloud. In: Proc. 26th Symp. on Parallel Algorithms and Architectures (SPAA), pp. 2–11 (2014)
11. Seiden, S.S.: On the online bin packing problem. J. ACM 49, 640–671 (2002)

Pal^k is Linear Recognizable Online

Dmitry Kosolobov, Mikhail Rubinchik, and Arseny M. Shur

Ural Federal University, Ekaterinburg, Russia
dkosolobov@mail.ru, mikhail.rubinchik@gmail.com, arseny.shur@urfu.ru

Abstract. Given a language L that is online recognizable in linear time and space, we construct a linear time and space online recognition algorithm for the language $L \cdot \mathrm{Pal}$, where Pal is the language of all nonempty palindromes. Hence for every fixed positive k, Pal^k is online recognizable in linear time and space. Thus we solve an open problem posed by Galil and Seiferas in 1978.

1 Introduction

In the last decades the study of palindromes constituted a notable branch in formal language theory. Recall that a string $w = a_1 \cdots a_n$ is a *palindrome* if it is equal to $\overleftarrow{w} = a_n \cdots a_1$. There is a bunch of papers on palindromes in strings. Some of these papers contain the study of strings "rich" in palindromes (see, e.g., [7]), some other present solutions to algorithmic problems like finding the longest prefix-palindrome [12] or counting distinct subpalindromes [9].

For languages constructed by means of palindromes, an efficient recognition algorithm is often not straightforward. In this paper we develop a useful tool for construction of acceptors for such languages. Before stating our results, we recall some notation and known facts.

The language of nonempty palindromes over a fixed alphabet is denoted by Pal. Let $\mathrm{Pal}_{ev} = \{w \in \mathrm{Pal}: |w| \text{ is even}\}$, $\mathrm{Pal}_{>1} = \{w \in \mathrm{Pal}: |w| > 1\}$, $\mathrm{Pal}^k = \{w_1 \cdots w_k: w_1, \ldots, w_k \in \mathrm{Pal}\}$. Given a function $f: \mathbb{N} \to \mathbb{N}$ and a language L, an algorithm *recognizes L in $f(n)$ time and space* if for any string w of length n, the algorithm decides whether $w \in L$ using at most $f(n)$ time and at most $f(n)$ additional space. An algorithm recognizes a given language *online* if the algorithm processes the input string sequentially from left to right and decides whether to accept each prefix after reading the rightmost letter of that prefix.

It is well known that every context-free language can be recognized by relatively slow Valiant's algorithm (see [13]). According to [11], there are still no examples of context-free languages that cannot be recognized in linear time on a RAM computer. Some "palindromic" languages were considered as candidates to such "hard" context-free languages.

At some point, it was conjectured that the languages Pal_{ev}^* and $\mathrm{Pal}_{>1}^*$, where $*$ is a Kleene star, cannot be recognized in $O(n)$ (see [8, Section 6]). But a linear algorithm for the former was given in [8] and for the latter in [6]. The recognition of Pal^k appeared to be a more complicated problem. Linear algorithms for the cases $k = 1, 2, 3, 4$ were given in [6]. Their modified versions can

G.F. Italiano et al. (Eds.): SOFSEM 2015, LNCS 8939, pp. 289–301, 2015.

be found in [3, Section 8]. In [6] and [3] it was conjectured that there exists a linear time recognition algorithm for Pal^k for arbitrary k. In this paper we present such an algorithm. Moreover, our algorithm is online. The main contribution is the following result.

Theorem. *Suppose a given language L is online recognizable in $f(n)$ time and space, for some function $f : \mathbb{N} \to \mathbb{N}$. Then the language $L \cdot \mathrm{Pal}$ can be recognized online in $f(n) + cn$ time and space for some constant $c > 0$ independent of L.*

Corollary. *For arbitrary k, Pal^k is online recognizable in $O(kn)$ time and space.*

Note that the related problem of finding the minimal k such that a given string belongs to Pal^k can be solved online in $O(n \log n)$ time [5], and it is not known whether a linear algorithm exists.

The paper is organized as follows. Section 2 contains necessary combinatorial properties of palindromes; similar properties were considered, e.g., in [2]. In Sect. 3 we describe an auxiliary data structure used in the main algorithm. An online recognition algorithm for Pal^k with $O(kn \log n)$ working time is given in Sect. 4. Finally, in Sect. 5 we speed up this algorithm to obtain the main result. Proofs omitted due to space constraints are included in the full version [10].

2   Basic Properties of Palindromes

A *string of length n* over the alphabet Σ is a map $\{1, 2, \ldots, n\} \mapsto \Sigma$. The length of w is denoted by $|w|$ and the empty string by ε. We write $w[i]$ for the ith letter of w and $w[i..j]$ for $w[i]w[i+1] \ldots w[j]$. Let $w[i..i-1] = \varepsilon$ for any i. A string u is a *substring* of w if $u = w[i..j]$ for some i and j. The pair (i, j) is not necessarily unique; we say that i specifies an *occurrence* of u in w. A string can have many occurrences in another string. A substring $w[1..j]$ (resp., $w[i..n]$) is a *prefix* [resp. *suffix*] of w. An integer p is a *period* of w if $w[i] = w[i+p]$ for $i = 1, \ldots, |w|-p$.

A substring [resp. suffix, prefix] of a given string is called a *subpalindrome* [resp. *suffix-palindrome*, *prefix-palindrome*] if it is a palindrome. We write $w = (uv)^*u$ to state that $w = (uv)^k u$ for some nonnegative integer k. In particular, $u = (uv)^*u$, $uvu = (uv)^*u$. Two basic combinatorial lemmas are quite useful.

Lemma 1. *Suppose p is a period of a nonempty palindrome w; then there are palindromes u and v such that $|uv| = p$, $v \neq \varepsilon$, and $w = (uv)^*u$.*

Lemma 2. *Suppose w is a palindrome and u is its proper suffix-palindrome or prefix-palindrome; then the number $|w|-|u|$ is a period of w.*

A string is *primitive* if it is not a power of a shorter string. Denote by p the minimal period of a palindrome w. By Lemma 1, we obtain palindromes u, v such that $w = (uv)^*u$, $v \neq \varepsilon$, and $|uv| = p$. The string uv is primitive. The representation $(uv)^*u$ is called *canonical decomposition* of w. Let $w[i..j]$ be a subpalindrome of the string w. The number $(i+j)/2$ is the *center* of $w[i..j]$.

The center is integer [half-integer] if the subpalindrome has an odd [resp., even] length. For any integer n, shl (w, n) denotes the string $w[t+1..|w|]w[1..t]$, where $t = n \bmod |w|$.

Lemma 3. *Suppose $(xy)^*x$ is a canonical decomposition of w and u is a subpalindrome of w such that $|u| \geq |xy|-1$; then the center of u coincides with the center of some x or y from the decomposition.*

3 Palindromic Iterator

Let $w[i..j]$ be a subpalindrome of a string w. The number $\lfloor (j-i+1)/2 \rfloor$ is the *radius* of $w[i..j]$. Let $\mathcal{C} = \{c > 0 : 2c \text{ is an integer}\}$ be the set of all possible centers for subpalindromes. *Palindromic iterator* is the data structure containing a string *text* and supporting the following operations on it:

1. append$_i(a)$ appends the letter a to the end;
2. maxPal returns the center of the longest suffix-palindrome;
3. rad(x) returns the radius of the longest subpalindrome with the center x;
4. nextPal(x) returns the center of the longest proper suffix-palindrome of the suffix-palindrome with the center x.

Example 1. Let *text* $= aabacabaa$. Then maxPal $= 5$. Values of rad and nextPal are listed in the following table (the symbol "−" means undefined value):

$text$		a		a		b		a		c		a		b		a		a	
x	0.5	1	1.5	2	2.5	3	3.5	4	4.5	5	5.5	6	6.5	7	7.5	8	8.5	9	9.5
rad(x)	0	0	1	0	0	1	0	0	0	4	0	0	0	1	0	0	1	0	0
nextPal(x)	−	−	−	−	−	−	−	−	−	8.5	−	−	−	−	−	−	9	9.5	−

A *fractional array* of length n is an array with n elements indexed by the numbers $\{x \in \mathcal{C} : 0 < x \leq \frac{n}{2}\}$. Fractional arrays can be easily implemented using ordinary arrays of double size. Let refl$(x, y) = y + (y - x)$ be the function returning the position symmetric to x with respect to y.

Proposition 1. *Palindromic iterator can be implemented such that* append$_i$ *requires amortized $O(1)$ time and all other operations require $O(1)$ time.*

Proof. Our implementation uses a variable s containing the center of the longest suffix-palindrome of *text*, and a fractional array r of length $2s$ such that for each $i \in \mathcal{C}$, $0 < i \leq s$, the number $r[i]$ is the radius of the longest subpalindrome centered at i. Obviously, maxPal $= s$. Let us describe rad(x). If $x \leq s$, rad$(x) = r[x]$. If $x > s$, then each palindrome with the center x has a counterpart with the center refl(x, s). On the other hand, rad$(x) \leq |text|-\lfloor x \rfloor$, implying rad$(x) = \min\{r[\text{refl}(x, s)], |text|-\lfloor x \rfloor\}$. To implement nextPal and append$_i$, we need additional structures.

We define an array $lend[0..|text|-1]$ and a fractional array $nodes[\frac{1}{2}..|text|+\frac{1}{2}]$ to store information about maximal subpalindromes of *text*. Thus, $lend[i]$ contains centers of some maximal subpalindromes of the form $text[i+1..j]$. Precisely,

$lend[i] = \{x \in C : x < s$ and $\lceil x \rceil - \mathsf{rad}(x) = i+1\}$. Each center x is also considered as an element of a biconnected list with the fields $x.\mathrm{next}$ and $x.\mathrm{prev}$ pointing at other centers. We call such elements nodes and store in the array $nodes$. The following *invariant of palindromic iterator* holds.

> Let $c_0 < \ldots < c_k$ be the centers of all suffix-palindromes of $text$. For each $j \in \overline{0, k-1}$, $nodes[c_j].\mathrm{next} = c_{j+1}$ and $nodes[c_{j+1}].\mathrm{prev} = c_j$.

Clearly, $c_0 = s$, $c_k = |text| + \frac{1}{2}$. Let $\mathsf{link}(x)$ and $\mathsf{unlink}(x)$ denote the operations of linking x to the end of the list and removing x from the list, respectively. Obviously, $\mathsf{nextPal}(x) = nodes[x].\mathrm{next}$. The following pseudocode of append_i uses the three-operand **for** loop like in the C language.

```
1: function appendᵢ(a)
2:     for (s₀ ← s; s < |text| + 1; s ← s + ½) do
3:         r[s] ← min(r[refl(s, s₀)], |text| − ⌊s⌋);              ▷ fill r
4:         if ⌊s⌋ + r[s] = |text| and text[⌈s⌉−r[s]−1] = a then
5:             r[s] ← r[s] + 1;      ▷ here s is the center of the longest suffix-pal.
6:             break;
7:         lend[⌈s⌉−r[s]−1] ← lend[⌈s⌉−r[s]−1] ∪ {s};              ▷ fill lend
8:     text ← text · a;
9:     link(nodes[|text|]); link(nodes[|text|+½]);      ▷ adding trivial suffix-pals.
10:    for each x in lend[⌈s⌉ − rad(s)] do
11:        unlink(nodes[refl(x, s)]);      ▷ removing invalid centers from the list
```

The code in lines 2–8 is a version of the main loop of Manacher's algorithm [12]; see also [3, Chapter 8]. The array $lend$ is filled simultaneously with r. Let us show that the invariant is preserved.

Suppose that a symbol is added to $text$ and the value of s is updated. Denote by S the set of centers $x > s$ such that the longest subpalindrome centered at x has lost its status of suffix-palindrome on this iteration. Once we linked the one-letter and empty suffix-palindromes to the list, it remains to remove the elements of S from it. Let $t = \lceil s \rceil - \mathsf{rad}(s)$. Since $text[t..|text|]$ is a palindrome, we have $lend[t] = \{\mathsf{refl}(x, s) : x \in S\}$. Thus, lines 10–11 unlink S from the list.

Since append_i links exactly two nodes to the list, any sequence of n calls to append_i performs at most $2n$ unlinks in the loop 10–11. Further, any such sequence performs at most $2n$ iterations of the loop 2–8 because each iteration increases s by $\frac{1}{2}$ and $s \leq |text|$. Thus, append_i works in the amortized $O(1)$ time.

4   Palindromic Engine

Palindromic engine is the data structure containing a string $text$, bit arrays m and res of length $|text|+1$, and supporting a procedure $\mathsf{append}(a, b)$ such that

1. $\mathsf{append}(a, b)$ appends the letter a to $text$, sets $m[|text|]$ to b, and calculates $res[|text|]$;
2. m is filled by append except for the bit $m[0]$ which is set initially;
3. $res[i] = 1$ iff there is $j \in \mathbb{N}$ such that $0 \leq j < i$, $m[j] = 1$, and $text[j+1..i] \in$ Pal (thus $res[0]$ is always zero).

Lemma 4. *Let L be a language. Suppose that for any $i \in \overline{0, |text|}$, $m[i] = 1$ iff $text[1..i] \in L$; then for any $i \in \overline{0, |text|}$, $res[i] = 1$ iff $text[1..i] \in L \cdot \text{Pal}$.*

Let f, g be functions of integer argument. We say that a palindromic engine works in $f(n)$ time and $g(n)$ space if any sequence of n calls to append on empty engine requires at most $f(n)$ time and $g(n)$ space.

Proposition 2. *Suppose a palindromic engine works in $f(n)$ time and space, and a language L is online recognizable in $g(n)$ time and space; then the language $L \cdot \text{Pal}$ is online recognizable in $f(n) + g(n) + O(n)$ time and space.*

We use the palindromic iterator in our implementation of palindromic engine. Let $\text{len}(x)$ be the function returning length of the longest subpalindrome centered at x, i.e., $\text{len}(x) = 2 \cdot \text{rad}(x) + \lfloor x \rfloor - \lfloor x - \frac{1}{2} \rfloor$. The operations of bitwise "or", "and", "shift" are denoted by or, and, shl respectively. Let $x \overset{\text{or}}{\leftarrow} y$ be short for $x \leftarrow (x \text{ or } y)$. The naive $O(n^2)$ time implementation is as follows; to improve it, we have to decrease the number of suffix-palindromes to loop through.

```
1: function append(a, b)
2:     append_i(a); n ← |text|; res[n] ← 0; m[n] ← b;
3:     for (x ← maxPal; x ≠ n+½; x ← nextPal(x)) do
4:         res[n] ←ᵒʳ m[n−len(x)];           ▷ loop through all suffix-palindromes
```

A nonempty string w is *cubic* if its minimal period p is at most $|w|/3$. A subpalindrome $t = w[i..j]$ is *leading in* w if any period p of any longer subpalindrome $w[i'..j]$ satisfies $2p > |t|$. For example, the only cubic subpalindrome of $w = aababababa$ is $w[2..8] = abababa$, and the only non-leading subpalindrome is $w[4..8] = ababa$. In general, a non-leading subpalindrome t has a period $p \leq |t|/2$ and then $t = (uv)^k u$ with $k \geq 2$ (Lemma 1); moreover, $(uv)^{k+1}u$ is a subpalindrome also, so t is a suffix of some leading cubic subpalindrome.

Lemma 5. *Let $s = w[i..j]$ be a leading subpalindrome of w, with the canonical decomposition $(uv)^*u$, and $t = w[i'..j]$ be the longest proper suffix-palindrome of s that is leading in w. Then $t = u$ if $s = uvu$, and $t = uvu$ otherwise.*

Lemma 6. *A string of length n has at most $\log_{\frac{3}{2}} n$ leading suffix-palindromes.*

To obtain a faster algorithm, we loop through leading suffix-palindromes only. Informally, to take into account other suffix-palindromes, we gather the corresponding bits of m into an additional bit array z described below.

For every $i \in \overline{0, |text|}$, let j_i be the maximal number j' such that $text[i+1..j']$ is a leading subpalindrome. Since any empty subpalindrome is leading, j_i is well defined. Let p_i be the minimal period of $text[i+1..j_i]$. Denote by d_i the length of the longest proper suffix-palindrome of $text[i+1..j_i]$ such that $text[j_i-d_i+1..j_i]$ is leading in $text$. By Lemma 5, $d_i = \min\{(j_i-i)-p_i, p_i+((j_i-i) \bmod p_i)\}$. The array z is maintained to support the following invariant:

$$z[i] = m[i] \text{ or } m[i+p_i] \text{ or } \ldots \text{ or } m[j_i-d_i-2p_i] \text{ or } m[j_i-d_i-p_i] \text{ for all } i \in \overline{0, |text|}.$$

Proposition 3. *The palindromic engine can be implemented to work in* $O(n \log n)$ *time and* $O(n)$ *space.*

Proof. Consider the following implementation of the function append.

```
1: function append(a, b)
2:     append_i(a); n ← |text|; res[n] ← 0; m[n] ← b; z[n] ← b; d ← 0;
3:     for (x ← maxPal; x ≠ n+½; x ← n−(d−1)/2) do   ▷ for leading suf-pal
4:         p ← len(x) − len(nextPal(x));            ▷ min period of processed suf-pal
5:         d ← min(p+(len(x) mod p), len(x)−p);     ▷ lenght of next leading s-pal
6:         if 3p > len(x) then                       ▷ processed suf-pal is not cubic
7:             z[n−len(x)] ← m[n−len(x)];
8:         else z[n−len(x)] ←ᵒʳ m[n−d−p];            ▷ processed suf-pal is cubic
9:         res[n] ←ᵒʳ z[n−len(x)];
```

Let $w_0, \dots, w_k$ be all leading suffix-palindromes of $text$ and $|w_0| > \dots > |w_k|$. We show by induction that the values taken by x are the centers of $w_0, \dots, w_k$ (in this order). In the first iteration $x = \mathsf{maxPal}$ is the center of w_0. Let x be the center of w_i. The minimal period p of w_i is calculated in line 4 according to Lemmas 1 and 2. By Lemma 5, the value assigned to d in line 5 is $|w_{i+1}|$. Thus, the third operand in line 3 sets x to the center of w_{i+1} for the next iteration.

Let x and $(uv)^*u$ be, respectively, the center and the canonical decomposition of w_i. Denote by w any suffix-palindrome such that $|w_i| \geq |w| > |w_{i+1}|$. By Lemma 3, $w = (uv)^*u$. If the invariant of z is preserved, the assignment in line 9 is equivalent to the sequence of assignments $res[n] \overset{\text{or}}{\leftarrow} m[n-|w|]$ for all such w. Since i runs from 0 to k, finally one gets $res[n] \overset{\text{or}}{\leftarrow} m[n-|w|]$ for all suffix-palindromes w, thus providing that the engine works correctly. To finish the proof, let us show that our implementation preserves the invariant on-the-fly, setting the correct value of $z[n-|w_i|]$ in lines 7, 8 just before it is used in line 9.

As in the pseudocode presented above, denote by n the length of $text$ with the letter c appended. For any $j \in \overline{0, n-1}$, the bit $z[j]$ is changed iff $text[j+1..n]$ is a leading suffix-palindrome. Assume that $w_i = text[j+1..n]$ is a leading suffix-palindrome and x is its center. If w_i is not cubic, line 7 gives the correct value of $z[j]$, because $n-d-p = j$. Suppose w_i is cubic. Let $(uv)^*u$ be a canonical decomposition of w_i. Then $w' = text[i+1..n-|vu|]$ is a leading subpalindrome. Indeed, $w' = (uv)^*u$ and $|w'| \geq |uvuvu|$. For some $i' \leq i$, suppose that $text[i'+1..n-|uv|]$ is a leading subpalindrome, p is its minimal period, and $2p < |w'|$; then since $p \geq |uv|$, we have, by Lemmas 1 and 3, that either $2p > |w'|$ or $|uv|$ divides p. Hence $text[i'+1..n-|uv|] = (uv)^*u$. Thus $i' = i$ because $text[i+1..n]$ is leading. Since w' is leading, we restore the invariant for $z[n-|w_i|]$ in line 8.

Since the number of iterations of the **for** cycle equals the number of leading suffix-palindromes of $text$, it is $O(\log n)$ by Lemma 6. This gives us the required time bound; the space bound is obvious.

5   Linear Algorithm

Consider the *word-RAM* model with $\beta+1$ bits in the machine word, where the bits are numbered starting with 0 (the least significant bit). A standard assumption is $\beta > \log |text|$. For a bit array $s[0..n]$ and integers i_0, i_1 such that

$0 \leq i_1 - i_0 \leq \beta$, we write $x \leftarrow s[\overrightarrow{i_0..i_1}]$ to get the number x whose jth bit, for any $j \in \overline{0,\beta}$, equals $s[i_0+j]$ if $0 \leq i_0+j \leq \min\{n, i_1\}$ and 0 otherwise. Similarly, $x \leftarrow s[\overleftarrow{i_0..i_1}]$ defines x with a jth bit equal to $s[i_1-j]$ if $\max\{0, i_0\} \leq i_1-j \leq n$ and to 0 otherwise. We write $s[\overrightarrow{i_0..i_1}] \leftarrow x$ and $s[\overleftarrow{i_0..i_1}] \leftarrow x$ for the inverse operations. A bit array is called *forward* [*backward*] if each read/write operation for $s[\overrightarrow{i_0..i_1}]$ [resp. $s[\overleftarrow{i_0..i_1}]$] takes $O(1)$ time. Forward [backward] arrays can be implemented on arrays of machine words with the aid of bitwise shifts.

Processing a string of length n, we can read/write a group of $\log n$ elements of forward or backward array in a constant time. In this section we speed up palindromic engine using bitwise operations on groups of $\log n$ bits. This sort of optimization is often referred to as four Russians' trick (see [1]). Note that there is a simpler algorithm recognizing Pal^k in $O(kn \log n)$ time, but it cannot be sped up in this fashion.

In the sequel n denotes $|text|$. As above, our palindromic engine contains a palindromic iterator and a bit array z. Arrays $m[0..n]$ and $z[0..n]$ are backward, while slightly extended array $res[0..n+\beta]$ is forward.

5.1   Idea of the Algorithm

We say that a call to append is *predictable* if it retains the value of maxPal (or, in other words, extends the longest suffix-palindrome). For a predictable call, we know from symmetry which suffix-palindromes will be extended. This crucial observation allows us to fill $res[n..n+\beta]$ in advance so that in the next β calls we need only few changes of res provided that these calls are predictable.

Let $text = vs$ at some point, where s is the longest suffix-palindrome. The number of subsequent calls preserving maxPal is at most $|v| = n-\text{len}(\text{maxPal})$: this is the case if we add $\overleftarrow{v}$. Consider those calls. Let $c_0 < \dots < c_k$ be the list of centers of all suffix-palindromes of $text$. Let $i \in \overline{1,k}$. After some predictable call c_i can vanish from this list. Let p_i be the number of predictable calls that retain c_i on the list. Then $p_i = \text{rad}(\text{refl}(c_i, \text{maxPal})) - \text{rad}(c_i)$ (in Figure 1 $p_1 = 5-3 = 2$).

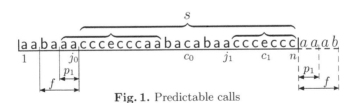

Fig. 1. Predictable calls

Let $j_i = n-\text{len}(c_i)$. If the operation $res[\overrightarrow{n..n+\beta}] \overset{\text{or}}{\leftarrow} m[\overleftarrow{j_i-p_i..j_i}]$ is performed for some $i \in \overline{1,k-1}$, we do not need to consider the suffix-palindrome with the center c_i during the next β predictable calls. Similarly, if $res[\overrightarrow{n..n+\beta}] \overset{\text{or}}{\leftarrow} m[\overleftarrow{j_0-\beta..j_0}]$ or $(m[\overleftarrow{j_k-p_k..j_k-1}] \text{ shl } 1)$ is performed, we do not consider the centers c_0 and c_k (a shift appears because the empty suffix-palindrome is ignored). The algorithm is roughly as follows. When the assignments above are performed, each of the next β predictable calls just adds two suffix-palindromes (one-letter and empty) and performs the corresponding assignments for them. When an unpredictable call or the $(\beta{+}1)$st predictable call occurs, we make new assignments in the current position and use array z to reduce the number of suffix-palindromes to loop through. Let us consider details.

5.2   Algorithm

We add to the engine an integer variable f such that $0 \le f \le \min\{\beta, n - \mathsf{len}(\mathsf{maxPal})\}$. The value of $\overrightarrow{res[n..n{+}f]}$ is called the *prediction*. Let us describe it. The centers c_i and the numbers p_i are defined in Sect. 5.1. Let $\mathrm{pr}\colon \{c_0, \ldots, c_k\} \to \mathbb{N}_0$ be the mapping defined by $\mathrm{pr}(c_0) = f$ and $\mathrm{pr}(c_i) = \min\{p_i, f\}$ for $i > 0$. Obviously, $\mathrm{pr}(c_i)$ is computable in $O(1)$ time. According to Sect. 5.1, the following value, called *f-prediction*, takes care of the palindromes with the centers $c_0, \ldots, c_k$ during all the time when they are suffix-palindromes:

$$m[\overleftarrow{j_0{-}\mathrm{pr}(c_0)..j_0}] \text{ or } \cdots \text{ or } m[\overleftarrow{j_{k-1}{-}\mathrm{pr}(c_{k-1})..j_{k-1}}] \text{ or } (m[\overleftarrow{j_k{-}\mathrm{pr}(c_k)..j_k{-}1}] \text{ shl } 1).$$

The prediction calculated by our algorithm will sometimes deviate from the f-prediction, but in a way that guarantees condition 3 of the definition of palindromic engine. Now we describe the nature of this deviation.

Let $c \in \mathcal{C}$ and $c > n{+}\frac{1}{2}$. Denote $c' = \mathsf{refl}(c, \mathsf{maxPal})$. Suppose $c' > 0$ and $\lceil c \rceil - \mathsf{rad}(c') \le n{+}1$ (see Figure 2). Let r be a positive integer such that $r \le \mathsf{rad}(c'){+}1$ and $\lceil c \rceil - r \le n$. The values c and r are chosen so that after a number of predictable calls *text* will contain a suffix-palindrome with the center c and the radius $r{-}1$. Then $res[\lceil c \rceil{+}r{-}1] = 1$ if $m[\lceil c \rceil{-}r] = 1$. We call the value $g = m[\lceil c \rceil{-}r]$ shl $(\lceil c \rceil{+}r{-}1{-}n)]$ an *additional prediction*. The assignment $res[n..n{+}f] \overset{\mathrm{or}}{\Leftarrow} g$ performs disjunction of the bits $res[\lceil c \rceil{+}r{-}1]$ and $m[\lceil c \rceil - r]$ (we suppose $\lceil c \rceil{+}r{-}1 \le n{+}f$). Setting this bit to 1 is not harmful: if there will be no unpredictable calls before the position $\lceil c \rceil{+}r{-}1$, then this bit will be set to 1 when updating the f-prediction on the $\lfloor c \rfloor$th iteration. Additional predictions appear as a byproduct of the linear-time implementation of the engine.

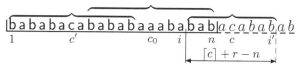

Fig. 2. Additional prediction; $c_0 = \mathsf{maxPal}$, $c' = \mathsf{refl}(c, c_0)$, $i = c - r$, $i' = c + r - 1$

We define the prediction through the *main invariant of palindromic engine*: $\overrightarrow{res[n..n{+}f]}$ equals the bitwise "or" of the f-prediction and some additional predictions. Such a definition guarantees that $res[n] = 1$ iff $m[j] = 1$ and

$text[j+1..n] \in$ Pal for some j, $0 \leq j < n$. Thus, the goal of append(a, b) is to preserve the main invariant. Our implementation of append(a, b) consists of three steps:

1. call append$_i(a)$ to extend $text$ (and increment n); then assign b to $m[n]$;
2. if maxPal remains the same and $f > 0$, decrement f and perform $\overrightarrow{res[n..n+f]} \overset{\text{or}}{\leftarrow} \overleftarrow{m[n-1-\text{pr}(n)..n-1]}$ or $(m[n-\text{pr}(n+\frac{1}{2})..n-1]$ shl $1)$;
3. otherwise, assign $f \leftarrow \min\{\beta, n-\text{len}(\text{maxPal})\}$ and recalculate the prediction $\overrightarrow{res[n..n+f]}$.

The operations of step 2 correspond to a predictable call and obviously preserve the main invariant. In the sequel we only consider step 3; step 1 is supposed to be performed: a is appended to $text$, n is incremented, and $m[n] = b$.

5.3 Prediction Recalculation

Recall that $c_0 < \ldots < c_k$ are the centers of suffix-palindromes, $j_i = n - \text{len}(c_i)$. First, clear the prediction: $\overrightarrow{res[n..n+f]} \leftarrow 0$. To get the f-prediction, it suffices to assign $\overrightarrow{res[n..n+f]} \overset{\text{or}}{\leftarrow} m[j_i - \text{pr}(c_i)..j_i]$ for $i = 0, \ldots, k-1$ and $\overrightarrow{res[n..n+f]} \overset{\text{or}}{\leftarrow} \overleftarrow{m[j_k - \text{pr}(c_k)..j_k - 1]}$ shl 1. But our algorithm processes leading suffix-palindromes only, and the bits of m that correspond to non-leading suffix-palindromes are accumulated in a certain fast accessible form in the array z. For simplicity, we process the empty suffix separately.

Let $i_0 < \ldots < i_h$ be integers such that $c_{i_0} < \ldots < c_{i_h}$ are the centers of all leading suffix-palindromes, $r \in \overline{0, h-1}$ and $s = i_{r+1} - i_r - 1 > 0$. Denote by w the suffix-palindrome centered at c_{i_r}. Let $(uv)^*u$ be the canonical decomposition of w. It follows from Lemma 3 that $c_{i_r+1}, \ldots, c_{i_r+s}$ are the centers of $(uv)^{s+1}u, \ldots, (uv)^2u$, $c_{i_r+s+1} = c_{i_{r+1}}$ is the center of uvu, and $w = (uv)^{s+2}u$. Then w is cubic. The converse is also true, i.e., if $w = (uv)^{s+2}u$ is a cubic suffix-palindrome, then $(uv)^{s+1}u, \ldots, (uv)^2u$ are non-leading suffix-palindromes, and uvu is a leading suffix-palindrome. So, non-leading suffix-palindromes are grouped into series following cubic leading suffix-palindromes.

Recall that the palindromic iterator allows one, in $O(1)$ time, to 1) get c_{i+1} from c_i; 2) find the minimal period of a suffix-palindrome; 3) using Lemma 5, get $c_{i_{r+1}}$ from c_{i_r}. The prediction recalculation involves the following steps:

1. accumulate some blocks of bits from m into z (see below);
2. for all $r \in \overline{0, h-1}$, assign $\overrightarrow{res[n..n+f]} \overset{\text{or}}{\leftarrow} m[j_{i_r} - \text{pr}(c_{i_r})..j_{i_r}]$;
3. for all $r \in \overline{1, h-1}$, if c_{i_r} is the center of a cubic suffix-palindrome and $\text{len}(c_{i_r}) \leq 2\beta$, assign $\overrightarrow{res[n..n+f]} \overset{\text{or}}{\leftarrow} \overleftarrow{m[j_{i_r+s} - \text{pr}(c_{i_r+s})..j_{i_r+s}]}$ for $s = 1, 2, \ldots, i_{r+1} - i_r - 1$;
4. for all $r \in \overline{0, h-1}$, if c_{i_r} is the center of a cubic suffix-palindrome and either $\text{len}(c_{i_r}) > 2\beta$ or $c_{i_r} = c_0$, perform the assignments of step 3 in $O(1)$ time with the aid of the array z.

Thus, "short" and "long" non-leading suffix-palindromes are processed separately (resp., on step 3 and step 4). Steps 1 and 4 require further explanation.

5.4   Content of z and Prediction of Long Suffix-Palindromes

Let w be a cubic leading suffix-palindrome such that $|w| > 2\beta$ or $|w| =$ len(maxPal). Suppose $(uv)^*u$ is the canonical decomposition of w. Then $p = |uv|$ is the minimal period of w. Denote the centers of suffix-palindromes $w, \ldots, uvu, u$ by $c_1, c_2, \ldots, c_k$ respectively. Let us describe the behavior of those suffix-palindromes in predictable calls.

Let t be the longest suffix of $text$ with the period p (t is not necessarily a palindrome). Then $|t| = |w| + \mathsf{rad}(\mathsf{refl}(c_k, c_1)) - \mathsf{rad}(c_k)$ is computable in $O(1)$ time. Since w is leading and cubic, $|t| < |w| + p$. In a predictable call to append, the suffix t extends if $text[n] = text[n{-}p]$, and breaks otherwise. Suppose t extended to ta. The suffix-palindromes centered at $c_2, \ldots, c_k$ also extended, while w extends iff $|w| < |t|$. Thus, in a series of such extensions of t the set of centers loses its first element during each p steps. Suppose t broke. Now the palindromes centered at $c_2, \ldots, c_k$ broke, while w can extend provided that $w = t$.

Example 2. Let $text = baaaabaaa$. Then $\mathsf{maxPal} = 6$; $w = aaa$ is a leading cubic suffix-palindrome; $w = (uv)^*u$ for $u = \varepsilon$ and $v = a$; $t = w$. Suffix-palindromes aaa, aa, a, ε have the centers $c_1 = 8, c_2 = 8.5, c_3 = 9, c_4 = 9.5$ respectively. After the predictable call to append, $text = baaaabaaaa$, t is extended, and w (with the center c_1) broke. After the second predictable call, $text = baaaabaaaab$, t is broken, and only c_2 remains the center of a suffix-palindrome.

Consider the first f predictable calls. Let q be the maximal number such that the suffix t of period p extends over the first q of these calls. Since w is "long", i.e., $|w| > 2\beta$ or w is the longest suffix-palindrome, and $f \le \beta$, one can be obtain q in $O(1)$ time: $q = \min\{f, \mathsf{rad}(\mathsf{refl}(c_k, \mathsf{maxPal})) - \mathsf{rad}(c_k)\}$. If $q < f$, the $(q{+}1)$st predictable call breaks the suffix of period p; as a result, at most one palindrome $w' = (uv)^*u$ extends to a suffix-palindrome at this moment (cf. Example 2). The length of w' in the initial text equals $|t|-q$, implying $(|t|-q-|u|) \bmod p = 0$. To process w', we perform $res[\overrightarrow{n..n{+}f}] \overset{\text{or}}{\underset{}{\leftarrow}} m[\overleftarrow{j-\mathsf{pr}(c_i)..j}]$ for $j = n-|w'|$, $c_i = n-(|w'|-1)/2$. To process other palindromes $(uv)^*u$, we consider z.

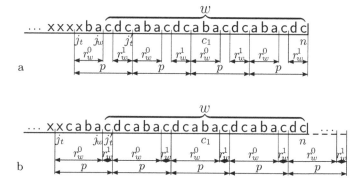

Fig. 3. Series of palindromes with a common period p. The cases presented are (a) $p > \beta{+}1$ $(= r_w^0 + r_w^1 = 5)$ and (b) $\beta{+}1 \ge p$ $(= r_w^0 + r_w^1 = 6)$

Denote $j_t = n - |t|$, $j'_t = j_t + p - 1$, and $j_w = n - |w|$, see Figure 3 a, b. We store the information about the series of palindromes $(uv)^*u$ in the block $z[j_t...j'_t]$ of length $p = |uv|$. For any $j \geq 0$, $i_j = j'_t - ((j + j'_t - j_w) \bmod p)$. Thus, $i_0 = j_w$, $i_1 = j_w - 1$ if $j_w \neq j_t$, and $i_1 = j'_t$ otherwise. Hence while j increases, i_j cyclically shifts left inside the range j_t, j'_t. We fill the block $z[j_t..j'_t]$ such that each of its bits is responsible for the whole series of suffix-palindromes with the period p.

$$\forall j \in \overline{0,\beta}:\ z[i_j] = m[i_j] \text{ or } m[i_j + p] \text{ or } \dots \text{ or } m[i_j + lp] \text{ for } l = \lfloor (n - i_j)/p \rfloor\ . \quad (1)$$

Let $r_w^0 = \min\{\beta, j_w - j_t\} + 1$, $r_w^1 = \min\{\beta + 1 - r_w^0, j'_t - j_w\}$. Clearly, $r_w^0 + r_w^1 = \min\{\beta + 1, p\}$. Hence, i_j in (1) runs through the ranges $[j_w - r_w^0 + 1..j_w]$ and $[j'_t - r_w^1 + 1..j'_t]$. Let $d = (1 \text{ shl } (q+1)) - 1$; thus, d is the bit mask consisting of $q+1$ ones. Suppose $\beta + 1 < p$ (see Figure 3,a). To recalculate the prediction, it suffices to assign $res[\overrightarrow{n..n+q}] \overset{\text{or}}{\leftarrow} d \text{ and}(z[\overleftarrow{j_w - r_w^0 + 1..j_w}] \text{ or } (z[\overleftarrow{j'_t - r_w^1 + 1..j'_t}] \text{ shl } r_w^0))$. Suppose $\beta + 1 \geq p$ (see Fig. 3,b). Let $k = \lceil q/p \rceil$. To recalculate the prediction, it suffices to perform the following:

$$
\begin{aligned}
res[\overrightarrow{n..n+q}] &\overset{\text{or}}{\leftarrow} d \text{ and}(z[\overleftarrow{j_t..j_w}] \text{ or } (z[\overleftarrow{j_w+1..j'_t}] \text{ shl } r_w^0)), \\
res[\overrightarrow{n..n+q}] &\overset{\text{or}}{\leftarrow} d \text{ and}((z[\overleftarrow{j_t..j_w}] \text{ or } (z[\overleftarrow{j_w+1..j'_t}] \text{ shl } r_w^0)) \text{ shl } p), \quad (2) \\
&\dots \\
res[\overrightarrow{n..n+q}] &\overset{\text{or}}{\leftarrow} d \text{ and}((z[\overleftarrow{j_t..j_w}] \text{ or } (z[\overleftarrow{j_w+1..j'_t}] \text{ shl } r_w^0)) \text{ shl } (kp))\ .
\end{aligned}
$$

To perform these assignments in $O(1)$ time, we use a precomputed array g of length β such that $g[i] = \sum_{j=0}^{\lfloor \beta/i \rfloor} 2^{ij}$ is the bit mask containing ones separated by $i-1$ zeroes. Then the sequence of assignments (2) is equivalent to the operation $res[\overrightarrow{n..n+q}] \overset{\text{or}}{\leftarrow} d \text{ and}((z[\overleftarrow{j_t..j_w}] \text{ or } (z[\overleftarrow{j_w+1..j'_t}] \text{ shl } r_w^0)) \cdot g[p])$.

Along with the f-prediction, the described method can produce additional predictions. Indeed, suppose we processed a cubic leading suffix-palindrome $w = (uv)^*u$. If $q > |v|$, the position $n + (|v|+1)/2$ is the center of the suffix-palindrome v after $|v|$ predictable calls. However, the corresponding assignment $res[n+|v|] \overset{\text{or}}{\leftarrow} m[n]$ is performed much earlier: calculating the prediction in the nth call of append, we accumulate the bit $m[n]$ in the array z (see (1)) and then use it in updating $res[n..n+q]$. The assignment $res[n+|v|+1] \overset{\text{or}}{\leftarrow} m[n-1]$ is performed at the same moment but corresponds to the $(|v|+1)$st predictable call, and so on. If $q > |vuv|$, we have the same situation with the suffix-palindrome vuv after $|vuv|$ calls. All these premature assignments are not necessary but bring no trouble.

Lemma 7. *Given the array z, the prediction recalculation requires $O(l + \min\{2\beta, s\})$ time, where l is the number of leading suffix-palindromes and s is the length of the second largest leading suffix-palindrome.*

Proof. The above analysis shows that each of steps 2, 4 takes $O(1)$ time per series of palindromes with a common period. Step 3 takes $O(\min\{2\beta, s\})$ time.

5.5   Recalculation of the Array Z and the Time Bounds

Lemma 8. *Recalculation of z requires $O(l+(n-n_0))$ time, where l is the number of leading suffix-palindromes and n_0 is the length of text at the moment of the previous recalculation.*

Lemma 9. *After an unpredictable call to* append, *k successive predictable calls require $O(k)$ time in total.*

Proof. A predictable call without recalculation takes $O(1)$ time. The number of recalculations is $\lfloor k/\beta \rfloor$. Since the number of leading suffix-palindromes is $O(\log n)$ by Lemma 6, it follows from Lemmas 7, 8 that the recalculation takes $O(\log n + \min\{2\beta, O(n)\}) + O(\log n + O(\beta)) = O(\beta)$ time, whence the result.

Lemma 10. *An unpredictable call requires $O(\mathsf{maxPal}-\mathsf{maxPal}_0 + n-n_0)$ time, where maxPal_0 is the center of the longest suffix-palindrome and n_0 is the length of text at the moment of the previous unpredictable call.*

Proof. By Lemmas 7, 8, the prediction recalculation takes $O(l + \min\{2\beta, s\}) + O(l + (n - n_0))$ time, where l is the number of leading suffix-palindromes and s is the length of the second longest leading suffix-palindrome. Since $l \leq s$, it remains to show that $s = O(\mathsf{maxPal}-\mathsf{maxPal}_0)$. See [10] for details.

Proposition 4. *The palindromic engine can be implemented to work in $O(n)$ time and space.*

Proof. The correctness of the implementation described in Sect. 5.2, 5.3 was proved in Sect. 5.2–5.4. It remains to prove the time bound. Consider the sequence of n calls to append. Let $n_1 < n_2 < \ldots < n_k$ be the numbers of all unpredictable calls to append and $\mathsf{maxPal}_1 < \mathsf{maxPal}_2 < \ldots < \mathsf{maxPal}_k$ be the centers of the longest suffix-palindromes just before each of these calls. By Lemma 10, all these calls require $O(1 + (\mathsf{maxPal}_2-\mathsf{maxPal}_1) + (n_2-n_1) + (\mathsf{maxPal}_3-\mathsf{maxPal}_2) + (n_3-n_2) + \ldots + (\mathsf{maxPal}_k-\mathsf{maxPal}_{k-1}) + (n_k-n_{k-1})) = O(n)$ time. A reference to Lemma 9 ends the proof.

Proposition 4 together with Proposition 2 implies the main theorem.

6   Conclusion

In the RAM model considered in this paper all operations are supposed to be constant-time. This is the so called *unit-cost RAM*. Our algorithm heavily relies on multiplication and modulo operations, and we do not know whether it can be modified to use only addition, subtraction, and bitwise operations.

It was conjectured that there exists a context-free language that can not be recognized in linear time by a unit-cost RAM machine. This paper shows that a popular idea to use palindromes in the construction of such a language is quite likely to fail. For some discussion on this problem, see [11].

Acknowledgements. This research was supported under the Agreement 02.A03.21.0006 of 27.08.2013 between the Ministry of Education and Science of the Russian Federation and Ural Federal University.

References

1. Arlazarov, V., Dinic, E., Kronrod, M., Faradzev, I.: On economical construction of the transitive closure of a directed graph. Dokl. Akad. Nauk. 194(11), 1209–1210 (1970) (in Russian)
2. Breslauer, D.: Efficient String Matching on Coded Texts. In: Galil, Z., Ukkonen, E. (eds.) CPM 1995. LNCS, vol. 937, pp. 27–40. Springer, Heidelberg (1995)
3. Crochemore, M., Rytter, W.: Jewels of Stringology. World Scientific Publ. (2002)
4. Crochemore, M., Rytter, W.: Squares, cubes, and time-space efficient string searching. Algorithmica 13(5), 405–425 (1995)
5. Fici, G., Gagie, T., Kärkkäinen, J., Kempa, D.: A Subquadratic Algorithm for Minimum Palindromic Factorization. J. Discrete Algorithms 28, 41–48 (2014)
6. Galil, Z., Seiferas, J.: A Linear-Time On-Line Recognition Algorithm for "Palstar". J. ACM 25(1), 102–111 (1978)
7. Glen, A., Justin, J., Widmer, S., Zamboni, L.Q.: Palindromic richness. European J. Combinatorics 30(2), 510–531 (2009)
8. Knuth, D.E., Morris, J.H., Pratt, V.R.: Fast pattern matching in strings. SIAM J. Comput. 6, 323–350 (1977)
9. Kosolobov, D., Rubinchik, M., Shur, A.M.: Finding Distinct Subpalindromes Online. In: Proc. Prague Stringology Conference, pp. 63–69 (2013)
10. Kosolobov, D., Rubinchik, M., Shur, A.M.: Pal^k Is Linear Recognizable Online. CoRR, arXiv:1404.5244.v2 [cs.FL] (2014)
11. Lee, L.: Fast context-free grammar parsing requires fast boolean matrix multiplication. J. ACM 49(1), 1–15 (2002)
12. Manacher, G.: A new linear-time on-line algorithm finding the smallest initial palindrome of a string. J. ACM 22(3), 346–351 (1975)
13. Valiant, L.G.: General context-free recognition in less than cubic time. Journal of Computer and System Sciences 10(2), 308–314 (1975)

Two Grammatical Equivalents
of Flip-Pushdown Automata*

Peter Kostolányi

Department of Computer Science, Faculty of Mathematics, Physics and Informatics,
Comenius University in Bratislava, Mlynská dolina, 842 48 Bratislava, Slovakia
kostolanyi@fmph.uniba.sk

Abstract. Flip-pushdown automata, introduced by Sarkar [7], are pushdown automata with an additional ability to reverse the contents of their pushdown, and with the most interesting setting arising when the number of such flips is limited by a constant. Two characterizations of flip-pushdown automata (with a limited number of flips) in terms of grammars are presented in this paper. First, the model is characterized by context-free grammars with an extra ability to generate reversals, which are called *reversal-generating context-free grammars* (RGCFG). Next, a model of parallel word production called *parallel interleaving grammar system* (PIGS) is introduced, for which the equivalence with flip-pushdown automata is proved, linking flip-pushdown automata to parallelism. The characterization in terms of PIGS is used to prove that flip-pushdown automata (with a limited number of flips) are weaker than ET0L systems, which solves an open problem of Holzer and Kutrib [2].

Keywords: flip-pushdown automaton, reversal-generating context-free grammar, RGCFG, parallel interleaving grammar system, PIGS.

1 Introduction

Nondeterministic flip-pushdown automata (NFPDA) are an extension of ordinary (nondeterministic) pushdown automata, given an additional ability to flip the pushdown store, i.e., to reverse its contents. This flipping operation allows to read the pushdown store at both of its ends and, in particular, to use it as a dequeue.

Since dequeue automata can be used to simulate Turing machines, it is obvious that when the number of pushdown flips is unlimited, flip-pushdown automata and Turing machines are equal in their computational power [7]. Thus, the research focus has been mainly on flip-pushdown automata with the number of flips limited by a constant. Flip-pushdown automata, limited in this way, are known to be more powerful than ordinary pushdown automata, while retaining virtually all of their pleasant properties, which makes the study of flip-pushdown automata particularly interesting [2].

---

* This work has been supported by the grants VEGA 1/0979/12 and UK/322/2014.

G.F. Italiano et al. (Eds.): SOFSEM 2015, LNCS 8939, pp. 302–313, 2015.

Flip-pushdown automata have been introduced by Sarkar in [7]. Since then, many of their properties have been resolved. Most of these results have been obtained by Holzer and Kutrib in [2] and [3].

However, up to now, there has been no type of grammar proved to be equivalent to NFPDA with a limited number of flips. In this paper, we introduce such grammars. First, we show that (limited) NFPDA are equivalent to an extension of context-free grammars that we call *reversal-generating context-free grammars* (RGCFG). Such grammars have an ability to generate reversals along with ordinary symbols, and are related to the flip-pushdown input-reversal technique introduced by Holzer and Kutrib [2]. Next, we show that (limited) NFPDA are equivalent to systems of parallel grammars that we call *parallel interleaving grammar systems* (PIGS). PIGS appear to be a natural way of describing parallel word productions, so the equivalence of NFPDA and PIGS establishes a strong link between NFPDA and parallelism.

The viewpoint of grammars can make the reasoning about NFPDA-languages much easier, as we demonstrate by providing an example application. In particular, Holzer and Kutrib have posed in [2] an open problem of the relation between (limited) NFPDA and E0L systems and between (limited) NFPDA and ET0L systems, and in both cases they have conjectured incomparability. The former problem has been solved by Ďuriš and Košta in [1], where they have proved that NFPDA and E0L are in fact incomparable. In this paper, we solve the latter problem by showing that the strict inclusion holds, i.e., NFPDA with a limited number of flips are strictly weaker than ET0L systems.

2   Definitions

In this section, we shall review the formal definition of NFPDA and the corresponding families of languages, and briefly survey some of the basic results obtained so far.

Definition 1. *A (nondeterministic) flip-pushdown automaton (NFPDA) is a tuple $A = (K, \Sigma, \Gamma, \delta, \Delta, q_0, Z_0, F)$, where K is a finite set of states, Σ is an input alphabet, Γ is a pushdown alphabet, δ is an ordinary transition function from $K \times (\Sigma \cup \{\varepsilon\}) \times \Gamma$ to finite subsets of $K \times \Gamma^*$, Δ is a flip transition function from K to subsets of K, q_0 in K is an initial state, Z_0 in Γ is a bottom-of-pushdown symbol, and $F \subseteq K$ is a set of accepting states.*

A *configuration* of the NFPDA A is defined similarly as for ordinary PDA, i.e., it is a triple (q, w, s), where q in K is a state, w in Σ^* is an unread part of the input word, and s in Γ^* is a content of the pushdown (written from the bottom of the pushdown). A *computation step* of the NFPDA A is a relation $\vdash_A$ on its configurations defined as follows: for p, q in K, a in $\Sigma \cup \{\varepsilon\}$, u in Σ^*, s, t in Γ^*, and Z in Γ, we define $(p, au, sZ) \vdash_A (q, u, st)$ if (q, t) is in $\delta(p, a, Z)$ (ordinary transitions) and $(p, u, Z_0 s) \vdash_A (q, u, Z_0 s^R)$ if q is in $\Delta(p)$ (flip transitions). Observe that the flip transition function does not depend neither on the input, nor on the contents of the pushdown. However, it may be easily seen that this definition is equivalent to the definition with a flip transition function Δ going from $K \times (\Sigma \cup \{\varepsilon\}) \times \Gamma$ to K. If A is understood from the context, we shall write $\vdash$ instead of $\vdash_A$. The *language* $L(A)$ accepted by A *by a final state* and the language $N(A)$ accepted by A *by empty pushdown* are defined as usual: $L(A) = \{w \in \Sigma^* \mid \exists q \in F, s \in \Gamma^* : (q_0, w, Z_0) \vdash^* (q, \varepsilon, s)\}$, and $N(A) = \{w \in \Sigma^* \mid \exists q \in K : (q_0, w, Z_0) \vdash^* (q, \varepsilon, \varepsilon)\}$. We say that the NFPDA A *operates in at most (exactly) k flips*, if in every its computation, the pushdown store is reversed at most (exactly) k times. This can be easily turned into a syntactic definition as well.

In [2], it is proved that NFPDA accepting by a final state and NFPDA accepting by empty pushdown are equivalent, while the simulations involved do not change the number of flips performed. Moreover, it is proved there that NFPDA with at most k flips are equivalent to NFPDA with *exactly* k flips. In [7], it is observed that NFPDA with unrestricted number of pushdown flips are equivalent to Turing machines in their computational power.

Definition 2. *We denote the family of languages accepted (either by a final state or by empty pushdown) by some NFPDA operating in at most k flips by* $\mathscr{L}(NFPDA_k)$. *Further, we define*

$$\mathscr{L}(NFPDA_{fin}) = \bigcup_{k=0}^{\infty} \mathscr{L}(NFPDA_k),$$

and denote the family of languages accepted by arbitrary NFPDA by $\mathscr{L}(NFPDA)$ *(it is already shown that* $\mathscr{L}(NFPDA) = \mathscr{L}(RE)$*).*

In the previous works [7], [2], and [3], these families of languages have been defined in a slightly different manner. However, both definitions can be clearly seen to be equivalent. In [2], it has been proved that the families $\mathscr{L}(\text{NFPDA}_k)$ form an infinite hierarchy, i.e.,

$$\mathscr{L}(\text{CF}) = \mathscr{L}(\text{NFPDA}_0) \subsetneq \mathscr{L}(\text{NFPDA}_1) \subsetneq \mathscr{L}(\text{NFPDA}_2) \subsetneq \dots$$

Finally, let us state the important *Flip-pushdown input-reversal theorem*, introduced by Holzer and Kutrib in [2] (in a slightly different form).

Theorem 1 (Holzer, Kutrib [2]). *Let k be in $\mathbb{N}$. A language L is accepted by empty pushdown by a NFPDA $A_1 = (K, \Sigma, \Gamma, \delta, \Delta, q_0, Z_0, \emptyset)$ operating in $k+1$ pushdown flips iff the language*

$$L_R = \{uv^R \mid (q_0, u, Z_0) \vdash^*_{A_1} (q_1, \varepsilon, Z_0 s) \text{ with } k \text{ flips}, q_2 \in \Delta(q_1),$$
$$\text{and } (q_2, v, Z_0 s^R) \vdash^*_{A_1} (q_3, \varepsilon, \varepsilon) \text{ without any flip}\}$$

is accepted by empty pushdown by some NFPDA A_2 operating in k pushdown flips. The same statement holds for NFPDA accepting by a final state.

3   Reversal-Generating Context-Free Grammars

In this section, we define an extension of context-free grammars that has an additional ability to generate reversals. We shall call such context-free grammars *reversal-generating* and we shall prove that they are equivalent to NFPDA. Our definition of these grammars will be closely related to the *flip-pushdown input-reversal technique* of Holzer and Kutrib [2]. In fact, word production by reversal-generating grammars can be viewed as an inverse of this technique. While in the flip-pushdown input-reversal technique one gets a word from a NFPDA-language and transforms this word by a series of reversals into a word from a certain context-free language, a reversal-generating grammar generates a context-free language with special reversal symbols. By interpreting these reversal symbols in an appropriate order, one actually performs an inverse of the flip-pushdown input-reversal technique, and obtains a word from some NFPDA-language.

Definition 3. *A* reversal-generating context-free grammar (RGCFG) *is a five-tuple* $G = (N, T, P, \sigma, \circledR)$, *where* (N, T, P, σ) *is a context-free grammar, and* $\circledR$ *is a special reversal symbol belonging to* T.

A *derivation step* of G is defined as for the context-free grammar $G' = (N, T, P, \sigma)$. The only difference is in the definition of the generated language. We define $L(G) = \{\varrho(w) \mid w \in L(G')\}$, where $\varrho : T^* \to (T - \{\circledR\})^*$ is the reversal-interpreting function defined inductively by

$$\varrho(w) = \begin{cases} w & \text{for } w \text{ without an occurrence of } \circledR, \\ u\varrho(v^R) & \text{for } w = u\circledR v, \ u \text{ without an occurrence of } \circledR, \text{ and } v \text{ in } T^*. \end{cases}$$

That is, reversal symbols are interpreted in the left-to-right order. We illustrate this by the following example.

Example 1. Let us consider a RGCFG $G = (N, T, P, \sigma, \circledR)$, such that (N, T, P, σ) generates a language consisting of words having the form $u\circledR v\circledR x\circledR y\circledR z$. Then, for every such word, the language $L(G)$ contains the word

$$\varrho(u\circledR v\circledR x\circledR y\circledR z) = u\varrho(z^R\circledR y^R\circledR x^R\circledR v^R) = uz^R\varrho(v\circledR x\circledR y) =$$
$$= uz^R v\varrho(y^R\circledR x^R) = uz^R vy^R\varrho(x) = uz^R vy^R x.$$

The following proposition can be proved easily by induction, so we omit its proof.

Proposition 1. *Let* $G = (N, T, P, \sigma, \circledR)$ *be a RGCFG and let us denote the context-free grammar* (N, T, P, σ) *by* G'. *Then,*

$$L(G) = \{w_1 w_{2n}^R w_2 w_{2n-1}^R \ldots w_n w_{n+1}^R \mid w_1 \circledR w_2 \circledR \ldots \circledR w_{2n} \in L(G')\} \cup$$
$$\cup \ \{w_1 w_{2n+1}^R w_2 w_{2n}^R \ldots w_n w_{n+2}^R w_{n+1} \mid w_1 \circledR w_2 \circledR \ldots \circledR w_{2n+1} \in L(G')\}.$$

Similarly as in the case of NFPDA, we shall be interested mainly in reversal-generating grammars generating a limited number of reversal symbols. We shall call RGCFG generating at most k reversal symbols k-*reversal-generating*. It can be observed that this can be easily turned into a syntactic constraint, and that the condition of generating at most k reversal symbols is equivalent to the condition of generating *exactly* k reversal symbols. Further, a *reversal-aware normal form* of k-reversal generating CFGs can be considered, such that every nonterminal is aware of which reversals it produces in a terminal word. This can be formalized by taking a set of nonterminal symbols of the form $N = N' \times 2^{\{1,\dots,k\}}$, where the second projection of each nonterminal is a (possibly empty) set of indices, corresponding to reversal symbols it generates in a terminal word. Obviously, these sets always consist of contiguous numbers. A proof that this is indeed a normal form is easy, and left to the reader.

Definition 4. *We denote the family of languages generated by k-reversal generating context-free grammars by $\mathscr{L}(RGCFG_k)$. Furthermore, we define*

$$\mathscr{L}(RGCFG_{fin}) = \bigcup_{k=0}^{\infty} \mathscr{L}(RGCFG_k).$$

We denote the family of languages generated by unrestricted reversal-generating context-free grammars by $\mathscr{L}(RGCFG)$.

Finally, we may proceed to the main theorem of this section, asserting that NFPDA and RGCFG are equivalent.

Theorem 2. *For all k in $\mathbb{N}$, the identity $\mathscr{L}(NFPDA_k) = \mathscr{L}(RGCFG_k)$ holds.*

Proof. First, let G be a RGCFG producing k reversal symbols. An equivalent NFPDA A performing k pushdown flips can be defined similarly as in the standard simulation of context-free grammars on pushdown automata (see, e.g., [4]). Let us initialize P_{sim} to be the set of production rules of G. The automaton A first makes its pushdown store contain the word $Z_0\sigma$, where Z_0 is a fixed bottom-of-pushdown symbol (different from the terminals and nonterminals of G), and σ is an initial nonterminal of G. Next, the following procedure is repeated: If a nonterminal of G is on the top of the pushdown, read nothing from the input and rewrite the nonterminal on the pushdown using some production rule in P_{sim}. If ℞ is on the top of the pushdown, erase it, perform a flip, and reverse the right-hand sides of production rules in P_{sim}. If any other terminal is on the top of the pushdown, erase it, and read the same symbol from the input. If it is not possible, A gets stuck. The automaton accepts if Z_0 is on the top of the pushdown (and the whole input is read). Clearly, $L(A) = L(G)$.

Now, let us prove the remaining inclusion. Let A be a NFPDA performing k flips. Obviously, by a minor change of its transition function, it is possible to obtain a NFPDA A', which behaves exactly like A, except that before every pushdown flip, it is forced to read some new special symbol # (and which cannot accept the input if some #-transition is not followed by a pushdown flip). Clearly, $L(A')$ consists of words $u = u_1 \# u_2 \# \dots \# u_{k+1}$, such that $u_1 u_2 \dots u_{k+1}$ can be accepted by A with k flips performed at the positions marked in u by symbols #. Now, by applying the flip-pushdown input-reversal technique [2] (see also Theorem 1 of the present paper) k times, one obtains the language

$$L' = \{u_1 \# u_3 \# \ldots \# u_k \# u_{k+1}^R \# u_{k-1}^R \# \ldots \# u_2^R \mid u_1 \# u_2 \# \ldots \# u_{k+1} \in L(A')\},$$

if k is odd, and the language

$$L' = \{u_1 \# u_3 \# \ldots \# u_{k+1} \# u_k^R \# u_{k-2}^R \# \ldots \# u_2^R \mid u_1 \# u_2 \# \ldots \# u_{k+1} \in L(A')\},$$

if k is even (this can be easily proved by induction). The Flip-pushdown input-reversal theorem implies that the language L' is context-free. Consider a context-free grammar G' generating L', and define a RGCFG G to be the same as G', but generating ⑧ instead of $\#$. Then, it follows by Proposition 1 that

$$L(G) = \{u_1 u_2 \ldots u_{k+1} \mid u_1 \# u_2 \# \ldots \# u_{k+1} \in L(A')\} = L(A).$$

The theorem is proved.   □

Corollary 1. $\mathscr{L}(NFPDA_{fin}) = \mathscr{L}(RGCFG_{fin})$.

4   Parallel Interleaving Grammar Systems

In this section, we shall introduce systems of parallel grammars, which we shall call *parallel interleaving grammar systems* (PIGS). Roughly said, languages generated by PIGS consist of words $u_1 v_1 u_2 v_2 \ldots u_m v_m$, where $u_1 \# u_2 \# \ldots \# u_m$ is generated by one context-free grammar, and $v_1 \# v_2 \# \ldots \# v_m$ is generated by another one. Here, $\#$ is a special *switch symbol* marking points in which the grammars interleave. The number of switch symbols generated need not be the same for both grammars, but the principle sketched above can be clearly generalized to hold in this setting as well.

Thus, we have two context-free grammars interleaving each other. However, these two grammars need not generate their words from scratch, but some sequential precomputation may take place. This means that first, an initial sentential form is precomputed sequentially, and this is then used by both grammars to start the generative process from. We shall prove that if the language of sequentially precomputed sentential forms is regular, then PIGS (with some restriction on the number of switch symbols) are equivalent to NFPDA (with a restriction on the number of pushdown flips).

To sum up, the generative process of PIGS can be divided into three stages. First, a sequential precomputation takes place, resulting in some sentential form. Next, this sentential form is used as an axiom by two context-free grammars, both of which (asynchronously) produce a terminal word with special switch symbols $\#$. Finally, these two words are combined into the final output by the following procedure: start by copying the word generated by the first grammar. When a switch symbol is encountered, do not copy it to the output, but continue by copying the word generated by the second grammar, etc.

Definition 5. *A parallel interleaving grammar system with two context-free grammars and with a regular set of axioms (PIGS(2,CF,Reg)) is a six-tuple* $G = (N, T, P_1, P_2, \#, I)$, *where* N *is a finite set of nonterminals,* T *is a finite set of terminals,* $N \cap T = \emptyset$, $P_1, P_2 \subseteq N \times (N \cup T)^*$ *are two finite sets of context-free production rules,* $\#$ *in* T *is a switch symbol, and* $I \subseteq (N \cup T)^*$ *is a regular language of initial sentential forms (axioms).*

Remark 1. In order to make the above definition strictly finitary, the regular set I may be replaced, e.g., by a regular grammar generating I.

Remark 2. The notation PIGS(2,CF,Reg) has been chosen in regard to possible future generalizations: PIGS consisting of more than two grammars can be studied, and some other family of grammars, and/or axiom sets can be considered. We believe that these generalizations are worth of research interest.

A *derivation step of the first grammar* of G is a binary relation $\Rightarrow_{G,1}$ (or simply $\Rightarrow_1$, if G is understood) on $(N \cup T)^*$ defined as follows: $u \Rightarrow_{G,1} v$ iff there are words u_1, u_2, x in $(N \cup T)^*$ and a nonterminal ξ in N, such that $u = u_1 \xi u_2$, $v = u_1 x u_2$, and $\xi \to x$ is in P_1. A *derivation step of the second grammar* of G, $\Rightarrow_{G,2}$ (or $\Rightarrow_2$, if G is understood), is defined similarly. The *language generated by the first grammar* of G from the axiom x is defined by

$$L(G, 1, x) = \{ w \in T^* \mid x \Rightarrow_1^* w \},$$

and we make an analogous definition for the second grammar of G as well. Let us define a *locally regulated shuffle* $w_1 \shuffle_\# w_2$ of words w_1, w_2 to be the output $\varphi(w_1, w_2)$ of the word combining function φ defined by

$$\varphi(u_1 \# u_2 \# \dots \# u_i, v_1 \# v_2 \# \dots \# v_j) = u_1 v_1 u_2 v_2 \dots u_j v_j u_{j+1} \dots u_i$$

for $i \geq j$, and by

$$\varphi(u_1 \# u_2 \# \dots \# u_i, v_1 \# v_2 \# \dots \# v_j) = u_1 v_1 u_2 v_2 \dots u_i v_i v_{i+1} \dots v_j$$

for $i \leq j$. For languages L_1, L_2, let us define

$$L_1 \shuffle_\# L_2 = \{ w_1 \shuffle_\# w_2 \mid w_1 \in L_1, w_2 \in L_2 \}.$$

Then, we define the *language generated* by G by

$$L(G) = \bigcup_{x \in I} L(G, 1, x) \shuffle_\# L(G, 2, x).$$

Remark 3. Every language L, generated by some PIGS(2,CF,Reg), can be expressed by $L = \bigcup_{w \in R} \tau_1(w) \shuffle_\# \tau_2(w)$, where R is a regular language, and τ_1, τ_2 are context-free substitutions.

Remark 4. The operation $\shuffle_\#$ can be regarded as an analogy of *shuffles on trajectories*, studied in [5], with a local control instead of a global one. Thus, we believe it is worth of research attention.

Similarly as in the case of NFPDA and RGCFG, we shall be interested mostly in PIGS generating a limited number of switch symbols. We shall call a PIGS (i,j)-*switch-generating*, if its first grammar generates i switch symbols, and its second grammar generates j switch symbols (it can be easily seen that this is equivalent to the condition of generating *at most* i resp. j switch symbols). By a direct analogy with reversal-aware RGCFGs, a *switch-aware normal form* of (i,j)-switch-generating PIGS can be defined (cf. Section 3). However, two sets of indices – one for each grammar – have to be remembered here for every nonterminal.

Definition 6. *We denote the family of languages generated by (i,j)-switch generating PIGS(2,CF,Reg) by $\mathscr{L}(PIGS_{(i,j)}(2, CF, Reg))$. Further, we define*

$$\mathscr{L}(PIGS_{fin}(2, CF, Reg)) = \bigcup_{i,j \geq 0} \mathscr{L}(PIGS_{(i,j)}(2, CF, Reg)),$$

and we denote the family of languages generated by unrestricted PIGS(2,CF,Reg) by $\mathscr{L}(PIGS(2, CF, Reg))$.

Now, we may present the main result of this paper, characterizing languages accepted by NFPDA in terms of PIGS.

Theorem 3. *For all $k \geq 1$, $\mathscr{L}(NFPDA_k) = \mathscr{L}(PIGS_{\lceil \frac{k-1}{2} \rceil, \lfloor \frac{k-1}{2} \rfloor}(2, CF, Reg))$.*

Proof. We shall use Theorem 2, i.e., we shall prove our statement by showing that the identity

$$\mathscr{L}(RGCFG_k) = \mathscr{L}(PIGS_{\lceil \frac{k-1}{2} \rceil, \lfloor \frac{k-1}{2} \rfloor}(2, CF, Reg))$$

holds for all $k \geq 1$. In the rest of the proof, we shall assume that k is odd. The proof for the case when k is even is analogous.

First, let $G = (N, T, P, \sigma, \circledR)$ be a k-reversal-generating RGCFG in the reversal-aware normal form. We shall construct a $((k-1)/2, (k-1)/2)$-switch-generating PIGS(2,CF,Reg) $G' = (N', T', P'_1, P'_2, \#, I')$, such that $L(G') = L(G)$. We shall call nonterminals, from which the middle (i.e., the $(k+1)/2$-st) reversal symbol is generated, *middle nonterminals*. Obviously, the initial nonterminal σ is middle, and there is at most one middle nonterminal in each sentential form.

Now, for each word w generated by G, consider a derivation such that production rules from middle nonterminals are used first, followed by all other rules. The first stage of the derivation has a form $\sigma \Rightarrow^* u \circledR v$, where u, v are in $(N \cup T)^*$, and it follows by Proposition 1 that $u \Rightarrow^* w_1 \circledR \ldots \circledR w_{(k+1)/2} =: x$, $v \Rightarrow^* w_{(k+3)/2} \circledR \ldots \circledR w_{k+1} =: y$, where $w = w_1 w_{k+1}^R \ldots w_{(k+1)/2} w_{(k+3)/2}^R$. Thus, it is obviously sufficient to construct G' so that its first grammar generates x, and its second grammar generates y^R (in both cases with $\circledR$ replaced by $\#$), for all words x, y as above (and nothing else is generated by G').

In order to do this, let us first observe that a regular language R and homomorphisms h_1, h_2 do exist, such that

$$\{(h_1(z), h_2(z)) \mid z \in R\}$$

is exactly the set of all pairs (u, v^R), such that G generates uⓇⓋ in the first stage of some derivation, where the reversal symbol between u and v is middle. The language R may be, for instance, defined to be a language over the alphabet of production rules of G, consisting of all valid chains of production rules from middle nonterminals, together forming a complete first stage of some derivation.

Suppose that the alphabet Σ_R of R and $N \cup T$ are disjoint. Now, set $I' = R$, and define P'_1 to contain rules simulating h_1, i.e., rewriting all symbols c from Σ_R by $h_1(c)$, and, furthermore, all production rules of G with Ⓡ replaced by #. Formally, $P'_1 = \{c \to h(h_1(c)) \mid c \in \Sigma_R\} \cup \{\xi \to h(x) \mid \xi \to x \in P\}$, where h is a homomorphism such that $h(Ⓡ) = \#$ and $h(c) = c$ for c in $(N \cup T) - \{Ⓡ\}$. Furthermore, define P'_2 to contain rules simulating h_2 and rules of G with their right side reversed (and with Ⓡ replaced by #). This can be formally written as $P'_2 = \{c \to h(h_2(x)) \mid c \in \Sigma_R\} \cup \{\xi \to h(x)^R \mid \xi \to x \in P\}$. Finally, let us set $N' = N \cup \Sigma_R$, $T' = (T \cup \{\#\}) - \{Ⓡ\}$, and the PIGS G' is completely defined. By what has been noted above, it is clear that $L(G') = L(G)$.

Now, we shall prove the remaining inclusion. Suppose that we are given a $((k-1)/2, (k-1)/2)$-switch-generating PIGS(2,CF,Reg) $G = (N, T, P_1, P_2, \#, I)$. We shall construct a k-reversal-generating RGCFG $G' = (N', T', P', \sigma', Ⓡ)$, such that $L(G') = L(G)$. Let $G_I = (N_I, T_I, P_I, \sigma_I)$ be a regular grammar generating I, such that N_I and $N \times \{1, 2\} \cup T \cup \{Ⓡ\}$ are disjoint. Let us define G' as follows: $N' = N_I \cup N \times \{1, 2\}$, $T' = (T \cup \{Ⓡ\}) - \{\#\}$, $\sigma' = \sigma_I$, and

$$P' = \{\xi \to h_1(x)\eta h_2(x)^R \mid x \in T_I^*, \eta \in N_I, \xi \to x\eta \in P_I\} \cup$$
$$\cup \{\xi \to h_1(x)Ⓡh_2(x)^R \mid x \in T_I^*, \xi \to x \in P_I\} \cup$$
$$\cup \{(\xi, 1) \to h_1(x) \mid \xi \to x \in P_1\} \cup \{(\xi, 2) \to h_2(x)^R \mid \xi \to x \in P_2\},$$

where $h_1, h_2 : (N \cup T)^* \to (N' \cup T')^*$ are homomorphisms defined by $h_1(\xi) = (\xi, 1), h_2(\xi) = (\xi, 2)$ for ξ in N, $h_1(\#) = h_2(\#) = Ⓡ$ and $h_1(c) = h_2(c) = c$ for c in $T - \{\#\}$. It is obvious that the context-free grammar (N', T', P', σ') generates words $w_1Ⓡ \ldots Ⓡw_{(k+1)/2}Ⓡw_{(k+3)/2}^R Ⓡ \ldots Ⓡw_{k+1}^R$ such that the first grammar of G generates $w_1\# \ldots \#w_{(k+1)/2}$, and the second grammar generates $w_{k+1}\# \ldots \#w_{(k+3)/2}$. Then, $L(G') = L(G)$ follows directly by Proposition 1. □

Corollary 2. $\mathscr{L}(NFPDA_{fin}) = \mathscr{L}(PIGS_{fin}(2, CF, Reg))$.

Proof. The inclusion $\mathscr{L}(NFPDA_{fin}) \subseteq \mathscr{L}(PIGS_{fin}(2, CF, Reg))$ follows directly by Theorem 3. The remaining inclusion holds, since for all i, j in $\mathbb{N}$, we have $\mathscr{L}(PIGS_{i,j}(2, CF, Reg)) \subseteq \mathscr{L}(PIGS_{\max\{i,j\},\max\{i,j\}}(2, CF, Reg))$, which, by Theorem 3, equals $\mathscr{L}(NFPDA_{2\max\{i,j\}+1})$. □

5   A Relation to ET0L Systems

In this section, we shall present an example application of the characterization of NFPDA in terms of PIGS: We shall prove that flip-pushdown automata (with a constant number of flips) are strictly weaker than ET0L systems (for the definition, see, e.g., [6]).

The problem of the relation between flip-pushdown automata and ET0L systems has been posed by Holzer and Kutrib in [2]. There, they have conjectured that both NFPDA and E0L, and NFPDA and ET0L are incomparable. Ďuriš and Košta have already confirmed the former conjecture [1]. In this section, we shall prove that the latter conjecture does not hold, i.e., NFPDA are strictly weaker than ET0L systems.

Theorem 4. $\mathscr{L}(NFPDA_{fin}) \subsetneq \mathscr{L}(ET0L)$.

Proof. It has been already known that $\mathscr{L}(\mathrm{NFPDA}_{fin}) \not\supseteq \mathscr{L}(\mathrm{ET0L})$. This can be proved, e.g., by showing that the language $L = \{a^n b^n c^n \mid n \in \mathbb{N}\}$ is in $\mathscr{L}(\mathrm{ET0L})$, but not in $\mathscr{L}(\mathrm{NFPDA}_{fin})$. For more information, see [2].

Thus, it remains to prove that $\mathscr{L}(\mathrm{NFPDA}_{fin}) \subseteq \mathscr{L}(\mathrm{ET0L})$. We shall show this by applying Theorem 3, i.e., we shall prove that for all nonnegative integers i, j and every (i, j)-switch-generating PIGS(2,CF,Reg) $G = (N, T, P_1, P_2, \#, I)$, there is an ET0L system $\mathscr{S} = (V, \mathscr{P}, x, \Sigma)$, such that $L(\mathscr{S}) = L(G)$. Without loss of generality, we may suppose that $i = j$, that G is in the switch-aware normal form, and that $I \subseteq N^*$. Further, let $G_I = (N_I, T_I, P_I, \sigma_I)$ be a regular grammar generating I, such that $P_I \subseteq N_I \times (T_I N_I \cup \{\varepsilon\})$, $N_I \cap (N \cup T) = \emptyset$, and $T_I \subseteq N$. Finally, we shall assume that G generates switch symbols only by rules of the form $\alpha \to \#$, with α in N.

Since $\mathscr{L}(\mathrm{CF}) \subseteq \mathscr{L}(\mathrm{ET0L})$, it is possible to simulate both context-free grammars of G by an ET0L-system. Thus, the only problem is with interleaving. To overcome this, we have to generate regular axioms of G already appropriately interleaved. The only serious problem here is when a nonterminal from an axiom generates a switch symbol (in one or both of the grammars of G), since then it can generate symbols in two different contiguous parts of the final word, and it may produce some other nonterminals that generate symbols in even other contiguous parts of the final word (between the two contiguous parts it generates symbols in directly). In that case, we shall split the nonterminal into two parts that will be rewritten always in parallel. These parts may be split even further.

Besides this, it is important to mark each symbol with the number (1 or 2) of the grammar of G it corresponds to. Then, after generating the interleaved and marked axioms, it is possible to simulate both grammars on these axioms (with the presence of split nonterminals described above).

Now, we shall describe the formal construction. In what follows, h_1 and h_2 are homomorphisms, such that $h_1(\xi) = (\xi, 1)$ and $h_2(\xi) = (\xi, 2)$ for ξ in $N \cup N_I$, and $h_1(c) = h_2(c) = c$ for c in T. Next, for ξ in N, we define $\zeta_1(\xi)$ to be (s, t) if ξ generates switch symbols s to t in the first grammar of G, and to be $(0, 0)$ if it generates none. The notation $\zeta_2(\xi)$ has a similar meaning for the second grammar of G. The derivation of $\mathscr{S}$ will proceed in four phases, with the right order enforced by symbols Π_1, Π_2, Π_3, and Π_4 in $V - \Sigma$. In the s-th phase, the symbol Π_s is present in the sentential form. Moreover, every table of production rules designated for the s-th phase contains rules $\Pi_s \to \Pi_s$ and $\Pi_t \to F$ for $t \neq s$, where F in $V - \Sigma$ is a special *fail* symbol. The only rule from F in each table is $F \to F$. Thus, the sentential form containing F cannot be terminated.

The axiom x of $\mathscr{S}$ will be $h_1(\sigma_I)\$_1 h_2(\sigma_I)\$_{i+2}\$_2\$_{i+3}\ldots\$_{i+1}\$_{2i+2}\Pi_1$, where $\$_s$ is a symbol in $V - \Sigma$ denoting the end of the s-th contiguous part of the word being generated. These symbols will be erased in the end.

In the first phase, appropriately interleaved axioms of G are generated. For each rule $\alpha \to x\beta$ of G_I, there is one separate table $P_{\alpha\to x\beta}$ in $\mathscr{P}$. If $\zeta_1(x) = (0,0)$, then there is a rule $h_1(\alpha) \to h_1(x)h_1(\beta)$ in $P_{\alpha\to x\beta}$, and similarly for $\zeta_2(x) = (0,0)$. If $\zeta_1(x) = (s,t) \neq (0,0)$, then there are rules $h_1(\alpha) \to (h_1(x), left)$, $\$_k \to L_k R_k \$_k$ for $s < k \le t$ and $\$_{t+1} \to (h_1(x), right)h_1(\beta)\$_{t+1}$ in $P_{\alpha\to x\beta}$, and similarly for the second grammar (these rules correspond to the splitting of nonterminals). Symbols L_k and R_k are placeholders for the further subdivision of split nonterminals. It follows from our assumption of all switch-producing rules having the form $\alpha \to \#$ that all these symbols will be rewritten sometimes. In addition, there are rules $h_1(\xi) \to F$ and $h_2(\xi) \to F$ for all ξ in N_I, $\xi \neq \alpha$ in this table. This is in order to assure that in the first phase, only the tables of rules corresponding to the nonterminal of G_I present in the sentential form are used. Moreover, there are rules $\Pi_1 \to \Pi_1$ and $\Pi_s \to F$ for $s > 1$ in this table. For all other symbols y, there is a rule $y \to y$.

For each rule $\alpha \to \varepsilon$ of G_I, there is one separate table $P_{\alpha\to\varepsilon}$ in $\mathscr{P}$. This contains rules $h_1(\alpha) \to \varepsilon$, $h_2(\alpha) \to \varepsilon$. Further, for all ξ in N_I, $\xi \neq \alpha$, there are rules $h_1(\xi) \to F$ and $h_2(\xi) \to F$. Since this table is used at the end of the first phase, there is a rule $\Pi_1 \to \Pi_2$. Similarly as above, rules $\Pi_s \to F$ for $s > 1$ are present, and for all other symbols y, there is a rule $y \to y$.

In the second phase, split (switch-producing) nonterminals are rewritten. We shall only describe the tables corresponding to the first grammar of G, since the tables for the second grammar are analogous. Let ξ in N be a nonterminal, such that $\zeta_1(\xi) \neq (0,0)$. Then, for each rule $\xi \to u\eta v$ in P_1 with u, v in $(N \cup T)^*$ and $\zeta_1(\eta) = \zeta_1(\xi)$ (note that the only switch-generating nonterminal on the right side is η), there is a table $P^{(1)}_{\xi\to u\eta v}$ in $\mathscr{P}$ containing the rules $(h_1(\xi), left) \to h_1(u)(h_1(\eta), left)$, $(h_1(\xi), right) \to (h_1(\eta), right)h_1(v)$, $\Pi_2 \to \Pi_2$, $\Pi_s \to F$ for $s \neq 2$, and $y \to y$ for all other symbols y. This table can be used also if (split) ξ is not present – in that case, the sentential form is not changed.

Now, let us consider rules with two switch-producing nonterminals on the right side, i.e., $\xi \to u\alpha v\beta w$ in P_1, where $\zeta_1(\xi) = (s,t) \neq (0,0)$, $\zeta_1(\alpha) = (s,k)$, $\zeta_1(\beta) = (k+1,t)$ for some $s \le k < t$ and u, v, w in $(N \cup T)^*$. To every such rule, there is a table $P^{(1)}_{\xi\to u\alpha v\beta w}$ in $\mathscr{P}$ containing the following production rules: $(h_1(\xi), left) \to h_1(u)(h_1(\alpha), left)$, $L_{k+1} \to (h_1(\alpha), right)h_1(v)$, $R_{k+1} \to (h_1(\beta), left)$, and $(h_1(\xi), right) \to (h_1(\beta), right)h_1(w)$. Further, for every non-terminal $\eta \neq \xi$ generating the k-th and the $(k+1)$-st switch symbol, this table contains the rules $(h_1(\eta), left) \to F$ and $(h_1(\eta), right) \to F$. This is in order to assure that if in some terminable derivation, L_{k+1} or R_{k+1} is rewritten using this table, then this derivation step is valid, i.e., $(h_1(\xi), left)$ and $(h_1(\xi), right)$ were present. However, when neither L_{k+1} nor R_{k+1} is present in the sentential form, this table can be used (with no effect). Finally, the table contains rules $\Pi_2 \to \Pi_2$, $\Pi_s \to F$ for $s \neq 2$, and $y \to y$ for all other symbols y. The case of more than two switch-producing nonterminals on the right side is analogous.

For all rules of the type $\alpha \to \#$ in P_1, there is a table $P^{(1)}_{\alpha \to \#}$ in $\mathscr{P}$, with rules $(h_1(\alpha), left) \to \varepsilon$, $(h_1(\alpha), right) \to \varepsilon$, $\Pi_2 \to \Pi_2$, $\Pi_s \to F$ for $s \neq 2$, and $y \to y$ for all other symbols y.

A table used for finalizing the second phase has rules $\Pi_2 \to \Pi_3$, $\Pi_s \to F$ for $s \neq 2$, and $y \to y$ for all other symbols y. This table may be used also if the second phase is not finished yet, however, as we shall see, the sentential form cannot be terminated in that case (F will be produced in the third phase).

In the third phase, the rest of nonterminals is rewritten. For each rule $\xi \to u$ in P_1, where $\zeta_1(\xi) = (0,0)$ and u is in $(N \cup T)^*$, there is a table $P^{(1)}_{\xi \to u}$ in $\mathscr{P}$, with rules $h_1(\xi) \to h_1(u)$, and $h_1(\xi) \to h_1(\xi)$. Further, for all (split) switch-producing nonterminals η and all symbols R_k and L_k, it contains rules that rewrite them to F. Finally, it contains rules $\Pi_3 \to \Pi_3$, $\Pi_s \to F$ for $s \neq 3$, and $y \to y$ for all other symbols y. Similarly for rules $\xi \to u$ in P_2.

A table for finalizing the third phase is similar to the table for finalizing the second phase.

Finally, there is only one table designated to the fourth phase. For each nonterminal, it contains rules rewriting it to F (that is, the third phase has to be finished). Moreover, it contains rules $\$_k \to \varepsilon$ for all k, $1 \leq k \leq 2i + 2$, $\Pi_4 \to \varepsilon$, $\Pi_s \to F$ for $s \neq 4$, and $y \to y$ for all other symbols y. $\square$

Acknowledgements. Many thanks go to Branislav Rovan and Pavel Labath, for the insightful comments they have made on the preliminary version of this paper.

References

1. Ďuriš, P., Košta, M.: Flip-Pushdown Automata with k Pushdown Reversals and E0L Systems are Incomparable. Inform. Process. Lett. 114, 417–420 (2014)
2. Holzer, M., Kutrib, M.: Flip-Pushdown Automata: $k + 1$ Pushdown Reversals Are Better than k. In: Baeten, J.C.M., Lenstra, J.K., Parrow, J., Woeginger, G.J. (eds.) ICALP 2003. LNCS, vol. 2719, pp. 490–501. Springer, Heidelberg (2003)
3. Holzer, M., Kutrib, M.: Flip-Pushdown Automata: Nondeterminism is Better than Determinism. In: Ésik, Z., Fülöp, Z. (eds.) DLT 2003. LNCS, vol. 2710, pp. 361–372. Springer, Heidelberg (2003)
4. Hopcroft, J.E., Motwani, R., Ullman, J.D.: Introduction to Automata Theory, Languages, and Computation, 2nd edn. Addison-Wesley, Reading (2001)
5. Mateescu, A., Rozenberg, G., Salomaa, A.: Shuffle on trajectories: Syntactic constraints. Theor. Comput. Sci. 197, 1–56 (1998)
6. Rozenberg, G., Salomaa, A.: The Mathematical Theory of L Systems. Academic Press, London (1980)
7. Sarkar, P.: Pushdown automaton with the ability to flip its stack. Report No. 81, Electronic Colloquium on Computational Complexity (2001)

On the Hierarchy Classes
of Finite Ultrametric Automata

Rihards Krišlauks and Kaspars Balodis

University of Latvia Faculty of Computing,
Raiņa bulvāris 19, Riga, LV-1459, Latvia
kbalodis@gmail.com

Abstract. This paper explores the language classes that arise with respect to the head count of a finite ultrametric automaton. First we prove
that in the one-way setting there is a language that can be recognized
by a one-head ultrametric finite automaton and cannot be recognized
by any k-head non-deterministic finite automaton. Then we prove that
in the two-way setting the class of languages recognized by ultrametric finite k-head automata is a proper subclass of the class of languages
recognized by $(k + 1)$-head automata. Ultrametric finite automata are
similar to probabilistic and quantum automata and have only just recently been introduced by Freivalds. We introduce ultrametric Turing
machines and ultrametric multi-register machines to assist in proving
the results.

1 Introduction

Ultrametric finite automata and ultrametric Turing machines were first introduced by [3]. This development has been followed by several papers in which
various aspects of these machines are studied in depth. [1] have studied the
descriptional complexity of ultrametric automata. They showed that ultrametric automata can achieve an exponential advantage in terms of the number of
states required when compared with equivalent deterministic automata. [8] have
studied the reversal complexity of ultrametric Turing machines.

Ultrametric machines are similar to probabilistic machines, with the
difference that for ultrametric machines it is not necessary for amplitudes (which
are the equivalent of probabilities in probabilistic automata) to be within the
range of 0 and 1. Instead, general p-adic numbers are used. We should note that
in [12], a similar generalization of probabilistic automata was introduced, where
"probabilities" can be arbitrarily large numbers, and the acceptance criterion
is whether the probability to be in an accepting state is greater than a given
threshold. Furthermore, it was shown that this generalization is in fact equivalent
to probabilistic automata. However, unlike the concepts used for these pseudo-
probabilistic machines, the definition of ultrametric machines uses the concept
of a p-adic norm.

It can be argued that the definition introduced by Freivalds is natural, because
in 1916, Alexander Ostrowski proved that any non-trivial absolute value on the

G.F. Italiano et al. (Eds.): SOFSEM 2015, LNCS 8939, pp. 314–326, 2015.

rational numbers Q is equivalent to either the usual real absolute value or a p-adic absolute value. This result demonstrates that using p-adic numbers is not merely one of many possibilities to generalize the definition of deterministic algorithms, but rather the only remaining possibility not yet explored [3]. Additionally, useful properties have been proven for the definition of ultrametric machines—[1] proved that the language class recognized by regulated p-adic machines coincides with the class of regular languages.

In this paper, we address the question of the language hierarchy associated with the number of heads of ultrametric multi-head automata. Several results can be found in the literature that consider deterministic, nondeterministic, and probabilistic finite automata in both—two-way and one-way—cases [5,9,11,13]. This paper examines whether similar results regarding the separation in classes with respect to the head count can be achieved for ultrametric multi-head finite automata. Other results consider the relationships of the language classes recognized by ultrametric and classical automata.

2   *P*-Adic Numbers

p-adic numbers are discussed in more detail in [10]. The use of p-adics in other sciences can be seen in [7,2]. Here, we only restate the definition of the p-adic absolute value.

For every non-zero rational number α there exists a unique prime factorization $\alpha = \pm 2^{\alpha_2} 3^{\alpha_3} 5^{\alpha_5} 7^{\alpha_7} \cdots$ where $\alpha_i \in \mathbb{Z}$.

Definition 1. *The p-adic absolute value (also called the p-**norm**) of a rational number* $\alpha = \pm 2^{\alpha_2} 3^{\alpha_3} 5^{\alpha_5} 7^{\alpha_7} \cdots$ *is*

$$\|\alpha\|_p = \begin{cases} p^{-\alpha_p}, & \text{if } \alpha \neq 0 \\ 0, & \text{if } \alpha = 0. \end{cases}$$

3   One-Way Multi-head Automata

3.1   Definitions

We extend the definition of ultrametric automata given in [1] by adding rejecting states.

Definition 2. *A finite one-way p-ultrametric one-head automaton ($1u_pfa$ or $1u_pfa(1)$) is a sextuple $\langle S, \Sigma, s_0, \delta, Q_A, Q_R \rangle$ where*

- *S is a finite set—the set of states,*
- *Σ is a finite set ($\$ \notin \Sigma$)—input alphabet,*
- *$s_0 : S \to \mathbb{Q}_p$ is the initial amplitude distribution,*
- *$\delta : (\Sigma \cup \{\$\}) \times S \times S \to \mathbb{Q}_p$ is the transition function,*
- *$Q_A, Q_R \subseteq S$ are the sets of accepting and rejecting states, respectively.*

The automaton works as follows: At every timestep, each of its states has an associated p-adic number called its amplitude. The automaton starts with an initial amplitude distribution s_0. It subsequently proceeds by processing the input word's $w = w_1 \ldots w_n$ symbols one at a time. The amplitude distribution after processing the i-th symbol is denoted as s_i, with $s_i(y) = \sum_{x \in S} s_{i-1}(x) \cdot \delta(w_i, x, y)$ for every $y \in S$. After the n-th symbol, the end marker \$ is similarly processed, obtaining the final amplitude distribution s_{n+1}. If the sum of the p-norms of final amplitudes over accepting states is greater than the sum of final amplitudes over rejecting states, i.e. if $\sum_{x \in Q_A} \|s_{n+1}(x)\|_p > \sum_{x \in Q_R} \|s_{n+1}(x)\|_p$, then the word w is said to be accepted, otherwise—rejected.

A two-way k-head finite automaton consists of an input tape containing the input word on which the heads of the automaton can move freely in both directions, not crossing the endmarkers. The tape is read-only. We use the standard definition for the two-way k-head non-deterministic finite automaton:

Definition 3 ([5]). *A two-way non-deterministic k-head finite automaton (2nfa (k)) is a sextuple $\langle S, \Sigma, k, s_0, \delta, F \rangle$, where*

- *S is a finite set—the set of states,*
- *Σ is a finite set ($\triangleright, \triangleleft \notin \Sigma$)—the input alphabet ($\triangleright$ and $\triangleleft$ are the left and right endmarkers, respectively),*
- *$k \geq 1$ is the number of heads,*
- *$s_0 \in S$ is the starting state,*
- *$\delta : S \times (\Sigma \cup \{\triangleright, \triangleleft\})^k \to 2^{S \times \{-1,0,1\}^k}$ is the partial transition function. Whenever $(s', (d_1, \ldots, d_k)) \in \delta(s, (a_1, \ldots, a_k))$ is defined, then $d_i \in \{0,1\}$ if $a_i = \triangleright$, and $d_i \in \{-1,0\}$ if $a_i = \triangleleft$, for $1 \leq i \leq k$,*
- *$F \subseteq S$ is the set of accepting states.*

If for any state and k-tuple of symbols the transition function δ is either undefined or singleton, then the automaton is said to be deterministic ($2dfa(k)$). If the heads of the automaton never move left, then the automaton is defined to be one-way. Nondeterministic and deterministic one-way k-head automata are denoted by $1nfa(k)$ and $1dfa(k)$, respectively.

3.2   Relation to Classical Automata

Strict hierarchies of classes have been shown for both one-way multi-head deterministic and nondeterministic automata with regard to the head count of the automata [5,13]. In 1978, [13] used the language

$$L'_k = \left\{ w_1 \$ w_2 \$ \ldots \$ w_{2k} \,\middle|\, w_i \in \{a, b\}^* \wedge w_i = w_{2k+1-i} \text{ for all } 1 \leq i \leq k \right\}$$

to prove the separation of the class of languages that can be recognized by a $1dfa(k)$ from the class that can be recognized by a $1dfa(k+1)$.

We will consider a similar language, L_k.

Theorem 1. *For every $k \geq 1 \in \mathbb{N}$, there exists a language L_k such that:*

(1) for every prime p there exists a $1u_pfa(1)$ that recognizes L_k,

(2) L_k cannot be recognized by any $1nfa(k)$.

Proof. Let $n = \binom{k}{2} + 1$. The sought language is

$$L_k = \{w_1 1 w_2 1 \ldots 1 w_{2n} | w_i \in \{0^m | m \geq 1\} \wedge w_i = w_{2n-i+1}\}.$$

We will now prove that L_k satisfies the points of our theorem.

(1) We show that for an arbitrary language L_k, a $1u_pfa(1)$ can be built for every prime number p. The automaton starts in n different starting states $q_{1,1,1}, q_{1,2,1}, \ldots, q_{1,n,1}$ with amplitude 1. Each of these states begins a computational path that is intended to accumulate amplitude in one of n different rejecting states $q_{2n,1,2}, q_{2n,2,2}, \ldots, q_{2n,n,2}$. Every branch contains two kinds of states—states of the 1st group $q_{i,j,1}$ are responsible for generating amplitudes, and states of the 2nd group $q_{i,j,2}$ are intended for amplitude accumulation, $i \in [1, 2n], j \in [1, n]$.

If 0 is read from the input and the automaton is in one of the 1st group states $q_{i,j,1}$, where $i \leq n$, then the amplitude of the state remains the same and with amplitude 1 the automaton goes to a 2nd group state, $q_{i,j,2}$. By doing so, the state's accumulated amplitude is added to $q_{i,j,2}$. If 0 is read in a 2nd group state $q_{i,j,2}$, the state's amplitude remains the same. If 1 is read in a 1st group state $q_{i,j,1}$, where $i < n$, then the automaton with amplitude $j + 1$ transitions to $q_{i+1,j,1}$, thereby transitioning there with amplitude $(j+1) \cdot |q_{i+1,j,1}|$ (by $|q_i|$, we denote the amplitude of the state q_i). In contrast, if 0 is read in the 1st group state $q_{i,j,1}$, where $i > n$, the amplitude of the state remains unchanged and the transition to $q_{i,j,2}$ is made with amplitude -1. If 1 is read in the 1st group state $q_{i,j,1}$, where $i \geq n$, the transition to $q_{i+1,j,1}$ is made with amplitude $-(j+1)$. If 1 is read in a 2nd group state, a transition is made from $q_{i,j,1}$ to $q_{i+1,j,1}$, with amplitude 1. The exception is the last column of states, $q_{2n,j,1}$ and $q_{2n,j,2}$, which are responsible for reading in the last block of the word. In this case, the transition if 1 is read is not defined. A schematic representation of the described automaton is presented in Fig. 1.

As a result, if a word $0^{a_1} 1 0^{a_2} 1 0^{a_3} 1 \ldots 1 0^{a_{2n}}$ was read, then each of the rejecting states $q_{2n,j,2}$ has accumulated an amplitude equal to
$$a_1 + a_2 \cdot (j+1) + a_3 \cdot (j+1)^2 + \cdots + a_n \cdot (j+1)^{n-1} - a_{n+1} \cdot (j+1)^{n-1} - a_{n+2} \cdot (j+1)^{n-2}$$
$$- \cdots - a_{2n},$$
which is equal to 0 if the word belongs to the language; i.e. if

$$a_1 = a_{2n} \wedge a_2 = a_{2n-1} \wedge \ldots \wedge a_n = a_{n+1}.$$

It follows that a word is in L_k iff the following equations hold:

$$\begin{cases} a_1 + a_2 \cdot 2 + a_3 \cdot 2^2 + \cdots + a_n \cdot 2^{n-1} - a_{n+1} \cdot 2^{n-1} - a_{n+2} \cdot 2^{n-2} - \cdots - a_{2n} \\ = 0 \\ a_1 + a_2 \cdot 3 + a_3 \cdot 3^2 + \cdots + a_n \cdot 3^{n-1} - a_{n+1} \cdot 3^{n-1} - a_{n+2} \cdot 3^{n-2} - \cdots - a_{2n} \\ = 0 \\ \cdots \\ a_1 + a_2 \cdot (n+1) + a_3 \cdot (n+1)^2 + \cdots + a_n \cdot (n+1)^{n-1} - a_{n+1} \cdot (n+1)^{n-1} \\ \qquad\qquad\qquad - a_{n+2} \cdot (n+1)^{n-2} - \cdots - a_{2n} = 0 \end{cases}$$

rewriting

$$\begin{cases} (a_1 - a_{2n}) + 2 \cdot (a_2 - a_{2n-1}) + 2^2 \cdot (a_3 - a_{2n-2}) + \cdots + 2^{n-1} \cdot (a_n - a_{n+1}) \\ = 0 \\ (a_1 - a_{2n}) + 3 \cdot (a_2 - a_{2n-1}) + 3^2 \cdot (a_3 - a_{2n-2}) + \cdots + 3^{n-1} \cdot (a_n - a_{n+1}) \\ = 0 \\ \cdots \\ (a_1 - a_{2n}) + (n+1) \cdot (a_2 - a_{2n-1}) + (n+1)^2 \cdot (a_3 - a_{2n-2}) + \cdots \\ \qquad\qquad\qquad + (n+1)^{n-1} \cdot (a_n - a_{n+1}) = 0 \end{cases}$$

We see that the coefficients of the system form a Vandermonde matrix. Therefore, its determinant is non-zero, and since the given system is homogeneous, only the trivial solution exists.

However, if the word does not belong to L_k, then no more than 4 lines can hold true. However, even in this case at least one line will exist that is not equal to 0. Otherwise the system would have a nontrivial solution. Therefore, a word belongs to L_k iff the sum of the final amplitude norms of the rejecting states is greater than 0.

By adding an accepting state with a sufficiently small norm of the amplitude (decreasing as the length of the word increases) it is possible to make the automaton accept the language L_k.

(2) Proven by [4], a proof for a similar language can also be found in [13]. The idea of this proof relies on the fact that any two heads that have been used to compare a pair cannot be used to compare another pair. This implies that if the number of block pairs in a word n is greater than the number of pairs of heads $\binom{k}{2}$, then the language cannot be recognized with k heads. □

4   Two-Way Automata

4.1   Definitions

Ultrametric multi-head automata are defined by generalizing the definition of ultrametric one-head one-way automata in a natural way. The definition of multi-head automata due to [5] is used as well.

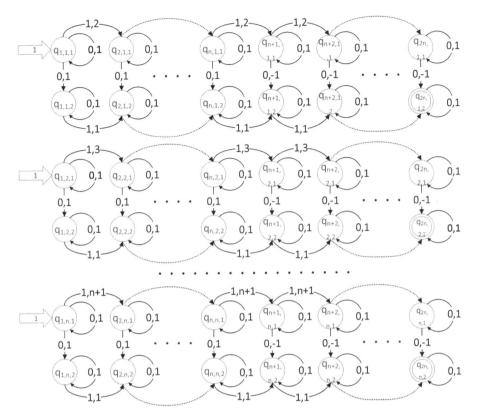

Fig. 1. Automaton for recognizing $0^n 10^m 10^h 1 \cdots 10^h 10^m 10^n$. Double-circled states are rejecting. Large arrows with labels in them show the amplitude distribution when the automaton starts. Small labelled arrows show transitions. A label (a, b) indicates that if the automaton reads a, transition with amplitude b should be made.

Definition 4. *A finite k-head two-way p-ultrametric automaton ($2u_pfa(k)$) is a septuple $\langle S, \Sigma, k, s_0, \delta, Q_A, Q_R \rangle$ where*

- *S is a finite set of states,*
- *Σ is a finite set ($\triangleright, \triangleleft \notin \Sigma$)—the input alphabet ($\triangleright$ and $\triangleleft$ are the left and right endmarkers, respectively),*
- *$k \geq 1$ is the number of heads,*
- *$s_0 : S \to \mathbb{Q}_p$ is the initial distribution of amplitudes,*
- *$\delta : S \times (\Sigma \cup \{\triangleright, \triangleleft\})^k \times S \times \{-1, 0, 1\}^k \to \mathbb{Q}_p$ is the partial transition function. Whenever $\delta(s, (a_1, \ldots, a_k), s', (d_1, \ldots, d_k))$ is defined and not equal to 0, then $d_i \in \{0, 1\}$ if $a_i = \triangleright$, and $d_i \in \{-1, 0\}$ if $a_i = \triangleleft$, for $1 \leq i \leq k$,*
- *$Q_A, Q_R \subseteq S$ are the sets of accepting and rejecting states, respectively.*

$2u_pfa(k)$ works in a similar way as $1u_pfa$, with the exception that the automaton now has k heads that can move freely, as in a $2nfa(k)$. In contrast to

$1u_pfa$, amplitudes are now given for a pair consisting of the positions of the heads and a state. The amplitude of the state $y \in S$ with heads in positions $(p_1, \ldots, p_k) \in \{0, \ldots, |w| + 1\}^k$ on the word $w = a_1 \ldots a_n$ after the i-th operation is given by

$$s_i(y, (p_1, \ldots, p_k)) = \sum_{\substack{x \in S, p_1', \ldots, p_k' \in \{0,1,\ldots,n+1\}: \\ \delta\left(x, \left(a_{p_1'}, \ldots, a_{p_k'}\right), y, (p_1 - p_1', \ldots, p_k - p_k')\right) \text{ is defined}}} s_{i-1}\left(x, (p_1', \ldots, p_k')\right) \cdot \delta\left(x, \left(a_{p_1'}, \ldots, a_{p_k'}\right), y,\right.$$

$$(p_1 - p_1', \ldots, p_k - p_k'))$$

Similarly as before, the acceptance of a word is determined by comparing the final amplitude p-norm sum of the accepting and rejecting states. That is, the word is accepted iff

$$\sum_{x \in Q_A} \left\| \sum_{\substack{i \in \mathbb{N}, \\ (p_1, \ldots, p_k) \in F_i(x)}} s_i(x, (p_1, \ldots, p_k)) \right\|_p > \sum_{x \in Q_R} \left\| \sum_{\substack{i \in \mathbb{N}, \\ (p_1, \ldots, p_k) \in F_i(x)}} s_i(x, (p_1, \ldots, p_k)) \right\|_p$$

where $(p_1, \ldots, p_k) \in F_i(x)$ iff the automaton with some amplitude halts in the i-th step in the state x with the heads in positions $p_1, \ldots, p_k$.

The class of languages recognized by $2u_pfa(k)$ is denoted $2U_pFA(k)$.

Ultrametric Turing machines are defined to be used as a device to assist in proving results regarding multi-head automata classes. We modify the definition for Turing machine by [6].

Definition 5. *A p-ultrametric Turing machine with k work tapes ($u_ptm(k)$ or simply u_ptm) is an octuple $M = \langle Q, \Sigma, b, \Gamma, q_0, \delta, Q_A, Q_R \rangle$ where:*

- *Q is a nonempty set of states,*
- *Σ is a nonempty set—the input alphabet,*
- *$b \notin \Sigma$ is the "empty" symbol,*
- *$\Gamma \supseteq \Sigma \cup \{b\}$ is the working alphabet,*
- *$q_0 : Q \to \mathbb{Q}_p$ is the initial amplitude distribution,*
- *$\delta : Q \times \Gamma^k \times Q \times (\Gamma \times \{-1, 0, 1\})^k \to \mathbb{Q}_p$ is a partially-defined transition function where -1 denotes moving the head to the left, 1 denotes moving the head to the right, 0 denotes not moving the head, and $\mathbb{Q}_p$ is the amplitude of the transition,*
- *$Q_A, Q_R \subseteq S$ are the sets of accepting and rejecting states, respectively.*

The class of languages recognized by a u_ptm is denoted U_pTM.

Similarly, as with ultrametric multi-head automata, the machine with a given amplitude is in one of its possible configurations. However, now the configuration consists of the state of the finite control, the position of the k heads, and the contents of all k tapes. The amplitude with which the machine is in one of its configurations in the i-th step is computed analogously as in the case with ultrametric multi-head automata. The criteria for acceptance are analogous to those of ultrametric multi-head automata.

Ultrametric multi-register automata are also used as an intermediate device for proofs.

Definition 6. *A finite p-ultrametric k register automaton (also referred to as a machine) ($2u_pra(k)$) consists of a p-ultrametric finite control and k registers that can hold natural numbers. The automaton begins with the input number in the first register with the specified starting amplitude distribution in some of its states . A predicate $\overset{?}{=} 0$ and the operations $+1$ or -1 can be applied to a register. Each of the transitions of the finite control can have a predicate or an operation associated with it that is triggered when the transition is made. Acceptance criteria are analogous to those of ultrametric automata.*

4.2   Simulation of Ultrametric Automata by Ultrametric Turing Machines

In this subsection, we will show how to simulate an ultrametric multi-head automaton by an ultrametric Turing machine. The techniques used here are similar to the techniques used by [9] for probabilistic automata.

Without loss of generality, we can assume that the simulated automaton has at most 2 transitions from each state and input symbol. We will show how to construct a p-ultrametric 2-tape Turing machine that, having received the description of a k-head p-ultrametric automaton as an input, will accept exactly the same words as the automaton. We will consider only automata recognizing unary languages, and we will show how to encode the description of the automaton as a word in the form $1^{2^n}, n \in \mathbb{N}$. Furthermore, we will require that the amplitudes of the simulated automaton are p-constructible sets.

Definition 7. *A set S of p-adic numbers is p-constructible if there exists a p-ultrametric Turing machine that, having received a description of a number $x \in S$ as an input, reaches a marked state q_u with amplitude x.*

A $2u_pfa(k)$ can be described with a binary sequence denoting, in turn: the number of states, the number of heads, the transitions between the states, and their respective amplitudes.

The simulation is performed as follows: The $2u_pfa(k)$ $\mathcal{A}$ with transitions from a p-constructible set is simulated by a $u_ptm(2)$ $\mathcal{T}$. $\mathcal{T}$ receives the description of $\mathcal{A}$ in unary alphabet as a word in the form $1^{2^n}, n \in \mathbb{N}$. $\mathcal{T}$ starts by reading the input word and deterministically (making transitions with amplitude 1) writes n in binary on the second tape, where n is the description of the automaton $\mathcal{A}$. Additionally, a space of size $O(k \cdot \log m)$ is reserved (where m is the length of the input word on which the automaton must be simulated) for k counters denoting the positions of the heads of the automaton $\mathcal{A}$. From this point forward, only the second tape will be used (we will refer to it as the work tape).

While processing the input word, the automaton $\mathcal{A}$ can be in different configurations in parallel with different amplitudes. If $\mathcal{A}$ is in a configuration with an amplitude a, $\mathcal{T}$ will simulate it by being in a configuration in which the content of the work tape corresponds to the respective configuration of $\mathcal{A}$ with the same amplitude a.

To show that $\mathcal{T}$ can simulate $\mathcal{A}$ in such a way, we must show that every transition of $\mathcal{A}$ can be realized by $\mathcal{T}$. If $\mathcal{A}$ has a transition from q_1 to q_2 with an amplitude a_1, and from q_1 to q_3 with an amplitude a_2, then $\mathcal{T}$ being in a configuration corresponding to q_1 can make transitions to configurations corresponding to q_2 and q_3 with amplitudes a_1 and a_2, respectively. This simulation is accomplished by $\mathcal{T}$ branching into two branches, and in both of them writing on a special place on the tape d_1 or d_2, respectively, with an amplitude 1, where d_1 and d_2 are the descriptions of the transitions and the amplitudes a_1 and a_2, respectively. This is done deterministically with amplitude 1. Next, a subroutine is called that transitions to a marked state q_u with an amplitude a_1 or a_2. (Because all transitions of $\mathcal{A}$ are p-constructible, there exists such a subroutine.) Subsequently, $\mathcal{T}$ changes the work tape so that it corresponds to the respective transition (again, this is done deterministically with amplitude 1). Because the only transition that is accomplished with an amplitude other than 1 is the transition to the state q_u, after this procedure $\mathcal{T}$ is in a configuration corresponding to q_2 with amplitude a_1, and in a configuration q_3 with amplitude a_2.

4.3   Multi-head Automata

[11] has proven that for finite deterministic and nondeterministic automata, the language class that can be recognized using k heads is a proper subclass to the class of languages that can be recognized if $k+1$ heads are allowed. Using similar methods, [9] has proven the same for finite probabilistic automata. We prove here that the same holds for finite ultrametric automata.

Definition 8. *By $\widehat{C}$, we denote the subset of a language class C containing only the words in the form $1^{2^n}, n \in \mathbb{N}$, more precisely*

$$\widehat{C} = \left\{ L \in C | \forall x \in L \, \exists n \in \mathbb{N} : x = 1^{2^n} \right\}$$

Theorem 2. *For every natural number k and prime p:*

$$2\widehat{U_p FA}(k) \subsetneq \widehat{U_p TM}.$$

Proof. We will construct a special p-ultrametric Turing machine with 2 tapes and log-space space complexity called $\mathcal{T}$. We will show that its recognized language cannot be recognized by a p-ultrametric automata with k heads for any k.

Similarly as in the previous section, we will construct $\mathcal{T}$ so that it simulates a $2u_p fa(k)$ $\mathcal{A}$ given in its input. The input word on which $\mathcal{A}$ will be simulated will be the input of $\mathcal{T}$, i.e. the description of $\mathcal{A}$.

More precisely, $\mathcal{T}$ 1st tape contains $1^m, m \in \mathbb{N}$, $\mathcal{T}$ checks whether the word is in the form $1^{2^n}, n \in \mathbb{N}$ (if not, the word is rejected), and by taking up to $O(log(n))$ space, writes n's binary representation on the 2nd tape (we will refer to it as the work tape). It then checks whether n's binary representation is syntactically a valid $2u_p fa(k)$; if not, the word is rejected. Then, $\mathcal{T}$ designates a space on the work tape to be used for k counters that will be used to represent $\mathcal{A}$'s head positions. Since the counter values can be in $\{0, \ldots, 2^n + 1\}$, $O(k \cdot log(2^n)) \sim O(k \cdot n)$ space is required. All previous actions are performed deterministically. Next, $\mathcal{T}$ is run on the input string similarly as it is shown in the previous section. (The head of the first tape is not used anymore; instead, a check is performed to determine whether or not the counters corresponding to head positions are inside word boundaries.) Consequently, $\mathcal{T}$ is with respective amplitudes in all of the possible configurations of $\mathcal{A}$ after processing the word. Afterwards, $\mathcal{T}$ checks whether the contents of the work tape suggest that $\mathcal{A}$ is in an accepting state, and halts in a rejecting state if $\mathcal{A}$ would have accepted the word; $\mathcal{T}$ halts in an accepting state if $\mathcal{A}$ would have rejected the word. Therefore, $\mathcal{T}$ yields the opposite result to that of $\mathcal{A}$ with the same amplitudes.

Let us consider the language $L(\mathcal{T})$ recognized by $\mathcal{T}$. We can see that for every $2u_p fa(k)$ denoted by $\mathcal{J}$, there exists a word w such that it is either in $L(\mathcal{T})$ but not in $L(\mathcal{J})$, or it is in $L(\mathcal{J})$ but not in $L(\mathcal{T})$; namely, $\mathcal{J}$'s specification.   □

Similarly, as in [9] and [11], in the following proofs we will use the function $f_k : \{1^{2^n} | n \in \mathbb{N}\} \to \{1^{2^n} | n \in \mathbb{N}\}$, where $f_k(1^{2^n}) = 1^{2^{k \cdot n}}$.

When f_k is applied to a language, we refer to the following function: $f_k(L) = \{f_k(x) | x \in L\}$.

Lemma 1. *For every language $L \in \widehat{U_p TM}$ that is recognized by a 2-tape $u_p tm$ in logarithmic space, there exists a natural number u such that: $f_u(L) \in 2\widehat{U_p FA}(3)$.*

Proof. We will show how a $u_p tm$ denoted by $\mathcal{T}$ that recognizes L can be transformed into a $u_p tm$ called $\mathcal{T}'$, which can then be replaced by a p-ultrametric 3 register machine. From this, it easily follows that there exists a $2u_p fa(3)$ that recognizes a "stretched" variant of L, where stretching is done by f_u.

We will construct $\mathcal{T}'$ so that it simulates $\mathcal{T}$. $\mathcal{T}'$ will hold the following information on its work tape:

- the binary information of the input word,
- $\mathcal{T}$ head position on the 1st tape (we will call it the input tape),
- $\mathcal{T}$ 2nd tape contents (we will call it the work tape) (requires $O(log n)$ space),
- $\mathcal{T}$ head position on the work tape.

The simulation of $\mathcal{T}$ on $\mathcal{T}'$'s work tape is less complicated than in the case of ultrametric automata, since $\mathcal{T}'$ can use the original finite control of $\mathcal{T}$, and the transitions are not required to be simulated on the tape. To simulate $\mathcal{T}$, $\mathcal{T}'$ uses only the work tape.

We can see that given $\mathcal{T}'$, a corresponding p-ultrametric 3 register machine can be constructed. If the input is of the form 1^{2^n}, then the register machine starts with n in the first register and with 0 in the remaining registers. The contents of the work tape of $\mathcal{T}'$ can be simulated by manipulating the registers of the first two stored sub-words, v and h^{rev}, and by using the third as an auxiliary register. To do this, we use operations "add 1" and "divide by 2", which can be carried out by using the auxiliary register.

Since a position of a multi-head automaton directly corresponds to a number in a register, and the simulated Turing machine has *log*-space space complexity, a 3 register machine can be replaced with a p-ultrametric 3 head automaton. However, since its heads cannot cross word boundaries and therefore cannot simulate arbitrarily large numbers, the input words must be sufficiently long. This is achieved by selecting a large enough u.   □

Lemma 2. *For all languages $L \in \widehat{U_p TM}$ and all $u, v \geq 1, u, v \in \mathbb{N}$: $f_u(L) \in 2\widehat{U_p FA}(v) \Rightarrow L \in 2\widehat{U_p FA}(u \cdot v)$.*

Proof. Let the $2u_p fa(v)$ in the premise be $\mathcal{A}$ and the $2u_p fa(u \cdot v)$ in the conclusion–$\mathcal{A}'$.

Consider the operation of $\mathcal{A}$ on a word $f_u(l), l = 1^{2^n} \in L, n \in \mathbb{N}$. The position of each of v heads of $\mathcal{A}$ can be described with an integer $h_i \in [0, 2^{u \cdot n} - 1]$. h_i can be written in base 2^n with u digits. As the position of each head of $\mathcal{A}'$ can be described with a digit in base 2^n, each head of $\mathcal{A}$ can be simulated with u heads of $\mathcal{A}'$. As each movement of a head of $\mathcal{A}$ corresponds to movement of heads of $\mathcal{A}'$, the respective transitions can be accomplished with equal amplitudes, and the accepting amplitudes of the words remain the same.

Note that this simulation is performed analogously as for deterministic automata in [11] and for probabilistic automata in [9].   □

Lemma 3. *For every language $L \in \widehat{U_p TM}$ and every $u > v > 1, u, v \in \mathbb{N}$:*

$$f_{u+1}(L) \in 2\widehat{U_p FA}(v) \Rightarrow f_u(L) \in 2\widehat{U_p FA}(v + 1).$$

Proof. Let the $2u_p fa(v)$ in the premise be $\mathcal{A}$ and the $2u_p fa(v + 1)$ in the conclusion–$\mathcal{A}'$.

Consider the operation of $\mathcal{A}$ on a word $f_{u+1}(l), l = 1^{2^n} \in L, n \in \mathbb{N}$.

The position of each of v heads of $\mathcal{A}$ on the input word can be described with an integer $h_i \in [0, 2^{(u+1) \cdot n} + 1]$.

We will simulate the position h_i in the automaton $\mathcal{A}'$ with a head g_i and an additional number $x_i \in [0, 2^n]$, so that $h_i = g_i + x_i \cdot 2^{u \cdot n}$. It is evident that the values in the necessary interval can be denoted this way:

$$2^{u \cdot n} + (2^n - 1) \cdot 2^{u \cdot n} = 2^{u \cdot n} \cdot (1 + (2^n - 1)) = 2^{u \cdot n + n} = 2^{(u+1) \cdot n}.$$

All v numbers x_i are coded with the $(v+1)$-th head of $\mathcal{A}'$, similarly to [11]. This can be accomplished if sufficient space exists on the tape of $\mathcal{A}'$, specifically if $(2^n)^v < 2^{u \cdot n}$, which holds as $v < u$.

Again, as each movement of a head of $\mathcal{A}$ corresponds to the movement of heads of $\mathcal{A}'$, the respective transitions can be accomplished with equal amplitudes, and the accepting amplitudes of the words remain the same.   □

The result concerning the superiority of a $k + 1$ head over k heads follows from the previous lemmas and Theorem 2.

Theorem 3. *For all $k \geq 2 \in \mathbb{N}$:*

$$2\widehat{U_p FA}(k) \subsetneqq 2U_p\widehat{FA(k+1)}.$$

Proof. We prove from the contrary by showing that if there exists such $h \geq 2$ that $2\widehat{U_p FA}(h) = 2U_p\widehat{FA(h+1)}$, it implies $2U_p\widehat{FA(h \cdot (h+1))} = \widehat{U_p TM}$, which contradicts 2.

Take $L \in \widehat{U_p TM}$ for some prime p. Lemma 1 implies that there exists $m \in \mathbb{N}$ such that $f_m(L) \in 2\widehat{U_p FA}(3)$. Consequently, $f_m(L) \in 2\widehat{U_p FA}(h)$. Lemma 3 implies that if $m > h+1$, then $f_{m-1}(L) \in 2U_p\widehat{FA(h+1)} = 2\widehat{U_p FA}(h)$. Reduce m by 1 and repeat until we get $f_m(L) \in 2\widehat{U_p FA}(h)$ and $m = h+1$. Lemma 2 implies that if $f_m(L) \in 2\widehat{U_p FA}(h)$, then $L \in 2U_p\widehat{FA(h \cdot m)} = 2U_p\widehat{FA(h \cdot (h+1))}$. Contradiction with Theorem 2. $\square$

Corollary 1. *Since $2U_p FA(k+1)$ is a superset of $2U_p FA(k)$ and we proved that there exists a language that can be recognized with $k+1$ heads and not by k, the head hierarchy result holds for languages in multi-letter alphabets as well:*

$$2U_p FA(k) \subsetneqq 2U_p FA(k+1).$$

References

1. Balodis, K., Beriņa, A., Cīpola, K., Dimitrijevs, M., Iraids, J., Jēriņš, K., Kacs, V., Kalējs, J., Krišlauks, R., Lukstiņš, K., Raumanis, R., Scegulnaja, I., Somova, N., Vanaga, A., Freivalds, R.: On the state complexity of ultrametric finite automata. In: SOFSEM 2013: Theory and Practice of Computer Science. vol. 2, pp. 1–9 (2013)
2. Dragovich, B., Dragovich, A.: A p-adic model of DNA sequence and genetic code. p-Adic Numbers, Ultrametric Analysis, and Applications 1(1), 34–41 (2009)
3. Freivalds, R.: Ultrametric automata and Turing machines. In: Voronkov, A. (ed.) Turing-100. EPiC Series, vol. 10, pp. 98–112. EasyChair (2012)
4. Freivalds, R.: Language recognition using finite probabilistic multitape and multi-head automata. Problemy Peredachi Informatsii 15(3), 99–106 (1979) (in Russian)
5. Holzer, M., Kutrib, M., Malcher, A.: Multi-head finite automata: Characterizations, concepts and open problems. Electronic Proceedings in Theoretical Computer Science 1, 93–107 (2009), http://dx.doi.org/10.4204/EPTCS.1.9
6. Hopcroft, J.E., Ullman, J.D.: Introduction to Automata Theory, Languages and Computation. Addison-Wesley (1979)
7. Kozyrev, S.V.: Ultrametric analysis and interbasin kinetics, pp. 121–128. American Institute of Physics (2006)
8. Krišlauks, R., Rukšāne, I., Balodis, K., Kucevalovs, I., Freivalds, R., Nāgele, I.: Ultrametric Turing machines with limited reversal complexity. In: SOFSEM 2013: Theory and Practice of Computer Science. vol. 2, pp. 87–94 (2013)
9. Macarie, I.: Multihead two-way probabilistic finite automata. In: Baeza-Yates, R., Poblete, P.V., Goles, E. (eds.) LATIN 1995. LNCS, vol. 911, pp. 371–385. Springer, Heidelberg (1995), http://dx.doi.org/10.1007/3-540-59175-3_103
10. Madore, D.A.: A first introduction to p-adic numbers. Online (2000), http://www.madore.org/~david/math/padics.eps

11. Monien, B.: Two-way multihead automata over a one-letter alphabet. RAIRO - Theoretical Informatics and Applications - Informatique Thorique et Applications 14(1), 67–82 (1980)
12. Turakainen, P.: Generalized automata and stochastic languages. Proceedings of The American Mathematical Society 21, 303–309 (1969)
13. Yao, A.C., Rivest, R.L.: k + 1 heads are better than k. vol. 25, pp. 337–340. ACM, New York (1978), http://doi.acm.org/10.1145/322063.322076

Nash-Williams-type and Chvátal-type Conditions in One-Conflict Graphs

Christian Laforest[1] and Benjamin Momège[2,*]

[1] LIMOS, CNRS, UMR 6158 – Université Blaise Pascal, Clermont-Ferrand
Campus des Cézeaux, 24 Avenue des Landais, 63173 Aubière Cedex, France
laforest@isima.fr
[2] Univ. Nice Sophia Antipolis, CNRS, I3S, UMR 7271, 06900 Sophia Antipolis,
INRIA, France
benjamin.momege@inria.fr

Abstract. Nash-Williams and Chvátal conditions (1969 and 1972) are well known and classical sufficient conditions for a graph to contain a Hamiltonian cycle. In this paper, we add constraints, called *conflicts*. A conflict is a pair of edges of the graph that cannot be both in a same Hamiltonian path or cycle. Given a graph G and a set of conflicts, we try to determine whether G contains such a Hamiltonian path or cycle without conflict. We focus in this paper on graphs in which each vertex is part of at most one conflict, called *one-conflict graphs*. We propose Nash-Williams-type and Chvátal-type results in this context.

Keywords: graph, conflict, Hamiltonian, path, cycle.

1 Introduction

These last decades many works have been done to prove sufficient conditions for a graph to have a Hamiltonian cycle. The most classical ones in a graph with n vertices are the following: Dirac's condition [3] (the degree of each vertex is at least $\frac{n}{2}$), Ore's condition [10] (for any non adjacent vertices u and v, degree of u plus degree of v is at least n), Bondy-Chvátal's condition (based on closure of graphs), Nash-Williams' condition (every k-regular graph on $2k+1$ vertices has a Hamiltonian cycle) and Chvátal's condition (based on the degree sequences). For more results in this area, see the recent survey [8]. More recently, several papers were devoted to the introduction of *conflicts* into graphs. A conflict is a pair of edges of the graph and, as they are in conflict, they cannot be both in a structure as a path or a tree. Conflicts are useful to model situations in which it is forbidden to use two incompatible objects (because of their nature, functions, etc.) in a same structure. In [11,5] the authors investigate the problem to find a path without conflict between two vertices in a graph with a given set of restrictive type of conflicts (the two edges of a conflict share a same vertex). For these graphs, the problem of finding two-factors[1] is considered in [4] and

---

[*] B. Momège has a PhD grant from CNRS and région Auvergne.
[1] A subgraph such that for any vertex its in-degree and its out-degree is exactly one.

a dichotomy between tractable and intractable instances is also given. In [6], the authors study the problem of spanning tree without conflict. All the known results show that adding conflicts to a graph considerably increase the complexity of problems.

In a very recent paper [7] we tried to extend the classical Dirac's, Ore's and Bondy-Chvátal's results to graphs in which each vertex is part of at most one conflict (called *one-conflict graphs*). We shown that it is not possible in all cases. We then proposed sufficient conditions (inspired by Dirac's and Ore's ones) for a one-conflict graph to contain a Hamiltonian path or cycle without conflict and Bondy-Chvátal-type conditions.

In this paper we try to extend the classical Nash-Williams' condition and Chvátal's condition to one-conflict graphs. We show that it is not possible in all cases. We then propose sufficient Nash-Williams-type and Chvátal-type conditions for a one-conflict graph to contain a Hamiltonian path or cycle without conflict.

2   Preliminaries, Notations and Definitions

In this paper, we only consider undirected, unweighted and simple graphs. We refer to [1] for definitions and undefined notations. The vertex set of a graph G is denoted by V_G and its edge set by E_G. If $|V_G| = n$, the graph G is called an n-vertex graph. An edge between u and v in a graph G is denoted by uv. The two endpoints of an edge are said to be adjacent to each other. The complete graph K_n is a graph with n vertices in which every vertex is adjacent to every other. A path in G consists of a sequence of distinct vertices with each two consecutive vertices in the sequence adjacent to each other in the graph. A cycle in G consists of a sequence of distinct vertices (except the starting and ending vertex) starting and ending at the same vertex, with each two consecutive vertices in the sequence adjacent to each other in the graph. A path (or a cycle) of G is *Hamiltonian* if it contains all the vertices of G. The length of a path or a cycle is the number of edges it contains. For example, in an n-vertex graph the length of a Hamiltonian path is $n-1$ and the length of a Hamiltonian cycle is n. A *matching* in a graph is a set of edges without common vertices. An edge and a vertex on that edge are called *incident*. The *degree* of a vertex v of G is the number of edges incident to v. It is denoted by $deg_G(v)$. The *minimum degree* of G is denoted by $\delta(G)$ (*i.e.* $\delta(G) = \min_{v \in V_G} deg_G(v)$). A graph where each vertex has degree k is called a k-*regular graph*. If G is a graph, a *conflict* in G is a pair $\{e_1, e_2\}$ of distinct edges of G. We denote by $(G, Conf)$ a graph G with a set of conflicts $Conf$. A *path without conflict* P in $(G, Conf)$ is a path P in G such that for any e, e' of P, $\{e, e'\} \notin Conf$ (similarly for Hamiltonian path without conflict, cycle without conflict and Hamiltonian cycle without conflict).

In this paper we only consider graph with conflicts such that each vertex is not involved in more than one conflict. We call such graphs *one-conflict graphs*. From now, $(G, Conf)$ is an n-vertex one-conflict graph.

3   Nash-Williams-type Conditions in One-Conflict Graphs

Let $k \geq 0$ be an integer. In 1969, Nash-Williams proves the following result:

Theorem 1 (Nash-Williams). *[9] Every k-regular graph on $2k + 1$ vertices has a Hamiltonian cycle.*

Remark 2. *Let $\alpha, \beta \in \mathbb{N}$. If G is an α-regular graph on $2\beta + 1$ vertices, we have:*

$$2|E_G| = \sum_{v \in V_G} deg_G(v) = \alpha(2\beta + 1)$$

and therefore α is necessarily divisible by two. Thus, for a k-regular graph on $2k + 1$-vertices, k is divisible by two.

Remark 3. *Theorem 1 is "tight", that is to say, it becomes false if we replace $2k + 1$ vertices with $2k + 2$ vertices or k-regular with $(k - 1)$-regular $(k \geq 1)$.*

Proof. If we replace $2k + 1$ vertices with $2k + 2$ vertices. The disjoint union of two complete graphs on $k + 1$ vertices is a k-regular graph on $2k + 2$ vertices that is not connected and therefore admits no Hamiltonian cycle.

If we replace k-regular with $(k - 1)$-regular. Consider the disjoint union of a complete graph on $k + 1$ vertices deprived of the edges of a perfect matching (it exists because $k+1$ is even by Remark 2) and a complete graph on k vertices. It is a $(k-1)$-regular graph on $2k+1$ vertices that is not connected and therefore admits no Hamiltonian cycle.   □

Remark 4. *In general, Theorem 1 is false for one-conflits graphs. For example, if $k = 2$, the cycle on 5 vertices with a conflict is a k-regular graph on $2k + 1$ vertices that contains no Hamiltonian cycle without conflict.*

In this section, we show the following result:

Theorem 5. *Every k-regular one-conflict graph on $2k + 1$ vertices has a Hamiltonian path without conflict.*

If $k = 0, 1$ or 2 the result is obvious. Indeed, if $k = 0$, a 0-regular one-conflict graph on 1 vertices is a single vertex. If $k = 1$ there is no a 1-regular one-conflict graph on 3 vertices (as for all k odd) from Remark 2. If $k = 2$, a 2-regular one-conflict graph on 5 vertices is a cycle with at most one conflict and simply remove an edge (of the conflict, if it exists) to obtain a Hamiltonian path. We now assume that k is greater than or equal to 3.

Lemma 6. *Every k-regular one-conflict graph on $2k + 1$ vertices $(G, Conf)$ has a path without conflict of length at least k.*

Proof. Suppose the length i of a longest path without conflict

$$P = v_0, \ldots, v_i$$

in $(G, Conf)$ is strictly less than k (i.e. $i < k$). We remove one edge outside P from each conflict. We denote by G' the graph obtained. Each vertex of G' is of degree greater than or equal to $k - 1$. A path in G' is then a path without conflict in $(G, Conf)$. Therefore, in G', the vertices v_0 and v_i cannot be adjacent to a vertex outside P. As the degrees $deg_{G'}(v_0)$ and $deg_{G'}(v_1)$ are greater than or equal to $k - 1$ we have $v_0 v_i \in E_{G'}$. Thus,

$$C = v_0, \ldots, v_i, v_0$$

is a cycle without conflict of length $i + 1$ in $(G, Conf)$.

Now consider any vertex v_j of the cycle C. Since $deg_G(v_j) = k$, the vertex v_j is adjacent to a vertex v outside the cycle in G. If there is no edge of the cycle C in conflict with vv_j then

$$v, v_j, \ldots, v_i, v_0, \ldots, v_{j-1}$$

(only the $k + 1$ first vertex if $j = 0$) is an path without conflict of length $i + 1$ in $(G, Conf)$, which contradicts the maximality of i. There is, therefore, an edge ab of C in conflict with vv_j. We then have $a \neq v_j$ or $b \neq v_j$. By symmetry we may assume that $a \neq v_j$. As a cannot be in more than one conflict, we can replace the vertex v_j chosen initially by a to execute the above operation and get an path without conflict of length $i + 1$ in $(G, Conf)$. In all cases, the maximality of i is contradicted. So the initial assumption is false and there is a path without conflict of length at least k. □

Lemma 7. *Let $P = v_0, \ldots, v_i$ be a path without conflict of length i in a k-regular one-conflict graph on $2k + 1$ vertices $(G, Conf)$ that cannot be extended by adding a vertex adjacent to one of its endpoints. If $k \leq i \leq 2k - 3$, $(G, Conf)$ has a cycle without conflict of length $i + 1$.*

Proof. We remove one edge outside P from each conflict. We denote by G' the graph obtained. Each vertex of G' is of degree greater than or equal to $k - 1$. In G', the vertices v_0 and v_i are in no conflict and are of degree greater than or equal to $k - 1$. They each are adjacent to $k - 1$ vertices of P (because otherwise we could extend P by adding an adjacent vertex of one of its endpoints).

There is $0 \leq j \leq i - 1$ such that v_0 is adjacent to v_{j+1} and v_i is adjacent to v_j. Indeed, if for any vertex v_{j+1} adjacent to v_0, v_j is not adjacent to v_i, at least $k - 1$ vertices of P are not adjacent to v_i. Thus, v_i is adjacent to at most $i - (k - 1)$ vertices of P and as

$$i - (k - 1) \leq 2k - 3 - (k - 1) = k - 2 < k - 1,$$

this is impossible.

Now, if $j = 0$ or $j = i - 1$,

$$v_0, \ldots, v_p, v_0$$

is a cycle without conflict of length $i + 1$ in $(G, Conf)$ and if $0 < j < i - 1$,

$$v_0, \ldots, v_j, v_i, \ldots, v_{j+1}, v_0$$

is a cycle without conflict of length $i + 1$ in $(G, Conf)$. □

Lemma 8. *For $k \leq i \leq 2k - 3$, any cycle without conflict $C = v_0, \ldots, v_i, v_0$ of length $i + 1$ in a k-regular one-conflict graph on $2k + 1$ vertices $(G, Conf)$ can give rise to a path without conflict of length $i + 1$ in $(G, Conf)$.*

Proof. If $i = k$. As the degree of the vertices of $(G, Conf)$ is equal to k, each of the k vertices outside of the cycle C is connected to at least one vertex of the cycle. There are therefore at least k edges between the vertices of the cycle and the vertices outside. As the cycle has $k + 1$ edges, at most $\lfloor \frac{k+1}{2} \rfloor$ of them may be involved in conflicts (two adjacent edges of C cannot both being part of the set of edges involved in distinct conflicts). As $k > \lfloor \frac{k+1}{2} \rfloor$ ($k \geq 3$), there is an edge $v_j v$ between a vertex v_j of C and a vertex v outside C which is in conflict with no edge of C. We add vv_j to C and remove an edge of C containing v_j to obtain a path without conflict length $i + 1$ in $(G, Conf)$.

If $i > k$. Take a vertex v outside the cycle. As the degree of v is equal to k it is adjacent to at least two vertices v_j and v_l of C. Since v is not in more than one conflict, one of the edges vv_j or vv_l is not in conflict with the edges of C. For example if it is vv_j, we add vv_j to C and remove one edge of C containing v_j to obtain a path without conflict of length $i + 1$ in $(G, Conf)$.   □

In a k-regular one-conflict graph on $2k+1$ vertices $(G, Conf)$, any path without conflict of length i with $k \leq i \leq 2k - 3$, can give rise to a path without conflict of length $i + 1$ in the following way:

- If possible add a vertex that is adjacent to one of its endpoints,
- Otherwise construct a cycle without conflict of length $i + 1$ using the technique presented in Lemma 7 and then construct a path without conflict of length $i + 1$ from this cycle using the technique presented in Lemma 8.

Lemma 6 shows that there is a path without conflict of length at least k in $(G, Conf)$ and using iteratively the above operation we get the following result:

Lemma 9. *Every k-regular one-conflict graph on $2k + 1$ vertices $(G, Conf)$ has a path without conflict of length at least $2k - 2$.*

We show now that there is a path without conflict of length at least $2k - 1$ in every k-regular one-conflict graph on $2k + 1$ vertices $(G, Conf)$.

Lemma 10. *Every k-regular one-conflict graph on $2k+1$ vertices $(G, Conf)$ has a path without conflict of length at least $2k - 1$.*

Proof. Let

$$P = v_0, \ldots, v_{2k-2}$$

be a path without conflict of length $2k - 2$ in $(G, Conf)$. Let v and v' be the vertices outside P. We remove one edge outside P from each conflict. We denote by G' the graph obtained. We will show that there is a Hamiltonian path in G' which will be a Hamiltonian path without conflict in $(G, Conf)$. To do this, suppose that there is no path of length greater than $2k - 2$ in G'. We note that for each vertex x of G':

$$deg_{G'}(x) = k - 1 \quad \text{or} \quad deg_{G'}(x) = k.$$

If $vv' \in E_{G'}$:

In this case v and v' cannot be adjacent to v_0, to v_1, to v_{2k-3}, and to v_{2k-2} in G'. Thus, v and v' are each adjacent to at least $k - 2$ vertices of the set

$$\{v_2, v_3, \ldots, v_{2k-4}\}$$

of $2k - 5$ vertices. As v (resp. v') cannot be adjacent to two consecutive vertices of P in G', it is adjacent to the subset of vertices of even index

$$\{v_2, v_4, \ldots, v_{2k-4}\}$$

of the previous set. So

$$v_0, v_1, v_2, v, v', v_4, \ldots, v_{2k-2}$$

is a path without conflict of length at least $2k - 1$ and $(G, Conf)$ which contradicts the assumption that there is no path of length greater than $2k - 2$ in G'.

We therefore have $vv' \notin E_{G'}$. In this case, v cannot be adjacent to v_1, to v_{2k-2}, and to two consecutive vertices of P in G'. So the only possible solution is $deg_{G'}(v) = k - 1$ and v is adjacent to the vertices with odd index of P in G. Similarly for v'.

Furthermore, we can replace a vertex x of even index different to 0 and $2k - 2$ ($k - 2$ possible vertices) by v to obtain a new path without conflict of length $2k - 2$ in $(G, Conf)$, a new graph G' and an edge $xv' \notin E_{G'}$. For the same reasons as above we see that x must be adjacent to the vertices of odd index of P in G. Similarly for all the $k - 2$ vertices of different even index different to 0 and $2k - 2$.

The vertex v_2 is then adjacent to v_1, v, v' and to the $k - 2$ vertices of even index different to 0 and $2k - 2$ in G. So its degree is strictly greater than k which contradicts the assumption on the degree of the vertices of G.

Finally, assuming that there is no path without conflict of length greater than $2k - 2$ in $(G, Conf)$ leads to a contradiction.   □

We show now that there is a Hamiltonian path without conflict in every k-regular one-conflict graph on $2k + 1$ vertices $(G, Conf)$.

Proof (of Theorem 5). Let

$$P = v_0, \ldots, v_{2k-1}$$

be a path without conflict of length $2k - 1$ in $(G, Conf)$. Let v be the vertex outside P. We remove one edge outside P from each conflict. We denote by G'

the graph obtained. We will show that there is a Hamiltonian path in G' which will be a Hamiltonian path without conflict in $(G, Conf)$. To do this, assume the contrary.

We denote by $N_{G'}(x)$ the set of vertices adjacent to the vertex x of G' i.e. $N_{G'}(x) := \{y \in V_{G'} \mid xy \in E_{G'}\}$. We note that every vertex x of G' satisfies:

$$deg_{G'}(x) = |N_{G'}(x)| = k - 1 \quad \text{or} \quad deg_{G'}(x) = k.$$

We have $v_0 \notin N_{G'}(v)$ and $v_{2k-1} \notin N_{G'}(v)$. If $v_0 v_{2k-1} \in E_{G'}$, then there is a cycle $v_0, \ldots, v_{2k-1}, v_0$ of length $2k$ in G' with a vertex adjacent to v and so there is a Hamiltonian path in G'. So $v_0 v_{2k-1} \notin E_{G'}$. If $|N_{G'}(v)| = k$ then v is connected to two vertices adjacent in P and then there is a Hamiltonian path in G'. Finally, $|N_{G'}(v)| = k - 1$ and either the differences between the indices of two consecutive vertices of $N_{G'}(v)$ ordered by increasing index are all 2 i.e.

$$N_{G'}(v) = \{v_1, v_3, \ldots, v_{2k-3}\}$$

or

$$N_{G'}(v) = \{v_2, v_4, \ldots, v_{2k-2}\},$$

either the differences between the indices of two consecutive vertices of $N_{G'}(v)$ ordered by increasing index are all 2 except one that is 3 i.e.

$$N_{G'}(v) = \{v_1, v_3, \ldots, v_i\} \cup \{v_{i+3}, v_{i+5}, \ldots, v_{2k-2}\}.$$

If the differences between the indices of two consecutive vertices of $N_{G'}(v)$ ordered by increasing index are all 2. By symmetry we may assume that

$$N_{G'}(v) = \{v_1, v_3, \ldots, v_{2k-3}\}.$$

In this case

$$N_{G'}(v_0) = N_{G'}(v).$$

Indeed, if v_0 is adjacent to $v_l \in \{v_2, v_4, \ldots, v_{2k-2}\}$ then

$$v_l, v_0, \ldots, v_{l-1}, v, v_{l+1}, \ldots, v_{2k-1}$$

is a Hamiltonian path in G'. Now, we want to show that there are two adjacent vertices in

$$X := \{v_2, v_4, \ldots, v_{2k-2}\}.$$

To do this, assume the contrary. We see that for all x in X we have $N_{G'}(x) \subseteq N_{G'}(v) \cup \{v_{2k-1}\}$. Let $X' := X \setminus \{v_{2k-2}\}$. For all v_i in X' we have $v_i v_{2k-1} \notin E_{G'}$. Indeed, if $v_i v_{2k-1} \in E_{G'}$ then

$$v_0, \ldots, v_{i-1}, v, v_{i+1}, \ldots, v_{2k-1}, v_i$$

is a Hamiltonian path in G'. So there are at least $k - 1$ edges between each vertex in X and $N_{G'}(v)$ and therefore $|X'|(k-1) = (k-2)(k-1)$ edges between X' and $N_{G'}(v)$, $k - 2$ edges between v_{2k-2} and $N_{G'}(v)$, $|N_{G'}(v)| = k - 1$ edges between v_0 and $N_{G'}(v)$ and $|N_{G'}(v)| = k - 1$ edges between v and $N_{G'}(v)$ and as $deg_{G'}(y) \leq k$ for all $y \in N_{G'}(v)$ we have:

$$(k-1)k \geq \sum_{y \in N_{G'}(v)} deg_{G'}(y) \geq (k+1)(k-1) - 1 > (k-1)k.$$

This is a contradiction, and therefore there are two adjacent vertices v_i and v_{i+j} in X. Then

$$v_0, \ldots v_{i-1}, v, v_{i+j-1}, v_i, v_{i+j} \ldots, v_{2k-1}$$

is a Hamiltonian path in G'. This contradicts the initial assumption that there is no Hamiltonian path in G'.

So the differences between the indices of two consecutive vertices of $N_{G'}(v)$ ordered by increasing index are all 2 except one that is 3 and we have

$$N_{G'}(v) = \{v_1, v_3, \ldots, v_i\} \cup \{v_{i+3}, v_{i+5}, \ldots, v_{2k-2}\}.$$

In this case

$$N_{G'}(v_0) = N_{G'}(v).$$

Indeed, if v_0 is adjacent to v_{j+1} for $v_j \in N_{G'}(v)$ $(j \neq 2k-2)$ then

$$v, v_j, \ldots, v_0, v_{j+1}, \ldots, v_{2k-1}$$

is a Hamiltonian path in G', and if v_0 is adjacent to v_{i+2}, then

$$v_{i+1}, v_{i+2}, v_0, \ldots, v_i, v_{i+3}, \ldots v_{2k-1}$$

is Hamiltonian path in G'. Similarly, we can show that

$$N_{G'}(v_{2k-1}) = N_{G'}(v).$$

Let S be the set of the $k - 3$ vertices of P located between two vertices of $N_{G'}(v)$ whose indices differ by 2 *i.e.*

$$S := \{v_1, v_2, \ldots, v_{2k-2}\} \setminus (\{v_1, v_3, \ldots, v_i\} \cup \{v_{i+3}, v_{i+5}, \ldots, v_{2k-2}\} \cup \{v_{i+1}, v_{i+2}\}).$$

Let $x \in S$. We can easily see that if $N_{G'}(x) \neq N_{G'}(v)$ there is a Hamiltonian path in G'. So we have

$$N_{G'}(x) = N_{G'}(v) = N_{G'}(v_0) = N_{G'}(v_{2k-1}).$$

The vertex v_i is adjacent to v, v_0, v_{i+1}, v_{2k-1} and to the $k - 3$ vertices of S and therefore we have $deg_{G'}(v_i) \geq k + 1$. This contradicts the initial hypothesis that G is a k-regular graph.

Thus, assuming that there is no Hamiltonian path in G' leads to a contradiction and therefore there is a Hamiltonian path in G'. Finally, Theorem 5 is true. $\qquad \square$

4   Chvátal-type Conditions in One-Conflict Graphs

Definition 1. *The degree sequence $(d_1, \ldots, d_n)$ of a graph is the sequence of the degrees of the vertices, in increasing order, with repetitions as needed:*

$$d_1 \leq \cdots \leq d_n.$$

We recall the result of Chvátal (1972) for "classical" graphs (*i.e.* without conflict):

Theorem 11 (Chvátal). *[2] Let $(G, \mathcal{C}onf)$ be is an n-vertex one-conflict graph with $n \geq 3$. If the degree sequence $(d_1, \ldots, d_n)$ of an n-vertex graph satisfies*

$$\forall i < \frac{n}{2}, \ d_i \leq i \Rightarrow d_{n-i} \geq n - i$$

then it has a Hamiltonian cycle.

This result becomes false when we consider conflicts. Indeed, let G_1 be the graph obtained from the disjoint union of K_1 and K_{n-1} by adding two edges between K_1 and K_{n-1}. A Hamiltonian cycle in G_1 necessarily contains these two edges. If $\mathcal{C}onf$ contains the conflict composed of these two edges, $(G_1, \mathcal{C}onf)$ satisfies the conditions of Theorem 11 but contains no Hamiltonian cycle without conflict.

A second counterexample for $n \geq 4$ is the graph G_2 obtained from the disjoint union of K_2 (with $V_{K_2} = \{x, y\}$) and K_{n-2} (with $V_{K_{n-2}} = \{v_1, \ldots, v_{n-2}\}$) by adding the three edges xv_1, xv_2 and yv_2. A Hamiltonian cycle in G_2 necessarily contains xv_1 and yv_2. If $\mathcal{C}onf$ contains the conflict composed of these two edges, $(G_2, \mathcal{C}onf)$ satisfies the conditions of Theorem 11 but contains no Hamiltonian cycle without conflict.

We start with a useful lemma:

Lemma 12. *We denote by G' a graph obtained from an n-vertex one-conflict graph $(G, \mathcal{C}onf)$ by removing in G one edge from each conflict. The degree sequence $(d'_1, \ldots, d'_n)$ of G' satisfies*

$$\forall i \in \{1, \ldots, n\}, \ d_i \leq d'_i + 1.$$

Proof. By definition we have:

$$\forall i \in \{1, \ldots, n\}, \ \exists v_i \in V_{G'} = V_G \ s.t. \ deg_{G'}(v_i) = d'_i.$$

As each vertex is not involved in more than one conflict we have:

$$\forall i \in \{1, \ldots, n\}, \ deg_G(v_i) \leq deg_{G'}(v_i) + 1 = d'_i + 1$$

and as

$$d'_1 \leq \cdots \leq d'_n$$

we have

$$\forall i \in \{1, \ldots, n\}, \ deg_G(v_1), \ldots, deg_G(v_i) \leq d'_i + 1.$$

So for all $i \in \{1, \ldots, n\}$, G contains at least i vertices of degree less than or equal to $d'_i + 1$. Thus, we have

$$\forall i \in \{1, \ldots, n\}, \ d_i \leq d'_i + 1.$$

$\square$

We deduce from Theorem 11 and Lemma 12 the following result:

Theorem 13. *Let $(G, Conf)$ be is an n-vertex one-conflict graph. If the degree sequence $(d_1, \ldots, d_n)$ of G satisfies*

$$\forall i < \frac{n}{2}, \ d_i \leq i + 1 \Rightarrow d_{n-i} \geq n - i + 1$$

then $(G, Conf)$ has a Hamiltonian cycle without conflict.

Proof. Suppose G satisfies the conditions. We remove one edge from each conflict. We denote by G' the graph obtained. We will show that there is a Hamiltonian cycle in G' which will be a Hamiltonian cycle without conflict in $(G, Conf)$. We denote by

$$d'_1 \leq \cdots \leq d'_n$$

the degree sequence of G'. We will show that G' satisfies the condition of Theorem 11 *i.e.*:

$$\forall i < \frac{n}{2}, \ d'_i \leq i \Rightarrow d'_{n-i} \geq n - i.$$

Suppose that there is $i < \frac{n}{2}$ such that $d'_i \leq i$. By Lemma 12, this implies

$$d_i \leq i + 1.$$

By definition of G this implies

$$d_{n-i} \geq n - i + 1$$

and we use again Lemma 12 to obtain

$$d'_{n-i} \geq n - i$$

and finally G' satisfies the condition of Theorem 11. $\square$

By adding to a graph a universal vertex adjacent to all other vertices and then applying Theorem 11 to this new graph, Chvátal also proved in [2] the following result for Hamiltonian paths:

Theorem 14 (Chvátal). *[2] If the degree sequence $(d_1, \ldots, d_n)$ of an n-vertex graph satisfies*

$$\forall i \leq \frac{n}{2}, \ d_i \leq i - 1 \Rightarrow d_{n+1-i} \geq n - i$$

then it has a Hamiltonian path.

We deduce from Theorem 14 the following result:

Theorem 15. *Let* $(G, \mathcal{C}onf)$ *be is an n-vertex one-conflict graph. If the degree sequence* $(d_1, \ldots, d_n)$ *of* G *satisfies*

$$\forall i \leq \frac{n}{2}, \ d_i \leq i \Rightarrow d_{n+1-i} \geq n+1-i$$

then $(G, \mathcal{C}onf)$ *has a Hamiltonian path without conflict.*

Proof. Suppose G satisfies the conditions. We remove one edge from each conflict. We denote by G' the graph obtained. We will show that there is a Hamiltonian path in G' which will be a Hamiltonian path without conflict in $(G, \mathcal{C}onf)$. We denote by

$$d_1' \leq \cdots \leq d_n'$$

the degree sequence of G'. We will show that G' satisfies the condition of Theorem 14 *i.e.*:

$$\forall i \leq \frac{n}{2}, \ d_i' \leq i-1 \Rightarrow d_{n+1-i}' \geq n-i$$

Suppose that there is $i \leq \frac{n}{2}$ such that $d_i' \leq i-1$. By Lemma 12, this implies

$$d_i \leq i.$$

By definition of G this implies

$$d_{n+1-i} \geq n+1-i$$

and we use again Lemma 12 to obtain

$$d_{n+1-i}' \geq n-i$$

and finally G' satisfies the condition of Theorem 14.   □

Remark 16. *In our paper [7] we proved that an n-vertex one-conflict graph* $(G, \mathcal{C}onf)$ *s.t.* $\delta(G) \geq \frac{n}{2}$ *contains a Hamiltonian path without conflict. When n is odd, this result is a simple corollary of Theorem 15. Indeed, if* $n = 2p+1$ *and* $\delta(G) \geq \frac{n}{2}$ *we have* $\delta(G) \geq p+1$ *thus the degree sequence* $(d_1, \ldots, d_n)$ *of* G *satisfies :*

$$\forall i \in \{1, \ldots, n\}, \ d_i \geq p+1$$

and finally :

$$\forall i \leq \frac{n}{2}, \ d_i \geq p+1 > \frac{n}{2} \geq i.$$

5   Conclusion and Perspectives

In this paper we give sufficient conditions for a one-conflict graph to contain a Hamiltonian path or cycle without conflict. As the classical Nash-Williams' condition and Chvátal's condition cannot always be generalized, we extended and proved similar conditions. Our perspectives now are to investigate system with more conflicts. We already have partial results.

Acknowledgements. We thank Mamadou M. Kanté and anonymous referees for helping us to improve a first version of this work.

References

1. Bondy, J.A., Murty, U.S.R.: Graph Theory. Springer London Ltd. (2010)
2. Chvátal, V.: On hamilton's ideals. J. Combinatorial Theory (B) 12, 163–168 (1972)
3. Dirac, G.A.: Some theorems on abstract graphs. Proc. London Math. Soc. 2, 69–81 (1952)
4. Dvořák, Z.: Two-factors in orientated graphs with forbidden transitions. Discrete Mathematics 309(1), 104–112 (2009)
5. Kanté, M.M., Laforest, C., Momège, B.: An exact algorithm to check the existence of (elementary) paths and a generalisation of the cut problem in graphs with forbidden transitions. In: van Emde Boas, P., Groen, F.C.A., Italiano, G.F., Nawrocki, J., Sack, H. (eds.) SOFSEM 2013. LNCS, vol. 7741, pp. 257–267. Springer, Heidelberg (2013)
6. Kanté, M.M., Laforest, C., Momège, B.: Trees in Graphs with Conflict Edges or Forbidden Transitions. In: Chan, T.-H.H., Lau, L.C., Trevisan, L. (eds.) TAMC 2013. LNCS, vol. 7876, pp. 343–354. Springer, Heidelberg (2013)
7. Laforest, C., Momège, B.: Hamiltonian conditions in one-conflict graphs. Accepted at IWOCA (2014)
8. Li, H.: Generalizations of Dirac's theorem in hamiltonian graph theory - a survey. Discrete Mathematics 313(19), 2034–2053 (2013)
9. Nash-Williams, C.St.J.A.: Valency sequences which force graphs to have hamiltonian circuits. In: University of Waterloo Research Report. Waterloo, Ontario: University of Waterloo (1969)
10. Ore, Ø.: Note on Hamiltonian circuits. American Mathematical Monthly (67), 55 (1960)
11. Szeider, S.: Finding paths in graphs avoiding forbidden transitions. Discrete Applied Mathematics 126(2-3), 261–273 (2003)

Optimal State Reductions of Automata with Partially Specified Behaviors

Nelma Moreira[1,*], Giovanni Pighizzini[2,**], and Rogério Reis[1,*]

[1] Centro de Matemática e Faculdade de Ciências da Universidade do Porto, Portugal
{nam,rvr}@dcc.fc.up.pt
[2] Dipartimento di Informatica, Università degli Studi di Milano, Italy
pighizzini@di.unimi.it

Abstract. Nondeterministic finite automata with *don't care* states, namely states which neither accept nor reject, are considered. A characterization of deterministic automata compatible with such a device is obtained. Furthermore, an optimal state bound for the smallest compatible deterministic automata is provided. Finally, it is proved that the problem of minimizing nondeterministic and deterministic *don't care* automata is NP-complete.

1 Introduction

Finite state automata are well-known and widely investigated language acceptors. On each input string x, the behavior of a finite automaton is an answer *yes/no* to the question of the membership of x to the accepted language. In some situations, however, we could have some input sequences for which the answer of the automaton is not interesting, or even situations where the automaton does not need to consider all possible strings over the input alphabet. For example, an automaton could receive its input from another machine or program, which produces only sequences in a special form, thus excluding all the other sequences which are definable over the input alphabet. We give a couple of trivial but immediate examples over the alphabet $\{-, 0, 1, \ldots, 9\}$. If the inputs of the automaton represent numbers in decimal notation produced by a (correct) program, the automaton cannot expect sequences starting by 0 (with the only exception of the sequence 0) as 00123, sequences starting by -0, and sequences containing the symbol − after the leftmost position, as 4-9-2014. On the other hand, if the inputs would represent calendar dates, the last string will be a valid input, while a string as -1234 will be invalid (unless a strange and counterintuitive format is used).

In these cases, we do not need to define the behavior of the automaton, namely acceptance or rejection, on the strings which are not interesting or will never

---

* Authors partially funded by the European Regional Development Fund through the programme COMPETE and by the Portuguese Government through the FCT under projects PEst-C/MAT/UI0144/2013 and FCOMP-01-0124-FEDER-020486.
** Author partially supported by MIUR under the project PRIN "Automi e Linguaggi Formali: Aspetti Matematici e Applicativi", code H41J12000190001.

G.F. Italiano et al. (Eds.): SOFSEM 2015, LNCS 8939, pp. 339–351, 2015.

appear as input. This suggests us the idea of studying finite automata with three kinds of states: accepting states, rejecting states, and *don't care* states. We call these models automata with *don't care* states or, shortly, *don't care* automata. A quite natural problem we consider in the paper is the state reduction of these models. Of course, to perform this reduction, we can arbitrarily accept or reject strings on the which the behavior of the automaton is not specified.

This idea is not completely new, if fact, in digital systems design, Moore automata (or equivalently Mealy automata) are used to specify several kinds of algorithms, protocols and processes which then are used in sequential circuits synthesis. Usually, the automata are incomplete (laking either outputs or transitions from some inputs), and the elimination of redundant states reduces the size of the logic needed to be implemented, tested or verified. However, the standard algorithm for minimizing deterministic complete automata is not enough for incomplete ones. The first algorithm for the exact solution was described by Paull and Unger [11], and Pfleeger [13] proved that the minimization of incomplete deterministic Moore machines is a NP-complete problem. Since then many other exact and heuristic algorithms have been proposed, some considering that the initial machine is nondeterministic [14,8,12,9,3]. The standard Paull and Unger approach is based on the identification of sets of compatible states and the obtention of a minimal closed cover. The use of *don't care* states has been also considered for different purposes in the case of automata on infinite words [4].

In this paper, we mainly investigate *nondeterministic* automata with *don't care* states (dcNFA). Given a such a device A, we are interested in finding a smallest deterministic finite automaton (DFA) B which is "compatible" with it, in the sense that all the strings accepted by A are also accepted by B and all the strings rejected by A are also rejected by B, while on the remaining strings B can have an arbitrary behavior. This problem can be reformulated as a *separation problem*: given two regular languages L_1 and L_2, find a language L with minimal state complexity that *separates* L_1 and L_2, i.e. such that $L_1 \subseteq L \subseteq L_2^c$ (where L^c is the complement of L). In the context of model checking, this version of the problem was considered by Chen et al. [1], but there the general Paull and Unger algorithm was used.

Here we obtain a precise characterization of the DFAs which are compatible with a given dcNFA. This result is useful to obtain an upper bound for the number of states of the smallest compatible DFAs. We also show that this bound is tight. We also study computational complexity aspects. To this respect, we show that the problem of obtaining a smallest DFA compatible with a given dcNFA is NP-complete, and it remains NP-complete if the given *don't care* automaton is deterministic. The paper concludes with some considerations concerning dcNFAs over one-letter alphabets.

Due to the lack of space, some of the proofs are omitted from this version of the paper.

2   Automata with *don't care* States

Given an alphabet Σ, we consider the usual notions of deterministic finite automata (DFAs) and nondeterministic finite automata (NFAs) (with multiple initial states). Given an automaton A, we denote the language accepted by it as $\mathcal{L}(A)$. We also assume that the reader is familiar with the notion of *minimal* DFA. We now introduce the main notion we are interested in.

Definition 1. *A don't care nondeterministic finite automaton (dcNFA) A is a tuple $\langle Q, \Sigma, \delta, I, F^\oplus, F^\ominus \rangle$, where $A^\oplus = \langle Q, \Sigma, \delta, I, F^\oplus \rangle$ and $A^\ominus = \langle Q, \Sigma, \delta, I, F^\ominus \rangle$ are two NFAs such that $\mathcal{L}(A^\oplus) \cap \mathcal{L}(A^\ominus) = \emptyset$. A state $q \in Q$ is called an accepting (rejecting) state if $q \in F^\oplus$ ($q \in F^\ominus$, respectively). If $q \notin F^\oplus \cup F^\ominus$ then q is called a* don't care *state. Associated to A there are the two languages $\mathcal{L}^\oplus(A) = \mathcal{L}(A^\oplus)$ and $\mathcal{L}^\ominus(A) = \mathcal{L}(A^\ominus)$ called the accepted* and the *rejected language by A, respectively.*

The automaton A is a don't care *deterministic finite automaton (dcDFA) if the set I consists exactly of one element i and δ is a partial function from Q to Q, namely, for each $q \in Q$, $a \in \Sigma$, $\delta(q, a)$ contains at most one element.*

Notice that given a dcNFA A, its accepted (rejected) language consists of all words having a computation path from an initial state to an accepting (a rejecting, resp.) state. Hence, if all the states of A are reachable from the initial state, then the sets $F^\oplus$ and $F^\ominus$ must be disjoint. As usual, in the deterministic case we will denote a dcNFA as $\langle Q, \Sigma, \delta, i, F^\oplus, F^\ominus \rangle$ and we will write $p = \delta(q, a)$ instead of $p \in \delta(q, a)$, when $\delta(q, a)$ is defined. The function δ can be made total, in a standard way, by inserting an extra state, called *trap* or *dead state*. However, while in DFAs this state is rejecting, according to our definition in the case of dcNFAs this state should be a *don't care* state. Hence, in a dc-NFA with a partial transition function, a string $x \in \Sigma^*$ leading to an undefined transition is neither accepted nor rejected.

In this paper, given a *don't care* automaton A we are interested in finding automata that agree with A on its accepted and rejected languages. This leads to the following definition.

Definition 2. *Let A be a dcNFA. A language L is said to be compatible with A whenever $\mathcal{L}^\oplus(A) \subseteq L$ and $\mathcal{L}^\ominus(A) \subseteq L^c$. An NFA (or DFA) B is compatible with A when $\mathcal{L}(B)$ is compatible with A.*

Since, as already observed, unspecified transitions have different meanings for DFAs (rejection) and for dcDFAs (*don't care* condition), while counting the number of the states in the case of DFAs we will add the trap state when the transition function is not total, while in the case of dcDFAs we will never add any extra state.

Example 3. Consider the dcDFA A represented in Figure 1. We label each accepting state with a $\oplus$ and each rejecting state with a $\ominus$, leaving *don't care* states unlabeled. Hence, $F^\ominus = \{s_5\}$ and $F^\oplus = \{s_0, s_3\}$. Trivially, $\mathcal{L}^\oplus(A) = (a^3 b^3)^*(\varepsilon + a^3)$

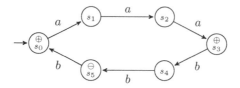

Fig. 1. The dcDFA A of Example 3

and $\mathcal{L}^{\ominus}(A) = (a^3b^3)^\star(a^3b^2)$. The language $L = (a^3b^3)^\star(\varepsilon + a + a^2 + a^3)$ is compatible with A. Notice that L is accepted by a DFA with the same transition graph of A (plus an implicit trap state) and with set of final states $\{s_0, s_1, s_2, s_3\}$. In the next sections, we will present several smaller DFAs compatible with A.

Let $G = (V, E)$ be an undirected graph. We recall that each complete subgraph of G is called a *clique*. We also say that a subset $\alpha \subseteq V$ forms a clique if the subgraph of G induced by α, namely the graph $(\alpha, E \cap (\alpha \times \alpha))$, is a clique. Furthermore, a clique $\alpha \subseteq V$ is maximal if any other subset of V which properly contains α does not form a clique. A *clique covering* of a graph G is a set of cliques such that every vertex of G belongs at least to one clique.

A self-verifying automaton (SVFA) A is a dcNFA where it is required that for each input string there exists at least one computation ending in an accepting or in a rejecting state, i.e., $\mathcal{L}^{\oplus}(A) = (\mathcal{L}^{\ominus}(A))^c$. This implies that the only language compatible with A is its accepted language $\mathcal{L}^{\oplus}(A)$. Hence, each SVFA can be transformed into an equivalent (and unique) minimal DFA. In [7] an optimal bound for the number of states of the minimal DFA equivalent to any given SVFA has been obtained. The authors associate with each n-state self-verifying automaton a graph with n vertices and prove that the state set of the minimum DFA equivalent to the given SVFA should be isomorphic to the set of the maximal cliques of such a graph. In the next sections, we use a similar approach for obtaining minimal DFAs compatible with a given automata with *don't care* states.

3 Conversion into Compatible Deterministic Automata

In this section we study how to convert any given dcNFA A into compatible DFAs. In particular we are interested in finding a minimal DFA compatible with A. As we will see, it is possible to have several minimal nonisomorphic smallest compatible DFAs.

Let us suppose that all the states of $A = \langle Q, \Sigma, \delta, I, F^{\oplus}, F^{\ominus} \rangle$ are reachable from the initial state. For each $q \in Q$, we denote by $L_q^{\oplus}$ and $L_q^{\ominus}$, respectively, the set of strings accepted and the set of strings rejected starting from q, that is, $L_q^{\oplus} = \{x \in \Sigma^* \mid \delta(q, x) \cap F^{\oplus} \neq \emptyset\}$ and $L_q^{\ominus} = \{x \in \Sigma^* \mid \delta(q, x) \cap F^{\ominus} \neq \emptyset\}$. Using the fact that q is reachable, it can be immediately verified that those two languages are disjoint. For the same reason, applying the subset construction to A, it turns out that $L_\alpha^{\oplus} \cap L_\alpha^{\ominus} = \emptyset$ for each subset $\alpha \subseteq Q$ whose states are all reachable by a same string, where $L_\alpha^\times = \bigcup_{q \in \alpha} L_q^\times$, for $\times \in \{\oplus, \ominus\}$. So, by suitable marking accepting and rejecting states, from A we can get the *subset dcDFA* A_s (with only reachable states) with $L^\times(A_s) = L^\times(A)$, for $\times \in \{\oplus, \ominus\}$.

As in [7], to study the structure of DFAs which are compatible with A, we introduce a *compatibility relation* on the state set Q. Intuitively, two states p, q of A are compatible if and only if two computations starting from p and q cannot give contradictory answers on the same string. Formally:

Definition 4. *Two states p, q of A are* compatible *if and only if*

$$(L_p^\oplus \cup L_q^\oplus) \cap (L_p^\ominus \cup L_q^\ominus) = \emptyset.$$

The compatibility graph *of A is the undirected graph whose vertex set is Q, and which contains the edge $\{p, q\}$ if and only if states p and q are compatible.*

It follows from the above discussion that if α is a state of the automaton A_s, then all states p, q in the set α must be compatible. Hence, each reachable state of A_s is represented by a clique in the compatibility graph.

In the case of SVFAs, it was proved that if for two reachable subsets $\alpha, \beta \subseteq Q$ of the subset automaton the set $\alpha \cup \beta$ is a clique of the compatibility graph then α and β are equivalent [7]. In our case, since the automaton A_s deriving from the subset construction could contain *don't care* states, we cannot properly define a similar equivalence over A_s states. However, we can prove the following result which will allow us to characterize DFAs that are compatible with A in terms of functions mapping states into cliques of the compatibility graph.

Theorem 5. *A DFA $A' = \langle Q', \Sigma, \delta', i', F' \rangle$ is compatible with a given dcNFA $A = \langle Q, \Sigma, \delta, I, F^\oplus, F^\ominus \rangle$ if and only if there is a function $\phi : Q' \to 2^Q$ such that:*

1. *$I \subseteq \phi(i')$,*
2. *for $q \in Q'$, $a \in \Sigma$, $\delta(\phi(q), a) \subseteq \phi(\delta'(q, a))$,*
3. *for $q \in Q'$, $\phi(q) \cap F^\oplus \neq \emptyset$ implies $q \in F'$ and $\phi(q) \cap F^\ominus \neq \emptyset$ implies $q \notin F'$.*

Furthermore, if A' is compatible with A then:

4. *for each $x \in \Sigma^*$, $\delta(I, x) \subseteq \phi(\delta'(i', x))$,*
5. *the set $\phi(Q')$ is a clique covering of the compatibility graph of A.*

Proof. First, let us suppose that A' is compatible with A. For each $q \in Q'$, we define

$$\phi(q) = \{p \in Q \mid \exists x \in \Sigma^* \text{ s.t. } q = \delta'(i', x) \text{ and } p \in \delta(I, x)\}.$$

By considering the empty string, we observe that $I \subseteq \phi(i')$, proving 1. Now, given $a \in \Sigma$, let $q' = \delta'(q, a)$. To prove 2 we show that $p' \in \delta(\phi(q), a)$ implies $p' \in \phi(q')$. To this aim, let us consider $p \in \phi(q)$ such that $p' \in \delta(p, a)$. By the definition of ϕ, there is a string $x \in \Sigma^*$ such that $q = \delta'(i', x)$ and $p \in \delta(I, x)$. Hence, $q' = \delta'(q, a) = \delta'(i', xa)$ and $p' \in \delta(p, a) \subseteq \delta(I, xa)$. According to the definition of ϕ this implies $p' \in \phi(q')$. Finally, the condition 3 follows immediately from our choice of ϕ.

To prove the converse, first of all it is useful to derive 4 from 1 and 2. We use an induction on the length of the string x. The basis $x = \epsilon$ is trivial. Now, let us consider a nonempty string $x = ya$ with $y \in \Sigma^*$ and $a \in \Sigma$, and suppose condition 4 true for y. Given $p \in \delta(I, x)$, there is a state $p' \in \delta(I, y)$ such that $p \in \delta(p', a)$. Furthermore, $q = \delta'(q', a)$ where $q' = \delta'(i', y)$. From the induction hypothesis we get that $p' \in \phi(\delta'(i', y)) = \phi(q')$ and, by condition 2, $\delta(\phi(q'), a) \subseteq \phi(\delta'(q', a))$ and, by putting all together, we complete the proof of 4:

$$p \in \delta(p', a) \subseteq \delta(\phi(q'), a) \subseteq \phi(\delta'(q', a)) = \phi(\delta'(i', x)).$$

Now, given $x \in \mathcal{L}^{\oplus}(A)$, let $p \in \delta(I, x) \cap F^{\oplus}$. Since $p \in \phi(\delta'(I, x))$, by condition 3 x should be accepted by A'. In a similar way, if $x \in \mathcal{L}^{\ominus}(A)$ then x should be rejected by A'. Hence we conclude that A' is compatible with A.

Concerning the second part of the theorem, we already proved 4. To prove 5, first we show that, for each $q \in Q'$, the set $\phi(q)$ is a clique of the compatibility graph of A, namely, each two states $p, r \in \phi(q)$ are compatible. Let $x, u \in \Sigma^*$ such that $q = \delta'(i', x) = \delta'(i', u)$, $p \in \delta(I, x)$, and $r \in \delta(I, u)$. By contradiction, suppose p and r not compatible. Then, there is $z \in \Sigma^*$ such that, without loss of generality, $\delta(p, z) \in F^{\oplus}$ and $\delta(r, z) \in F^{\ominus}$. It follows that z distinguishes strings x and u, which contradicts the fact that these strings lead to the same state in the automaton A'. This allows us to conclude that p, r must be compatible and, hence, $\phi(q)$ is a clique. Furthermore, since all the states of A are reachable, as a consequence of 4 for each $p \in Q$ there is a state $q \in Q'$ such that $p \in \phi(q)$. This completes the proof of 5.   □

Using Theorem 5, we now derive a "pseudo-subset construction" which allows to find some DFAs compatible with A. We remind the reader that we suppose that all the states of A are reachable from the initial state. Then we define a DFA $A' = \langle Q', \Sigma, \delta', i', F' \rangle$ as follows:

- Q' is the set of *all maximal cliques* of the compatibility graph of A; in the following, given a maximal clique $\alpha \subseteq Q$, we use the same name α to denote the corresponding state in Q';
- i' is a clique that includes the set I of initial states of A;
- for $\alpha \in Q'$, $\sigma \in \Sigma$, $\delta'(\alpha, \sigma)$ is a state $\beta \in Q'$ such that $\delta(\alpha, \sigma) \subseteq \beta$;
- the set F' of final states is a subset of Q' that contains those states α s.t. $\alpha \cap F^{\oplus} \neq \emptyset$ and does not contain those states α s.t. $\alpha \cap F^{\ominus} \neq \emptyset$, namely, each state of Q' that contains a state from $F^{\oplus}$ is marked as final, each state that contains a state from $F^{\ominus}$ is marked as nonfinal, while each one of the remaining states can be freely marked either as final or as nonfinal.

The above definition leaves some degrees of freedom, which allow to obtain different DFAs. For any possible choice, it can be immediately verified that the function $\phi : Q' \to 2^Q$ defined as $\phi(\alpha) = \alpha$ satisfies the conditions of Theorem 5. Hence, it turns out that each DFA A', defined as above, is compatible with A.

Example 6. Let us consider the dcDFA A of Example 3 (Figure 1). Its compatibility graph is depicted in Figure 2 (left). Applying the above construction we obtain 4 different DFAs, which are summarized in the Figure 2 (right). We have two choices for the initial state and two choices for the transition from state $\{s_1, s_2, s_5\}$ on b. These choices are represented by dotted arrows.

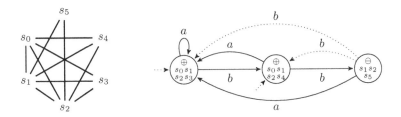

Fig. 2. The compatibility graph of the dcDFA A in Fig. 1 and four compatible DFAs

In the previous construction, we used the covering of the compatibility graph defined by maximal cliques. In general, we could also use a different covering, provided that the trivial function ϕ mapping each clique of the considered cover in itself satisfies the conditions 1, 2, and 3 of Theorem 5. For instance, further DFAs, compatible with the dcDFA A of Example 6, are depicted in Figure 3. We

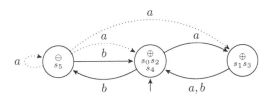

Fig. 3. More DFAs compatible with the dcDFA A in Figure 2

can observe that in this example the compatibility graph of A cannot be covered using less than 3 cliques. Hence, there are no DFAs compatible with A with less than 3 states, the number of maximal cliques in the compatibility graph. However, in general the situation can be different, as illustrated in the next example.

Example 7. Let us consider the dcDFA A depicted in the upper part of Figure 4 with its compatibility graph, which contains 4 maximal cliques. This graph has the following two coverings consisting each one of two cliques: $\{\{s_0, s_1\}, \{s_2, s_3\}\}$ and $\{\{s_0, s_3\}, \{s_1, s_2\}\}$. For these coverings we obtain two DFAs which are compatible with A (see also Figure 4). Since these DFAs have only two states and each DFA consisting only of one state cannot be compatible with A, it turns out that they are the smallest DFAs which are compatible with A. In Figure 5

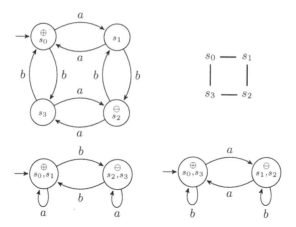

Fig. 4. The dcDFA A of Example 7 with its compatibility graph, and two compatible DFAs

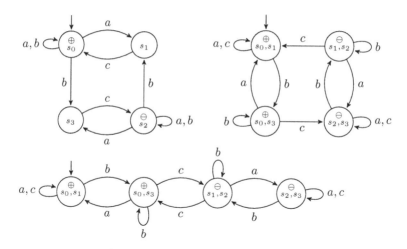

Fig. 5. The dcNFA $\widehat{A}$ of Example 7 (top left), with two compatible DFAs (top right and bottom)

it is depicted a dcNFA $\widehat{A}$ having the same compatibility graph as A, with two compatible DFAs whose states correspond to all maximal cliques of that graph.

We observe that in the automaton $\widehat{A}$ the strings a, b, bca, and bcb lead to the set of states $\{s_0, s_1\}$, $\{s_0, s_3\}$, $\{s_2, s_3\}$, and $\{s_1, s_2\}$, respectively. Hence, observing the compatibility graph and using condition 4 of Theorem 5 we can conclude that in this example *all maximal cliques* of the compatibility graph are necessary. Hence, each DFA compatible with $\widehat{A}$ should have at least 4 states.

Example 7 shows that we can have different dcNFAs A and $\widehat{A}$ with the same compatibility graph but with smallest compatible DFAs of different sizes.

The following theorem summarizes the situation, providing bounds for such a size in terms of cliques of the compatibility graph:

Theorem 8. *For each dcNFA A, there exists a compatible DFA whose number of states is bounded by the number of maximal cliques in the compatibility graph of A. Furthermore, each DFA compatible with A should have at least as many states as the smallest number of cliques covering the compatibility graph of A.*

4   State Complexity

In this section, we study descriptional complexity aspects. First we state an upper bound for the number of states of smallest DFAs compatible with a given dcNFA, showing that it can be effectively reached, i.e. it is tight. The arguments are adapted from those used for SVFAs [7].

Theorem 9. *For each integer $n \geq 2$ and each n-state dcNFA there exists a compatible DFA with at most $f(n)$ states, where*

$$f(n) = \begin{cases} 3^{\lfloor n/3 \rfloor}, & \text{if } n \equiv 0 \ (\text{mod } 3), \\ 4 \cdot 3^{\lfloor n/3 \rfloor - 1}, & \text{if } n \equiv 1 \ (\text{mod } 3), \\ 2 \cdot 3^{\lfloor n/3 \rfloor}, & \text{if } n \equiv 2 \ (\text{mod } 3). \end{cases}$$

Furthermore this bound can be effectively reached.

Proof. The upper bound immediately derives from Theorem 8 and from a result by Moon and Moser [10] stating that the maximum number of maximal cliques in a graph with n vertices is given by the function $f(n)$. The lower bound is a consequence of Theorem 10 in [7], where for each integer n an n-state SVFA A_n with multiple initial states such that the smallest equivalent DFA requires $f(n)$ states was provided. (See Figure 6 for the case of n multiple of 3.) Since SVFAs with multiple initial states are a special case of dcNFAs, the claimed result follows. $\square$

The optimality proof in Theorem 9 is a consequence of the optimality of the same bound for SVFAs with multiple initial states. Since the optimal bound in the case of SVFAs with a single initial state is slightly different $(1 + f(n-1))$, one could ask what happens in the case of dcNFAs with a *single initial state*. We are going to prove that in this case the optimal bound remains that of Theorem 9.

To this aim, for each n we consider an automaton A'_n, obtained by modifying the automaton A_n used to give the optimality in Theorem 9, as follows. We start from the same set of states of A_n and from the same transition graph. One of the initial states of A_n is chosen as the initial state of A'_n. Furthermore, we add a transition on a new input symbol c from a selected state of A_n to all the states that in A_n are initial. In this way each time the automaton A'_n makes a transition on the letter c, it is able to simulate a computation of A_n on a factor $w \in \{a, b\}^*$. We show that each DFA compatible with A'_n requires $f(n)$ states, where f is the function given in Theorem 9, by considering the following general lemma:

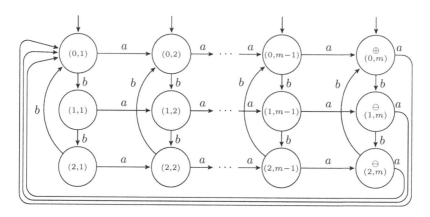

Fig. 6. Automaton A_n of Theorem 9 in the case of n multiple of 3

Lemma 10. *Let Σ be an alphabet and $c \notin \Sigma$ be an extra symbol. Given a dcNFA A over Σ, a nonempty set $K \subseteq (\Sigma \cup \{c\})^*$, two languages $J', J'' \subseteq \Sigma^*$, consider the languages $L' = Kc\mathcal{L}^{\oplus}(A) \cup J'$ and $L'' = Kc\mathcal{L}^{\ominus}(A) \cup J''$. If $L' \cap L'' = \emptyset$ then each DFA accepting a language L, with $L' \subseteq L \subseteq L''^c$ should have at least as many states as a smallest DFA compatible with A.*

Theorem 11. *For each integer n there is an n-state dcNFA with a unique initial state such that the smallest compatible DFA requires $f(n)$ states, where $f(n)$ is the function given in Theorem 9.*

After considering the restriction to dcNFAs having only one initial state, we further restrict to the case of deterministic transitions where, clearly, the bound of Theorem 9 can be reduced. In fact, given an n-state dcDFA A we can just arbitrarily mark each *don't care* state as accepting or rejecting in order to obtain a compatible DFA with the same number of states. Furthermore, if the set of *don't care* states of A is empty and A is minimal then we clearly cannot obtain a smaller compatible DFA. Hence, in the deterministic case n is a tight bound.

We can also observe that if A contains a *don't care* trap state then a compatible DFA can be always obtained by moving each transition leading to the trap state to an arbitrarily chosen state and by arbitrarily choosing final states among the remaining *don't care* states. Hence, the resulting DFA contains $n - 1$ states. For each n this bound cannot be further reduced. Consider in fact the n-state automaton consisting of a loop of $n - 1$ states accepting the language $(ab^{n-2})^*$ and rejecting all the strings in $(ab^{n-2})^*ab^k$ with $0 \leq k < n - 2$, plus a *don't care* trap state. Clearly, each two states on the loop are incompatible. Hence, they belong to different cliques of the compatibility graph. By Theorem 8, we conclude that each compatible DFA should have at least $n - 1$ states.

5 Time Complexity

In this section we shortly study time complexity of the reductions of dcDFAs and dcNFAs to minimal compatible DFAs. In both cases we prove NP-completeness. Our starting point is the following problem, which has been proved to be NP-complete by Pfleeger [13]:

> Given an "incomplete" DFA A and $k > 0$, is there a way to assign a state to each unspecified transition so that the resulting complete automaton has a minimal equivalent DFA with at most k states?

A clarification is necessary to explain the meaning of "incomplete" in this context. As already mentioned in the Section 2, reaching an undefined transition in a DFA is conventionally interpreted as the definitive rejection of the input and, hence, undefined transitions can be made defined by introducing a trap state, which is not final and, so, rejecting. In the above mentioned problem, an undefined transition will never be reached (e.g., because some restrictions on the form of possible input words) and, hence, it represents a *don't care* condition. Hence, an "incomplete" DFA $A = \langle Q, \Sigma, \delta, i, F \rangle$ in the previous problem, can be transformed in a complete dcDFA by adding a trap state q_t which is the only *don't care* state, and by choosing F as the set of accepting states and $Q - F$ as the set of rejecting states. From this discussion we immediately obtain the following result.

Theorem 12. *The problem of deciding if, given a dcDFA A and an integer $k > 0$, there exists a compatible DFA with at most k states is NP-hard.*

We note that the same result could also be deduced using NP-completeness of the inference of a DFA from a finite set of words [6]. We can also easily prove that the problem belongs to NP. However, we can do better, by proving that the problem is in NP even if A is nondeterministic. This allows us to obtain the main result of this section:

Theorem 13. *The problem of deciding if, given a dcNFA A and an integer $k > 0$, there exists a compatible DFA with at most k states is NP-complete.*

Proof. To show that the problem belongs to NP, we observe that in polynomial time it is possible to nondeterministically generate a DFA B with at most k states and verify if it is compatible with A. More into details, compatibility is verified by checking if $\mathcal{L}^{\oplus}(A) \subseteq \mathcal{L}(B)$ and $\mathcal{L}^{\ominus}(A) \subseteq (\mathcal{L}(B))^c$. To do that, from A and B we build a product (nondeterministic) automaton and verify that for each reachable state (p, q), when the component p is an accepting state of A then the component q is a final state of B and when the component p is a rejecting state of A then the component q is a nonfinal state of B.

NP-hardness follows from Theorem 12. □

6   The Unary Case

As well-known, in the unary case (namely the case of languages and automata defined over a one letter alphabet, which in the following we assume to be $\Sigma = \{a\}$) many state bounds are lower than in the general case. In this section we shortly present some considerations in this respect for dcNFAs. First of all, using standard results on diophantine equations, we can prove the following lemma:

Lemma 14. *A unary dcNFA cannot accept a string $a^{m'}$ along a path containing a state belonging to a loop of length ℓ' and reject another string $a^{m''}$ along a path containing a state belonging to a loop of length ℓ'', if ℓ' and ℓ'' are relatively prime.*

The maximal state gap between n-state unary NFAs and equivalent DFAs is $e^{\Theta(\sqrt{n \ln n})}$ [2]. Using Lemma 14, it is possible to show that the same gap cannot be reached starting from unary dcNFAs and compatible DFAs. This happens even in the case of SVFAs. However, it has been shown that the state gap between unary n-state SVFAs and DFAs grows at least as $e^{\Omega(\sqrt[3]{n \ln^2 n})}$ [5]. Hence, there is at least the same gap from dcNFAs to DFAs.

Theorem 15. *For each sufficiently large integer n there is a dcNFA with at most n states such that each compatible DFA requires at least $e^{\Omega(\sqrt[3]{n \ln^2 n})}$.*

References

1. Chen, Y.-F., Farzan, A., Clarke, E.M., Tsay, Y.-K., Wang, B.-Y.: Learning minimal separating dFA's for compositional verification. In: Kowalewski, S., Philippou, A. (eds.) TACAS 2009. LNCS, vol. 5505, pp. 31–45. Springer, Heidelberg (2009)
2. Chrobak, M.: Finite automata and unary languages. Theor. Comput. Sci. 47(3), 149–158 (1986)
3. Damiani, M.: The state reduction of nondeterministic finite-state machines. IEEE Trans. CAD 16(11), 1278–1291 (1997)
4. Eisinger, J., Klaedtke, F.: Don't care words with an application to the automata-based approach for real addition. Formal Methods in System Design 33(1-3), 85–115 (2008), http://dx.doi.org/10.1007/s10703-008-0057-6
5. Geffert, V., Pighizzini, G.: Pairs of complementary unary languages with "balanced" nondeterministic automata. Algorithmica 63(3), 571–587 (2012)
6. Gold, E.M.: Complexity of automaton identification from given data. Inf. Contr. 37(3), 302–320 (1978)
7. Jirásková, G., Pighizzini, G.: Optimal simulation of self-verifying automata by deterministic automata. Inf. Comput. 209(3), 528–535 (2011)
8. Kam, T., Villa, T., Brayton, R., Sangiovanni-Vincentelli, A.: A fully implicit algorithm for exact state minimization. In: Proc. ACM/IEEE Design Automation Conf., pp. 684–690 (1994)
9. Kam, T., Villa, T., Brayton, R., Sangiovanni-Vincentelli, A.: Theory and algorithms for state minimization of nondeterministic FSMs. IEEE Trans. CAD 16(11), 1311–1322 (1997)
10. Moon, J., Moser, L.: On cliques in graphs. Israel J. Math 3, 23–28 (1965)

11. Paull, M.C., Unger, S.H.: Minimizing the number of states in incompletely specified sequential switching functions. IRE Trans. on Elect. Comput. 3, 356–367 (1959)
12. Pena, J.M., Oliveira, A.L.: A new algorithm for exact reduction of incompletely specified finite state machines. IEEE Trans. CAD 18(11), 1619–1632 (1999)
13. Pfleeger, C.P.: State reduction in incompletely specified finite-state machines. IEEE Trans. Comput. 22(C), 1099–1102 (1973)
14. Rho, J.K., Hachtel, G., Somenzi, F., Jacoby, R.: Exact and heuristic algorithms for the minimization of incompletely specified state machines. IEEE Trans. CAD 13, 167–177 (1994)

Quantum Pushdown Automata
with a Garbage Tape

Masaki Nakanishi

Faculty of Education, Art and Science, Yamagata University,
Yamagata, 990–8560, Japan
masaki@cs.e.yamagata-u.ac.jp

Abstract. Several kinds of quantum pushdown automaton models have
been proposed, and their computational power is investigated intensively. However, for some quantum pushdown automaton models, it is
not known whether quantum models are at least as powerful as classical
counterparts or not. This is due to the reversibility restriction. In this paper, we introduce a new quantum pushdown automaton model that has
a garbage tape. This model can overcome the reversibility restriction by
exploiting the garbage tape to store popped symbols. We show that the
proposed model can simulate any quantum pushdown automaton with a
classical stack as well as any probabilistic pushdown automaton. We also
show that our model can solve a certain promise problem exactly while
deterministic pushdown automata cannot. These results imply that our
model is strictly more powerful than classical counterparts in the setting
of exact, one-sided error and non-deterministic computation.

Keywords: quantum pushdown automata, deterministic pushdown automata, quantum computation models.

1 Introduction

One important question in quantum computing is whether a computational gap
exists between models that are allowed to use quantum effects and models that
are not. Several types of quantum computation models have been proposed,
including quantum finite automata, quantum counter automata, and quantum
pushdown automata. Quantum finite automata are the simplest model of quantum computation, and have been investigated intensively[3–5, 7, 8, 13–15, 17, 22–
24, 26–28]. Several quantum automata augmented with additional computational
resources have also been proposed, including quantum counter automata and
quantum pushdown automata [6, 12, 16–18, 20, 23, 29, 30].

It might be a surprising result that some of simple quantum computation
models can be less powerful than classical counterparts[15, 29, 30] due to the
reversibility restriction. Thus, it is a natural question what kinds of restrictions make quantum models less powerful than classical counterparts, and what
kinds of computational resources make quantum models more powerful. Motivated by those questions, quantum pushdown automata have been investigated.

G.F. Italiano et al. (Eds.): SOFSEM 2015, LNCS 8939, pp. 352–363, 2015.

Quantum pushdown automata were first proposed in [17], but their model is the generalized quantum pushdown automata whose evolution does not have to be unitary. Then Golovkins proposed quantum pushdown automata including unitarity criteria[12], and he showed that quantum pushdown automata can recognize every regular language and some non-context-free languages. However, it is still open whether Golovkins's model of quantum pushdown automata are more powerful than probabilistic pushdown automata or not. In [18], it is shown that a certain promise problem can be solved exactly by Golovkins's model of quantum pushdown automata while it cannot be solved by deterministic pushdown automata. However, it is not known whether Golovkins's model can simulate any deterministic pushdown automaton or not. This is because quantum computation models must be reversible while pop operation deletes the stack-top symbol, which is not a reversible operation. In [20], a quantum pushdown automaton model that has a classical stack is proposed, and it is shown that the model is strictly more powerful than classical counterparts in the setting of one-sided error as well as non-deterministic computation.

The above mentioned results are for the models whose state transitions are described by unitary operators. It is known that by allowing more general operators such as trace preserving completely positive (TPCP) maps, quantum finite automata can simulate classical counterparts as well as several quantum finite automata mentioned above[13, 14]. These results were generalized and it was shown how to define general quantum operators for other models in [28]. For counter automata and pushdown automata, it is also known that generalized quantum models (i.e., the models that can use TPCP maps) can simulate classical counterparts[23, 24].

In this paper, we focus on the restricted quantum computation models (i.e., the models whose state transitions are described by unitary operators) rather than the general models (i.e., the models whose state transitions are described by TPCP maps). As mentioned above, it is known that the generalized quantum computation models can simulate classical counterparts and sometimes can be strictly more powerful than classical counterparts. Nevertheless, studying restricted models is important. That is, our goal is to investigate what kinds of restriction makes quantum models less powerful and under what kinds of restrictions quantum models are still more powerful than classical counterparts. This could lead to understand the source of the power of quantum computation in architecturally restricted models such as quantum automata.

Motivated by these discussions, we introduce a new model of quantum pushdown automata, called quantum pushdown automata with a garbage tape. This model has a garbage tape on which popped symbols are stored, and thus, we can pop the stack-top symbol preserving reversibility. The garbage tape is a write-only memory, and thus, classical pushdown automata cannot exploit it. A quantum computation model that has a write-only memory was proposed in [25]. The model uses a write-only memory in order to control interference between distinct computation paths. In our model, the write-only garbage tape is restricted to store popped symbols. Also the similar notion of garbage tapes were proposed in [8, 22]. In those models, a garbage tape is used to make transitions reversible. Our model is constructed so as to take advantages of both of a write-only tape and a garbage tape.

Another motivation is that it is expected that investigating quantum pushdown automata reveals how last-in first-out manner of memory access affects (or limits) quantum computation. However, for this purpose, Golovkins's model[12] is too restrictive on pop operation, i.e., we can pop a stack-top symbol only if we can delete stack-top symbol preserving reversibility. Thus, we cannot identify from which the impossibilities come from, reversibility or last-in first-out manner of memory access. In contrast, our model is useful for this purpose since pop operations can always be executed preserving reversibility.

In this paper, we show that the proposed model can simulate any quantum pushdown automaton with a classical stack, which is proposed in [20], as well as any classical pushdown automaton. It is known that quantum pushdown automata with a classical stack are strictly more powerful than classical counterparts in the setting of one-sided error and non-deterministic computation[20]. Thus, so is our model. We also show that our model can solve a certain promise problem exactly while deterministic pushdown automata cannot. This implies that our model is strictly more powerful than classical counterparts also in the setting of exact computation. It is a common technique to apply the pumping lemma (or Ogden's lemma[21], which is a generalization of the pumping lemma) in order to show that a language is not context-free, i.e., pushdown automata cannot recognize the language. However, our problem is a promise problem. Thus, we cannot apply the pumping lemma to our case.[1] In [2], the pumping lemma is proved through the analysis of pushdown automata. We modify their notion of *full state*, and use it to show the impossibility by directly analyzing time evolution of pushdown automata. This is a new technique to prove that a certain promise problem cannot be computed by pushdown automata.

2   Preliminaries

A quantum pushdown automaton with a garbage tape (QPAG) has an input tape, a stack and a garbage tape. A QPAG also has a finite state control. The input tape contains a classical input string, and its tape head is implemented by qubits that represent the position of the tape head. The stack and the garbage tape are implemented by qubits that represent contents of the stack and the garbage tape, respectively. The finite state control is also implemented by qubits that represents the current state. A QPAG reads the stack top symbol and the input symbol pointed by the input tape-head, and then evolves as follows: The tape head can move to the right or stay at the same position, the finite state control moves to the next state, and a stack symbol is pushed to the stack or

---

[1] As far as the author knows, [18] is the only exception in which the pumping lemma is used for a promise problem. The technique in [18] can be applied only to the limited cases. For OBDD models, an impossibility proof for a partial function, which is a function counterpart of promise problems, was shown recently in [1].

popped from the stack. When we pop a symbol from the stack, the popped symbol is written on the garbage tape, moving the garbage tape head to the right. This allows a QPAG to pop the stack top symbol preserving reversibility. We define QPAGs formally as follows.

Definition 1. *A quantum pushdown automaton with a garbage tape (QPAG) is defined as the following 7-tuple: $M = (Q, \Sigma, \Gamma, \delta, q_0, Q_{acc}, Q_{rej})$, where Q is a set of states, Σ is a set of input symbols including the left and the right endmarkers $\{\cent, \$\}$, respectively, Γ is a set of stack symbols including the bottom symbol Z, δ is a quantum state transition function $(\delta : (Q \times \Sigma \times \Gamma \times Q \times G \cup \{\varepsilon, pop\} \times \{0, 1\}) \longrightarrow \mathbb{C})$, where $G(\subseteq (\Gamma \backslash \{Z\})^+)$ is a finite set and $(\Gamma \backslash \{Z\})^+$ is the set of all nonempty strings of finite length from alphabet $\Gamma \backslash \{Z\}$, q_0 is the initial state, $Q_{acc} (\subseteq Q)$ is the set of accepting states, and $Q_{rej} (\subseteq Q)$ is the set of rejecting states, where $Q_{acc} \cap Q_{rej} = \emptyset$.* $\square$

$\delta(q, a, b, q', b', D) = \alpha$ means that the amplitude of the transition from q to q' updating the input tape head to D ($D = 1$ means 'right' and $D = 0$ means 'stay') and pushing b' to the stack (or popping the stack-top symbol if $b' = pop$) is α when reading input symbol a and stack symbol b. A configuration of a QPAG is (q, k, w_s, w_g), where $q \in Q$ is the current state of the finite state control, k is the position of the input tape head, and w_s and w_g are the strings on the stack and the garbage tape, respectively. We store a configuration of a QPAG in a quantum register, where a basis state is described as $|q, k, w_s, w_g\rangle$. For input string x, we define the time evolution operator U^x as follows:

$$U^x(|q, k, w_s b, w_g\rangle)$$
$$= \sum_{q' \in Q, b' \in G \cup \{\varepsilon, pop\}, D \in \{0,1\}} \delta(q, x(k), b, q', b', D) |q', k + D, w'_s, w'_g\rangle,$$

where $x(k)$ is the k-th input symbol of input x,

$$w'_s = \begin{cases} w_s bb' & \text{if } b' \neq pop \\ w_s & \text{if } b' = pop \end{cases}$$

$$\text{and } w'_g = \begin{cases} w_g & \text{if } b' \neq pop \\ w_g b & \text{if } b' = pop \end{cases} \text{ (b is the popped stack-top symbol).}$$

If U^x is unitary (for any input string x), then the corresponding QPAG is called well-formed. A well-formed QPAG is considered valid in terms of the quantum theory. We consider only well-formed QPAGs. For so-called well-formedness conditions, readers may refer to [19]. Let the initial quantum state and the initial position of the input tape head be q_0 and '0', respectively. We define $|\psi_0\rangle$ as $|\psi_0\rangle = |q_0, 0, Z, \varepsilon\rangle$. We also define E_{non}, E_{acc} and E_{rej} as follows:

$$E_{non} = span\{|q, k, w_s, w_g\rangle \,|\, q \notin Q_{acc} \text{ and } q \notin Q_{rej}\},$$
$$E_{acc} = span\{|q, k, w_s, w_g\rangle \,|\, q \in Q_{acc}\}, \quad E_{rej} = span\{|q, k, w_s, w_g\rangle \,|\, q \in Q_{rej}\}.$$

We define observable $\mathcal{O}$ as $\mathcal{O} = E_{non} \oplus E_{acc} \oplus E_{rej}$. For notational simplicity, we define the outcome of a measurement corresponding to E_j as j for $j \in \{non, acc, rej\}$. A QPAG computation proceeds as follows:

(a) U^x is applied to $|\psi_i\rangle$, and we obtain $|\psi_{i+1}\rangle = U^x |\psi_i\rangle$.

(b) $|\psi_{i+1}\rangle$ is measured with respect to $\mathcal{O}$. Let $|\phi_j\rangle$ be the projection of $|\psi_{i+1}\rangle$ to E_j. Then each outcome j is obtained with probability $|\,|\phi_j\rangle\,|^2$. Note that this measurement causes $|\psi_{i+1}\rangle$ to collapse to $\frac{1}{||\phi_j\rangle|}|\phi_j\rangle$, where j is the obtained outcome.

(c) If the outcome of the measurement is acc or rej, the automaton outputs the measurement result and halts. Otherwise, go to (a).

3   Simulation of QCPDAs

In this section, we show that a QPAG can simulate a quantum pushdown automaton with a classical stack (QCPDA). Since QCPDAs can simulate any probabilistic pushdown automata[20], QPAGs can simulate any probabilistic pushdown automata as well. For the definition of QCPDAs, readers may refer to [19], which is the technical report version of this paper, or may refer to [20]. A quantum pushdown automaton with a classical stack(QCPDA) is a quantum pushdown automaton whose classical stack operations are determined by measurement results. We can use the garbage tape so that if we measure the garbage tape, the stack contents will be identical among all the basis states contained in the resulting superposition. Therefore, we can simulate a QCPDA by a QPAG.

Theorem 1. *Let $M_{qc} = (Q, \Sigma, \Gamma, \delta, q_0, \sigma, Q_{acc}, Q_{rej})$ be a QCPDA. Then, there exists a QPAG M_q such that for any input, the acceptance probability of M_q is the same as that of M_{qc}.*

Proof. For a transition of M_{qc} from state q to q' moving the input tape head to D, we construct the corresponding transitions of M_q, which consist of three successive transitions, as follows: Note that the stack operation of M_{qc} is determined solely by the state q' to which it transits, denoted by $\sigma(q')$. We add two new states q_a and q_b to Q and also add $\sigma(q')$ to Γ. Then, we replace the original transition with the transition from q to q_a such that the stack operation is the same as the original transition ($\sigma(q')$), the direction of the tape head is D and the transition probability is also the same. We define the transition from q_a to q_b, whose probability is one, as a transition pushing the label $\sigma(q')$ to the stack, the input tape head staying at the same position. We also define the transition from q_b to q', whose probability is one, as a transition popping $\sigma(q')$ from the stack and moving it to the garbage tape, the input tape head staying at the same position. This records the history of stack operations in the garbage tape. Thus, if the history of stack operations are different between two computation paths, they do not interfere with each other since the contents of the garbage tape are different. This means that if we measure the garbage tape, the contents of the stack are identical between any basis states contained in the resulting superposition at any moment of computation. In other words, if we trace out the garbage tape, then, the stack configuration is not in a superposition but in a classical mixture of basis states. Thus, it can be regarded as a classical stack, and the resulting QPAG M_q simulates the original QCPDA M_{qc}.   □

It is known that QCPDAs can recognize a certain non-context-free language with one-sided error[20]. This means that QPAGs are strictly more powerful than classical pushdown automata in the setting of one-sided error as well as non-deterministic computation.

Corollary 1. *The class of languages recognized by one-sided error QPAGs properly includes the class of languages recognized by one-sided error probabilistic pushdown automaton as well as by non-deterministic pushdown automaton.* □

4 Possibility and Impossibility of Solving a Certain Promise Problem

We say that two strings, u and v, have even (resp. odd) distinctions, denoted by $u \overset{e}{\sim} v$ (resp. $u \overset{o}{\sim} v$), if $|u| = |v|$ and u and v are different at even (resp. odd) number of positions. For example, $1100 \overset{e}{\sim} 1111$ since the third and the fourth bits are different between the two strings, and $1000 \overset{o}{\sim} 1111$ since the second, the third and the fourth bits are different between the two strings. We define a promise problem, Problem I, as follows:

Problem I

- *An input is of the form $w_1 \# w_2 \# w_3$, where $w_1, w_2 \in \{a, b, c\}^n$ and $w_3 \in \{a, b, c, d\}^n$.*
- *Problem: Compute $w_1 \overset{e}{\sim} w_2^R$ xor $w_1 \overset{e}{\sim} w_3^R$.*   □

We show that QPAGs can solve Problem I exactly while deterministic pushdown automata cannot solve it. This result combined with Theorem 1 implies that QPAGs are strictly more powerful than classical pushdown automata in the setting of exact computation.

Theorem 2. *There exists a QPAG that solves Problem I exactly.*

Proof. We use the same technique as in Theorem 3.1 of [18]. We construct a QPAG, M, that solves Problem I as follows: We consider two sub-automata, M_1 and M_2, such that M_1 (resp. M_2) computes whether $w_1 \overset{e}{\sim} w_2^R$ (resp. $w_1 \overset{e}{\sim} w_3^R$), and run them in a superposition. It is straightforward to see that M_1 and M_2 can be implemented by reversible deterministic pushdown automata with a garbage tape, which is a special case of QPAGs, and we can construct M_1 and M_2 so that the contents of the garbage tape at the moment of reading the right-endmarker can be the same between the two sub-automata. Then, we utilize the algorithm in [9] (the improved Deutsch-Jozsa algorithm[11]) to compute the exclusive-or exactly using the two sub-automata as the oracle for Deutsch's problem[10].   □

In the following, we show that no deterministic pushdown automata can solve Problem I.

Theorem 3. *No deterministic pushdown automata can solve Problem I.*   □

We introduce several lemmas in order to prove Theorem 3. We divide w_1 into two segments $w_1 = w_{1L}w_{1R}$. Similarly, we divide w_2 and w_3 as $w_2 = w_{2L}w_{2R}$ and $w_3 = w_{3L}w_{3R}$, respectively. In the following discussion, we assume that there exists a deterministic pushdown automaton that solves Problem I. Let $h_{max}(k)$ be the maximum height of the stack over all w_1's at the moment of reading the k-th symbol of w_1. Note that stack height can increase at most $O(1)$ when reading each symbol[2]. Then, it is obvious that there is a constant, c, for which the following holds:

$$h_{max}(\frac{n}{c}) < \log_{|\Gamma|}\left(3^{\frac{c-1}{c}n}/(\#states \cdot n(n+1))\right),$$

where $\#states$ denotes the number of states of the finite state control, and $n = |w_1|$. We fix such a constant c, and also fix the length of w_{1L} to be n/c.

We say that pushdown automaton M is in a *state-configuration* of (q, a) if M is in the state q and the stack-top symbol is a. In other words, a state-configuration is a configuration of a pushdown automaton ignoring the position of the tape head and the stack contents except for the stack-top. The notion of a state-configuration is a modification of the notion of *full state* in [2]. Note that the tape head can be stationary at a transition. Thus, the stack height can increase (or decrease) multiple times with multiple transitions during reading one symbol. Let $h_b(i)$ and $c_b(i)$ denote the stack height and the state-configuration, respectively, immediately before reading the i-th symbol of the input. Also let $h_a(i)$ denote the set of stack heights that consists of the stack height after reading the i-th symbol and the stack heights during reading the $(i+1)$-th symbol with the tape head being stationary on the $(i+1)$-th symbol. For each $h \in h_a(i)$, let $c_a(i, h)$ be the corresponding state-configuration. We define the notations "$h_b(i) > h_a(j)$" and "$h_b(i) - h_a(j)$" as follows: $h_b(i) > h_a(j)$ iff $h_b(i) > min_{h' \in h_a(j)}h'$. $h_b(i) - h_a(j) = h_b(i) - min_{h' \in h_a(j)}h'$. A *zero-stack* pair is a pair (l, r) $(1 \le l < r \le n)$ such that $h_b(l) \in h_a(r)$ and $h_b(l) \not> h_a(t)$ for any t $(l < t < r)$. Then, we have the following lemma.

Lemma 1. *We fix w_1 arbitrarily. Let (i, j) be a zero-stack pair such that the maximum of $h_b(k) - h_a(l)$ for $i < k < l < j$ is $\omega(1)$. Then, for any zero-stack pair (i', j') $(1 \le i' < j' < i$ or $j < i' < j' \le n)$, the maximum of $h_b(k) - h_a(l)$ for $i' < k < l < j'$ is $O(1)$.*

Proof. We consider a zero-stack pair (i, j) $(1 \le i < j \le n)$ such that the maximum of $h_b(k) - h_a(l)$ for $i < k < l < j$ is $\omega(1)$. Let the maximum (resp. minimum) height of the stack during processing from the i-th symbol to the j-th symbol be h_{max} (resp. h_{min}). Note that $h_{max} - h_{min} > \omega(1)$. For each $h \in \{h_{min}, \ldots, h_{max}\}$, let ZS_h be the set of zero-stack pairs such that $ZS_h = \{(l, r)|h_b(l) = h, i \le l < r \le j\}$. Note that for at least a constant fraction of $\{h_{min}, \ldots, h_{max}\}$, ZS_h is nonempty. For each $h \in \{h_{min}, \ldots, h_{max}\}$, we choose at most one $(l_h, r_h) \in ZS_h$ such that $l_h < l_{h+1}$ and $r_{h+1} \le r_h$. It

---
[2] Note that, on the other hand, stack height may decrease more than $\omega(1)$ when reading each input symbol.

is obvious that we can have such (l_h, r_h)'s for at least a constant fraction of $\{h_{min}, \ldots, h_{max}\}$. Let $(c_a(k,h), t)$ be a pair of a state-configuration and an input symbol where $t \in \Sigma$ is the input symbol pointed by the tape head at the moment when the automaton is in the state-configuration $c_a(k, h)$ with the k and h. Then, there exists two distinct zero-stack pairs (l_{h_1}, r_{h_1}) and (l_{h_2}, r_{h_2}) $(h_1 < h_2)$ such that $c_b(l_{h_1}) = c_b(l_{h_2})$ and $(c_a(r_{h_1}, h_1), t) = (c_a(r_{h_2}, h_2), t)$ for some t since $|\Sigma|$ and the number of possible state-configurations are both $O(1)$ while we have $\omega(1)$ pairs of (l_h, r_h)'s. We divide w_1 as $w_1 = uvxyz$ where $u = w_1(1) \cdots w_1(l_{h_1} - 1)$, $v = w_1(l_{h_1}) \cdots w_1(l_{h_2} - 1)$, $x = w_1(l_{h_2}) \cdots w_1(r_{h_2})$, $y = w_1(r_{h_2} + 1) \cdots w_1(r_{h_1})$, and $z = w_1(r_{h_1} + 1) \cdots w_1(n)$, where $w_1(i)$ denotes the i-th symbol of w_1. Then, for any $i \geq 0$, the configuration after reading $uv^i xy^i z$ and the configuration after reading $uvxyz$ are the same, including the contents of the stack.

We assume that there exists two zero-stack pairs (i_1, j_1) and (i_2, j_2) $(1 \leq i_1 < j_1 < i_2 < j_2 \leq n)$ such that the maximum of $h_b(k) - h_a(l)$ for $i_1 < k < l < j_1$ and the maximum for $i_2 < k < l < j_2$ are both $\omega(1)$. Then, we can divide w_1 in two ways: $w_1 = u_k v_k x_k y_k z_k$ with (i_k, j_k) $(k \in \{1, 2\})$. It is obvious that there exist p and q such that $|u_1 v_1^p x_1 y_1^p z_1| = |u_2 v_2^q x_2 y_2^q z_2|$. Thus, there exist two inputs $u_1 v_1^p x_1 y_1^p z_1$ and $u_2 v_2^q x_2 y_2^q z_2$ for which the configurations after reading the two inputs are the same, including the contents of the stack. This implies that for any completion of the inputs, both of $u_1 v_1^p x_1 y_1^p z_1$ and $u_2 v_2^q x_2 y_2^q z_2$ leads to the same answer, which is a contradiction. □

Let w_{pref} be a string for which there is a zero-stack pair (i, j) and the maximum of $h_b(k) - h_a(l)$ for $i < k < l < j$ is $\omega(1)$ where $|w_{pref}| = c|w_{1L}|$ for some constant c $(0 < c < 1)$. If there is no such zero-stack pair for any long enough w_{pref}, we define w_{pref} to be an empty string. We fix such a w_{pref}. We define $a^+ = b, b^+ = c, c^+ = a$. For two strings $u, v \in \{a, b, c\}^n$, we say $u \leq v$ iff $[(u(k) = x) \longrightarrow (v(k) = x$ or $v(k) = x^+)]$, where $x \in \{a, b, c\}$ and $u(k)$ (resp. $v(k)$) represents the k-th symbol of u (resp. v). Let WL_{all} be the set of w_{1L}'s such that $WL_{all} = \{w_{pref} a^{|w_{1L}| - |w_{pref}| - k} b^k | 0 \leq k \leq |w_{1L}| - |w_{pref}| - 1\}$ $(= \{w_{pref} aaa \ldots aaa, w_{pref} aaa \ldots aab, w_{pref} aaa \ldots abb, w_{pref} aaa \ldots bbb, \ldots, w_{pref} abb \ldots bbb\})$. Note that for any two distinct $u, v \in WL$, $u \leq v$ or $v \leq u$. Then, we have the following lemma.

Lemma 2. *There exists $WL \subseteq WL_{all}$ satisfying the following conditions: (1) Any $w \in WL$ leads to the same state-configuration, say C_{WL}. (2) Given a constant c, after reading w, the stack contents between the top and the c-th from the top are the same among all $w \in WL$. (3) $|WL| = \Theta(n)$.* □

Proof. There exists a constant fraction of WL_{all}, which is WL, satisfying the first and the second conditions of the lemma since the number of possible state-configurations is a constant and the number of possible stack contents between the top and the c-th from the top is also a constant. It is obvious that $|WL| = \Theta(n)$ since $|WL_{all}| = |w_{1L}| - |w_{pref}| = \Theta(n)$. □

We consider the case that the following Condition I holds:

Condition I. *There exists $w_{1L} \in WL$ and w_{2L} such that for at least $1/(n+1)$ fraction of $\{w_{1R}\}$, stack height is less than $\log_{|\Gamma|}(3^{n-|w_{1L}|}/(\#states \cdot n(n+1)))$ at the moment of reading the last symbol of $w_{1L}w_{1R}w_{2L}$.* ☐

In this case, at the moment when stack height is less than $\log_{|\Gamma|}(3^{n-|w_{1L}|}/(\#states \cdot n(n+1)))$, the number of possible configurations (including stack contents and the position of the input tape head) is less than $\frac{1}{n+1}3^{n-|w_{1L}|}$, which means there exist at least two distinct partial inputs $w_{1L}w_{1R}w_{2L}$ and $w_{1L}w'_{1R}w_{2L}$ that result in the same configuration (including stack contents and the position of an input tape head) since $|\{w_{1R}\}| = 3^{n-|w_{1L}|}$. Thus both of $w_{1L}w_{1R}w_{2L}$ and $w_{1L}w'_{1R}w_{2L}$ lead to the same answer for any completion of the rest of the input. This is a contradiction. Thus, we can say the negation of Condition I holds. In this case, given w_2, at every step of processing w_2, for at most $1/(n+1)$ fraction of $\{w_{1R}\}$, stack height becomes less than $\log_{|\Gamma|}(3^{n-|w_{1L}|}/(\#states \cdot n(n+1)))$. Thus, for at most $n/(n+1)$ fraction of $\{w_{1R}\}$, stack height becomes less than $\log_{|\Gamma|}(3^{n-|w_{1L}|}/(\#states \cdot n(n+1)))$ while processing w_2; for at least $1/(n+1)$ fraction of $\{w_{1R}\}$, stack height is always more than or equal to $\log_{|\Gamma|}(3^{n-|w_{1L}|}/(\#states \cdot n(n+1)))$ while processing w_2. We consider the case that the following Condition II holds:

Condition II. *For any $w_{1L} \in WL$ and w_2, at least $1/(n+1)$ fraction of $\{w_{1R}\}$, stack height is always greater than or equal to $\log_{|\Gamma|}(3^{n-|w_{1L}|}/(\#states \cdot n(n+1)))$ while processing w_2.* ☐

We define w_{3L} as the prefix of w_3 such that stack height is always higher than $\hat{h} - O(1) (= \hat{h}')$ during reading $w_{1R}w_2w_{3L}$ and it becomes $\hat{h}'$ when reading the last symbol of w_{3L}, where $\hat{h}$ denotes the stack height after reading the last symbol of w_{1L}. If stack height is always higher than $\hat{h}'$ during reading w_3, we define $w_{3L} = w_3$.

Lemma 3. *We assume that there exists a deterministic pushdown automaton that solves Problem I. Then, there exist w_{1R}, k $(1 \le k \le n)$ and a set W_2 of w_2's such that, starting from C_{WL}, $w_{1R}w_2w_{3L}$ leads to the same state-configuration for all $w_2 \in W_2$ where $w_{3L} = d^k$, stack height is always greater than or equal to $\hat{h} - O(1)$, and $|W_2| = \Omega(\frac{1}{n^2}3^n)$, where C_{WL} is as in Lemma 2.*

Proof. Note that for each w_2, there are more than $\frac{1}{n+1}3^{|w_{1R}|}$ of w_{1R}'s for which stack height is always greater than or equal to $\log_{|\Gamma|}(3^{n-|w_{1L}|}/(\#states \cdot n(n+1)))$ while processing w_2 by Condition II. This means that for some w_{1R}, there are $\Omega(\frac{1}{n}3^n)$ of w_2's for which stack height is always greater than or equal to $\log_{|\Gamma|}(3^{n-|w_{1L}|}/(\#states \cdot n(n+1)))$ while processing w_2. By Lemma 1 and the fact that $\hat{h} < \log_{|\Gamma|}(3^{|w_{1R}|}/(\#states \cdot n(n+1)))$, the lemma follows immediately. ☐

We fix w_{1R}, k and W_2 as those in Lemma 3 in the following. For WL in Lemma 2, the following lemma holds.

Lemma 4. *We assume that there exists a deterministic pushdown automaton that solves Problem I. For w_{1R}, k and W_2 in Lemma 3, there exist $w_{1L} \in WL$ and two distinct partial inputs $w_{1L}w_{1R}w_2w_{3L}$ and $w_{1L}w_{1R}w_2'w_{3L}$ ($w_2, w_2' \in W_2$ and $w_{3L} = d^k$) such that $w_1 \overset{e}{\sim} w_2^R$ and $w_1 \overset{o}{\sim} w_2'^R$.*

Proof. Let $WL = \{w_{1L}^1, w_{1L}^2, \ldots, w_{1L}^m\}$ where $w_{1L}^i \le w_{1L}^{i+1}$. $W_{2,even}^1$ denotes the set of $w_2 \in \{a, b, c\}^n$ such that $w_{1L}^1 w_{1R} \overset{e}{\sim} w_2^R$. Also $W_{2,even}^2$ denotes the set of $w_2 \in W_2^1$ such that $w_{1L}^2 w_{1R} \overset{e}{\sim} w_2^R$. Similarly, $W_{2,even}^i$ denotes the set of $w_2 \in W_{2,even}^{i-1}$ such that $w_{1L}^i w_{1R} \overset{e}{\sim} w_2^R$. In other words, for all $w_2 \in W_{2,even}^i$ and $j \le i$, $w_{1L}^j w_{1R} \overset{e}{\sim} w_2^R$. We show that $|W_{2,even}^i| \le c|W_{2,even}^{i-1}|$ for some constant $c < 1$ in the following. We consider the positions at which w_{1L}^{i-1} and w_{1L}^i differ. We define the set of such positions to be D^i. Note that $w_{1L}^{i-1}(k) = a$ and $w_{1L}^i(k) = b$ for $k \in D^i$, where $w(k)$ represents the k-th symbol of w. We define $S = \{w_2 \in W_{2,even}^{i-1} | \exists k \in D^i \ w_2^R(k) = b \text{ or } w_2^R(k) = c.\}$, where $w_2(k)$ denotes the k-th symbol of w_2. It is obvious that $|S| \ge c_1|W_{2,even}^{i-1}|$ for some constant $c_1 < 1$. For $w_2 \in S$, let l be the largest position in D^i such that $w_2^R(l) = b$ or $w_2^R(l) = c$. We assume that $w_2^R(l) = b$ without loss of generality. We consider w_2' such that $w_2'^R(i) = w_2^R(i)$ for $i \ne l$ and $w_2'^R(i) = c$. It is obvious that w_2' is also in S. Then, either $w_{1L}^i w_{1R} \overset{o}{\sim} w_2^R$ or $w_{1L}^i w_{1R} \overset{o}{\sim} w_2'^R$ holds. This implies that a half of elements in S cannot belong to $W_{2,even}^i$. Thus, $|W_{2,even}^i| \le |W_{2,even}^{i-1}| - \frac{c_1}{2}|W_{2,even}^{i-1}| = c_2|W_{2,even}^{i-1}|$, where $c_2 = 1 - \frac{c_1}{2}$. Similar to $W_{2,even}^1$, we define $W_{2,odd}^1$, and then, similarly, it can be shown that $|W_{2,odd}^i| \le c|W_{2,odd}^{i-1}|$ for some constant $c < 1$. Therefore, $|W_{2,even}^i|$ and $|W_{2,odd}^i|$ can be smaller than $|W_2|$ for $i \in \Theta(n)$. The lemma follows. □

(Proof of Theorem 3)
We assume that there exists a classical deterministic pushdown automaton that solves Problem I. Then, by Lemma 4, we have two input strings, $w_a = w_{1L}w_{1R} w_2w_{3L}$ and $w_b = w_{1L}w_{1R}w_2'w_{3L}$ ($w_2, w_2' \in W_2$ and $w_{3L} = d^k$), such that $w_1 \overset{o}{\sim} w_2^R$ and $w_1 \overset{e}{\sim} w_2'^R$. We fix $w_{3R} = d^{n-k}$. Then, the answer only depends on the number of distinctions between w_1 and w_2^R (or $w_2'^R$). Thus, one is YES and the other is NO for w_a and w_b. However, the configurations (including the contents of the stack and the position of the input tape head) at the moment of reading the last symbol of w_{3L} are the same between w_a and w_b if $k \ne n$. On the other hand, if $k = n$, the state-configuration at the moment of reading the last symbol of w_a and w_b are the same. Thus, both of w_a and w_b lead to the same answer. This is a contradiction. □

5   Comparison between Quantum Pushdown Automata with and without a Garbage Tape–Concluding Remarks

In this paper, we showed that QPAGs are strictly more powerful than classical pushdown automata in the setting of exact, one-sided error and nondeterministic computation. In this section, we discuss comparison between quantum pushdown automata with and without a garbage tape. Our conjecture is that Problem I cannot be solved exactly by quantum pushdown automata without a garbage tape, which is Golovkins's model[12], since it seems to be impossible to compute $w_1 \overset{e}{\sim} w_2^R$ or $w_1 \overset{e}{\sim} w_3^R$ without a garbage tape. On the other hand, in the QPAG model, popped symbols are always stored in the garbage tape. Thus, if the contents of the garbage tape are different between two computation paths, they no longer interfere with each other. In other words, only the two computation paths that have the same contents in the garbage tape can interfere with each other. This might make the QPAG model less powerful than Golovkins's model. Therefore, we conjecture that the class of languages recognized by the two models are incomparable. We also conjecture that even the generalized quantum pushdown automata without a garbage tape constructed by the technique in [28] cannot solve Problem I. At least, the generalized model of quantum pushdown automata without a garbage tape cannot execute the algorithm in Theorem 2. This is because, although the garbage tape is in a superposition in the middle of the computation of the algorithm, the generalized quantum pushdown automaton cannot represent such a superposition without a garbage tape. Thus, our model and the generalized model without a garbage tape might also be incomparable.

Acknowledgments. This work was supported by JSPS KAKENHI Grant No. 24500003 and No. 24106009.

References

1. Ablayev, F., Gainutdinova, A., Khadiev, K., Yakaryılmaz, A.: Very narrow quantum OBDDs and width hierarchies for classical OBDDs. In: Proceedings of 16th International Workshop on Descriptional Complexity of Formal Systems (DCFS'14). pp. 53–64 (2014)
2. Amarilli, A., Jeanmougin, M.: A proof of the pumping lemma for context-free languages through pushdown automata, coRR, abs/1207.2819 (2012)
3. Ambainis, A., Freivalds, R.: 1-way quantum finite automata: strengths, weakness and generalizations. In: Proceedings of the 29th Symposium on Foundations of Computer Science (FOCS'98). pp. 332–341 (1998)
4. Ambainis, A., Watrous, J.: Two-way finite automata with quantum and classical states. Theoretical Computer Science 287(1), 299–311 (2002)
5. Ambainis, A., Yakaryılmaz, A.: Superiority of exact quantum automata for promise problems. Information Processing Letters 112(7), 289–291 (2012)
6. Bonner, R., Freivalds, R., Kravtsev, M.: Quantum versus probabilistic one-way finite automata with counter. In: Proceedings of the 28th Conference on Current Trends in Theory and Practice of Informatics (SOFSEM2001). pp. 181–190 (2001)
7. Brodsky, A., Pippenger, N.: Characterizations of 1-way quantum finite automata. SIAM Journal on Computing 31(5), 1456–1478 (2002)
8. Ciamarra, M.P.: Quantum reversibility and a new model of quantum automaton. In: Proceedings of the 13th International Symposium on Fundamentals of Computation Theory (FCT'01). pp. 376–379 (2001)
9. Cleve, R., Ekert, A., Macchiavello, C., Mosca, M.: Quantum algorithms revisited. Proceedings of the Royal Society A 454, 339–354 (1998)

10. Deutsch, D.: The Church-Turing principle and the universal quantum computer. Proceedings of the Royal Society A 400, 97–117 (1985)
11. Deutsch, D., Jozsa, R.: Rapid solution of problem by quantum computation. Proceedings of the Royal Society A 439, 553–558 (1992)
12. Golovkins, M.: Quantum pushdown automata. In: Proceedings of 27th Conference on Current Trends in Theory and Practice of Informatics (SOFSEM2000). pp. 336–346 (2000)
13. Hirvensalo, M.: Various aspects of finite quantum automata. In: Proceedings of Developments in Language Theory 2008 (DLT2008). pp. 21–33 (2008)
14. Hirvensalo, M.: Quantum automata with open time evolution. International Journal of Natural Computing Research (IJNCR) 1(1), 70–85 (2010)
15. Kondacs, A., Watrous, J.: On the power of quantum finite state automata. In: Proceedings of the 38th Symposium on Foundations of Computer Science (FOCS'97). pp. 66–75 (1997)
16. Kravtsev, M.: Quantum finite one-counter automata. In: Proceedings of 26th Conference on Current Trends in Theory and Practice of Informatics (SOFSEM1999). pp. 432–442 (1999)
17. Moore, C., Crutchfield, J.P.: Quantum automata and quantum grammars. Theoretical Computer Science 237(1–2), 275–306 (2000)
18. Murakami, Y., Nakanishi, M., Yamashita, S., Watanabe, K.: Quantum versus classical pushdown automata in exact computation. IPSJ Journal 46(10), 2471–2480 (2005)
19. Nakanishi, M.: Quantum pushdown automata with a garbage tape, arXiv:1402.3449 (2014)
20. Nakanishi, M., Hamaguchi, K., Kashiwabara, T.: Expressive power of quantum pushdown automata with classical stack operations under the perfect-soundness condition. IEICE Transactions on Information and Systems E89-D(3), 1120–1127 (2006)
21. Ogden, W.: A helpful result for proving inherent ambiguity. Mathematical Systems Theory 2(3), 191 – 194 (1968)
22. Paschen, K.: Quantum finite automata using ancilla qubits (2000), technical report, University of Karlsruhe, (2000)
    http:\\digbib.ubka.uni-karlsruhe.de\volltexte\1452000
23. Say, A.C.C., Yakaryılmaz, A.: Quantum counter automata. International Journal of Foundations of Computer Science 23(5), 1099–1116 (2012)
24. Yakaryılmaz, A.: Superiority of one-way and realtime quantum machines. RAIRO - Theoretical Informatics and Applications 46(04), 615–641 (2012)
25. Yakaryılmaz, A., Freivalds, R., Say, A.C.C., Agadzanyan, R.: Quantum computation with wirte-ony memory. Natural Computing 11(1), 81–94 (2012)
26. Yakaryılmaz, A., Say, A.C.C.: Efficient probability amplification in two-way quantum finite automata. Theoretical Computer Science 410(20), 1932–1941 (2009)
27. Yakaryılmaz, A., Say, A.C.C.: Succinctness of two-way probabilistic and quantum finite automata. Discrete Mathematics and Theoretical Computer Science 12(4), 19–40 (2010)
28. Yakaryılmaz, A., Say, A.C.C.: Unbounded-error quantum computation with small space bounds. Information and Computation 209(6), 873–892 (2011)
29. Yamasaki, T., Kobayashi, H., Imai, H.: Quantum versus deterministic counter automata. Theoretical Computer Science 334(1–3), 275–297 (2005)
30. Yamasaki, T., Kobayashi, H., Tokunaga, Y., Imai, H.: One-way probabilistic reversible and quantum one-counter automata. Theoretical Computer Science 289(2), 963–976 (2002)

Towards a Characterization of Leaf Powers by Clique Arrangements

Ragnar Nevries and Christian Rosenke

Department of Computer Science, University of Rostock
{ragnar.nevries,christian.rosenke}@uni-rostock.de

Abstract. In this paper, we use the new notion of clique arrangements to suggest that leaf powers are a natural special case of strongly chordal graphs. The clique arrangement $\mathcal{A}(G)$ of a chordal graph G is a directed graph that represents the intersections between maximal cliques of G by nodes and the mutual inclusion of these vertex subsets by arcs. Recently, strongly chordal graphs have been characterized as the graphs that have a clique arrangement without bad k-cycles for $k \geq 3$.

The class $\mathcal{L}_k$ of k-leaf powers consists of graphs $G = (V, E)$ that have a k-leaf root, that is, a tree T with leaf set V, where $xy \in E$ if and only if the T-distance between x and y is at most k. Structure and linear time recognition algorithms have been found for 2-, 3-, 4-, and, to some extent, 5-leaf powers, and it is known that the union of all k-leaf powers, that is, the graph class $\mathcal{L} = \bigcup_{k=2}^{\infty} \mathcal{L}_k$, forms a proper subclass of strongly chordal graphs. Despite that, no essential progress has been made lately.

In this paper, we characterize the subclass of strongly chordal graphs that have a clique arrangement without certain bad 2-cycles and show that $\mathcal{L}$ is contained in that class.

1 Introduction

Leaf powers have been introduced by Nishimura et al. [22] to model the problem of reconstructing phylogenetic trees. In particular, a given graph $G = (V, E)$ is called the k-leaf power of a tree T for some $k \geq 2$, if V is the set of leaves in T and any two distinct vertices $x, y \in V$ are adjacent, that is, $xy \in E$ if and only if the distance of x and y in T is at most k. For all $k \geq 2$, the class of graphs that are a k-leaf power of some tree, is simply called k-leaf powers and denoted by $\mathcal{L}_k$. The general problem, from a graph theoretic point of view, is to structurally characterize $\mathcal{L}_k$ for all fixed $k \geq 2$ and to provide efficient recognition algorithms.

On the other hand, if we push k to infinity, then it turns out that not every graph is a k-leaf power for some $k \geq 2$. In particular, a k-leaf power is, by definition, the subgraph of the kth power of a tree T induced by the leaves of T. Since trees are sun-free chordal and as taking graph powers and induced subgraphs do not destroy this property, it follows trivially that every k-leaf power, despite the value of k, is sun-free chordal. By Farber [16], strongly chordal graphs are exactly the sun-free chordal graphs and, hence, k-leaf powers are strongly chordal. But not every strongly chordal graph is a k-leaf power. In fact, we are

G.F. Italiano et al. (Eds.): SOFSEM 2015, LNCS 8939, pp. 364–376, 2015.

aware of exactly the one counterexample shown as G_7 in Figure 1 which has been found by Bibelnieks et al. [1]. It is reasonable to ask for a characterization of the strongly chordal graphs that are not a k-leaf power for any $k \geq 2$. This problem can equivalently be formulated as to describe the class $\mathcal{L} = \bigcup_{k=2}^{\infty} \mathcal{L}_k$, which we call leaf powers, for short.

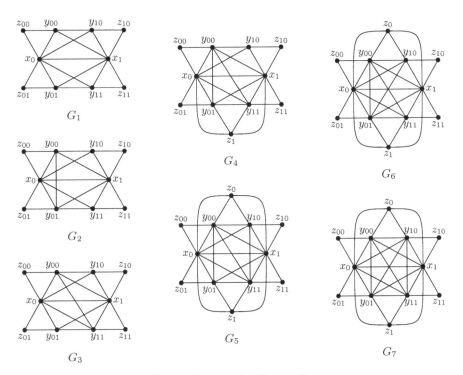

Fig. 1. The graphs $G_1, \ldots, G_7$

Recently, we introduced the clique arrangement in [20], a new data structure that is especially valuable for the analysis of strongly chordal graphs and, thus, also for the study of leaf powers. The clique arrangement $\mathcal{A}(G) = (\mathcal{X}, \mathcal{E})$ of a chordal graph G is a directed acyclic graph that has certain vertex subsets of G as a node set and describes the mutual inclusion of these sets by arcs. In particular, for every set $C_1, C_2, \ldots$ of maximal cliques of G there is a node $X = C_1 \cap C_2 \cap \ldots$ in $\mathcal{X}$ and two nodes $X, Z \in \mathcal{X}$ are joined by an arc $XZ \in \mathcal{E}$, if $X \subset Z$ and there is no $Y \in \mathcal{X}$ with $X \subset Y \subset Z$. In [20], we characterize strongly chordal graphs in terms of forbidden bad k-cycles in their clique arrangement for $k \geq 3$, and we show how to construct the clique arrangement of a strongly chordal graph in $O(n \log n)$ time. Moreover, it is known that the clique arrangements of ptolemaic graphs are even directed trees [24].

That strongly chordal graphs are characterized by clique arrangements with forbidden bad k-cycles, $k \geq 3$, and that ptolemaic graphs are precisely the graphs that have cycle-free clique arrangements justifies the question, if there is anything interesting between these two borders. As $\mathcal{L}$ can be found between strongly chordal graphs and ptolemaic graphs [3] it appears likely that the degree of acyclicity in clique arrangements of leaf powers is between forbidden bad k-cycles, $k \geq 3$, and the complete absence of cycles.

Thus, analyzing the clique arrangements of leaf powers and even more of G_7 can lead to new insights that will ultimately bring the solution to the characterization of $\mathcal{L}$. In this paper, we first introduce a new notion of bad 2-cycles that has in fact been inspired by the study of the clique arrangement of G_7. Secondly, we characterize the subclass of strongly chordal graphs that have no bad 2-cycles in their clique arrangement by the seven forbidden induced subgraphs depicted in Figure 1. Finally, we prove that $\mathcal{L}$ is a subset of our new graph class. Beside the fact that this starts to explain how maximal cliques intersect in leaf powers, this means also that for the first time a natural graph class has been identified that is a proper subclass of strongly chordal graphs but a superclass of leaf powers. So far, to the best of our knowledge, every known natural subclass of strongly chordal graphs has also been a proper subclass of $\mathcal{L}$.

Because of space limitations most of the proofs have been omitted. We refer to the technical report [21] for the missing details.

2   Previous Work

Leaf powers are a well studied graph class which still offers a large number of open questions that remained unsolved for several years. Obviously, a graph G is a 2-leaf power if and only if it is the disjoint union of cliques, that is, G does not contain an induced path of length 2. Dom et al. [14,15] prove that 3-leaf powers are exactly the chordal graphs that do not contain an induced bull, dart, or gem. Brandstädt et al. [4] contribute to the characterization of 3-leaf powers by showing that they are exactly the graphs that result from substituting cliques into the nodes of a tree. Moreover, they give a linear time algorithm to recognize 3-leaf powers building on their characterization.

A characterization of 4-leaf powers in terms of forbidden subgraphs is yet unknown. However, basic 4-leaf powers, the 4-leaf powers without true twins, are characterized as the chordal graphs that are free of eight forbidden subgraphs presented by Rautenbach [23]. The structure of basic 4-leaf powers has further been analyzed by Brandstädt et al. [8], who provide a nice characterization of the two-connected components of basic 4-leaf powers that leads to a linear time recognition algorithm even for 4-leaf powers.

For 5-leaf powers, a polynomial time recognition algorithm was given in [13]. However, no structural characterization is known, even for basic 5-leaf powers. Only for distance-hereditary basic 5-leaf powers a characterization in terms of chordal graphs that are free of 34 forbidden induced subgraphs has been discovered by Brandstädt et al. [6].

Except from the result in [10] that $\mathcal{L}_k \subseteq \mathcal{L}_{k+1}$ is not true for every k, there have not been any more essential advances in determining the structure of k-leaf powers for $k \geq 5$ since 2007. Instead, research has focused on generalizations of leaf powers such as simplicial powers [5], (k, ℓ)-leaf powers [9] and pairwise compatibility graphs [19,12].

Brandstädt et al. [3] show that $\mathcal{L}$ coincides with the class of fixed tolerance NeST (neighborhood subtree tolerance) graphs, a well-known graph class with an absolutely different motivation given by Bibelnieks et al. [1]. Naturally, characterizations and an efficient recognition algorithms for this class are also open questions today. However, by Brandstädt et al. [2,3], it is known that $\mathcal{L}$ is a superclass of ptolemaic graphs, that is, gem-free chordal graphs [18], and even a superclass of directed rooted path graphs, introduced by Gavril [17]. Hence, up to this point, it is known that $\mathcal{L}$ has to be searched in a relatively small gap between strongly chordal graphs and directed rooted path graphs.

3   Preliminaries

Several used graph classes are not explicitly defined due to space limitations. For a comprehensive survey on graph classes we would like to refer to [7].

Throughout this paper, all graphs $G = (V, E)$ are simple, without loops and, with the exception of clique arrangements, undirected. We usually denote the vertex set by V and the edge set by E, where the edges are also called arcs in a directed graph. We write $x-y$, respectively $x{\rightarrow}y$ in the directed case, for $xy \in E$ and $x|y$ for $xy \notin E$. For all $x \in V$ we let $N(x) = \{y \mid xy \in E\}$ denote the *open neighborhood* and $N[x] = N(x) \cup \{x\}$ the *closed neighborhood* of x in an undirected graph G. In a directed graph, $N_o(x) = \{y \mid xy \in E\}$ denotes the set of neighbors that are reachable from x by a single arc and $N_i(x) = \{y \mid yx \in E\}$ are the neighbors that reach x by a single arc. If $|N_i(x)| = 0$ then x is a *source* and if $|N_o(x)| = 0$ then x is a *sink*.

An *independent set* is a set of mutually nonadjacent vertices. A *clique* C is a set of mutually adjacent vertices and C is called maximal, if there is no clique C' with $C \subset C'$. The set of all maximal cliques of G is denoted by $\mathcal{C}(G)$.

A graph $F = (U, E_U)$ is an *induced subgraph* of a graph $G = (V, E)$, if $U \subseteq V$ and $E_U = \{xy \mid x, y \in U, xy \in E\}$.

A *(simple) path* in a graph G is a sequence $x_1, x_2 \ldots, x_k$ of non-repeating vertices in G, such that $x_i x_{i+1} \in E$ for all $i \in \{1, \ldots, k-1\}$. If E is clear from the context, then we denote the path by $x_1-x_2-\ldots-x_k$ in an undirected graph. In a directed graph, $x_1{\rightarrow}x_2{\rightarrow}\ldots{\rightarrow}x_k$ specifies a *directed path* and we say that x_1 *reaches* x_k. An arc $x{\rightarrow}y$ is called transitive, if x reaches y by a directed path that does not use the arc $x{\rightarrow}y$. The *distance* $d_G(x, y)$ between two vertices x, y of an (un-) directed graph G is the minimum number of edges in an (un-) directed path starting in x and ending in y. If the edge $x_k x_1$ is additionally present in E, then we talk of a *(simple) cycle* in G, and as for paths, an undirected cycle is denoted by $x_1-x_2-\ldots-x_k-x_1$. An undirected cycle is called *induced k-cycle C_k*, if G contains $x_i x_j$ if and only if $j = i+1$ or $i = k$ and $j = 1$.

A *tree* T is an undirected connected acyclic graph, that is, for all pairs x, y of vertices there exists a path $x- \ldots -y$, and T is free of cycles. Directed graphs are *acyclic*, if they are free of directed cycles.

A clique $X = \{x_0, \ldots, x_{k-1}\}$ and an independent set $Y = \{y_0, \ldots, y_{k-1}\}$ induce a k-*sun* in G, if every edge $x_i y_j$ between X and Y fulfills $i = j$ or $i + 1 = j$, where the indices are counted modulo k. By definition, a graph is *chordal* if and only if it does not contain induced k-cycles for any $k \geq 4$, and by Farber [16] a graph is *strongly chordal* if and only if it is chordal and does not contain induced k-suns for any $k \geq 3$.

A graph $G = (V, E)$ is the k-*leaf power* of a tree T for $k \geq 2$, if V is the set of leaves in T and $xy \in E$ if and only if $d_T(x, y) \leq k$ for all $x, y \in V$. In this case T is called a k-*leaf root* of G. Notice that k-leaf roots are not necessarily unique for given k-leaf powers. For all $k \geq 2$, the class $\mathcal{L}_k$ consists of all graphs that are a k-leaf power for some tree and $\mathcal{L} = \bigcup_{k=2}^{\infty} \mathcal{L}_k$ is the class of leaf powers.

The *clique arrangement* $\mathcal{A}(G) = (\mathcal{X}, \mathcal{E})$ of a chordal graph G, as introduced in [20], is a directed acyclic graph with node set

$$\mathcal{X} = \left\{ X \;\middle|\; X = \bigcap_{C \in \mathcal{C}} C \text{ with } \mathcal{C} \subseteq \mathcal{C}(G) \text{ and } X \neq \emptyset \right\},$$

containing a node for every intersection of a set of maximal cliques, and arc set

$$\mathcal{E} = \{XZ \mid X, Z \in \mathcal{X} \text{ with } X \subset Z \text{ and } \nexists Y \in \mathcal{X} : X \subset Y \subset Z\}$$

that describes their mutual inclusion. Notice that $\mathcal{A}(G)$ does not contain transitive arcs. Clearly, the set of sinks in $\mathcal{A}(G)$ corresponds exactly to $\mathcal{C}(G)$. Moreover, as the nodes represent the intersections of maximal cliques, it follows that the vertices of any particular node form a clique in G.

Although $\mathcal{A}(G)$ is acyclic by definition, we call the following structure a cycle for the lack of a better term. For any $k \in \mathbb{N}$, a k-*cycle* of $\mathcal{A}(G)$ is a set of nodes $S_0, \ldots, S_{k-1}, T_0, \ldots, T_{k-1}$ such that for all $i \in \{0, \ldots, k-1\}$ there is a directed path from S_i to T_i and a directed path from S_i to T_{i-1} (counted modulo k). Then $S_0, \ldots, S_{k-1}$ are called *starters* of the cycle and $T_0, \ldots, T_{k-1}$ are called *terminals* of the cycle. Note that $S_i \subseteq T_i \cap T_{i-1}$ for all $i \in \{0, \ldots, k-1\}$. In [20], we call a k-cycle *bad*, if $k \geq 3$ and for all $i, j \in \{0, \ldots, k-1\}$ there is a directed path from S_i to T_j, if and only if $j \in \{i, i-1\}$ (counted modulo k).

Theorem 1 (Nevries and Rosenke [20]). *Let G be a chordal graph and let $\mathcal{A}(G)$ be the clique arrangement of G. Then G is strongly chordal if and only if $\mathcal{A}(G)$ is free of bad k-cycles for all $k \geq 3$.*

In this paper we also apply the following property of clique arrangements for induced subgraphs:

Lemma 1. *If G is a chordal graph with clique arrangement $\mathcal{A}(G) = (\mathcal{X}, \mathcal{E})$ that occurs as an induced subgraph of a chordal graph G' with clique arrangement $\mathcal{A}(G') = (\mathcal{X}', \mathcal{E}')$, then there is a function $\phi : \mathcal{X} \to \mathcal{X}'$ that fulfills for all $X, Y \in \mathcal{X}$: (1) $X = Y \Leftrightarrow \phi(X) = \phi(Y)$, and (2) $\mathcal{A}(G)$ has a directed path from X to Y if and only if $\mathcal{A}(G')$ has a directed path from $\phi(X)$ to $\phi(Y)$.*

4   Bad 2-Cycles in Clique Arrangements and the Graphs Characterized by Their Exclusion

As shown in [20], strongly chordal graphs can be characterized by forbidden bad k-cycles in their clique arrangements, where $k \geq 3$. But as leaf powers are a proper subclass of strongly chordal graphs, this kind of cycles does not fully capture the structure that is forbidden in the clique arrangements of leaf powers. To cover the required stronger acyclicity we studied the clique arrangement of G_7, the only known example of a strongly chordal graph that is not a leaf power, and extracted the concept of bad 2-cycles, which are a natural continuation of bad k-cycles with $k \geq 3$. Accordingly, we call a 2-cycle *bad*, if for all $i, j \in \{0, 1\}$

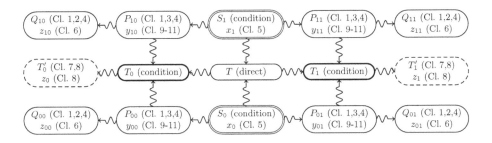

Fig. 2. Proof sketch: Based on the precondition of a bad 2-cycle with starters S_0, S_1 (double frames) and terminals T_0, T_1 (bold frames) we firstly show the existence of other nodes as P_{ij} and sinks as Q_{ij} that are arranged along directed paths, depicted as wavy arrows in the figure. Secondly, we identifiy vertices in the nodes that ultimately build the forbidden graphs $G_1, \ldots, G_7$. The numbers in the brackets identify the claims that show the respective objects.

there is a directed path from starter S_i to terminal T_j that does not contain a node X which fulfills $S_0 \cup S_1 \subseteq X \subseteq T_0 \cap T_1$. Notice that there may still be other paths from S_i to T_j containing nodes X as described in the definition. In particular, the node $T = T_0 \cap T_1$ is always a superset of $S_0 \cup S_1$ and thus, situated on certain paths from S_i to T_j. That a node X does not occur is, however, only required for one path from S_i to T_j. Figure 3 shows the seven clique arrangements of the graphs $G_1, \ldots, G_7$, which all have a bad 2-cycle.

In this section we show that, in terms of forbidden subgraphs, $G_1, \ldots, G_7$ exactly characterize the subclass of strongly chordal graphs that have a clique arrangement without bad 2-cycles. The following theorem provides the main argument of this paper:

Theorem 2. *Let $G = (V, E)$ be a strongly chordal graph with clique arrangement $\mathcal{A}(G) = (\mathcal{X}, \mathcal{E})$. The graph $\mathcal{A}(G)$ contains a bad 2-cycle if and only if G contains one of the graphs $G_1, \ldots, G_7$ as an induced subgraph.*

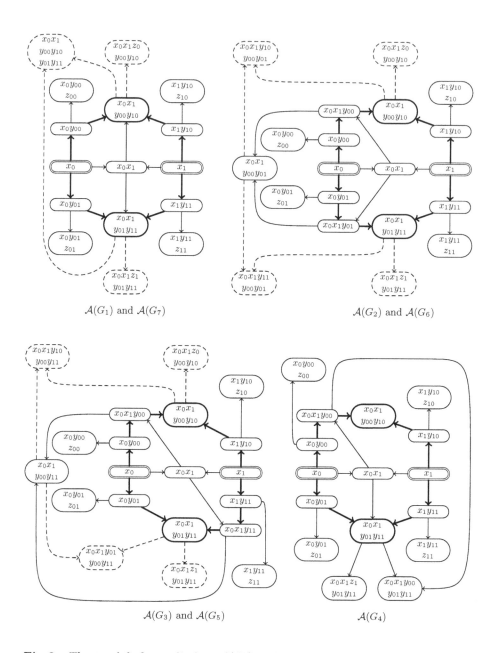

Fig. 3. The top left figure displays $\mathcal{A}(G_7)$ and, without dashed nodes and arcs, it shows $\mathcal{A}(G_1)$. Similarly, the top right figure presents $\mathcal{A}(G_6)$ or, without the dashed parts, $\mathcal{A}(G_2)$. The bottom left figure shows $\mathcal{A}(G_5)$ and, without dashed parts, $\mathcal{A}(G_3)$. The bottom right contains $\mathcal{A}(G_4)$. Bold arcs emphasize the bad 2-cycle, where starters are double framed and terminals bold framed.

Proof. The proof starts by showing the first direction, that is, if $\mathcal{A}(G)$ contains a bad 2-cycle, then G contains one of the graphs $G_1, \ldots, G_7$ as an induced subgraph. A sketch and overview for this direction can be found in Figure 2.

Among the bad 2-cycles of $\mathcal{A}(G)$ we select a cycle with starters S_0, S_1 and terminals T_0, T_1 that primarily minimizes the summed cardinalities of the terminals $|T_0| + |T_1|$ and secondarily maximizes the summed cardinalities of the starters $|S_0| + |S_1|$. Because T_0 and T_1 have a non-empty intersection there is a node $T = T_0 \cap T_1$, which by the way contains $S_0 \cup S_1$. In the following we provide a number of claims to support our arguments. We start by shaping the bad 2-cycle:

Claim 1. *For all $i, j \in \{0, 1\}$ there is a path B_{ij} from S_i to T_j that does not contain a node X with $S_0 \cup S_1 \subseteq X \subseteq T_0 \cap T_1$ (hence, B_{ij} does also not contain T) such that there is a node P_{ij} on B_{ij} with (1) $S_i \subseteq P_{ij} \subseteq T_j$, (2) $S_{1-i} \not\subseteq P_{ij}$, (3) $P_{ij} \not\subseteq T$, and (4) there exists a sink Q_{ij} in $\mathcal{A}(G)$ with $S_{1-i} \not\subseteq Q_{ij}$ that fulfills $P_{ij} = Q_{ij} \cap T_j$.*

In the following we refer to the nodes P_{ij} by the P-nodes and we call Q_{ij} the Q-nodes. The pure existence of the Q-nodes does not directly imply that they are different:

Claim 2. *For all $i, j, i', j' \in \{0, 1\}$ with $(i, j) \neq (i', j')$, the sinks Q_{ij} and $Q_{i'j'}$ differ.*

Hence, the Q-nodes represent distinct maximal cliques in G. For the pairwise intersection between the P-nodes, Claim 1 directly implies for all $i, j, j' \in \{0, 1\}$ that $P_{ij} \not\subseteq P_{(1-i)j'}$. We can now infer the following two additional statements about the intersections between the P-nodes and the intersections between the Q-nodes:

Claim 3. *For all $i \in \{0, 1\}$ it is true that $P_{i0} \cap P_{i1} = S_i$.*

Claim 4. *For all $i, i' \in \{0, 1\}$ it is true that $P_{0i} \cap P_{1i'}$ and $Q_{0i} \cap Q_{1i'}$ are subsets of T.*

We deduce that $P_{ij} \cap P_{i'j'} \subseteq T$ for all $i, j, i', j' \in \{0, 1\}$ with $(i, j) \neq (i', j')$. Following the construction of the P-nodes, we also know for all $i, j \in \{0, 1\}$ that the set $P'_{ij} = P_{ij} \setminus T$ is not empty.

Using the collected facts about the mentioned nodes on the bad 2-cycle, the next two claims start selecting vertices to construct one of the induced subgraphs $G_1, \ldots, G_7$:

Claim 5. *For all $i \in \{0, 1\}$, the starter S_i has a vertex x_i that is not contained in $Q_{(1-i)0} \cup Q_{(1-i)1}$.*

Claim 6. *For all $i, j \in \{0, 1\}$, there is a vertex $z_{ij} \in Q_{ij} \setminus P_{ij}$ with the following two properties: (1) for all $i', j' \in \{0, 1\}$ it is true that $z_{ij} = z_{i'j'} \iff (i, j) = (i', j')$ and (2) z_{ij} is neither adjacent to $x_{1-i}, z_{i(1-j)}, z_{(1-i)j}, z_{(1-i)(1-j)}$, nor to any vertex in $P'_{i(1-j)}$, in $P'_{(1-i)j}$ or in $P'_{(1-i)(1-j)}$.*

Depending on the edges between the six central vertices of $G_1, \ldots, G_7$, there exist up to two additional vertices in $G_4, \ldots, G_7$. This dependency is also visible in the clique arrangement. Consider the sets $V_0 = P_{00} \cup P_{01}$, $V_1 = P_{10} \cup P_{11}$, $D_0 = P_{00} \cup P_{11}$ and $D_1 = P_{01} \cup P_{10}$ and moreover, for all $i, j \in \{0, 1\}$ let $C_{ij} = V_i \cup D_j$. If one of the sets $C_{ij}, i, j \in \{0, 1\}$ induces a clique in G, then it follows that T_0 or T_1 are proper subsets of maximal cliques in G:

Claim 7. *For all $i, j \in \{0, 1\}$ and $k = (i + j + 1) \mod 2$, the node T_k is not a sink in $\mathcal{A}(G)$, if C_{ij} is a clique in G.*

In such a case, if C_{ij} is a clique, we select an additional vertex from a sink that is reachable from T_k:

Claim 8. *For all $i, j \in \{0, 1\}$ and $k = (i + j + 1) \mod 2$, the set C_{ij} being a clique in G implies the existence of a sink T'_k that is reachable from T_k and that contains a vertex $z_k \in T'_k \setminus (P_{0k} \cup P_{1k} \cup T_{1-k})$ such that (1) z_k is not one of the vertices $z_{1-k}, z_{00}, z_{01}, z_{10}, z_{11}$, (2) z_k is neither adjacent to $z_{1-k}, z_{0(1-k)}, z_{1(1-k)}$ nor to any vertex in $T_{1-k} \setminus T$, and (3) z_k is adjacent to at most one vertex of z_{0k} and z_{1k}.*

Object of the remainder of the proof is to select the central vertices y_{ij} from P'_{ij} for all $i, j \in \{0, 1\}$ to ultimately induce a forbidden subgraph. But before giving the strategy of how to select these four vertices, we briefly summarize the results gathered in the proof so far. By Claim 5, we know that there are vertices $x_0 \in S_0$ and $x_1 \in S_1$. As we select y_{ij} from P'_{ij} it follows from $P'_{ij} \subseteq T_j$, which is made sure in the construction of the P-nodes according to Claim 1, that $\{x_0, x_1, y_{00}, y_{10}\} \subseteq T_0$ and $\{x_0, x_1, y_{01}, y_{11}\} \subseteq T_1$ are cliques in G, regardless of the particular choice of $y_{00}, y_{01}, y_{10}, y_{11}$. Moreover, by Claim 6, there exists an independent set $\{z_{00}, z_{01}, z_{10}, z_{11}\}$ in G such that for all $i, j \in \{0, 1\}$, the vertex z_{ij} is adjacent to x_i but not to x_{1-i}. Again, since the vertices y_{ij} come from $P'_{ij} \subseteq P'_{ij}$, it follows that z_{ij} is adjacent to y_{ij} but not to any of $y_{i(1-j)}, y_{(1-i)j}, y_{(1-i)(1-j)}$. Finally, Claim 8 states that certain circumstances imply the existence of two non-adjacent vertices z_0 and z_1 in G that are both adjacent to x_0 and x_1 and such that for all $k \in \{0, 1\}$ it is true that z_k is adjacent to y_{0k} and y_{1k} but not adjacent to $y_{0(1-k)}, y_{1(1-k)}, z_{0(1-k)}$ and $z_{0(1-k)}$. The claim leaves it open, if z_k can be adjacent to either z_{0k} or z_{1k} and, consequently, we cope with this problem during the following vertex selection.

The main difference between the graphs G_1 to G_7 are the edges between the two cliques $\{x_0, x_1, y_{00}, y_{10}\}$ and $\{x_0, x_1, y_{01}, y_{11}\}$. In this proof, the possible edges $y_{00}y_{01}, y_{00}y_{11}, y_{01}y_{10}$ and $y_{10}y_{11}$ are reflected by the sets V_0, V_1, D_0 and D_1. In fact, each of these four edges has both endpoint in exactly one of the sets. To complete this direction of the proof, we therefore analyze the edges inside the sets V_0, V_1, D_0 and D_1. For the decision, which forbidden subgraph is induced in G, it suffices to know whether these sets are cliques. The following three claims perform this analysis for all necessary cases:

Claim 9. *If at most one of the sets V_0, V_1, D_0, D_1 is a clique in G, then G contains G_1, G_2 or G_3 as an induced subgraph.*

Claim 10. *If exactly two of the sets V_0, V_1, D_0, D_1 are cliques in G, then G contains G_2, G_3 or G_4 as an induced subgraph.*

Claim 11. *If at least three of the sets V_0, V_1, D_0, D_1 are cliques in G, then G contains at least one of $G_1, \ldots, G_7$ as an induced subgraph.*

The proof is completed by the converse direction, that is, if G contains one of $G_1, \ldots, G_7$ as an induced subgraph, then $\mathcal{A}(G)$ has a bad 2-cycle. We basically use Lemma 1. The clique arrangements of all graphs $G_1, \ldots, G_7$ contain bad 2-cycles with the same starters $S_0 = \{x_0\}$ and $S_1 = \{x_1\}$ and the same terminals $T_0 = \{x_0, x_1, y_{00}, y_{10}\}$ and $T_1 = \{x_0, x_1, y_{01}, y_{11}\}$. Moreover, there exist nodes $P_{ij} = \{x_i, y_{ij}\}$ and $Q_{ij} = \{x_i, y_{ij}, z_{ij}\}$ such that we have paths $S_i \to \ldots \to P_{ij} \to \ldots \to T_j$ and $P_{ij} \to \ldots \to Q_{ij}$ for all $i, j \in \{0, 1\}$. If G contains one of $G_1, \ldots, G_7$ as an induced subgraph, then there is a function ϕ, that maps these nodes to some nodes of $\mathcal{A}(G)$ such that $\phi(S_i) \to \ldots \to \phi(P_{ij}) \to \ldots \to \phi(T_j)$ and $\phi(P_{ij}) \to \ldots \to \phi(Q_{ij})$ for all $i, j \in \{0, 1\}$. Hence, the clique arrangement $\mathcal{A}(G)$ contains a 2-cycle with starters $\phi(S_0), \phi(S_1)$ and terminals $\phi(T_0), \phi(T_1)$.

Assume that this 2-cycle is not bad, that is, at least one of the four paths in $\mathcal{A}(G)$, without loss of generality say $\phi(S_0) \to \ldots \to \phi(T_0)$, contains a node X with $\phi(S_0) \cup \phi(S_1) \subseteq X \subseteq \phi(T_0) \cap \phi(T_1)$. If X is situated on the subpath $\phi(S_0) \to \ldots \to \phi(P_{00})$, then it follows that $X \subset Q_{00}$ and, hence, $x_1 - z_{00}$, a contradiction.

Hence, X is on the subpath $\phi(P_{00}) \to \ldots \to \phi(T_0)$. Here, $\phi(P_{00})$ is a subset of X and as $X \subseteq \phi(T_0) \cap \phi(T_1)$ by definition, $\phi(P_{00})$ is also a subset of $\phi(T_1)$. This means that $y_{00} \in \phi(T_1)$, which implies $y_{00} - y_{01}$ and $y_{00} - y_{11}$. Consequently, we are in the case were the induced subgraph in G is one of $G_4, \ldots, G_7$. The clique arrangement of all these graphs contains a sink $T_1' = \{x_0, x_1, y_{01}, y_{11}, z_1\}$ that is reached from T_1. In $\mathcal{A}(G)$, we have $\phi(T_1) \to \ldots \to \phi(T_1')$, thus, $\phi(P_{00}) \subset \phi(T_1')$, which finally means that $y_{00} - z_1$, a contradiction.

Hence, X does not exist and $\mathcal{A}(G)$ contains a bad 2-cycle with starters $\phi(S_0), \phi(S_1)$ and terminals $\phi(T_0), \phi(T_1)$.   □

5   New Forbidden Induced Subgraphs for Leaf Powers

The firstly known strongly chordal graph that is not in $\mathcal{L}$, namely G_7, has been found by Bibelnieks et al. [1]. In fact, they were looking for a strongly chordal graph that is not a fixed tolerance NeST graph, but by Brandstädt et al. [3], we know that $\mathcal{L}$ and this class are equal. Notwithstanding all efforts, no counterexample structurally different from G_7 has been discovered for some years and it was even assumed that G_7 is the smallest possible strongly chordal graph without a leaf root.

To show that G_7 is not in $\mathcal{L}$, Bibelnieks et al. [1] use a lemma of Broin et al. [11]. The basic idea of the proof of this lemma is to show for certain pairs of edges $x_1 y_1$ and $x_2 y_2$ in G that the path between x_1 and y_1 is disjoint from the path between x_2 and y_2 in every leaf root of G. In particular, this happens, if vertices a, b exist in G with $x_1, y_1 \in N(a) \setminus N[b]$ and $x_2, y_2 \in N(b) \setminus N[a]$.

The graph G_7 has a cycle $x_0-y_{00}-y_{10}-x_1-y_{11}-y_{01}-x_0$, where the condition is fulfilled for many pairs of edges in the cycle. It follows that every leaf root of G_7 would have a cycle, which is a contradiction.

In this section, we show that $G_1, \ldots, G_6$ are also not in $\mathcal{L}$. Interestingly, every of these six graphs is smaller than G_7. For our proof, we generalize the argument of Bibelnieks et al. [1] for pairs of edges $x_1 y_1$ and $x_2 y_2$ that correspond to disjoint paths in leaf roots. The following Lemma provides three corresponding conditions:

Lemma 2. *Let $G = (V, E)$ be a k-leaf power of a tree T for some $k \geq 2$ and let $x_1 y_1$ and $x_2 y_2$ be two edges of G on distinct vertices $x_1, y_1, x_2, y_2 \in V$. The paths $x_1 - \ldots - y_1$ and $x_2 - \ldots - y_2$ in T are disjoint, that is, do not share any node, if at least one of the following conditions holds:*

1. *At most one of the edges $x_1 x_2, x_1 y_2, y_1 x_2, y_1 y_2$ is in E.*
2. *There is a vertex $a \in V$ such that $x_1, y_1 \in N(a)$, and $x_2, y_2 \notin N[a]$, and $|N(x_1) \cap \{x_2, y_2\}| \leq 1$, and $|N(y_1) \cap \{x_2, y_2\}| \leq 1$.*
3. *There are distinct vertices $a, b \in V$ such that $x_1, y_1 \in N(a) \setminus N[b]$, and $x_2, y_2 \in N(b) \setminus N[a]$.*

Based on this, we can find a cycle $x_0-y_{00}-y_{10}-x_1-y_{11}-y_{01}-x_0$ in every graph from $G_1, \ldots, G_7$ such that many pairs of edges in the cycle fulfill at least one of the three conditions. The following theorem states that this is never compatible with the existence of a leaf root.

Theorem 3. *The graphs $G_1, \ldots, G_7$ are not in $\mathcal{L}$.*

This implies that $G_1, \ldots, G_7$ are forbidden induced subgraphs for $\mathcal{L}$. As these seven graphs precisely characterize the strongly chordal graphs that have no bad 2-cycle in their clique arrangement, it follows that the maximal cliques of a leaf power cannot intersect in the way as described by bad 2-cycles. This is formalized in the following corollary:

Corollary 4. *If G is a graph in $\mathcal{L}$, then the clique arrangement $\mathcal{A}(G)$ does not contain a bad 2-cycle.*

6   Conclusion

In this paper, we were able to identify and characterize the proper subclass of strongly chordal graphs that have a clique arrangement without bad 2-cycles. This class naturally stands between strongly chordal graphs, which have clique arrangements without bad k-cycles for $k \geq 3$, and ptolemaic graphs, whose clique arrangements are entirely free of cycles. By providing the first case for a natural subclass of strongly chordal graphs that is a superclass of $\mathcal{L}$, we essentially tighten the gap for the location of $\mathcal{L}$ in the hierarchy of chordal graphs. It remains for future work to find a complete characterization of $\mathcal{L}$ in terms of forbidden subgraphs or forbidden cyclic structures in the clique arrangements.

References

1. Bibelnieks, E., Dearing, P.M.: Neighborhood Subtree Tolerance Graphs. Discrete Applied Mathematics 43, 13–26 (1993)
2. Brandstädt, A., Hundt, C.: Ptolemaic Graphs and Interval Graphs are Leaf Powers. In: Laber, E.S., Bornstein, C., Nogueira, L.T., Faria, L. (eds.) LATIN 2008. LNCS, vol. 4957, pp. 479–491. Springer, Heidelberg (2008)
3. Brandstädt, A., Hundt, C., Mancini, F., Wagner, P.: Rooted Directed Path Graphs are Leaf Powers. Discrete Mathematics 310(4), 897–910 (2009)
4. Brandstdt, A., Le, V.B.: Structure and Linear Time Recognition of 3-Leaf Powers. Information Processing Letters 98, 133–138 (2006)
5. Brandstädt, A., Le, V.B.: Simplicial powers of graphs. In: Yang, B., Du, D.-Z., Wang, C.A. (eds.) COCOA 2008. LNCS, vol. 5165, pp. 160–170. Springer, Heidelberg (2008)
6. Brandstädt, A., Le, V.B., Rautenbach, D.: Distance-hereditary 5-Leaf Powers. Discrete Mathematics 309, 3843–3852 (2009)
7. Brandstädt, A., Le, V.B., Spinrad, J.P.: Graph Classes: A Survey. SIAM Monographs on Discrete Mathematics and Applications (1999)
8. Brandstdt, A., Le, V.B., Sritharan, R.: Structure and Linear Time Recognition of 4-Leaf Powers. ACM Transactions on Algorithms 5(1) (2008)
9. Brandstädt, A., Wagner, P.: On (k,ℓ)-Leaf Powers. In: Kučera, L., Kučera, A. (eds.) MFCS 2007. LNCS, vol. 4708, pp. 525–535. Springer, Heidelberg (2007)
10. Brandstädt, A., Wagner, P.: On k- Versus $(k + 1)$-Leaf Powers. In: Yang, B., Du, D.-Z., Wang, C.A. (eds.) COCOA 2008. LNCS, vol. 5165, pp. 171–179. Springer, Heidelberg (2008)
11. Broin, M.W., Lowe, T.J.: A Dynamic Programming Algorithm for Covering Problems with (Greedy) Totally Balanced Constraint Matrices. SIAM Journal on Algebraic Discrete Methods 7(3), 348–357 (1986)
12. Calamoneri, T., Montefusco, E., Petreschi, R., Sinaimeri, B.: Exploring Pairwise Compatibility Graphs. Theoretical Computer Science 468, 23–36 (2013)
13. Chang, M.-S., Ko, M.-T.: The 3-Steiner Root Problem. In: Brandstädt, A., Kratsch, D., Müller, H. (eds.) WG 2007. LNCS, vol. 4769, pp. 109–120. Springer, Heidelberg (2007)
14. Dom, M., Guo, J., Hffner, F., Niedermeier, R.: Error Compensation in Leaf Root Problems. Algorithmica 44(4), 363–381 (2006)
15. Dom, M., Guo, J., Hüffner, F., Niedermeier, R.: Extending the Tractability Border for Closest Leaf Powers. In: Kratsch, D. (ed.) WG 2005. LNCS, vol. 3787, pp. 397–408. Springer, Heidelberg (2005)
16. Farber, M.: Characterizations of Strongly Chordal Graphs. Discrete Mathematics 43, 173–189 (1983)
17. Gavril, F.: A Recognition Algorithm for the Intersection Graphs of Directed Paths in Directed Trees. Discrete Mathematics 13(3), 237–249 (1974)
18. Howorka, E.: A Characterization of Ptolemaic Graphs. Journal of Graph Theory 5, 323–331 (1981)
19. Kearney, P., Munro, J.I., Phillips, D.: Efficient Generation of Uniform Samples from Phylogenetic Trees. In: Benson, G., Page, R.D.M. (eds.) WABI 2003. LNCS (LNBI), vol. 2812, pp. 177–189. Springer, Heidelberg (2003)
20. Nevries, R., Rosenke, C.: Characterizing and Computing the Structure of Clique Intersections in Strongly Chordal Graphs. In: Brandstädt, A., Jansen, K., Reischuk, R. (eds.) WG 2013. LNCS, vol. 8165, pp. 382–393. Springer, Heidelberg (2013)

21. Nevries, R., Rosenke, C.: Towards a Characterization of Leaf Powers by Clique Arrangements (2014), http://arxiv.org/
22. Nishimura, N., Ragde, P., Thilikos, D.M.: On Graph Powers for Leaf Labeled Trees. Journal of Algorithms 42(1), 69–108 (2002)
23. Rautenbach, D.: Some Remarks about Leaf Roots. Discrete Mathematics 306(13), 1456–1461 (2006)
24. Uehara, R., Uno, Y.: Laminar Structure of Ptolemaic Graphs and its Applications. In: Deng, X., Du, D.-Z. (eds.) ISAAC 2005. LNCS, vol. 3827, pp. 186–195. Springer, Heidelberg (2005)

Filling Logarithmic Gaps in Distributed Complexity for Global Problems[*]

Hiroaki Ookawa and Taisuke Izumi

Nagoya Institute of Technology, Gokiso-cho, Showa-ku, Nagoya, Aichi,
466-8555, Japan
cht15031@nitech.jp, t-izumi@nitech.ac.jp

Abstract. Communication complexity theory is a powerful tool to show time complexity lower bounds of distributed algorithms for global problems such as minimum spanning tree (MST) and shortest path. While it often leads to nearly-tight lower bounds for many problems, polylogarithmic complexity gaps still lie between the currently best upper and lower bounds. In this paper, we propose a new approach for filling the gaps. Using this approach, we achieve tighter deterministic lower bounds for MST and shortest paths. Specifically, for those problems, we show the deterministic $\Omega(\sqrt{n})$-round lower bound for graphs of $O(n^\epsilon)$ hop-count diameter, and the deterministic $\Omega(\sqrt{n/\log n})$ lower bound for graphs of $O(\log n)$ hop-count diameter. The main idea of our approach is to utilize the two-party communication complexity lower bound for a function we call *permutation identity*, which is newly introduced in this paper.

1 Introduction

In distributed computing theory, many graph problems are naturally treated as the problems in networks, where each vertex represents a computing entity (node) and each edge does a communication link between two nodes. The theory of *distributed graph algorithms* has been developed so far for the efficient in-network computation of graph problems. A crucial factor of distributed graph algorithms is *locality*. Local algorithms require each node to compute its output only by the interaction to the nodes within a bounded distance smaller than the diameter of the network. In other words, local algorithms must terminate within $o(D)$ rounds, where D is the hop-count diameter of the network. There are a number of problems allowing local solutions: Maximal matchings, colorings, independent sets, and so on. On the other hand, some of other graph problems (e.g., minimum spanning tree, shortest path, minimum cut, and so on) are known to have no local solution. They are called *global problems*. By the definition, the (worst-case) run of any algorithm for global problems inherently takes $\Omega(D)$ rounds.

For both local and global problems, the time complexity is one of the central measures to evaluate distributed algorithms. In this paper, we focus on the

---

[*] This work is supported in part by KAKENHI No. 25106507 and No. 25289114.

G.F. Italiano et al. (Eds.): SOFSEM 2015, LNCS 8939, pp. 377–388, 2015.

distributed complexity of two well-known global problems: Minimum spanning tree (MST) and shortest s-t path. As we stated above, these problems have trivial $\Omega(D)$-round lower bounds. If the communication bandwidth of each link is not bounded, every global problem has an optimal-time algorithm with $O(D)$ rounds: A process aggregates the whole information of the network, and computes the result locally. However the assumption of so rich bandwidth is far from real systems, and thus the challenge of global problems is to solve them in the environment with a limited bandwidth. Theoretically, such environments are called *CONGEST model*, where processes work under the round-based synchrony, and each link can transfer $O(\log n)$-bit messages per one round.

A seminal result about the lower bounds for global problems is the one by Das Sarma et al. [1], which exhibits that many problems, including MST and shortest s-t path, are more expensive tasks. Precisely, it shows that p $\Omega(\sqrt{n}/\log n + D)$-round lower bounds hold for many global problems even if D is small (i.e., $D = O(\log n)$). The core of this result is a general framework to obtain the lower bounds based on the reduction from two-party communication complexity by Yao [15]. Two-party communication complexity is a theory to reveal the amount of communication to compute a global function whose inputs are distributed among two players. The reduction framework in [1] induces the hardness of MST and shortest s-t path from the two-party communication complexity of *set-disjointness* function. While the framework is a powerful tool to bound the time complexity of global problems, all the bounds obtained by that approach have the form of $\Omega(f(n)/(m\log n))$, where $f(n)$ is the amount of information inherently exchanged among the network to solve the target problem, m is the number of the links where the information must be transferred, and $\log n$ factor is the bandwidth of each link (that is, $m \log n$ is the amount of information transmittable within a round). Unfortunately, that form does not strictly match the known corresponding upper bounds, which typically have the form of $O(f(n)\mathrm{polylog}(n)/m)$. That is, for many global problems, the currently best bounds still have (poly)logarithmic gaps.

The objective of this paper is to close those gaps. For that goal, we introduce a new two-party function called *permutation identity*, whose deterministic communication complexity is slightly more expensive than set-disjointness, and show new reductions using it on the top of the framework by Das Sarma et al. [1]. Specifically, for MST and shortest s-t path, we show the deterministic $\Omega(\sqrt{n})$-round lower bound for graphs of $O(n^\epsilon)$ hop-count diameter, and the deterministic $\Omega(\sqrt{n/\log n})$ lower bound for graphs of $O(\log n)$ hop-count diameter. The comparison with the prior work are shown in Table 1. As far as we consider the complexity of *deterministic* and *exact* computation, our bounds beat the currently best ones.

2   Related Work

The paper by Das Sarma et al. [1] is the first one explicitly considering the distributed verification problem, which has given a general framework to lead lower

Table 1. Comparison with the prior work. SP (resp. MST) means the shortest s-t path problem (resp. the minimum spanning tree problem). The phrase "$\alpha(n)$-approximation" implies that $\alpha(n)$-approximation is hard for any non-trivial function $\alpha(n)$.

paper	bound	problem	comments
Garay et al. [4]	$O(\sqrt{n}\log^* n + D)$	MST	deterministic
Nanongkai [11]	$O(\sqrt{n}D^{1/4} + D)$	SP	$(1 + o(1))$-approximation single-source SP
Das Sarma et al.[1]	$\Omega(\sqrt{\frac{n}{\log n}})$	SP,MST	randomized $\alpha(n)$-approximation $D = O(n^\epsilon)$ $(\epsilon < 1/2)$
Das Sarma et al.[1]	$\Omega(\frac{\sqrt{n}}{\log n})$	SP, MST	randomized $\alpha(n)$-approximation $(D = \Theta(\log n))$
This paper	$\Omega(\sqrt{n})$	SP, MST	deterministic $D = O(n^\epsilon)$ $(\epsilon < 1/2)$
This paper	$\Omega(\sqrt{\frac{n}{\log n}})$	SP, MST	deterministic $D = O(\log n)$

bounds and approximation hardness for a vast class of problems. It is used in several following papers to obtain the complexity for a number of graph problems: Weighted/unweighted diameter and all-pair shortest paths [13,6,8,9], minimum cuts [5,11], distance sketches [8], weighted single-source shortest paths [8,11], fast random walks [12], and so on. While the framework by Das Sarma et al. [1] pointed out a general relationship interconnecting the communication complexity theory and distributed complexity theory, the construction of the worst-case instances used in the framework is much inspired by the earlier papers leading the time lower bound for the distributed MST construction [14,10,3]. Dinitz et al. [2] shows a bit complexity result for non graph-theoretic problems based on communication complexity theory.

3 Preliminaries

3.1 Round-Based Distributed Systems

A distributed system consists of n nodes interconnected with communication links. We model it by a weighted graph $G = (V, E, w)$, where V is the set of nodes, $E \subseteq V \times V$ is the set of links (edges), and $w : E \to \mathbb{R}$ is a weight function. The hop-count diameter of G (i.e., the diameter of the unweighted graph (V, E)) is denoted by D. Executions of the system proceed with a sequence of consecutive rounds. In each round, each process sends a (possibly different) message to each neighbor, and within the round, all messages are received. After receiving the messages, the process performs local computation. Throughout this paper, we restrict the number of bits transmittable through any communication link per one round to $O(\log n)$ bits. This is known as the CONGEST model.

3.2 Distributed MST and Single-Source Shortest Paths

In this paper we consider two popular graph problems: Minimum spanning tree (MST) and shortest s-t path. The distributed minimum spanning tree problem requires the system to find the MST of the (weighted) network. After the computation by distributed MST algorithms, each node must identify the incident edges constituting the MST. In the shortest s-t path problem, the algorithm takes two input nodes s and t, and computes a shortest path between them. After the computation, each node on the computed path must identify the incident edge toward s and the distance from s.

3.3 Two-Party Communication Complexity

Communication complexity, which is first introduced by Yao [15], reveals the amount of communication to compute a global function whose inputs are distributed in the network. The most successful scenario in communication complexity is *two-party* communication complexity, where two players, called Alice and Bob, respectively have their inputs $x, y \in U$ (where U is the domain of inputs), and compute a global function $f : U \times U \to \{0, 1\}$. The communication complexity of a two-party protocol is the number of one-bit messages exchanged by the protocol for the worst case input (if the protocol is randomized, it is defined as the expected number of bits exchanged for the worst-case input). One of the most popular functions in two-party communication complexity is *set-disjointness*, which is the function over two k-bit 0-1 vectors $x, y \in {0, 1}^k$ and return value one if and only if there exists a common position $i \in [0, k-1]$ such that i-th bits of x and y are one.

While the known best lower bounds for MST and shortest s-t path is obtained by using the communication complexity of set-disjointness, it seems difficult to extend that proof for a stronger bound we will prove. Thus in this paper, we introduce a new function called *permutation identity*, which is defined as follows:

Definition 1. *Let* $\pi_A, \pi_B : [1, N] \to [1, N]$ *be permutations over* $[1, N]$. *The permutation identity function* $ident_N$ *is defined as follows:*

$$ident_N(\pi_A, \pi_B) = \begin{cases} 1 \text{ if } \forall i \in [1, N] : \pi_A \circ \pi_B(i) = i, \\ 0 \text{ otherwise,} \end{cases}$$

where $\pi_A \circ \pi_B$ *means the composition of* π_A *and* π_B, *that is,* $\pi_A \circ \pi_B(i) = \pi_A(\pi_B(i))$.

Theorem 1. *The deterministic communication complexity of two-party permutation identity over* $[1, N]$ *is* $\Omega(N \log N)$ *bits.*

While we omit the proof of this theorem due to lack of space, it is almost the same as that for *equality* function, which is found in a standard textbook of communication complexity [7]. We also show a fundamental lemma for the permutation identity function, which is used in the following sections.

Lemma 1. *Let π_A and π_B be permutations over $[1, N]$. If $\pi_A \circ \pi_B$ is not identical, there exists $i \in [1, N]$ such that $\pi_A \circ \pi_B(i) < i$ holds.*

4   General Framework for the Reduction

The proof basically follows the framework by Das Sarma et al. [1]. The core of this framework is the reduction from two-party computation via a hard instance for distributed computation. In this section, we introduce the framework which is slightly modified for our proof.

4.1   Graph Construction

The graph we construct is denoted by $G(N, M)$, where N and M are design parameters of the graph. For simplicity of the argument, throughout the paper, we assume that $M + 1$ is a power of 2, i.e., $M = 2^p - 1$ for some nonnegative integer p. Note that the assumption is not essential and it is not difficult to remove it. The graph is built by the following steps:

1. Prepare N paths of length M, each of which is denoted by P_i ($1 \leq i \leq N$). The nodes constituting P_i are identified by $v_i^0, v_i^1, \cdots, v_i^M$ from left to right.
2. Add edges (v_i^0, v_j^1) and $(v_i^{(M-1)}, v_j^M)$ for any $i, j \in [1, N]$.
3. Add edges $(v_i^0, v_{(i+1)}^0)$ and $(v_i^M, v_{(i+1)}^M)$ for any $i \in [1, N-1]$.
4. Construct a complete binary tree $T(M)$ with $M + 1$ leaf. where each leaf is identified by $u^0, u^1, \cdots, u^M$ from left to right.
5. Add edges (u^i, v_j^i) for any $i \in [0, M]$ and $j \in [1, N]$.

The weight of each edge depends on concrete reductions, which is determined later. Note that the number n of nodes in $G(N, M)$ is $\Theta(NM)$, and its diameter is $D = O(\log n)$. We also define the sets of nodes $A = \{u^0\} \cup \{v_i^0, v_i^1 | i \in [1, N]\}$ and $B = \{u^M\} \cup \{v_i^{(M-1)}, v_i^M | i \in [1, N]\}$. The whole construction is illustrated in Figure 1. For this graph, we can show the following theorem.

Theorem 2 (Das Sarma et al. [1]). *Let $\mathcal{A}$ be any algorithm running on the graph $G(N, M)$ with an arbitrary edge-weight function, and $r < M$ be an arbitrary value. Then there exists a two-party protocol satisfying the following three properties:*

- *At the beginning of the protocol, Alice (resp. Bob) knows the whole topological information of $G(N, M)$ except for the subgraph induced by B (resp. A),*

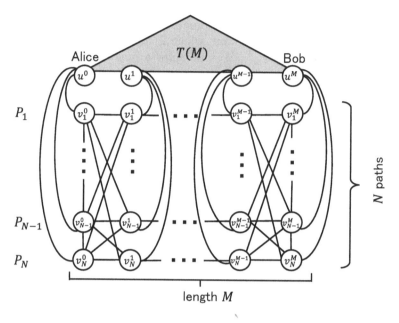

Fig. 1. Construction of $G(N, M)$

- after the run of the protocol, Alice and Bob output the internal states of the processes in A and B at round $\lceil (r-3)/2 \rceil$ in the execution of $\mathcal{A}$ on $G(N, M)$, respectively, and
- the protocol consumes at most $O(r(\log MN)^2)$-bit communication.

While the graph used in this paper is a slightly modified version of the original construction in [1], the theorem above is proved in the almost same way. So we just quote it without the proof.

4.2 Networked Two-Party Computation

To obtain time lower bounds for distributed algorithms, we use a variation of the two-party computation problem in distributed settings. We assume that Alice and Bob are placed at two nodes in a network of n nodes, and have inputs $x \in U$ and $y \in U$ for two-party function $f : U \times U \to \{0, 1\}$, respectively. It is also assumed that each node in the network (including ones other than Alice and Bob) knows everything (i.e., the complete knowledge of the network topology) except for the inputs held by Alice and Bob. Then all nodes must work cooperatively for outputting the value of $f(x, y)$ as fast as possible. In what follows, we call this problem setting the *networked two-party computation* (and the networked permutation identity problem if $f = ident_N$). Note that the measurement of the networked two-party computation is not the amount of communication, but the number of rounds.

Obviously the time complexity of networked two-party computation problems relies on the target function f and the topology of the network. A useful consequence from Theorem 2 is that we can transform the communication lower bound for any two-party computation into the time lower bound for its networked version. In the original version by Das Sarma et al. [1], the transformation from two-party set-disjointness is considered. Here we derive the similar fact from two-party permutation identify function (the proof is omitted for lack of space):

Theorem 3. *Let $M = N/\log N$. For any deterministic algorithm $\mathcal{A}$ solving the networked permutation identity over $[1, N]$ in $G(N, M)$, its worst-case running time is $\Omega(\sqrt{n/\log n})$ rounds.*

4.3   Lower Bound for MST

We show the reduction from networked permutation identity to MST. In this reduction we construct an instance of the MST problem by virtually assigning some weight to each edge in $G(N, M)$ for $M = N/\log N$ to encode an instance (π_A, π_B) of permutation identity over $[1, N]$. After the construction of the MST, Alice and Bob can determine the identity of $\pi_A \circ \pi_B$ from the computed MST. Let $L(\pi_A, \pi_B)$ be the instance of the MST problem corresponding to the permutation identity instance (π_A, π_B), which is constructed by defining edge-weight function w as follows:

1. For any $i \in [1, N]$ and $j \in [1, M-1]$, $w(u^j, v_i^j) = 100NM$.
2. For any $i \in [1, N-1]$, $w(v_i^0, v_{i+1}^0) = 100NM$ and $w(v_i^M, v_{i+1}^M) = 100NM$.
3. For any $i \in [1, N]$, $w(u^0, v_i^0) = 2i$ and $w(u^M, v_i^M) = 2i - 1$.
4. For any $i, j \in [1, N]$, $w(v_i^0, v_j^1) = 1$ if $\pi_A(j) = i$. Otherwise $w(v_i^0, v_j^1) = 100NM$. Similarly, for any $i, j \in [1, N]$, $w(v_i^{M-1}, v_j^M) = 1$ if $\pi_B(j) = i$. Otherwise $w(v_i^{M-1}, v_j^M) = 100NM$.
5. All other edges have weight one.

The construction of $L(\pi_A, \pi_B)$ is illustrated in Figure 2. Let $E_A = \{(u^0, v_i^0) | i \in [1, N]\}$ and $E_B = \{(u^M, v_i^M) | i \in [1, N]\}$. The following lemma is the core of the reduction.

Lemma 2. *The MST of $L(\pi_A, \pi_B)$ contains no edge in E_A if and only if $\pi_A \circ \pi_B$ is identical.*

Proof. Let P_i' be the path consisting of the nodes $v_{\pi_A(\pi_B(i))}^0, v_{\pi_B(i)}^1, v_{\pi_B(i)}^2, \cdots,$ $v_{\pi_B(i)}^{M-1}, v_i^M$. Following the standard greedy algorithm for constructing the MST, every edge with weight one is contained in the MST. Thus, the components $P_1', P_2', \cdots, P_N'$ and $T(M)$ are MST fragments. A component P_i' is merged with $T(M)$ by choosing either $(u^0, v_{\pi_A(\pi_B(i))}^0)$ or (u^M, v_i^M) (all other edges merging them are too heavy (i.e., $100NM$) and never chosen as a MST edge). If $\pi_A \circ \pi_B$ is identical, $\pi_A(\pi_B(i)) = i$ holds. Thus we have $w(u^0, v_{\pi_A(\pi_B(i))}^0) = 2i$ and $w(u^M, v_i^M) = 2i - 1$ for any $i \in [1, N]$. This implies that P_i' is merged with $T(M)$

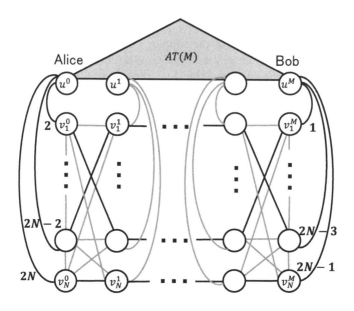

Fig. 2. An example of $L(\pi_A, \pi_B)$. Every unlabeled edge has weight one. All the edges with weight $100NM$ are grayed out.

by edge (u^M, v_i^M) (Figure 3). On the other hand, if $\pi_A \circ \pi_B$ is not identical, from Lemma 1, there exists at least one i satisfying $\pi_A \circ \pi_B(i) < i$. Then for such i we have $w(u^0, v^0_{\pi_A(\pi_B(i))}) \leq 2(i-1)$ and $w(u^M, v_i^M) = 2i - 1$. Thus P'_i and $T(N)$ is merged with edge $w(u^0, v^0_{\pi_A(\pi_B(i))}) \in E_A$ (Figure 4). The lemma is proved   □

Lemma 3. *If an algorithm $\mathcal{A}$ solves the MST problem in $L(\pi_A, \pi_B)$ within r rounds, there exists an algorithm solving the networked permutation identity over $[1, N]$ in $G(N, M)$ within $O(r)$ rounds.*

Proof. At the rounds one and two, each node sets up the instance $L(\pi_A, \pi_B)$ of the MST problem according to the input (π_A, π_B). Then the system runs the MST algorithm $\mathcal{A}$. From Lemma 2, no edge in E_A is not included in the constructed MST if $\pi_A \circ \pi_B$ is identical. Then, after the construction of the MST, each node v_i^0 ($i \in [1, N]$) sends to u^0 the information that no incident edge is contained in the MST. By this information, u^0 can determine whether $\pi_A \circ \pi_B$ is identical or not. That is, the networked permutation identity is solved in $G(N, M)$ within $O(r)$ rounds.   □

Combining Theorem 3 and Lemma 3, we have the main theorem below.

Theorem 4. *Any deterministic algorithm solving the MST problem, its worst-case running time is $\Omega(\sqrt{n/\log n})$ rounds.*

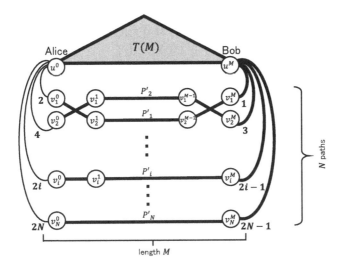

Fig. 3. Graph $L(\pi_A, \pi_B)$ when $\pi_A \circ \pi_B$ is identical

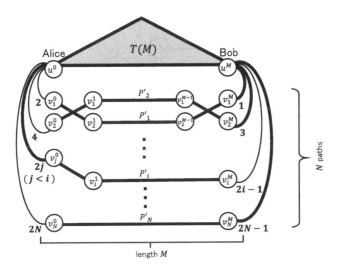

Fig. 4. Graph $L(\pi_A, \pi_B)$ when $\pi_A \circ \pi_B$ is not identical

4.4 Lower Bound for Shortest s-t Path

The argument in this section is almost the same as Section 4. We construct a graph $L'(\pi_A, \pi_B)$ by fixing a weight function w for the network $G(N, M)$ for $M = N \log N$. The weight function w is defined as follows:

1. For any $i \in [1, N]$ and $j \in [0, M]$, $w(u^j, v_i^j) = 100NM$.
2. For any $i \in [1, N-1]$, $w(v_i^0, v_{i+1}^0) = 1$ and $w(v_i^M, v_{i+1}^M) = 1$.
3. For any $i, j \in [1, N]$, $w(v_i^0, v_j^1) = 100NM$ if $\pi_A(j) = i$. Otherwise $w(v_i^0, v_j^1) = 100NM$. Similarly, for any $i, j \in [1, N]$, $w(v_i^{M-1}, v_j^M) = 1$ if $\pi_B(j) = i$. Otherwise $w(v_i^{M-1}, v_j^M) = 100NM$.
4. For any $i \in [1, N]$ and $j \in [1, M-1]$, $w(v_i^j, v_i^{j+1}) = 1$.
5. Every edge in $T(M)$ has weight $100NM$.

We also define $s = v_1^0$ and $t = v_N^M$. Then, we have the following lemma:

Lemma 4. *In graph $L'(\pi_A, \pi_B)$, the length of the shortest s-t path is $N + M - 1$ if and only if $\pi_A \circ \pi_B$ is identical.*

Proof. The path $v_1^0, v_2^0, \cdots v_N^0, v_N^1, v_N^2, \cdots V_N^{M-1}, v_N^M$ is the s-t path of length $N + M - 1$. We first show that this is the shortest path if $\pi_A \circ \pi_B$ is identical. Since the length of the shortest path between s and t is at most $N + M - 1$, it contains no edge with weight $100NM$. Thus we omit those edges. Then, if $\pi_A \circ \pi_B$ is identical, v_i^0 and v_i^M are connected by a path of length M. Thus, the graph (where all isolated nodes in $T(M)$ are removed) becomes a subdivision of a ladder graph. It is not difficult to see that the shortest path between s and t is $N + M - 1$ (Figure 5).

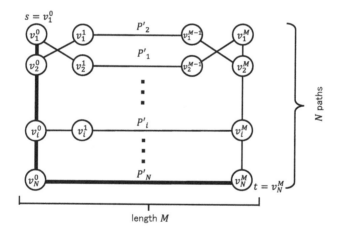

Fig. 5. Graph $L'(\pi_A, \pi_B)$ when $\pi_A \circ \pi_B$ is identical

We next consider the case where $\pi_A \circ \pi_B$ is not identical. From Lemma 1, there exists i satisfying $\pi_A \circ \pi_B(i) < i$. Then, we have an s-t path $v_1^0, v_2^0, \cdots v_{\pi_A(\pi_B(i))}^0, v_{\pi_B(i)}^1, v_{\pi_B(i)}^2, \cdots, v_{\pi_B(i)}^{M-1}, v_i^M, v_{i+1}^M, \cdots, v_N^M$ of length less than $N + M - 1$ (Figure 6). The lemma is proved. □

Lemma 5. *If an algorithm $\mathcal{A}$ solves the shortest s-t path problem in $L'(\pi_A, \pi_B)$ within r rounds, there exists an algorithm solving the networked permutation identity over $[1, N]$ within $O(r)$ rounds.*

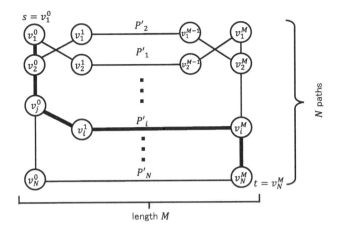

Fig. 6. Graph $L'(\pi_A, \pi_B)$ when $\pi_A \circ \pi_B$ is not identical ($\pi_A \circ \pi_B(i) = j$)

The proof is almost the same as that for Lemma 3, and thus we omit it. Finally we obtain the following theorem.

Theorem 5. *Any deterministic algorithm solving the shortest s-t path problem, its worst-case running time is $\Omega(\sqrt{n/\log n})$ rounds.*

5   Lower Bound for the Graphs with $O(n^\epsilon)$ Hop-Count Diameter

For the case of larger diameter graphs, we obtain stronger bounds by slightly modifying the framework graph $G(N, M)$. Since the fundamental idea has been proposed in the prior work [1], we state only the results in this paper. Theorem 4 and 5 are extended as follows:

Theorem 6. *Any deterministic algorithm solving the MST problem or the shortest s-t path problem, its worst-case running time is $\Omega(\sqrt{n/\log n})$ rounds for graphs with diameter $O(\log n)$. In addition, for graphs with diameter $O(n^\epsilon)$ $(0 < \epsilon < 1/2)$, the worst-case running time is $\Omega(\sqrt{n})$ rounds.*

6   Concluding Remarks

In this paper, we introduced a new function called *permutation identity*. By using the seminal reduction framework by Das Sarma et al.[1], we show the deterministic $\Omega(\sqrt{n/\log n})$-round lower bounds for MST and shortest s-t path. Furthermore, for graph with for graphs with $O(n^\epsilon)$ hop-count diameter, we obtained $\Omega(\sqrt{n})$ lower bounds.

An open problem is that we show the same lower bounds for randomized algorithms. Because of the analogy to the equality function, the permutation identity problem is unlikely to have the same lower bound (i.e., $\Omega(N \log N)$ bits) for randomized protocols. It is an interesting future work to explore a (two-party) function over permutations exhibiting $\Omega(N \log N)$-bit randomized lower bound.

References

1. Sarma, A.D., Holzer, S., Kor, L., Korman, A., Nanongkai, D., Pandurangan, G., Peleg, D., Wattenhofer, R.: Distributed verification and hardness of distributed approximation. In: Proc. of the 43rd Annual ACM Symposium on Theory of Computing, pp. 363–372 (2011)
2. Dinitz, Y., Moran, S., Rajsbaum, S.: Bit complexity of breaking and achieving symmetry in chains and rings. Journal of the ACM 55(1) (2008)
3. Elkin, M.: An unconditional lower bound on the hardness of approximation of distributed minimum spanning tree problem. In: Proc the 30th ACM Symposium on Theory of Computing(STOC), pp. 331–340 (2004)
4. Garay, J.A., Kutten, S., Peleg, D.: A sublinear time distributed algorithm for minimum-weight spanning trees. SIAM Journal on Computing 27(1), 302–316 (1998)
5. Ghaffari, M., Kuhn, F.: Distributed minimum cut approximation. In: Afek, Y. (ed.) DISC 2013. LNCS, vol. 8205, pp. 1–15. Springer, Heidelberg (2013)
6. Holzer, S., Wattenhofer, R.: Optimal distributed all pairs shortest paths and applications. In: Proc. of the 2012 ACM Symposium on Principles of Distributed Computing (PODC), pp. 355–364 (2012)
7. Kushilevitz, E., Nisan, N.: Communication Complexity. Cambridge University Press (1997)
8. Lenzen, C., Patt-Shamir, B.: Fast routing table construction using small messages: Extended abstract. In: Proc. of the 45th Annual ACM Symposium on Symposium on Theory of Computing (STOC), pp. 381–390 (2013)
9. Lenzen, C., Peleg, D.: Efficient distributed source detection with limited bandwidth. In: Proc. of the 2013 ACM Symposium on Principles of Distributed Computing (PODC), pp. 375–382 (2013)
10. Lotker, Z., Patt-Shamir, B., Peleg, D.: Distributed mst for constant diameter graphs. Distributed Computing 18(6), 453–460 (2006)
11. Nanongkai, D.: Distributed approximation algorithms for weighted shortest paths. In: Proc. of the 46th ACM Symposium on Theory of Computing (STOC), pp. 565–573 (2014)
12. Nanongkai, D., Das Sarma, A., Pandurangan, G.: A tight unconditional lower bound on distributed random walk computation. In: Proc. of the 30th Annual ACM SIGACT-SIGOPS Symposium on Principles of Distributed Computing (PODC), pp. 257–266 (2011)
13. Peleg, D., Roditty, L., Tal, E.: Distributed algorithms for network diameter and girth. In: Czumaj, A., Mehlhorn, K., Pitts, A., Wattenhofer, R. (eds.) ICALP 2012, Part II. LNCS, vol. 7392, pp. 660–672. Springer, Heidelberg (2012)
14. Peleg, D., Rubinovich, V.: A near-tight lower bound on the time complexity of distributed minimum-weight spanning tree construction. SIAM Journal on Computing 30(5), 1427–1442 (2000)
15. Yao, A.C.-C.: Some complexity questions related to distributive computing(preliminary report). In: Proc. of the 11th Annual ACM Symposium on Theory of Computing (STOC), pp. 209–213 (1979)

On Visibly Pushdown Trace Languages

Friedrich Otto

Fachbereich Elektrotechnik/Informatik, Universität Kassel,
34109 Kassel, Germany
otto@theory.informatik.uni-kassel.de

Abstract. We present a characterization of the class of (linearizations of) visibly pushdown trace languages in terms of cooperating distributed systems (CD-systems) of stateless deterministic restarting automata with window size one that use an external pushdown store for computing the global successor relation.

1 Introduction

Input-driven languages were already presented in 1983 by von Braunmühl and Verbeek [6], but they became the focus of much attention only after their reinvention by Alur and Madhusudan [2] under the name of *visibly pushdown languages*. This class of languages properly extends the regular languages by some important context-free concepts, but they are still sufficiently restricted to inherit many closure and decidability results from the regular languages. In particular, they have found applications in the modelling of questions concerning the program analysis of recursive programs (see, e.g., [17]). On the other hand, trace languages have been studied with regard to their ability to model concurrent processes (see, e.g., [7,9]), and by combining these two concepts a variant of the *visibly pushdown trace languages* is used in [5] to analyze concurrent recursive programs. In that paper the visibly pushdown trace languages are realized by so-called *concurrent visibly pushdown automata* (CVPA) that combine the concept of a visibly pushdown automaton with that of Zielonka's *asynchronous automaton* [18]. However, for the concurrent pushdown alphabets considered in [5], it is assumed that the push and pop operations of any process are independent of all the push and pop operations of any other process.

Here we study the class of formal languages that occur as linearizations of visibly pushdown trace languages without enforcing any such restriction, that is, the dependency relation D on a given pushdown alphabet Σ is completely independent of the actual partitioning of Σ into call, return, and internal symbols. Our main result gives a characterization of the class of (linearizations of) all visibly pushdown trace languages in terms of a cooperating distributed system (CD-system) of stateless deterministic restarting automata of a particularly simple type, the so-called VPD-CD-R(1)-systems.

In [11] CD-systems of stateless deterministic R(1)-automata were introduced. Although the restarting automata of this type are very restricted in their computational power, these systems accept a class of semi-linear languages that

G.F. Italiano et al. (Eds.): SOFSEM 2015, LNCS 8939, pp. 389–400, 2015.

contains all (linearization of) rational trace languages. In [13] these CD-systems were then extended by introducing an external pushdown store that is used to determine the successor of the current automaton. Essentially such a system can be interpreted as a traditional pushdown automaton, in which the operation of reading an input symbol has been replaced by a stateless deterministic R(1)-automaton. Hence, not the first symbol is necessarily read, but some symbol that can be reached by this automaton by moving across a prefix of the current input word. Other variants of pushdown automata that do not simply read their input sequentially from left to right have been studied before. For example, in [3] pushdown automata are considered that can reverse their input. It is shown in [13] that these types of CD-systems yield an automata-theoretical characterization for the class of (linearizations of) context-free trace languages (see also [14] for an extended presentation).

Here we combine the idea of the above CD-systems with an external pushdown with the idea of a *visibly pushdown automaton*. We will see that the class of languages accepted by these VPD-CD-R(1)-systems properly contains (the linearizations of) all visibly pushdown trace languages. In fact, a subclass of these CD-systems will be described that characterizes (the linearizations of) all visibly pushdown trace languages. We also present a deterministic class of our CD-systems, but we will see that the class of languages they accept is incomparable under inclusion to (the linearizations of) all visibly pushdown trace languages.

This paper is structured as follows. In Section 2 we restate the necessary notions and notation on traces and trace languages, and in the next section we describe the CD-systems of stateless deterministic R(1)-automata and present the announced VPD-CD-R(1)-systems. Section 4 presents our characterization result, and in Section 5 we discuss the deterministic variant of our CD-systems.

2   Visibly Pushdown Trace Languages

Let Σ be a finite alphabet, and let D be a binary relation on Σ that is reflexive and symmetric. Then D is a *dependency relation* on Σ, and $I_D = (\Sigma \times \Sigma) \smallsetminus D$ is the corresponding *independence relation*. Obviously, the relation I_D is irreflexive and symmetric. The dependency relation D induces a binary relation $\equiv_D$ on Σ^* that is defined as the smallest congruence relation (with respect to the operation of concatenation) that contains the set of pairs $\{\, (ab, ba) \mid (a, b) \in I_D \,\}$. For $w \in \Sigma^*$, the congruence class of w mod $\equiv_D$ is denoted by $[w]_D$, that is, $[w]_D = \{\, z \in \Sigma^* \mid w \equiv_D z \,\}$. These congruence classes are called *traces*, and the factor monoid $M(D) = \Sigma^*/\equiv_D$ is a *trace monoid* (see, e.g., [7]). By φ_D we denote the morphism $\varphi_D : \Sigma^* \to M(D)$ that is defined by $w \mapsto [w]_D$ for all words $w \in \Sigma^*$.

The underlying idea is the following. The letters of Σ are interpreted as actions, and $(a, b) \in I_D$ expresses the fact that actions a and b are independent, which means that they can be executed in parallel, or when realized on a single processor, then they can be executed in any order. In contrast, if $(c, d) \in D$, then the actions c and d are dependent on each other, which means that the sequences of actions cd and dc will have different effects.

For a subset S of the trace monoid $M(D)$, the set of words $L = \varphi_D^{-1}(S) \subseteq \Sigma^*$ is called a *linearization* of S. Now a language $L \subseteq \Sigma^*$ is called (the linearization of) a *rational trace language*, if there exists a dependency relation D on Σ and a regular language $R \subseteq \Sigma^*$ such that $L = \varphi_D^{-1}(\varphi_D(R)) = \bigcup_{w \in R} [w]_D$. By $\mathcal{LRAT}$ we denote the set of linearizations of all rational trace languages.

Analogously, a language $L \subseteq \Sigma^*$ is the *linearization* of a *context-free trace language*, if there exist a dependency relation D on Σ and a context-free language $R \subseteq \Sigma^*$ such that $L = \varphi_D^{-1}(\varphi_D(R)) = \bigcup_{w \in R} [w]_D$ [1,4]. By $\mathcal{LCF}$ we denote the set of all linearizations of context-free trace languages.

Here we are interested in a particular subclass of $\mathcal{LCF}$, the class $\mathcal{LVPD}$ of linearizations of all visibly pushdown trace languages. Following the presentation in [2], a language L is a *visibly pushdown language* if it is accepted by a *visibly pushdown automaton* (or VPDA for short). A VPDA is a restricted type of pushdown automaton the input alphabet Σ of which is a *pushdown alphabet*, that is, it is partitioned into three disjoint subsets Σ_c of *call symbols*, Σ_r of *return symbols*, and Σ_{int} of *internal symbols*. A VPDA M over $\Sigma = \Sigma_c \,\dot\cup\, \Sigma_r \,\dot\cup\, \Sigma_{int}$ is defined by a 5-tuple $M = (Q, Q_{in}, \Gamma, \delta, Q_{fin})$, where

- Q is a finite set of states, $Q_{in} \subseteq Q$ is the subset of initial states, and $Q_{fin} \subseteq Q$ is the subset of final states,
- Γ is a finite stack alphabet not containing the bottom marker $\bot$,
- and $\delta \subseteq (Q \times \Sigma_c \times Q \times \Gamma) \cup (Q \times \Sigma_r \times (\Gamma \cup \{\bot\}) \times Q) \cup (Q \times \Sigma_{int} \times Q)$ is a transition relation.

A transition of the form (q, a, q', A), where $q, q' \in Q$, $a \in \Sigma_c$, and $A \in \Gamma$, is a *push transition*. On reading the letter a, M can push A onto its pushdown and switch from state q to state q'. Observe that this action is not influenced by the content of the pushdown. A transition of the form (q, a, A, q'), where $q, q' \in Q$, $a \in \Sigma_r$, and $A \in \Gamma \cup \{\bot\}$, is a *pop transition*. On reading the letter a, M can remove the symbol A from the top of its pushdown and switch from state q to q', if $A \in \Gamma$; if, however, $A = \bot$, then on reading a, M can switch from state q to q' without changing the content of its pushdown, if $\bot$ is the topmost symbol on the pushdown. Finally, a transition of the form (q, a, q'), where $q, q' \in Q$ and $a \in \Sigma_{int}$, is an *internal transition*. On reading the letter a, M can switch from state q to q' without looking at the pushdown at all.

Obviously, the content of the pushdown is always a word of the form $\bot \alpha$ for some $\alpha \in \Gamma^*$. A word $w \in \Sigma^*$ is *accepted* by M, if there exists a computation of M that starts from an initial configuration of the form $(q_0, w, \bot)$, where $q_0 \in Q_{in}$, and that ends with an accepting configuration of the form $(q_1, \varepsilon, \bot \alpha)$ for some $q_1 \in Q_{fin}$ and a word $\alpha \in \Gamma^*$. By $L(M)$ we denote the language accepted by M, and VPL denotes the class of all visibly pushdown languages.

It is known that REG $\subsetneq$ VPL $\subsetneq$ CFL holds, that VPL is closed under union, intersection, complementation, product, and Kleene star, and that each VPDA can effectively be converted into an equivalent *deterministic* VPDA [2].

Definition 1. Let $\Sigma = \Sigma_c \,\dot\cup\, \Sigma_r \,\dot\cup\, \Sigma_{int}$ be a pushdown alphabet. A language $L \subseteq \Sigma^*$ is the linearization of a visibly pushdown trace language, *if there exist* *a dependency relation D on Σ and a visibly pushdown language $R \subseteq \Sigma^*$ such that $L = \varphi_D^{-1}(\varphi_D(R)) = \bigcup_{w \in R}[w]_D$. By $\mathcal{LVPD}(D)$ we denote the set of all linearizations of visibly pushdown trace languages obtained from (Σ, D), and $\mathcal{LVPD}$ is the set of all linearizations of visibly pushdown trace languages.*

We illustrate this definition by a simple example.

Example 1. Let $R = \{ a^n(bc)^n \mid n \geq 0 \}$, where $\Sigma_c = \{a\}$, $\Sigma_r = \{c\}$, and $\Sigma_{int} = \{b\}$. Then it is easily seen that R is a visibly pushdown language over Σ. Let $D = \{(a,a),(a,b),(a,c),(b,a),(b,b),(c,a),(c,c)\}$, that is, $xbcy \equiv_D xcby$ holds for all $x, y \in \Sigma^*$. Then $L = \varphi_D^{-1}(\varphi_D(R)) = \{ a^n v \mid n \geq 0, v \in \{b,c\}^*, |v|_b = |v|_c = n \}$, which is not even context-free.

3   CD-Systems of Stateless Deterministic R(1)-Automata

Stateless types of restarting automata were introduced in [8]. Here we are only interested in the most restricted form of them, the *stateless deterministic R-automaton* of window size one. A *stateless deterministic R(1)-automaton* is a one-tape machine that is described by a 5-tuple $M = (\Sigma, \text{¢}, \$, 1, \delta)$, where Σ is a finite alphabet, the symbols $\text{¢}, \$ \notin \Sigma$ serve as markers for the left and right border of the work space, respectively, the size of the *read/write window* is one, and $\delta : \Sigma \cup \{\text{¢}, \$\} \to \{\text{MVR}, \text{Accept}, \varepsilon\}$ is the (partial) *transition function*. There are three types of transition steps: *move-right steps* (MVR), which shift the window one step to the right, combined *rewrite/restart steps* (denoted by ε), which delete the content a of the window, thereby shortening the tape, and place the window over the left end of the tape, and *accept steps* (Accept), which cause the automaton to halt and accept. Finally, we use the notation $\delta(a) = \emptyset$ to express the fact that the function δ is undefined for the symbol a. Some additional restrictions apply in that the sentinels ¢ and $\$ must not be deleted, and that the window must not move right on seeing the $\$-symbol.

A *configuration* of M is described by a pair (α, β), where either $\alpha = \varepsilon$ (the empty word) and $\beta \in \{\text{¢}\} \cdot \Sigma^* \cdot \{\$\}$ or $\alpha \in \{\text{¢}\} \cdot \Sigma^*$ and $\beta \in \Sigma^* \cdot \{\$\}$; here $\alpha\beta$ is the current content of the tape, and it is understood that the head scans the first symbol of β. A *restarting configuration* is of the form $(\varepsilon, \text{¢}w\$)$, where $w \in \Sigma^*$; to simplify the notation such a configuration is usually written as $\text{¢}w\$.

If $M = (\Sigma, \text{¢}, \$, 1, \delta)$ is a stateless deterministic R(1)-automaton, then we can partition its alphabet Σ into four disjoint subalphabets:

(1) $\Sigma_M = \{ a \in \Sigma \mid \delta(a) = \text{MVR} \}$,   (3) $\Sigma_A = \{ a \in \Sigma \mid \delta(a) = \text{Accept} \}$,
(2) $\Sigma_\varepsilon = \{ a \in \Sigma \mid \delta(a) = \varepsilon \}$,   (4) $\Sigma_\emptyset = \{ a \in \Sigma \mid \delta(a) = \emptyset \}$.

To exclude some trivial cases we assume in the following that $\delta(\text{¢}) = \text{MVR}$ holds for each stateless deterministic R(1)-automaton considered.

In [10] CD-systems of restarting automata were introduced. Here we study a restricted variant of the *pushdown CD-system of stateless deterministic R(1)-automata* (PD-CD-R(1)-system) of [13,14]. Such a system consists of a CD-system of stateless deterministic R(1)-automata and an external pushdown store. Formally, it is defined as a tuple $\mathcal{M} = (I, \Sigma, (M_i, \sigma_i)_{i \in I}, \Gamma, \bot, I_0, \delta)$, where

- I is a finite set of indices, and Σ is a finite input alphabet,
- for all $i \in I$, $M_i = (\Sigma, \text{¢}, \$, 1, \delta_i)$ is a stateless deterministic R(1)-automaton on Σ, and $\sigma_i \subseteq I$ is a non-empty set of possible successors for M_i,
- Γ is a finite pushdown alphabet, and $\bot \notin \Gamma$ is the bottom marker of the pushdown store,
- $I_0 \subseteq I$ is the set of initial indices, and
- $\delta : (I \times \Sigma \times (\Gamma \cup \{\bot\})) \to 2^{I \times (\Gamma \cup \{\bot\})^*}$ is the successor relation.

A *configuration* of $\mathcal{M}$ is a triple $(i, \omega, \bot\alpha)$, where $i \in I$, $\omega \in \text{¢} \cdot \Sigma^* \cdot \$ \cup \{\text{Accept}\}$, and $\alpha \in \Gamma^*$. It describes the situation that the component automaton M_i has just been activated, $\text{¢}w\$$ is the corresponding restarting configuration, and $\bot\alpha$ is the current content of the pushdown with the last symbol of α at the top. An *initial configuration* of $\mathcal{M}$ on input $w \in \Sigma^*$ has the form $(i_0, \text{¢}w\$, \bot)$ for any $i_0 \in I_0$, and an *accepting configuration* has the form $(i, \text{Accept}, \bot)$ for any $i \in I$.

The *single-step computation relation* $\Rightarrow_{\mathcal{M}}$ that $\mathcal{M}$ induces on the set of configurations is defined by the following three rules, where $i \in I$, $w \in \Sigma^*$, $\alpha \in \bot \cdot \Gamma^*$, $A \in \Gamma$, and, for each $i \in I$, $\Sigma_{\text{M}}^{(i)}$, $\Sigma_{\varepsilon}^{(i)}$, and $\Sigma_{\text{A}}^{(i)}$ are the subsets of Σ according to the above definition that correspond to the automaton M_i:

(1) $(i, \text{¢}w\$, \alpha A) \Rightarrow_{\mathcal{M}} (j, \text{¢}w'\$, \alpha\eta)$ if $\exists u \in {\Sigma_{\text{M}}^{(i)}}^*, a \in \Sigma_{\varepsilon}^{(i)}, v \in \Sigma^*$ such that $w = uav, w' = uv$, and $(j, \eta) \in \delta(i, a, A)$;

(2) $(i, \text{¢}w\$, \bot) \Rightarrow_{\mathcal{M}} (j, \text{¢}w'\$, \bot\eta)$ if $\exists u \in {\Sigma_{\text{M}}^{(i)}}^*, a \in \Sigma_{\varepsilon}^{(i)}, v \in \Sigma^*$ such that $w = uav, w' = uv$, and $(j, \bot\eta) \in \delta(i, a, \bot)$;

(3) $(i, \text{¢}w\$, \alpha) \Rightarrow_{\mathcal{M}} (i, \text{Accept}, \alpha)$ if $\exists u \in {\Sigma_{\text{M}}^{(i)}}^*, a \in \Sigma_{\text{A}}^{(i)}, v \in \Sigma^*$ such that $w = uav$, or $w \in {\Sigma_{\text{M}}^{(i)}}^*$ and $\delta_i(\$) = \text{Accept}$.

Notice that the content of the pushdown store is always of the form $\bot\alpha$ for some $\alpha \in \Gamma^*$. By $\Rightarrow_{\mathcal{M}}^*$ we denote the *computation relation* of $\mathcal{M}$, which is the reflexive and transitive closure of the relation $\Rightarrow_{\mathcal{M}}$. The language $L(\mathcal{M})$ accepted by $\mathcal{M}$ consists of all words for which $\mathcal{M}$ has an accepting computation, that is,

$$L(\mathcal{M}) = \{ w \in \Sigma^* \mid \exists i_0 \in I_0 \, \exists i \in I : (i_0, \text{¢}w\$, \bot) \Rightarrow_{\mathcal{M}}^* (i, \text{Accept}, \bot) \}.$$

Thus, the system $\mathcal{M}$ accepts if and when the currently active component M_i executes an accepting tail computation starting from the current restarting configuration $\text{¢}w'\$$, and the pushdown store just contains the bottom marker $\bot$.

Definition 2. A PD-CD-R(1)-*system* $\mathcal{M} = (I, \Sigma, (M_i, \sigma_i)_{i \in I}, \Gamma, \bot, I_0, \delta)$ *is in strong normal form if it satisfies the following conditions, where, for all $i \in I$, $\Sigma_{\text{M}}^{(i)}, \Sigma_{\varepsilon}^{(i)}, \Sigma_{\text{A}}^{(i)}, \Sigma_{\emptyset}^{(i)}$ is the partitioning of alphabet Σ for the automaton M_i:*

(1) $\exists i_+ \in I : \delta_{i_+}(\text{¢}) = \text{MVR}, \delta_{i_+}(\$) = \text{Accept}, \text{ and } \Sigma_{\emptyset}^{(i_+)} = \Sigma$;

(2) $\forall i \in I \smallsetminus \{i_+\} : \delta_i(\text{¢}) = \text{MVR}, |\Sigma_{\varepsilon}^{(i)}| = 1, \Sigma_{\text{A}}^{(i)} = \emptyset, \text{ and } \delta_i(\$) = \emptyset$.

Thus, if $\mathcal{M}$ is in strong normal form, then it has a unique component M_{i_+} that can execute accept instructions, but it only accepts the empty word, while all other components each delete a single kind of letter. In [13] it is shown that each PD-CD-R(1)-system $\mathcal{M}$ can effectively be converted into a PD-CD-R(1)-system $\mathcal{M}'$ in strong normal form such that $L(\mathcal{M}') = L(\mathcal{M})$. By $\mathcal{L}$(PD-CD-R(1)) we denote the class of languages that are accepted by PD-CD-R(1)-systems. In [13] a particular subclass of these systems is described that characterizes the class of linearizations of all context-free trace languages.

Now we introduce the restricted model of the PD-CD-R(1)-system announced above, where $\Sigma = \Sigma_c \,\dot\cup\, \Sigma_r \,\dot\cup\, \Sigma_{int}$ is a pushdown alphabet.

Definition 3. *A* visibly pushdown-CD-R(1)-system *(a* VPD-CD-R(1)-*system for short) over Σ is a* PD-CD-R(1)-*system* $\mathcal{M} = (I, \Sigma, (M_i, \sigma_i)_{i \in I}, \Gamma, \bot, I_0, \delta)$ *in strong normal form that satisfies the following additional restrictions:*

[**Push**] $\forall a \in \Sigma_c \,\forall i \in I: \text{ if } \Sigma_\varepsilon^{(i)} = \{a\}, \text{ then } \exists \Gamma(a,i) \subseteq \Gamma \text{ such that}$
$\delta(i, a, B) = \{ (j, BA) \mid j \in \sigma_i, A \in \Gamma(a, j) \} \text{ for all } B \in \Gamma \cup \{\bot\},$

that is, if component automaton M_i erases the input letter $a \in \Sigma_c$, then a symbol from the set $\Gamma(a, i)$ is pushed onto the pushdown store. This operation is independent of the current topmost symbol on the pushdown.

[**Pop**] $\forall a \in \Sigma_r \,\forall i \in I: \text{ if } \Sigma_\varepsilon^{(i)} = \{a\}, \text{ then } \exists \Gamma(a,i) \subseteq \Gamma \cup \{\bot\} \text{ such that}$
$- \delta(i, a, A) = \{ (j, \varepsilon) \mid j \in \sigma_i \} \quad \text{for all } A \in \Gamma(a, i) \smallsetminus \{\bot\},$
$- \delta(i, a, \bot) = \{ (j, \bot) \mid j \in \sigma_i \}, \text{ if } \bot \in \Gamma(a, i), \text{ and}$
$- \delta(i, a, B) = \emptyset \qquad\qquad \text{for all } B \in (\Gamma \cup \{\bot\}) \smallsetminus \Gamma(a, i),$

that is, if component automaton M_i erases the input letter $a \in \Sigma_r$, then a symbol from the set $\Gamma(a, i)$ must be popped from the pushdown store unless it only contains the bottom marker $\bot$.

[**Internal**] $\forall a \in \Sigma_{int} \,\forall i \in I: \text{ if } \Sigma_\varepsilon^{(i)} = \{a\}, \text{ then}$
$\delta(i, a, B) = \{ (j, B) \mid j \in \sigma_i \} \text{ for all } B \in \Gamma \cup \{\bot\},$

that is, if component automaton M_i erases the input letter $a \in \Sigma_{int}$, then the pushdown store is not used in the choice of the successor component.

An input word $w \in \Sigma^*$ is accepted by $\mathcal{M}$, if there exists a computation of the form $(i_0, \math12cent w\$, \bot) \Rightarrow_{\mathcal{M}}^* (i_+, \mathsf{Accept}, \bot\alpha)$ for some $i_0 \in I_0$ and $\alpha \in \Gamma^*$. As usual we denote the language of all input words that are accepted by $\mathcal{M}$ as $L(\mathcal{M})$.

Observe that it is not required that the pushdown only contains the bottom marker $\bot$ at the end of an accepting computation.

Example 2. Let $L = \{ a^n v \mid v \in \{b, c\}^*, |v|_b = |v|_c = n, n \geq 0 \}$, and let $\mathcal{M} = (I, \Sigma, (M_i, \sigma_i)_{i \in I}, \Gamma, \bot, I_0, \delta)$ be the VPD-CD-R(1)-system that is defined by $\Sigma_c = \{a\}$, $\Sigma_r = \{c\}$, and $\Sigma_{int} = \{b\}$, $I = \{a_1, a_2, b, c_1, c_2, +\}$, and $\Gamma = \{C, D\}$, the component automata are defined by the following transition functions:

$\delta_{a_1}(a) = \varepsilon, \quad \delta_b(b) = \varepsilon, \qquad \delta_{c_1}(c) = \varepsilon, \qquad \delta_{c_2}(c) = \varepsilon, \qquad \delta_+(\$) = \mathsf{Accept},$
$\delta_{a_2}(a) = \varepsilon, \quad \delta_b(c) = \mathsf{MVR}, \quad \delta_{c_1}(b) = \mathsf{MVR}, \quad \delta_{c_2}(b) = \mathsf{MVR},$

$\sigma_{a_1} = \{a_2, b\}$, $\sigma_{a_2} = \{a_2, b\}$, $\sigma_b = \{c_1, c_2\}$, $\sigma_{c_1} = \{+\}$, $\sigma_{c_2} = \{b\}$, $\sigma_+ = \{+\}$, and $I_0 = \{a_1, +\}$, and δ is defined as follows, where $A \in \Gamma \cup \{\bot\}$:

$\quad$(1) $\delta(a_1, a, A) = \{(a_2, AD), (b, AD)\}$, $\quad$(4) $\delta(c_1, c, D) = \{(+, \varepsilon)\}$,
$\quad$(2) $\delta(a_2, a, A) = \{(a_2, AC), (b, AC)\}$, $\quad$(5) $\delta(c_2, c, C) = \{(b, \varepsilon)\}$,
$\quad$(3) $\delta(b, b, A) = \{(c_1, A), (c_2, A)\}$,

and for all other tripels, δ yields the empty set. The letter a is the only one in Σ_c, and M_{a_1} and M_{a_2} are those component automata that can erase this letter. We have $\Gamma(a_1, a) = \{D\}$ and $\Gamma(a_2, a) = \{C\}$. The letter c is the only one in Σ_r, and M_{c_1} and M_{c_2} are those component automata that can erase this letter. We have $\Gamma(c_1, c) = \{D\}$ and $\Gamma(c_2, c) = \{C\}$.

The automaton M_+ just accepts the empty word, while M_{a_1} and M_{a_2} delete the first letter, if it is an a; otherwise, they get stuck. The automaton M_b reads across c's and deletes the first b it encounters, and analogously, M_{c_1} and M_{c_2} read across b's and delete the first c they encounter. Thus, we see from the successor sets that $\mathcal{M}$ can only accept certain words of the form $a^m v$ such that $v \in \{b, c\}^*$. In fact, it can be shown that $\mathcal{M}$ accepts iff $m = |v|_b = |v|_c$ holds. Hence, we see that $L(\mathcal{M}) = L$.

By $\mathcal{L}(\text{VPC-CD-R}(1))$ we denote the class of languages that are accepted by VPC-CD-R(1)-systems. When $\Sigma_{int} = \Sigma$, then a VPC-CD-R(1)-automaton over Σ does not use its pushdown at all. It now follows easily that VPC-CD-R(1)-systems can simulate the stl-det-local-CD-R(1)-systems of [11,12]. As the language L of Example 2 is not accepted by a system of that type, we have the following result.

Proposition 1. $\mathcal{L}(\text{stl-det-local-CD-R}(1)) \subsetneq \mathcal{L}(\text{VPC-CD-R}(1))$.

Further, VPD-CD-R(1)-systems accept all visibly pushdown languages, as from a given VPDA we can easily construct a VPD-CD-R(1)-system that accepts the same language

Proposition 2. VPL $\subsetneq \mathcal{L}(\text{VPD-CD-R}(1))$.

Let $\psi : \Sigma^* \to \mathbb{N}^n$ denote the *Parikh mapping*, where $\Sigma = \{a_1, \ldots, a_n\}$. Applying a construction similar to the one used in [14], we can derive the following result.

Theorem 1. *Each language* $L \in \mathcal{L}(\text{VPD-CD-R}(1))$ *contains a sublanguage* $E \in$ VPL *such that* $\psi(L) = \psi(E)$ *holds. In fact, a* VPDA *for* E *can be constructed effectively from a* VPD-CD-R(1)-*system for* L.

The context-free language $L = \{a^n b a^n \mid n \geq 1\}$ does not contain any sublanguage that is a visibly pushdown language and that is letter-equivalent to the language itself. Hence, Theorem 1 yields the following negative result.

Proposition 3. $L = \{a^n b a^n \mid n \geq 1\} \notin \mathcal{L}(\text{VPD-CD-R}(1))$.

As PD-CD-R(1)-systems accept all context-free languages, the following proper inclusion can be derived.

Proposition 4. $\mathcal{L}(\text{VPC-CD-R}(1)) \subsetneq \mathcal{L}(\text{PD-CD-R}(1))$.

While the language $L = \{a^n v \mid v \in \{b, c\}^*, |v|_b = |v|_c = n\}$ of Example 2 is accepted by a VPD-CD-R(1)-system, the intersection $L \cap (a^* \cdot b^* \cdot c^*)$ of L with the regular set $a^* \cdot b^* \cdot c^*$ is not even accepted by any PD-CD-R(1)-system (see [14], Prop. 3.11). Thus, we have the following closure and non-closure results.

Theorem 2. *The language class* $\mathcal{L}$(VPC-CD-R(1)) *is effectively closed under union, but it is neither closed under complement, nor under intersection, nor under intersection with regular sets.*

In the following we consider a subclass of the VPC-CD-R(1)-systems that characterizes the visibly pushdown trace languages.

4   Characterizing the Visibly Pushdown Trace Languages

We are interested in the (linearizations of) visibly pushdown trace languages. By a modification of the proof of Proposition 2, the following result can be shown.

Theorem 3. *If D is a dependency relation on a finite pushdown alphabet $\Sigma = \Sigma_c \mathbin{\dot\cup} \Sigma_r \mathbin{\dot\cup} \Sigma_{int}$, then $\mathcal{L}\mathcal{V}\mathcal{P}\mathcal{D}(D) \subseteq \mathcal{L}$(VPD-CD-R(1)).*

The language $L' = \{\, wa^m \mid |w|_a = |w|_b = |w|_c \geq 1, m \geq 1 \,\}$ on $\Sigma = \{a, b, c\}$ is accepted by a stl-det-local-CD-R(1)-system ([12], Example 4.15), and hence, it is accepted by a VPD-CD-R(1)-system by Proposition 1, but as shown in [14], Prop. 4.8, this language is not the linearization of any context-free trace language over Σ. Thus, we obtain the following result.

Corollary 1. $\mathcal{L}\mathcal{V}\mathcal{P}\mathcal{D} \subsetneq \mathcal{L}$(VPD-CD-R(1)).

We now concentrate on a restricted class of VPD-CD-R(1)-systems.

Definition 4. *Let $\mathcal{M} = (I, \Sigma, (M_i, \sigma_i)_{i \in I}, \Gamma, \bot, I_0, \delta)$ be a VPD-CD-R(1)-system that satisfies the following condition:*

$$(*) \qquad \forall i, j \in I : \Sigma_\varepsilon^{(i)} = \Sigma_\varepsilon^{(j)} \text{ implies that } \Sigma_M^{(i)} = \Sigma_M^{(j)},$$

that is, if two component automata erase the same letter, then they also read across the same subset of Σ. With $\mathcal{M}$ we associate the binary relation $I_\mathcal{M} = \bigcup_{i \in I}(\Sigma_M^{(i)} \times \Sigma_\varepsilon^{(i)})$, that is, $(a, b) \in I_\mathcal{M}$ iff there exists a component automaton M_i such that $\delta_i(a) = $ MVR and $\delta_i(b) = \varepsilon$. Further, let $D_\mathcal{M} = (\Sigma \times \Sigma) \setminus I_\mathcal{M}$.

Observe that the relation $I_\mathcal{M}$ is necessarily irreflexive, but it will in general not be symmetric. Now the following characterization can be established.

Theorem 4. *Let $\mathcal{M}$ be a VPD-CD-R(1)-system over Σ satisfying condition $(*)$ above. If the associated relation $I_\mathcal{M}$ is symmetric, then $D_\mathcal{M}$ is a dependency relation on Σ, and $L(\mathcal{M}) \in \mathcal{L}\mathcal{V}\mathcal{P}\mathcal{D}(D_\mathcal{M})$. In fact, from $\mathcal{M}$ one can construct a VPDA B over Σ such that $L(\mathcal{M}) = \bigcup_{u \in L(B)} [u]_{D_\mathcal{M}}$.*

The system $\mathcal{M}$ constructed in the proof of Theorem 3 satisfies property $(*)$, and the associated relation $I_\mathcal{M}$ coincides with the relation I_D, and hence, it is symmetric. Thus, Theorems 3 and 4 yield the following characterizations.

Corollary 2. *A language $L \subseteq \Sigma^*$ is the linearization of a visibly pushdown trace language if and only if there exists a VPD-CD-R(1)-system $\mathcal{M}$ satisfying condition $(*)$ such that the relation $I_\mathcal{M}$ is symmetric and $L = L(\mathcal{M})$.*

5   Deterministic VPD-CD-R(1)-Systems

It is known that, from a given VPDA M, a deterministic VPDA M' can be constructed such that $L(M') = L(M)$ holds [2]. Does a corresponding result hold for VPD-CD-R(1)-systems that accept linearizations of visibly pushdown trace languages?

In [15] a deterministic variant of the PD-CD-R(1)-systems was introduced and studied, the *deterministic pushdown CD-systems of* R(1)-*automata*, or det-PD-CD-R(1)-systems, and it was show that the language class $\mathcal{L}(\text{det-PD-CD-R}(1))$ is incomparable to the class of linearizations of context-free trace languages with respect to inclusion. Here we adjust the notion of a deterministic PD-CD-R(1)-system to obtain a deterministic variant of the VPD-CD-R(1)-system.

Definition 5. *A deterministic visibly pushdown-CD-R(1)-system (or det-VPD-CD-R(1)-system) over a pushdown alphabet* $\Sigma = \Sigma_c \,\dot{\cup}\, \Sigma_r \,\dot{\cup}\, \Sigma_{int}$ *is a* det-PD-CD-R(1)-*system* $\mathcal{M} = (I, \Sigma, (M_i, \sigma_i)_{i \in I}, \Gamma, \bot, i_0, \delta)$ *that satisfies the following additional restrictions:*

[Push] $\forall a \in \Sigma_c \, \forall i \in I :$ *if* $a \in \Sigma_{\varepsilon}^{(i)}$, *then* $\exists j \in \sigma_i \, \exists A \in \Gamma$ *such that*
$\delta(i, a, B) = (j, BA)$ *for all* $B \in \Gamma \cup \{\bot\}$,

that is, if component automaton M_i erases the input letter $a \in \Sigma_c$, then a uniquely determined symbol $A \in \Gamma$ is pushed onto the pushdown store. This operation is independent of the current topmost symbol on the pushdown.

[Pop] $\forall a \in \Sigma_r \, \forall i \in I :$ *if* $a \in \Sigma_{\varepsilon}^{(i)}$, *then* $\forall A \in \Gamma \cup \{\bot\} \, \exists j \in \sigma_i$ *such that*
$-\ \delta(i, a, A) = (j, \varepsilon), \quad$ *if* $A \in \Gamma$, *and*
$-\ \delta(i, a, \bot) = (j, \bot), \quad$ *if* $A = \bot$,

that is, if component automaton M_i erases the input letter $a \in \Sigma_r$, then a symbol must be popped from the pushdown store unless it only contains the bottom marker $\bot$.

[Internal] $\forall a \in \Sigma_{int} \, \forall i \in I :$ *if* $a \in \Sigma_{\varepsilon}^{(i)}$, *then* $\exists j \in \sigma_i$ *such that*
$\delta(i, a, B) = (j, B)$ *for all* $B \in \Gamma \cup \{\bot\}$,

that is, if component automaton M_i erases the input letter $a \in \Sigma_{int}$, then the pushdown store is not used in the choice of the successor component.

An input word $w \in \Sigma^*$ is accepted by $\mathcal{M}$, if there exists a computation of the form $(i_0, \mathorigin{\mathcal{c}} w\$, \bot) \Rightarrow_{\mathcal{M}}^* (i, \mathorigin{\mathcal{c}} x\$, \bot\alpha) \Rightarrow_{M_i}^* (i, \text{Accept}, \bot\alpha)$ *for some $i \in I$ and some $\alpha \in \Gamma^*$, that is, if the computation that starts with the initial configuration of $\mathcal{M}$ on input w ends with a component M_i that accepts the current tape contents $\mathorigin{\mathcal{c}} x\$. Observe that here it is not required that the given input is completely erased by an accepting computation.*

Again we illustrate this definition by a simple example, in which we consider a slight variation of the language of Example 2.

Example 3. Let $L_{Ex3} = \{\, a^n v \mid v \in \{b, c\}^*, |v|_b = |v|_c = n, n \geq 0$, and $|v_1|_b \geq |v_1|_c$ for each prefix v_1 of $v \,\}$. In analogy to the language considered in Example 2 it follows that this language is not context-free and that it is not accepted by any stl-det-local-CD-R(1)-system.

Let $\mathcal{M} = (I, \Sigma, (M_i, \sigma_i)_{i \in I}, \Gamma, \bot, 0, \delta)$ be defined as follows, where Σ is turned into a pushdown alphabet by taking $\Sigma_c = \{a\}$, $\Sigma_r = \{c\}$, and $\Sigma_{int} = \{b\}$:

- $I = \{0, 1, 2, 3, +\}$, and $\Gamma = \{C, D\}$,
- M_0, M_1, M_2, M_3, and M_+ are defined by the following transition functions:

$$\delta_0(\$) = \mathsf{Accept}, \delta_0(a) = \varepsilon, \delta_1(a) = \varepsilon, \delta_1(b) = \varepsilon,$$
$$\delta_+(\$) = \mathsf{Accept}, \delta_2(b) = \varepsilon, \delta_3(c) = \varepsilon, \delta_3(b) = \mathsf{MVR},$$

- $\sigma_0 = \{1\}$, $\sigma_1 = \{1, 3\}$, $\sigma_2 = \{3\}$, $\sigma_3 = \{2, +\}$, and
- δ is defined as follows, where $A \in \Gamma \cup \{\bot\}$:

$$
\begin{array}{ll}
(1)\ \delta(0, a, A) = (1, AD), & (4)\ \delta(2, b, A) = (3, A), \\
(2)\ \delta(1, a, A) = (1, AC), & (5)\ \delta(3, c, C) = (2, \varepsilon), \\
(3)\ \delta(1, b, A) = (3, A), & (6)\ \delta(3, c, D) = (+, \varepsilon).
\end{array}
$$

It is easily seen that $L(\mathcal{M}) = L_{Ex3}$ holds.

As each visibly pushdown language is accepted by a deterministic VPDA, and as a deterministic VPDA can easily be turned into a det-VPD-CD-R(1)-system for the same language, we have the following inclusion result.

Proposition 5. VPL $\subsetneq \mathcal{L}(\text{det-VPD-CD-R}(1))$.

This inclusion is proper as seen from the language L_{Ex3}. Further, Theorem 1 also holds for det-VPD-CD-R(1)-systems. Next we consider some of the closure properties of the class $\mathcal{L}(\text{det-VPD-CD-R}(1))$.

Theorem 5. $\mathcal{L}(\text{det-VPD-CD-R}(1))$ *is closed under complement. In fact, from a given* det-VPD-CD-R(1)-*system* $\mathcal{M}$ *for a language* L *over* Σ, *one can effectively construct a* det-VPD-CD-R(1)-*system* $\mathcal{M}^c$ *for the language* $L^c = \Sigma^* \smallsetminus L$.

Unfortunately, $\mathcal{L}(\text{det-VPD-CD-R}(1))$ is not closed under intersection with regular sets, as $L_{Ex3} \cap (a^* \cdot b^* \cdot c^*) = \{a^n b^n c^n \mid n \geq 0\}$, which does not contain any sublanguage that is a visibly pushdown language and that is letter-equivalent to $L_{Ex3} \cap (a^* \cdot b^* \cdot c^*)$. Hence, we obtain the following nonclosure results.

Corollary 3. $\mathcal{L}(\text{det-VPD-CD-R}(1))$ *is not closed under union or intersection.*

It follows that $\mathcal{L}(\text{det-VPD-CD-R}(1))$ is a proper subclass of $\mathcal{L}(\text{VPD-CD-R}(1))$ that is incomparable to the class $\mathcal{LVPD}$ with respect to inclusion.

6   Conclusion

We have introduced a class of pushdown CD systems of restarting automata that accepts a proper superclass of the linearizations of all visibly pushdown trace languages, and we have characterized the latter class in terms of a restricted type of pushdown CD systems. In addition, we have considered a deterministic type of these systems, but as it turned out they do not accept the linearizations

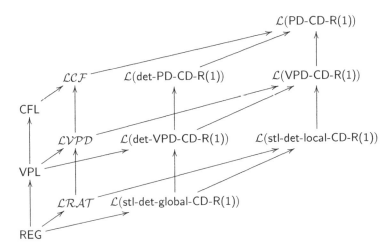

Fig. 1. Hierarchy of language classes accepted by various types of CD-systems of stl-det-R(1)-automata. Each arrow represents a proper inclusion, and classes that are not connected by a sequence of arrows are incomparable under inclusion.

of all visibly pushdown trace languages. The diagram in Figure 1 summarizes the hierarchy results that we have obtained for the classes of languages that are accepted by (deterministic and nondeterministic) VPD-CD-R(1)-systems.

The emptiness problem and the finiteness problem are decidable for context-free languages. From Theorem 1 we know that we can construct a visibly pushdown sublanguage E of $L(\mathcal{M})$ from a given (det-)VPD-CD-R(1)-system $\mathcal{M}$ such that E is letter-equivalent to $L(\mathcal{M})$. Hence, $L(\mathcal{M})$ is empty (finite) if and only if E is empty (finite). Further, $\mathcal{L}(\text{det-VPD-CD-R}(1))$ is effectively closed under complementation. As $L(\mathcal{M}) = \Sigma^*$ iff $\Sigma^* \smallsetminus L(\mathcal{M}) = \emptyset$, we immediately obtain that universality is decidable for these systems.

On the other hand, for stl-det-global-CD-R(1)-systems, the inclusion problem is undecidable. As the latter systems can be seen as a special type of det-VPD-CD-R(1)-systems, the inclusion problem is also undecidable for our systems. Actually, it is even undecidable whether $L(\mathcal{M})$ is contained in a given regular language, or whether it contains a given regular language (see [16], Cor. 5.4).

References

1. Aalbersberg, I., Rozenberg, G.: Theory of traces. Theoret. Comput. Sci. 60, 1–82 (1988)
2. Alur, R., Madhusudan, P.: Visibly pushdown languages. In: STOC 2004, pp. 202–211. ACM Press (2004)
3. Bordihn, H., Holzer, M., Kutrib, M.: Input reversals and iterated pushdown automata: a new characterization of Khabbaz geometric hierarchy of languages. In: Calude, C.S., Calude, E., Dinneen, M.J. (eds.) DLT 2004. LNCS, vol. 3340, pp. 102–113. Springer, Heidelberg (2004)

4. Bertoni, A., Mauri, G., Sabadini, N.: Membership problems for regular and context-free trace languages. Inform. Comput. 82, 135–150 (1989)
5. Bollig, B., Grindei, M.L., Habermehl, P.: Realizability of concurrent recursive programs. In: de Alfaro, L. (ed.) FOSSACS 2009. LNCS, vol. 5504, pp. 410–424. Springer, Heidelberg (2009)
6. Braunmühl, B., von Verbeek, R.: Input-driven languages are recognized in log n space. In: Karpinski, M. (ed.) FCT 1983. LNCS, vol. 158, pp. 40–51. Springer, Heidelberg (1983)
7. Diekert, V., Rozenberg, G.: The Book of Traces. World Scientific, Singapore (1995)
8. Kutrib, M., Messerschmidt, H., Otto, F.: On stateless two-pushdown automata and restarting automata. In: Csuhaj-Varjú, E., Ésik, Z. (eds.) Proc. Automata and Formal Languages, AFL 2008, pp. 257–268. Computer and Automation Research Institute, Hungarian Academy of Sciences (2008)
9. Mazurkiewicz, A.: Concurrent program schemes and their interpretations. DAIMI Rep. PB 78, Aarhus University, Aarhus (1977)
10. Messerschmidt, H., Otto, F.: Cooperating distributed systems of restarting automata. Intern. J. Found. Comput. Sci. 18, 1333–1342 (2007)
11. Nagy, B., Otto, F.: CD-systems of stateless deterministic R(1)-automata accept all rational trace languages. In: Dediu, A.-H., Fernau, H., Martín-Vide, C. (eds.) LATA 2010. LNCS, vol. 6031, pp. 463–474. Springer, Heidelberg (2010)
12. Nagy, B., Otto, F.: On CD-systems of stateless deterministic R-automata with window size one. JCSS 78, 780–806 (2012)
13. Nagy, B., Otto, F.: An automata-theoretical characterization of context-free trace languages. In: Černá, I., Gyimóthy, T., Hromkovič, J., Jefferey, K., Královič, R., Vukolić, M., Wolf, S. (eds.) SOFSEM 2011. LNCS, vol. 6543, pp. 406–417. Springer, Heidelberg (2011)
14. Nagy, B., Otto, F.: CD-systems of stateless deterministic R(1)-automata governed by an external pushdown store. RAIRO 45, 413–448 (2011)
15. Nagy, B., Otto, F.: Deterministic pushdown-CD-systems of stateless deterministic R(1)-automata. Acta Inform. 50, 229–255 (2013)
16. Nagy, B., Otto, F.: Globally deterministic CD-systems of stateless R-automata with window size 1. Intern. J. Comput. Math. 90, 1254–1277 (2013)
17. Srba, J.: Visibly pushdown automata: From language equivalence to simulation and bisimulation. In: Ésik, Z. (ed.) CSL 2006. LNCS, vol. 4207, pp. 89–103. Springer, Heidelberg (2006)
18. Zielonka, W.: Notes on finite asynchronous automata. RAIRO 21, 99–135 (1987)

Dominating an *s*-*t*-Cut in a Network

Ralf Rothenberger[1], Sascha Grau[2], and Michael Rossberg[2]

[1] Friedrich-Schiller-Universität Jena, Jena, Germany
ralf.rothenberger@uni-jena.de
[2] Technische Universität Ilmenau, Ilmenau, Germany
{first.lastname}@tu-ilmenau.de

Abstract. We study an optimization problem with applications in design and analysis of resilient communication networks: given two vertices s, t in a graph $G = (V, E)$, find a vertex set $X \subset V$ of minimum cardinality, such that X and its neighborhood constitute an s-t vertex separator. Although the problem naturally combines notions of graph connectivity and domination, its computational properties significantly differ from these relatives.

In particular, we show that on general graphs the problem cannot be approximated to within a factor of $2^{\log^{1-\delta} n}$, with $\delta = 1/\log\log^c n$ and arbitrary $c < 1/2$ (if P $\neq$ NP). This inapproximability result even applies if the subgraph induced by a solution set has the additional constraint of being connected. Furthermore, we give a $2\sqrt{n}$-approximation algorithm and study the problem on graphs with bounded node degree. With Δ being the maximum degree of nodes $V \setminus \{s, t\}$, we identify a $(\Delta + 1)$ approximation algorithm.

Keywords: graph theory, approximation algorithms, inapproximability.

1 Introduction

In recent years, the development of secure overlay networks has strongly advanced (e.g. [1,2,3]). As a consequence, we are approaching a situation, where the effort an attacker needs to spend on identifying worthwile targets may exceed the costs of mounting the actual attack. This is especially true, since huge botnets, which are able to conduct massive denial-of-service attacks, can be cheaply rent on the internet. In contrast, while actively observing a network node will reliably reveal its communication partners, it might be connected to a risk of detection, a risk of failure or a considerable amount of resources necessary to obtain the involved nodes' addresses.

Motivated by these facts, we study a problem that we term as CUT DOMINATION: given a graph G and a pair of particular nodes s and t, we seek to select a node set X of minimum cardinality such that the nodes in X and their neighborhood constitute an s-t-vertex-separator.

This problem is posed to an attacker possessing knowledge about the network topology, but not about actual addresses of the participants (needed to mount the attack). Examples for such settings are virtual private networks with dynamic

G.F. Italiano et al. (Eds.): SOFSEM 2015, LNCS 8939, pp. 401–411, 2015.
© Springer-Verlag Berlin Heidelberg 2015

routing [1], where the data itself is encrypted and authenticated, but denial-of-service attacks may still cause a severe thread.

A related theoretical problem was first described in [4,5], where, given graph G and a budget $x \leq n$, the number of node pairs rendered unreachable by removing x nodes and their respective neighborhood was to be maximized. The derived decision problem was shown to be NP-complete, but needs to be solved in order to build more resilient networks.

CUT DOMINATION differs from this problem as only paths between a specific node pair s, t are to be disrupted. One can imagine s and t as important and well-equipped communication partners, leaving the intermediate network nodes as easier targets for an attack. Although the related network formation problem is easy to solve (connect s and t by as many isolated, parallel paths as possible), the properties of CUT DOMINATION are interesting in their own respect and might provide insight into the practically-motivated problem described in [4,5].

Our contributions are the following: After introducing a problem formalization in Section 2, we show in Section 3 that generally the CUT DOMINATION problem cannot be approximated to within a factor of $2^{\log^{1-\delta} n}$, with $\delta = 1/\log \log^c n$ and arbitrary $c < 1/2$ (if P $\neq$ NP). This result also holds if the observed node set has to be connected (CONNECTED CUT DOMINATION) and for the respective weighted variant (WEIGHTED CUT DOMINATION). In Section 4 we give a $2\sqrt{n}$-approximation algorithm for WEIGHTED CUT DOMINATION, which can also be used to approximate CONNECTED CUT DOMINATION to within a ratio of $n^{2/3}$ of an optimal solution. Finally, in Section 5, we show a $(\Delta + 1)$-approximation algorithm for WEIGHTED CUT DOMINATION, with Δ denoting the maximum degree of nodes $V \setminus \{s, t\}$. Since CUT DOMINATION is a special case of WEIGHTED CUT DOMINATION, all upper bound results for the weighted variant also apply to the unweighted version.

2   Problem Definition and Notation

To formalize the studied optimization problem, we first introduce necessary notation: For an undirected graph $G = (V, E)$ and a node $u \in V$, let the *inclusive neighborhood of u* be $N^+(u) = \{u\} \cup \{v \in V \mid \{u, v\} \in E\}$. Analogously, for a set $U \subseteq V$, let $N^+(U) = \bigcup_{u \in U} N^+(u)$ be the *inclusive neighborhood of U*.

Furthermore, for an undirected graph $G = (V, E)$ and non-adjacent nodes $s, t \in V$, an *s-t-vertex-separator* is a node set $U \subseteq V \setminus \{s, t\}$ with the property that the removal of U from G disconnects s and t. It is a well-known result, that a minimum s-t-vertex-separator can be found in polynomial time [6]. Sometimes, such a set is also called an *s-t vertex cut*. In the same context, we define an *s-t-cut dominator* to be a set $U \subseteq V \setminus \{s, t\}$, so that $N^+(U) \setminus \{s, t\}$ is an *s-t-vertex-separator* of G. In other words, U dominates an s-t-vertex-separator.

Given a simple undirected graph $G = (V, E)$ and two non-adjacent nodes $s, t \in V$, the CUT DOMINATION problem consists of finding a minimum s-t-cut dominator. Furthermore, we define the CONNECTED CUT DOMINATION problem

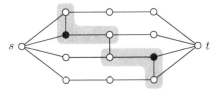 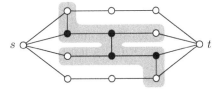

(a) Smallest *s-t*-cut dominator (black) and   (b) Smallest *connected s-t*-cut dominator
its dominated *s-t*-cut (grey)   (black) and its dominated *s-t*-cut (grey)

Fig. 1. Variants of cut domination

of finding a *connected s-t*-cut dominator of minimum cardinality. Examples for typical solutions are given in Figure 1.

Both problems admit a natural generalization by adding a weight function $w : V \to \mathbb{R}^+$ that assigns positive weights to the nodes of G. Trying to find a (connected) set $U \subseteq V \setminus \{s, t\}$ of *minimum weight* $w(U) = \sum_{v \in U} w(v)$ dominating an *s-t*-separator is called WEIGHTED (CONNECTED) CUT DOMINATION.

In the rest of the paper let the problem size $n = |V \setminus \{s, t\}|$, the number of nodes excluding s and t.

3   Inapproximability of CUT DOMINATION

We show an approximation-preserving polynomial-time reduction from RED-BLUE SET COVER to CUT DOMINATION.

The RED-BLUE SET COVER problem is a generalization of the SET COVER problem, where the universe U is partitioned into two subsets, a set R of red elements and a set B of blue elements. We are given a collection of sets $\mathcal{S} = \{S_1, S_2, \ldots, S_n\}$ over the universe U and have to find a subcollection $C \subseteq \mathcal{S}$ containing all blue elements while also containing a minimum number of red elements. Let $R(C) = \bigcup_{S_i \in C} S_i \cap R$ denote the set of red elements covered by the subcollection C.

Carr et al. showed in [7] that RED-BLUE SET COVER is $\mathcal{O}(2^{\log^{1-\delta} n})$-inapproximable with $\delta = 1/\log \log^c n$ for every constant $c < 1/2$, unless P=NP. This result even holds for RED-BLUE SET COVER with the additional constraint that every set $S_i \in \mathcal{S}$ only contains one blue and two red elements.

Theorem 1. CUT DOMINATION *is* $\mathcal{O}(2^{\log^{1-\delta} n})$-*inapproximable for every constant* $c < 1/2$, *with* $\delta = 1/\log \log^c n$, *if* P$\neq$NP.

Proof. We are given an instance $I = (\mathcal{S}, R, B)$ of RED-BLUE SET COVER with the constraint, that every set contains one blue and two red elements. W.l.o.g. we can assume, that every red element and every blue element is contained in at least one set $S \in \mathcal{S}$. Furthermore, we assume an arbitrary ordering of the sets in $\mathcal{S}$. We now build an instance $I' = (G = (V, E), s, t)$ of CUT DOMINATION with the following properties:

(1) Every feasible solution $C \subseteq S$ for I corresponds to a feasible solution $U \subseteq V \setminus \{s,t\}$ of size $|R(C)|$ for I'.
(2) For every feasible solution $U \subseteq V \setminus \{s,t\}$ for I', we can find a solution $C \subseteq S$ for I with $|R(C)| \leq |U|$.

Starting from an empty graph, we first create nodes s and t. Then we add a complete subgraph of 'red' nodes $V_R = \{v_r \mid r \in R\}$, each of them corresponding to one of the red elements in R. Afterwards, we construct two s-t-pathways, so called *b-connectors*, for each blue element $b \in B$ and connect them to some of the 'red' nodes. This is done in such a way, that all of the b-connectors have to be cut to disconnect s and t, while cutting them can be done by selecting pairs of 'red' nodes whose corresponding elements are in a set together with b. The construction of the b-connectors will be explained in greater detail now, for an arbitrary, but fixed blue element $b \in B$.

For every set $S_i \in S$ that contains b, we do the following. First, we create a pair of nodes u_l^i, u_k^i corresponding to the red elements r_l, r_k in S_i. Second, we add edges $\{u_l^i, v_l\}$ and $\{u_k^i, v_k\}$. Third, we connect both u_l^i and u_k^i to each node of the previously created pair for b. If there is no previously created pair for b, we connect u_l^i and u_k^i to s. After examining all sets, we connect both nodes of the lastly created pair to t. This gives us the first b-connector. An example of such a b-connector can be seen in Fig. 2. By repeating this procedure and creating node pairs w_l^i, w_k^i instead of u_l^i, u_k^i, we obtain the second b-connector.

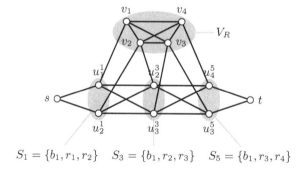

$$S_1 = \{b_1, r_1, r_2\} \quad S_3 = \{b_1, r_2, r_3\} \quad S_5 = \{b_1, r_3, r_4\}$$

Fig. 2. The first b_1-connector

We do this for all blue elements $b \in B$ to get the graph G. The construction can obviously be performed in polynomial time and creates a graph G with exactly $4|S| + |R|$ nodes, excluding s and t. Since every red element appears in at least one set of S and every set contains exactly two red elements, there can be at most $2|S|$ red elements. Thus, it holds that $|V(G) \setminus \{s,t\}| \leq 6|S|$.

The only thing left for us to show is, how to transform feasible solutions of the RED-BLUE SET COVER instance I to feasible solutions of the CUT DOMINATION instance I' and vice versa while containing their costs.

Given a solution $C \subseteq \mathcal{S}$ of I we determine the set $R(C)$ of red elements covered by C and take the nodes $U = \{v_r \in V_R \mid r \in R(C)\}$ as an s-t-cut dominating set in G. Note that U only consists of nodes from V_R and does not contain any node of a b-connector. By definition, it holds that $|R(C)| = |U|$. Choosing at least one of the $v_r \in V_R$, all nodes of V_R are dominated since V_R induces a complete subgraph. Therefore the only s-t paths left may lead over the b-connectors. Consider an arbitrary blue element $b \in B$. Since C is a solution to I it has to cover b. Consequently, there has to be a set $S_i \in C$ containing b. Furthermore, both the red elements r_k and r_l from S_i have to be contained in $R(C)$. This means, that v_{r_k} and v_{r_l} are in U. According to the construction, there are node pairs u_k^i, u_l^i and w_k^i, w_l^i in the b-connectors and edges $\{v_{r_k}, u_k^i\}$, $\{v_{r_l}, u_l^i\}, \{v_{r_k}, w_k^i\}$ and $\{v_{r_l}, w_l^i\}$. Due to the fact, that v_{r_k} and v_{r_l} are chosen, u_k^i, u_l^i, w_k^i and w_l^i are dominated, therefore cutting both b-connectors. Since this holds for every $b \in B$, all b-connectors are cut, thus separating s and t.

Consider now a set $U \subseteq V(G) \setminus \{s,t\}$ that dominates an s-t-separator of G. First, we show that we can choose a set $U' \subseteq V_R$ with $|U'| \leq |U|$ that also dominates an s-t-separator of G. To do so, we consider the b-connectors. Due to the construction there are only two ways to dominate a cut of a b-connector: either by choosing a node of the connector or by choosing two nodes $v_{r_k}, v_{r_l} \in V_R$, so that $\{b, r_k, r_l\} \in \mathcal{S}$. In the second case, both b-connectors are cut and we are done. Otherwise, there has to be at least one node from U on each of the two b-connectors. Instead of taking these two nodes, we can arbitrarily choose a set $S_j \in \mathcal{S}$ with $b \in S_j$ and take the nodes $v_{r_k}, v_{r_l} \in V_R$ corresponding to the red elements $r_k, r_l \in S_j$. By doing so we still cut both b-connectors and additionally dominate all nodes from V_R, if they were not already. We can do this for all $b \in B$ while still containing the size of the solution. This first step ensures that all b-connectors are cut by nodes in V_R. Afterwards, we eliminate all nodes from $V \setminus V_R$ from the solution to obtain U'.

We now define $R(U')$ to be the set of all red elements whose corresponding nodes are in U'. We can choose C as the collection of all sets $S \in \mathcal{S}$ that contain only red elements from $R(U')$. Since $U' \subseteq V_R$ and $R(C) \subseteq R(U')$ it now holds that $|R(C)| \leq |R(U')| = |U'|$. Our transformation ensures that for every blue element $b \in B$ there are nodes $v_{r_k}, v_{r_l} \in U'$ with $\{b, r_k, r_l\} \in \mathcal{S}$. Especially $r_k, r_l \in R(U')$ and therefore $\{b, r_k, r_l\} \in C$. Since this holds for every $b \in B$, C has to cover all blue elements. $\qquad\square$

WEIGHTED CUT DOMINATION has to be at least as difficult to approximate as the unweighted case. Hence, the inapproximability result also holds for the weighted variant of the problem. The s-t-cut dominating set that can be constructed from a RED-BLUE SET-COVER solution consists only of nodes from V_R, therefore it is connected. Furthermore, we transform a feasible solution of the constructed graph G to a solution of the same size that consists only of nodes from V_R. This is possible for every s-t-cut dominating set of G. Especially, it is possible for every *connected* s-t-cut dominating set of G. Consequently, Theorem 1 also holds for CONNECTED CUT DOMINATION.

4   A $2\sqrt{n}$-Approximation in General Graphs

We give a $2\sqrt{n}$-approximation algorithm for WEIGHTED CUT DOMINATION. Since the unweighted problem is a special case of the weighted variant, the approximability result holds for both.

Algorithm 1. Weighted Cut Domination Approximation (G, s, t, w)

1 **foreach** $w_i \in \{w(v) \mid v \in V \setminus \{s, t\}\}$ **do**
2      $U_1 \leftarrow \emptyset$;

3   **while** $\exists\, v \in V \setminus (U_1 \cup \{s, t\})$: $\frac{|N^+(v) \setminus (\{s,t\} \cup N^+(U_1))|}{w(v)} \geq \frac{\sqrt{n}}{w_i}$ **do**
4   choose such a node v arbitrarily;
5          $U_1 \leftarrow U_1 \cup \{v\}$;

6      $G' \leftarrow G \setminus (N^+(U_1) \setminus \{s, t\})$;
7   **if** s and t are in the same connected component H of G' **then**
8   **foreach** $v \in V \setminus \{s, t\}$ **do** $w'(v) \leftarrow \min \{w(u) \mid u \in N^+(v) \setminus \{s, t\}\}$
          $C \leftarrow$ min-vertex-cut(H, s, t, w');
9          $C_i \leftarrow \emptyset$;
10   **foreach** $v \in C$ **do**
11             $v' \leftarrow \text{argmin}\{w(u) \mid u \in N^+(v) \setminus \{s, t\}\}$;
12             $C_i \leftarrow C_i \cup \{v'\}$;
13         $C_i \leftarrow C_i \cup U_1$;

14   **else**
15         $C_i \leftarrow U_1$;

16 **return** $\text{argmin}_{X \in \{C_i \mid 1 \leq i \leq n\}}\{w(X)\}$;

Algorithm 1 proceeds as follows: the weights appearing in the graph are, one by one, considered as maximum weight of a vertex from an optimal solution. For every weight w_i, we compute a candidate solution C_i as follows.

First, we greedily choose nodes v, which dominate at least $w(v)\sqrt{n}/w_i$ currently undominated nodes, including themselves, and consider their inclusive neighborhoods as dominated. This is repeated until either an s-t-cut is dominated or no appropriate node is left. The result of this greedy selection is a set U_1. Second, we consider the induced subgraph G' of currently undominated nodes. If s and t are not connected in G', the candidate solution for w_i is the set $C_i = U_1$ and the algorithm continues with w_{i+1}. Otherwise we consider the connected component H of G' which contains s and t. The nodes of $V \setminus \{s, t\}$ are then assigned new weights w', so that $w'(v) := \min \{w(u) \mid u \in N^+(v) \setminus \{s, t\}\}$. The new weights represent the cost to dominate these nodes. Now we compute a minimum s-t-cut C in (H, w'). Third, we choose a minimum-weight neighbor in G for each node $v \in C$ arbitrarily. This gives us a set C_i of weight at most $w'(C)$. The candidate solution for w_i is the set $C_i = U_1 \cup C_i$.

After calculating the sets C_i for $1 \le i \le n$, the set with minimum weight $w(C_i)$ is returned.

Theorem 2. *Algorithm 1 is a $\sqrt{n} \cdot (1 + w_{max}/OPT)$ approximation algorithm, where OPT denotes the weight of an optimal solution and w_{max} denotes the maximum weight of a node from this optimal solution. Especially, this is at most $2\sqrt{n}$.*

Proof. Consider an optimal solution $V_{opt} \subseteq V \setminus \{s, t\}$. Now let

$$w_{max} = \max \{w(v) \mid v \in V_{opt}\}.$$

It follows that: (i) $w(v) \le w_{max}$ for all $v \in V_{opt}$ and
(ii) $w_{max} \le w(V_{opt})$.
Since Algorithm 1 does one round for each node's weight, there must be a round $1 \le j \le n$ with $w_j = w_{max}$. Consider the respective run of the algorithm's main loop.

To bound the weight of U_1, we take a look at the nodes $v^1, v^2, \ldots v^{|U_1|}$ of U_1 in the order in which they are included in U_1 by the algorithm. Let the set $U_1^k = \{v^1, \ldots, v^k\}$ be the set U_1 after the k-th round of the greedy selection and $U_1^0 = \emptyset$. Let the set of newly dominated nodes for v^k be $N^k := N^+(v^k) \setminus (\{s, t\} \cup N^+(U_1^{k-1}))$. These sets are pairwise disjoint.

Since every node $v^k \in U_1$ dominated at least $|N^k| \ge w(v^k)\sqrt{n}/w_{max}$ new nodes, it holds that

$$w(U_1) = \sum_{k=1}^{|U_1|} w(v^k) \le \sum_{k=1}^{|U_1|} |N^k|\, w_{max}/\sqrt{n}$$
$$\le \frac{w_{max}}{\sqrt{n}} n = \sqrt{n}\, w_{max}.$$

After the greedy selection all $v \in V \setminus (U_1 \cup \{s, t\})$ fulfill

$$|N^+(v) \setminus (\{s, t\} \cup N^+(U_1))| < w(v)\sqrt{n}/w_{max}. \tag{1}$$

Now take a closer look at $G' = G \setminus (N^+(U_1) \setminus \{s, t\})$, the induced subgraph of currently undominated nodes. If s and t are cut in G', U_1 is an s-t-cut dominating set of weight at most $\sqrt{n}\, w(V_{opt})$ as desired. Let us now assume that this is not the case, i.e. there is a connected component H of G' which contains both s and t. We know that V_{opt} dominates an s-t-cut in G. Therefore, $V_{opt} \setminus U_1$ has to dominate an s-t-cut in H. It now holds that $(N^+(V_{opt} \setminus U_1) \setminus (N^+(U_1) \cup \{s, t\})) \cap V(H)$ is an s-t-cut in H. Therefore, the weight $w'(C)$ of the minimum w'-weight s-t-cut in H is at most $w'((N^+(V_{opt} \setminus U_1) \setminus (N^+(U_1) \cup \{s, t\})) \cap V(H))$. Furthermore, since for every node $v \in C$ there is a node $u \in N^+(v) \setminus \{s, t\}$ with $w(u) = w'(v)$, it holds that

$$w(C_j) \le w'(C). \tag{2}$$

This leads to

$$w(C_j) \le w'(C)$$

$$\leq w'((N^+(V_{opt} \setminus U_1) \setminus (N^+(U_1) \cup \{s,t\})) \cap V(H))$$

$$\leq \sum_{v \in V_{opt} \setminus U_1} \sum_{u \in N^+(v) \setminus (N^+(U_1) \cup \{s,t\})} w'(u)$$

$$\leq \sum_{v \in V_{opt} \setminus U_1} |N^+(v) \setminus (N^+(U_1) \cup \{s,t\})| w(v)$$

$$\overset{(1)}{<} \sum_{v \in V_{opt} \setminus U_1} \sqrt{n}\, w(v) \frac{w(v)}{w_{max}}$$

$$\overset{(i)}{\leq} \sum_{v \in V_{opt} \setminus U_1} \sqrt{n}\, w(v) \leq \sqrt{n}\, w(V_{opt}).$$

Hence, by uniting C_j and U_1, we obtain a set of nodes with weight at most $\sqrt{n}\, w(V_{opt}) + \sqrt{n}\, w_{max}$ in run j of the algorithm. Consequently, the algorithm returns a set of weight at most $\sqrt{n}(1 + w_{max}/w(V_{opt}))\, w(V_{opt})$.   □

In the unweighted version all weights are 1. Therefore, the following simplifications of the algorithm can be applied. First, one run of the algorithm's main loop will be sufficient, since we know $w_{max} = 1$. Second, the greedy procedure only chooses nodes dominating at least $\sqrt{n}$ new nodes. Third, the minimum-weight-function is not necessary. It suffices to calculate a minimum cardinality s-t-vertex-cut of G'.

So, the algorithm will degenerate to greedily choosing nodes which dominate at least $\sqrt{n}$ new nodes and computing a minimum s-t-vertex-cut in the resulting graph of undominated nodes. The approximation ratio of this algorithm is $\sqrt{n}\,(1 + \frac{1}{OPT})$, where OPT denotes the *size* of the optimal solution. Since OPT is at least one the simplified algorithm would give a $2\sqrt{n}$-approximation in the worst case. We can improve this ratio by adding a preprocessing step that enumerates all subsets $U \subseteq V$ up to a constant size $k \in \mathbb{N}$ and checks whether they dominate an s-t-cut. The first subset to do is an optimal solution. If none of the subsets dominates an s-t-cut, the optimal solution has to be of size at least $k + 1$. Therefore, the approximation ratio of the simplified algorithm with such a preprocessing step is at most $\sqrt{n}(1 + \frac{1}{k+1})$. The preprocessing needs $\mathcal{O}(kn^k(|V| + |E|))$ time, since for all $\sum_{i=1}^{k} \binom{n}{i}$ subsets it has to construct the graph G' of undominated nodes and test whether s and t are connected in G'. We state this observation in the following corollary.

Corollary 1. CUT DOMINATION *can be approximated with ratio* $\sqrt{n} \cdot (1 + \frac{1}{k+1})$ *for every constant* $k \in \mathbb{N}$.

A variation of Algorithm 1 can also be applied to approximate CONNECTED CUT DOMINATION. In particular, we can use it as a subroutine to solve one instance I_v of WEIGHTED CUT DOMINATION for each node $v \in V \setminus \{s,t\}$ and return a minimum weight result. The nodes of that result are then connected to v via shortest paths in $G \setminus \{s,t\}$. In instance I_v, the weight of a node is set to its hop distance to node v, thereby (over-)paying for intermediate nodes.

In addition, the greedy phase has to be adapted to start with $U_1 = \{v\}$ and choose nodes u that dominate at least $w(u)\sqrt{n}/w_{max}\sqrt{2w_{max}}$ new nodes. It can be shown that this algorithm achieves an approximation ratio of $\sqrt{n}\sqrt{OPT}$, which is at most $n^{2/3}$.

5   The Case of Bounded Vertex Degrees

We show that WEIGHTED CUT DOMINATION is $(\Delta + 1)$-approximable, if all but at most a logarithmic number of nodes are of degree Δ or less.

Theorem 3. *Let $W \subseteq V\backslash\{s,\ t\}$ with $|W| = \mathcal{O}(\log n)$. Then WEIGHTED CUT DOMINATION is $(\Delta_W + 1)$-approximable where Δ_W is the maximum degree of nodes from $V \backslash (\{s,t\} \cup W)$ in $G \backslash \{s,t\}$.*

Proof. Consider an algorithm that iterates all $2^{|W|}$ subsets $U \subseteq W$. For each $U \subseteq W$ the algorithm proceeds like one round of Algorithm 1, but with U taking the place of U_1, and calculates a candidate solution C_U. It then outputs the candidate solution of minimum weight.

Now we need to show, that this algorithm outputs a $(\Delta_W + 1)$-approximate solution. Let V_{opt} be a minimum weight s-t-cut dominating set of G. Furthermore, let $W_{opt} = V_{opt} \cap W$. Since $W_{opt} \subseteq W$, there is a round where $U = W_{opt}$. We know that every node from $V_{opt} \backslash W_{opt}$ is from $V \backslash (\{s,t\} \cup W)$ and therefore has a maximum degree of Δ_W in $G \backslash \{s,t\}$. It follows, that for every node $v \in V_{opt} \backslash W_{opt}$

$$\sum_{u\in N^+(v)\backslash\{s,t\}} w'(u) \leq (\Delta_W + 1)w(v). \tag{3}$$

Let us now consider the induced subgraph $G_{W_{opt}}$ of nodes which are not dominated by W_{opt}. We know that V_{opt} dominates an s-t-vertex-cut of G. Therefore, $V_{opt} \backslash W_{opt}$ has to dominate an s-t-vertex-cut of $G_{W_{opt}}$. It now holds, that $N^+(V_{opt} \backslash W_{opt}) \backslash (N^+(W_{opt}) \cup \{s,t\})$ is an s-t-cut of $G_{W_{opt}}$. Since C is a minimum s-t-cut of $G_{W_{opt}}$ according to w', it is also true that

$$w'(C) \leq w'(N^+(V_{opt} \backslash W_{opt}) \backslash (N^+(W_{opt}) \cup \{s,t\})). \tag{4}$$

Therefore, for the s-t-cut dominating set $C_{W_{opt}}$ constructed from C, it holds that

$$
\begin{aligned}
w(C_{W_{opt}}) &\overset{(2)}{\leq} w'(C) \\
&\overset{(4)}{\leq} w'(N^+(V_{opt} \backslash W_{opt}) \backslash (N^+(W_{opt}) \cup \{s,t\})) \\
&\leq \sum_{v\in V_{opt}\backslash W_{opt}} \sum_{u\in N^+(v)\backslash\{s,t\}} w'(u) \\
&\overset{(3)}{\leq} (\Delta_W + 1)w(V_{opt} \backslash W_{opt}).
\end{aligned}
$$

Hence, it holds that $w(W_{opt} \cup C_{W_{opt}}) \leq w(W_{opt}) + (\Delta_W + 1)w(V_{opt} \setminus W_{opt}) = (\Delta_W + 1)w(V_{opt}) - \Delta_W w(W_{opt})$. We obtain a $(\Delta_W + 1)$-approximation in the worst case and an upper bound for the size of the algorithm's solution. Since the algorithm computes $2^{|W|}$ induced subgraphs and minimum weight s-t-vertex-cuts, its running time is $\mathcal{O}(2^{|W|}(|V| + |E| + \sqrt{|V|}|E|))$, which is polynomial in n if and only if $|W| = \mathcal{O}(\log n)$. $\qquad\square$

Theorem 3 is especially relevant in practical applications, since communication overlay networks usually are of constant or logarithmic degree for scalability reasons.

As the minimum s-t-vertex-cut provides an upper bound for the minimum s-t-cut-dominating set, CUT DOMINATION can be solved in polynomial time for all graphs with a minimum s-t-vertex-cut of constant size. In particular, this includes all graphs, where s and t have degrees bounded by a constant.

6   Conclusion

Although a minimum s-t-vertex-separator can be found in polynomial time, we showed that it is much more complex to efficiently *dominate* any s-t-vertex separator. In particular, we proved that the CUT DOMINATION problem is not approximable to within a factor of $2^{\log^{1-\delta} n}$, with $\delta = 1/\log\log^c n$ and arbitrary $c < 1/2$ (if P $\neq$ NP) by reducing from RED-BLUE SET COVER. Thus, its inapproximability is higher than that of DOMINATING SET, the problem of finding a smallest set of nodes dominating *all* nodes of a graph [8] (again, if P $\neq$ NP).

On the positive side, we were able to show that WEIGHTED CUT DOMINATION is $2\sqrt{n}$-approximable in general graphs and $(\Delta+1)$-approximable in graphs with maximum degree Δ. In practice, the case of bounded node degrees is of special interest, since common overlay networks feature at most logarithmic degrees. The obtained (in-)approximability results are similar to the best known results for RED-BLUE SET COVER [7,9], which is believed to be a canonical representative from the class of optimization problems with superpolylogarithmic but potentially subpolynomial approximability. Closing the gap between approximability and inapproximability of WEIGHTED CUT DOMINATION by showing stronger inapproximability or approximability results, as well as investigating inapproximability for graphs of bounded node degree, remains for future research.

CONNECTED CUT DOMINATION is also of special interest, as in computer networks attacks sometimes spread from one node to another. We showed that the inapproximability result carries on to CONNECTED CUT DOMINATION. An approximation algorithm similar to Algorithm 1 achieves an approximation ratio of $\sqrt{n}\sqrt{OPT}$. It remains open, whether the approximation ratio can be lowered to the same ratio as for CUT DOMINATION or which ratio is achievable for WEIGHTED CONNECTED CUT DOMINATION.

Another matter of interest is the relation of CUT DOMINATION to the original problem described in [4,5]. It is a goal for future work to show similar approximability bounds for that problem.

References

1. Bollapragada, V., Khalid, M., Wainner, S.: IPSec VPN Design. Cisco Press (2005)
2. Vasserman, E., Jansen, R., Tyra, J., Hopper, N., Kim, Y.: Membership-concealing overlay networks. In: 16th ACM CCS, pp. 390–399. ACM (2009)
3. Clarke, I., Sandberg, O., Toseland, M., Verendel, V.: Private communication through a network of trusted connections: The dark freenet. Network (2010)
4. Rossberg, M., Girlich, F., Schaefer, G.: Analyzing and Improving the Resistance of Overlays against Bandwidth Exhaustion Attacks. In: International Workshop on Reliable Networks Design and Modeling (RNDM) (2012)
5. Girlich, F., Rossberg, M., Schaefer, G.: On the Resistance of Overlay Networks against Bandwidth Exhaustion Attacks. Accepted for Telecommunication Systems Journal (Special Issue) (2014)
6. Ahuja, R.K., Magnanti, T.L., Orlin, J.B.: Network Flows: Theory, Algorithms, and Applications. Prentice-Hall, Inc., Upper Saddle River (1993)
7. Carr, R.D., Doddi, S., Konjevod, G., Marathe, M.: On the red-blue set cover problem. In: ACM-SIAM Symposium on Discrete Algorithms, pp. 345–353 (2000)
8. Alon, N., Moshkovitz, D., Safra, S.: Algorithmic construction of sets for k-restrictions. ACM Trans. Algorithms 2(2), 153–177 (2006)
9. Peleg, D.: Approximation algorithms for the label-covermax and red-blue set cover problems. Journal of Discrete Algorithms 5(1), 55–64 (2007)

Lower Bounds for Linear Decision Trees with Bounded Weights

Kei Uchizawa[1] and Eiji Takimoto[2]

[1] Graduate School of Science and Engineering, Yamagata University, 4-3-16 Jonan, Yonezawa,Yamagata, 992-8510, Japan
uchizawa@yz.yamagata-u.ac.jp
[2] Department of Informatics, Graduate School of Information Science and Electrical Engineering, Kyushu University, 744 Motooka, Nishi-ku, Fukuoka 819-0395, Japan
eiji@inf.kyushu-u.ac.jp

Abstract. In this paper, we consider a linear decision tree such that a linear threshold function at each internal node has a bounded weight: the sum of the absolute values of its integer weights is at most w. We prove that if a Boolean function f is computable by such a linear decision tree of size (i.e., the number of leaves) s and rank r, then f is also computable by a depth-2 threshold circuit containing at most $s(2w+1)^r$ threshold gates with weight at most $(2w+1)^{r+1}$ in the bottom level. Combining a known lower bound on the size of depth-2 threshold circuits, we obtain a $2^{\Omega(n/\log w)}$ lower bound on the size of linear decision trees computing the Inner-Product function modulo 2, which improves on the previous bound $2^{\sqrt{n}}$ if $w = 2^{o(\sqrt{n})}$.

1 Introduction

A binary decision tree is one of the basic models of computation. In a standard decision tree, each internal node is labeled with a Boolean variable and each leaf with 0 or 1. For a given input assignment to the Boolean variables, a path from the root to a leaf is naturally defined according to the values of the variables, and the decision tree returns as output the label (0 or 1) of the leaf in the path. Thus, a decision tree represents a Boolean function. In this paper, we consider a *linear* decision tree in which a classification rule at each internal node is given by a linear threshold function, where a linear threshold function g is defined by integer weights $w_1, w_2, \ldots, w_n$ and threshold t, and for every input $\boldsymbol{x} \in \{0,1\}^n$, the output of g is given by $g(\boldsymbol{x}) = \text{sign}(\sum_{i=1}^n w_i x_i - t)$.

For a linear decision tree T, its depth is defined to be the length of the longest path from the root to a leaf in T, and its size is to be the number of leaves in T. The depth and size are natural complexity measures for decision trees since these correspond to the time and space complexity from the viewpoint of parallel computation, and several lower bound results are obtained for each measure. Nisan proved that any linear decision tree computing Inner-Product function IP_n of $2n$ variables has depth $\Omega(n/\log n)$ by showing that a liner threshold function has small communication complexity [15]. Gröger and Turán obtained

G.F. Italiano et al. (Eds.): SOFSEM 2015, LNCS 8939, pp. 412–422, 2015.

a linear lower bound on the depth of linear decision trees computing IP_n by an adversary argument [8]. Turán and Vatan generalized the Gröger and Turán's result to $2^{\Omega(n/r)}$ in terms of rank r of linear decision trees [17], where the rank of a binary tree T is the maximal depth of a complete binary tree embedded in T (see Section 2 for the precise definition). Since a lower bound on the depth is that on the size, the lower bounds above imply ones on the size of linear decision trees computing IP_n. In particular, the result in [17] immediately implies that any linear decision tree computing IP_n has size $2^{\Omega(\sqrt{n})}$ regardless of its rank, as follows. If a tree satisfies $r \leq \sqrt{n}$, then their result gives the desired lower bound; otherwise, we obtain the bound from the definition of the rank: If a tree has rank $r > \sqrt{n}$, then the tree must have size $s > 2^{\sqrt{n}}$. In the paper [18], we directly investigated the size complexity of linear decision trees, and obtained a weaker lower bound which is applicable to a wide class of Boolean functions (having high communication complexity) .

On the other hand, a simple counting argument gives a stronger lower bound on the size. More precisely, we can show that there exists an n-variable Boolean function that requires size $2^n/poly(n)$, since the number of the n-variable Boolean functions computable by linear threshold functions is bounded by $2^{O(n^2)}$ [12,13], and the binary trees of size s have at most 4^s different structures (more precisely, it equals to the $(s-1)$-th Catalan number), and thus linear decision trees of size s can compute at most $(2^{n^2})^s \cdot 4^s$ Boolean functions. To the best of our knowledge, we do not know any explicitly defined Boolean function that requires size $2^n/poly(n)$ for its linear decision tree representation.

In this paper, to tighten the gap between $2^{\Omega(\sqrt{n})}$ and $2^n/poly(n)$, we restrict ourselves to the case where the weights of each linear threshold function are bounded: the weight vector $(w_1, w_2, \ldots, w_n)$ should satisfy $\sum_i |w_i| \leq w$ for some integer w. Our main result is to prove a $2^{\Omega(n/\log w)}$ lower bound on the size of such linear decision trees computing IP_n. Our lower bound improves on Turán and Vatan's one if $w = 2^{o(\sqrt{n})}$. We note that the computational power of a linear threshold function is well-studied in terms of the weight. In particular, it is known that every linear threshold function has a representation where each weight has magnitude at most $2^{O(n \log n)}$ (see, for example, Corollary 2.3 in [13]), and this was shown to be tight in [9].

Informally, we show in the proof that any given linear threshold function can be converted to a depth-2 threshold circuit, which immediately implies the desired lower bound, since strong lower bounds against depth-2 threshold circuits are known [7]. To be more specific, the proof proceeds in the following three steps. At the first step, we use an argument of Blum [2] to show that any linear decision tree T of size s, weight w and rank r can be converted into a linear decision list L of size s, weight w and term r. At the next step, in a similar way to a conversion method in [11], we convert L into a depth-3 threshold circuit C in which the top gate is a threshold gate, the middle layer contains s AND gates with fan-in r, and the bottom layer contains threshold gates with bounded weight w. At the final step, using the idea in Beigel et al. [1], we complete the proof by collapsing the middle and bottom layers of C to a single layer containing at most $s(2w+1)^r$ threshold gates each of which has bounded weight $(2w+1)^{r+1}$.

We note that the steps of our proof do not depend on any explicit property of IP_n, and hence we can apply them to linear decision trees computing any function. Consequently, we can also obtain a similar lower bound for linear decision trees computing a Boolean function for which a lower bound on the size of depth-2 threshold circuits is known, such as $ORT_{p,n}$ which determines the orthogonality of two given 1-dimensional homogeneous linear subspaces of $\mathbb{F}_p^n$ [7].

The rest of the paper is organized as follows. In Section 2, we define some terms on linear decision trees, linear decision lists, and threshold circuits. In Section 3, we give the lower bound for linear decision trees. In Section 4, we conclude with some remarks.

2   Definitions

In this section, we give definitions of linear decision trees, linear decision lists, and threshold circuits.

2.1   Linear Decision Trees and Lists

Let g be a *linear threshold function* with n inputs, weights $w_1, w_2, \ldots, w_n$ and a threshold t. Then the output $g(\boldsymbol{x})$ of g for every input $\boldsymbol{x} = (x_1, x_2, \ldots, x_n) \in \{0,1\}^n$ is given as follows:

$$g(\boldsymbol{x}) = \text{sign}\left(\sum_{i=1}^n w_i x_i - t\right)$$

where, for any number η, $\text{sign}(\eta) = 1$ if $\eta \geq 0$ and $\text{sign}(\eta) = 0$, otherwise. We assume throughout the paper that the weights and threshold of every threshold function are integers. We define the *weight* w of a threshold function g as the sum of the absolute values of the weights $w_1, w_2, \ldots, w_n$ of g for the n input variables $x_1, x_2, \ldots, x_n$.

A *linear decision tree* T is a binary decision tree in which a classification rule at each internal node is given by a linear threshold function and each leaf is labelled by zero or one. Given an input $\boldsymbol{x} \in \{0,1\}^n$ to T, the output $T(\boldsymbol{x})$ of T is determined by the following procedure starting from the root until reaching a leaf: if the output of the linear threshold function at the current node is zero for $\boldsymbol{x}$, then go to the left child; otherwise go the right. If the leaf reached is labeled by $z \in \{0,1\}$, then $T(\boldsymbol{x}) = z$. The *size* s of T is defined to be the number of leaves in T. We say that T has weight w if the linear threshold function at every internal node of T has weight w. The *rank* of a linear decision tree T is inductively defined as follows: if T consists of a single leaf, then $rank(T) = 0$; otherwise,

$$rank(T) = \begin{cases} rank(T_l) + 1 & \text{if } rank(T_l) = rank(T_r); \\ \max\{rank(T_l), rank(T_r)\} & \text{otherwise,} \end{cases} \quad (1)$$

where T_l and T_r denote the left and right subtrees of the root, respectively.

For a positive integer m, a *linear decision list L of length m* is a sequence of pairs $(S_0, z_0), (S_1, z_1), \ldots, (S_{m-2}, z_{m-2}), (S_{m-1} = \emptyset, z_{m-1})$, where, for each j, $0 \leq j \leq m - 1$, S_j is a set of linear threshold functions, and $z_j \in \{0, 1\}$. Given an input $\boldsymbol{x} \in \{0, 1\}^n$ to L, the output $L(\boldsymbol{x})$ is defined to be such $z_{j'}$ that every linear threshold function in $S_{j'}$ outputs one for $\boldsymbol{x}$, while an output of at least one of the linear threshold functions in S_j is zero for every j, $0 \leq j \leq j' - 1$. We say that L has *term k* if $\#S_j \leq k$ holds for every j, $0 \leq j \leq m - 1$, where $\#S_j$ indicates the cardinality of S_j, and that L has *weight w* if every linear threshold function in $S_0 \cup S_2 \cup \ldots \cup S_{m-1}$ has weight w.

Clearly, a linear decision tree is a generalization of an ordinary decision tree such that a classification rules at each internal node is given by a single Boolean variable. Similarly, a linear decision list is a generalization of an ordinary decision list such that each internal node has a set of Boolean literals. The following two facts are well-known for the ordinary decision trees and lists:

Fact 1 ([2]). *An ordinary decision tree T of size s can be simulated by an ordinary decision list of length s and term $rank(T)$.*

Fact 2 ([4]). *If an ordinary decision tree T has size s then $rank(T) \leq \log s$.*

Since Boolean literals for ordinary decision trees and lists play same role as threshold functions for linear decision trees and lists, we immediately have the following lemma from Fact 1 and 2:

Lemma 1. *Any linear decision tree T of size s and weight w can be simulated by a linear decision list of length s, term $rank(T)$ and weight w. Moreover, it also can be simulated by a linear decision list of length s, term at most $\log s$ and weight w.*

We will use this lemma to obtain our main result.

2.2   Threshold Circuits

A *threshold gate* with an arbitrary number n of inputs computes a linear threshold function with n inputs. We say that a threshold gate g has *weight w* if the corresponding linear threshold function of g has weight w. A *threshold circuit C* is a combinatorial circuit of threshold gates. In this paper, we only consider threshold circuits of depth 2: a circuit consists of two layers such that the bottom level contains a number of gates, and the second level does a single gate, called *top gate*. The *size s* of a depth-2 threshold circuit is defined to be the number of threshold gates in the bottom level of the circuit.

To obtain the lower bounds for linear decision trees, we consider two Boolean functions IP_n and $ORT_{p,n}$, defined as follows.

IP_n is defined to be a Boolean function of $2n$ variables such that, for every pair of inputs $x = (x_1, x_2, \cdots, x_n) \in \{0,1\}^n$ and $y = (y_1, y_2, \cdots, y_n) \in \{0,1\}^n$,

$$IP_n(x, y) = x_1 y_1 \oplus x_2 y_2 \oplus \ldots \oplus x_n y_n,$$

where $\oplus$ denotes the XOR function.

Let $\mathbb{F}_p$ be the prime field of characteristic p, and $\mathbb{F}_p^n$ be the n-dimensional vector space of $\mathbb{F}_p$. Let $\mathbb{P}_{n-1}(\mathbb{F}_p)$ be the $(n-1)$-dimensional projective space, where the elements in $\mathbb{P}_{n-1}(\mathbb{F}_p)$ are 1-dimensional linear subspaces of $\mathbb{F}_p^n$. Then, $ORT_{p,n}$ is defined to be a Boolean function of $2n\lceil \log n \rceil$ variables such that, for every pair of inputs $q \in \{0, 1, \ldots, p-1\}^n$ and $q' \in \{0, 1, \ldots, q-1\}^n$,

$$ORT_{p,n}(q, q') = \begin{cases} 0 & \text{if } q \text{ is orthogonal to } q'; \\ 1 & \text{otherwise.} \end{cases}$$

Forster *et al.* derived the following lower bounds on the size of threshold circuits for IP_n and $ORT_{p,n}$ [7].

Lemma 2 ([7]). *Suppose a Boolean function f is computed by a depth-2 threshold circuit whose bottom level contains z gates. If f is IP_n, and each of the z gates has weight at most w, then*

$$z \geq \frac{2^{n/2} - 1}{w + 1}.$$

Moreover, if f is $ORT_{p,n}$, and each of the z gates has weight at most w, then

$$z \geq \frac{p^{n/2} - 1}{w + 1}.$$

3   Lower Bounds for Linear Decision Trees

In Section 3.1, we present a technical lemma, and give results obtained from the lemma. In Section 3.2, we prove the lemma.

3.1   Our Results

The following lemma shows a relationship between linear decision trees and depth-2 threshold circuits. We will give a proof of the lemma in the next section.

Lemma 3. *Suppose a linear decision tree T of size s, weight w and rank r computes a Boolean function $f : \{0,1\}^n \to \{0,1\}$. Then, f is also computable by a depth-2 threshold circuit such that its bottom level contains at most $s(2w+1)^r$ gates, each of which has weight at most $(2w+1)^{r+1}$.*

Combining Lemma 2 with Lemma 3, we immediately obtain our main result for linear decision trees, as follows.

Theorem 1. *Let T be a linear decision tree of weight w and rank r. If T computes IP_n, the size of T is at least*

$$\frac{2^{n/2} - 1}{(2w + 1)^r((2w + 1)^{r+1} + 1)}.$$

If T computes $ORT_{p,n}$, the size of T is at least

$$\frac{p^{n/2} - 1}{(2w + 1)^r((2w + 1)^{r+1} + 1)}.$$

Proof. We only give a proof for IP_n. Let T be a linear decision tree computing IP_n, and let s, w and r be the size, weight and rank of T. By Lemma 3, we obtain from T a depth-2 threshold circuit C that computes IP_n and has size

$$z \le s(2w + 1)^r \tag{2}$$

and weight at most $(2w + 1)^{r+1}$. On the other hand, Lemma 2 implies that

$$\frac{2^{n/2} - 1}{(2w + 1)^{r+1} + 1} \le z. \tag{3}$$

Therefore, we have the claim from (2) and (3).   □

By Lemma 1 and Theorem 3.2, we can obtain lower bounds on the size of linear decision trees regardless of its rank.

Corollary 1. *Let T be a linear decision tree of weight w. If T computes IP_n, the size of T is at least*

$$\left(\frac{2^{n/2} - 1}{2}\right)^{1/(10 \log w)}.$$

If T computes $ORT_{p,n}$, the size of T is at least

$$\left(\frac{p^{n/2} - 1}{2}\right)^{1/(10 \log w)}.$$

3.2 Proof of Lemma 3

In this section, we prove Lemma 3. Let T be an arbitrary linear decision tree computing a Boolean function $f : \{0, 1\}^n \to \{0, 1\}$, and be s, w and r be be size, weight and rank of T, respectively. We construct the desired depth-2 threshold circuit C^*.

Firstly, we convert T to a linear decision list by applying Lemma 1: we obtain from T a linear decision list L that computes f and has length s, term r and weight w. We denote by $(S_0, z_0), (S_1, z_1), \ldots, (S_{s-1}, z_{s-1})$ the sequence of pairs composing L; and, for each j, $0 \le j \le s-1$, let $a_0^{(j)}, a_1^{(j)}, \ldots, a_{\#S_j-1}^{(j)}$ be the threshold functions in S_j. Furthermore, for every k, $0 \le k \le \#S_j - 1$, we denote by $w_{k,1}^{(j)}, w_{k,2}^{(j)}, \ldots, w_{k,n}^{(j)}$ the weights of $a_k^{(j)}$ for $x_1, x_2, \ldots, x_n$, respectively; and denote by $t_k^{(j)}$ the threshold of $a_k^{(j)}$, that is, we have

$$a_k^{(j)}(\boldsymbol{x}) = \mathrm{sign}\left(\sum_{i=1}^n w_{k,i}^{(j)} x_i - t_k^{(j)}\right).$$

Secondly, we convert L to a depth-3 intermediate circuit C. The top gate g of C has threshold zero, and receives outputs of s circuits $C_0, C_1, \ldots, C_{s-1}$ of depth-2 given as follows: For each j, $0 \le j \le s-1$, the circuit C_j computes AND of the outputs of threshold gates $g_0^{(j)}, g_1^{(j)}, \ldots, g_{\#S_j-1}^{(j)}$, where $g_k^{(j)}$ computes $a_k^{(j)}$ for each k, $0 \le k \le \#S_j - 1$. Note that the fan-ins of the top gates of $C_0, C_1, \ldots, C_{s-1}$ equal to $\#S_0, \#S_1, \ldots, \#S_{s-1}$, respectively. In addition, we have $\#S_j \le r$ for every j, $0 \le j \le s-1$. For each j, $0 \le j \le s-1$, we then define the weight w_j of g for the output of C_j as

$$w_j = \begin{cases} 2^{(s-1)-j} & \text{if } z_j = 1; \\ -2^{(s-1)-j} & \text{otherwise.} \end{cases} \tag{4}$$

Eq. (4) implies that, for any index j', we have

$$\sum_{j=j'+1}^{s-1} |w_j| < |w_{j'}|, \tag{5}$$

Since g has threshold zero, (5) implies that the output of $C_{j'}$ dominates the outputs of $C_{j'+1}, C_{j'+2}, \ldots, C_s$, that is, for every $\boldsymbol{x} \in \{0,1\}^n$, $C(\boldsymbol{x})$ equals to the output of $C_{j'}$ with the least index j' satisfying

$$C_j(\boldsymbol{x}) = 0$$

for every $1 \le j \le j'-1$ and

$$C_{j'}(\boldsymbol{x}) = 1.$$

Therefore, C clearly simulates L, and hence computes f.

Finally, we obtain the desired depth-2 circuit C^* from C. The top gate g^* of C^* has threshold zero. For the bottom level of C^*, we employ a standard idea to construct a certain set of threshold gates from C_j for each j, $0 \le j \le s-1$ so that the gates somehow compute the same function as C_j does. (See, for example, [1] or [10].) We below give the construction for completeness.

Let j, $0 \le j \le s-1$, be an arbitrary index. Recall that $w_{k,1}^{(j)}, w_{k,2}^{(j)}, \ldots, w_{k,n}^{(j)}$ are the weights of $a_k^{(j)}$ (and hence of $g_k^{(j)}$) for $x_1, x_2, \ldots, x_n$. We define a function $p^{(j)}$ as follows: For every $\boldsymbol{x} = (x_1, x_2, \ldots, x_n) \in \{0,1\}^n$,

$$p^{(j)}(\boldsymbol{x}) = \sum_{k=0}^{\#S_j-1} (2w+1)^k \left(w + \sum_{i=1}^n w_{k,i}^{(j)} x_i\right); \tag{6}$$

in other words, $p(x)$ is a $(2w + 1)$-nary integer of $\#S_j$ digits such that the $(k+1)$-st digit corresponds to the value

$$w + \sum_{i=1}^{n} w_{k,i}^{(j)} x_i.$$

Note that we have

$$0 \leq w + \sum_{i=1}^{n} w_{k,i}^{(j)} x_i \leq 2w + 1$$

and hence

$$0 \leq p^{(j)}(x) \leq (2w + 1)^{\#S_j} \leq (2w + 1)^r.$$

We denote by m the integer $(2w + 1)^r$ hereafter.

The value $p^{(j)}(x)$ uniquely determines the outputs of $g_0^{(j)}, g_1^{(j)}, \ldots, g_{\#S_j-1}^{(j)}$ as in the following claim:

Claim. For every k, $0 \leq k \leq \#S_j - 1$, we define $p_k^{(j)}$ as

$$p_k^{(j)}(x) \equiv \left\lfloor \frac{p^{(j)}(x)}{(2w + 1)^k} \right\rfloor \quad \text{mod } 2w + 1$$

for every $x \in \{0, 1\}^n$. Then $g_k^{(j)}(x) = 1$ if and only if $w + t_k^{(j)} \leq p_k^{(j)}$.

Proof. Let k, $0 \leq k \leq \#S_j - 1$ be an arbitrary index. By Eq. (6), we have

$$\left\lfloor \frac{p^{(j)}(x)}{(2w + 1)^k} \right\rfloor = \sum_{k'=k}^{\#S_j - 1} (2w + 1)^{k'-k} \left(w + \sum_{i=1}^{n} w_{k',i}^{(j)} x_i \right)$$

$$= \left(w + \sum_{i=1}^{n} w_{k,i}^{(j)} x_i \right) + \sum_{k'=k+1}^{\#S_j - 1} (2w+1)^{k'-k} \left(w + \sum_{i=1}^{n} w_{k',i}^{(j)} x_i \right) \quad (7)$$

Since the second term of the right hand side of Eq. (7) is a multiple of $2w + 1$, we clearly have

$$p_k^{(j)}(x) \equiv w + \sum_{i=1}^{n} w_{k,i}^{(j)} x_i \quad \text{mod } 2w + 1.$$

Thus, $w + t_k^{(j)} \leq p_k^{(j)}$ if and only if

$$t_k^{(j)} \leq \sum_{i=1}^{n} w_{k,i}^{(j)} x_i,$$

and hence the claim follows.   □

For each integer l, $0 \leq l \leq m$, we say that l *satisfies* C_j if both $p^{(j)}(x) = l$ and $C_j(x) = 1$ hold for some $x \in \{0, 1\}^n$.

By an elementary transformation from Eq. (6), we have

$$p^{(j)}(\boldsymbol{x}) = \sum_{i=1}^{n} w_i^{(j)} x_i + \sum_{k=0}^{\#S_j - 1} (2w + 1)^k w, \tag{8}$$

where

$$w_i^{(j)} = \sum_{k=0}^{\#S_j - 1} (2w + 1)^k w_{k,i}^{(j)}.$$

For each integer l, $0 \le l \le m$, we construct a threshold gate $h_l^{(j)}$ with threshold l and weight $w_i^{(j)}$ for each i, $1 \le i \le n$: For every $\boldsymbol{x} = (x_1, x_2, \dots, x_n) \in \{0,1\}^n$,

$$h_l^{(j)}(\boldsymbol{x}) = \mathrm{sign}(p_j(\boldsymbol{x}) - l).$$

Eq. (8) implies that $h_l^{(j)}$ is a linear threshold function. In addition, we have by construction that $h_l^{(j)}(\boldsymbol{x}) = 1$ for every l, $0 \le l \le p^{(j)}(\boldsymbol{x})$, while $h_l^{(j)}(\boldsymbol{x}) = 0$ for every l, $p^{(j)}(\boldsymbol{x}) + 1 \le l \le m$.

We now define the weight $w_0^{(j)}$ of g^* for the output of $h_0^{(j)}$ as

$$w_0^{(j)} = \begin{cases} w_j & \text{if the integer zero satisfies } C_j; \\ 0 & \text{otherwise.} \end{cases} \tag{9}$$

Inductively, for each integer l, $1 \le l \le m$, the top gate g^* receives the output of $h_l^{(j)}$ with weight $w_l^{(j)}$ defined as

$$w_l^{(j)} = \begin{cases} w_j & \text{if the value } l - 1 \text{ does not satisfy } C_j, \text{ but } l \text{ does;} \\ -w_j & \text{if the value } l - 1 \text{ satisfies } C_j, \text{ but } l \text{ does not;} \\ 0 & \text{otherwise.} \end{cases} \tag{10}$$

Eq. (10) then clearly imply that, for every $\boldsymbol{x} \in \{0,1\}^n$,

$$\sum_{l=0}^{m} w_l^{(j)} h_l^{(j)}(\boldsymbol{x}) = \begin{cases} w_j & \text{if } p^{(j)}(\boldsymbol{x}) \text{ satisfies } C_j; \\ 0 & \text{otherwise.} \end{cases} \tag{11}$$

Thus, using $h_0^{(j)}, h_1^{(j)}, \dots, h_m^{(j)}$, we can simulates the contributions of C_j to the top gate g^*.

Consequently, we have obtained a depth-2 threshold circuit C^* that simulates C, and hence C^* computes f. The bottom level of C^* contains at most $sm = s(2w + 1)^r$ gates. Furthermore, Eq. (8) implies that, for an arbitrary j, $0 \le j \le s - 1$, each bottom-level gate constructed from C_j has weight

$$\sum_{i=1}^{n} \sum_{k=0}^{\#S_j - 1} (2w + 1)^k |w_{k,i}^{(j)}| = \sum_{k=0}^{\#S_j - 1} (2w + 1)^k \left(|w_{k,1}^{(j)}| + |w_{k,2}^{(j)}| + \cdots + |w_{k,n}^{(j)}| \right)$$

$$\le \sum_{k=0}^{\#S_j - 1} (2w + 1)^k \cdot w$$

$$\le (2w + 1)^{r+1}$$

as desired.

4   Conclusions

In this paper, we derive lower bounds on the size of a linear decision tree by simulating a linear decision tree by depth-2 threshold circuit. Our lower bound gives a better bound than the previous one for IP_n if the weight w of trees satisfy $w = 2^{o(\sqrt{n})}$. Note that we can apply Lemma 2 to any linear decision trees computing a Boolean function f, and hence it yields such lower bounds if f has a lower bound for depth-2 threshold circuits.

Acknowledgments. This research is supported by MEXT KAKENHI Grant No. 24106010 and JSPS KAKENHI Grant No. 23300003 and 25330005.

References

1. Beigel, R., Reingold, N., Spielman, D.: PP is closed under intersection. Journal of Computer and System Sciences 50(2), 191–202 (1995)
2. Blum, A.: Rank-r decision trees are a subclass of r-decision lists. Information Processing Letters 42(4), 183–185 (1992)
3. Dobkin, D.P., Lipton, R.J.: On the complexity of computations under varying sets of primitives. Journal of Computer and System Sciences 3, 1–8 (1982)
4. Ehrenfeucht, A., Haussler, D.: Learning decision trees from random examples. Information and Computation 82(3), 231–246 (1989)
5. Erickson, J.: Lower bounds for linear satisfiability problems. In: Proceedings of the 6th Annual ACM-SIAM Symposium on Discrete Algorithms, pp. 388–395 (1995)
6. Fleischer, R.: Decision trees: Old and new results. Information and Computation 152, 44–61 (1999)
7. Forster, J., Krause, M., Lokam, S.V., Mubarakzjanov, R., Schmitt, N., Simon, H.U.: Relations between communication complexity, linear arrangements, and computational complexity. In: Hariharan, R., Mukund, M., Vinay, V. (eds.) FSTTCS 2001. LNCS, vol. 2245, pp. 171–182. Springer, Heidelberg (2001)
8. Gröger, H.D., Turán, G.: On linear decision trees computing boolean functions. In: Leach Albert, J., Monien, B., Rodríguez-Artalejo, M. (eds.) ICALP 1991. LNCS, vol. 510, pp. 707–718. Springer, Heidelberg (1991)
9. Håstad, J.: On the size of weights for threshold gates. SIAM Journal on Discrete Mathematics, 484–492 (1994)
10. Hofmeister, T.: A Note on the Simulation of Exponential Threshold Weights. In: Cai, J.-Y., Wong, C.K. (eds.) COCOON 1996. LNCS, vol. 1090, pp. 136–141. Springer, Heidelberg (1996)
11. Klivans, A.R., Servedio, R.A.: Learning DNF in time $2^{n^{1/3} poly(\log n)}$. Journal of Computer and System Sciences 68(2), 303–318 (2004)
12. Parberry, I.: Circuit Complexity and Neural Networks. MIT Press, Cambridge (1994)
13. Siu, K.Y., Roychowdhury, V., Kailath, T.: Discrete Neural Computation; A Theoretical Foundation. Prentice-Hall, Inc., Upper Saddle River (1995)
14. Steele, J.M., Yao, A.C.: Lower bounds for algebraic decision trees. Journal of Algorithms 18, 86–91 (1979)
15. Nisan, N.: The communication complexity of threshold gates. In: Proceedings of "Combinatorics, Paul Erdös is Eighty", pp. 301–315 (1999)

16. Klivans, A.R., Servedio, R.A.: Learning DNF in time $2^{\tilde{O}(n^{1/3})}$. Journal of Computer and System Sciences 68(2), 303–318 (2004)
17. Turán, G., Vatan, F.: Linear decision lists and partitioning algorithms for the construction of neural networks. Foundations of Computational Mathematics, 414–423 (1997)
18. Uchizawa, K., Takimoto, E.: Lower Bounds for Linear Decision Trees via an Energy Complexity Argument. In: Murlak, F., Sankowski, P. (eds.) MFCS 2011. LNCS, vol. 6907, pp. 568–579. Springer, Heidelberg (2011)

A Model-Driven Approach to Generate External DSLs from Object-Oriented APIs

Valerio Cosentino, Massimo Tisi, and Javier Luis Cánovas Izquierdo

AtlanMod team (Inria, Mines Nantes, LINA), Nantes, France
{firstname.lastname}@inria.fr

Abstract. Developers in modern general-purpose programming languages create reusable code libraries by encapsulating them in Applications Programming Interfaces (APIs). Domain-specific languages (DSLs) can be developed as an alternative method for code abstraction and distribution, sometimes preferable to APIs because of their expressivity and tailored development environment. However the cost of implementing a fully functional development environment for a DSL is generally higher. In this paper we propose DSLit, a prototype-tool that, given an existing API, reduces the cost of developing a corresponding DSL by analyzing the API, automatically generating a semantically equivalent DSL with its complete development environment, and allowing for user customization. To build this bridge between the API and DSL technical spaces we make use of existing Model-Driven Engineering (MDE) techniques, further promoting the vision of MDE as a unifying technical space.

1 Introduction

Modern General-purpose Programming Languages (GPLs) provide facilities for program abstraction and reuse, to foster the development of distributable libraries. Code in a programming library is encapsulated behind Application Programming Interfaces (APIs), that are used in user programs by the mechanisms provided by the GPL (e.g., function call or class inheritance). Sometimes library developers prefer to provide their users with a Domain-Specific Language (DSL), instead of (or in addition to) an API. APIs and DSLs can be seen as alternative methods to access the library functionalities, and are characterized by specific advantages. Programs written in the DSL can be more expressive, maintainable, concise and readable than corresponding GPL programs using the API (e.g., by avoiding the user to write some boilerplate code) [1,2]. On the other side, APIs allow for natural integration in complex programs written in their native language (or in other languages when coupled with suitable interface bindings).

Literature distinguishes DSLs in internal and external [3]. Internal DSLs are created by embedding DSL constructs into an existing GPL, which acts as host language. Although the internal approach allows DSLs to be easily developed [4], the corresponding tooling relies on the existing support for the host language, which limits the domain-specific assistance [5]. External DSLs instead are characterized by a separate syntax and specific development facilities. An important advantage of this approach

G.F. Italiano et al. (Eds.): SOFSEM 2015, LNCS 8939, pp. 423–435, 2015.

is that the domain-specific development environment can be tailored to ease coding in the DSL:

- Static validation can be enriched to enforce semantic constraints hidden in the API. Thus some runtime errors can be avoided at compile time.
- Features like syntax highlighting, code completion, outlining, folding, can be tailored to the DSL.
- The DSL interpretation/compilation step can be designed to automatically optimize the DSL code execution.

While, depending on the case, DSL or API (or both) may be the preferable solution [6], the development cost of a DSL, especially if external, is in general much higher. Users have to define the DSL (i.e., abstract and concrete syntaxes, and semantics) and develop the domain-specific environment (e.g., syntax highlighting, code assistance, folding, outline view) which are tedious and time-consuming tasks.

In this paper we propose a method to automatically analyze an existing object-oriented API and generate an external DSL out of it. Our approach leverages model-driven techniques to analyze and represent APIs at high-level of abstraction (i.e., as metamodels) which are later used to automatically generate the DSL components and the corresponding tooling, including parser, compiler and development environment. Developers can influence the DSL generation by editing the model-based API representation and by specifying design choices about the structure of the DSL to generate.

We provide a proof-of-concept implementation of the method in the DSLit tool, that is able to analyze Java APIs and generate external textual DSLs using the Xtext framework [7]. DSLit is currently able to deal with two API categories that we describe. The first category is called *Plain Old Data* (POD[1]) and indicates simple APIs that have the purpose of creating and maintaining a data structure. Usually such APIs are composed by classes made exclusively of getters, setters and constructors. The second category is called *Fluent* and contains those APIs that rely on chaining method calls. The return value of these method calls is an object representing the context of the keyword, and it is used to structure the language, defining which keywords can follow other keywords. For APIs not included in these categories, we also provide a fallback category, called *SimpleJava* based on a subset of Java which includes statements and declarations.

While currently limited in scope, the DSLit prototype, freely available at the project website[2], demonstrates the feasibility and usefulness of the approach.

The paper is structured as follows. Section 2 presents concrete examples to motivate our approach. Section 3 describes the conceptual framework applied to obtain a DSL from a Java API, while Section 4 presents the implementation of the prototype tool and the solution of the running cases. Section 5 lists the related work and Section 6 finalizes the paper and outlines some further work.

---

[1] http://en.wikipedia.org/wiki/Plain_old_data_structure
[2] http://www.emn.fr/z-info/atlanmod/index.php/DSLit

2   Motivating Examples

While APIs have proven to be a flexible means to encapsulate and reuse program logic, their usage can be cumbersome. A typical example is Java Swing, an API based on the Abstract Window Toolkit (AWT[3]) for the development of graphical user interfaces for Java applications. Several DSLs have been developed to allow more concise and readable interface specifications with respect to Swing-like code. An example is JavaFX[4], a framework specifically tailored to create rich internet applications (RIA) that includes the so-called JavaFX Script, which is a DSL enabling the fast definition of user interfaces. Figures 1a and 2b compare two equivalent chunks of code written in Java Swing and JavaFX Script, respectively.

Java Swing	JavaFX Script	DSLit
| ```
JFrame jFrame1 = new JFrame();
jFrame1.setTitle("My Java Application");
jFrame1.setSize(500, 300);
JLabel jLabel1 = new JLabel();
jFrame1.add(jLabel1);
jLabel1.setText("Hello World!");
jFrame1.setVisible(true);
``` | ```
Frame {
    title: "My Java Application"
    width: 500
    height: 300
    content: Label {
        text: "Hello World!"
    }
    visible: true
}
``` | ```
JFrame {
    title: "My Java Application"
    size: 500, 300
    JLabel {
        text: "Hello World!"
    }
    visible: true
}
``` |
| (a) | (b) | (c) |

Fig. 1. A Swing example using (a) Java code, (b) JavaFX and (c) DSLit

Both examples specify the creation of a frame including a title and a label and the JavaFX Script version is remarkably more concise and readable. Developers in JavaFX are free from writing most of the boilerplate code and can use a declarative language specifically adapted to the creation of user interfaces. However, JavaFX Script was not developed as a DSL for targeting the Java Swing API but as a DSL to develop user interfaces rapidly and effectively. Thus, JavaFX Script code does not compile to the corresponding Java Swing code and it incorporates extra features such as declarative animation or mutation triggers.

As can be seen, a clear correspondence can be drawn among the DSL constructs and the Java API calls. For instance the "title:" element corresponds to a call to setTitle(). In this case DSL and API seem to lay at the same abstraction level, thus theoretically allowing for a purely syntactic translation. Note that this example does not show how to handle user interface event handlers, which can execute arbitrary actions, and are therefore generally written in GPL code.

The snippet in the Figure 1c is written in the DSL obtained with DSLit by analyzing the Swing API. The snippet shows only a few lexical differences with the JavaFX version. In this sense, the constructs of the automatically-generated DSL mimic the structure defined in the API, e.g., there is a DSL element for each API method. In addition, a compiler is also generated by DSLit that translates this snippet to the program in the Figure 1a.

As we will show, the conciseness of the previous DSL comes from the particular containment structure of the Swing API. As a significantly different example we show

---

[3] http://java.sun.com/products/jdk/awt
[4] http://docs.oracle.com/javafx
```

```
                    jRTF                                              DSLit
rtf().section(                                 rtf section {
  p( "first paragraph" ),                        p { "first paragraph" },
  p(                                             p {
    tab(),                                         tab,
    " second par ",                                " second par ",
    bold( "with something in bold" ),              bold{ "with something in bold" },
    " and ",                                       " and ",
    italic( underline( "italic underline" ) )      italic{ underline{ "italic underline" } }
  )                                              }
).out( out );                                  } out { out }
                   (a)                                               (b)
```

Fig. 2. An excerpt of Java code using the jRTF Fluent API and the corresponding automatically-generated DSL

in Figure 2a a program using jRTF[5], a Fluent API to construct Rich Text Format (RTF) documents by Java. According to M. Fowler[6], a Fluent API is an implementation of an object-oriented API that aims to provide more readable code. It is normally implemented by using method chaining to relay the instruction context of subsequent calls (but the Fluent paradigm is not limited to method chaining). Fluent APIs are becoming a very popular way to implement internal DSLs in Java. The jRTF example in Figure 2 shows how a method chain in the Fluent API can closely resemble a DSL.

Figure 2b shows the DSL automatically generated from jRTF by DSLit. Differently from the Swing case, the DSL in Figure 2 provides very little syntactic simplification w.r.t. its corresponding Java code. However, even in this case, the generation allows, for instance, generating an environment that has a domain-specific outline representing the RTF document structure, and it can be augmented with static checking capabilities (that are poor in the fluent API version).

In this paper we present a method that, given an API, generates an equivalent DSL. Our current application of this approach supports APIs fitting in one of the two categories previously defined plus a fallback category which resembles Java-like languages. We provide DSLit, a tool that generates such DSL, together with its development environment and a Java compiler, providing the following benefits:

- The generated DSL development environment has features like syntax highlighting or code completion that are tailored to the API domain.
- Semantic constraints can be made explicit and static validation can be enriched by parameterizing the generation process. E.g., the DSL for Swing can be customized so that labels are always created in a single container frame, and the frame name is a mandatory attribute (avoiding at compile time some common mistakes in interface development).
- The DSL compiler can be manually improved to optimize the resulting API code (e.g., reordering DSL definition elements to get optimal performance).

---

[5] http://code.google.com/p/jrtf/

[6] http://martinfowler.com/bliki/FluentInterface.html

3   From API to DSL

Figure 3 gives an overview of the linguistic architecture of our approach, spanning over three technical spaces (TS)[7]: (1) the API TS, in which API objects (i.e., in memory) conform to the set of defined API classes, (2) the Model TS (we refer to the MOF/Ecore TS), in which models conform to metamodels, and (3) the Grammar TS, in which programs conform to the grammar (e.g., a GPL).

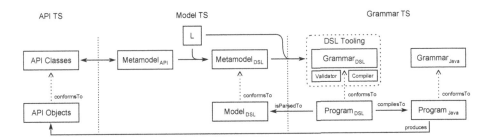

Fig. 3. Technical spaces and bridges

The process for obtaining a DSL from an API is split into three steps: (1) extracting an API metamodel for the set of class definitions in the API, (2) computing a DSL metamodel from the API metamodel, and (3) generating the grammar and the DSL tooling from the DSL metamodel. The steps are detailed in the following subsections.

3.1   API Classes to API Metamodel

The first step generates an API metamodel by applying a bridge between the API and Model TSs. This bridge maps API class definitions into metamodel elements. Thus Java classes are mapped into metaclasses, while attributes and methods are mapped into metaclass attributes and references/operations, respectively. The mapping is not trivial because of the semantic differences between class definitions and metaclasses, but it is well studied in works such as [9] and [10].

When applied to big APIs, this step may generate very large metamodels. For instance in Figure 1 the bridge would create a metaclass for each class and interface of the Swing API. Our approach provides a customization mechanism that allows filtering out API elements (e.g., classes, methods, attributes) in order to influence the construction of the DSL metamodel. Filtered elements will typically include technical classes that are out of the developer's interest or do not belong to the level of abstraction of the DSL. For instance, in the Swing example the developer may be interested only in the JFrame and JLabel class, with all their ancestors in the inheritance hierarchy.

---

[7] The concept of Technical Space (TS) is introduced in [8]. It is defined according to a conformance relationship that associates artifacts (e.g. program) with meta-artifacts (e.g. grammar). Bridges can be defined to transfer artifacts from one to another TS.

Once the API metamodel is generated, developers can leverage in metamodel techniques to make explicit semantics in the DSL that are hidden in Java. An example is the semantics of references in metamodels: references can have containment semantics and multiplicity constraints that are implicit in Java attributes. To this aim, developers may manually enrich the API metamodel to exploit these aspects in the generated DSL, which can be statically checked on the DSL code. For instance, by adding a containment property to the reference between JFrame and JLabel the resulting DSL may automatically check that a label is not contained in two distinct frames.

3.2   API Metamodel to DSL Metamodel

API classes represent an internal abstraction mechanism of object-oriented languages, i.e., an *in-language abstraction*. The API metamodel we obtained in the previous step is an artifact describing this in-language abstraction. The purpose of the second step is to transform the in-language abstraction in a *linguistic abstraction*, i.e., an abstraction defined by language constructs. We perform the transformation between in-language abstraction and linguistic abstraction within the Model TS as a model transformation, thus creating the DSL metamodel.

Lifting the abstraction to the linguistic level is not a trivial step, as the logic to apply is strongly dependent on the structure of the DSL the user wants to obtain. For instance in the Swing DSL we want to generate language concepts for classes (e.g. JFrame, JLabel) and attributes (e.g. title, size) of the API. Conversely, in the RTF example the language structure contains a concept for each method of the API.

The linguistic abstraction of a DSL contains: (a) the domain concepts, which are extracted from the API metamodel; and (b) its structure and capabilities, which define how the concepts can be defined, linked and composed (e.g., which concepts become *Statements*, whether the DSL is going to use *Blocks*, etc.). While the former is domain-specific, the latter can be considered domain-independent and be reused in different DSLs. For instance, in our first example the domain contains the elements JFrame and JLabel (and their attributes) while the structure of programs in the Swing DSL may be composed by a sequence of statements initializing the JFrame and JLabel attributes.

In order to generate the DSL metamodel we built a template system which receives as inputs: (a) the API metamodel, and (b) a template defining the structure and capabilities of the languages. Our approach currently provides three templates for the categories considered in DSLit but more templates can also be plugged in.

3.3   DSL Metamodel to DSL Environment

The last step is a bridge between the Model TS and the Grammar TS which produces the needed artefacts for the DSL. The Model TS already contains several well-known tools that help in generating the components of an external textual DSL environment (e.g., Xtext). Therefore, this step is devoted to generate the input artifacts for these tools, including: (1) the mapping of metamodel elements (i.e., the abstract syntax definition of the DSL) into the grammar rules of the concrete syntax, (2) development environment (e.g., validators, type system, etc.) and (3) compiler.

The generation process is parameterized by the DSL metamodel and the template chosen. While the DSL metamodel provides domain-specific information (e.g., concepts, attributes, references) and basic semantics (e.g., cardinalities, containment, etc.), the template drives the grammar structure of the resulting DSL and also the development environment and compiler.

The resulting compiler is able to transform DSL programs in their corresponding Java programs. As shown in Figure 3 the compiler is an artefact of the Grammar TS, and the execution of the compiled Java program produces the set of API objects in the API TS (plus possibly other objects). Most tools create also a parser towards the Model TS that extracts from programs the corresponding instance of the DSL metamodel.

The next section explains how we implement the approach and illustrates in detail how the motivating examples are addressed.

4   DSLit

As a proof of concept of the described approach, we have implemented DSLit, a prototype DSL generator integrated in the Eclipse platform. The current prototype contains three DSL templates that aim to address the APIs that fall under the categories previously introduced, respectively *POD*, *Fluent* and *SimpleJava*.

Once one of the provided templates is selected, our tool is able to generate a domain-specific development environment, using Xtext. Currently DSLit supports the generation of a Proposal Provider and Validator components of the environment. In future work we plan to investigate the domain-specific customization of other components.

4.1   Grammar Generation

In this section, we describe the DSL templates included in DSLit, covering three possible representations of the information contained within the API metamodel.

POD DSL Generator. DSLit provides an ad-hoc POD DSL Generator. The generator can be applied to any API, but it only considers their POD part (setters, getters and constructors) for the definition of the DSL. This generator has been applied to a Swing subset and generates the DSL in Figure 1c.

In the following we briefly describe the transformation logic for the generation of the DSL grammar model (conforming to the Xtext metamodel) from the API metamodel. The full code of the transformation is available at the paper website.

The POD DSL extracted out of the API metamodel takes into account only attributes and references contained in the classes defined in the metamodel. All the classes, attributes and references are mapped to grammar rules in Xtext and their names will appear as terminals in the grammar. Each class is transformed into a rule that contains, wrapped into braces, the features that correspond to the attributes and references for that class. Each of those features expands to the rule that represents the type of the corresponding attribute or reference, while its cardinality depends on the cardinality of the corresponding attribute or reference. In particular, a multi valued attribute is expressed as a feature list, an optional attribute is expressed as an optional feature and finally a single valued attribute is expressed as a single feature. Finally, since we know the semantics of the POD template, our tool is able to append additional grammar rules in order to avoid identifying the root class of the API metamodel. Thus, it adds to the template the rules *Grammar* and *Element*. The former is the grammar root rule and contains a list of *Elements*, while the latter includes as alternatives all the rules that correspond to the metaclasses.

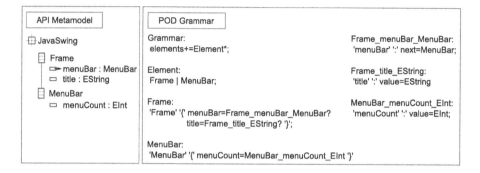

Fig. 4. Example of the POD DSL for Swing

In Figure 4, an excerpt of the generation of the Java Swing DSL is shown. The classes *Frame, MenuBar* are mapped to the corresponding grammar rules (generated according to the schema *className-attrOrRefName-AttrOrRefTypeName*). The rule *Frame* contains as terminal the name of the related class and two optional features *menuBar* and *title* that represent respectively the reference and the attribute embedded in the class *Frame*. They expand in two rules *Frame_menuBar_MenuBar* and *Frame_title_EString*. The former contains the name of the reference *menuBar* as terminal and the feature *next* (inserted by the generator) that expand in the rule *MenuBar*; the latter contains the name of the attribute *title* as terminal and the feature *value* (inserted as well by the generator), since the *title* is defined as a primitive type. The remaining rules *MenuBar* and *MenuBar_menuCount_EInt* follow the same mapping strategy.

Fluent DSL Generator. The Fluent DSL generator transforms a fluent API into an equivalent external DSL. The generator included in DSLit is able to handle simple fluent APIs like the jRTF (Figure 2).

The API metamodel generated from a Fluent API is composed by operations defined over the classes of the metamodel. For each class, a grammar rule is created. It contains a set of alternatives that are the grammar rules corresponding to the operations for that class. Inside these rules, each operation parameter is mapped to a feature that, depending on the parameter cardinality, can be optional or a list. In addition, an optional feature *next* is defined, that expands to the rule representing the return type of the operation.

It is important to note that in the Fluent DSL, all the names of the operations defined in the classes of the metamodel appear as terminal in the grammar.

Figure 5 shows the generation of a Fluent DSL for a small subset of the jRTF API. The classes *Rtf* and *RtfPara* are mapped to the corresponding rules; while the operations

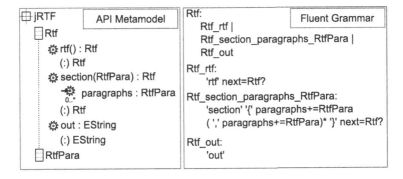

Fig. 5. Example of the generation of a Fluent DSL

rtf( ), *section(RtfPara)* and *out( )* are mapped to rules which names follow the schema: *className-operationName-parameterNames-returnType*.

In particular, *Rtf_rtf* contains as terminal the name of the operation *rtf( )* and an optional feature *next*, that represents the return type *Rtf* of the operation *rtf( )*. *Rtf_section _paragraphs_RtfPara* contains the terminal *section*, a list of paragraphs that expand the rule *RtfPara* (not shown in the example) and the optional feature *next*. Finally, *Rtf_out* contains only the terminal *out*, since the operation *out( )* has neither parameters nor a return type that is a class of the input API metamodel.

SimpleJava DSL Generator. In the case in which the API does not suit one of the previous templates, the user has still the possibility to generate a fallback DSL. Since no assumptions can be made on the structure of the API and on its intended use, the generated DSL needs to offer similar capabilities to the Java language. The SimpleJava DSL generator produces a DSL that resembles Java but is restricted to the use of the analyzed API. The generated DSL in this case has the purpose of being a starting point for DSL development, since 1) the DSL developer can easily add domain-specific features to the generated environment, 2) moving the domain information to the linguistic level makes it more suitable to automated analysis.

The SimpleJava template (an excerpt is shown in Figure6) defines typical Java concepts, e.g. *Declaration* and *Assignment*. The transformation *API2Grammar* expands the SimpleJava template with the information contained within the API metamodel. It is important to note that names of classes, attributes and references in the API metamodel are inserted in the template respectively as alternatives of the grammar rules *Type* and *Attribute*.

As mentioned above, the SimpleJava template does not make any assumption on the structure of the API. Therefore, it is applicable to any API, including those for which other templates such as Plain Old Data structure, and Fluent DSL are applicable. Figure 7 revisits the Swing and jRTF examples: the DSL code samples of Figure 1 and 2 are now expressed in textual syntaxes derived using the SimpleJava template. This figure further illustrates how SimpleJava works.

```
Program:                                     VariableExpression:
    statements+=Statement*;                      var=ReferenceVar ('.' attr=Attribute)?;
Statement:                                   ReferenceVar:
    Declaration | Assignment;                    value=[Value];
Declaration:                                 Literal:
    var=Var;                                     FloatLiteral | StringLiteral | ...;
Assignment:                                  Var:
    left=VariableExpression '=' right=Expression;    name=ID ':' type=Type;
Expression:                                  Attribute:
    VariableExpression | Literal;                name=__AllAttributeNames__;

                                             Type:
                                                 name=__AllClassNames__;
```

Fig. 6. Excerpt of the DSL Template for the SimpleJava DSL generator

SimpleJava for Swing   SimpleJava for jRTF

```
                              r : Rtf                       b : Bold
f : Frame                     s : Section                   b.contents = "text in bold"
f.title = "My Java Application"   r.section = s              p2.contents = b
f.width = 500                 p1 : RtfPara                   p2.contents = " and "
f.height = 300                p1.contents = "first paragraph"   i : Italic
l : Label                     p2 : RtfPara                   u : Underline
l.text = "Hello World!"       t : Tab                        u.contents = "italic underline"
f.content = l                 p2.contents = t                i.contents = u
f.visible = true              p2.contents = " second par "   p2.contents = i
```

Fig. 7. SimpleJava template applied to Swing and jRTF

4.2 Development-Environment Generation

Once the DSL grammar is generated, Xtext is able to produce several artifacts composing a DSL development environment. In particular, it offers 1) a *proposal provider* that provides a list of accessible keywords according to the current terminal of the grammar (i.e. content assist) and 2) a *validator* performing static analysis during editing. However, since we know the semantics of the DSL template that has been used to generate the grammar, we can automatically derive improved versions of such environment components by mixing the DSL domain-independent part, that comes from the template structure, and the DSL domain-specific part, that is inferred from the API metamodel.

In our prototype, these improved components are generated for the POD and SimpleJava DSLs and can be can be eventually redefined by the developer if needed. For the Fluent DSL, since the proposal provider and validator are embedded in the structure of the grammar (i.e., how the feature of a grammar rule expands in other rules), we rely instead on the Xtext default components.

4.3 Compiler Generation

Since the semantics of the DSL template is well-defined, a DSL instance can be transformed into its equivalent in Java. For instance concepts like *Declaration*, *Assignment* and *Statement* in the SimpleJava template (Figure6) have a one-to-one correspondence with Java programming language's constructs. Xtext provides the capability to generate a model-representation of the DSL grammar according to the Xtext metamodel. Such a DSL model is then transformed to a Java model leveraging on MoDisco[8] that is in turn translated to a Java readable file using Acceleo[9], a model-to-text transformation tool.

5 Related Work

The work presented in [11] about Framework-Specific Modeling Languages studies how to identify and extract domain-specific knowledge from APIs. Some of the ideas in that work inspired our research. Works such as [12] investigate current analysis techniques to understand APIs and extract some usage patterns. Such studies could complement ours in identifying specific API features and therefore improve our process.

Integration between the model and the API TS has been considered in works such as [9,10,13], where approaches to define bridges between these two TSs are presented. However, none of them enables the generation of a DSL from an API definition nor the selection of an appropriate structure for the resulting DSL.

Existing approaches such as METABORG [14], SugarJ [15] and Helvetia[10] enable the definition of internal DSLs and the corresponding generation of the domain-specific environment in the host language. Instead, our approach targets external DSLs. Furthermore ours can automatically generate a DSL definition for an API, which could be used as input for the aforementioned approaches.

Some GPLs with flexible concrete syntaxes like Haskell or Ruby enable direct definitions of DSLs directly using GPL syntax. An example of such a DSL is given in [16]. One problem with this kind of approaches is that DSL concrete syntax must be a subset of GPL concrete syntax (i.e., DSL syntax must be valid GPL code). Another problem is that the corresponding API has to be defined specifically so that GPL syntax may be used directly as a DSL. Therefore, compromises must be made on both API and DSL.

In [17] the authors evaluate 10 different approaches to implement DSLs, concluding that embedded DSLs are the simplest to implement. Our approach could be considered as an additional approach with which a significant amount of DSL customization is attainable at a comparatively low cost: (a) it provides a specific textual syntax not limited by a host GPL and (b) it kickstarts a DSL tooling ready to be used for the DSL.

The approach presented in [18] provides abstractions for repeatedly used patterns. Instead, we target on abstracting API calls into DSL constructs. However, both approaches could be combined so that simple DSL construct could be mapped to complex patterns of API usage.

---

[8] http://www.eclipse.org/gmt/modisco/technologies/J2SE5/
[9] http://www.eclipse.org/acceleo
[10] http://scg.unibe.ch/research/helvetia

6   Conclusion and Future Work

In this paper we have presented an approach to automatically generate an external DSL out of an object-oriented API by transforming in-language abstraction into linguistic abstraction. The generation process has been presented as a bridge between the involved technical spaces (i.e., API TS, Model TS, and Grammar TS) and uses a template mechanism, which allows customization of the resulting DSL. Our approach has been implemented on the Eclipse platform, as a plugin called DSLit, which generates an Xtext-based textual DSL out of Java-based APIs. The current prototype incorporates two templates that generate specific DSL structures (POD and Fluent) as well as a fallback template covering a subset of Java (SimpleJava).

In future work we plan to study how our method could cope with APIs that allow custom code extension (e.g., providing implementations of interfaces or abstract classes). We would also like to define more templates allowing for different types of DSLs, which in turn will need a deeper study on API characterization. Another possibility to explore is the generation of DSL interpreters instead of compilers. They would for instance make it possible to load and execute DSL code at runtime. Finally, since a GPL allows interleaving calls to distinct APIs, one open question to study is how the generated DSLs may be combined in order to achieve the same kind of interleaving achievable with a GPL.

References

1. Lédeczi, A., Bakay, A., Maróti, M., Völgyesi, P., Nordstrom, G., Sprinkle, J., Karsai, G.: Composing domain-specific design environments. Computer 34(11), 44–51 (2001)
2. Kelly, S., Tolvanen, J.P.: Domain-Specific Modeling: Enabling Full Code Generation. Wiley-IEEE Computer (2008)
3. Fowler, M.: Domain-Specific Languages. Addison-Wesley (2010)
4. Sánchez Cuadrado, J., García Molina, J.: A model-based approach to families of embedded domain-specific languages. IEEE Trans. Softw. Eng. 35(6), 825–840 (2009)
5. Sánchez Cuadrado, J., Cánovas Izquierdo, J.L., García Molina, J.: Comparison Between Internal and External DSLs via RubyTL and Gra2MoL. In: Formal and Practical Aspects of Domain-Specific Languages: Recent Developments, pp. 109–131 (2013)
6. Mernik, M., Heering, J., Sloane, A.M.: When and how to develop domain-specific languages. ACM Comput. Surv. 37(4), 316–344 (2005)
7. Eysholdt, M., Behrens, H.: Xtext: implement your language faster than the quick and dirty way. In: SPLASH, pp. 307–309 (2010)
8. Kurtev, I., Bézivin, J., Aksit, M.: Technological Spaces: an Initial Appraisal. In: DOA, pp. 1–6 (2002)
9. Cánovas Izquierdo, J.L., Jouault, F., Cabot, J., García Molina, J.: API2MoL: Automating the building of bridges between APIs and Model-Driven Engineering. Inform. Software Tech. 54(0), 257–273 (2012)
10. Cuadrado, J.S., Guerra, E., de Lara, J.: The program is the model: Enabling transformations@ run. time. In: SLE, pp. 104–123 (2013)
11. Antkiewicz, M., Czarnecki, K., Stephan, M.: Engineering of Framework-Specific Modeling Languages. IEEE Trans. Softw. Eng. 35(6), 795–824 (2009)
12. Robillard, M.P., Bodden, E., Kawrykow, D., Mezini, M., Ratchford, T.: Automated API Property Inference Techniques. IEEE Trans. Softw. Eng., 613–637 (2012)

13. Song, H., Huang, G., Chauvel, F., Xiong, Y., Hu, Z., Sun, Y., Mei, H.: Supporting runtime software architecture: A bidirectional-transformation-based approach. J. Syst. Software 84(5), 711–723 (2011)
14. Bravenboer, M., Visser, E.: Concrete syntax for objects: domain-specific language embedding and assimilation without restrictions. In: SPLASH, pp. 365–383 (2004)
15. Erdweg, S., Rendel, T., Kastner, C., Ostermann, K.: SugarJ: Library-based Syntactic Language Extensibility. In: OOPSLA, pp. 391–406 (2011)
16. Cunningham, H.C.: A little language for surveys: constructing an internal dsl in ruby. In: ACMSE, pp. 282–287 (2008)
17. Kosar, T., Barrientos, P.A., Mernik, M., et al.: A preliminary study on various implementation approaches of domain-specific language. Inform. Software Tech. 50(5), 390–405 (2008)
18. Chodarev, S., Pietriková, E., Kollár, J.: Towards Automated Program Abstraction and Language Enrichment. In: SLATE, pp. 51–64 (2013)

Function Based Requirements Engineering and Design –Towards Efficient and Transparent Plant Engineering

Florian Himmler

University of Erlangen-Nuremberg, Nuremberg, Germany
evosoft GmbH, Nuremberg, Germany
`florian.himmler@fau.de, florian.himmler@evosoft.com`

Abstract. Industrial enterprises are facing an increasing complexity of their manufactured products. This goes along with a high number of fast moving and volatile customer requirements. In order to support engineering organizations in increasing their profitability, this paper presents major success factors for efficient requirements engineering and design in plant engineering. In this paper a *Function Based Requirements Engineering and Design* concept is introduced, which helps plant engineering companies achieve a more efficient requirement analysis and solution concept engineering. By splitting up the requirements engineering process in three layers – *Problem Layer*, *Abstraction Layer* and *Solution Layer* – the concept fosters an increase in reusability of requirements and features, better traceability as well as higher transparency throughout the whole process. We were able to prove the feasibility of the concept based on a case study involving the engineering of baggage handling systems.

Keywords: plant engineering, requirements engineering, function based engineering, industrial engineering.

1 Introduction

Today, industrial enterprises in plant engineering are facing a rising complexity of their manufactured products and manufacturing processes [1,2]. Especially the European plant engineering companies are exposed to increasing competitive pressure, especially from the Asian market [3–5]. These changing basic conditions require that companies enhance their planning efficiency as well as their planning quality [6].

Prior research on this topic has shown that the most important levers to reach these goals are currently:

- The development of standardization and modularization strategies to improve interdisciplinary cooperation and the project comprehensive reuse of artifacts [3,5].
- The development of requirements engineering (RE) concepts and processes which improve the requirements elicitation at the beginning of a project and help to build the right product during the project [4,7].

G.F. Italiano et al. (Eds.): SOFSEM 2015, LNCS 8939, pp. 436–448, 2015.

In this paper we will introduce a novel concept for a function based requirements and systems engineering concept for the plant engineering domain. This concept can be applied in order to improve the following aspects:

- Improved coverage of the requirements of engineered plants ("Build the right product").
- Improved efficiency during the early phases of the engineering process (i.e. basic engineering).
- Better communication with the customer regarding the engineered plant and its features.
- Full transparency and traceability throughout the whole engineering process.

We will first provide an overview of work related to this topic. This will be followed by a description of the traditional requirements management approach usually employed in plant engineering. We will then introduce our integrated *Function Based Requirements Engineering and Design* approach. Finally, we will evaluate the concept based on a case study and give a brief outlook on future research needed in order to further extend this topic.

2   Related Work

RE is a long established information systems and software engineering discipline with a high number of publications available. Most of these concepts focus on defining guidelines and establishing good practices on how to perform the different RE tasks, such as stakeholder analyses, requirements elicitation, documentation, prioritization etc. [8]. Current problems relating to RE in general include a lack of information (about the needs and requirements of customers and their relevance), heterogeneous understanding of requirements, the high degree of necessary coordination and communication, limited tool support and limited guidance and adoption strategies (especially in complex domains) [9–11].

Despite its economic importance, research is still not focusing on the handling of customer requirements and the requirements based solution design and validation in context of the engineering of industrial plants. In [12], major challenges (authority requirements, aging, communication, knowledge transfer, representation of requirements and tools) were identified regarding RE in the energy plant engineering domain. Other authors describe concepts about how to define requirements at different levels of abstraction, such as system level, function level, and software level [13].

Three independent market studies performed recently highlight the urgency for developing plant engineering specific RE and design concepts. In [14], especially large scale plant engineering companies state that they see a high potential in making the RE and requirements management process more efficient. Others expect the optimization of the requirements management process in combination with an optimization of the solution design engineering as the main topic for future activities [15]. The third study found, that only one third of the companies follow a structured approach to gather customer requirements and integrate them into their solution portfolio. This leads to *"complexity and diverging customer require-ments [being] key challenges, driving the need for a structured approach"* [16].

3 Traditional Requirements Management in Plant Engineering

In this section we will provide an overview of the traditional RE and design approach usually applied in plant engineering, and point out potential issues emerging from such an approach.

Today, especially within the plant engineering domain, a typical RE process is split up in a *Problem Layer*, which defines the problem to be solved for a customer, and a *Solution Layer, w*here the solution to the problem is designed (Figure 1).

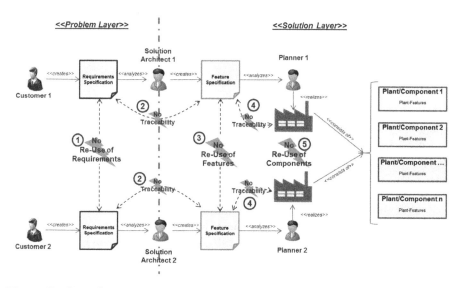

Fig. 1. Traditional requirements engineering and design approach in plant engineering

The process usually starts with a *Customer* creating a *Requirements Specifi-cation* for a plant that the *Customer* wants to purchase. This specification is then handed over to a plant engineering company, which is supposed to realize the plant for the *Customer*. A *Solution Architect* of the plant engineering company then an-alyzes this *Requirements Specification* document in order to obtain an overview of what the Customer needs. Based on his interpretation of the *Customer's* needs, the *Solution Architect* then defines the concept as to how the plant could be realized to comply with the *Customer's needs*. This concept is documented and described in a *Feature Specification* document and handed over to a *Planner* for further process-ing. The *Planner* then analyzes the *Feature Specification* document and engineers the plant according to the concept described in the document. The engineered plant itself is an assembly of different sub-plants, machines and components. If a second *Customer* is also requesting a plant at the same time, an identical process is started in parallel. Due to the complexity of such engineering projects, the roles of the second request are usually handled by different persons than for the first *Customer*. Communication and design matching in the different phases of the parallel processes usually doesn't take place.

During such RE and design processes as described above, multiple issues may emerge, especially in bigger companies that are running multiple customer projects in parallel:

- **No Reuse of *Requirements*:** *Requirements* are typically written in a natural language style with each *Customer* having different ways of expressing himself. This may lead to situations where two *Requirements* of two different *Customers* have the same semantic meaning but a completely different text. Situations such as these are not identified, resulting in redundantly designing individual solutions to address the same *Requirements*.

- **No Traceability between *Requirements* and *Features*:** The *Solution Architect* creates a solution concept to satisfy the *Customer's* needs and documents it in a *Feature Specification*. Since *Requirements Specifications* are rarely split into atomic *Requirements,* there is usually no traceability between a single *Requirement* and the associated *Feature* to address this *Requirement*. Transparency towards the *Customer* (e.g. which *Requirements* can/cannot be fulfilled) requires a significant level of manual effort.

- **No Reuse of *Features*:** There is no systematic reuse of solutions (*Features*) for the same *Requirements*. The same *Requirement* defined by two *Customers* may lead to two different solutions as they are handled by different *Solution Architects*.

- **No Traceability between the Engineered *Plants* and their *Features*:** The *Planner* analyzes the *Feature Specification* and engineers the real *Plant* according to this specification. Engineered *Plants* usually consist of a number of *Sub-Plants/Sub-Components*. Significant manual effort is needed to analyze which parts of the *Feature Specification* are realized by which *Sub-Plant/Sub-Component*.

- **No Reuse of Components:** There is no systematic reuse of *Sub-Plants/ Sub-Components* used to realize a *Plant*. Similar *Features* defined in different projects and engineered by different *Planners* may result in completely different, individually engineered solutions using different *Sub-Plants/Sub-Components*.

- **Limited Reporting and Analytics Capabilities:** The capabilities to analyze the RE and design processes are limited. Regarding the *Problem Layer*, reports about *Requirements,* which are often requested by *Customers* but cannot be fulfilled by the company, are hard to realize. *Solution Layer* reports such as *Plants/Components* which are part of the *Customer's* portfolio, but rarely requested by a *Customer,* also require significant manual effort.

- **Know-How Required between the *Problem Layer* and *Solution Layer*:** Every request by a *Customer* requires a *Solution Architect* to analyze the *Requirements Specification* and design a solution to it. This even applies to cases where the requested plants are of limited complexity. Since *Solution Architects* are highly specialized, this results in high engineering costs for every plant.

Some of these issues mentioned above could be mitigated to a certain extent at the process level. When defining close collaboration and communication processes between *Solution Architects* and *Planners* over multiple projects, the reuse ratios regarding *Features* and engineered *Plants* could be improved. However, the higher the number of parallel projects being processed in a company, the harder it will be to establish such processes.

Another approach to mitigate some of these issues would be to use formal templates for defining requirements or use cases, as proposed by [17]. However, this would only promote a better and unique interpretation of requirements – the majority of the issues, especially regarding traceability and reuse, would still exist.

4   Function Based Requirements Management

In this chapter we will introduce our novel *Function Based Requirements Engineering and Design* concept for industrial plants. We will first describe the basic concept behind the approach, followed by the potential benefits resulting from it.

4.1   Basic Concept

The concept described here has been developed based on domain expert knowledge and extensive practical experience gained in various customer projects of an IT consulting company. Such a concept could solve the issues and problems identified in the previous section. The foundation for the concept is the function based standardization framework described by [18]. By applying our new concept, the RE and design process can be split up into three different layers (Figure 2):

- **Problem Layer:** In this layer the customer defines the problem that needs to be solved using a customer-specific language and wording.
- **Abstraction Layer:** The problem that has been defined by a customer in the *Problem Layer* is mapped to an abstract, standardized representation of the problem.
- **Solution Layer:** In this layer, a standardized set of *Functions* and *Features* is defined, which can be used to compose a solution that can satisfy the customer' s *Requirements*.

The problem defined by the *Customer* in the *Problem Layer* is still documented in a *Requirement Specification* document. This document is then split into single *Requirements,* which ideally fulfill the characteristics of good *Requirements*: unitary, complete, consistent, atomic, traceable, current, unambiguous, and verifiable [19,20]. This split could either be performed by the *Customer*

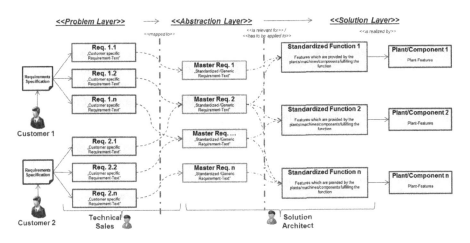

Fig. 2. Function Based Requirements Engineering and Design concept overview

himself or by the *Solution Architect* of the engineering company. The result is a list of customer-specific *Requirements*. Due to the fact that *Customers* use their own language and wording to define *Requirements*, situations can occur where two *Customers* semantically have the same *Requirement* regarding the system to be designed, but phrase it in a totally different way, therefore making it hard to recognize the similarity between them (e.g. *Must be able to operate with normal German supply voltage"* vs. *Must be able to be connected to 230 V power outlets"*).

This problem is solved by introducing a set of *Master Requirements* within the *Abstraction Layer*. Such *Master Requirements* are *Requirements* that are defined in a standardized and generic way – and should fulfill the characteristics of good *Requirements*. They always comprise two elements: a textual representation of the requirement and, when applicable, a corresponding formula, which allows automated evaluations and validations (e.g. Text: *"The Voltage must be $<<value>> V$"*, Formula: *"Voltage $= <<value>> V$"*). Each customer-specific Requirement has to be mapped to exactly one *Master Requirement* (while one *Master Requirement* can be mapped to any number of customer-specific *Requirements*). This task is performed by a *Technical Sales* person in the engineering company talking directly with the *Customer*. The complexity of having *Customer*-specific languages and wordings can be eliminated by applying this concept. If there is no suitable *Master Requirement* to address a customer-specific *Requirement*, a new *Master Requirement* needs to be created. The result is a list of *Master Requirements*, which is a standardized representation of the customer-specific *Requirements*.

A solution now needs to be engineered to address these *Master Requirements*. The *Solution Layer* becomes relevant at this point. The solution portfolio of the engineering company is represented by a set of standardized *Functions* that can be used and combined in order to create specific solutions for customers. Each of the *Functions* is defined as a set of characteristics that describe the capabilities of the same (e.g. *"Maximum speed"*, *Maximum weight"*, *"Voltage"* etc.). Each of the *Functions* has a number of components assigned to it, which are able to realize the function. This structure follows the *Function Based Engineering*, a framework which has been introduced as a standardization and modularization strategy for the engineering of industrial plants in [18]. Each of the *Functions* in the solution portfolio has a link to those *Master Requirements*, which are relevant for the *Function* or have to be applied for a particular *Feature* of the *Function* (e.g. if a *Function* has a specific voltage in its characteristics, then the corresponding *Master Requirements* need to be assigned to this *Function*). This assignment of *Master Requirements* to the relevant *Functions* is performed by the *Solution Architect* of the engineering company as part of project-independent activities. This task requires extensive engineering know-how.

As soon as the relationships between the *Master Requirements* and the standardized *Functions* are defined, a solution can be engineered to address the customer's *Requirements*. Since these *Requirements* are already represented as a list of *Master Requirements*, and each of the *Master Requirements* is related to the relevant *Functions*, it can be easily evaluated and validated as to which (combination) of the *Functions* from the *Solution Layer* is able to fulfill the *Customers'* needs.

The left part of Figure 3 shows such an example for a mapping between customer specific *Requirements* (*Req. 1.1* to *Req. 1.3*) to their corresponding representation as *Master Requirements* (*"Master Req. 1"* to *"Master Req. 3"*). In this example, the *Requirements* are linked using an AND logic operation ("□"), which means that all of them need to be fulfilled by a potential solution. Alternatively, *Requirements* can also be linked using OR and XOR logic operations.

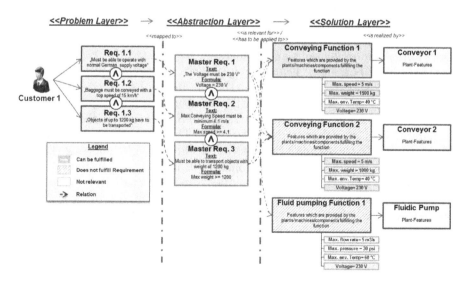

Fig. 3. Mapping between *Problem Layer*, *Abstraction Layer* and *Solution Layer*

The right part of Figure 3 shows the validation of the *Master Requirements* towards the *Functions* (*Conveying Function 1*, *Conveying Function 2* and *Fluid Pumping Function 1*) available in the *Solution Layer*. The only *Function* which could fulfill the *Requirements* is the *Conveying Function 1* as it can satisfy all of the specified *Master Requirements*. The *Fluid Pumping Function 1* is not a suitable solution as it does not have a *"Max. Weight"* characteristic, and therefore cannot satisfy *Master Requirement 3*. *Conveying Function 2* is also not suitable as its *"Max. Weight"* characteristic of *1000 kg* does not match the *Master Requirement 3*, which requires a *"Max. Weight"* of at least *1200 kg*.

4.2   Benefits of the Concept

By applying the *Function Based Requirements Engineering and Design* concept as described, most of the issues associated with traditional RE in plant engineering can be solved. In particular, the following advantages can be achieved with this concept:

- **Standardized Representation of *Requirements*:** By using *Master Requirements* between the *Problem Layer* and the *Solution Layer*, a set of standardized requirements is introduced leading to a higher degree of standardization during the whole RE process. This results in a higher cross-project reuse ratio with regard to the *Requirements* used – as well as to the solutions that are engineered.
- **Elimination of Customer-Specific Wording:** Customer-specific wording is eliminated by mapping each customer-specific *Requirement* to a *Master Requirement*. This results in a unique/non-ambiguous interpretation of the *Requirements*.
- **Early Pre-determination of Practicability and Automated Solution Proposals:** Each *Master Requirement* comprises a text-based as well as a formula-based representation. The formula representation could be used for an automated solution proposal by validating the *Requirements* formulas against the characteristics of the standardized *Functions* in the *Solution Layer*. This approach means that it can be identified as to whether a solution requested by a *Customer* can be fulfilled or not at a very early stage.
- **Automated *Requirements* Validation:** Analogous to the automated generation of solution proposals, for every solution engineered in the *Solution Layer* it is possible to automatically validate whether the customers' *Requirements* are fulfilled.
- **Full Traceability:** Due to the relationship between customer-specific *Requirements*, *Master Requirements* and the standardized *Functions*, full traceability is possible at the *Requirements* level from the *Problem Layer* to the *Solution Layer* and vice versa.
- **Identification of Unused *Functions*:** Due to the full traceability, it is possible to identify how often each of the standardized *Functions* is used in the *Solution Layer*. This could help to reassess the solution portfolio of the engineering company, e.g. by eliminating unused functions.

- **Identification of Frequently Requested *Requirements* that cannot be Fulfilled:** Due to the full traceability, it is possible to identify those *Requirements* of the *Customers* that cannot be fulfilled with the solution portfolio of the engineering company. Mapping to the *Master Requirements* makes it easy to identify those *Requirements* that are frequently requested by multiple *Customers,* but have never (or rarely) been fulfilled.
- **Improved Reuse Ratio within the *Solution Layer*:** By employing *Master Requirements*, a higher reuse ratio can be achieved regarding the solution engineered in the *Solution Layer*. Once all customer *Requirements* are mapped to *Master Requirements*, it can be analyzed as to how similar customer requests have been solved in the past, in order to reuse these solutions (or parts of them).
- **Reduced Engineering Costs:** Using the traditional approach, each customer request requires a *Solution Architect* for further processing. By applying our new approach, the highly qualified *Solution Architect* can focus on his tasks that are independent of any particular project: perform the engineering of a new standardized *Function* (i.e. to define the *Solution Layer*), define *Master Requirements* and relate them to the relevant standardized *Functions*. Project-specific tasks, such as communicating to the customer and creating a first solution concept, can be handed over to a *Technical Sales* person in the future. This would reduce the engineering costs for every plant requested by a *Customer*.

5   Case Study

In this section, we will describe a case study, which we performed in cooperation with an international IT consulting company involved in plant engineering projects. The goal of the case study was to prove the feasibility of the *Function Based Requirements Engineering and Design* described above by validating the following statements:

- *Statement 1 (S1): Individual customer Requirements can be mapped to common Master Requirements in order to reduce complexity and identify Requirements with the same semantic.*
- *Statement 2 (S2): It is possible to define unique Master Requirements by using a standardized text-based representation in conjunction with a formula-based one.*
- *Statement 3 (S3): Mapping from Master Requirements to standardized Functions in the Solution Layer facilitates efficient solution engineering based on the customers' Requirements.*

In this case study we initially analyzed 52 different projects that the participating company had executed within the last years. These projects involved the engineering of small airport baggage handling systems (i.e. max. 4 million passengers per year, max. 1500 bags/hour). The first step was to identify those projects where the customer provided a detailed *Requirements Specification* during its request for proposal. This reduced the number of relevant projects down to six. In 46 of the projects, no structured *Requirements Specification* was provided (in these cases the *Requirements* were identified in verbal face-to-face communication between a sales person and the *Customer*). The next step was to analyze the *Requirements Specifications* provided within these six remaining projects. The size of the documents ranged from 7 to 135 pages, resulting in completely different structures and level of details within the documents. However, it was noticeable that only few of the *Requirements* defined in the document were able to fulfill the criteria for good *Requirements* as described above. We then started splitting up the *Requirements Specifications* into single and unique *Requirements*. In order to limit the number of *Requirements* to a manageable amount, we focused on *Requirements* for belt conveyors regarding the baggage which needs to be handled, the dimensions of the conveyors, the generated noise level, the conveying speed and the surrounding conditions. Overall we extracted 53 customer *Requirements* regarding these categories from the *Requirements Specifications*. We then analyzed these *Requirements* in detail and defined the respective *Master Requirements* in order to create the initial *Abstraction Layer*. We found that those 53 customer *Requirements* can be abstracted and generalized by using 14 different *Master Requirements*. This corresponds to a *Requirements* abstraction rate of 73% (Table 1).

Table 1. Number of *Requirements* identified during the case study

	Number of Requirements per category					
	Conveyor dimensions	Baggage specification	Noise level	Speed	Surrounding conditions	Sum
Customer Requirements	15	23	5	5	5	*53*
Master Requirements	6	4	1	1	2	*14*

The standardized *Functions* determining the *Solution Layer* for this case study has already been defined in an earlier case study relating to *Function Based Engineering* [18] and were reused for our purpose. Overall, 30 different belt conveying functions (e.g. *Straight Conveying 1200 mm, Curved Conveying 90°* etc.) and their corresponding *Features* had been defined there, which we were able to reuse in the context of this case study. In the next step, we were able to map each of the *Master Requirements* defined earlier to one or more *Features*, which were assigned to the standardized *Functions*.

Reflecting the case study and referring to the statements defined above, it can be stated that we were able to confirm statements S1 and S2. Up until now, statement S3 could only be partially confirmed. Regarding statement S1, we were able to define an abstracted set of *Master Requirements* where each of the customers' *Requirements* could be mapped, thus reducing the complexity of the domain. We further specified each of the *Master Requirements* using a text-based as well as formula-based representation (Statement S2). Mapping from *Requirements* de-

Table 2. Sample of mapping customer *Requirement* to *Master Requirement* during case study

Category	Customer Requirement	➡	Master Requirement
Speed	**Customer 1:** Collecting band will have a 0.5 m/s speed, and since for each baggage there'll be allocated 2 m slots resulting in a 900 baggage/hour capacity. **Customer 2:** Components for low-speed conveyors shall be used for all conveyors with speeds up to 85.m.p.m **Customer 3:** The containers shall be moved at a minimum of 0.3 m/s by the conveyors. **Customer 4:** Conveying speed: 0.4 m/s	➡	**Text-based representation:** Maximum conveying speed must be minimum <<value>> m/s **Formula-based representation:** Max. speed >= <<value>> m/s

fined by the *Customers* to standardized *Master Requirements* for *Requirements* regarding the speed of the conveyors is shown as example in Table 2.

Regarding statement S3, we were able to map every *Master Requirement* to the corresponding *Features* of the standardized *Functions*. However, we were not able to fully prove the improved efficiency of the solution design. To confirm this statement to its full extent, it would be necessary to apply our concept within a new engineering project from the very beginning. However, since we were able to automatically reduce the number of potential conveying functions available for configuration of the plants by matching their *Features* to the *Master Requirements*, it can be assumed that it will also have a positive effect on the overall efficiency of the solution design process.

During the case study, we additionally identified a couple of key lessons learned:

- As already stated earlier, only a few customers (6 out of 52) provided structured *Requirements Specifications*. For those *Customers* not having such a structured overview of the *Requirements*, the existence of the *Master Requirements* could help the *Technical Sales* person to perform a structured *Requirement* elicitation from the *Customer*

- The process of splitting up the *Requirements Specifications* of the *Customers* into single *Requirements* already has a first positive impact. Since each *Requirement* must be interpreted and mapped to *Master Requirements*, it is necessary that the *Requirements* are completely understood. Existing ambiguities are solved in cooperation with the *Customer*, ensuring that the correct solutions are engineered. **Example:** *Requirement*: *"Luggage size to be transported: 900 x 700 x 500mm."*; **Ambiguity**: Which of the numbers is for width, length or height?

- A frequent source of inconsistency between *Requirements* of different *Customers* is that different units are used. In the *Abstraction Layer*, each *Master Requirement* needs to have a standardized unit of measure. When linking a customer *Requirement* with a different unit of measure to a *Master Require-*

ment, the value must be converted. Conversion tables could additionally be saved with the *Master Requirements*. **Example**: ***Dimensions***: millimeter (mm) vs. meter (m) =>standardized as millimeter.

- The initial effort to define the *Abstraction Layer* (i.e. the *Master Requirements*) and the mapping to the *Solution Layer* (i.e. the standardized *Functions*) is a time consuming task, which needs to be performed prior to the first *Customer* projects. However, this is an activity that is only required once, and it should have a fast payback time.

6 Conclusion

The aim of this article was to introduce a novel *Function Based Requirements Engineering and Design* concept for the plant engineering domain. We have shown the problems plant engineering companies are facing with traditional approaches to RE. We then introduced a concept to solve the problems by introducing an *Abstraction Layer* with *Master Requirements* between the *Requirements* in the *Problem Layer* and the standardized *Functions* in the *Solution Layer*. We were able to prove the feasibility of our concept by analyzing real customer projects, in this particular case the engineering of baggage handling systems. Using these projects, three statements have been validated and partially confirmed. However, this concept is currently still in an early phase and further research has to prove that it can be implemented in practice and can achieve the goals that have been set. For evaluation purposes, we recommend that prototypes and case studies are performed in cooperation with other plant engineering companies – preferably using projects from various domains.

The *Function Based Requirements Engineering and Design* described in this article has the potential to significantly improve the efficiency and transparency of designing customer-specific plant solutions (i.e. the basic engineering). In order to further establish the concept, some additional research is necessary regarding this topic. On the one hand, additional case studies should be carried out in order to further prove the applicability of the concept. On the other hand, the concept should be integrated in an IT infrastructure in order to verify to what extent it is possible to automate process steps such as the mapping of customer *Requirements* to *Master Requirements*, the automated and *Requirements*-based design proposal etc. Additionally, an evaluation should be performed to determine to what extent the framework could be transferred to other domains in addition to plant engineering (e.g. software engineering). These remain challenges that need to be addressed by future research.

References

1. Hicks, C., Earl, C.F., McGovern, T.: An analysis of company structure and business processes in the capital goods industry in the UK. IEEE Trans. Eng. Manag. 47, 414–423 (2000)
2. Aerni, A.: Modularization and Standardization in Plant and System Engineering. Digit. Bus. + Eng., pp. V–VIII (2004)

3. VDMA: Was macht den Großanlagenbau robust für die Zukunft–Erfolgsfaktor Wettbewerbsfähigkeit (2011)
4. VDMA: Großanlagenbau packt neue Herausforderungen an - Beiträge zum Industrieanlagenbau (2012)
5. VDMA: Staying competitive in a volatile environment (2014)
6. Hicks, C., McGovern, T., Earl, C.: Supply chain management: A strategic issue in engineer to order manufacturing. Int. J. Prod. Econ. 65, 179–190 (2000)
7. Little, A.D.: Ergebnissicherung im Anlagenbau (2003)
8. Finkelstein, A.: Requirements engineering: a review and research agenda. In: Proceedings of 1st Asia-Pacific Software Engineering Conference, pp. 10–19. IEEE Comput. Soc. Press (1994)
9. Schneider, L., Hajji, K., Schirbaum, A., Basten, D.: Knowledge Creation in Requirements Engineering – A Systematic Literature Review. In: Proc. 11th Int. Conf. Wirtschaftsinformatik, pp. 1829–1843 (2013)
10. Alves, V., Niu, N., Alves, C., Valença, G.: Requirements engineering for software product lines: A systematic literature review. Inf. Softw. Technol. 52, 806–820 (2010)
11. Liu, L., Li, T., Peng, F.: Why Requirements Engineering Fails: A Survey Report from China. In: 2010 18th IEEE Int. Requir. Eng. Conf., pp. 317–322 (2010)
12. Raatikainen, M., Mannisto, T., Tommila, T., Valkonen, J.: Challenges of requirements engineering – A case study in nuclear energy domain. In: 2011 IEEE 19th Int. Requir. Eng. Conf., pp. 253–258 (2011)
13. Buhne, S., Halmans, G., Pohl, K., Weber, M., Kleinwechter, H., Wierczoch, T.: Defining requirements at different levels of abstraction. In: Proceedings. 12th IEEE International Requirements Engineering Conference, pp. 325–326. IEEE (2004)
14. VDMA: Modularisierung und Standardisierungsansätze im Anlagenbau – Mythos oder Realität (2014)
15. VDMA, VDI, ID-Consult: Der Einfluss von modularen Produktbaukästen (MPB) auf den Unternehmenserfolg (2014)
16. Berger, R.: How to leverage modular product kits for growth and globalization Study (2012)
17. Toro, A.D., Jiménez, B.B., Cortés, A.R., Bonilla, M.T.: A Requirements Elicitation Approach Based in Templates and Patterns. In: Work. em Eng. Requisitos, pp. 17–29 (1999)
18. Himmler, F., Loy, H., Ostapovski, V., Amberg, M.: Function Based Engineering - A Standardization Framework for the Plant Engineering Domain. In: Kundisch, D., Suhl, L., Beckmann, L. (eds.) Tagungsband Multikonferenz Wirtschaftsinformatik 2014 (MKWI 2014), Paderborn, pp. 404–416 (2014)
19. Davis, A., Weidner, M.: Software Requirements: Objects, Functions and States. Prentice Hall (1993)
20. IEEE: IEEE Recommended Practice for Software Requirements Specifications (1998)

Software Developer Activity as a Source for Identifying Hidden Source Code Dependencies

Martin Konôpka and Mária Bieliková

Slovak University of Technology, Faculty of Informatics
and Information Technologies, Ilkovičova 2, 842 16 Bratislava, Slovakia
{martin_konopka,maria.bielikova}@stuba.sk

Abstract. Connections between source code components are important to know in the whole software life. Traditionally, we use syntactic analysis to identify source code dependencies which may not be sufficient in cases of dynamically typed programming languages, loosely coupled components or when multiple programming languages are combined. We aim at using developer activity as a source for identifying implicit source code dependencies, to enrich or supplement explicitly stated dependencies in the source code. We propose a method for identification of implicit dependencies from activity logs in IDE, mainly of switching between source code files in addition to usually used logs of copy-pasting code fragments and commits. We experimentally evaluated our method using data of students' activity working on five projects. We compared implicit dependencies with explicit ones including manual evaluation of their significance. Our results show that implicit dependencies based on developer activity partially reflect explicit dependencies and so may supplement them in cases of their unavailability. In addition, implicit dependencies extend existing dependency graph with new significant connections applicable in software development and maintenance.

Keywords: software component, dependency, source code, developer activity, dependency graph, implicit dependency, implicit feedback.

1 Introduction

Source code dependencies traditionally reflect explicit statements in the source code and are identified with syntactic analysis of source code contents. As a dependency we understand oriented connection between two source code components of selected granularity, namely instance or type reference, inheritance relationship or call reference.

Identified dependencies are used to form a dependency matrix or an oriented graph of interconnected software components to study organization and hierarchy of software components and their attributes [7]. Dependencies are also sourced for identifying problematic places, possibly code smells and complexity of the web of software components, which is important for maintenance activities on evolving software in particular.

G.F. Italiano et al. (Eds.): SOFSEM 2015, LNCS 8939, pp. 449–462, 2015.

Source code dependencies allow software developers to learn about how the existing source code works and how it is composed, e.g., how it will be affected by an introduced change or how much effort will be required for refactoring. Both adding new functionality and changing existing functionality require developers to know about the dependencies before making a change in the source code.

Traditional approaches use a syntactic analysis for identifying source code dependencies. Other approach is to employ developer activity as a source for identifying source code dependencies, but mostly logs of copy-pasting code fragments and commits to identify hidden dependencies in source code are used. We propose a method for identification of source code dependencies that extends existing works with utilizing data of developer's navigation in source code space (logs of switching between source code components). Based on the source for identification, we distinguish identified source code dependencies as *implicit*, i.e., identified from developer activity as an implicit feedback related to the source code, in addition to the traditional *explicit* dependencies reflecting explicit statements in the source code. With our method we broaden the space of known source code dependencies, thus extending a dependency graph with new edges relevant for the development or other evaluations of the source code.

For the identification of the implicit source code dependencies we use developer activity recorded in an integrated development environment (IDE) and commit logs from a revision control system (RCS) [14]. Our work is inspired by the research project PerConIK – Personalized Conveying of Information and Knowledge (perconik.fiit.stuba.sk) [2] with its goal to bring new software metrics based on evaluating data of developer activity and context of software development. Infrastructure of this project [3] provides us with data collected in software house and university environment (student team projects), which we use for evaluation of our method.

2 Related Work

Software product and its source code result from software developer activity. This motivates current research to look for how software attributes (mostly maintainability) [4] are affected by activities performed during the development together with the context which developers had resided in. Developers are often disrupted at work, switch between multiple tasks or take over another developer's task. Because of that, various tools for navigation in the source code were proposed, notably dependency graph of software components [7] and task-related tools, e.g., for source code and developer recommendation [1,6,13].

We may infer programming sessions [6] from developer activity monitored during the software development and use them to describe tasks which developers had worked on with task contexts [1,15], i.e., source code artifacts relevant to the currently solved task by developer. Developers visit places in the source code related to the task more often during the work on that particular task [4,8]. From the recorded data we may then reconstruct what the developer was working on [6,13] or what particular developer specializes in [11].

There are multiple sources and types of data that we can gather when monitoring developers during the processes of software development [14], for example: source code files and their contents from source code repositories [13]; development tasks from issue tracking systems; developer activity in an IDE [1,3,7]; developer activity outside of an IDE, in operating system or even events in real life.

It is important to not affect monitored developers during their activities [8], being it a similar problem of gathering implicit feedback on the Web [2]. Several issues arise in the design and development of the infrastructure for a system for gathering mentioned data of developer activity together – scalability and efficient processing online among them. One of the already proposed solutions is developed within the project PerConIK – Personalized Conveying of Information and Knowledge which considers software repository as a web of software components and applies "webification" of software development [2,3], i.e., employing methods and techniques from Web engineering to identify new information about software development and propose new software metrics.

Traditional source code metrics rely on a syntactic analysis of source code, omitting the information from development process. Basic example is the *lines of code* metric which evaluates the size of source code but not the time spent working on the measured source code. Similarly, traditional dependency graph of software components is created with identified references of source code components [7], helping developers with software development and maintenance. Authors in [17] also successfully applied network algorithms on identified dependencies to predict problems in software design.

Dependencies identified with syntactic analysis of source code are *explicit* since they reflect explicit statements in the source code [7]. We identify following main problems of the explicit dependencies and of their identification:

- Explicit dependencies do not capture cross-language connections in source code, e.g. in Web projects developed in combination of HTML and C# language.
- Explicit dependencies do not capture connections with configuration files, schema template files or runtime dependencies.
- Syntactic analysis of source code is not trivial or even possible for dynamically typed languages, e.g., JavaScript, Ruby.
- Explicit dependencies do not reflect sources of solutions in source code, developer's inspiration and places required to check when particular component is changed.

We see possibilities of employing developer activity as a source for identification of source code dependencies, inspired by the task context approaches [1,6,13]. Source code does not contain information about the developer's intents, inspirations and decisions for implemented solutions, what may suitably extend existing dependency graph. Moreover, because developer activity is not language-dependent, we may identify dependencies across different programming languages and also dependencies with configuration files or others which are currently not covered by explicit dependencies.

3   Implicit Source Code Dependencies

Software developer interacts with source code components during the development and maintenance, and so implicitly reveals task-related dependencies hidden in the source code. Selected example situations are:

- Developer studies existing code and navigates between dependent components to understand implemented logic (navigation paths) [8].
- Developer implements functionality in multiple components at the time.
- Developer copy-pastes a code fragment from existing component further changes of the original implementation may lead to inconsistency [4].
- Developer configures source code components by creating or maintaining external configuration files

In all these situations developer's navigation and activity performed in the source code implicitly reveals dependent components. When we do not take into account contents of source code files, developer activity also reveals dependencies on configuration files or on components implemented in different programming language. The idea behind the implicit dependencies is similar to the identification of task contexts, i.e., the developer works with software components that are related with each other for the task completion, and is based on empirical observations of developer activity from two sources – activities recorded in an IDE, e.g., custom extension for Microsoft Visual Studio or Eclipse [3], and commit history in a RCS, e.g., Microsoft Team Foundation Server or Git [12]. We chose to use these low-level logs of activities with source code components from all available types of logs [14] provided by the project PerConIK [2]:

- navigation in the source code - open, close and switch-to a source code file in an IDE - time-related activities changing currently opened file;
- copy-paste code fragment between two source code files; and
- commit (or check-in) of a collection of source code files to a RCS.

Although we do not force choice of the granularity for source code components (e.g., line of code, method, class or library), but for our experiments and implementation we consider source code files as components.

Our method (see Figure 1) consists of steps for converting raw logs of developer activity in an IDE and from a RCS to the format used by our method, continued by the identification of implicit dependencies, their weighting and validation and finally construction of a dependency graph.

Logs of activity in   Filtering of logs for   Identification of   Weighting and   Dependency graph   Graph with
IDE and RCS   identification   implicit   validation   construction   implicit
                                    dependencies   dependencies

Fig. 1. Overview of method for identification of implicit source code dependencies

3.1   Identification of Implicit Dependencies

We define source code dependencies as oriented connections between pairs of software components. Dependencies are weighted according to their significance and, considering software evolution, are valid for the particular time. Let S be the set of source code components of the selected granularity, then the space of dependencies in the source code is $D = S \times S \times T \times R$ in the time T with the weights R. Note that the explicit dependencies are valid in time while explicit statements are present in the source code.

Based on the types of activity logs used for identification of the implicit dependencies we define three specialized types of implicit dependencies $D_{imp} \subseteq D$:

- time-related implicit dependencies $D_{imp,time}$,
- content-related implicit dependencies $D_{imp,content}$,
- commit-related implicit dependencies $D_{imp,commit}$.

The most common activity performed by developer during the development is the navigation in the source code space. Ttime-related activities with source code components in an IDE are described with the tuples *(source, target, operation, timestamp)* containing the *source* and the *target* component of the operation, *type* of performed operation (e.g., open, close or switch-to file) and *timestamp* when the activity occurred. We reconstruct developer activity from these logs to create time-related implicit dependencies $d_{imp,time}$ (1) between the software components s_1 and s_2, which occurred at the time t, when developer spent time span in the target component (the *time window* property) before making next operation. The weight w is determined by the weighting function using the *time window* property.

$$d_{imp,time} = (s_1, s_2, t, w, time\ window) \tag{1}$$

Content-related activities of copying and pasting code are described with the tuples *(target, code operation, content, timestamp)*, containing the *target* component where the *code operation* was performed (copy, cut or paste) with the code *content*, and when the operation happened. Final copy-paste operation is logged with at least two actions, i.e., copying the code from the source code component X and pasting it into the source code component Y. Because of that we reconstruct the clipboard stack to identify content-related implicit dependencies $d_{imp,content}$ (2) where the *content* property contains the copy-pasted code fragment and may be used for determining the weight w.

$$d_{imp,content} = (s_1, s_2, t, w, content) \tag{2}$$

The last type of implicit dependencies are identified from commit operations. Developers tend to submit changes in a collection of source code components as a solution for the particular task, thus the changed components are implicitly connected with each other. For each pair of the changed components (s_1, s_2) in the same commit we create dependency $d_{imp,commit}$ (3) with *total count* of all committed components and weight w.

$$d_{imp,commit} = (s_1, s_2, t, w, total\ count) \tag{3}$$

3.2   Weighting of Implicit Dependencies

Each specialized type of implicit dependency extends general D_{imp} with extra property, being it *time window*, *content* or *count*. We use these properties to determine the significance of dependencies with the weighting function *weight* differently for each specialized type (4), ranging from insignificant to fully significant dependency.

$$weight : D_{imp} \rightarrow \langle 0, 1 \rangle \tag{4}$$

Time-related implicit dependencies are weighted according to the time spent in the visited component, i.e., significance of visiting (opening, switching-to) that component for the developer. The weighting function may be specified for the particular implementation. In our case we chose the weighting function to be as shown in Figure 2. To eliminate mistakes in developer's navigation in source code space, the dependency becomes fully significant after the selected threshold a. But after the threshold b the dependency is becoming irrelevant (the threshold c). After experiments we chose the thresholds to be $a = 10\ seconds$, $b = 10\ minutes$ and $c = 15\ minutes$ for our method.

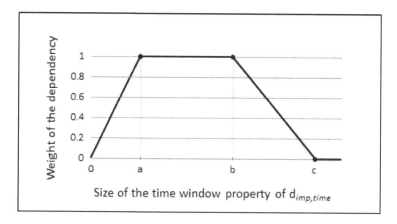

Fig. 2. The weighting function for time-related implicit dependencies $d_{imp,time}$ with threshold parameters a, b and c

Content-related implicit dependencies are identified from the copy-paste operations, thus their significance may correspond to the amount of the code copied or its contents. However, for simplicity we chose to weight every $d_{imp,content}$ with constant of *1*.

Commit-related implicit dependencies are weighted to the total count of committed components, as in (5), to promote smaller, fine-grained, commits solving single tasks. The lower the count of changed components in a commit, the more dependent they are together and contrariwise.

$$weight\,(d_{imp,commit}) = \frac{1}{count} \tag{5}$$

3.3   Validation of Implicit Dependencies

Even though changes in the source code invoke changes of its explicit dependencies, we are not able to similarly validate implicit dependencies over time. To model their relevance we validate them to the selected time using a forgetting function [9]. Developer interacts with the source code components mostly based on the task they currently solve, thus when the contents change over time or the task is finished, the developer's interactions lose their significance. Because of that the definition for validation functions of implicit dependencies D_{imp} (6) is similar to of explicit dependencies D_{exp} (7).

$$validity_{imp} : D_{imp} \times T \to \langle 0, 1 \rangle \tag{6}$$

$$validity_{exp} : D_{exp} \times T \to \{0, 1\} \tag{7}$$

Explicit dependencies are valid or not according to existence of explicit statements to the selected time, hence the validity selected to 0 or 1. Implicit dependencies are valid according to the forgetting function to the selected time t (8), which we chose to use from [16] with parameters set to $a = 1$ and $b = 0.08$:

$$y = ae^{(-b\sqrt{t})} \tag{8}$$

3.4   Dependency Graph

After the identification of implicit dependencies, we extend existing dependency graph $G(V, E)$ of V for the vertices of software components (source code files), and E for edges of aggregated dependencies. We differentiate between explicit and implicit edges in the graph because of differences in their meaning, i.e., $E = E_{exp} \cup E_{imp}$:

- Explicit dependencies represent statements in the source code, e.g., references, call hierarchy, inheritance.
- Implicit dependencies represent how developers interacted with source code during the development, e.g., their inspirations.

Both explicit and implicit edges of dependency graph are weighted with sum of weights of aggregated explicit or implicit dependencies of the corresponding type (9), possibly validated to the selected time. Resulting weight of implicit edge represents significance of aggregated implicit dependencies over time with single value.

$$weight\,(e_{imp,\,s_1,s_2}) = \sum_{d_{imp}\in D_{imp,s_1,s_2}} validity\,(d_{imp})\,w \qquad (9)$$

4   Evaluation

We propose implicit source code dependencies to be mainly applied during the processes of development and maintenance. To evaluate contribution of implicit dependencies we performed two experiments to show that:

- implicit dependencies reflect explicit dependencies of components which developer worked on during the task,
- implicit dependencies enrich dependency graph with new significant connections usable during the maintenance.

The PerConIK project is developed in cooperation with a medium sized software company. Before deploying our method in real work environment, we opted for evaluating it using data of five on-going student software projects provided also by the PerConIK project. These projects were developed by students of master courses Software Engineering or Information Systems. All projects were developed in C#/.NET and other Microsoft technologies in time span of one academic year:

- *Project A* – development of a web/desktop application by 3 developers,
- *Project B* – development of class libraries by 1 developer,
- *Project C* – development of class libraries by 1 developer,
- *Project D* – development of a web/desktop application by the team of 7 developers,
- *Project E* – development of a web application by the team of 7 developers.

Table 1 shows total numbers of activity logs from these projects, total number of explicit edges in their dependency graphs and also results of the selected source code metrics – *lines of code, maintainability index* and *cyclomatic complexity*. In our evaluation we use source code files for components used in the method. Because all the evaluated projects were developed using Microsoft technologies, we employed the Code Map functionality of Microsoft Visual Studio to identify explicit dependencies using the built-in reference recognition and the Code Metrics to evaluate other metrics.

Table 1. Results of the source code metrics for the software projects used for our evaluation, total numbers of explicit edges in their dependency graphs and numbers of activity logs recorded in 1 year of their development (LOC – lines of code, MI – maintainability index, CC – cyclomatic complexity)

Project	LOC	MI	CC	$|E_{exp}|$	No. of activity logs
A	2,402	82	1,285	232	15,215
B	4,384	74	2,164	274	8,584
C	4,993	81	3,213	528	24,213
D	5,342	81	1,839	270	55,877
E	3,721	81	1,779	189	48,717

Table 2. Total numbers of edges for identified implicit dependencies in dependency graphs for the evaluated software projects with the same vertices set

Project	$|E_{imp}|$ for thresholds			
	0	1	2	3
A	640	423	200	124
B	507	339	141	88
C	792	524	224	124
D	797	556	285	201
E	755	464	246	164

4.1 Reflection of Explicit Dependencies

We expect that implicit dependencies reflect explicit ones based on the existence of the developer's task context. We compared existing explicit dependencies of evaluated projects with identified implicit dependencies by comparing the E_{exp} and E_{imp} sets of dependency graphs. Table 2 shows total numbers of identified implicit edges in dependency graphs constrained by threshold for edge weights ranging from 0 to 3, when considering only edges between the files that appear in the explicit dependency graphs.

For evaluating the overlap of implicit and explicit edges we used weight thresholds of 1 and 2 to filter out less significant implicit edges. Higher values may have been also used but not for the dataset of size of ours. Table 3 shows how much of the identified implicit edges are explicit as well and how many explicit edges were identified with implicit edges (last column). We found out that up to 54% of all the identified implicit dependencies are common with explicit dependencies. We also analyzed whether the rest of the implicit dependencies are significant or not (see Section 4.2).

Table 3. Overlaps of implicit and explicit edges in dependency graphs with selected thresholds to filter out less significant implicit edges

Project	E_{imp} threshold	$\|E_{exp} \cap E_{imp}\|$	$\dfrac{\|E_{exp} \cap E_{imp}\|}{\|E_{imp}\|}$	$\dfrac{\|E_{exp} \cap E_{imp}\|}{\|E_{exp}\|}$
A	1	173	40.90%	74.57%
	2	108	54.00%	46.55%
B	1	140	41.30%	51.09%
	2	70	49.65%	25.55%
C	1	210	40.08%	39.77%
	2	120	53.57%	22,73%
D	1	191	34.35%	70.74%
	2	123	43.16%	45.56%
E	1	149	32.11%	78.84%
	2	108	43.90%	57.14%

Secondly, identified implicit dependencies overlapped up to 78.84% of explicit dependencies between files included in the space of implicit dependencies. That means that we are able to partially compensate inability of identification of explicit dependencies in cases where source code analysis is not possible but monitoring of developer activity is. Such example is usage of multiple programming language in the same software project, e.g., when simply combining HTML, JavaScript and CSS.

4.2   Significance of Implicit Dependencies

For the second experiment we expected that implicit dependencies provide new and significant information about connected software components, e.g., when particular component in a source code file relies on settings in a configuration file or when components are loosely coupled. Our task was to discuss identified implicit dependencies with the developers and decide whether they reflect connections in source code usable in the maintenance or not. As a significant connection of source code files, i.e., significant implicit dependency, we understood: *If the components in the source code file A are changed, the contents of the file B should be checked or changed as well.*

We were able to perform this experiment on the first four projects only and we chose to validate implicit dependencies with the weight threshold of 2. We chose this threshold to evaluate as most of the implicit edges as possible while still keeping the number of edges relatively low and the experiment bearable for developers. Developers manually checked each dependency and decided its significance, hence evaluating all the identified edges would have been too difficult. In this experiment the counts of implicit edges were higher than in the first experiment because we also evaluated dependencies between files which were not included in the explicit dependency graphs, e.g., webpages, configuration files, etc. To simplify the evaluation process, we generated dependency graphs in DGML format (for Microsoft Visual Studio) with implicit edges only, excluding the explicit ones. Developers were able to switch between the graph and the file, thus ensure in their decision of keeping or removing the dependency from the graph. In the end we compared the results with the original files. We achieved precision of more than 75% for the evaluated projects (Table 4).

Table 4. Evaluation of significance of implicit dependencies with the weight threshold of 2

Project	No. of implicit dependencies		Precision
	Original	Significant	
A	180	138	76.67%
B	112	103	91.96%
C	257	203	78.99%
D	634	576	90.85%

During the experiment we led discussion with the participated developers and received positive feedback for ability to identify dependencies between loosely coupled components across layers of the Model-View-Controller pattern. Even more, developers reminded reasons why the files were dependent during their work with them.

We highly appreciate the ability to identify dependencies just from logs of switching between source code files in the IDE. This is important when looking for dependencies on configuration files, schema template files or dynamically resolved dependencies. As an example, these situations were correctly identified with our method from developer activity in the IDE (for Web projects in ASP.NET MVC):

- Dependencies between C# classes and the XML configuration files, e.g., key-value settings, database connection strings, web service definitions.
- The View layer components (HTML webpages) displaying contents of the Model layer components (C#), e.g., webpage displaying data of a data class in table view.
- Dependencies between the View layer components (HTML webpages) on the Controller layer components (C#), e.g., when linking to a controller action.
- Dependencies between JavaScript source code files and C# files.
- Correct pairings of the View layer (HTML) with its code-behind (C#).
- Transitive dependencies on class inheritance hierarchies through interfaces.

5 Conclusion

Knowledge about dependencies of software components is utilized mostly during the development and maintenance processes, helping software developers with navigation and study of the existing source code. However, identification of explicit dependencies does not provide information about all connections in the source code. Moreover, in case of dynamically typed languages, it is sometimes even impossible to identify dependencies at all. Because of that we proposed the identification of implicit dependencies from developer activity to enrich existing dependency graph with new significant edges, or to supplement explicit dependencies in case of their unavailability.

For the evaluation of our method we used data gathered in the course of student (team) software projects. While we see natural difference between behavior of students and professional developers (work habits and schedule, experience), the evaluated projects were of relatively large size considering school projects and served as a basis for next step in our research, which aims at evaluating our method in real work environment. In our first experiment we showed overlap of both explicit and implicit edges in dependency graphs, thus possibilities of supplementing explicit dependencies with implicit ones. In our second experiment we attempted to manually evaluate significance of identified implicit dependencies. We achieved positive results, with correctly identifying hidden dependencies in the source code, and also cross-language dependencies.

Achieved results allowed us to deploy our monitoring infrastructure to a medium size software company. The infrastructure is aimed at recording implicit feedback of software developers and annotating source code with information tags created manually by the developers (during code reviews) or automatically based on source code analysis and developer activity analysis [3]. In June 2014 we have started to record activity data from two teams of total 25 developers working on web information systems development in this software company. Just before the deployment of our developed infrastructure within the PerConIK project for recording implicit feedback of software developers we tested the infrastructure extensively.

First impression of developers on the dependencies enriched by implicit dependencies was very positive including examples of such dependencies identified even by hand in existing software repositories. While the straightforward application of implicit dependencies is to visualize them in the form of a graph, we discussed our method with professional software developers and received valuable feedback to simply provide prioritized list of software components to be checked for the selected component upon developer's request. We plan to continue in experimental evaluation with dataset based on professional developers' work.

Acknowledgment. This work was partially supported by grants No.VG 1/0752/14 and it is the partial result of the Research and Development Operational Programme for the project Research of methods for acquisition, analysis and personalized conveying of information and knowledge, ITMS 26240220039, co-funded by the ERDF.

References

1. Antunes, B., Cordeiro, J., Gomez, P.: An Approach to Context-based Recommendation in Software Development. In: Proc. of the 6th ACM Conf. on Recommendation Systems, pp. 171–178. ACM (2012)
2. Bieliková, M., Návrat, P., Chudá, D., Polášck, I., Barla, M., Tvarožek, J., Tvarožek, M.: Webification of Software Development: General Outline and the Case of Enterprise Application Development. In: AWERProcedia Information Technology and Computer Science: 3rd World Conf. on Information Technology, vol. 3, pp. 1157–1162 (2013)
3. Bieliková, M., Polášek, I., Barla, M., Kuric, E., Rástočný, K., Tvarožek, J., Lacko, P.: Platform Independent Software Development Monitoring: Design of an Architecture. In: Geffert, V., Preneel, B., Rovan, B., Štuller, J., Tjoa, A.M. (eds.) SOFSEM 2014. LNCS, vol. 8327, pp. 126–137. Springer, Heidelberg (2014)
4. Bird, C., Nagappan, N., Gall, H., et al.: Putting It All Together: Using Sociotechnical Networks to Predict Failures. In: 20th Int. Symposium on Software Reliability Engineering, pp. 109–119. IEEE CS Press (2009)
5. Boehm, B.W., Brown, J.R., Lipow, M.: Quantitative Evaluation of Software Quality. In: Proc. of the 2nd Int. Conf. on Program Comprehension, pp. 592–605. IEEE CS Press (1976)
6. Coman, I.D., Sillitti, A.: Automated Identification of Tasks in Development Sessions. In: Proc. of 16th IEEE Int. Conf. on Program Comprehension, pp. 212–217. IEEE CS Press (2008)
7. Counsell, S., Hassoun, Y., Loizou, G., et al.: Common Refactorings, a Dependency Graph and Some Code Smells: An Empirical Study of Java OSS. In: Proc. of the ACM/IEEE Int. Symp. on Empirical Software Engineering, pp. 288–296. ACM (2006)
8. DeLine, R., Czerwinski, M., Robertson, G.: Easing Program Comprehension by Sharing Navigation Data. In: Proc. of the 2005 IEEE Symp. on Visual Languages and Human-Centric Computing, pp. 241–248. IEEE CS Press (2005)
9. Ebbinghaus, H.: Memory: A Contribution to Experimental Psychology. Ruger, H.A., Bussenius, C.E. (trans.) Teachers College, New York (1885/1913)
10. Fenton, N.E., Pfleeger, S.L.: Software Metrics: A Rigorous and Practical Approach, 2nd edn. PWS Pub. Co., Boston (1998)
11. Fritz, T., Murphy, G.C., Hill, E.: Does a Programmer's Activity Indicate Knowledge of Code? In: Proc. of 6th Joint Meeting of the European Software Eng. Conf. and the ACM SIGSOFT Symp. on The Foundations of Software Eng., pp. 341–350. ACM (2007)
12. Kalliamvakou, E., Gousios, G., Spinellis, D., et al.: Measuring Developer Contribution from Software Repository Data. In: Proc. of the 4th Mediterranean Conf. on Information Systems, pp. 600–611 (2008)
13. Kersten, M., Murphy, G.C.: Using Task Context to Improve Programmer Productivity. In: Proc. of 14th ACM SIGSOFT Int. Symp. on Foundations of Software Eng., pp. 1–11. ACM (2006)
14. Polášek, I., Ruttkay-Nedecký, I., Ruttkay-Nedecký, P., Tóth, T., Černík, A., Dušek, P.: Information and Knowledge within Software Projects and Their Graphical Representation for Collaborative Programming. Acta Polytechnica Hungarica 10(2), 173–192 (2013) ISSN: 1785-8860
15. Robillard, M.P., Murphy, G.C.: Automatically Inferring Concern Code from Program Investigation Activities. In: Proc. of 18th IEEE Int. Conf. on Automated Software Engineering, pp. 225–234. IEEE CS Press (2003)

16. White, K.G.: Forgetting Functions. Animal Learning & Behavior 29(3), 193–207 (2001)
17. Zimmermann, T., Nagappan, N.: Predicting Defects Using Network Analysis on Dependency Graphs. In: Proc. of 30th Int. Conf. on Software Engineering, pp. 531–540. ACM (2008)

Discovering Motifs in Real-World Social Networks

Lotte Romijn[1], Breanndán Ó Nualláin[1,2], and Leen Torenvliet[2]

[1] Amsterdam University College
lotteromijn@student.auc.nl, bon@science.uva.nl
[2] ILLC, University of Amsterdam
L.Torenvliet@uva.nl

Abstract. We built a framework for analyzing the contents of large social networks, based on the approximate counting technique developed by Gonen and Shavitt. Our toolbox was used on data from a large forum—boards.ie—the most prominent community website in Ireland. For the purpose of this experiment, we were granted access to 10 years of forum data. This is the first time the approximate counting technique is tested on real-world, social network data.

Keywords: approximate counting, software development, social networks, big data.

1 Introduction

Many real-world systems are complex networks and consist of a large number of highly connected interacting components. Examples are the World Wide Web, Internet, neural and social networks. Complex networks can be represented as graphs. These graphs contain characteristic patterns and substructures, such as cycles or triangles. Such patterns are called network motifs, subgraphs or templates. There are several algorithmic procedures to count or detect network motifs of size $O(\log n)$. Counting and detecting motifs is a method of identifying functional properties of a network. The frequency of certain motifs indicates how nodes behave in the network. The term "motif" was coined by Milo et al., who subsequently found motifs in biochemical, neurobiological, ecological and engineering networks [11]. A problem with counting patterns in graphs is that the general problem is known to be #P-hard[13], and therefore no efficient (*i.e.,* polynomial time) algorithms are known. Attempts to count motifs in networks must either be limited to networks of a modest size, so exponential algorithms finish in reasonable time, or give an approximate answer. As networks of interest are emerging that are large by nature, the latter approach seems the way to go.

Approximate counting of motifs has not been attempted on large networks (more than a few thousand nodes) [16]. In the experiment which this paper reports, we set up and tested a framework that is capable of analyzing large, real-world, social-media networks, by transforming them into graphs and approximately counting motifs in these networks.

G.F. Italiano et al. (Eds.): SOFSEM 2015, LNCS 8939, pp. 463–474, 2015.

This paper is organized as follows. We start with an overview of related work, then we discuss the approximate counting algorithm and parameter settings that lead to acceptable results. Then, we discuss the dataset on which the experiments were run. Section 6 contains the results of our analysis and our findings based on this, and the final section contains directions of further results and expansion of the algorithms.

2   Related Work

Networks with similar global topology can have varying local structures. In fact, local motifs are increasingly considered to be the small building blocks which are responsible for local functions in a network. Milo *et al.* [11] found network motifs in biochemical, neurobiological, ecological and engineering networks. An example of functional properties of motifs was illustrated by Becchetti *et al.* [5]. They showed that the local number of triangles in large-scale Web graphs is an indication of spamming activity. Przulj *et al.* [12] uses the term graphlet to denote a connected network with a small number of nodes.

The search for motifs in networks focuses on either induced or non-induced motifs [8]. Induced motifs have an additional restriction: an induced motif is a subset of vertices that contains all the edges between those vertices that are present in the original network. In general, searching non-induced motifs is more informative because a vertex in a network could have functions not associated with all of its adjacent edges [8]. Motif *detection* is equivalent to the subgraph isomorphism problem, a well-known problem in Graph Theory, which is NP-complete [7]. The exact solution can be found by enumerating all possible combinations of vertices that together form the size of the motif, and checking whether the edges present correspond to the edges in the subgraph. Ullmann described an exponential algorithm for subgraph isomorphism which takes polynomial time for a fixed choice of a subgraph [14]. Counting the number of motifs of a particular vertex builds upon the subgraph isomorphism problem. Counting motifs amounts to enumerating how many subgraphs can be found in which a particular vertex is included. Finding non-induced motifs grows rapidly in computation time with input size, and has not been attempted on large-scale, real-world networks [16]. Reducing this computation time represents a major research challenge.

Brute-force search for a particular motif requires the enumeration of all possibilities. For instance, finding all triangles in a network requires finding every pair of edges with the same vertex as one of their end vertices, and checking whether there is an edge connecting the other end vertices of these two edges. Without any form of approximation, the most efficient way to solve this problem uses matrix multiplication, which is of order $O(n^3)$ if a textbook method is used. Without matrix multiplication, a naive algorithm takes computation time of $O(n^5)$. For larger networks testing such naive algorithms is problematic.

Counting the number of a certain motif is #P-hard. #P-problems are of the form "compute the value of a function $f(x)$," where $f(x)$ is the number of possible solutions to the corresponding NP-decision problem [13]. They are at least as hard as the NP-problem, since solving the decision problem entails finding out if this number is nonzero. There are a few existing algorithms for counting and detecting non-induced motifs. These techniques go back to Larry Stockmeyer's (1983) theorem for approximate counting. He proved that for every #P-problem there is a randomized approximate algorithm that determines the count, using an NP-oracle. [13] This means that for a particular instance a of P and $\epsilon > 0$, the algorithm returns the count C with a high probability such that $(1 - \epsilon)P(a) \leq C \leq (1 + \epsilon)P(a)$. The randomized algorithm is in principle an (ϵ, δ)-approximation method.

3   Approximate Counting Algorithm

3.1   Color Coding

The approximate counting algorithm makes use of the color coding technique introduced by Alon *et al.* [2], used there to detect simple paths, cycles and bounded treewidth subgraphs. (See, *e.g.*, Bodlaender and Koster [6]). Recently, the color coding technique has been used to detect signaling pathways in PPI-networks [1].

The color coding technique is based on random assignments of colors to the vertices of an input graph. It can detect specific subgraphs efficiently by only considering specific color assignments, in time proportional to a polynomial function of the input $n = \|V\|$. If the assignment of colors is repeated sufficiently many times the method will find a specific occurrence of the motif of size $O(\log(n))$ with high probability. Multiple algorithms use elements of, or are entirely based on, the color coding technique [3,1]. Arvind and Raman [3] used color coding for counting the number of subgraphs isomorphic to a bounded treewidth graph. Alon *et al.* [1] described a polynomial time algorithm for approximating the number of non-induced occurrences of trees and bounded treewidth subgraphs with $k \in O(\log n)$ vertices. In 2007, Hüffner, Wernicke and Zichner [10] presented various algorithmic improvements for color coding that lead to savings in time and memory consumption.

Other methods have been explored to approximate the number of motifs, such as the exploitation of subgraph symmetries by Grochow and Kellis [9]. It could happen that a subgraph H can be mapped to a given subset G of a graph multiple times. Eliminating these subgraph symmetries significantly decreases computation time. However, the running time of the algorithm still increases exponentially with the size of the motif.

Zhao *et al.* [16] have recently shown that using color coding in addition to parallel programming can find motifs in networks with millions of nodes. They have combined parallelization of color coding with stream based partitioning. Their "ParSE" algorithm was tested on large-scale, synthetically generated, social contact networks for urban regions.

In this paper, color coding is also employed to count motifs. Gonen and Shavitt's algorithm for counting simple paths will be explained in detail, together with its implementation in the Python programming language and performance on the forum data set. By testing the algorithm on the forum data, Gonen and Shavitt's simple path algorithm is applied for the first time to a real-world, social network.

3.2 Simple Path Algorithm

Gonen and Shavitt's algorithm to find simple paths uses the color coding technique by Alon, Yuster and Zwick [2]. It approximates the number of paths of length $k - 1$, where k is the number of colors in the color set. The input is the graph G, a vertex $v \in V$, the path length $k - 1$, fault tolerance ϵ, and error probability δ.

1: $t = \log(\frac{1}{\delta})$; $s = \frac{4k^k}{\epsilon^2 k!}$;
2: **for** $j = 1$ to t **do**
3:     **for** $i = 1$ to s **do**
4:        Color each vertex of F independently and uniformly
5:        at random with one of the k colors
6:        **for all** $u \in V$ **do**
7:           $C_i(u, \emptyset) = 1$
8:        **end for**
9:        **for all** $l \in [k]$ **do**
10:           $C_i(v, l) = \begin{cases} 1 \text{ if } col(v) = l \\ 0 \text{ otherwise} \end{cases}$
11:        **end for**
12:        **for all** $S \subseteq [k]$ s.t. $\|S\| > 1$ **do**
13:           $C_i(v, S) = \sum_{u \in N(v)} C_i(u, S \backslash col(v))$
14:        **end for**
15:        $P_i(v, [k]) = \sum_{l=1}^{k} \sum_{(S_1, S_2) \in A_{l,v}} \sum_{u \in N(v)} C_i(v, S_1) C_i(u, S_2),$
16:        where $A_{l,v} = \{(S_i, S_j) \mid S_i \subseteq [k], S_j \subseteq [k],$
17:                  $S_i \cap S_j = \emptyset, \|S_i\| = l, \|S_j\| = k - l\}$
18:        Let $X_i^v = P_i(v, [k])$
19:     **end for**
20:     Let $Y_j^v = \frac{\sum_{i=1}^{s} X_i^v}{s}$
21: **end for**
22: Let Z^v be the median of $Y_1^v \ldots, Y_t^v$
23: Return $Z^v.k^k/k!$

This algorithm is an (ϵ, δ)-approximation for counting simple paths of length $k - 1$ containing vertex v, "simple" meaning that there are no repeated vertices in the path. $P_i(v, S)$ is the number of colorful paths (*i.e.*, paths on which all nodes have a distinct color) containing v using colors in S at the ith coloring. $C_i(v, S)$ is the number of colorful paths for which one of the endpoints is v using colors in S at the ith coloring. The algorithm finds an approximation of the number of paths within $[(1 - \epsilon)r, (1 + \epsilon)r]$, where r is the actual number of paths in the graph, with a probability of at least $1 - 2\delta$.

The estimator used in this algorithm is also called "median of means" and it can be shown, using Chebyshev's inequality and Chernoff bounds, that the expected value of the number of colorful paths (*i.e.*, the number of paths times $k!/k^k$) can be approximated arbitrarily closely using a limited number of iterations.

When experimenting with the algorithm on known graphs, two problems with the original pseudocode of Gonen and Shavitt became apparent. First of all, when v is a node on the path, the colorful paths containing v are counted "in both directions", in other words twice. Second, when v is an endpoint, one of the sets in the partition is the empty set and therefore $C_i(u, \{\}) = 1$ for all neighbors of v. To get results that are both theoretically correct and experimentally acceptable, we had to adapt the value of P_i as follows.

$$P_i(v, [k]) = C_i(v, k) + \frac{1}{2} \sum_{l=2}^{k-1} \sum_{(S_1.S_2) \in A_{l,v}} \sum_{u \in N(v)} C_i(v, S_1).C_i(u, S_2)$$

This adaptation in computing P_i influences the complexity of computing P_i only by a constant, leaving the complexity of the algorithm $O((2e)^k \|E\|. \log(\frac{1}{\delta})/\epsilon^2)$, as in the Gonen-Shavitt case.

4   About the Implementation

To implement and test the simple path algorithm in Python, we made use of the packages NetworkX (http://networkx.github.io/) for generating graphs of vertices and edges to which weights and labels can be associated, and Numpy (http://www.numpy.org/) that provides a library of mathematical functions for performing computations on large arrays.

To achieve a better performance we have made use an efficient bitwise representation of color sets. The sets of colors required are all "small" sets with cardinality certainly less than 32. Such sets can efficiently be represented as 32-bit integers using a bitwise representation where, for example, the set $\{6, 4, 3, 1\}$ is represented by the integer $2^6 + 2^4 + 2^3 + 2^1 = 90$. The counterparts to the necessary set operations can then be efficiently implemented as combinations of logical bit operations.

To test the implementation, we generated complete and random graphs with less than 20 nodes in which the number of paths of a given length can be computed exactly using built-in NexworkX functions and compared the results with the results of the color coding algorithm. An example of the measured results is in Figure 1.

For obvious reasons, exact counting cannot be done on large graphs. However, both calculations and tests suggested that the number of iterations needed to achieve acceptable accuracy in the color coding case could be reduced significantly, resulting in a significant reduction in computation time (See Figure 2).

A series of further tests led to fine tuning of the parameters of the algorithm for the analysis of the real data set.

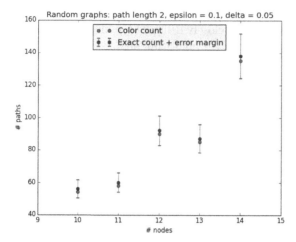

Fig. 1. Simple path counts of length 2 for random graphs with 2/3 possible edges

t	s	Number of iterations	Mean	Standard deviation
5	20	100	107.166375	6.722599292
5	90	450	107.85075	3.584042681
5	180	900	108.01275	2.277940394
5	270	1350	107.99725	1.998800875
5	360	1800	108.150938	1.535474752
5	540	2700	108.042375	1.245173181
5	720	3600	108.0345	1.199928865
5	900	4500	107.92755	1.023959715
5	1080	5400	108.06	0.980441546
5	1260	6300	107.981786	0.87689016
5	1440	7200	108.090234	0.878974341
5	1620	8100	108.073917	0.747840553
5	1800	9000	107.9613	0.64225233

Fig. 2. Accuracy of the algorithm for varying s and t

5   About the Data Set

The real-world, complex networks for which the numbers of simple paths have been counted are generated from an Irish forum data set. This data set was put online in 2008 for the "boards.ie SIOC Data Competition." (SIOC stands for Semantically-Interlinked Online Communities Project). The complete data set contains ten years of Irish online life from Irelands largest community website "boards.ie" over the years 1998-2007. Since the foundation of the website in 1998, over 36 million posts have been made and the current posting rate is around 17,000 a day (retrieved from http://www.boards.ie/content/about-us). The data set is a large collection of RDF-files (Resource Description Framework), in which each file contains a post in a thread on a forum. The RDF-files have a tree-like structure, corresponding to the way in which the board's website contains various forums, each forum contains multiple threads and each thread contains board posts that are chronological replies to each other. In this project, useful information from the data set is extracted by parsing the RDF-files. Such information involves the topic of the post (title), the username of the person who posted, his FOAF-person profile, and the thread which contains the posts. The Python package that is used for this purpose is `rdflib` (`https://github.com/RDFLib/rdflib/`), which can query and extract certain elements from an RDF-file. From there, we generate graphs that represent the structure of the data set by using the NetworkX Python package. For the main analysis of this project, graphs are constructed in which nodes represent users with accounts on the `boards` website, and edges the connections between users if they posted in the same threads. For an initial analysis, the data of the years 1998-2000 were used. The distribution of simple paths in these graphs were compared to artificial data, such as randomly generated graphs, preferential attachment graphs, and small-world graphs. While testing the algorithm on the data sets, the algorithm was revised and further optimized. The results of the motif counts were analyzed and compared, and further analysis of these substructures was used to yield conclusions about this forum data set.

6   Results and Conclusions

6.1   Analysis

Having tested, corrected and fine-tuned our software, it was time to run tests on the SIOC data and compare characteristics of these data with artificially created networks of different sorts. Due to space limitation, only a small fraction of the results obtained can be presented here.

First we give a logarithmic representation of the means of all nodes per path length. We obtain a linear plot, which indicates the mean of the counts grows exponentially in path length. The plot shows a standard deviation close to the means, indicating a large tail of the distribution. Similar results were obtained for 1999 and 2000.

Figure 4 shows the relationship between the number of paths of length 3 vs length 2 and 4 vs 7 of the same node for the 1998 data set. The relationship is linear and was found for each pair of length up till length 9.

The distribution of paths in the `boards` forum data was compared with random, preferential attachment and small-world graphs. For the graphs of 1998, 1999 and 2000 the distribution of the number of paths of a specific length is plotted. In Figures 5 and 6 the x-axis represents the number of paths of specific length that were counted of which a given node is a member. The y-axis shows the relative number of occurrences of this number of paths for all nodes in the graph in Figure 5. In Figure 6, the y-axis shows the relative occurrence of this number of paths for 150 graphs generated by graph models in NetworkX, in which each time the number of paths for a random node was determined. (These graphs have the same number of nodes and edges as the

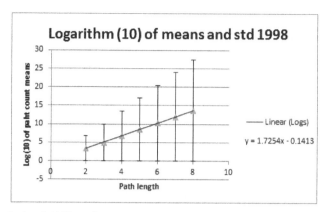

Fig. 3. Analysis of 1998 path counts. Logarithm of means and standard deviations

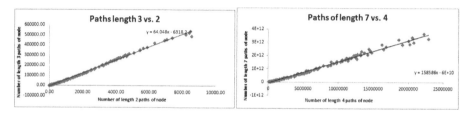

Fig. 4. Relationship between path lengths in 1998 data

1998 data.) Based on the minimum and maximum number of paths, the data is segmented into 50 bins. The data was fitted to all valid parametric probability distributions in Matlab, using the function `Allfitdist` (retrieved from `http://www.mathworks.nl/matlabcentral/fileexchange/`). The four best fitting probability density functions are displayed in the plot. Below the results are shown for the **boards** graph of 1998. Since the relationship between different path lengths is linear (Figure 4), the main analysis below is performed with path length 3. The graphs in Figure 6 are, respectively, generated randomly, then scale-free according to the preferential attachment model by Barabási and

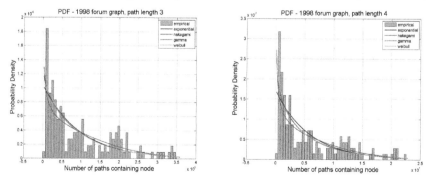

Fig. 5. Paths of length 3 and 4 in the 1998 data and fitted probability density functions

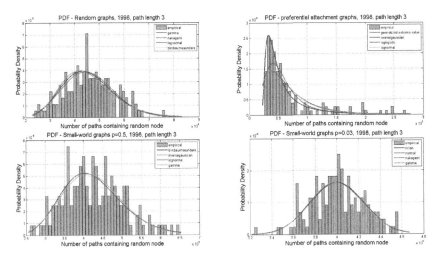

Fig. 6. Path counts in 150 randomly generated graphs according to different models

Albert [4] and then twice according to the small-world model of Watts and Strogatz [15] with rewiring probabilities 0.5 and 0.03. Other probabilities were also tested and showed similar results.

A final plot compares path counts to degree.

6.2 Conclusions

Concerning the Algorithms. The most important conclusion to draw from the experiments is that the algorithms of Gonen & Shavitt and our implementation of them indeed work. Except from a minor glitch in the pseudocode, mentioned above, and the fact that in practice much fewer iterations of the randomized algorithm were necessary than were required by the theory everything went more or less smoothly. This again supports the practice of testing implementations against data for which the count is known. Nonetheless, there were some performance issues which support the consideration to port the implementation to a lower-level language, such as C, before making the tools more widely available. Meanwhile however, extensions as discussed in the next section, and tests on even larger networks seem in order.

Concerning the Experiments. The histograms of the number of paths (Figure 5) show that the distribution of the number of paths is broader than for random graphs with the same number of nodes and edges. The peak for path length 3 of the 1998 data is at 1.5×10^4 paths, while for random graphs with the same number of nodes the peak is at 4.5×10^4 (≈ 3 times larger than the SIOC graph). On the other hand, the tails of path distributions of the SIOC graphs are much larger for all path lengths. Again, for path length 3 the largest number of paths of the SIOC data is around 3.5×10^5, while for random graphs it is around 2.3×10^4 (≈ 15 times smaller than the SIOC graph). Similar results were obtained for larger path lengths. This suggests that there are a few members extremely active on the **boards** website and are connected with other nodes through many paths. For instance, when a node is a member of a large number of paths of length 3 (4 nodes), it suggests that the node itself posts prolifically and has many threads in common with other users (prolific poster) and it could have posted in a thread in which an extremely active user has posted (extremely active neighbor). However, from Figure 7 it can be concluded that counting the number of paths a node is a member of and its degree are highly correlated. This suggests that prolific users are generally in more paths. The two outliers in this correlation could be explained by considering neighbors; there is one node with low degree (degree 3) and a high number of paths (173000). This means that it is probably connected to a very prolific user, or even to two. There is one node with a higher degree (degree 45) and a low number of paths (8022).

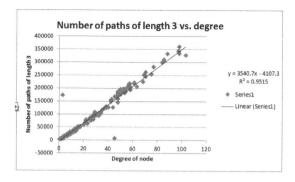

Fig. 7. paths of length 3 vs degree

Comparing the distribution of counts against those of artificially generated graphs revealed that the distribution of the **boards** data graph is similar to that of preferential attachment graphs with the same amount of nodes and edges. The small-world model [15] with varying rewiring probabilities does not show similar path-count distributions as the SIOC data. In the simple path counts it is seen that the distribution of the counts looks similar to preferential attachment networks. Each new individual on the **boards** website posting in a thread has a higher probability of posting in a thread in which a prolific poster has posted earlier. Hence, it has a higher probability of forming an edge with that person. This mechanism is essentially how a preferential attachment network is dynamically generated. In a preferential attachment graph, every new node connects to already existing nodes and has a higher probability of connecting to "popular" nodes in the graph. However, the preferential attachment model differs from the real-world SIOC graph in several aspects. Firstly, in a preferential attachment graph, a new node has a fixed number of starting edges. Each new node connects to existing nodes with the same amount of edges, while in the SIOC data a new individual on a forum could actively engage in many threads, post new questions and/or be an observer. Also, the preferential attachment model does not take into account that new nodes can make new threads. In addition, the degree distribution of a scale-free network (that can be generated by the preferential attachment model) generally follows a power-law distribution. This was not observed for the SIOC data. In the SIOC data, the number of nodes with higher degree decreases less rapidly. However, it is likely that the preferential attachment process causes the similarity in distribution of path counts between the preferential attachment and the SIOC data graphs.

7   Further Research

In this project we created a toolbox for analyzing real-world, social networks of considerable size. Our approach was concentrated on simple paths, but the algorithms can easily be extended to researching other motifs and graph properties. The toolbox we created can be expanded in several directions:

- Algorithmically: the algorithms and code developed so far can still be optimized in several ways. For instance, the approach taken allows for numerous variations of parallelization. Not only can different parts of the graph be explored simultaneously, as was done on artifical data in [16] because of the limited size of the structures searched for, but also the iterations sketched above can be parallelized, with which a speedup is expected or, equivalently, the ability to tackle even larger graphs.
- Semantically: Though the motifs considered so far are interesting, they do not represent the only properties one would extract from social networks. For instance a question of interest is: which users form cliques interested in the same subset of subjects? Is the opinion of users about subjects related to their geographic location (IP-number)? Which (opposing) coalitions are formed? Simple paths are just a start. Yet, the triangles mentioned in Section 1 are simple paths of length two, of which the endpoints are neighbors. Circles are like paths of arbitrary length, and cliques are circles in which all points are neighbors.
- Dynamically: So far only static graphs have been considered. However, we have data spanning a long period of time. We would make the toolbox adequate to deal with development of structures over time. For example time decay could be added to the preferential attachment model to account for recency of threads.

We will continue to pursue this research.

Acknowledgement. We thank Tom Murphy of `boards.ie` and John G. Breslin, University College Galway, for providing access to the data from the SIOC project and are grateful to the anonymous referees for useful comments.

References

1. Alon, N., Dao, P., Hajirasouliha, I., Hormozdiari, F., Sahinalp, S.C.: Biomolecular network motif counting and discovery by color coding. Bioinformatics 24(13), 241–249 (2008)
2. Alon, N., Yuster, R., Zwick, U.: Color-coding. Journal of the ACM 42(4), 844–856 (1995)
3. Arvind, V., Raman, V.: Approximation algorithms for some parameterized counting problems. In: Bose, P., Morin, P. (eds.) ISAAC 2002. LNCS, vol. 2518, pp. 453–464. Springer, Heidelberg (2002)
4. Barabási, A.L.: Scale-free networks: A decade and beyond. Science 325(5935), 412–413 (2009)
5. Becchetti, L., Boldi, P., Castillio, C., Gionis, A.: Efficcient semi-streaming algorithms for local triangle counting in massive graphs. In: Proceedings of the 14th ACM SIGKDD International Conference on Knowledge Discovery and Data Mining, pp. 16–24 (2008)
6. Bodlaender, H., Koster, A.: Combinatorial optimization on graphs of bounded treewidth. The Computer Journal 51(3), 255–269 (2008)
7. Garey, M.R., Johnson, D.S.: Computers and Intractability: A Guide to the Theory of NP-Completeness. W.H. Freeman and Company, New York (1979)
8. Gonen, M., Shavitt, Y.: Approximating the number of network motifs. In: Avrachenkov, K., Donato, D., Litvak, N. (eds.) WAW 2009. LNCS, vol. 5427, pp. 13–24. Springer, Heidelberg (2009)
9. Grochow, J., Kellis, M.: Network motif discovery using subgraph enumeration and symmetry-breaking. In: Speed, T., Huang, H. (eds.) RECOMB 2007. LNCS (LNBI), vol. 4453, pp. 92–106. Springer, Heidelberg (2007)
10. Hüffner, F., Wernicke, S., Zickner, T.: Algorithm engineering for color-coding with applications to signaling pathway detection. Algorithmica 52, 114–132 (2007)
11. Milo, R., Shen-Orr, S., Itzkovitz, S., Cashtan, N., Chklovskii, D., Alon, U.: Network motifs: simple building blocks of complex networks. Science 298(5594), 824–827 (2002)
12. Przulj, N., Corneil, D., Jurisica, I.: Modelling interactome: Scale-free or geometric. Bioinformatics 150, 216–231 (2005)
13. Stockmeyer, L.: The complexity of approximate counting. In: Proceedings of the 15th Annual ACM Symposium on Theory of Computing, pp. 118–126 (1983)
14. Ullmann, J.: An algorithm for subgraph isomorphism. Journal of the Association for Computing Machinery 22(1), 31–42 (1976)
15. Watts, D., Strogatz, S.: Collective dynamics of "small-world" networks. Nature 393, 440–442 (1998)
16. Zhao, Z., Khan, M., Kumar, V., Marathe, M.: Subgraph enumeration in large social contact networks using parallel color coding and streaming. In: Proceedings of the 39th Conference on Parallel Processing, pp. 594–603 (2010)

Exploiting Semantic Activity Labels to Facilitate Consistent Specialization of Abstract Process Activities

Andreas Bögl[1], Michael Karlinger[2], and Christoph Schütz[2], Michael Schrefl[2], and Gustav Pomberger[3]

[1] Software Competence Center Hagenberg GmbH, Hagenberg, Austria
andreas.boegl@scch.at
[2] Department of Business Informatics – Data & Knowledge Engineering,
Johannes Kepler University, Linz, Austria
{karlinger,schuetz,schrefl}@dke.uni-linz.ac.at
[3] Department of Business Informatics – Software Engineering,
Johannes Kepler University, Linz, Austria
gustav.pomberger@jku.at

Abstract. Designing business processes from scratch is an intricate and challenging task for process modellers. For this reason, the reuse of process patterns has become an integral part of process modelling in order to deal with recurring design issues in a given domain when modelling new business processes and variants thereof. The specialization of abstract process activities remains a key issue in process pattern reuse. Depending on the intended purpose of process pattern reuse, the specialization of abstract process activities typically ranges from the substitution of abstract process activities with sub-processes to the substitution of activity labels with specialized labels. The specialization of abstract process activities through label specialization has been hardly investigated so far in the business process community. The approach presented in this paper achieves consistent specialization of abstract process activities by ensuring consistent specialization of activity labels through exploitation of semantic activity labels as introduced in previous work. Semantic activity labels encode the linguistic meaning of process activities and thereby facilitate the establishment of consistency criteria based on the implicit semantics captured by activity labels.

Keywords: semantic activity labels, activity label specialization, consistent reuse of process patterns.

1 Introduction

Traditional business process management relies on business process models which prescribe a sequence of process activities that are performed in order to meet an organization's business objectives. The modelling of such business processes is supported by various languages such as *Event-driven Process Chains* (EPCs) [7] or

G.F. Italiano et al. (Eds.): SOFSEM 2015, LNCS 8939, pp. 475–485, 2015.

the *Business Process Model and Notation* (BPMN) [14]. Process patterns represent core business processes and allow for the description and reuse of best-practice process knowledge in organizations rather than defining business processes either from scratch or through updates of existing business process models (cf. [21,22]). Typically, process patterns are represented by process model templates which mainly consist of abstract process activities, the labels of which lack domain-specific expressiveness (see also [2,15]). Empirical studies show that activity labels are the key for human model readers to understand process models [12] as *"the use of informative and unambiguous labels improves an overall understanding of a process model"* [10]. As a consequence, the reuse of process patterns faces the issue of substituting abstract process activity labels with more informative labels, leading to what we call *activity specialization by label specialization*. In order to illustrate the need for activity specialization by label specialization, consider a process pattern for the definition of development processes which consists of, amongst others, abstract process activity *"Define Requirements"*. When reusing this process pattern for modelling a domain-specific development process, e.g., a software development project, the modeller replaces the *"Define Requirements"* activity label with a more expressive label, e.g., *"Define Software Requirements With Customer"* or *"Define Component Requirements for Architecture Design"*, in order to accomplish a higher informative value for human model readers.

The reuse of process patterns requires a mechanism for checking the consistency of the resulting specialized process model with respect to the original process patterns. Violations of consistency criteria may have significant impact on the correctness of propagated best-practice knowledge, thereby posing an impediment to controlled process evolution in organizations. While consistency along the behavioural dimension of process models has been thoroughly investigated, e.g., in data-centric business process modelling [17,1], consistent specialization of abstract process activities by label specialization still is an open research issue.

Consistent label specialization has to cope with two critical challenges. First, empirical studies show that the current labelling practice of process activities is conducted rather arbitrarily [13], which inherently causes a potential threat for understanding process models by human model readers as well as information systems. Second, consistent specialization requires a verification of the linguistic meaning of activity labels at any level of abstraction that these labels appear. In order to deal with this first issue, we proposed in previous work [3] a comprehensive lexical and semantic labelling framework. This framework introduces, on the one hand, lexical labels to facilitate unambiguous formulation of activity labels by text clauses and, on the other hand, the refactoring of semantic activity labels to facilitate linguistic encoding of activity labels. As a key feature, the framework covers the analysis tasks of automated label refactoring to accomplish human understandability and the automated semantic annotation of process activities to accomplish interpretability by information systems. In order to deal with the second issue, this paper extends the lexical and semantic labelling framework with consistency criteria for label specialization.

In particular, the proposed consistency criteria exploit semantic activity labels annotated to process activities.

The remainder of this paper is organized as follows. Section 2 presents the annotation of process activities with semantic activity labels. For illustration purposes, the presented approach towards consistent activity specialization employs, without loss of generality, a more compact formalization of semantic activity labels with respect to previous work [3]. Section 3 presents the main contribution of this paper. Section 4 provides an application example in order to illustrate the main benefits of the proposed approach for the process modeller. Section 5 discusses the state of the art on the specialization of process activities.

2   Annotation of Process Activities with Semantic Activity Labels

The formulation of activity labels by process modellers typically includes the description of tasks or actions (e.g. *"Verify"* or *"Identify"*) and process objects (e.g. *"Order"* or *"Requirements"*) which together constitute the main building blocks for activity labels. Accounting for the linguistic meaning of activity labels, however, requires the consideration of the product of the meaning of the words that compose the activity labels (cf. [11, p. 68 et seq.]). In other words, linguistic encoding of activity labels involves the description of task/process objects on the one side and the description of relationships between used words on the other side. In previous work [3], we introduced semantic activity labels to facilitate linguistic encoding of process activity labels into a semantic representation that abstracts from any lexical representation. A semantic activity label embodies a set of process items, i.e., process objects and tasks and a set of relationships between them. The process items and relationships encapsulated by semantic activity labels originate from a *process knowledge base* to accomplish controlled evolution and specification of process activity labels. The formalism for process knowledge bases employed in this paper (Definition 1) is a simplification with respect to previous work [3] which allows for a more focussed discussion of the specialization aspects of semantic activity labels.

Definition 1 (Process Knowledge Base).
A process knowledge base $\mathbb{P} = (I, \leq_I, R, \tau, \leq_R)$ *consists of*
- *the set of* process items *I;*
- *the partial order* $\leq_I$ *over I, called* process item hierarchy. *The set of all sub-process items of a process item $i \in I$, denoted by $sub(i)$, is given by* $sub(i) = \{i' \in I \mid i' \leq_I i\}$;
- *the set of* relationships *R;*
- *the function* $\tau : R \mapsto I \times I$ *which assigns to each relationship $r \in R$ a pair of process items $(i, i') \in I \times I$, called the* relationship type *of r;*
- *the partial order* $\leq_R$ *over R, called* relationship hierarchy, *which requires that* $\forall (r, r') \in R \times R$, *if $r \leq_R r'$ then $i \leq_I \bar{i}$ and $i' \leq_I \bar{i}'$, where $(i, i') = \tau(r)$ and $(\bar{i}, \bar{i}') = \tau(r')$.*

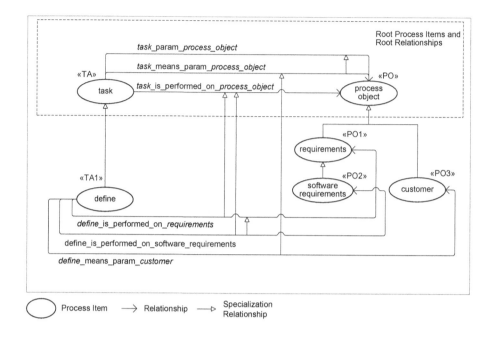

Fig. 1. Running example of a process knowledge base $\mathbb{P}_1$

Figure 1 illustrates a process knowledge base $\mathbb{P}_1$ according to Definition 1. By convention, the labels annotated to process items in Figure 1 are abbreviations and are used for textual descriptions of respective process items throughout the remainder of this paper. The process items ≪TA≫ and ≪PO≫ serve as root process items for the specification of tasks and process objects, respectively. A task represents an action which is performed on a process object, which is expressed by root relationship < is_performed_on > (≪TA≫, ≪PO≫). A process object represents a real or an abstract thing being of interest within a process domain. In addition to < is_performed_on > relationships between tasks and process objects, process knowledge base $\mathbb{P}_1$ considers parameter relationships. A parameter denotes a process object which might be used for defining intentional aspects relevant for executing a task on a process object. The specification of a parameter is derived from the root relationship < param > (≪TA≫, ≪PO≫). The illustrated process knowledge base states one intentional aspect named *means* between a task and a process object expressed by root relationship < means_param > (≪TA≫, ≪PO≫) which represents a specialization of the < param > (≪TA≫, ≪PO≫) relationship (see [3] for further parameter types). A means parameter refers to a process object which enables task execution, i.e., a process object that is a means to complete the task. For example, the activity *"Define Software Requirements with Customer"* indicates *"Customer"* as means parameter.

Definition 2 (Semantic Activity Label). *Let* $\mathbb{P} = (I, \leq_I, R, \tau, \leq_R)$ *be a process knowledge base. A semantic activity label* $l = (I_l, R_l)$ *over* $\mathbb{P}$ *consists of a* *set of process items $I_l \subseteq I$, and a set of relationships $R_l \subseteq R$. It is required that* $\forall r \in R_l : \{i, i'\} \subseteq I_l$ *where* $(i, i') = \tau(r)$.

Given process knowledge base $\mathbb{P}_1$ (Fig. 1), semantic activity label l_A describing process activity *"Define Requirements"* consists of

- the process item set $I_{l_A} = \{\ll\text{TA1}\gg, \ll\text{PO1}\gg\}$, the process items of which are the machine-readable interpretations of the *"Define"* task and the *"Requirements"* process object, and
- the relationship set $R_{l_A} = \{<\text{is\_performed\_on}> (\ll\text{TA1}\gg, \ll\text{PO1}\gg)\}$

and semantic activity label l_S describing process activity *"Define Software Requirements With Customer"* consists of

- the process item set $I_{l_S} = \{\ll\text{TA1}\gg, \ll\text{PO2}\gg, \ll\text{PO3}\gg\}$, the process items of which are the machine-readable interpretations of the task *"Define"* and the process objects *"Software Requirements"* and *"Customer"*, and
- the relationship set $R_{l_S} = \{<\text{is\_performed\_on}> (\ll\text{TA1}\gg, \ll\text{PO2}\gg), <\text{means\_parameter}> (\ll\text{TA1}\gg, \ll\text{PO3}\gg)\}$

The exploitation of the expressive power of semantic activity labels requires their annotation to process activities. For this purpose, we introduce a *process activity repository* as follows.

Definition 3 (Process Activity Repository). *A process activity repository* $\mathbb{R}_A = (\mathbb{P}, \mathcal{A}, \mathcal{L}, \lambda)$ *consists of*
- *a process knowledge base* $\mathbb{P} = (I, \leq_I, R, \tau, \leq_R)$,
- *a set of process activities* $\mathcal{A} = \{A_1, \ldots, A_n\}$,
- *a set of semantic activity labels* $\mathcal{L}$ *over* $\mathbb{P}$,
- *function* $\lambda : \mathcal{A} \mapsto \mathcal{L}$ *which annotates each process activity* $A \in \mathcal{A}$ *with exactly one semantic activity label* $l_S \in \mathcal{L}$.

3   Consistent Specialization of Process Activities by Label Specialization

This section provides an answer to the question *when exactly a process activity S is considered a consistent specialization of a process activity A*. The key property that determines consistent specialization of activity labels concerns the observable differences between the activity labels of A and S. More precisely, we say that S is an observation-consistent specialization of A if and only if the linguistic meaning of A is observable in S. This means that all process items and relationships associated with A are observable in the process items and relationships associated with S. The specialization of process activities based on observable differences between the activities' labels is what we call *process activity specialization by label specialization*. Process activity specialization by label specialization requires a reasoning over observable differences and correspondences between semantic activity labels annotated to the respective process activities. In this context, a pair of semantic activity labels l_A and l_S is considered equivalent if both consist of the same set of process items, i.e., $I_{l_A} = I_{l_S}$ and, moreover, both consist of the same set of relationships, i.e., $R_{l_A} = R_{l_S}$. Based on the notion of equivalence of semantic activity labels, we introduce predicate $ocs(l_S, l_A)$ which evaluates to true if and only if l_S is an observation-consistent specialization of l_A.

Definition 4 (Specialization of Semantic Activity Labels). *Let* $\mathbb{P} = (I, \leq_I, R, \tau, \leq_R)$ *be a process knowledge base and let* $\mathcal{L}$ *be a set of semantic activity labels over* $\mathbb{P}$. *Semantic activity label* $l_S = (I_{l_S}, R_{l_S})$ *is defined to be an observation-consistent specialization of semantic activity label* $l_A = (I_{l_A}, R_{l_A})$, *denoted by predicate* $ocs_l(l_S, l_A)$, *iff there exists a mapping* $\mathfrak{A} : I_{l_A} \mapsto I_{l_S}$ *such that*

1. $\mathfrak{A}$ *is total and injective, and*
2. $\forall i \in I_{l_A} : \mathfrak{A}(i) \leq_I i$, *and*
3. $\forall r \in R_{l_A} : \exists! r' \in R_{l_S}$ *such that*
    i. $r' \leq_R r$, *and*
    ii. $(\mathfrak{A}(\bar{i}), \mathfrak{A}(\bar{j})) = \tau(r')$ *where* $(\bar{i}, \bar{j}) = \tau(r)$

Definition 4 requires additional comments. Mapping function $\mathfrak{A} : I_{l_A} \mapsto I_{l_S}$ is a *total* and *injective* function (Requirement 1) and thus ensures that every process item $a \in I_{l_A}$ is mapped to a distinct process item $s \in I_{l_S}$. If process item a maps to process item s then it is required that s is a specialization of a according to $\mathbb{P}$ (Requirement 2). Consequently, Requirements 1 and 2 ensure that a more general process item in l_A is substituted only with a more special process item in l_S. Requirement 3 addresses consistency criteria regarding the relationships of R_{l_A} and R_{l_S}. Observation-consistent specialization requires that for any relationship r in R_{l_A} there is a single relationship r' in R_{l_S} such that r' is a specialization of r according to $\mathbb{P}$ (Requirement 3.i), and the process items of r are mapped to the process items of r' according to $\mathfrak{A}$ (Requirement 3.ii).

For purposes of illustration of observation-consistent label specialization, recall from the running example the specialization of process activity *"Define Requirements"* annotated with semantic activity label l_A to process activity *"Define Software Requirements With Customer"* annotated with semantic activity label l_S. Semantic activity label l_S is an observation-consistent specialization of l_A because there exists an injective mapping function $\mathfrak{A}$ from concepts ≪TA1≫ and ≪PO1≫ of semantic activity label l_A to the concepts ≪TA1≫ and ≪PO2≫ in l_S, respectively, and the relationship between ≪TA1≫ and ≪PO1≫ of l_A corresponds to the relationship between ≪TA1≫ and ≪PO2≫ of l_S according to Definition 4.

Based on the specialization of semantic activity labels, we finally define process activity specialization by label specialization as follows. A process activity S is a specialization of another process activity A if the semantic activity label l_S annotated to S is an observation-consistent specialization of the semantic activity label l_A annotated to A. Definition 5 provides a formalization of the notion of activity specialization by label specialization.

Definition 5 (Activity Specialization by Label Specialization). *Let* $\mathbb{R}_A = (\mathbb{P}, \mathcal{A}, \mathcal{L}, \lambda)$ *be a process activity repository. Given process activities* $\{A, S\} \subseteq \mathcal{A}$, *then process activity S is defined to be a specialization of A, denoted by predicate $ocs_a(S, A)$, iff $ocs_l(\lambda(S), \lambda(A))$.*

4 Application: User Support for Specializing Abstract Process Activities

This section illustrates the potential benefits of consistent reuse of process patterns to process modellers. For this purpose, Figure 2 shows a reuse example using BPMN. Figure 2a illustrates process pattern *Plan Development Project* as a BPMN model which serves as process template for the design of domain specific development projects (e.g., software or hardware development projects). The process pattern consists of the three abstract process activities *"Define Requirements"*, *"Design Architecture"*, and *"Define Development Plan"* as well as the elementary process activity *"Check Development Risk"*. In contrast to abstract process activities, elementary process activities already represent executable process activities that need not be further specialized for the purpose

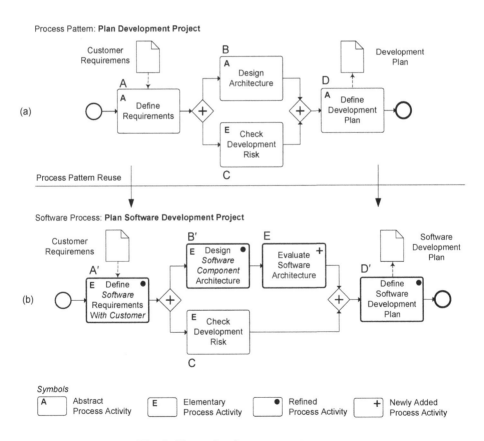

Fig. 2. Example of process pattern reuse

of describing domain-specific processes. Figure 2b shows a domain-specific development process for software projects which results from the reuse of the previously described process patterns. Then, one may observe that the original process activity *"Define Requirements"* has been substituted with *"Define Software Requirements With Customer"*, *"Design Architecture"* with *"Design Software Component Architecture"*, and *"Define Development Plan"* with *"Define Software Development Plan"*.

A process activity repository $\mathbb{R}_A$ presents the key to assisting process modellers with the reuse of process patterns by providing a hierarchical arrangement of process activities, also referred to for short as process activity hierarchy. A process activity hierarchy is a complete lattice structure $(\mathbb{R}_A, \mathbf{S})$ where $\mathbf{S} \subseteq \mathbb{R}_A \times \mathbb{R}_A$ is the set of all specialization relationships between the activities in $\mathbb{R}_A$ according to Definition 5. In other words, a process hierarchy results from inferring observation-consistent specialization relationships between semantic activity labels annotated to respective process activities in $\mathbb{R}_A$.

For the purpose of illustrating a process activity hierarchy, consider the example shown in Figure 3. Figure 3b illustrates a process activity hierarchy over the process activities shown in Figure 3a. A directed arc leading from an activity A to another activity S means that S is a specialization of A under observation-consistent label specialization.

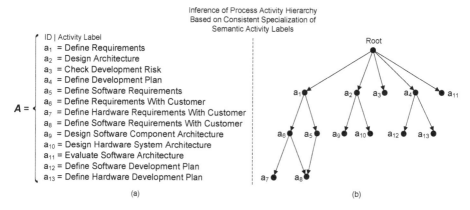

(a)

(b)

Fig. 3. (a) Set of process activities and (b) hierarchical arrangement over process activities

In the course of reusing process pattern *Plan Development Project*, a process modeller can explore the process activity hierarchy for desired substitutions. For example, if abstract process activity *"Design Architecture"* should be substituted with a domain-specific process activity then process modeller can choose one of the process activity specializations associated with *"Design Architecture"*. In the example shown in Figure 3, a process modeller may select either *"Design Hardware System Architecture"* or *"Design Software Component Architecture"*, thereby ensuring the consistent specialization of process activities by specializing their labels. In case of non-availability of the desired process activities, the process modeller can populate the process activity repository with new process activities. Afterwards, a hierarchical update procedure automatically infers specialization relationships between all process activities and, therefore, consistent reuse in a subsequent step is ensured.

5   Related Work

Existing work on the specialization of process activities mainly relates to business process model decomposition and abstraction (e.g., [6,8,9,18]). Some of the proposed approaches [8,18] exploit activity labels to address activity specialization while other approaches [9,16,6] assume the existence of an activity ontology which provides prescribed decomposition relations between process activities. To our knowledge, the approach presented in this paper is the first theoretical approach to support consistent specialization of process activities by specializing their activity labels.

Another field of research that relates to the work presented in this paper is the work on the comparison and alignment of process models. One of the main issues in that area represents the identification of one-to-one correspondences between process activities of different process models [19]. Most approaches suggest the computation of lexical or linguistic similarity measures between activity labels to infer correspondences between process activities [20,5,4]. The proposed similarity measures consider activity labels as compositions of words or noun phrases only and neglect implicit meaning of relations between involved words. Moreover, similarity measures do not facilitate inference of process activities on different levels of abstraction. Consequently, it is impossible to check whether abstract process activities have been specialized in a consistent fashion in the course of process patterns reuse.

6   Conclusion and Future Work

This paper introduces the specialization of process activities by specialization of their labels, which represents an orthogonal approach to the decomposition of process activities into sub-processes. Thereby, we contribute to the identification of process activities on different levels of abstraction from a semantic point of view, which represents a key issue in the field of establishing alignments between process models. This paper presents the theoretical foundations for a lexical and semantic labelling framework. In order to facilitate a comprehensive evaluation, future work will provide an implementation of a lexical and semantic labelling framework based on the concepts presented in this paper.

From the results in this paper, future research will evolve in two main directions. The first direction will be devoted to the development of refinement and extension operations for activity labels. These operations should assist process designers in preserving consistent activity specialization during the overall reuse process. The second direction will exploit semantic activity labels for the purpose of finding frequent process patterns in existing process descriptions. In particular, semantic activity labels will be used to identify semantically correct one-to-one correspondences between process models which serve as input for an envisioned framework for finding frequent process patterns.

Acknowledgement. This publication has been written within the project Vertical Model Integration (VMI) 4.0. The project VMI 4.0 is supported within the program Regionale Wettbewerbsfähigkeit OÖ 2007-2013 by the European Fund for Regional Development as well as the State of Upper Austria.

References

1. van der Aalst, W.M.P., Basten, T.: Inheritance of workflows: an approach to tackling problems related to change. Theor. Comput. Sci. 270(1-2), 125–203 (2002)
2. Becker, J., Delfmann, P., Knackstedt, R.: Adaptive reference modeling: Integrating configurative and generic adaptation techniques for information models. In: Becker, J., Delfmann, P. (eds.) Reference Modeling, pp. 27–58. Physica-Verlag HD (2007), http://dx.doi.org/10.1007/978-3-7908-1966-3_2
3. Bögl, A., Karlinger, M., Schrefl, M., Pomberger, G.: EPCs annotated with lexical and semantic labels to bridge the gap between human understandability and machine interpretability. In: Smolnik, S., F.T, Thomas, O. (eds.) Semantic Technologies for Business and Information Systems Engineering, pp. 214–241. IGI Global (2012), NOTE: The published version of this book chapter unfortunately contains numerous typesetting errors caused by the publisher. An error-free version is available at the authors' website http://www.dke.jku.at/research/publications
4. Dijkman, R., Dumas, M., van Dongen, B., Käärik, R., Mendling, J.: Similarity of business process models: Metrics and evaluation. Inf. Syst. 36(2), 498–516 (2011), http://dx.doi.org/10.1016/j.is.2010.09.006
5. Ehrig, M., Koschmider, A., Oberweis, A.: Measuring similarity between semantic business process models. In: Roddick, J.F., Hinze, A. (eds.) Proceedings of the Fourth Asia-Pacific Conference on Conceptual Modelling (APCCM 2007). CRPIT, vol. 67, pp. 71–80 (2007)
6. Greco, G., Guzzo, A., Pontieri, L., Saccá, D.: An ontology-driven process modeling framework. In: Galindo, F., Takizawa, M., Traunmüller, R. (eds.) DEXA 2004. LNCS, vol. 3180, pp. 13–23. Springer, Heidelberg (2004)
7. Keller, G., Nüttgens, M., Scheer, A.W.: Semantische Prozeßmodellierung auf der Grundlage Ereignisgesteuerter Prozeßketten (EPK). Tech. Rep. 89, Universität des Saarlandes, Germany, Saarbrücken, Germany (January 1992)
8. Koschmider, A., Blanchard, E.: User assistance for business process model decomposition. In: First IEEE International Conference on Research Challenges in Information Science, pp. 445–454 (2007)
9. Lausen, G.: Modeling and analysis of the behavior of information systems. IEEE Trans. Software Eng. 14(11), 1610–1620 (1988)
10. Leopold, H., Smirnov, S., Mendling, J.: On labeling quality in business process models. In: Proceedings of the 8th GI-Workshop Geschftsprozessmanagement mit Ereignisgesteuerten Prozessketten (EPK), Berlin, Germany (2009)
11. Lyons, J.: Linguistic Semantics: An Introduction. Cambridge University Press (1995)
12. Mendling, J., Reijers, H.A., Cardoso, J.: What makes process models understandable? In: Alonso, G., Dadam, P., Rosemann, M. (eds.) BPM 2007. LNCS, vol. 4714, pp. 48–63. Springer, Heidelberg (2007)

13. Mendling, J., Reijers, H.A., Recker, J.: Activity labeling in process modeling: Empirical insights and recommendations. Inf. Syst. 35(4), 467–482 (2010)
14. OMG: Business Process Model and Notation (BPMN), Version 2.0, OMG Document Number: formal/2011-01-03, http://www.omg.org/spec/bpmn/2.0/ (last visited September 29, 2014)
15. Reinhartz-Berger, I., Soffer, P., Sturm, A.: Extending the adaptability of reference models. IEEE Transactions on Systems, Man, and Cybernetics, Part A 40(5), 1045–1056 (2010)
16. Schrefl, M.: Behavior modelling by stepwise refining behavior diagrams. In: Proceedings of the 9th International Conference on Entity-Relationship Approach (ER 1990), pp. 113–128 (1990)
17. Schrefl, M., Stumptner, M.: Behavior-consistent specialization of object life cycles. ACM Trans. Softw. Eng. Methodol. 11(1), 92–148 (2002)
18. Smirnov, S., Reijers, H.A., Weske, M.: From fine-grained to abstract process models: A semantic approach. Inf. Syst. 37(8), 784–797 (2012)
19. Weidlich, M., Barros, A., Mendling, J., Weske, M.: Vertical alignment of process models – how can we get there? In: Halpin, T., Krogstie, J., Nurcan, S., Proper, E., Schmidt, R., Soffer, P., Ukor, R. (eds.) Enterprise, Business-Process and Information Systems Modeling. LNBIP, vol. 29, pp. 71–84. Springer, Heidelberg (2009)
20. Weidlich, M., Dijkman, R., Mendling, J.: The iCoP framework: Identification of correspondences between process models. In: Pernici, B. (ed.) CAiSE 2010. LNCS, vol. 6051, pp. 483–498. Springer, Heidelberg (2010)
21. Weidlich, M., Mendling, J., Weske, M.: A foundational approach for managing process variability. In: Mouratidis, H., Rolland, C. (eds.) CAiSE 2011. LNCS, vol. 6741, pp. 267–282. Springer, Heidelberg (2011)
22. Weidlich, M., Weske, M.: Structural and behavioural commonalities of process variants. In: Gierds, C., Sürmeli, J. (eds.) ZEUS 2010. CEUR Workshop Proceedings, vol. 563, pp. 41–48. CEUR-WS.org (2010)

Efficient Similarity Search by Combining Indexing and Caching Strategies[*]

Nieves R. Brisaboa[1], Ana Cerdeira-Pena[1], Veronica Gil-Costa[3],
Mauricio Marin[2], and Oscar Pedreira[1]

[1] Database Lab., Facultade de Informática, Universidade da Coruña, Spain
{brisaboa,acerdeira,opedreira}@udc.es
[2] CITIAPS, DIINF, University of Santiago, Chile
mauricio.marin@usach.cl
[3] DCC, National University of San Luis, Argentina
gvcosta@unsl.edu.ar

Abstract. A critical issue in large scale search engines is to efficiently handle sudden peaks of incoming query traffic. Research in metric spaces has addressed this problem from the point of view of creating caches that provide information to, if possible, exactly/approximately answer a query very quickly without needing to further process an index. However, one of the problems of that approach is that, if the cache is not able to provide an answer, the distances computed up to that moment are wasted, and the search must proceed through the index structure. In this paper we present an index structure that serves a twofold role: that of a cache and an index in the same structure. In this way, if we are not able to provide a quick approximate answer for the query, the distances computed up to that moment are used to query the index. We present an experimental evaluation of the performance obtained with our structure.

1 Introduction

New applications for search engines demand the use of data more complex than plain-text. Metric spaces have proven useful and practical for performing similarity search on very-large collections of complex objects. In this case, queries are objects of the same type of those stored in the database where, for example, one is interested in retrieving the k most similar objects to a given query. The similarity between any two objects is calculated by an application-dependent *distance function*, which is usually expensive to compute. The database is indexed using pre-computed distances to reduce comparisons during the search.

One of the critical issues in large scale search engines is efficiently handling sudden peaks in incoming query traffic. Typically, a large search engine is composed of one or more *front-service* (FS) machines and a collection of P processors

[*] Partially funded by: MICIN ref. TIN2009-14560-C03-02 (PGE & FEDER), Xunta de Galicia ref. GRC2013/053 (FEDER), and CDTI-MINECO-Axencia Galega de Innovación EXP 00064563/ITC-20133062 for authors in UDC[1]. FONDEF IDeA CA12i10314 for M. Marín. Mincyt-Conicyt CH1204 for V. Gil-Costa.

G.F. Italiano et al. (Eds.): SOFSEM 2015, LNCS 8939, pp. 486–497, 2015.

forming a distributed memory system. The Front-Service is in charge of receiving and sending queries to processors for results calculation. Each processor is seen as a *search node* which is in charge of a fraction of the whole object collection. Efficient search is supported by an index data structure that is distributed onto the P processors and parallel query processing is performed by sending the query to a number of processors. For systems under heavy query traffic it is critical to reduce the number of computations and yet to maintain an efficient throughput (number of queries entirely solved per unit time).

Research in metric-space similarity search has mainly focused on optimizing the execution of single queries. In multiple query settings, where query arrival rate can drastically change in intensity, and query content can become dynamically skewed in unpredictable ways, a relevant question is how we can make current queries benefit from previous query results so that they can be answered with approximate results. The underlying assumption is that approximate answers can be computed with much less computing cycles than regular answers so that servers are able to cope with drastic increase in incoming query traffic.

Caching query results is a feasible solution, and strategies such as QCache and RCache [1] have been proposed. However, they fail to reduce overall computing cycles as they treat independently caching and indexing. Namely, incoming queries that do not benefit from the cache are redirected to the metric index so query answers are computed from scratch. Our experiments show that the cost of accessing the cache in previous strategies can be even more expensive (or similar) than computing the answer for a query from the index itself.

To illustrate the above claim we performed experiments under the setting described in Section 4. We used a query log on the following three cases: (1) each query is sent to a RCache [1] and if the cache fails to produce the top-k results, the query is solved with an M-Tree [2] index, (2) each query is sent to a QCache [1], (a variation of the RCache) and if the cache fails to produce the top-k results, the query is sent to the M-Tree index and, (3) each query is directly solved with the M-Tree index. In each case we computed the running time and the number of distance evaluations required to process the full query log. Figure 1 shows the results normalized to 1. Cache sizes were set to 1%,

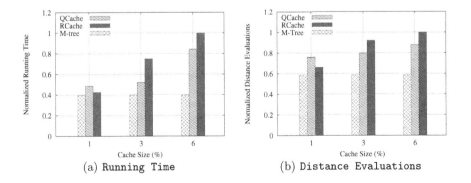

(a) **Running Time**   (b) **Distance Evaluations**

Fig. 1. Performance achieved by different cache strategies

3% and 6% of the number of queries, which produces cache hit ratios of 15%, 22% and 33% respectively. The results show that the running time and distance evaluations when using caches are a significant part of the search cost.

In this paper we propose a strategy which contains a cache embedded in the index so that computing cycles are not wasted when cache contents are not able to produce good approximate results. In such a case, previous computations for the query are used to continue the traversal of the index in order to produce approximate query results as fast as possible.

The remaining of this paper is organized as follows. Section 2 reviews related work on similarity search and caching. In Section 3 we present a new index structure that combines indexing and caching strategies. Section 4 presents the results obtained in the experimental evaluation of our structure. Finally, Section 5 summarizes the main conclusions from our work.

2   Related Work

A *metric space* (U, d) is composed of a universe of objects U and a *metric*, a function $d : U \times U \to R^+$ that measures the dissimilarity between any two objects and that holds the properties of strictly positiveness ($d(x, y) > 0$ and if $d(x, y) = 0$ then $x = y$), symmetry ($d(x, y) = d(y, x)$), and the triangle inequality ($d(x, z) \leq d(x, y) + d(y, z)$). The *database* or *collection* of objects is a finite subset $X \subseteq U$, with size $n = |X|$.

There are two main queries of interest: (a) *range search*, $R_X(q, r)$, that retrieves all the objects $u \in X$ within a search radius r of the query q, and (b) *k-nearest neighbors search*, $kNN_X(q)$, that retrieves the set k most similar objects to q. Given a query $q \in U$, the goal is to retrieve the most similar objects to q with the minimum number of object comparisons.

Many metric index structures have been proposed and studied (see [3,4]). The proposals of this paper make use of one of those structures, the *List of Clusters* (LC) [5], which has shown to outperform well-known alternative metric-space indexes [6]. LC partitions the collection into a set of disjoint clusters as follows. We first choose a *cluster center* $c \in X$ and a radius r_c. The *cluster ball* (c, r_c) contains the subset of elements of X at distance at most r_c from c. We define $I_{X,c,r_c} = \{u \in X - \{c\}, d(c, u) \leq r_c\}$ as the cluster of *internal* elements which lie inside (c, r_c), and $E_{X,c,r_c} = \{u \in X, d(c, u) > r_c\}$ as the *external* elements. The clustering process is recursively applied in E. As shown in [5] a good policy for selecting the next center is to choose the object in the collection that maximizes the sum of distances to previous centers.

Given a query $R_x(q, r)$, q is sequentially compared with the cluster centers of the LC. Given a center c, we exhaustively scan its cluster I (that is, we compare q with the objects $u \in I$) if the query ball (q, r) intersects the cluster ball (c, r_c). The search then continues with the next cluster in LC. At any point of the search, the search stops if the query ball (q, r) is totally and strictly contained in the cluster ball (c, r_c), since the construction process ensures that all the elements that are inside the query ball (q, r) have been inserted in I (as shown in line

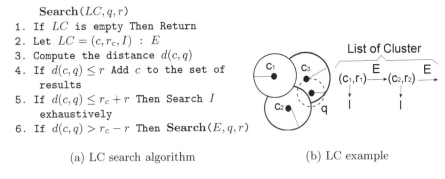

Search(LC, q, r)
1. If LC is empty Then Return
2. Let $LC = (c, r_c, I)$: E
3. Compute the distance $d(c, q)$
4. If $d(c, q) \leq r$ Add c to the set of
   results
5. If $d(c, q) \leq r_c + r$ Then Search I
   exhaustively
6. If $d(c, q) > r_c - r$ Then **Search**(E, q, r)

(a) LC search algorithm   (b) LC example

Fig. 2. The List of Clusters (LC) strategy

6 of Figure 2a). Figure 2b shows three clusters and a query $R_X(q, r)$. In this example, q has to be compared with the objects in the clusters with centers c_2 and c_3, but the cluster with center c_1 is directly discarded.

The selection of effective pivot /cluster centers for metric indexes has been deeply studied. One of the existing proposals for center selection is SSS [7,8] that selects a new object as a center if it is far enough from those already selected. Being M the maximum distance between any two objects in the space, and α a parameter such that $0 \leq \alpha \leq 1$, an object in the collection is selected as a new center if its distance to the previously selected centers is greater than $M \times \alpha$. In [9] it has been shown that the most effective pivots for a given object of the database are the nearest and furthest pivots.

2.1 Parallel Processing for List of Clusters

We assume a parallel architecture in which a *front-service* receives queries and evenly distributes their processing onto the processors. The work in [6] studied various forms of parallelization of the LC strategy concluding that a global indexing strategy called *GG*, which stands for Global Index and Global Centers, achieves the best performance.

The GG strategy builds a LC and distributes it uniformly at random the clusters of the LC onto the processors. Upon reception of a query q, the broker sends it to a circularly selected processor. This processor becomes the *ranker* for that query. It calculates the *query plan*, that is, the list of clusters that intersect (q, r). To this end, it broadcasts the query to all processors and they calculate in parallel a fraction $1/P$ of the query plan. Then they send their n_q/P pieces of the global plan to the ranker, which merges them to get the global plan with clusters sorted in construction order. The then ranker sends the query q and its plan to the processor i containing the first cluster to be visited. This processor i goes directly to the GG clusters that intersect with q, compares q against the objects stored in them, and returns to the ranker those within (q, r). The remaining part of the query plan is passed to the next processor j and so on, till completing the processing of the query.

2.2   Metric Space Cache

A *metric space cache* $\mathcal{C}$ consists of a set of past queries with their respective results. Let $q_i \in \mathcal{C}$, if the query along with all results in $kNN_X(q_i, k)$ are in the cache. Also $o_i \in \mathcal{C}$ denotes that the object $o_i \in X$ is stored in the cache and thus belongs to at least one set $kNN_X(q_i, k)$ associated with a cached query q_i. Let r_q denote the radius of the smallest hyper-sphere centered in q which contains all objects in $kNN_X(q, k)$. The *safe radius* s_q [1,10] of the query q with respect to a query $q_i \in \mathcal{C}$ is the radius r_{q_i} minus the distance from q to q_i, namely $s_q(q_i) = r_{q_i} - d(q, q_i)$. Every cached query q_i gives complete knowledge of the space up to distance r_{q_i} from q_i. If q is inside the hypersphere centered in q_i with radius r_{q_i}, then as long as we restrict ourselves to look inside this hypersphere, we have complete knowledge of the $k' \leq k$ nearest neighbours of q.

Thus, if the safe radius $s_q(q_i)$ of a query $q \in U$ with respect to a query $q_i \in \mathcal{C}$ is a positive value, then every object in the range query $R_X(q, s_q(q_i))$ is also in the cache $\mathcal{C}$ and thus can be solved over the cache with the range query $R_{\mathcal{C}}(q, s_q(q_i))$. Furthermore, the k' objects in $R_{\mathcal{C}}(q, s_q(q_i))$ are also the k' nearest neighbours of q in the whole database X.

In [1,10,11] two different metric-space cache algorithms were presented: RCache (Result Cache) and QCache (Query Cache). RCache uses a hash table $\mathcal{H}$ where, for each query $q_i \in \mathcal{C}$, it stores tuples of the form $(q_i, kNN(q_i, k))$, being the object q_i the hash key. If the query is not in $\mathcal{C}$, then it attempts to give an approximate answer. To search for an approximated answer, RCache uses a metric-space index $\mathcal{M}$ to perform a $kANN_{\mathcal{C}}(q, k)$ search of the k closest objects to q_i which are currently stored in $\mathcal{C}$. QCache builds the metric index $\mathcal{M}$ over the query objects instead of indexing every single object returned by the queries in the cache, as the RCache algorithm does. This reduces by a factor of k the number of indexed objects. The main idea is to search sets of suitable cached queries first and then to use the cached results of those queries to find an approximate answer. According to the experimental results reported in [1,10], the quality of the approximate results returned by both algorithms are comparable.

In [12] a caching metric-space index called D-File was proposed. D-File uses a hash table with entries $[o_1, o_2, d(o_1, o_2)]$, where o_1 and o_2 are the objects identifier and the third component is the computed distance between them. D-File is kept in main memory in order to reduce the number of distance computations performed over a second index like the M-tree [2]. This goal is achieved when the distance $d(o_1, o_2)$ is in the D-File, or by obtaining a lower or upper bound of $d(o_1, o_2)$ and thereby improving the pruning over the M-tree. However, as shown in [13], D-File suffers from a too high internal processing cost because of the hash table. Recently, SnakeTable [13] was specifically proposed for scenarios in which queries are received in streams of very similar queries.

3   Combining Indexing and Caching

One of the problems of previous proposals in caching for searching in metric spaces is the cost of processing the cache in terms of distance computations. If the cache succeeds in providing an answer, the distance computations to process the cache save the distance computations needed to solve the query with an index. However, if the cache fails, those distance computations for processing the cache are wasted and add up to the overall search cost.

In this section we present a variant of the list of clusters that combines indexing and caching policies into the same structure so, if the cache cannot provide an exact or approximate answer for the query, at least the distance computations needed to process the cache are not wasted, since they would be necessary anyway in order to solve the query in the index.

3.1   Index Structure and Construction

The index structure is that of a LC in which cluster centers are not selected among the objects in the database, but among the queries received in search time. Therefore, there is no index built until the system starts receiving queries. This implies that the search cost will be higher for the first queries. To avoid this in a real scenario, we can make use of an additional index at the beginning of the process that would be dropped when the new index structure has stabilized.

Since queries are dynamic, it is not possible to follow the center selection policy of LC. We use SSS [7], that adapts to the dynamic nature of the queries, and guarantees that the centers will be well distributed in the space. Therefore, if the distance from a new query q to existing cluster centers is greater than $M \times \alpha$, q will become a new cluster center and the index will be restructured accordingly. The cluster corresponding to the new cluster center q may be empty if no object in the collection is closer to q than to any other past query used as a cluster center. In this case, the cluster would be removed. As in LC, each object belongs to, and only to, the cluster formed by its closest cluster center.

Reorganizing the index when a new query is selected as a cluster center has a cost in terms of distance computations. However, SSS guarantees that the number of pivots will stabilize at some point, so the cost of that reorganization will be amortized among all processed queries.

The reason for choosing the cluster centers among the queries received by the system is that, in this way, they will better cover the portion of the space defined by the queries, which is not necessarily the same space defined by the objects in the database. This would benefit the algorithms for range and kNN search since the cluster centers would be more similar to future queries.

In addition, we keep additional information in each cluster. Instead of storing only the cluster center, the list of objects belonging to the cluster, and the covering radius, we keep the distances from the center to each of the objects in the cluster, as proposed in [14]. In this way, the cluster centers also play the role of a pivot during the search. As shown in [9], the nearest pivot is the most promising for each object, so using the cluster center as a pivot for each of the objects belonging to that cluster should be the most effective choice. In addition, keeping the distances from the cluster center to each object in the cluster will also make possible to return approximate results to queries sufficiently similar to one of the pivots.

3.2   Index as a Cache for Approximate Search

The main motivation for this index structure is that it can be used as a classical index or as a cache that may help to quickly provide approximate answers to queries when the system is receiving a huge number of queries. When a new query is submitted to the system, the first step consists in comparing the query object with the cluster centers. Since these objects are past queries, they reflect the space defined by the queries, and it is probable that new queries are equal or similar to some of the cluster centers. The higher the number of past queries used as cluster centers, the more the chances that a new query is sufficiently similar to a past query used as a cluster center.

If $d(q, c_i) = 0$ for some c_i, the approximate answer to q will be extracted by just using the information contained in C_i. This may leave some objects out of the answer, but it would not require additional distance computations.

The case in which a new query is exactly equal to a past query used as a cluster center will happen very few times. However, it is still possible to provide an approximate result if q is sufficiently close to some cluster center c_i. Notice that by selecting the cluster centers with SSS, all of them are at least at a distance $M \times \alpha$. Therefore, the covering radius of each cluster will be at most $\frac{M \times \alpha}{2}$. We consider that q is sufficiently similar to the cluster center c_i if:

$$d(q, c_i) \leq \frac{M \times \alpha}{2} \times \rho$$

where $0 \leq \rho \leq 1$. Therefore, ρ determines how close a query has to be to a past query in order to return an approximate answer. Since the structure of the index does not depend on ρ, this parameter can be easily changed during the search phase depending on the processing demands on the system.

3.3   Searching

Algorithm 1 shows the pseudocode of the algorithm for range search. Given $R(q, r)$, and being I the index, the search process proceeds as follows:

- In a first step, the query q is compared with the center of each of the clusters of I (lines 2 and 3). This comparison allows us to determine if q will become a new cluster center or not. If the $d(q, p_i) > M \times \alpha$ for all pivots p_i, q becomes a new cluster center (lines 4 and 5). In this case, it is necessary to restructure the index so the objects that are closer to this new pivot are assigned to its cluster. The procedure $AddNewCluster(I, q)$ carries out this restructuring.

  Notice that it is possible to carry out this reorganization without comparing q with all objects in the database, since the distances from each object x to its cluster center provides us with lower bounds on $d(q, x)$ so it may not be necessary to compute $d(q, x)$.
- After comparing the query q with each of the cluster centers, we can use this information and the index as a cache (lines 6-8). First, if $d(q, c_i) = 0$ for some p_i, we can answer the query with the information contained in the cluster

```
 1  Algorithm: RangeSearch (I, q, r)
    Data: I: Index structure; q: query object; r: search radius;
 2  for i = 1 to I.size do
 3  │    d[i] = d(q, p_i) ;
 4  if ∀i, d[i] ≥ M × α then
 5  │    AddNewCluster(I,q);
 6  else
 7  │    if ∃p_i / d[i] ≤ (M×α)/2 × ρ then
 8  │    │    ApproximateRangeSearch(I.c_i, q, r);
 9  │    else
10  │    │    for i = 1 to I.size do
11  │    │    │    if d[i] − cr_i ≤ r then
12  │    │    │    │    foreach x_j ∈ I.C_i do
13  │    │    │    │    │    if |d[i] − d(c_i, x_j)| ≤ r then
14  │    │    │    │    │    │    if d(q, x_j) < r then
15  │    │    │    │    │    │    │    Result ← Result ∪ {x_j}
```

Algorithm 1. Pseudocode for range search with approximate search

formed by c_i. Even if an exact match does not occur, the information of the index allows us to provide an approximate answer. If $d(q, c_i) \leq \frac{M \times \alpha}{2} \times \rho$ for some c_i, we will also build the answer to q from the information contained in the cluster formed by c_i. In that case, we restrict the search to the cluster of c_i. Since the index stores the distances $d(x, c_i)$ for all the objects in the cluster, we can use these distances to obtain lower bounds for $d(q, x)$ and discard some objects x without comparing them with q.

– If cannot provide an approximate answer, the search continues exploiting the rest of information in the index by combining cluster and pivot criteria to discard the objects in the clusters (lines 9-15). Notice that at this point we already have the distance from q to the cluster centers. Those clusters that do not intersect the query ball (q, r) are directly discarded from the result. For those clusters that cannot be completely discarded, we use the distances from the center to the objects x in the cluster to obtain lower bounds on $d(q, x)$. In those cases in which these lower bounds do not allow us to directly discard x, we finally have to compute the distance $d(q, x)$.

With this algorithm, if the structure is not able to provide an approximate answer for a query (thus acting like a cache), the distances computed up to that moment are not wasted, since they are the same needed to search using the index. As in proposals in which the cache is a component separated from the index, in the proposed algorithm it is also possible to activate/deactivate the use of the cache in function of the processing requirements of the system. Actually, the parameter ρ allows us to control the degree to which the cache will be used (and in consequence, the quality of the approximate answers).

Although we have only explained the algorithm for range search, adapting it for the case of kNN search is straightforward. Given a new query q, its list of k nearest neighbors can be initialized with the first objects it is compared with.

Then, the search proceeds as a range search, but updating the range at each step if necessary as the distance from q to its current k^{th} nearest neighbor.

4 Experimental Results

In this section we provide experimental results on different aspects of our structure. We implemented it using the Metric Spaces Library [15]. We used two collections from the library, widely used in the state of the art, namely English, a dictionary containing $69,069$ words, and Nasa, a collection of $40,151$ images represented by feature vectors of dimension 20. The edit distance was used for English, and the Euclidean distance for Nasa.

For each collection, 90% of the objects were used as the database, and the remaining 10% were used as base queries. The query sets were created from the base queries in order to reflect the typical human behavior in real search engines. To do so, the base queries were replicated following the same distribution obtained from a set of real queries obtained from a web search engine. In this way, the queries have a biased distribution that matches that we would obtain in a real system. Following this procedure, we generated files with $10,000$ queries for each collection. The query range for English was set to 2, and for Nasa we used a range that retrieves an average of 0.01% of the collection for each query.

In a first set of experiments, we analyzed how the structure performance evolves when the first queries are submitted to the system. To do this, we ran the $10,000$ queries for each collection with an initially empty index. In order to focus on the cost of solving queries and updating the index, the parameter ρ was set to 0 in these experiments when the cache was used, which means that the cache is used only when a new query is exactly equal to a past query. Figures 3 and 4 show for each collection the number of distances needed to answer the range queries with and without using the cache part of the algorithm (RWC and RWOC respectively), the number of distances needed to update the index with and without cache (UWC and UWOC respectively), and the sum of these two numbers (TWOC and TWC). The results are shown in terms of the number of queries received by the system (from 0 to $10,000$). As we can see in the results, the cost of solving a query decreases in all cases as the index gets more information. In the case of Nasa, which needs a smaller number of centers, we can see how the cost of updating the index quickly decreases as the number of past queries increases.

We conducted experiments to analyze the performance when we use the structure as a cache and an index at the same time, providing approximate but quick answers when possible. In order to leave the cost of updating the index out of the results so it does not interfere with the purpose of these experiments, we ran the $10,000$ queries twice for each collection: the first one allows the index to build and stabilize (which has been analyzed in the previous experiments), and the second one allows us to measure search performance when very few changes in the index happen. Figure 5 shows the average number of distance computations to solve range queries and 4-NN queries for different values of the parameter ρ. The results reflect only the performance in the second phase. The figure shows

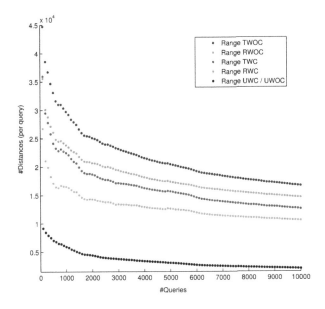

Fig. 3. Search and update cost for range queries in English

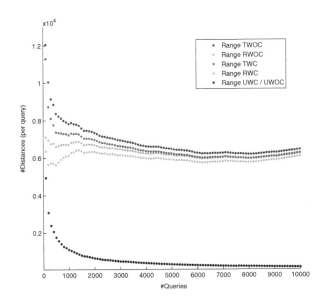

Fig. 4. Search and update cost for range queries in Nasa

for each value of ρ the cost of solving the query using the cache (English-TWC and Nasa-TWC) and without using it (English-TWOC and Nasa-TWOC). As the value of ρ grows, the requirements for a query to be solved using only information of its closest cluster are lower, which results in less distance computations.

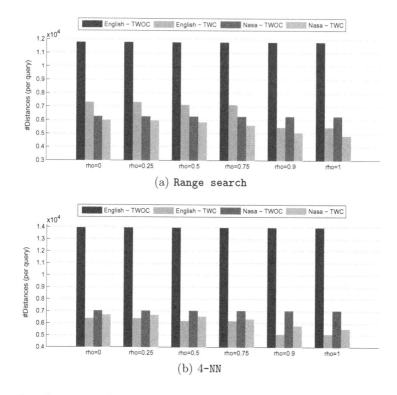

Fig. 5. Cost for range and 4-NN search using the structure as a cache and as an index

Even for the smallest values of ρ, the improvement with respect to the version that does not use the cache is very significant.

5   Conclusions

We have presented a metric structure for similarity search that allows us to use it as a cache or a classic metric index. One of the main differences with the original List of Clusters structure is that cluster centers are selected among past queries instead of among the objects in the collection, and that they are selected with a criterion that ensures they will be distributed in the space. In addition, each cluster stores the distances from the center to each of its objects. Given a new query, we provide an approximate result if it is equal or very similar to one of the past queries we keep in the index. If we cannot provide an approximate result, the distances computations needed to examine past queries are used to continue the search in the index, and to prune part of the search space.

The main advantage of this metric structure is that it can work as a cache and an index at the same time. Since the cluster centers are past queries, the comparison of each query with them allows us to provide an approximate but quick result. Even if the first step cannot return an approximate answer, those distance computations carried out to that moment are not wasted work, since they would be necessary anyway to query the index. Therefore, the main difference with previous proposals is that the cache is integrated with the index, not requiring any additional processing.

Some lines for future work are still open. We are working on replacement policy to dynamically update the past queries kept in the index as cluster centers. This can be important if the distribution of the queries received by the system are very skewed in certain and short periods of time.

References

1. Falchi, F., Lucchese, C., Orlando, S., Perego, R., Rabitti, F.: Caching content-based queries for robust and efficient image retrieval. In: Procs. of EDBT, pp. 780–790 (2009)
2. Ciaccia, P., Patella, M., Zezula, P.: M-tree: An efficient access method for similarity search in metric spaces. In: Procs. of VLDB, pp. 426–435 (1997)
3. Chávez, E., Navarro, G., Baeza-Yates, R., Marroquín, J.L.: Searching in metric spaces. ACM Computing Surveys 33, 273–321 (2001)
4. Zezula, P., Amato, G., Dohnal, V., Batko, M.: Similarity search. The metric space approach. Advances in Database Systems, vol. 32. Springer (2006)
5. Chavez, E., Navarro, G.: A compact space decomposition for effective metric indexing. Pattern Recognition Letters 26(9), 1363–1376 (2005)
6. Gil-Costa, V., Marin, M., Reyes, N.: Parallel query processing on distributed clustering indexes. Journal of Discrete Algorithms 7(1), 3–17 (2009)
7. Pedreira, O., Brisaboa, N.R.: Spatial selection of sparse pivots for similarity search in metric spaces. In: van Leeuwen, J., Italiano, G.F., van der Hoek, W., Meinel, C., Sack, H., Plášil, F. (eds.) SOFSEM 2007. LNCS, vol. 4362, pp. 434–445. Springer, Heidelberg (2007)
8. Bustos, B., Pedreira, O., Brisaboa, N.: A dynamic pivot selection technique for similarity search in metric spaces. In: Procs. of SISAP, pp. 105–112. IEEE Press (2008)
9. Ares, L.G., Brisaboa, N.R., Esteller, M.F., Pedreira, O., Ángeles, S.: Places: Optimal pivots to minimize the index size for metric access methods. In: Procs. of SISAP, pp. 74–80. IEEE Press (2009)
10. Falchi, F., Lucchese, C., Orlando, S., Perego, R., Rabitti, F.: A metric cache for similarity search. In: Procs. of LSDS-IR, pp. 43–50 (2008)
11. Falchi, F., Lucchese, C., Orlando, S., Perego, R., Rabitti, F.: Similarity caching in large-scale image retrieval. Information Processing and Management (2011)
12. Skopal, T., Lokoc, J., Bustos, B.: D-cache: Universal distance cache for metric access methods. Transactions on Knowledge and Data Engineering 99 (2011)
13. Barrios, J., Bustos, B., Skopal, T.: Snake table: A dynamic pivot table for streams of k-nn searches. In: Navarro, G., Pestov, V. (eds.) SISAP 2012. LNCS, vol. 7404, pp. 25–39. Springer, Heidelberg (2012)
14. Marin, M., Gil-Costa, V., Bonacic, C.: A search engine index for multimedia content. In: Luque, E., Margalef, T., Benítez, D. (eds.) Euro-Par 2008. LNCS, vol. 5168, pp. 866–875. Springer, Heidelberg (2008)
15. Figueroa, K., Navarro, G., Chávez, E.: Metric spaces library (2007), http://www.sisap.org/Metric_Space_Library.html

Retrieving Documents Related to Database Queries

Vladimir Soares Catão, Marcus Costa Sampaio, and Ulrich Schiel

Systems and Computing Department, Federal University of Campina Grande, Brazil
vladimir@copin.ufcg.edu.br
{sampaio,ulrich}@dsc.ufcg.edu

Abstract. Databases and documents are commonly isolated from each other, controlled by Database Management Systems (DBMS) and Information Retrieval Systems (IRS), respectively. However, both systems are likely to store data about the same entities, a strong argument in favor of their integration. We propose a DBMS-IRS integration approach that uses terms in DBMS queries as keywords to IRS searches, retrieving documents strongly related to the queries. The IRS keywords are built "expanding" an initial set of user-provided keywords, with top-ranked terms found in a query result: the terms are ranked based on a measure of term diffusion over the query result. Our experiments show the effectiveness of the approach in two different domains, in comparison to other DBMS-IRS integration methods, as well as to other term-ranking methods.

Keywords: information integration, DBMS-IRS integration, query expansion, term-ranking methods.

1 Introduction

Databases and documents are the basic sources of information in the majority of organizations, controlled by Database Management Systems (DBMS) and Information Retrieval Systems (IRS), respectively. While databases follow strict rules for data organization, documents are based on free text. Databases are queried by means of formal query languages (like SQL), whereas documents are normally retrieved through keyword lists. It is understood that database queries are exact; on the contrary, document retrieval is inexact by nature, frequently retrieving much irrelevant documents.

These differences reinforces the isolation of databases and documents inside organizations; more precisely, DBMS and IRS usually do not communicate with each other. However, there are still many DBMS-IRS integration opportunities [1,2].

The "dataspace" [3], for example, is one of the approaches that envisions a transparent access to data stored in either a DBMS or an IRS. Nonetheless, many authors still do not see it as a "ready-to-use" technology [4]. We can also

G.F. Italiano et al. (Eds.): SOFSEM 2015, LNCS 8939, pp. 498–510, 2015.
© Springer-Verlag Berlin Heidelberg 2015

point other integration approaches, such as structured queries over documents [5,6], keyword searches over databases [7], and XML retrieval [8].

Our approach takes a different rationale, using the fact that, in a given organization, the DBMS and the IRS will probably keep different information about the same entities. Example: find contracts, business proposals, or presentation slides – side IRS – regarding last month company's clients – side DBMS –; or recover emails with user complaints – side IRS – related to products made on company's division XYZ – side DBMS.

With that in mind, we propose DBFIRe (**DataBases** **f**or **I**nformation **R**etrieval), an approach that borrows ideas from query expansion [9] in IR. Documents are retrieved through a keyword search formed by top-ranked terms[1] extracted from a database query result, "expanding" an initial set of keywords provided by the user; the user keywords act as a rough description of the information need. Besides using the query result as expansion corpus, DBFIRe also offers a new term-ranking method based on the diffusion of a term in the query result: as we assume database queries return exact responses, the more the term is spread over the query result, the better its ranking.

Two other approaches [10,11] also build keyword searches from database queries: in [11], entire column contents are ranked, instead of isolated terms as DBFIRe does, whereas in [10] the keywords are taken from the query body. Expansion could also be done using other term-ranking approaches, such as [12–14]. Thus, our experiments investigated the following hypotheses:

- Could we get the same results using only the original user keywords, that is, without expanding those keywords?
- How does DBFIRe behave compared with similar approaches [10,11]?
- What is the performance of our term-ranking method compared to other methods, like [12–14]?

In all tests with two different domains, DBFIRe achieved significantly better results.

The rest of the paper is structured as follows: we review related work in Section 2 and detail DBFIRe itself in Section 3; Section 4 presents the experiments we ran for DBFIRe validation; and Section 5 concludes the paper, with some suggestions of further improvements to the method.

2   Related Work

In this section, we detail approaches that use DBMS queries to feed IRS searches, as well as concepts about query expansion, showing other term-ranking methods that inspired DBFIRe.

---

[1] Regarding the DBMS world, we use "term"; regarding the IRS world, we use "keyword".

2.1 SCORE (Symbiotic Context Oriented Information Retrieval)

Instead of ranking individual terms, SCORE [11,15] ranks the whole content of tuple elements[2] (consider a tuple element as the intersection between a row and a column among the tuples of the query result). Thus, considering a query over a single database table, each element is ranked according to its higher frequency in the query result and its lower presence in the rest of the table, as detailed in (1):

$$W(A,t) = N_Q(A,t) \, log(\frac{1 + |T| - |Q(T)|}{1 + N_T(A,t) - N_Q(A,t)}) \tag{1}$$

In this equation, consider table T with $|T|$ rows, and a query over $T - Q(T)$ – with $|Q(T)|$ rows. $W(A,t)$ is the weight of tuple element t over column A, $N_Q(A,t)$ is the number of times the element t appears in column A of $Q(T)$ and $N_T(A,t)$ is the number of times the element t appears in column A of T. SCORE returns the best scored N elements with the higher values of W, where N is a tuning parameter. Though the formulation targets a single table, it can be easily extended for database queries made of joins between different tables.

Note that all terms in the top N elements are sent to the IRS. This decreases the quality of document retrieval, as we will see in the experiments detailed in Section 4.

2.2 SEMEX (SEMantic EXplorer)

Suggesting documents related to a database query is only a portion of SEMEX functionalities. Actually, its focus is Personal Information Management (PIM) [16]; the part of the system which is indeed related to DBFIRe is detailed in [10].

Unlike SCORE, SEMEX disregards the query result, looking for terms on the query body only. SEMEX sees queries as graphs: literals in WHERE clauses and names of tables label nodes; attribute-value pairs and join predicates label edges. After a number of traversals, SEMEX suggests node and edge labels that best summarize the graph.

For example, consider the following query:

SELECT * FROM papers
WHERE keywords LIKE %information% AND keywords
LIKE %retrieval% AND year='2014'

The query selects papers whose keywords contain the words "information" and "retrieval" in a specific year. This query is modelled by the graph in Figure 1.

Given the graph, SEMEX suggests the following keyword search:

papers information retrieval 2014

---

[2] The main paper about SCORE [11] uses "terms" when referring to tuple elements.

See that not all graph elements are suggested: in this case, edge labels were considered "distraction" to the IRS. Indeed, it is a good description of the information need; perhaps a human could come with the same keywords.

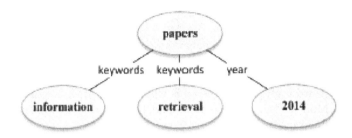

Fig. 1. Example query for SEMEX

However, by not using the query result, the method neglects an important source of useful terms.

2.3   Query Expansion

The main purpose of query expansion in IR is to give a better description of the user's information need. It starts from an initial set of keywords, which is augmented by terms in some expansion source. Despite the availability of a great number of approaches, our focus is on automatic expansion methods, based on a first round of keyword search, called pseudo-relevance feedback (PRF) methods [9].

Considering each tuple composed of a bag of words, PRF environments are very similar to ours: think of the documents in the first round of keyword search as equivalent to the tuples of the database query result. In this case, expansion methods can be easily adapted.

PRF ranks terms inside top-k documents returned by the first keyword search. The definite keyword search is built with the original user keywords and expanded with the best n ranked terms; the documents returned to the user are those from the expanded search. Parameters k and n must be manually tuned.

In this paper, we compared DBFIRe with three widely used PRF methods, each one with its own score function, which ranks the best n terms for expansion. One of them is called KLD (Kullback-Leibler Divergence [13]). KLD defines its term score as in (2):

$$score(t) = p_R(t) \log(\frac{p_R(t)}{p_C(t)})$$

(2)

The score quantifies the divergence between the probability of finding term t among top-k documents (P_R) and the probability of finding t in the document collection (P_C).

Another method also measures differences among term distributions; it is known as DFR (Divergence from Randomness [12]). DFR term score is detailed in (3):

$$score\,(t) = tf_k \log_2 \left(\frac{1 + P_n}{P_n} \right) + \log_2(1 + P_n) \qquad (3)$$

In (3), tf_k means the term frequency among the top-k documents, and P_n is the frequency of the term on the whole collection divided by the number of documents in the collection.

Finally, the third approach we tested is what we call RM, because it is based on Relevance Models [14]. As with KLD and DFR, terms are ranked based on its score as defined in (4):

$$score\,(t) = \sum_{d \in R} p(d)p(t|d) \prod_{q \in Q} p(q|d) \qquad (4)$$

In this case, R is the set of top-k documents, $p(d)$ is the probability of document d appearing in R (normally estimated by its rank), $p(t|d)$ is the probability of term t occurring in document d, Q is the set of the user keywords, and $p(q|d)$ is the probability of keyword q occurring in document d.

Though the basic idea in RM is to rank terms based only on term probabilities over the expansion source, documents that do not contain at least one of the user keywords, will not contribute to the term score; that is the so-called "zero-probability" problem. The authors suggest smoothing this effect, using also the term probability over the whole collection [17]. Therefore, to get $p(q|d)$ we should follow (5):

$$p\,(q\,|\,d) = \lambda \frac{\#(q,d)}{|d|} + (1 - \lambda)\frac{\#(q,C)}{|C|} \qquad (5)$$

In this case, $\#(q,\,d)$ and $\#(q,\,C)$ are the number of occurrences of q in document d and in the whole collection C, respectively; likewise, $|d|$ and $|C|$ are the size in words of document d and collection C; at last, the formula is smoothed by parameter λ, which is set at 0.2.

The ideas of query expansion inspired our own method, which will we show in the next section.

3   DBFIRe: *DataBases for Information Retrieval*

DBFIRe starts from an initial database query and a set of keywords provided by the user; the keywords are a brief summary of his/her information need. These keywords are then expanded with top ranked terms found in the query result; the new set of keywords is sent to the IRS in order to retrieve related documents.

However, how should we rank the terms? Should we choose just any expansion method? Note that all methods detailed in Section 2 use some heuristic regarding term occurrences over the whole document collection. If we consider database queries producing exact answers, perhaps we could focus exclusively on the query result as expansion source, and forget how the terms appear in the collection as a whole.

Therefore, one possibility would be using the pure term frequencies over the query result; in fact, as we will see ahead, this already gives a good advantage over traditional methods. However, if we also consider the presence of the term over tuple elements of the query result, we can do even better.

We believe the joint measure of both frequencies gives a better estimation of how much spread is a term over the database query result. Therefore, each term should be ranked based on its already mentioned *term frequency* – *tf(t)*, which is the number of occurrences of *t* along the whole query result, and also on the term's *element frequency* – *ef(t)*, which stands for the number of tuple elements in which *t* occurs. This ranking formula is shown in (6), where $s(t)$ is the overall score of term t.

$$s\left(t\right) = tf\left(t\right)\sqrt{ef(t)} \qquad (6)$$

The computation of *ef(t)* considers only the *presence* of a term in a tuple element, no matter how many times it may occur in that element. Moreover, once we will always have less tuple elements than terms, a small increase on its element frequency –*ef(t)* – may cause a stronger impact on $s(t)$; therefore, we smooth this effect by applying the square root to the *ef(t)* component.

Though useful to measure the diffusion of a term over the query result, $s(t)$ can also give higher weights to stopword terms, that is, terms that don't help expansion. Thus, common stopwords [18] are discarded.

Lastly, it is also possible to tune the amount of overhead for DBFIRe. Similar to other expansion methods, DBFIRe has two parameters that can be adjusted manually: the number of added terms (a parameter called n), as well as the number of tuples that should be processed (parameter k).

3.1   Term Weighting

A known effect in expansion methods is the so-called "query drift" [19]: as more terms are added, the IRS tends to retrieve more documents related to these terms instead of related to the user keywords. To avoid this, expansion terms should be given lower weights than the user keywords, but keeping the relative differences of their scores $s(t)$.

Thus, inspired by the Rocchio's beta [20], the weight of each term in DBFIRe normalizes its score $s(t)$ with respect to the maximum value of s, but limited to a user defined parameter (called β); β should be set in the interval $(0, 1)$. In DBFIRe, we weight terms as in (7):

$$weight\left(t\right) = \beta\frac{s(t)}{\max(s)} \qquad (7)$$

Parameter β tunes the importance that the expansion should have: the higher β, the higher the weight of top-ranked terms. Additionally, we assure the user keywords are specially treated assigning them the maximum weight (that is, 1.0). Though the best value of β will probably differ from domain to domain, we think a value that can be generally applied is $\beta = 0.5$, half way between the whole interval (0, 1).

3.2   An Example

Consider the online movie database, available at IMDB[3] (**Internet Movie Data Base**), and suppose we are interested in movies directed by Francis Ford Coppola; relevant documents for the keyword search could be reviews or general comments about the movies on the query, for example. We show some tuples of the query result in Table 1, highlighting non-stopword terms with at least two occurrences.

Table 1. Fragment of the query *Francis Ford Coppola movies*

Title	Plot
the **godfather** (1972)	vito **corleone**, head of the **corleone mafia family**, passes the **family** businesses to his son, michael ...
the **godfather**: part ii (1974)	the continuing saga of the **corleone** crime **family** ...
the **godfather**: part iii (1990)	final instalment of mario puzzo's **mafia** trilogy ...

With these data, we get the terms shown in Table 2, presented in descending order of their score s. We focused only on the highlighted terms of Table 1, showing their term frequencies, element frequencies and the corresponding scores s.

Table 2. Frequencies and score for highlighted terms in Table 1

t	$tf(t)$	$ef(t)$	$s(t)$
godfather	3	3	5.19
corleone	3	2	4.24
family	3	2	4.24
mafia	2	2	2.82

The definite keywords sent to the IRS are shown below, along with their weights according to (7). In this case, we set $k = 3$, $n = 4$ and $\beta = 0.5$, and considered "Francis Ford Coppola movies" as initial user keywords.

---

[3] http://imdb.com

1.0 francis 1.0 ford 1.0 coppola 1.0 movies 0.5 godfather
0.4 corleone 0.4 family 0.27 mafia

Though this is just a toy example, we wanted to stress the main idea of
DBFIRe, rewarding terms that appear in more than one row or column of the
query result. For example, despite "godfather" has the same term frequency than
"corleone", its score is higher due to a greater element frequency.

We can also see the effect of smoothing through the square root of $ef(t)$:
without smoothing, the score of "godfather" would be 50% higher than the
score of "corleone"; using smoothing, this difference goes around 22%. Other
smoothing functions could be used, but the square root gave the more "behaved"
results.[4]

4   Experiments

We judged the quality of DBFIRe regarding the relevance of the documents
returned. This is equivalent to the evaluation of an IRS, in which we usually use
test collections [21]: a set of documents, a set of topics, and pairs of relevance
assessments, determining which document is relevant for which topic. We used
Indri[5] as IRS and MAP (Mean Average Precision) [21] as quality indicator. We
simulated the user keywords through the *<title>* fields of each collection topic,
which represents our baseline, that is, a keyword search without expansion.

However, we have a strict requirement: the test collection must allow the usage
of a related database. Therefore, we focused on two test collections created under
INEX (INitiative for the Evaluation of XML retrieval) workshops [22]: the 2011
Data Centric [23] and the 2013 Linked Data [24], using the ad-hoc search task
of each one.

4.1   Environment Setup for Data Centric Collection

The collection is composed of XML documents derived from IMDB plain files.
There is no direct associated database, but it is possible to build a database
processing the XML files. In our case, XML markups identifying entities such as
movies, actors or locations became database tables whose data is filled through
the attribute/value pairs in each XML file. In Figure 2 we see a fragment of an
XML file converted into database tables.

In order to avoid a bias due to a database created as an exact copy of the data
in the document collection, the actual collection was appended with "dummy"
files, available in the TREC 2005 Robust Track test collection [25].

After loading the data, we created SQL queries for each collection topic, according to the each topic's description. Queries and database schema are available online.[6]

---

[4] The same example ("godfather" versus "corleone") smoothed through $log(ef(t))$
would give a score more than 50% higher to "godfather".

[5] http://www.lemurproject.org/indri

[6] https://sites.google.com/a/copin.ufcg.edu.br/sofsem-2015/

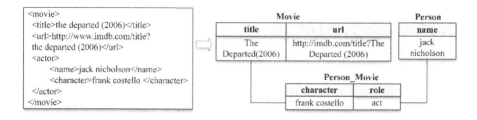

Fig. 2. Loading an XML file into the database (Data Centric collection)

4.2   Environment Setup for Linked Data Collection

In this case, documents are composed of Wikipedia articles, while the structured content is provided by YAGO [26] and DBpedia [27] ontologies. The queries were run over both ontologies, through a language similar to SQL: SPARQL.[7]

For the first 50 Jeopardy topics of the collection [24], we created SPARQL queries, returning triples in the form <resource, property, value>: a resource identifies an article on Wikipedia, a property corresponds to an attribute of that resource, and the last component is the value of the property for the resource. Resembling a database query result, we converted the triples into tuples, with resources and values as tuple elements, and properties naming tuple columns; Figure 3 shows an example of how we made this conversion. As with the Data Centric collection, all queries we created are available online.[6]

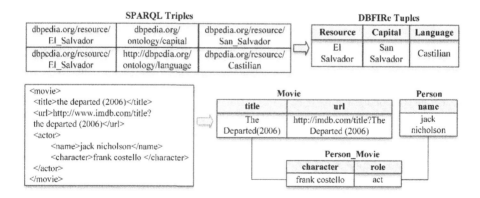

Fig. 3. Converting SPARQL triples into DBFIRe tuples (Linked Data collection)

---

[7] http://www.w3.org/TR/sparql11-query

4.3 Results

The benchmarks for each test collection investigated the three hypothesis stated in the beginning of the paper: we compared DBFIRe with the baseline (a keyword search without expansion), with other DBMS/IRS integration methods (SCORE and SEMEX), as well as with different expansion methods (KLD, DFR and RM). We also wanted to verify whether our term ranking could be better than ranking terms through their term frequencies only, a method we called TF.

All expansion methods (KLD, DFR, RM and TF) used the same expansion framework of DBFIRe: tuples as expansion corpus, and the same values for k, n and β (we set $\beta = 0.5$ in all experiments).

The benchmarks in Data Centric collection and Linked Data collection appear in Table 3 and Table 4, respectively. The tables also show the relative differences of each method compared to DBFIRe. We tested the methods for different combinations of parameters k and n, with most results verified at the 0.05 level through Wilcoxon signed rank tests [21]. However, we had three cases where the differences were verified only at the 0.1 level. These cases appear shadowed in Table 3 and Table 4.

Table 3. Comparisons for Data Centric test collection

k and n	DBFIRe	BASE LINE	SCORE	SEMEX	KLD	DFR	RM	TF
$k=10$, $n=10$	0.3702	0.3119 18.6%	0.1871 97.8%	0.2125 74.2%	0.3199 15.7%	0.3168 16.8%	0.3230 14.6%	0.3378 9.6%
$k=20$, $n=10$	0.3599	0.3119 15.3%	0.1871 92.3%	0.2125 69.3%	0.3182 13.1%	0.3126 15.1%	0.3268 10.1%	0.3403 5.7%
$k=10$, $n=20$	0.3483	0.3119 11.6%	0.1871 86.1%	0.2125 63.9%	0.3324 4.7%	0.3297 5.6%	0.3104 12.2%	0.3343 4.1%
$k=20$, $n=20$	0.3533	0.3119 13.2%	0.1871 88.8%	0.2125 66.3%	0.3334 5.9%	0.3275 7.8%	0.3163 11.7%	0.3337 5.8%

Table 4. Comparisons for Linked Data test collection

k and n	DBFIRe	BASE LINE	SCORE	SEMEX	KLD	DFR	RM	TF
$k=10$, $n=10$	0.3229	0.2923 10.4%	0.1286 >100%	0.1946 65.9%	0.3091 4.4%	0.3029 6.6%	0.3041 6.2%	0.3169 1.9%
$k=20$, $n=10$	0.3233	0.2923 10.6%	0.1286 >100%	0.1946 66.1%	0.3082 4.8%	0.2999 7.8%	0.3055 5.8%	0.3191 1.3%
$k=10$, $n=20$	0.3203	0.2923 9.5%	0.1286 >100%	0.1946 64.5%	0.3020 6.0%	0.2904 10.2%	0.3023 5.9%	0.3114 2.7%
$k=20$, $n=20$	0.3225	0.2923 10.3%	0.1286 >100%	0.1946 65.7%	0.3005 7.2%	0.2884 11.8%	0.3048 5.7%	0.3139 2.7%

4.4 Discussion

The first conclusion from the benchmarks is that DBFIRe is very effective when considering the baseline and other integration methods (SCORE and SEMEX).

Regarding SCORE, the usage of the whole content of tuple elements increases the chance of the query drift effect, once tuple elements may contain both "good" and "bad" terms; besides, SCORE neglects cues for the information need (such as the user keywords, for example). This explains why SCORE had the worst performance among all methods.

With respect to SEMEX, its approximation of the user information need is not enough to good retrieval results: it does show some improvement compared to SCORE, but its performance is way below the baseline, for example.

DBFIRe also significantly beats the baseline. That is, expanding user keywords with individual terms of the query result is much better than using the keywords alone. However, why not just *any* term-ranking method?

When comparing with traditional ranking methods (KLD, DFR and RM), we had significant differences in most tests among all different combinations of parameters k and n; only two of them in the Linked Data collection had a p-value above the usual "standard" 0.05 level. However, none of these methods was able to beat the baseline: in both test collections, the differences were not statistically significant, with p-values above 0.2 in all cases. Once all these methods use some heuristic for term frequencies over the whole collection, this result reinforces the idea of expanding the user keywords based in term distributions exclusively over the query result.

Nevertheless, which measure of term diffusion should we use? Ranking terms based only on their term frequencies does give good results, as we see by the numbers of method TF. However, using it together with a second measure of term spread (the tuple element frequency, used by DBFIRe) always presents a better result, though we had only slight differences in the Linked Data collection.

5   Conclusions and Future Work

The paper presented a method for retrieving documents associated to database queries. The method contains two innovations: the use of database query results as reliable sources of additional keywords for in information retrieval, and a new term-ranking method, which takes advantage of the diffusion of terms in database query results. These innovations can be a useful resource to DBMS/IRS integration inside organizations. Furthermore, it is easy to implement, either at application level or at DBMS level.

Still, some improvements are possible. One of them is to avoid the manual tuning of the weight of expansion terms for the IRS keyword search: we could automatically adjust DBFIRe tuning parameter, using the frequencies in the query result of the initial user keywords. The more user keywords in the query result, the more it should be "trusted", using a higher value for the tuning parameter.

Similarly, the amount of user keywords on the query result could also rank the tuples to be used for expansion. In a DBMS view, all tuples are equally relevant, but some tuples may present more useful terms for expansion than others may. Ranking tuples based on the diffusion of the user keywords, could provide a more useful expansion corpus.

References

1. Weikum, G.: DB & IR: both sides now. In: Proceedings of the 2007 ACM SIGMOD International Conference on Management of Data - SIGMOD 2007, pp. 25–30. ACM, New York (2007)
2. Chaudhuri, S., Ramakrishnan, R., Weikum, G.: Integrating DB and IR technologies: what is the sound of one hand clapping? In: Proceedings of the Second Biennial Conference on Innovative Data Systems Research - CIDR 2005, pp. 1–12. VLDB Foundation (2005)
3. Halevy, A., Franklin, M., Maier, D.: Principles of dataspace systems. In: Proceedings of the twenty-fifth ACM SIGMOD-SIGACT-SIGART Symposium on Principles of Database Systems - PODS 2006, pp. 1–9. ACM, New York (2006)
4. Mirza, H.T., Chen, L., Chen, G.: Practicability of dataspace systems. International Journal of Digital Content Technology and its Applications 4(3), 233–243 (2010)
5. Cafarella, M.J., Christopher, R., Suciu, D., Etzioni, O., Banko, M.: Structured querying of Web text: a technical challenge. In: Proceedings of the 3rd Biennial Conference on Innovative Data Systems Research - CIDR 2007, pp. 225–234. VLDB Foundation (2007)
6. Jain, A., Doan, A., Gravano, L.: SQL queries over unstructured text databases. In: Proceedings of the 23rd IEEE International Conference on Data Engineering - ICDE 2007, pp. 1255–1257. IEEE Computer Society, Washington-DC (2007)
7. Yu, J.X., Qin, L., Chang, L.: Keyword search in relational databases: a survey. IEEE Data Eng. Bull. 33(1), 67–78 (2010)
8. Luk, R.W.P., Leong, H.V., Dillon, T.S., Chan, A.T.S., Croft, W.B.: A survey in indexing and searching XML documents. J. Assoc. Inf. Sci. Technol. 53(6), 415–437 (2002)
9. Carpineto, C., Romano, G.: A survey of automatic query expansion in information retrieval. ACM Comp. Surv. 44(1), 1–50 (2012)
10. Liu, J., Dong, X., Halevy, A.: Answering structured queries on unstructured data. In: Proceedings of the 9th International Workshop on the Web and Databases - WebDB 2006, Chicago, USA, pp. 25–30 (2006)
11. Roy, P., Mohania, M., Bamba, B., Raman, S.: Towards automatic association of relevant unstructured content with structured query results. In: Proceedings of the fourteenth ACM Conference on Information and Knowledge Management - CIKM 2005, pp. 405–412. ACM, New York (2005)
12. Amati, G., Rijsbergen, C.J.V.: Probabilistic models of information retrieval based on measuring the divergence from randomness. ACM Trans. on Inf. Syst. 20(4), 357–389 (2002)
13. Carpineto, C., Mori, R., Romano, G., Bigi, B.: An information-theoretic approach to automatic query expansion. ACM Trans. on Inf. Syst. 19(1), 1–27 (2001)
14. Lavrenko, V., Croft, W.B.: Relevance-based language models. In: Proceedings of the 24th Annual International ACM SIGIR Conference on Research and Development in Information Retrieval - SIGIR 2001, pp. 120–127. ACM, New York (2001)
15. Roy, P., Mohania, M.: SCORE: symbiotic context oriented information retrieval. In: Pro-ceedings of the Joint 9th Asia-Pacific Web and 8th International Conference on Web-Age Information Management Conference on Advances in Data and Web Management - AP-Web/WAIM 2007, Huang Shan, China, pp. 30–38 (2007)
16. Dong, X.L., Halevy, A.: A platform for personal information management and integration. In: Proceedings of the Second Biennial Conference on Innovative Data Systems Research - CIDR 2005, pp. 119–130. VLDB Foundation (2005)

17. Lavrenko, V., Allan, J.: Real-time query expansion in relevance models. Internal Report 473, Center for Intelligent Information Retrieval - CIIR, University of Massachusetts (2006)
18. Fox, C.: Lexical analysis and stoplists. In: Frakes, W.B., Baeza-Yates, R. (eds.) Information Retrieval: Data Structures and Algorithms, pp. 102–130. Prentice Hall, USA (1992)
19. Mitra, M., Singhal, A., Buckley, C.: Improving automatic query expansion. In: Proceedings of the 21st Annual International ACM SIGIR Conference on Research and Development in Information Retrieval, SIGIR 1998, pp. 206–214. ACM, New York (1998)
20. Rocchio, J.J.: Relevance feedback in information retrieval. In: Salton, G. (ed.) SMART Retrieval System - Experiments in Automatic Document Processing, pp. 313–323. Prentice Hall, USA (1971)
21. Sanderson, M.: Test collection based evaluation of information retrieval systems. Found. Trend. Inf. Ret. 4(4), 247–375 (2010)
22. Lalmas, M., Tombros, A.: INEX 2002 - 2006: Understanding XML retrieval evaluation. In: Thanos, C., Borri, F., Candela, L. (eds.) Digital Libraries: Research and Development. LNCS, vol. 4877, pp. 187–196. Springer, Heidelberg (2007)
23. Wang, Q., Ramírez, G., Marx, M., Theobald, M., Kamps, J.: Overview of the INEX 2011 Data-Centric track. In: Geva, S., Kamps, J., Schenkel, R. (eds.) INEX 2011. LNCS, vol. 7424, pp. 118–137. Springer, Heidelberg (2012)
24. Bellot, P., et al.: Overview of INEX 2013. In: Forner, P., Müller, H., Paredes, R., Rosso, P., Stein, B. (eds.) CLEF 2013. LNCS, vol. 8138, pp. 269–281. Springer, Heidelberg (2013)
25. Voorhees, E.M.: The TREC Robust Retrieval Track. SIGIR Forum 39(1), 11–20 (2005)
26. Hoffart, J., Suchanek, F.M., Berberich, K., Weikum, G.: YAGO2: A spatially and temporally enhanced knowledge base from Wikipedia. Artif. Intell. 194, 28–61 (2013)
27. Lehmann, J., Isele, R., Jakob, M., Jentzsch, A., Kontokostas, D.: DBpedia – a large-scale, multilingual knowledge base extracted from Wikipedia. Semant. Web J. (in press)

Advantages of Dependency Parsing
for Free Word Order Natural Languages

Seyed Amin Mirlohi Falavarjani and Gholamreza Ghassem-Sani

Sharif University of Technology, Department of Computer Engineering,
Tehran, Iran
aminmir@ce.sharif.edu, sani@sharif.edu

Abstract. An important reason to prefer dependency parsing over
classical phrased based methods, especially for languages such as Persian, with the property of being "free word order", is that this particular property has a negative impact on the accuracy of conventional
parsing methods. In Persian, some words such as adverbs can freely be
moved within a sentence without affecting its correctness or meaning. In
this paper, we illustrate the robustness of dependency parsing against
this particular problem by training two well-known dependency parsers,
namely *MST Parser* and *Malt Parser*, using a Persian dependency corpus
called Dadegan. We divided the corpus into two separate parts including
only projective sentences and only non-projective sentences, which are
corelated with the free word order property. As our results show, *MST
Parsing* is not only more tolerant than *Malt Parsing* against the free
word order problem, but it is also in general a more accurate technique.

Keywords: data mining, knowledge discovery and machine learning,
knowledge modeling and processing.

1   Introduction

Working on dependency based grammars has been an interesting research avenue in the recent decade, as a suitable alternative for phrase based grammars
of languages such as Persian and Czech, which are regarded as free word order
languages. A phrase structure parse tree represents hierarchal relations between
words in a constituent. However, a dependency parse tree represents relations between individual words. Dependency parsing has become widely used in different
applications such as temporal relation extraction and machine translation [7].

Persian is an Indo-European language with a writing system that is based on
the Arabic script. In contrast with English, in which the order of main constituents
is SVO (Subject-Verb-Object), Persian is mainly a SOV language. Besides, in Persian, both subjects and/or objects may be omitted. Furthermore, some constituents
such as adverbs can be placed anywhere in the sentence. That is why Persian is regarded as a Free Word Order Language. This phenomenon allows different phrase
structures for the same sentence. As a result of this divergence, it may be difficult
for a phrase-based parser to learn the structure of Persian sentences. As we later

G.F. Italiano et al. (Eds.): SOFSEM 2015, LNCS 8939, pp. 511–518, 2015.

show, the dependency parsing is more suitable to overcome this problem. Figure 1 shows the phrase structure of sentence "the basket is on the table". Note that Persian sentences are written from right to left.

Sentence			
Verb Phrase			
	Preposition Phrase		
Verb	Noun	Preposition	Noun
است .	میز	روی	سبد
is	table	on	basket

Fig. 1. A phrase structure for "The basket is on the table" in Persian

Dependency parse trees can be divided into projective and non-projective trees. A projective dependency tree is shown in Figure 2. In the projective trees the edges do not cross each other and a word and its dependents can form a substring of the sentence, but in non-projective trees, there are crossing edges. A non-projective dependency tree is shown in Figure 3. In English, projective trees can cover most sentences, but in languages like Persian non-projective trees are more frequent.

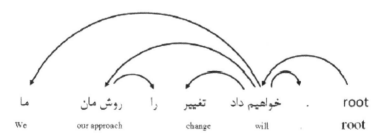

Fig. 2. A projective dependency tree for "We will change our approach" in Persian

In this paper, we present the evaluations of two of most successful dependency parsing methods, namely *Malt Parser* and *MST Parser* for Persian. We also show that one of these methods is more suitable for projective sentences and the other is more suitable for non-projective sentences, which correspond to the free word order property.

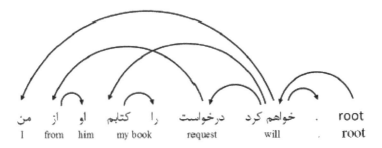

Fig. 3. A non-projective dependency tree for "I will request my book from him" in Persian

The paper is structured as follows. Section 2 briefly describes the background of parsing techniques. In Section 3, there is a description of *Malt Parser* and *MST Parser* followed by introducing a Persian dependency Treebank. Section 4 presents the results of our experiment. Finally, Section 5 includes our conclusions.

2   Background

There have been several researches on phrase based approaches for grammar induction. One of most successful grammar induction algorithms is the constituent context model (CCM) [4]. CCM is a distributional based method, which is based on a simple idea: the sequences of words or tags that construct the same constituents appear in analogous contexts. A constituent is a group of words or tags that constructs a single unit within a pars tree. A context is a pair of words that surround a constituent [9]. Figure 4 shows an example of a constituent and its context. Parent based CCM (PCCM) is a history based method, which improved CCM especially for Persian. PCCM employs the parent's information of each context and constituent to avoid divergence in the likelihood space [8].

There are compelling reasons to switch our attention from phrase structure grammar to dependency grammar. Persian is a so-called free word order natural language where constituents can be moved freely around each clause. These scrambled clauses are the most important reason to use a dependency grammar. The methods such as CCM that learn from frequency of constituents and contexts patterns have a lower accuracy in free word order languages such as Persian.

On the other hand, dependency grammars represent information by a number of head-dependent arcs. There is also an extra node called root, with no head. In a complete sentence, the governing verb is a dependent of this root node, and other words are directly or indirectly dependents of the governing verb. A dependency grammar is more robust against word movements within the sentences due to its word to word interactions. The parsers of dependency relations can be divided into two groups: 1) graph based parsers, and 2) transition based parsers.

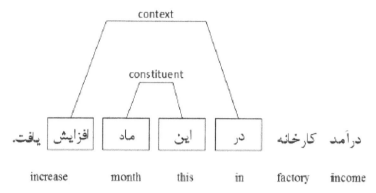

Fig. 4. A constituent and a context for "The factory income was increased last month" in Persian

In graph based approaches, every graph has a score and the graph with the highest score is equivalent to a maximum spanning tree. *MST Parser* is a graph based parser. On the other hand transition based approaches are based on certain transition systems. A transition system consists of some configurations and certain transitions between these configurations. *Malt Parser* is a transition based parser. *MST Parser* and *Malt Parser* are explained in more details in the next section.

3   Experimental Frameworks

In this section, we briefly describe the so-called *Malt Parsing* and *MST Parsing* methods. We also introduce the dependency Treebank that has been used for doing experiments on the mentioned parsing methods.

3.1   Dependency Parsing

Two of most successful dependency parsing methods are the so-called *Malt Parsing* and *MST Parsing* [11], [7]. *Malt Parsing* is a multilingual data-driven parsing method. This parsing method does not require any grammar and one can develop a parser for any new language with a Treebank consisting of dependency relations between words. The parsing method relies on a transition system that consists of a number of configurations and transitions between these configurations. The transition system is based on a simple idea: beginning from an initial configuration and after a number of valid transitions, the system obtains a dependency tree.

Malt Parser is based on three components:

- The parsing algorithm
- A Feature model
- A machine learning approach

Any transition-based parsing algorithm such as *Malt Parsing* needs to work with the following structures:

- A stack consisting of partially processed tokens.
- An input buffer consisting of remaining input tokens.
- An arc list consisting of previously recognized arcs [11].

The parser constructs the dependency structure of the input sentence incrementally by checking the existence of a dependency relation between the top of its stack and the first token of the remaining input buffer. The feature model contains some lexical features, part-of-speech tags, and some dependency based features. The parser also requires a machine learning algorithm to get a mapping from the feature model to one of the actions: *left-arc*, *right-arc*, *reduce*, and *shift*. *Malt Parser* uses a support vector machine algorithm to predict appropriate action in each configuration [2].

Another successful dependency parsing method is the so-called *MST Parsing*. *MST Parsing* is a graph-based approach. The parser extracts the spanning tree with the maximum score out of a complete graph. In this graph, the score of each arc is calculated by the dot product of a feature vector extracted from the input sentence and a weight vector obtained during the training phase. The complete graph contains all possible dependency arcs. The score of a dependency tree is the sum of scores of all the dependency arcs of the tree. The maximum spanning tree (*MST*) is a tree that maximizes the sum of all arc scores and contains all vertices. *MST Parser* searches the whole space of all spanning trees using the so-called Chu-Liu-Edmonds algorithm [7]. The main advantage of graph-based approaches is their simplicity.

3.2   A Persian Treebank

Dadegan is a public domain Persian dependency Treebank consisting of 29,982 sentences manually annotated with syntactic roles and morpho-syntatctic features. One valuable feature of this Treebank is that there are approximately

Table 1. Statistics about the persian dependency treebank

Number of Sentences	29,982
Number of Words	498,081
Number of Distinct Words	37,618
Percent of Sentences with Length 2 to 15	38
Percent of sentences with length higher than 15	62
Number of Sentences with Internal Punctuation	7341
Number of Sentences with Internal Verb	14830
Number of Sentences with Internal Punctuation and Verb contemporary	5677

5000 distinct verb lemmas in its sentences. The corpus is annotated with part of speech tags, person, number, and the tense-mood-aspect of words of sentences in the CoNLL-format. In this Treebank, 39.24% of the words are tagged as nouns, 12.62% as verbs, 11.64% as prepositions, and 7.39% as adjectives. Table 1 shows some statistics about the Dadegan Treebank [12].

There are two representations for objects accompanied by the accusative case maker. In the first representation, the accusative case maker is the head and in the other representation, it is the dependent of the direct object of the main verb of the sentence. In our experiments, we have used the first representation.

4   Persian Parsing

Our main goal of focusing our attention on dependency parsing for Persian is that Persian has the property of being "free word order". This particular property has a negative impact on the accuracy of conventional parsing methods [3,9]. In our experiments, we illustrate the extent of the data-driven dependency parsers success to overcome this problem. For this goal, we divided the Dadegan corpus into two separate parts including only projective sentences and only non-projective sentences, which are correlated with the free word order property. Then, we trained the parsers on each part of the Treebank. We used 90% of the data for training and the rest for test. Our features were gold standard POS tags taken from the Dadegan Treebank. We used the same data and features for both parsers.

To apply *Malt Parsing* to Persian, we selected the Arc-eager algorithm [10]. Due to the using of the *Reduce* action and not deleting the dependent in the *Right-arc* action, this algorithm seems to be more suitable than other methods for free word order languages. This algorithm can select four actions in every configuration. *Left-arc* and *Right-arc* actions add a dependency arc between the word on the top of stack and the first word of the input buffer. *Reduce* pops the stack and *shift* removes the first word of the input buffer and pushes it onto the stack. Arc-eager algorithm is a linear-time algorithm [11].

To apply *MST Parsing*, we used the non-projective and the order2 settings. The non-projective setting allows the parser to select non-projective arcs in addition to projective ones. With the order2 settings, the parser can find the interactions between three words in addition to the interactions between two words. This can increase the accuracy of the parser for free word order languages.

The accuracy obtained for projective sentences by the *Malt Parser* is 87% and by the *MST Parser* is 84%. Therefore, the *Malt Parser* is more accurate than the *MST Parser* on the sentences without any non-projective arcs. The accuracy obtained for non-projective sentences by the *Malt Parser* is 73% and by the *MST Parser* is 77%. Thus, the *MST Parser* is more accurate than the *Malt Parser* on the sentences with non-projective arcs, because the *Malt Parser* finds relations between the words on the top of the stack and the first word of the input buffer, which are sequential words in most cases. Therefore, when dependents of a word are not sequential and there is a space between them, the accuracy of *Malt Parser* will be reduced. Although Malt Parsing has obtained a higher accuracy on a Persian dependency Treebank called Uppsala [13], the accuracy of MST Parsing has been higher on Dadegan, because there are much more non-projective sentences in Dadegan. The results are given in Table 2.

Table 2. Accuracy of parsers on projective and non-projective sentences

	All Sentences	ProjectiveSentences	Non-projective Sentences
MST Parser	81	84	77
Malt Parser	78	87	73

5   Conclusions

In this paper, we examined the ability of the dependency parsers to overcome the free word order problem in Persian. We used the *MST Parser* as a graph based parser and the *Malt Parser* as a transition based one. Our experiments showed that the *Malt Parser* is more accurate than the *MST Parser* for sentences without any non-projective arcs, and reversely the *MST Parser* is more accurate for sentences with non-projective arcs. Thus, we can conclude that the *Malt Parsing* is a more suitable approach for languages like English with less non-projective sentences and *MST Parsing* is more appropriate for languages with the free word order property. One possible future work to improve the performance of dependency parsers can be the usage of semantic information. Semantic information can help the parser to find the correct relation between words.

References

1. Agirre, E., Kepa, B., Koldo, G., Nivre, J.: Improving dependency parsing with semantic classes. In: Association for Computational Linguistics: Human Language Technologies: short papers, pp 699–703 (2011)
2. Cortes, C., Vapnik, V.: Support-Vector Networks. Machine Learning, pp. 273–297 (1995)
3. Heshaam, F., Ghassem-Sani, G.: Unsupervised grammar induction using history based approach. In: Computer Speech & Language, pp. 644–658 (2006)
4. Klein, D., Manning, C.D.: Natural language grammar induction using a constituent-context model. In: Advances in Neural Information Processing Systems, pp. 35–42 (2001)
5. Klein, D., Manning, C.D.: Corpus-based induction of syntactic structure: Models of dependency and constituency. In: Association for Computational Linguistics (2004)
6. Koo, T., Carreras, X., Collins, M.: Simple semi-supervised dependency parsing. In: ACL, pp. 595–603 (2008)
7. McDonald, R., Pereira, F., Ribarov, K., Hajic, J.: Non-projective dependency parsing using spanning tree algorithms. In: Human Language Technology and Empirical Methods in Natural Language Processing (HLT/EMNLP), pp. 523–530 (2005)

8. Mesgar, M., Ghasem-Sani, G.: History Based Unsupervised Data Oriented Parsing. In: RANLP, pp. 453–459 (2013)
9. Mirroshandel, S.A., Ghassem-Sani, G.: Unsupervised Grammar Induction Using a Parent Based Constituent Context Model. In: ECAI, pp. 293–297 (2008)
10. Nivre, J.: Dependency grammar and dependency parsing. Technical Report MSI (2005)
11. Nivre, J., Hall, J., Nilsson, J.: Maltparser: A data-driven parser-generator for dependency parsing. In: Language Resources and Evaluation (LREC), Genoa, Italy, pp. 2216–2219 (2006)
12. Rasooli, M.S., Kouhestani, M., Moloodi, A.: Development of a Persian Syntactic Dependency Treebank. In: North American Chapter of the Association for Computational Linguistics: Human Language Technologies, pp. 306–314 (2013)
13. Seraji, M., Beatam, M., Nivre, J.: Dependency parsers for Persian. In: Asian Language Resources, COLING, pp. 35–43 (2012)

Detecting Identical Entities
in the Semantic Web Data

Michal Holub, Ondrej Proksa, and Mária Bieliková

Institute of Informatics and Software Engineering,
Faculty of Informatics and Information Technologies, Slovak University of Technology,
Ilkovičova 2, 842 16, Bratislava, Slovakia
{michal.holub,maria.bielikova}@stuba.sk, ondrej.proksa@gmail.com

Abstract. Large amount of entities published by various sources inevitably introduces inaccuracies, mainly duplicated information. These can even be found within a single dataset. In this paper we propose a method for automatic discovery of identity relationship between two entities (also known as instance matching) in a dataset represented as a graph (e.g. in the Linked Data Cloud). Our method can be used for cleaning existing datasets from duplicates, validating of existing identity relationships between entities within a dataset, or for connecting different datasets using the *owl:sameAs* relationship. Our method is based on the analysis of sub-graphs formed by entities, their properties and existing relationships between them. It can learn a common similarity threshold for particular dataset, so it is adaptable to its different properties. We evaluated our method by conducting several experiments on data from the domains of public administration and digital libraries.

Keywords: duplicates, identity, similarity, relationship, semantic web, owl:sameAs, Linked Data, web of data.

1 Introduction

The Web has shifted from a group of pages into an interconnected network of information processable by automatic agents. Many web pages contain structured data in formats such as XML or RDFa. New methods for extracting structured information from unstructured data emerge [2,16]. Facts about real-world entities are grouped into various datasets and published on the Web. Moreover, datasets are connected with each other using links and shared vocabularies (schemas), thus forming the Linked Data Cloud.

These datasets have the form of graphs with vertices representing entities and values, and edges representing relationships. They capture semantics usable for various adaptation and personalization tasks. Linked Data can help with content-based recommendations, answering near natural language search queries, filtering relevant information, or adapting web pages to specific users.

Linked Data can also be used in detection of similarity between objects, relationship discovery, or semantic enrichment of existing web pages. For all these

G.F. Italiano et al. (Eds.): SOFSEM 2015, LNCS 8939, pp. 519–530, 2015.

tasks we need to have a good representation [7]. The main problem linked with representation is the discovery of similarity and identity between entities [18].

An imminent problem of large datasets is that they usually contain duplicates and it is a challenge to find them. This is also referred to as the data linking problem where the goal is to find equivalent resources on the Semantic Web [5].

Marking entities as duplicates on the Semantic Web is usually done using a *owl:sameAs* relationship. However, a study [6] proved, that *owl:sameAs* is often wrongly used to connect e.g. two people or other objects with identical names, although their other attributes are not identical and they represent different real-world entities. Another problem is that sometimes *owl:sameAs* is deliberately or mistakenly used to connect very similar, although not identical entities.

In this paper we address the problem of identifying if two entities are the same with satisfactory precision. Our main goal is to discover whether two entities in one dataset refer to the identical real-world object, a problem which is also known as *instance matching*. We use graph algorithms and specific rules on various attributes, results of which we combine in a specific way tailored for each dataset in order to determine if two given entities are the same.

In our experiments we focused on finding duplicate people and companies in various datasets. Our main contribution is that the proposed method can automatically adapt to varying properties of datasets.

2   Related Work

Because the Linked Data datasets use various ontologies and schemas to describe their content, there is a problem of ontology diversity, which could cause that identical entities are not connected. Graph algorithms are used to address this problem in [20]. Graph patterns in the form of sub-graphs with identical vertices and edges are defined. At first, the authors integrate two datasets and detect the graph patterns in them. Then, ontology alignment on each of the graph patterns is performed. Finally, it aggregates similar ontology classes and properties.

Detecting duplicates in XML data was examined in [9]. The authors proposed to combine not only the information within elements, but also the information about how the data is structured. Their solution uses Bayesian network to determine the degree of similarity of two entities.

Analyzing sub-graphs of compared entities is also used in the domain of ontology matching problem [3,15]. They usually compare two similar datasets (describing the same domain, very similar entities and relationships but described using different schemas with different names). The prior knowledge that the two compared datasets are very close to each other makes the comparison easier, unless we are comparing one very rich dataset (lots of entities with plethora of attributes, richly connected through various relationships) with very sparse one (only few entities' attributes and basic relationships).

This approach can be used to map more ontologies to each other, and thus discovering identical entities in datasets which use these ontologies. Problem with this approach is that it requires manual verification as the precision is not high enough. This can be difficult and time consuming, especially for large datasets.

Another approach using graph algorithms is described in [8]. The proposed method finds relationships among entities in DBpedia. Its main idea is to split the graph into its components using the breadth-first search. Then, the shortest path between two entities is computed. The disadvantage is not considering the type of the relationship. However, we can use the presence of such relationship as an indicator that the two entities are identical.

When performing the instance matching the usual approach is to 1) quickly scan the whole dataset to find entities for comparison, and 2) measure the similarity between pairs of entities found in the previous step [17,12].

This approach is also used in [1]. The authors propose a method for instance matching using class hierarchy of instances. For selecting candidates of potentially identical entities they use small number of characteristic attributes (e.g. name or title). Then they search for entities which have similar values of these attributes. The challenge here is automatic selection of characteristic attributes, as well as the similarity threshold for the values of attributes.

Selecting potential candidates allows the algorithm to scale on large datasets. Otherwise, we would have to compare each possible pair of entities, resulting in quadratic complexity. On the other hand, if this step is executed with low precision, we may omit some duplicates, so on smaller datasets it is desirable to compare as many pairs of entities as possible.

A suitable similarity measure has to be applied on the attributes of compared entities. Each pair of attributes can be compared using different similarity measure according to the data type, e.g. text or numeric similarity. There can also be more complex measures, comparing e.g. geo data or address records [15]. We need to know the semantics of the attributes to correctly select the metrics.

Current approaches use either attribute values to compare pairs of entities [13], or graph algorithms to compute the similarity based on entity's relationships with its neighbors [11]. Instances can also be matched based on the representation of their metadata [14]. However, various datasets need different approaches according to their properties. In many datasets it is vital to use a combination of approaches with weights tailored for a particular domain.

3 Detecting Similarity between Entities

We propose a method for finding if two given entities are identical. It can compare each pair of entities within a dataset (or possibly more datasets). The method was designed to work on datasets which can be represented using graphs, such as (but not limited to) RDF representation, the basis of the Linked Data Cloud.

Our method is based on a hypothesis that the matching of entities is reflected in the similarity between the sub-graphs composed of classes and properties of the individual entities. For a given pair of entities we compute how similar they are. Then, we determine a common similarity threshold, above which we mark the entities to be identical.

We use matching of features of entities, as well as properties of the sub-graphs they are part of. Our method computes the similarity between entities as

a weighted sum of these four partial parameters: 1) similarity of entities' properties, 2) graph distance between entities, 3) graph distance between neighboring
entities, and 4) similarity of entities' relationships:

$$SGN = \frac{SNP \times w_{SNP} + SNR \times w_{SNR} + ND \cdot w_{ND} + DRN \times w_{DRN}}{w_{SNP} + w_{SNR} + w_{ND} + w_{DRN}} \quad (1)$$

where

SGN - similarity of graph nodes, final similarity between two entities
SNP - similarity of entities' properties
SNR - similarity of relationships between entities
ND - graph distance between entities
DRN - average graph distance between adjacent entities
w_i - weight of a particular component

The resulting values of the similarity, as well as the results produced by each
component, are from the interval $[0, 1]$. Similarity $= 1.0$ means that the two
entities are 100% similar (i.e. identical), whereas similarity $= 0.0$ means that the
two entities are not similar at all. All components of the eq. 1 have associated
weights, which determine the component's contribution to the total similarity.

In some datasets (e.g. in the domain of public administration data) it turns
out that the entities' properties are more important than relationships between
them (the records contain a lot of attributes, but few connections). On the other
hand, in domains such as social networks or digital libraries, there are many
relationships between entities, so the weights of related similarity components
should be higher. In our approach the weights can be automatically trained, so
that our method can adapt itself to a particular problem domain. They can also
be assigned manually by a domain expert.

As we already mentioned, we compute the similarity of two entities A, B from
the interval $SGN_{A,B} \in [0, 1]$. To determine whether these entities are identical
we need to find a threshold of similarity (denoted S) from the same interval.
The threshold S should be determined for each domain differently, as it should
reflect dataset's properties.

3.1   Similarity of Properties between Entities (SNP)

When computing the similarity of properties (attributes) of two entities, we
need to use suitable similarity measure according to the data types used. This is
illustrated in Tab. 1 representing people in a dataset of owners of companies in
Slovakia, which provides example properties used when comparing two entities.
As we can see, not all types of properties can be compared using textual similarity. Example in the first row depicts that sometimes the attribute labelled *name*
contains not only the name of a person (entity A), but also an abbreviation of
related organization (entity B). Row 2 demonstrates that names and addresses
may be mixed together.

Table 1. Comparison of example properties between various pairs of entities

Property label	Entity A	Entity B
Name	Jozef Turanovsky	Jozef Turanovsky - UNIP
Name, Address	Juraj Siroky Strme vrsky 137	Mgr. Juraj Siroky 137 Strmy vrsok 137
Address	Zombova 19 040 23 Kosice - Sidlisko KVP	Lomonosovova 30 040 01 Kosice - Juh
District	Kosice II	Kosice 4

If the entities are in the same dataset (which is the primary concern of this paper) or if their properties are described using the same vocabulary (ontology) we can easily determine which properties to compare. Otherwise, we need to apply some matching algorithm to determine appropriate attributes for comparison. For the similarity of properties between entities A and B we define:

$PR_x = \{prop_1, prop_2, \ldots prop_n\}$ - denotes a set of properties of entity x
$PR_y = \{prop_1, prop_2, \ldots prop_m\}$ - denotes a set of properties of entity y
$PR_{x,y} = PR_x \cap PR_y$ - denotes a set of properties common to both entities x and y

Similarity of properties (sig. SNP) between entities is defined as follows:

$$SNP = \frac{SP_0 \times w_{P_0} + SP_1 \times w_{P_1} + \ldots + SP_k \times w_{P_k}}{w_{P_0} + w_{P_1} + \ldots + w_{P_k}} \qquad (2)$$

where

$SP_0, SP_1, \ldots, SP_n$ - similarity between common properties (text, numeric, etc.)
$w_{P_0}, w_{P_1}, \ldots, w_{P_n}$ - weight of each similarity

Again, we determine the component's importance using weights which we train using machine learning. The eq. 2 reflects the weighted average of the individual similarities between the properties. Sometimes, it is possible that some properties will be ignored (we set their weight to zero). We do not consider properties unique to some entity, only those common to both of the compared entities.

We compare the properties of entities using textual similarity and numerical similarity. Textual similarity of properties is the average between the Levenshtein distance and 3-gram similarity of the compared properties, as these metrics are widely used and their good performance has been confirmed. Numerical similarity is defined the normalized numerical distance, which we compute as follows:

$$ndist(num_A, num_B) = \frac{MAX_{num} - \mid num_A - num_B \mid}{MAX_{num}} \qquad (3)$$

where

$ndis(num_A, num_B)$ - normalized distance between the numbers, $ndis \in [0, 1]$
MAX_{num} - maximum value of the numerical properties in the domain

Table 2. Comparison of names of equal properties between DBpedia and YAGO

Property	Name in DBpedia	Name in YAGO
School	*almaMater*	*school*
Date of birth	*birthDate*	*wasBornOnDate*
Nationality	*nationality*	*nationality*
Place of birth	*birthPlace*	*wasBornIn*
Title	*title*	*title*

Linking Properties from Different Vocabularies

Existing approaches usually consider the fact that the entities are described using the same vocabulary, so the names of properties will match. A problem occurs when the properties are defined using different schemas. Tab. 2 presents a comparison of several names of properties in DBpedia and YAGO datasets. As we can see, some of the names do not match, although they describe the same fact. To solve this problem we also use semantic similarity of properties' names. The semantic similarity between words is not easily computable as in general we do not know exact meaning of the words. Also, we often do not have complete information on their context. We use WordNet in order to determine if two words are in the same synset. Once at least one of these similarities is above the given threshold, we can say that the names represent the same property, so that we can compare the values using previously described metrics.

In order to link the names of the properties we take all the properties of the entity A. Then, we find the most similar property of the entity B according to its name. When the entities come from different datasets, it is not always possible to connect all the properties. Subsequently, we consider only the properties we are able to link when computing the overall similarity of properties (SNP).

3.2   Distance between the Entities (ND)

When calculating the distance between two entities in a graph we use the algorithm of breadth-first search. Our intention is to find the shortest path between the entities. We consider every edge to have the length of 1. The final distance is normalized to the interval $[0, 1]$. Normalization is based on the minimum and maximum distance in the given dataset.

Maximum and minimum distances for the normalization must be set for each dataset separately in the preprocessing phase. We experimented in the domain of public administration data, where the minimum distance between organizations is at least 2. Also, we conducted experiments on data from digital libraries, where the minimum and maximum distances between papers' authors are different (for more details see evaluation).

3.3   Average Distance between Adjacent Entities (DRN)

The average distance between the two entities is defined as the normalized average distance computed from the smallest distances to the neighboring entities. We
compute DRN of entities A and B using their direct neighbors (see Fig. 1).

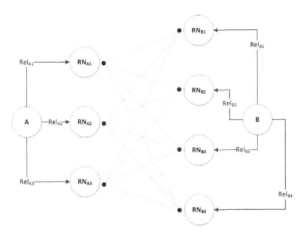

Fig. 1. Relationships nodes for two entities

In Fig. 1 entity A has three neighboring entities - RN_{A1}, RN_{A2} and RN_{A3}.
Entity B has four neighboring entities - RN_{B1}, RN_{B2}, RN_{B3} and RN_{B4}. For
each neighboring entity of A we find the closest neighboring entity of B and use
their distance. The value of DRN between A and B is equal to the average of
these distances. Finally, we normalize the average distance using maximum and
minimum average distances according to eq. 4.

$$DRN(A,\ B) = \frac{MAX_{drn} - avg\_dist(A,\ B)}{MAX_{drn} - MIN_{drn}} \tag{4}$$

$$avg\_dist(A,\ B) = \frac{\sum \forall RN_{Bj} \in RN_B\ :\ min(RN_{Ai},\ RN_{Bj})}{|\ RN_A\ |} \tag{5}$$

where

$DRN(A,\ B)$ - normalized average distance between A and B, $DRN(A\ B) \in [0,1]$
$avg\_dist(A,\ B)$ - average distance between entities A and B
MIN_{drn} - minimum average distance between entities
MAX_{drn} - maximum average distance between entities

3.4   Similarity of Relationships between Entities (SRN)

The last component of the similarity computation is the similarity of entities'
relationships. For both entities A and B we divide their neighboring entities

according to the type of the relationship between them. For each type we express the Jaccard coefficient between sets of adjacent entities of A and adjacent entities of B and we compute the average of the Jaccard coefficients:

$$SNR(A, B) = \frac{j(RNA_{R1}, RNB_{R1}) + \ldots + j(RNA_{Rn}, RNB_{Rn})}{|R|} \qquad (6)$$

where

R - list of the types of relationships for which the neighboring entities of A and B form non-empty sets

$j(R_A R_B)$ - Jaccard coefficient between sets of neighboring entities of A and B

4   Evaluation

We have conducted several experiments with various datasets in order to evaluate specific parts of our method, from finding duplicate organizations to finding namesakes of authors of research papers.

4.1   Finding Identical Organizations

In this experiment we used public data[1] with information about cities, villages, regions and organizations in Slovakia. It contains a lot of duplicates and our goal was to identify identical organizations which were wrongly labelled as diverse, thus helping to clean the dataset.

Throughout the time, organisations may change their names, cease to exist and be started again elsewhere, etc. However, each organization has assigned its unique ID number, which persists event if the organization changes some of its properties (e.g. name, owner). We used this ID for evaluation. For each organization X we found two other organizations which we used to compute the similarities: 1) one identical organization A with the same ID number, and 2) one random organization B that has a different ID number.

We expected A to be the most similar organization to X and B to be the least similar, according to our method. Note that have not used the unique ID when computing the similarity, we only used it for selecting the entities and evaluating the performance of our method.

For setting up the weights of components of our method we used machine learning, particularly support vector machines. We trained on the 80% and tested on 20% of the data. As a threshold (S) for two entities to be identical we used the value of $S = 0.5$, which we set manually based on our observations of the available data. This value could also be trained if enough test data was present. We used precision, recall, accuracy and F1 score as standard metrics of performance.

The dataset used for this experiment was composed by a graph of 13,820 vertices, they had 119,484 properties and were connected using 131,683 relationships. We compared the names and addresses of organizations. As a baseline we used textual similarity computed using Levenshtein distance and 3-grams.

---

[1] http://naseobce.sk

The results are summarized in Tab. 3. We used two setups: the first one involved computing the SGN value using all similarity components, in the second one we only computed the value of SNP (the similarity of properties). We have performed the training and testing on 5,000 and 10,000 examples, respectively.

Using only the similarity of properties resulted in the best overall F1 score because the entities' properties are the most dominant in this dataset. Including other components has worsen the results. The baseline method was better when finding identical organizations, but it also produced more false positives.

Table 3. Results of experiment with identical organization identification

Method	Size	Presicion	Recall	Accuracy	F1 score
SNP	5,000	0.98981	0.98812	0.99014	0.99012
SNP	10,000	0.99240	0.99275	0.99257	0.99257
SGN	5,000	0.95532	0.82688	0.89409	0.88647
SGN	10,000	0.96147	0.93057	0.94663	0.94576
3gramLev	5,000	1.00000	0.22811	0.61405	0.37148
3gramLev	10,000	1.00000	0.19275	0.59637	0.32320

4.2 Finding Namesakes of the Authors in DBLP

In the second experiment we used data from DBLP digital library [10]. In this dataset the authors are not represented as separate entities (e.g. using unique identificators), but only using their full names. No other information is available. It is not clear when two records represent the same person or they are namesakes. In this experiment we used our method to identify identical and diverse authors.

For each author we found the number of his namesakes using these steps:

1. Set the threshold to $S = 0.5$, a value we manually chose for this experiment.
2. For all papers transform all its authors to new temporary authors. Each temporal author was assigned to only one paper he wrote.
3. For each temporary author find an identical temporary author using our method. Merge found authors to one entity. Repeat until no further temporary author could be merged.

In order to evaluate the results we created an application which showed authors and their papers. We asked 5 researchers to combine the authors with the same name into one entity if all records represented one physical person. Thus, we obtained a golden standard for evaluation of our method. The manually processed dataset contained information about 100 distinct authors from three research communities, each author had his associated papers. The researchers were from the same communities and thus were able to correctly assess the authors. The number of papers varied from 3 to 20 papers per author.

For each of the 100 authors we found his namesakes analyzing the whole DBLP dataset (containing more than 1,300,000 author records). We then evaluated the precision of our method based on the number of namesakes it found compared with the expected number of namesakes. We normalized these values to the interval $[0, 1]$ and we calculated the weighted average, where the weights are the numbers of publications for all namesakes of the author. Normalized precision for one author was calculated using the normalization interval, where the minimum value was the original number of namesakes, and the maximum value was the total number of papers (because one author can be divided into at most as many namesakes as the number of papers he wrote). This way we obtained the weighted mean precision of 96.35%.

We compared our results with a method based on detecting communities of authors according to co-authorship of research papers [19]. It assumes that researchers who authored papers together are always the same people. However, author and his namesake would probably not have a common colleague, so their publications and co-authors will form two separate clusters. Our method outperformed this baseline by 12%.

We also compared the number of namesakes our method found with the number from the golden standard. We derived the precision as the ratio of correctly computed counts of namesakes to the number of all authors. Here, our result was 80.2%, whereas the baseline method achieved only 66.34%.

It turns out that considering only co-authors and shared publications is not always the best indicator for namesake detection. Our method achieved better results because it combines graph patterns better representing the real situation.

5   Conclusion

We proposed a method for identifying duplicates in existing data and finding identity relations between entities. It can be used to clean a dataset as well as to verify existing relationships denoting duplicate entities (e.g. using *owl:sameAs* link). Since we are able to process any graph data, our method can also be applied on the Linked Data datasets.

Our method is proposed as universal, its main contribution is combining similarity of attributes with the similarity of sub-graph neighbourhoods of compared entities. This can be helpful when finding e.g. namesakes of a person according to various other objects he is linked to. The method can automatically adapt its components (using weights) to reflect the properties of a particular dataset.

One important step in our method is setting the similarity threshold (S) which influences the accuracy. When the threshold is set too low it will result in high recall but low precision. Otherwise, setting the threshold too high will produce high precision but very low recall.

We analyzed this on a dataset from the project Annota [4], which is a social bookmarking and annotation tool. It is primarily aimed at bookmarking research papers available on the Web (with special support of metadata extraction from digital libraries such as ACM DL[2]). Annota's RDF dataset contains over 71,000 research papers, 390,000 authors, 9,000 publications and 6,400 publishers.

---

[2] http://dl.acm.org

For each author we randomly chose 5 other authors thus forming candidate pairs for duplicates. For each pair we calculated their similarity using our method. We observed that when setting the threshold $S < 0.5$ we found duplicates in around 70% of the cases. When the threshold was $S = 0.5$ the share of duplicates fell to around 32% and setting the threshold to $S = 0.6$ resulted in only around 2% of duplicates. Because for each author we selected very few candidates compared to all authors in the dataset, the probability of choosing real duplicate was insignificantly small.

This feature can be used to automatically adjust the threshold for an unknown dataset: randomly pair each entity with a small set of other entities, then calculate the similarity between each pair of entities. Start with a low similarity threshold and increment it iteratively until the number of duplicates found is close to zero. For very large datasets this process should be performed only on their subsets for performance sake.

The applications of our method are multiple: 1) it can be used to find duplicates in an existing dataset in order to clean the data, 2) it can be used to verify the existing identity relations, and 3) it can be used in the process of creating a new dataset so that we do not include the same entities twice.

Our method could also be used to connect a newly created dataset to the Linked Data cloud, which is vital for enriching the Semantic Web. Entities marked as identical can be connected using *owl:sameAs* link. In the future work we plan to evaluate this in experiments by linking Annota dataset to existing RDF datasets available on the Web.

Acknowledgement. This work was partially supported by grants No. VG 1/0646/15 and APVV-0208-10.

References

1. Araujo, S., Tran, D.T., de Vries, A.P., Schwabe, D.: SERIMI: Class-based Disambiguation for Effective Instance Matching over Heterogeneous Web Data. In: Proc. of 15th Int. Workshop on the Web and Databases, WebDB 2012, pp. 25–30 (2012)
2. Auer, S., Bizer, C., Kobilarov, G., Lehmann, J., Cyganiak, R., Ives, Z.: DBpedia: A Nucleus for a Web of Open Data. In: Aberer, K., et al. (eds.) ASWC 2007 and ISWC 2007. LNCS, vol. 4825, pp. 722–735. Springer, Heidelberg (2007)
3. Aumueller, D., Do, H., Massmann, S., Rahm, E.: Schema and Ontology Matching with COMA++. In: Proc. of 2005 ACM SIGMOD Int. Conf. on Management of Data, pp. 906–908. ACM Press (2005)
4. Holub, M., Móro, R., Ševcech, J., Lipták, M., Bieliková, M.: Annota: Towards Enriching Scientific Publications with Semantics and User Annotations. D-Lib Magazine, Vol. 20, No. 11/12 (2014)
5. Ferrara, A., Nikolov, A., Scharffe, F.: Data Linking for the Semantic Web. Int. Journal on Semantic Web and Information Systems 7(3), 46–76 (2011)
6. Halpin, H., Hayes, P.J., McCusker, J.P., McGuinness, D.L., Thompson, H.S.: When owl:sameAs Isn't the Same: An Analysis of Identity in Linked Data. In: Patel-Schneider, P.F., Pan, Y., Hitzler, P., Mika, P., Zhang, L., Pan, J.Z., Horrocks, I., Glimm, B. (eds.) ISWC 2010, Part I. LNCS, vol. 6496, pp. 305–320. Springer, Heidelberg (2010)

7. Harth, A., Hose, K., Schenkel, R.: Database Techniques for Linked Data Management. In: Proc. of 2012 ACM SIGMOD Int. Conf. on Management of Data, pp. 597–600. ACM Press (2012)
8. Lehmann, J., Schüppel, J., Auer, S.: Discovering Unknown Connections - the DBpedia Relationship Finder. In: Proc. of 1st Conf. on Social Semantic Web, CSSW, vol. 113, pp. 99–110 (2007)
9. Leitão, L., Calado, P., Herschel, M.: Efficient and Effective Duplicate Detection in Hierarchical Data. IEEE Trans. on Knowledge and Data Engineering 25(5), 1028–1041 (2013)
10. Ley, M.: The DBLP Computer Science Bibliography: Evolution, Research Issues, Perspectives. In: Laender, A.H.F., Oliveira, A.L. (eds.) SPIRE 2002. LNCS, vol. 2476, pp. 1–10. Springer, Heidelberg (2002)
11. Melnik, S., Garcia-Molina, H., Rahm, E.: Similarity Flooding: A Versatile Graph Matching Algorithm and its Application to Schema Matching. In: Proc. of 18th Int. Conf. on Data Engineering, pp. 117–128. IEEE CS (2002)
12. Ngomo, A.N., Auer, S.: LIMES: A Time-efficient Approach for Large-scale Link Discovery on the Web of Data. In: Proc. of 22nd Int. Joint Conf. on Artificial Intelligence, pp. 2312–2317. AAAI Press (2011)
13. Nikolov, A., d'Aquin, M., Motta, E.: Unsupervised Learning of Link Discovery Configuration. In: Simperl, E., Cimiano, P., Polleres, A., Corcho, O., Presutti, V. (eds.) ESWC 2012. LNCS, vol. 7295, pp. 119–133. Springer, Heidelberg (2012)
14. Shvaiko, P., Euzenat, J.: A Survey of Schema-based Matching Approaches. In: Spaccapietra, S. (ed.) Journal on Data Semantics IV. LNCS, vol. 3730, pp. 146–171. Springer, Heidelberg (2005)
15. Shvaiko, P., Euzenat, J.: Ontology Matching: State of the Art and Future Challenges. IEEE Trans. on Knowledge and Data Engineering 25(1), 158–176 (2013)
16. Suchanek, F.M., Kasneci, G., Weikum, G.: Yago: A Core of Semantic Knowledge. In: Proc. of 16th Int. Conf. on World Wide Web, pp. 697–706. ACM Press (2007)
17. Volz, J., Bizer, C., Gaedke, M., Kobilarov, G.: Silk - A Link Discovery Framework for the Web of Data. In: Proc. of the Linked Data on the Web Workshop (LDOW2009), CEUR Workshop Proceedings, vol. 538 (2009)
18. Weikum, G., Theobald, M.: From Information to Knowledge: Harvesting Entities and Relationships from Web Sources. In: Proc. of 29th ACM SIGMOD-SIGACT-SIGART Symposium on Principles of Database Systems, pp. 65–76. ACM Press (2010)
19. Zaïane, O.R., Chen, J., Goebel, R.: Mining Research Communities in Bibliographical Data. In: Zhang, H., et al. (eds.) WebKDD 2007. LNCS, vol. 5439, pp. 59–76. Springer, Heidelberg (2009)
20. Zhao, L., Ichsie, R.: Graph-based Ontology Analysis in the Linked Open Data. In: Proc. of 8th Int. Conf. on Semantic Systems, pp. 56–63. ACM Press (2012)

Conducting a Web Browsing Behaviour
Study – An Educational Scenario

Martin Labaj and Mária Bieliková

Slovak University of Technology in Bratislava,
Faculty of Informatics and Information Technologies,
Ilkovičova 2, 842 16 Bratislava, Slovakia
{martin.labaj,maria.bielikova}@stuba.sk

Abstract. Web browsing behaviour is a matter of study in several fields – from web usage mining, to its applications in adaptive and personalized systems. Current web browsers allow for parallel browsing – opening multiple web pages at once and switching between them. To capture such behaviour, client-side observations are typically performed, where attracting and retaining enough participants poses a challenge. In this paper, we describe a study based on an experiment on logging the parallel browsing behaviour, both in an adaptive web-based educational system and on the open Web, while using the educational system as a tool for recruiting and motivating the participants. We focus on how various types of users (here students), including their personality information, participated in the experiment regarding churn and their observed behaviour. The paper concludes with "lessons learned" important to consider when planning and performing similar studies.

Keywords: web browsing study, tabbed web browsing, educational systems, churn.

1 Introduction and Related Work

As more and more aspects of everyday life are now being carried out using the Web, if not completely on the Web, observing and analysing web browsing behaviour comes into importance. By understanding how users browse websites, either within a single web-based system, or across heterogeneous systems on the open Web in general, we can possibly improve any personalization and adaptation feature that can be based on user behaviour (i.e. implicit feedback), such as:

- Domain modelling – creating links between pages, links between domain terms, or assigning domain terms to content based on user movement across the Web.
- User modelling – discovering user's current interests and predicting future ones from their visits to pages.
- Recommender systems – observing patterns in users' visits from a page to other pages and recommending relevant pages to similar users.

An important aspect of how users browse the Web stems from modern web browsers. These not only allow opening multiple windows at once, but also allow opening multiple web pages within a single window using tabs. Such behaviour is called parallel

G.F. Italiano et al. (Eds.): SOFSEM 2015, LNCS 8939, pp. 531–542, 2015.

browsing [1,8] or tabbed browsing [4,9,18]. Since multiple web pages are accessible to the user without having to load or reload them, the tabbing has changed the traditional web usage mining approaches [18].

Traditionally, server-side log-based web usage mining considers *page load* events, but with the parallel browsing, the user can switch contexts by navigating between opened tabs with various pages attributed to different user tasks without generating these page load events that could be visible to server-side logging. The user can also open a page in various ways. A link can be opened into the same tab, replacing the currently opened page, or in a new tab, leaving the source page opened and branching the browsing action tree. Therefore, even when the page load event is observed in the server-side logging, it can represent differing browsing actions when we take parallel browsing into account. Various inferences can be made from traditionally logged actions, e.g. when the user loads two different pages by clicking a link from the same page, the page must have still existed for the second click and therefore the first click must have been branching [7,18]; or tasks in the browsing can be discovered and then browsing behaviour can be estimated [1]. However, in order to *fully* capture and analyse the user browsing behaviour including using multiple pages (tabs) and switching between them, and to allow for the aforementioned personalization and adaptation features based on user behaviour, one must use *client-side* tracking and observe the actions on the user computers in their browsers.

A single web-based system can easily track its users on the client-side by including client-side-executed scripts into pages served by this system and aggregating user actions across multiple pages (activating/deactivating single pages) to reconstruct the browsing tree. We previously used such approach [10] to capture user browsing within adaptive web-based educational system ALEF (Adaptive Learning Framework).

Users, however, do browse in tabs across heterogeneous web systems in various situations [4], ranging from comparing pages against each other, keeping frequently used ones at hand, creating bookmarks or todo lists, to simply multitasking. To observe such behaviour, the tracking must be done either in all pages, e.g. by using an adaptive proxy that injects scripts into web pages passing through it, or by observing the user's browser directly via a browser extension. Except when monitoring connections in a school, in a workplace, etc., where an intercepting proxy [12] or browser can be configured authoritatively, active user participation is required to install modified network settings or a browser extension [4] to participate in such experiments.

Studies of browsing behaviour therefore often face a choice, either:

- *Passively use server-side* data that are easily observable from all visitors, but those do not provide details on advanced features of their browsing behaviour, such as parallel browsing. Or:
- *Actively monitor the participants with client-side software*, but attracting enough participants to install the logging software voluntarily (in order to observe their natural behaviour) and retaining these participants usually becomes a challenge.

In this paper, we describe a study where we observed users' browsing behaviour both within adaptive web-based educational system ALEF (Adaptive Learning Framework) [17] and on the open Web while using ALEF as a tool to find participants and motivate them to stay in the study, i.e. to observe the web browsing behaviour to maximum extent, while actually reliably finding and retaining participants. We focus on how students participated in the experiment.

Our aim is to understand how users participate in a voluntary long-term browsing behaviour study that requires installation of a logging browser extension which is, in fact, always a privacy intrusion to some degree, and how long they remain participating. By observing user features that can be observed independently from the study, such as participant demographics elicited through surveys, their academic performance (in the case of student users), or their actions in other web systems (such as in an adaptive learning system), we could predict how would users join before even starting the study and for example estimate required population to be acquired. If we could predict how long would they participate (their churn rate) and what causes them to stay/leave, we could obtain more results from more participants in similar studies.

The rest of this paper is organized as follows: first, we describe the browsing behaviour logging infrastructure and explain the experiment setup, including motivation and recruitment of the participants via ALEF. Next, the resulting dataset is described, followed by analysis of churn and user behaviour during the experiment. In the last part, we outline future work and implications of this study.

2 Study on Logging the Browsing Behaviour

2.1 Experiment Setup

For the purpose of this experiment, we implemented parallel browsing tracking as an extension in the Brumo platform (Browser-based User Modelling and Personalization Framework[1]) [15], which provides infrastructure for browser extensions, including client-server communication and storage, and allows distributing the extensions. We capture user actions in separate tabs, such as loading the page, bringing the page into focus or hiding it, and then combine these actions into single-timeline stream using reconstruction algorithm we described in [10]. We created an infrastructure, where the events are observed using the extension, sent to server and analysed [11]. The output is a browsing tree describing how a given user clicked on each single link or typed-in URLs (into the same tab or a new tab), how they switched between tabs or out of the browser – effectively reconstructing entire user session, allowing for analysis of switch frequencies between given resources, or user browsing styles.

We set up experiment with participants who used educational system ALEF – bachelor students of the Principles of software engineering course – instructing them to broaden their knowledge about topics presented in the system and look for appropriate external sources (URLs related to given content) on the open Web, attaching them to corresponding learning objects. This task was also motivated competitively using global user score and leaderboards – students were awarded score points for links attached to the content [6].

Submitted external sources were scored according to their novelty (repeated URLs were penalized), access level (sources attached as public and signed with own username were worth more points than anonymous sources), and finally according to their quality

---

[1] http://brumo.fiit.stuba.sk

and relevance to the given learning object. The last criterion was evaluated by a domain expert, who rated the sources in three levels: appropriate (accepted), neutral, rejected, which were rewarded with more points, rewarded with the default amount of points, or penalized with negative amount of points, respectively. On top of that, whenever a student achieved a reward level amounting to five top-rated (signed and approved) sources, an additional content (a recapitulating question for a final exam) was unlocked for the given student – motivating the students to browse for quality links.

Most importantly, we asked the students to let us see their browsing behaviour while looking for these external sources and in order to do that, the students could insert the external sources only when the Brumo extension with browser tracking was present. The main phase of the experiment (inserting the external sources) ran for more than a week and data collection continued voluntarily for a year among users who have kept the extension installed after the learning system experiment.

2.2 Dataset

For studying various user approaches in browsing style and participation in the study, we created a dataset consisting of 249 users. The users are structured into three groups:

1. 80 users: bachelor students from ALEF system who chose to participate in the browsing study. These are students who were learning the course content in ALEF and as other ALEF students, they were given the opportunity to install the extension and then submit external sources that rewarded points as a motivation. These students chose to install the extension in order to participate. Then they could leave the extension installed for some time, remaining in the study. This group serves as a *positive* sample and allows exploring which features influences joining the study and the length of participation.
2. 144 users: bachelor students from ALEF system who chose not to participate in the browsing study. These students were also exposed to the same motivation as the group (1), but chose not to participate. However, except the browsing behaviour on the open Web (which is naturally not available, since these students did not install the logging extension), other features about their activity in the learning system, personality, etc. (see below) were still observed for these users, therefore this group serves as a *negative* example and allows exploring which student/user properties have impact on not participating.
3. 25 users: older students (mostly master study) in the browsing study who did not come from the ALEF system. These users started using the logging extension through other means, for example by attending a student research seminar where the extension was propagated. The number of these users is rather low, because the logging extension was freshly deployed and there are little additional features known about these users apart from their browsing behaviour.

Additional to browsing actions within and outside of the educational system, we computed/included churn data (when the user joined, left, etc.), activity within the experiment (external sources statistics), activity within the educational system outside of the experiment (question-answer learning object answering), and study performance.

Two types of demographic traits were elicited: Felder and Silverman learning style [5] was obtained via a questionnaire within ALEF, and Big Five personality traits were obtained via professional assessment. Table 1 provides more detail about the data collected in the dataset.

Table 1. Dataset composition

Feature		Source	Explanation
A	Churn	Brumo extension, partially inferred	Date and time joined and left (started/stopped using the extension), participated in experiment, is still active
B	Browsing actions	Brumo extension, ALEF	Browsing behaviour within ALEF, outside of ALEF, during experiment, outside of experiment, total browsing
C	External sources	ALEF	External sources submitted, categorized as approved, rejected, and deleted
D	Learning activty	ALEF	Number of shown and number of answered question-answer learning objects; portion of views skipped without answering (tendency to "cheat" into viewing question-answer pairs without having to rate); portion of views rated with default value (similar)
E	Study	Course	Academic performance
F	Learning style	Questionnaire (ALEF)	Dimensions (active/reflective, sensing/intuitive, verbal/visual, sequential/global)
G	Personality traits	Professional assessment	Traits (openness to experience, conscientiousness, extraversion, agreeableness, neuroticism) – value and percentile

Since several of these attributes depend on participation, either by doing some voluntary activity in the educational system, or filling out the questionnaire, or taking the personality traits assessment, not all of these data are available for each user. Dataset features coverage is shown in Figure 1.

2.3 User Participation and Browsing Style

First, due to the nature of the continuing experiment, we were interested in churn rate (or retention conversely). Determining the churn is a task of finding when the user is likely to leave. The term comes from the telecommunications field, where one is interested in when and why is the user going to switch to competing service provider, but the churn is commonly explored in the field of adaptive web-based systems. For example, in a user-generated content community, rate of participation and feedback from other users can be associated with length of membership [14]. In another example, a length of participation in a community-based question answering system can be predicted using classifiers based on questions, answers, gratification and answerer demographic features [3]. Perhaps closer to our study is an example of content discovery system, where users

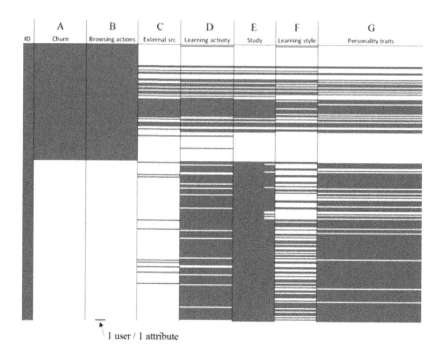

Fig. 1. Dataset sparsity: first column represents participations (by their ID), columns 2 to 8 represent particular features (according to Table 1); grey – attribute is present, white – attribute is missing (user did not participate in the source activity). First 105 users participated in the browsing study (groups 1+3), rest of them, 144, did not participate (group 2).

view recommended Web pages. Time spent, visit features and content features can be associated with number of sessions made by the user [2].

If we could predict the churn, we can, for example, increase or add user motivation during the experiment to prevent them leaving and obtain better data, or even predict how many participants we need in the beginning given the expected dropout through the course of experiment. We believe that simple actions like notification of the teacher or stimulated message to the student including gamification measures currently more and more employed in domain of educational systems can further improve students' engagement and so their results. There is, however, a rather important difference to traditional churn prediction tasks. In setups such as ours, the user is being observed in their natural activity on the Web and it is even desirable for the purpose of obtaining unbiased data to interfere with this activity as little as possible. Actions on the web therefore do not correspond with the user satisfaction or engagement in the experiment. In our case, we have multiple external features relevant to the user – study performance in the course, activity in ALEF, external source activity, or personal traits which could be possibly used universally.

A potential candidate for relevance to churn is the activity in ALEF related to question-answer learning objects, because the motivation is similar [16]. The user is presented with

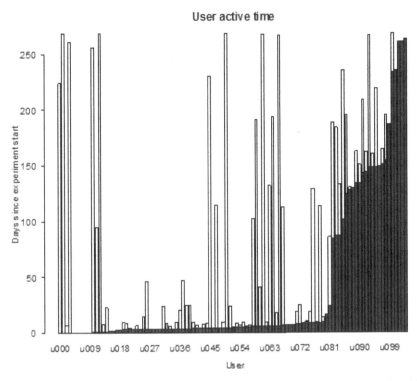

Fig. 2. User participation in the browsing study (groups 1 and 3). Dark bars represents time (days) since the start of the study until the user has joined. Light bars are the length of the participation (time until the user has left).

a pair consisting of a question and a student-provided answer and has to judge the correctness of the answer. The user is motivated by getting acquainted with potential questions for upcoming exams and recapitulating own knowledge. In the external sources gathering experiment, the students were rewarded for multiple quality links with a potential exam questions, so there is similar motivation in both of these activities.

Users can, however, "cheat" out of the work of rating the answer correctness by using an option to skip a question and get another one multiple times in a row, or by answering with the default correctness value. While some rate of skips and default answers is natural, for some users, rates as high as 90 % skipped questions were observed. This could relate to the user's approach to participation in other kinds of activity – e.g., if a user is only interested in the score and reward levels for the external sources, he/she could install the logging extension for a very brief period of time or even into a separate browser, submit prepared sources (to achieve a score for the attached sources) and stop participating.

The overview of how users joined and left the study is shown in Figure 2. The overview of relations found in the dataset is shown in Figure 3. Correlations within segments of features (group of features describing on property of subjects, e.g. their learning style) delimited by horizontal and vertical lines are expected, since some of the features are inferred from others in the same segment, or they may be related on each other, for example, the ALEF experiment ran from the beginning of the dataset for some time, therefore time of joining is related to the user having participated in the study from the ALEF experiment (users who joined sooner), or having participated in the study independently (users who joined later).

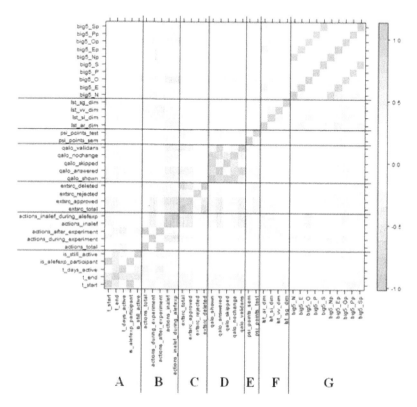

Fig. 3. Associations between various features of users found in the experiment (Pearson pairwise correlation). Horizontal and vertical lines mark feature segments of the same type. The notation identifies features according to their segment explanations, e.g., personality dimensions, learning style dimensions, or browsing actions (see Table 1).

The question-answer learning object activity described above correlates both to whether the user has participated in the study at all (0.297, p-value = 0.002) and how actively they participated (0.483, p-value ≈ 0), suggesting that these activities with similar underlying motivations (allowing the students to practice exam-like questions) attracted similar user behaviour. Therefore, *if we base user motivation in an experiment starting from an educational system on similar mechanics as another activity in the system with known usage, even when the activity is very different to the experiment, we can predict user participation to a degree.*

In the logs themselves, we indeed found a behaviour we can call "*short-participa-ting*". It is similar to the skipping and default-rating the question-answer learning objects – some users allowed tracking only for the minimal time, when they installed or enabled the tracking, looked up and attached several external sources to the learning content (inserting the sources into the system was the only activity that strictly required the presence of tracking) and stopped the tracking altogether. In some cases, only the inserting took place with the tracking. This can be induced by realising that the tracking is needed to insert the link only after having the external links already found, in that case, clearer communication (within the constraint of this being an uncontrolled experiment) may help.

In some cases, students mentioned that they use a different browser for their daily browsing activity and they used the two browsers for which the tracking is implemented only to insert sources looked up elsewhere. Therefore, in browsing behaviour studies, it is important to *cover as many various browsers as feasible technically and labour-wise*. Due to differences between browsers in creating extensions, integrating with the browser and pages, maintaining their compatibility through new browser versions, etc., covering every possible browser is infeasible. Often only one browser is observed.

If a study is concerned with a smaller number of participants (such as when follow-up in-person interviews are planned), they can be recruited with explicit requirement of everyday usage of a given supported browser, such as in [4]. However, in cases similar to our where we motivated users to participate, temporary switching the browser can hinder the goal of observing the natural user behaviour on a large scale. Note that we supported two common browsers (Chrome and Firefox), such as some other studies do [19], and yet there was a relatively significant portion of users who commented on using other browsers (Safari, Opera). On the other hand, supporting two and more browsers creates a possibility for users to intentionally install the extension into that browser they do not use naturally, cases of this behaviour were observed in this study.

We then looked on the *continuing participation* in the study. On one hand, using a "stealth" extension (one that does not present any specific features directly to the user, although it is not hidden in any way in the standard extension-listing user interfaces in the browser) to track browsing has an effect of observing the natural behaviour of the users and some participants will continue to use it because it does not interfere with their browser experience. On the other hand, this very same mechanism of the initial motivation (e.g. user score, questions) in an experiment and no subsequent user features of the tracking can mean that we lose participants, e.g., when the user reinstalls or switches the browsers. In interviews with selected participants, several of those who left the study at some time after the experiment (they did not short-participate as above) mentioned that no user features were the reason for deliberately uninstalling the tracking.

It seems that *the best approach is combining the initial motivation (as described here with ALEF experiment) to gather the initial user base together with later features available in the tracking software (allowing the users [of educational system] to see their browsing history, manage tabs, etc.) to retain the users for ongoing data collection.* Instead of having the logging stay quietly behind, making the logging software attractive to participants can not only help keep users in the study, but possibly also attract new users who did not receive or did not respond to the initial motivation such as that from the educational system we describe here.

Regarding the browsing behaviour within the adaptive learning system, a common type of browsing paths observed in the experiment was a *"loop"*, which is a sequence of tabbing actions that:

- starts in a learning object,
- follows user switches through several pages and
- returns to the same learning object.

Such feature could be, for example, used for discovering external sources complementing the content that the user was previously browsing. A method for web content enrichment was created based on the logged browsing behaviour [13].

3   Conclusions and Future Work

In this paper, we presented findings and *lessons learned* of conducting a study on open Web browsing behaviour of students who use an educational system, which served as a motivation tool for engagement in the study. Our study was based on the dataset consisting of 249 users structured into three groups according to activities they performed within the educational system ALEF. We explored user participation and found a correlation with participation in the system activity. We summed up our observations as recommendations for conducting browsing studies. These include using other activity in an adaptive web based system (here educational system) as a basis for predicting the participation in another study, avoiding "short-participating" when covering single, or on the other hand, multiple web browsers, and continuing the participation motivation in some form after the initial drafting motivation. By caring for these aspects, a browsing study can gather more participants, observe as much as possible of their natural behaviour (both in terms of length of the observed part and in its quality).

In spite of the fact that the resulting dataset is based on enough participants to consider their behaviour and traits, the number is insufficient for cross-validation of a predictive model. It, however, includes diverse set of attributes describing various aspects of the user from personality traits, to performance, to activity in an educational system, and activity on the open Web, which helps to get inside into browsing behaviour of study participants. Along the lines of such breadth oriented study, a qualitative experiment observing physical users could complement these findings.

Research described here was so far focused on the user side of the problem, i.e. how users participated and browsed. Another aspect, the item based view, on which we plan focusing now, is important for leveraging these findings in domain modelling, recommender systems, etc. – how browsing styles differ depending on the item. We have shown previously that the tabbing actions differ in learning environment based on the type of learning object (explanations, exercises, and questions) [10]. Various browsing styles could perhaps differentiate the quality of web pages, or help us reveal relations between objects, such as subsumed-by (one object logically follows after another, especially in, but not limited to, an educational environment), or related-to (one object explains concepts present in another object).

Acknowledgement. This work was partially supported by grants No. VG 1/0646/15, KEGA 009STU-4/2014 and it is the partial result of the Research and Development Operational Programme for the project "University Science Park of STU Bratislava", ITMS 26240220084, co-funded by the European Regional Development Fund.

References

1. Bonnin, G., Brun, A., Boyer, A.: Towards Tabbing Aware Recommendations. In: Proc. of the First Int. Conf. on Intelligent Interactive Technologies and Multimedia - IITM 2010, pp. 316–323. ACM Press, New York (2010)
2. Dave, K.S., Vaingankar, V., Kolar, S., Varma, V.: Timespent Based Models for Predicting User Retention. In: Proc. of the 22nd Int. Conf. on World Wide Web, WWW 2013, pp. 331–342 (2013)
3. Dror, G., Pelleg, D., Rokhlenko, O., Szpektor, I.: Churn Prediction in New Users of Yahoo! Answers. In: Proc. of the 21st Int. Conf. Companion on World Wide Web - WWW 2012 Companion, pp. 829–834. ACM Press, New York (2012)
4. Dubroy, P., Balakrishnan, R.: A Study of Tabbed Browsing Among Mozilla Firefox Users. In: Proc. of the 28th Int. Conf. on Human Factors in Computing Systems, CHI 2010, pp. 673–682. ACM Press, New York (2010)
5. Felder, R., Silverman, L.: Learning and Teaching Style in Engineering Education Richard. Eng. Educ. 78(7), 674–681 (1988)
6. Filipčík, R., Bieliková, M.: Motivating Learners in Adaptive Educational System by Dynamic Score and Personalized Activity Stream. In: Proc. of 9th Int. Workshop on Semantic and Social Media Adaptation and Personalization – SMAP 2014. IEEE (to appear, 2014)
7. Huang, J., Lin, T., White, R.W.: No Search Result Left Behind: Branching Behavior with Browser Tabs. In: Proc. of the 5th ACM Int. Conf. on Web Search and Data Mining - WSDM 2012, pp. 203–212. ACM Press, New York (2012)
8. Huang, J., White, R.W.: Parallel browsing behavior on the web. In: Proc. of the 21st Int. Conf. on Hypertext and Hypermedia - HT 2010, pp. 13–17. ACM Press, New York (2010)
9. Chierichetti, F., Kumar, R., Tomkins, A.: Stochastic Models for Tabbed Browsing. In: Proceedings of the 19th Int. Conf. on World Wide Web - WWW 2010, pp. 241–250. ACM Press, New York (2010)
10. Labaj, M., Bieliková, M.: Modeling parallel web browsing behavior for web-based educational systems. In: Proc. of 10th Int. Conf. on Emerging eLearning Technologies and Applications – ICETA 2012, pp. 229–234. IEEE (2012)
11. Labaj, M., Bieliková, M.: Tabbed Browsing Behavior as a Source for User Modeling. In: Carberry, S., Weibelzahl, S., Micarelli, A., Semeraro, G. (eds.) UMAP 2013. LNCS, vol. 7899, pp. 388–391. Springer, Heidelberg (2013)
12. Meiss, M., Duncan, J., Gonçalves, B., Ramasco, J.J., Menczer, F.: What's in a Session: Tracking Individual Behavior on theWeb. In: Proc. of the 20th Conf. on Hypertext and Hypermedia - HT 2009, p. 173. ACM Press, New York (2009)
13. Račko, M.: Automatic Web Content Enrichment Using Parallel Web Browsing. In: Proc. of 10th Student Reasearch Conference in Informatics and Information Technologies -IIT.SRC 2014, STU Bratislava, pp. 173–178 (2014)
14. Sarkar, C., Wohn, D.Y., Lampe, C.: Predicting Length of Membership in Online Community "Everything2" Using Feedback. In: Proc. of the Conf. on Computer Supported Cooperative Work Companion, CSCW 2012, pp. 207–210. ACM Press, New York (2012)
15. Šajgalík, M., Barla, M., Bieliková, M.: Efficient Representation of the Lifelong Web Browsing User Characteristics. In: Proc. of the 2nd Workshop on LifeLong User Modelling, in Conjunction with UMAP 2013, pp. 21–30 CEUR-WS (2013)

16. Šimko, J., Šimko, M., Bieliková, M., Ševcech, J., Burger, R.: Classsourcing: Crowd-Based Validation of Question-Answer Learning Objects. In: Bădică, C., Nguyen, N.T., Brezovan, M. (eds.) ICCCI 2013. LNCS, vol. 8083, pp. 62–71. Springer, Heidelberg (2013)
17. Šimko, M., Barla, M., Bieliková, M.: ALEF: A framework for adaptive web-based learning 2.0. In: Reynolds, N., Turcsányi-Szabó, M. (eds.) KCKS 2010. IFIP AICT, vol. 324, pp. 367–378. Springer, Heidelberg (2010)
18. Viermetz, M., Stolz, C., Gedov, V., Skubacz, M.: Relevance and Impact of Tabbed Browsing Behavior on Web Usage Mining. In: Proc. of IEEE/WIC/ACM Int. Conf. on Web Intelligence - WI 2006, pp. 262–269. IEEE (2006)
19. Von der Weth, C., Hauswirth, M.: DOBBS: Towards a Comprehensive Dataset to Study the Browsing Behavior of Online Users. In: Proc. of the 2013 IEEE/WIC/ACM Int. Joint Conf. on Web Intelligence (WI) and Intelligent Agent Technologies (IAT) WI-AT 2013, pp. 51–56. IEEE (2013)

A Uniform Programmning Language
for Implementing XML Standards

Pavel Labath[1] and Joachim Niehren[2]

[1] Commenius University, Bratislava
labath@dcs.fmph.uniba.sk
[2] Inria, Lille

Abstract. We propose X-Fun, a core language for implementing various XML standards in a uniform manner. X-Fun is a higher-order functional programming language for transforming data trees based on node selection queries. It can support the XML data model and XPATH queries as a special case. We present a lean operational semantics of X-Fun based on a typed lambda calculus that enables its in-memory implementation on top of any chosen path query evaluator. We also discuss compilers from XSLT, XQUERY and XPROC into X-Fun which cover the many details of these standardized languages. As a result, we obtain in-memory implementations of all these XML standards with large coverage and high efficiency in a uniform manner from SAXON's XPATH implementation.

Keywords: XML transformations, database queries, functional programming languages, compilers.

1   Introduction

A major drawback of query-based functional languages with data trees so far is that they either have low coverage in theory and practice or no lean operational semantics. Theory driven languages are often based on some kind of macro tree transducers [3,5,11], which have low coverage, in that they are not closed under function composition [4] and thus not Turing complete (for instance type checking is decidable [12]). The W3C standardised languages XQUERY [13] and XSLT [7], in contrast, have large coverage in practice (string operations, data joins, arithmetics, aggregation, etc.) and in theory, since they are closed by function composition and indeed Turing complete [8]. The definitions of these standards, however, consist of hundreds of pages of informal descriptions. They neither explain how to a build a compiler in a principled manner nor can they be used as a basis for formal analysis.

A second drawback is the tower of languages approach, adopted for standardised XML processing languages. What happened in the case of XML was the development of a separate language for each class of use cases, which all host the XPATH language for querying data trees based on node navigation. XSLT serves for use cases with recursive document transformations such as HTML publishing, while XQUERY was developed for use cases in which XML databases are

G.F. Italiano et al. (Eds.): SOFSEM 2015, LNCS 8939, pp. 543–554, 2015.

queried. Since the combination of both is needed in most larger applications, the XML pipeline language XPROC [16,17,18] was developed and standardised again by the W3C. This resulted in yet another functional programming language for processing data trees based on XPATH.

For resolving the above two drawbacks, the question is whether there exists a uniform core language for processing data trees that can cover the different XML standards in a principled manner. It should have a lean and formal operational semantics, support node selection queries as with XPATH and it should be sufficiently expressive in order to serve as a core language for implementing XQUERY, XSLT, and XPROC in a uniform manner.

Related work. An indicator for the existence of a uniform core language for XML processing is that the omnipresent Saxon system [14] implements XSLT and XQUERY on a common platform. However, there is no formal description of this platform as a programming language, and it does not support the XML pipeline language XPROC so far. Instead, the existing implementations of XPROC, CAL-ABASH [16] and QUIXPROC [18], are based on Saxon's XPATH engine directly.

The recent work from Castagna et al. [2] gives further hope that our question will find a positive answer. They present an XPATH-based functional programming language with a lean formal model based on the lambda calculus, which thus satisfies our first two conditions above and can serve as a core language for implementing a subset of XQUERY 3.0. We believe that relevant parts of XSLT and XPROC can also be compiled into this language, even though this is not shown there. The coverage, however, will remain limited, in particular on the XPATH core (priority is given to strengthening type systems). Therefore, our last requirement is not satisfied.

Contributions. In this paper, we present the first positive answer to the above question based on X-Fun. This is a new purely functional programming language. X-Fun is a higher-order language and it supports the evaluation of path-based queries that select nodes in data trees. The path queries are mapped to X-Fun expressions, whose values can be computed dynamically. In contrast to most previous interfaces between databases and programming languages, we overload variables of path queries with variables of X-Fun. In this manner, the variables in path queries are always bound to tree nodes, before the path query is evaluated itself. We note in particular, that path queries are not simply mapped to X-Fun expressions of type string.

The formal model of the operational semantics of X-Fun is a lambda calculus with a parallel call-by-value reduction strategy. Parallel evaluation is possible due to the absence of imperative data structures. The main novelty in X-Fun admission of tree nodes as values of type **node**. Which precise nodes are admitted depends on a tree store. New nodes can be created dynamically by adding new trees to the tree store. The same tree can be added twice to the store but with different nodes. How nodes are represented internally can be freely chosen by the X-Fun implementation and is hidden from the programmer.

X-Fun can serve as a uniform core language for implementing XQUERY, XSLT and XPROC. In order to do so, we have developed compilers of all three languages into X-Fun. We also discuss how to implement X-Fun in an in-memory fashion on top of any in-memory XPATH evaluator. Based on our compilers, we thus obtain new in-memory implementations of XQUERY, XSLT and XPROC with large coverage. Our implementation has very good efficiency and outperforms the most widely used XPROC implementation by a wide margin.

Outline. In Section 2 we introduce our general model of data trees, alongside its application to XML documents. The syntax and type system of the X-Fun language is introduced in Section 3. The applications of X-Fun to XML document transformation is studied in Section 4, where we discuss compilers from other XML processing languages into X-Fun. Section 5 contains our notes on the implementation of X-Fun and the results of our experiments.

2 Preliminaries

We introduce a general concept of data trees which will be used in the X-Fun language. We also show how to instantiate the trees to the XML data model.

2.1 Data Values and Data Trees

We fix a finite set *Char* whose elements will be called characters. A data value $"c_1 \cdots c_n"$ is a word of characters for $c_1, \ldots, c_n \in Char$. We define $String = Char^*$ to be the set of all data values, and $\texttt{nil}=""$ to be the empty data value.

Next, we will fix a natural number $k \geq 1$ and introduce data trees in which each node contains exactly k data values with characters in *Char*.

A *node label* is a k-tuple of data values, i.e., an element of $(String)^k$. The set of data trees $\mathcal{T}$ of label size k over *Char* is the least set that contains all pairs of node labels and sequences of data trees in $\mathcal{T}$. That is, it contains all unranked trees t with the abstract syntax $t ::= l(t_1, \ldots, t_n)$, where $n \geq 0$ and $l \in String^k$. It should be noticed that the set of node labels is infinite, but that each node label can be represented finitely.

2.2 XML Data Model

For XML data trees, we can fix $k = 4$ and *Char* the set of Unicode characters, and restrict ourselves to node labels of the following forms, where all v_i are data values:

("element", v_1, v_2, nil)	("attribute",	v_1, v_2, v_3)
("comment", nil, nil, v_3)	("processing-instruction",	v_1, nil, v_3)
("document", nil, nil, nil)	("text",	v_1, nil, nil)

An element ("element", v_1, v_2, nil) has three non-nil data values: its type "element", a name v_1 and a namespace v_2. An attribute has four data values: its type, a name v_1, a namespace v_2, and the attribute value v_3. A text node contains its type and its text value v_3. Besides these, there are comments, processing instructions and the rooting document node.

3   Language X-Fun

In this section, we introduce X-Fun, a new functional programming language for transforming data trees. X-Fun can be applied to all kinds of data trees with a suitable choice of its parameters. We will instantiate the case of data trees satisfying the XML data model concomitant with XPATH as a query language.

We start with introducing the types and values of X-Fun (Section 3.1). Then we explain how to map path queries to X-Fun values, by using particular X-Fun expressions with variables (Section 3.2). The general syntax of X-Fun expressions is given in Section 3.3. Some syntactic sugar and an example of an X-Fun program are given in Sections 3.5 and 3.6. Discussion of the typing rules for X-Fun's type system and the formal semantics of X-Fun can be found in the research report [10].

3.1   Types and Values

The X-Fun language supports higher-order values and expressions with the following types:

$$T ::= \mathbf{none} \mid \mathbf{node} \mid \mathbf{tree} \mid \mathbf{number} \mid \mathbf{bool} \mid \mathbf{char}$$
$$\mid T_1 \times \ldots \times T_n \mid [T] \mid T_1 \to T_2 \mid T_1 \cup T_2$$

A value of type **char** is an element of *Char*, a value of type **tree** is an element of $\mathcal{T}$. A value of type **number** is a floating point number, while the values of type **bool** are the Boolean values *true* and *false*. A value of type **node** will be a node of the graph of one of the trees stored by the environment of the X-Fun evaluator. The precise node identifiers chosen by the evaluator are left internal (to the mapping from trees to graphs).

As usual, we support list types $[T]$ which denote all lists of values of type T, product types $T_1 \times \ldots \times T_n$ whose values are all tuples of the values of types T_i, and function types $T_1 \to T_2$ whose values are all partial functions of values of type T_1 to values of type T_2. Besides these, we also support type unions in the obvious manner.

A data value $"c_1 \cdots c_n" \in String$ is considered as a list of characters of type **string** $= [\mathbf{char}]$. A node label is considered a k-tuple of strings, i.e., as a value of type **label** $= \mathbf{string}^k$. Hedges are considered as lists of trees of type **hedge** $= [\mathbf{tree}]$.

Since XPATH can return sequences of items of different types, we define the type **pathresult** as **node** $\cup$ **number** $\cup$ **string** $\cup$ **bool**. The result of evaluating a path expression will then be of type [**pathresult**]. To be able to specify path expressions, we define the type **path** as [**char** $\cup$ **pathresult** $\cup$ [**pathresult**]], i.e., as list of characters, individual items returned by a path expression, and whole sequences of those items.

3.2   XPath Queries as X-Fun Expressions

We will consider XPath expressions as values of our programming language. This is done in such a manner that the variables in XPATH expressions can be bound to values of the programming language. For instance, if we have an XPATH expression

```
$x//book[auth=$y]
```

then one might want to evaluate this expression while variable x is bound to a node of some tree and variable y to some data value. In X-Fun, the above query will be represented by the following expression of type **path**, where x is a variable of type **node** and y a variable of type **string**:

$$x :: '/' :: '/' :: 'b' :: 'o' :: 'o' :: 'k' :: '[' :: 'a' :: 'u' :: 't' :: 'h' :: '=' :: y :: ']' :: nil$$

The concrete syntax of X-Fun supports syntactic sugar for values of type **path**, so that the above expression can be defined as:

```
"$x//book[auth=$y]"
```

In order to enable the evaluation of path expressions, X-Fun supports a builtin function evalPath of type **path** → [**pathresult**]. In an implementation of X-Fun, this function can be mapped straightforwardly to existing XPATH evaluators.

3.3 Syntax of X-Fun Expressions

X-Fun is a purely functional programming language whose values subsume higher-order function, trees, strings, numbers and Boolean values. The evaluation strategy of X-Fun is fully parallel, which is possible since no imperative constructs are permitted.

The syntax of X-Fun programs E is given in Figure 1. All expressions of X-Fun are standard in functional programming languages, so we only briefly describe different kinds of subexpressions of X-Fun programs.

A variable x is evaluated to the value of the corresponding type. The constant expression c returns the respective constant, which can be a Boolean value,

Expressions

$E ::= x$
$\quad | \quad c$
$\quad | \quad E_1 :: E_2$
$\quad | \quad (E_1, \ldots, E_n), n \geq 2$
$\quad | \quad \textbf{match } E \ \{ \ P_1 \rightarrow E_1, \ \ldots, \ P_n \rightarrow E_n \ \}$
$\quad | \quad \textbf{fun } x : T_1 \rightarrow T_2 \ \{ \ E \ \}$
$\quad | \quad E_1(E_2)$
$\quad | \quad \textbf{try } E_1 \ \textbf{catch}(x) \ E_2$
$\quad | \quad \textbf{raise}(E)$

Patterns

$P ::= x : T$
$\quad | \quad !(E)$
$\quad | \quad P_1 :: P_2$
$\quad | \quad (P_1, \ldots, P_n), n \geq 2$

Fig. 1. Syntax of X-Fun's expressions

a number or a character from *Char*. The list constructor $E_1 :: E_2$ prepends an element to a list, while the tuple constructor $(E_1, \ldots, E_n)$ constructs tuples.

The match expression **match** $E \ \{ \ P_1 \rightarrow E_1, \ \ldots, \ P_n \rightarrow E_n \ \}$ selects one of the branches E_i based on the patterns P_i, which are matched against the value of E. The pattern $x : T$ captures a matched value of type T into a variable. The pattern $!(E)$ matches the value against the value of expression E. Here, the matching of functional values, or lists/tuples that contain functions is not permitted. Pattern $P_1 :: P_2$ matches a list if P_1 and P_2 match its head and tail, while the pattern $(P_1, \ldots, P_n)$ matches tuples.

A function expression **fun** $x : T_1 \rightarrow T_2 \ \{ \ E \ \}$ returns a new function, with the argument $x : T_1$ and the return value of type T_2 obtained by the evaluation of the function body E. The expression $E_1(E_2)$ applies a function to a value. X-Fun also supports exception handling, where exceptions are values of type **string**.

3.4   Builtin Operators

At the beginning of the evaluation, the environment contains bindings of the global variables given in Figure 2.

Parameters		Fixed	
Global variable TYPE		Global variable TYPE	
makeTree	**label** × **[tree]** → **tree**	nil	**[none]**
evalPath	**path** → **[pathresult]**	subtree	**node** → **tree**
less	**char** × **char** → **bool**	label	**node** → **label**
		addTree	**tree** → **node**

Fig. 2. Builtin operators of X-Fun

The first block contains three functions, whose semantics are parameters of the language, and depend on the query language and data model. For a label l and a sequence of trees h, the function application makeTree(l, h) returns the data tree $l(h)$, if $l(h)$ is a well-formed data tree (e.g., in the XML data model attributes cannot have children) and raises an exception otherwise. The function evalPath(p) evaluates a path expression p. Whenever p is not well-formed (e.g., with respect to the XPATH 3.0 specification) an error is raised. Note that path expressions are X-Fun values, which means they can be computed dynamically by the X-Fun program using information from the input data tree. We will also define functions evalPath$_T$, on top of evalPath, for $T = $ **[node]**, **[string]**, etc. These functions verify (using a **match** expression with a typecase) that the result of the path call is of type T and raise an exception otherwise.

The next four operators are generic and do not depend on the specific kind of data trees. The variable nil refers to the empty list. A function application subtree(v) returns the subtree rooted at node v, while a function application label(v) returns the label of the node. The function addTree returns the identifier of the root node of the tree, and is used for storing the graph of the tree in the environment. This function can be used to access nodes of newly generated trees by starting path navigation from their root.

3.5 Syntactic Sugar

In the X-Fun snippets in the rest of the paper we shall employ some syntactic shortcuts, which enable us to express more succinctly some X-Fun constructs:

List Concatenation. We shall use the binary operator $*$ to concatenate two lists.

Simplified Patterns. When the type of a capture variable can be deduced from the matched expression we shall omit the "$: T$" in the capture pattern. This happens when the **match** expression is used to decompose lists and tuples instead of doing a typecase. For example, we shall simply write **match** $E \{ h :: t \rightarrow E_1, e \rightarrow E_2 \}$ to get the head and tail of a list.

Let-Declarations. We shall use the syntax **let** $x_1 = E_1, \ldots, x_n = E_n$ **in** E instead of **match** $(E_1, \ldots, E_n) \{ (x_1, \ldots, x_n) \rightarrow E \}$ as a more familiar way to declare variables.

Tuple Arguments. We shall allow tuple arguments to functions to be written without an extra pair of parentheses. I.e., $f(a, b)$ instead of $f((a, b))$. This is unambiguous since tuples always have at least two members.

3.6 Example

In Figure 3 we illustrate a transformation that converts an address book into HTML. The address fields are assumed to be unordered in the input data tree, while the fields of the output HTML addresses should be published in the order name, street, city and, phone.

```
<addresses>                              <ol>
<address>                                  <li>
  <name>Jemal Antidze</name>                 <p>Jemal Antidze</p>
  <phone>99532 305972</phone>                <p>Tblissi</p>
  <city>Tblissi</city>                       <p>Phone: 99532 305972</p>
  <phone>99532 231231</phone>                <p>Phone: 99532 231231</p>
</address>                         ⇒       </li>
<address>                                  <li>
  <name>Joachim Niehren</name>               <p>Joachim Niehren</p>
  <city>Lille</city>                         <p>Rue Esquermoise</p>
  <street>Rue Esquermoise</street>           <p>Lille</p>
</address>                                 </li>
</addresses>                             </ol>
```

Fig. 3. Publication of an address book in HTML except for secret entries

An X-Fun program defining this transformation is given in Figure 4. Starting at the root it first locates all address records, and applies the function convert_address to each of them. For each address record, the program first extracts the values of the fields name, street, and city located at some children of x. These values are then bound to variables named alike and later output as text nodes. The example program uses the standard map function, which can be defined in X-Fun for every T and T' as follows

```
map_{T→T'} = fun  x:  (T → T') × [T] → [T']  { match  x  {
    (f, head::tail)  →  f(head)::map_{T→T'}(f, tail)
    other  →  nil
} }
```

and the functions element and text, which are wrappers around makeTree which facilitate creation of nodes of the correct kind.

```
fun  book  :  tree→tree  {
  let  bookroot = addTree(book)  in
  let  convert_address = fun  x  :  node→tree  {
    let  name = evalPath_{[node]}("$x/child::name/text()"),
         street = evalPath_{[node]}("$x/child::street/text()"),
         city = evalPath_{[node]}("$x/child::city/text()")  in
    element("li",
      element("p", map_{node→tree}(name, subtree)) ::
      element("p", map_{node→tree}(street, subtree)) ::
      element("p", map_{node→tree}(city, subtree)) ::
      map_{string→tree}(
        fun  x:  string → tree  {
            element("p", text("Phone: " * x)::nil)
        }, evalPath_{[string]}("data($x/child::phone)"))
    )
  } in
  element("ol", map_{node→tree}(convert_address,
        evalPath_{[node]}("$bookroot/descendant::address")))
}
```

Fig. 4. X-Fun program converting address books to HTML

4   Translations from Other XML Languages

In this section, we briefly sketch translations from the standard XML processing languages, XSLT XQUERY and XPROC.

A more thorough treatment of this topic can be found in the full version of the paper.[10]

By implementing these three compilers, we obtain a uniform implementation of the whole XML processing stack based on a single X-Fun evaluator.

XSLT. Each template in the XSLT stylesheet is translated to a function in X-Fun. Furthermore, for each mode, we produce an additional function which implements the selection of the correct template from the set of templates associated with that mode according to their match patterns. The `call-template` and `apply-templates` instructions are translated as calls to the template or mode functions respectively. In the `copy-of` instruction, the nodes returned by the XPATH expression are copied to the output using the subtree function and strings and numbers are converted to a new text node with a call to makeTree. The instructions constructing elements, attributes and other XML nodes translate to corresponding calls to makeTree. The `for-each` instruction translates to a call to map, where the list to map over is produced by a call to evalPath and the mapping function is the body of the `for-each` instruction. Other XSLT instructions like `if` and `choose` can be translated similarly.

XQUERY. The feature that most distinguishes XQUERY is the SQL-like FLWOR expression. It enables the programmer to create a stream of tuples using the **for** and **let** clauses, filter them with a **where** clause and then reorder them using the **order by** clause. There is no single expression in X-Fun which covers this functionality, but it is easy to build it piecewise. Using several evalPath calls we can construct the list of tuples which corresponds to the tuple stream of XQUERY. Sorting and filtering of a list are functions easily definable in a functional language, and the functionality of **where** and **order by** is translated to calls to these functions. The sort and filter conditions are given again by calls to evalPath with the appropriate XPATH expression. Translation of other XQUERY constructs like the **if** expressions and functions proceeds in a straight forward manner.

XPROC. By encapsulating each processing step in a function, X-Fun can easily express the multi-stage processing which is inherent in XPROC. The pipelines then become simple function compositions. XPROC steps which invoke XQUERY or XSLT processing are handled by defining a function whose body is the translation of the respective program. Simple XPROC steps like `split-sequence`, which splits a sequence of documents into two based on an XPATH criterion are defined as normal X-Fun functions and provided as a library. The pipeline them simply calls these functions to do the required processing. The rest of the constructs like choosing among alternative subpipelines (`choose`) or looping over documents in a sequence are compiled to **match** and map expressions in X-Fun.

5   Implementation and Experiments

We have implemented a proof-of-concept X-Fun language evaluator in the Java programming language. We have instantiated X-Fun with the XML data model, using standard Java libraries for manipulating XML trees. We have used XPATH as the path language, as implemented by SAXON. We have used standard techniques for implementing functional languages, using the heap to store the values and the environment of the program and a stack for representing recursive function calls. We reduce an expression in all possible positions in an arbitrary order.

We have attempted to interface our implementation with TATOO, a highly efficient evaluator of an XPATH fragment based on [1]. Unfortunately, the penalty of crossing the language barrier (TATOO is implemented in OCAML) shadowed all performance gains from a faster implementation, so we could not perform any significant experiments. To see the difference in performance in using a faster XPATH implementation, we would need to implement X-Fun in OCAML as well.

We have also implemented the compilers of XSLT and XQUERY into X-Fun. In order to support real-world XSLT and XQUERY, they need support for additional features, like modules and various optional attributes of expressions in these languages (e.g., grouping with the `group-starting-with` attribute, etc.). However, none of these limitations are fundamental and they are not implemented because of their volume. The supported fragment is wide enough to run all queries from the XMARK [15] benchmark.

We don't have an XPROC compiler implementation, but for the purposes of testing we have run X-Fun on manually translated programs.

5.1   Experiments

To evaluate the performance of our implementation, we have compared it with the leading industry tool, the SAXON XSLT and XQUERY processor. To compare our performance on XPROC pipelines, we have used CALABASH, the most frequently used XPROC processor, as baseline. The tests were run on a computer with an Intel Core i7 processor running at 2.8 GHz, with 4GB of RAM and a SATA hard drive, running 64-bit Linux operating system.

First, we have compared the running time of our implementation on XQUERY programs. We used the queries from the XMARK benchmark, and the results are in Figure 5. The tests show that the running time of both tools is comparable. X-Fun is faster in case of simple queries (Q6, Q7, Q15, which contain just a simple loop), while SAXON is faster on queries involving joins (e.g., Q8, Q9, Q11). On the rest of queries our implementation of X-Fun is at most 20% slower that the competition, which we consider a good result as Saxon is a highly optimised industry tool, while we have not spent much time optimising the performance of our X-Fun implementation.

For the XSLT test, we used a transformation publishing an address book to HTML. The transformation in question is a more elaborate version of the program in Figure 4, and it includes about 40 XPATH expressions. The tests show that SAXON is about 4 times faster than our tool (for example, 15.7 vs. 63 seconds on a 200 MB document) and that the time of both tools scales linearly with the document size.

In the XPROC comparison, we have a simple pipeline consisting of 4 steps. First, it selects subtrees from the input document, splits the resulting sequence into two based on the presence of some node. The documents from the two sequences are then joined into pairs and these pairs are concatenated to form a single document again. We have compared the performance of CALABASH with

Query	X-Fun	SAXON	Query	X-Fun	SAXON	Query	X-Fun	SAXON
Q1	13.5	10.9	Q8*	962	592	Q15	12.0	14.4
Q2	13.6	12.9	Q9*	1235	705	Q16	13.6	11.8
Q3	14.0	12.5	Q10	314	222	Q17	13.9	12.4
Q4	16.7	12.8	Q11*	650	410	Q18	14.4	12.5
Q5	17.2	13.8	Q12	595	317	Q19	20.8	15.4
Q6	11.5	13.6	Q13	20.5	11.6	Q20	13.8	12.0
Q7	11.4	12.5	Q14	14.5	12.8			

Fig. 5. Running time in seconds of X-Fun and SAXON on queries from the XMARK benchmark on a 500 MB document. The three queries marked with '*', due to their complexity, were run on a 300 MB document.

our implementation of the pipeline in X-Fun. Both implementations show linear scalability with respect to size of the input and the pipeline, as can be seen in Figures 6 and 7 (for scaling the pipeline size, we simply composed the described pipeline with itself). However, our own implementation is consistently at least two times faster, and for the larger pipelines the difference is even more apparent. While the relatively low processing speed per megabyte can be explained by the need to create many small documents (the element per megabyte density is much higher compared to the previous tests), it is surprising to see an implementation specifically designed for processing XPROC be outperformed by our unoptimised implementation of the pipeline steps.

Document size	X-Fun	CALABASH
2 MB	8.7 s	16.6 s
4 MB	15.3 s	32.6 s
6 MB	23.1 s	51.8 s
8 MB	39.5 s	78.7 s

Pipeline size	X-Fun	CALABASH
1	8.7 s	16.6 s
2	12 s	75.8 s
3	16 s	136.6 s
4	22 s	198.6 s

Fig. 6. Performance of X-Fun and CALABASH on a fixed pipeline with varying input tree size

Fig. 7. Performance of X-Fun and CALABASH on a 2 MB document with varying pipeline size

6   Conclusion and Future Work

We have presented X-Fun, a language for processing data trees and shown that can serve as a uniform programming language for XML processing and as a uniform core language for implementing XQUERY, XSLT, and XPROC on top of any existing XPATH evaluator. Our implementation based on SAXON's in-memory XPATH evaluator yields surprisingly efficient implementations of the three W3C standards, even there is a lot of space left for optimisation. We have obtained results which are a match for the SAXON's XQUERY and XSLT evaluators and in the case of XPROC, first results show that we are already faster than CALABASH.

Our prime objective in future is to build streaming implementations of X-Fun, and thus of XQUERY, XSLT, and XPROC. The main ideas behind it are described in a technical report [9]. These streaming implementation will serve in the tools called QUIXQUERY, QUIXSLT, and QUIXPROC. A first version of QUIXSLT is freely available for testing on our online demo machine [6] while streaming is not yet available for our current QUIXPROC implementation.

Acknowledgement. We would like to thank Guiseppe Castagna and Kim Nguyen for their helpful discussions about the type system of X-Fun.

References

1. Arroyuelo, D., et al.: Fast in-memory xpath search using compressed indexes. In: ICDE, pp. 417–428. IEEE (2010)
2. Castagna, G., Im, H., Nguyen, K., Benzaken, V.: A Core Calculus for XQuery 3.0 (2013) (unpublished manuscript), http://www.pps.univ-paris-diderot.fr/gc/papers/xqueryduce.pdf
3. Frisch, A., Nakano, K.: Streaming XML Transformation Using Term Rewriting. In: Programming Language Technologies for XML (PLAN-X), pp. 2–13 (2007)
4. Fülöp, Z., Vogler, H.: Syntax-Directed Semantics – Formal Models based on Tree Transducers. In: EATCS Monographs in Theoretical CS. Springer (1998)
5. Hakuta, S., Maneth, S., Nakano, K., Iwasaki, H.: XQuery Streaming by Forest Transducers. In: ICDE, pp. 952–963. IEEE (2014)
6. Innovimax, INRIA Lille: Quix tools suite, https://project.inria.fr/quix-tool-suite/
7. Kay, M.: XSL Transformations (XSLT) Version 3.0. W3C Last Call Working Draft (2013), http://www.w3.org/TR/xslt-30
8. Kepser, S.: A Simple Proof for the Turing-Completeness of XSLT and XQuery. In: Proceedings of the Extreme Markup Languages® (2004)
9. Labath, P., Niehren, J.: A Functional Language for Hyperstreaming XSLT. Research report (March 2013), http://hal.inria.fr/hal-00806343
10. Labath, P., Niehren, J.: A Uniform Programming Language for Implementing XML Standards. Research report (January 2015), http://hal.inria.fr/hal-00954692
11. Neumann, A., Seidl, H.: Locating matches of tree patterns in forests. In: Arvind, V., Sarukkai, S. (eds.) FST TCS 1998. LNCS, vol. 1530, pp. 134–146. Springer, Heidelberg (1998)
12. Maneth, S., Berlea, A., Perst, T., Seidl, H.: XML type checking with macro tree transducers. In: PODS 2005, pp. 283–294. ACM-Press (2005)
13. Robie, J., et al.: XQuery 3.0: An XML Query Language. W3C Proposed Recommendation (2013), http://www.w3.org/TR/xquery-30
14. Saxonica: SAXON 9.5: The XSLT and XQuery Processor, http://saxonica.com
15. Schmidt, A., et al.: XMark: A Benchmark for XML Data Management. In: In 28th International Conference on VLDB (2002)
16. Walsh, N.: XML Calabash, http://xmlcalabash.com
17. Walsh, N., et al.: XProc: An XML Pipeline Language. W3C Recommendation (2010), http://www.w3.org/TR/xproc
18. Zergaoui, M.: Innovimax: QuiXProc, https://project.inria.fr/quix-tool-suite/quixproc/

OntoSDM: An Approach to Improve Quality on Spatial Data Mining Algorithms

Carlos Roberto Valêncio[1], Diogo Lemos Guimarães[1], Geraldo F.D. Zafalon[1], Leandro A. Neves[1], and Angelo C. Colombini[2]

[1] Instituto de Biociências, Letras e Ciências Exatas, São José do Rio Preto, Brazil
[2] Universidade Federal de São Carlos, São Carlos, Brazil
`{valencio,leandro}@ibilce.unesp.br`, `{diogolg06,zafalon}@gmail.com`,
`accolombini@dc.ufscar.br`

Abstract. The increase in new electronic devices had generated a considerable increase in obtaining spatial data information; hence these data are becoming more and more widely used. As well as for conventional data, spatial data need to be analyzed so interesting information can be retrieved from them. Therefore, data clustering techniques can be used to extract clusters of a set of spatial data. However, current approaches do not consider the implicit semantics that exist between a region and an object's attributes. This paper presents an approach that enhances spatial data mining process, so they can use the semantic that exists within a region. A framework was developed, OntoSDM, which enables spatial data mining algorithms to communicate with ontologies in order to enhance the algorithm's result. The experiments demonstrated a semantically improved result, generating more interesting clusters, therefore reducing manual analysis work of an expert.

Keywords: data mining, ontology, context-aware.

1 Introduction

Nowadays, companies are obtaining and using more and more data and that brings different challenges related to efficiency [1]. Besides that, it's important to be able to extract and understand information from databases. Furthermore, data that are being stored had changed and not always are traditional data, such as strings and integers, but are also spatial data. Those data are becoming more popular due to the use of electronic devices, like smartphones and GPS [2].

The advancement of these new technologies provided a shift in the context of information extraction in databases [3]. New techniques needed to be considered, since spatial data has unique characteristics that distinguish them from traditional data and are also considered more complex [4]. Therefore, spatial data mining can be defined as the extensive use of statistical methods of pattern recognition technology, artificial intelligence, machine knowledge, etc. The main goal is to extract understandable, interesting and initially unknown information from spatial databases, such as data management bases, business database, or remote sensors [5].

G.F. Italiano et al. (Eds.): SOFSEM 2015, LNCS 8939, pp. 555–565, 2015.

There are many different techniques that allow extracting spatial information where each one have a different approach, for instance, there are spatial association rules techniques, classification and clustering methods. Regarding clustering algorithm, the goal is to partition the spatial objects in groups, called clusters, allowing finding pattern distributions in the data [6]. This technique can be applied to perform analysis of satellite images, on geographic information system, marketing strategies and so on [7]. Normally, the clusters are calculated considering a minimum distance from a point to other neighbors or can also consider an object's neighborhood [8]. Newer algorithms adds to the clustering formation the use of similarity measures, where besides density shall also be considered the non-spatial attributes, thus cluster in which points have a greater degree of similarity between them are created [9].

However, existing techniques do not consider that the points are in a geographical region and that region holds an implicit semantic knowledge. Normally, those knowledge are only considered when an expert analyses the results manually, and using the context provided by regions it can assert which clusters are relevant.

When analyzing Figure 1 it is possible to realize the importance of considering the region during clustering generation. Considering that, it's possible to create new algorithms that will address different approaches that were not considered before, for instance, one can consider a new cluster by having points that have some attributes that are more relevant for a specific region, or even disregard points for being in a region less interesting.

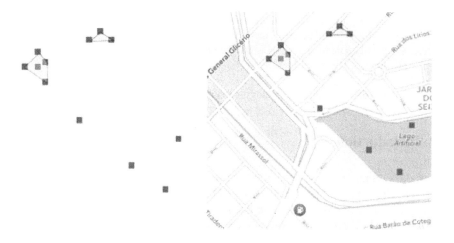

Fig. 1. Comparison between clusters without and with regions

Therefore, the developed work, OntoSDM, allows spatial data mining (SDM) algorithms to use the semantic relevance of a point in relation to their region. The semantic information is provided by a domain ontology created by the user, allowing enhancing the cluster generation process.

This paper is organized as follows: Section 2 presents the theoretical concepts needed to developed the work; in Section 3 the development stages of the approach are discussed;

in Section 4 the experiments and results are presented; and finally, in Section 5 the conclusion and future work are presented.

2   Concepts and Techniques

2.1   Spatial Clustering Methods

Spatial clustering techniques aim to group objects corresponding to the density and similarity between them. Accordingly, data in each cluster created have similar attributes among objects in the same cluster and as different as possible of others [6]. Researchers have developed several algorithms in this category where each one has a specific approach, and they can be organized in four methods: partitioning, hierarchical, density-based and grid-based methods [9].

The first, partitioning method, were used even before the term data mining became popular [8]. These algorithms [10, 11], normally need a parameter k as input so they can analyze the set of objects in a way to arrange them in k clusters. Algorithms belonging to the hierarchical method [11, 12] uses a tree structure called *dendrogram*, where the data from the database is recursively divide into subsets, which can be done using two approaches, "top-down" or "bottom-up".

Most partitioning algorithms can only find clusters in a spherical shape. On the other hand, density-based algorithm [13, 14] can generate clusters that have arbitrary shapes, and do so considering how dense is a point's surrounding prior to adding it to a cluster. Another advantages is that it is possible to eliminate isolated points, known as noises. One drawback to this method is the need of choosing good parameters to achieve a good result.

Finally, algorithms based on grid have better performance, reducing the processing time, moreover the number of objects do not directly impact the processing time. This is because, instead of considering each point individually, the whole area is divided into cells and if a particular cell contains a minimum number of points it is considered to be dense. The clusters are then formed by connecting the cells that were marked as dense.

2.2   Ontology

SDM algorithms usually have a manual analysis step, where an expert in the domain that is being analyzed will infer his knowledge in the algorithm's result. That allows to obtain a more interesting result, besides that it's possible to apply techniques that will improve future results.

Nowadays there is a trend in using ontologies techniques to express the context and infer knowledge about a specific domain [15]. Ontologies are normally chosen since they provide an efficient way to express a greater semantic representation due to its structure and inference capability [16]. There are many definitions for ontology applied to the computer field, one widely know defines it as a theoretical representation of a set of objects, properties of objects and the relationships between objects that are possible in a specific field of knowledge [17].

The use of ontologies has become quite popular for providing a common understanding, a shared knowledge and the ability to be able to share it with people and other applications for a variety of uses [18]. To ensure these efficient sharing and formalism, ontologies are usually based on the specifications of the semantic web [1].

Regarding this topic, it is important to understand the concepts of URI, universal resource identifier, which are a universal unique identifier for each resource in ontology. Furthermore, RDF, resource description framework, which is a data model that allows to describe resources using RDF triples that consists of a subject, a predicate and an object, as seen in Figure 2. New information is always added as a new RDF triple, by connecting them forming a graph structure.

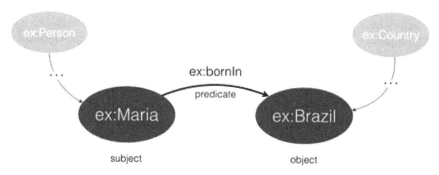

Fig. 2. A RDF triple example

As the first layer, RDF is limited in terms of semantic capability, since the main purpose of the layer is to ensure the basic framework structure for sharing information. The semantic concepts are added in the next layers as RDFS and OWL [19]. RDFS adds concepts related to classes and properties hierarchy, finally, OWL enhance the concepts introduced in RDFS adding complex structures based on description logic [18].

3   The OntoSDM Approach

This work aimed to develop a framework, able to extend SDM algorithms so they are able to use ontologies that represent the areas or regions of a domain that it is being analyzed. Thus it is considered a new attribute to a point, it's semantic weight.

The framework works as a middle tier between the triplestore repository and the SDM algorithm, Figure 3, display the representation of the architecture created. Furthermore, the developed approach is divided in two steps, preparation and execution.

The first stage is responsible to initialize and prepare the ontology. In this stage, the framework establishes a connection with the triplestore and obtain all defined classes in it. The user must then identify the ontology class that holds the region information. After defining the class, the user must specify weights for each predicate that are relating an ontology instance with other classes' instances described in the ontology. The weight range from -1 to 1, where -1 represents a lower relevance and 1 represents greater semantic relevance.

All the steps described earlier are executed during the preparation stage and the algorithm base has not being executed yet. The next stage concerns the logic and execution

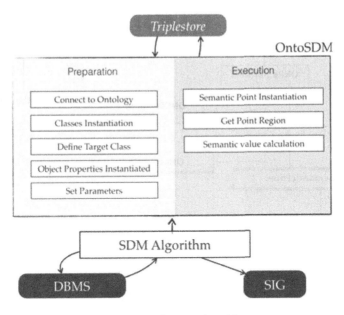

Fig. 3. OntoSDM framework architecture

stage, and it's executed during the execution of the SDM algorithm. Instead of the traditional point, that has only the coordinates information, the semantic point is initialized and a new attribute is considered, it's semantic weight. Figure 4 displays the simplified framework's model.

The two main classes are highlighted. The first, ONConection, is responsible for making the connection to the repository and perform all queries needed during the algorithm execution. The other class is the ONSemanticPoint, and it's responsible to hold the semantic point information and perform the semantic value calculation. To use the OntoSDM framework the SDM algorithm needs only to use the ONSemanticPoint class provided by the framework.

Firstly, the algorithm will obtain the spatial object, a point, from the database and will initialize it with the x and y coordinates. Since all ontology regions are known at the preparation stage, all information is transferred to classes, thus preventing to query the ontology unnecessarily. Therefore, it's possible to go through all regions and check if the point is within any known region.

If the point is within any known region each non-spatial attribute of the point it is then analyzed. First, the ontology is queried to verify if it has any information on the point's attributes regarding the region that the point is located. In case the ontology had any information, a list with all the object properties is returned and then it is checked whether the weights for the respective properties have been set by the user during the preparation stage. As stated above, the semantic weight ranges from -1 to 1, where a value of -1 is considered a low semantic relevance and 1 is a higher semantic relevance.

The formulas for calculating the semantic value are shown in (1), (2) and (3). The formula (1) indicates the value of A(p), where for a given point p which has, non-spatial

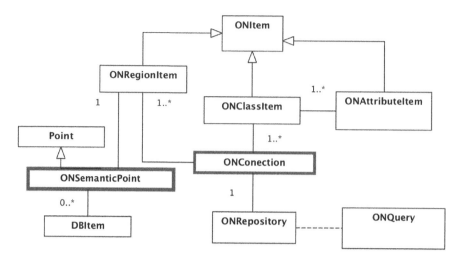

Fig. 4. OntoSDM class model illustration

attributes, o_i, such that, $i = 0 \ldots n$, the sum of non-spatial attributes that have a negative relevance value, o^-, is performed.

$$A(p) = \sum_{i=i}^{n} o_i^- \tag{1}$$

The value of $B(p)$ shown in formula (2) is obtained by summing the non-spatial attributes that have positive relevance, o^+, for a particular point p. Finally, the semantic value, $S(p)$ is calculated for the point p as shown in formula (3).

$$B(p) = \sum_{i=i}^{n} o_i^+ \tag{2}$$

$$S(p) = \frac{(A(p) + B(p))}{\max\{|A(p)|, B(p), 1\}} \tag{3}$$

This value is calculated to all points by checking in which region each one is and then getting all existing predicates and their respective weights.

With this approach, we introduce a new concept for geographical points that can be used in spatial data mining algorithms, it's semantic value that it's relative to the region that the point is located and it's non-spatial attributes. Lastly, the semantic point has all the necessary information, such as the point's region, list of non-spatial attributes, and especially the semantic value.

Therefore, it is needed now to consider what will be the approach and how the SDM algorithm will use the semantic information during its execution. They can use different strategies, such as disregarding the point immediately if it is a point that it's not consider to be semantic relevant in that region, or even, use the semantic value with other attributes to add a point to a new cluster that wouldn't be considered before and so on.

4   Experiments

For the experiments execution, the framework OntoSDM was applied in two SDM algorithms that have different approaches. The first, MRClustering, is a density-base algorithm based on VDBSCAN algorithm. The algorithm CHSMST+ is an extension from the graph-based algorithm CHSMST. The experiments were performed with the intention of proving that the approach can be used in different SDM clustering algorithms.

The experiments were then performed on a real database, which holds information on work accidents in the city of São José do Rio Preto, Brazil. The database has over 100 000 registered accidents where 30 000 are geo-referenced. The tests were applied using data sets ranging from 3 000 to 25 000 points, it was measured the execution time for each approach and the clusters formed were analyzed by a specialist.

Besides preparing the database, it was necessary to develop a domain ontology that represents the universe of work accidents in the city. Therefore, the ontology created for the experiments had the following structure:

- 4 regions (Commercial, Industrial, Residential and Hospitals);
- 50 objects involved, those represent objects that were related or cause the accident.
- 50 job occupation;
- 3 predicates that represent if the accident is considered to be frequent, common or rare.

4.1   Cluster Quality

The developed approach allows algorithms to choose how they want to use the semantic information provided by the OntoSDM. It is possible to disregard a point if it is not interesting or even include a new one if it's inside a region that makes it more relevant. The strategy used during the execution of the algorithm was to disregard a point if it is not considered semantically relevant to the region. The first experiments were aimed at validating the increase in quality of the results when compared with the algorithm with and without the developed approach for a given set of data.

Due to the strategy that it was used, disregarding non-relevant points, it was expected a reduction in the amount of generated clusters. This reduction provides better final results that facilitate further analysis that are made by an expert analyst. Table 1 shows the number of clusters obtained for the MRClustering algorithm. The reduction rates of clusters reached almost 25% with an average of 19% for all data sets.

Analyzing visually the result, it is possible to notice a general improvement. For instance, the industrial area of the city had 151 clusters and after applying the OntoSDM approach a total of 84 clusters was generated. In order to validate that only non-relevant points were disregarded, all cluster were analyzed. Following are shown some clusters that were not generated, in an industrial region, when using the OntoSDM approach.

- 9 clusters were removed since the points had "Industrial Machine" as object involved;
- 3 clusters had accidents that happened with "Saw" as object involved;
- 5 clusters where the job occupation was "Heavy machine operator";
- 12 cluster where the job occupation was "Welder".

Table 1. Quantity of cluster generated using the MRClustering algorithm

Points	Without OntoSDM	With OntoSDM
3.000	142 clusters	107 clusters
5.000	198 clusters	150 clusters
10.000	492 clusters	409 clusters
25.000	1 247 clusters	1 075 clusters

It is important to notice that the point is not disregarded if only one attribute is not relevant, but all attributes chosen by the users are taken into consideration to generate the semantic value, and only if this value is bellow the threshold is that the point is not considered relevant hence disregarded. This can be seen when analyzing the hospital regions, where even being described, in the ontology, that accidents inside this region that had the job occupation "Nursing Assistant" are considered frequent, 3 clusters were kept due to the other attributes.

Finally, how individual points are disregarded and not the cluster as a whole, it's possible to form more relevant and consistent clusters. Figure 5 shows the result by applying the algorithm CHSMST+. Each red point it's a work accident report, the blue regions represent the industrial areas of the city and the clusters are represented by the grey lines.

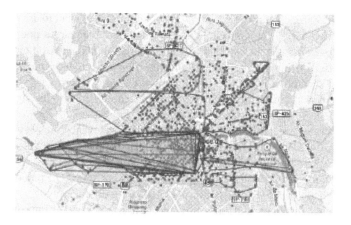

Fig. 5. Cluster Generated using the CHSMST+ Algorithm

As is possible to see, several clusters take a great area, almost crossing the entire city, although, when the OntoSDM approach is added the points that were in the industrial region are disregarded and that provides a more consistent and interested result, as can be seen in Figure 6.

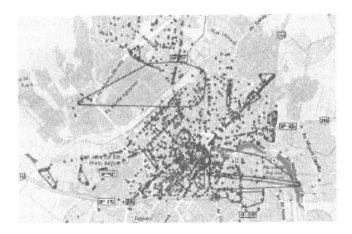

Fig. 6. Cluster Generated using The OntoSDM approach with the CHSMST+ Algorithm

5   Execution Time

The following experiments aimed to measure the impact caused by the OntoSDM approach when applied to a SDM algorithm. Therefore, the total running time of the algorithm was measured for each data set and it was then compared using both approaches, with and without using the OntoSDM. The execution times can be seen in Figure 7.

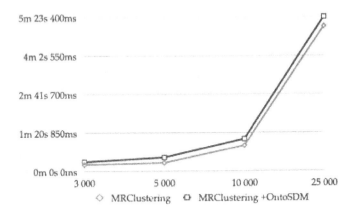

Fig. 7. Execution time

As expected, by using the OntoSDM, it caused an impact on the total execution time. To better analyze how the execution times are distributed, the 25 000 point data set were analyzed in more detail.

For this, the algorithm execution was divided into 4 stages: "Point instantiation", "Computing KdistSet and Eps", "Algorithm execution" and "Consolidation" as it's shown in Figure 8.

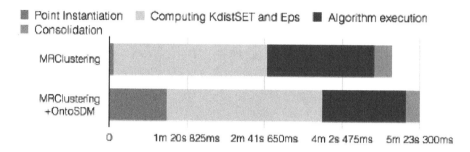

Fig. 8. Execution time detailed

As can be seen in the graph, the impact of the OntoSDM approach focuses on the instantiation stage. This is due to the fact that it needs to analyze each point that it's instantiate and calculate it's semantic value contained in the data set.

In the other hand, how the approach implemented disregard non-relevant points, those points do not need to be analyzed by following steps during the algorithm execution, such as the density and similarity calculation. Furthermore, another advantage is that the approach provided a reduction during the consolidation stage. That occurred since fewer clusters were generated preventing the algorithm to expend time processing and recording unnecessary clusters that will be discarded by an analyst in a future stage.

6   Conclusion

The developed approach achieved its objective by providing a novel approach that considers the relevance of a point's attributes in relation to the region that the point is located. With this data, algorithms can be adapted to use this information, allowing applying different strategies to get different results. The experiments showed a increase in the total time execution when compared with the same algorithm without using the OntoSDM approach, but the increase in time can be justified since the result presented a better quality in semantic.

In the future, it is intended to optimize the point instantiation stage and add other techniques to improve the results. For instance, manage situations when may have a region overlapping and enable the use of objects properties between different classes of the ontology.

References

1. Arvor, D., Durieux, L., Andrés, S., Laporte, M.-A.: Advances in Geographic Object-Based Image Analysis with ontologies: A review of main contributions and limitations from a remote sensing perspective. ISPRS J. Photogramm. Remote Sens. 82, 125–137 (2013)
2. Mennis, J., Guo, D.: Spatial data mining and geographic knowledge discovery—An introduction. Comput. Environ. Urban Syst. 33, 403–408 (2009)
3. Yueshun, H.: A study of spatial data mining architecture and technology. In: 2009 2nd IEEE Int. Conf. Comput. Sci. Inf. Technol., pp. 163–166 (2009)

4. Ji, M., Jin, F., Zhao, X., Ai, B., Li, T.: Mine geological hazard multi-dimensional spatial data warehouse construction research. In: 18th International Conference on Geoinformatics, pp. 1–5. IEEE (2010)
5. Hongfei, C., Xiaoyan, W.: Research on GIS-based spatial data mining. In: 2011 Int. Conf. Bus. Manag. Electron. Inf., pp. 351–354 (2011)
6. Han, J., Kamber, M.: Data mining: concepts and techniques. Morgan Kaufmann (2011)
7. Pattabiraman, V., et al.: A Novel Spatial clustering with Obstacles and Facilitators Co-straint Based on Edge Deduction and K- Mediods. In: Proceeding of 2009 International Conference on Computer Technology and Development, vol. 1, pp. 402–406. IEEE Computer Society, Kota Kinabalu (2009)
8. Bae, D.-H., Baek, J.-H., Oh, H.-K., Song, J.-W., Kim, S.-W.: SD-Miner: A spatial data mining system. In: 2009 IEEE Int. Conf. Netw. Infrastruct. Digit. Content, pp. 803–807 (2009)
9. Han, J., Kamber, M., Tung, A.K.H.: Spatial Clustering Methods in Data Mining: A Survey. Geogr. Data Min. Knowl. Discov. 1–29 (2001)
10. MacQueen, J.: Some Methods for Classification and Analysis of Multivariate Observations. In: Proc. fifth Berkeley Symp..., vol. 233, pp. 281–297 (1967)
11. Kaufman, L., Rousseeuw, P.: Finding groups in data: an introduction to cluster analysis (1990)
12. Karypis, G., Han, E.H., Kumar, V.: CHAMELEON: A Hierarchical Clustering Algorithm Using Dynamic Modeling. IEEE Comput (1999)
13. Ester, M., Kriegel, H., Sander, J., Xu, X.: A density-based algorithm for discovering clusters in large spatial databases with noise. In: KDD (1996)
14. Ankerst, M., Breunig, M.M., Kriegel, H.-P., Sander, J.: OPTICS: Ordering Points to Identify the Clustering Structure. In: Proceedings of the 1999 ACM SIGMOD International Conference on Management of Data, pp. 49–60. ACM, New York (1999)
15. Ongenae, F., Claeys, M., Dupont, T., Kerckhove, W., Verhoeve, P., Dhaene, T., De Turck, F.: A probabilistic ontology-based platform for self-learning context-aware healthcare applications. Expert Syst. Appl. 40, 7629–7646 (2013)
16. Ayres, R., Santos, M.P.: FOntGAR algorithm: Mining generalized association rules using fuzzy ontologies. Fuzzy Syst (FUZZ-IEEE),... 10–15 (2012)
17. Chandrasekaran, B., Josephson, J.R., Benjamins, V.R.: What are ontologies, and why do we need them? IEEE Intell. Syst. 14, 20–26 (1999)
18. Liao, S.-H., Chen, J.-L., Hsu, T.-Y.: Ontology-based data mining approach implemented for sport marketing. Expert Syst. Appl. 36, 11045–11056 (2009)
19. Allemang, D., Hendler, J.: Semantic web for the working ontologist: effective modeling in RDFS and OWL. Elsevier (2011)

Attribute-Based Encryption Optimized for Cloud Computing

Máté Horváth

Budapest University of Technology and Economics,
Department of Networked Systems and Services,
Laboratory of Cryptography and System Security (CrySyS Lab),
Budapest, Hungary
mhorvath@crysys.hu

Abstract. In this work, we aim to make attribute-based encryption (ABE) more suitable for access control to data stored in the cloud. For this purpose, we concentrate on giving to the encryptor full control over the access rights, providing feasible key management even in case of multiple independent authorities, and enabling viable user revocation, which is essential in practice. Our main result is an extension of the decentralized CP-ABE scheme of Lewko and Waters [6] with identity-based user revocation. Our revocation system is made feasible by removing the computational burden of a revocation event from the cloud service provider, at the expense of some permanent, yet acceptable overhead of the encryption and decryption algorithms run by the users. Thus, the computation overhead is distributed over a potentially large number of users, instead of putting it on a single party (e.g., a proxy server), which would easily lead to a performance bottleneck. The formal security proof of our scheme is given in the generic bilinear group and random oracle models.

Keywords: storage in clouds, access control, attribute-based encryption, multi-authority, user revocation.

1 Introduction

Recent trends show a shift from using companies' own data centres to outsourcing data storage to cloud service providers. Besides cost savings, flexibility is the main driving force for outsourcing data storage, although in the other hand it raises the issue of security, which leads us to the necessity of encryption. Traditional cryptosystems were designed to confidentially encode data to a target recipient (e.g. from Alice to Bob) and this seems to restrict the range of opportunities and flexibility offered by the cloud environment. Imagine the following scenario: some companies are cooperating on a cryptography project and from each, employees are working together on some tasks. Suppose that Alice wants to share some data of a subtask with those who are working on it, and with the managers of the project from the different companies. We see that encrypting this data with traditional techniques, causes that recipients must be

G.F. Italiano et al. (Eds.): SOFSEM 2015, LNCS 8939, pp. 566–577, 2015.

determined formerly, moreover either they has to share the same private key or several encrypted versions (with different keys) must be stored. These undermine the possible security, efficiency and the flexibility which the cloud should provide.

Attribute-based encryption (ABE) proposed by Sahai and Waters [12] is intended for one-to-many encryption in which ciphertexts are encrypted for those who are able to fulfil certain requirements. The most suitable variant for fine-grained access control in the cloud is called ciphertext-policy (CP-)ABE, in which ciphertexts are associated with access policies, determined by the encryptor and attributes describe the user, accordingly attributes are embedded in the users' secret keys. A ciphertext can be decrypted by someone if and only if, his attributes satisfy the access structure given in the ciphertext, thus data sharing is possible without prior knowledge of who will be the receiver preserving the flexibility of the cloud even after encryption.

Returning to the previous example, using CP-ABE Alice can encrypt with an access policy expressed by the following Boolean formula: "CRYPTOPROJECT" AND ("SUBTASK Y" OR "MANAGER"). Uploading the ciphertext to the cloud, it can be easily accessed by the employees of each company, but the data can be recovered only by those who own a set of attributes in their secret keys which satisfies the access policy (e.g. "CRYPTOPROJECT", "SUBTASK Y").

In spite of the promising properties, the adoption of CP-ABE requires further refinement. A crucial property of ABE systems is that they resist collusion attacks. In most cases (e.g. [2,14]) it is achieved by binding together the attribute secret keys of a specific user with a random number so that only those attributes can be used for decryption which contains the same random value as the others. As a result private keys must be issued by one central authority that would need to be in a position to verify all the attributes or credentials it issued for each user in the system. However even our example shows that attributes or credentials issued across different trust domains are essential and these have to be verified inside the different organisations (e.g. "MANAGER" attribute). To overcome this problem, we are going to make use of the results of Lewko and Waters [6] about decentralising CP-ABE.

The other relevant issue is user revocation. In everyday use, a tool for changing a user's rights is essential as unexpected events may occur and affect these. An occasion when someone has to be revoked can be dismissal or the revealing of malicious activity. Revocation is especially hard problem in ABE, since different users may hold the same functional secret keys related with the same attribute set (aside from randomization). We emphasise that user revocation is applied in *exceptional cases* like the above-mentioned, as all other cases can be handled simpler, with the proper use of attributes (e.g. an attribute can include its planned validity like "CRYPTOPROJECT2015").

Related Work. The concept of ABE was first proposed by Sahai and Waters [12] as a generalization of identity-based encryption. Bethencourt et al. [2] worked out the first ciphertext-policy ABE scheme in which the encryptor must decide who should or should not have access to the data that she encrypts (ciphertexts are associated with policies, and users' keys are associated with sets of descriptive attributes). This concept was further improved by Waters in [14].

The problem of building ABE systems with multiple authorities was first considered by Chase [3] with a solution that introduced the concept of using a global identifier (GID) for tying users' keys together. Her system relied on a central authority and was limited to expressing a strict AND policy over a pre-determined set of authorities. Decentralized ABE of Lewko and Waters [6] does not require any central authority and any party can become an authority while there is no requirement for any global coordination (different authorities need not even be aware of each other) other than the creation of an initial set of common reference parameters. With this it avoids placing absolute trust in a single designated entity, which must remain active and uncorrupted throughout the lifetime of the system. Several other multi-authority schemes (e.g. [10,13]) were shaped to the needs of cloud computing, although these lack for efficient user revocation.

Attribute revocation with the help of expiring attributes was proposed by Bethencourt et al. [2]. For single authority schemes Sahai et al. [11] introduced methods for secure delegation of tasks to third parties and user revocation through piecewise key generation. Ruj et al. [10], Wang et al. [13] and Yang et al. [15] show traditional attribute revocation (in multi-authority setting) causing serious computational overhead, because of the need for key re-generation and ciphertext re-encryption. A different approach is identity-based revocation, two types of which were applied to the scheme of Waters [14]. Liang et al. [9] gives the right of controlling the revoked set to a "system manager" while Li et al. [8], follow [5], from the field of broadcast encryption systems and give the revocation right directly to the encryptor. This later was further developed by Li et al. [7] achieving full security with the help of dual system encryption.

To the best of our knowledge no multi-authority system is integrated with identity-based user revocation and our work is the first in this direction.

Contribution. Based on [6] and [5] we propose a scheme that adds identity-based user revocation feature to distributed CP-ABE. With this extension, we achieve a scheme with multiple, independent attribute authorities, in which revocation of specific users (e.g. with ID_i) from the system with all of their attributes is possible without updates of attribute public and secret keys (neither periodically, nor after revocation event). We avoid re-encryption of all ciphertexts the access structures of which contain a subset of attributes of the revoked user. The revocation right can be given directly to the encryptor, just like the right to define the access structure which fits to the cloud computing scenario.

Organization. In section 2 we introduce the theoretical background that we use later and define the security of multi-authority CP-ABE schemes with ID-based revocation. In section 3 the details of our scheme can be found together with efficiency and security analysis. Directions for further research are proposed in the last section.

2   Background

We first briefly introduce bilinear maps, give formal definitions for access structures and relevant background on Linear Secret Sharing Schemes (LSSS). Then we give the algorithms and security definitions of Ciphertext Policy Attribute-Based Encryption with identity-based user revocation.

2.1   Bilinear Maps

We present the most important facts related to groups with efficiently computable bilinear maps.

Let $\mathbb{G}_0$ and $\mathbb{G}_1$ be two multiplicative cyclic groups of prime order p. Let g be a generator of $\mathbb{G}_0$ and e be a bilinear map (pairing), $e : \mathbb{G}_0 \times \mathbb{G}_0 \to \mathbb{G}_1$, with the following properties:

1. Bilinearity: $\forall u, v \in \mathbb{G}_1$ and $a, b \in \mathbb{Z}_p$, we have $e(u^a, v^b) = e(u, v)^{ab}$
2. Non-degeneracy: $e(g, g) \neq 1$.

We say that $\mathbb{G}_0$ is a bilinear group if the group operation in $\mathbb{G}_0$ and the bilinear map $e : \mathbb{G}_0 \times \mathbb{G}_0 \to \mathbb{G}_1$ are both efficiently computable. Notice that the map e is symmetric since $e(g^a, g^b) = e(g, g)^{ab} = e(g^b, g^a)$.

2.2   Access Structures

Definition 1 (Access Structure [1]). *Let $\{P_1, \ldots, P_n\}$ be a set of parties. A collection $\mathbb{A} \subseteq 2^{\{P_1, \ldots, P_n\}}$ is monotone if $\forall B, C :$ if $B \in \mathbb{A}$ and $B \subseteq C$ then $C \in \mathbb{A}$. An access structure (respectively, monotone access structure) is a collection (respectively, monotone collection) $\mathbb{A}$ of non-empty subsets of $\{P_1, \ldots, P_n\}$, i.e., $\mathbb{A} \subseteq 2^{\{P_1, \ldots, P_n\}} \setminus \{\emptyset\}$. The sets in $\mathbb{A}$ are called the authorized sets, and the sets not in $\mathbb{A}$ are called the unauthorized sets.*

In our case the access structure $\mathbb{A}$ will contain the authorized sets of attributes, furthermore we restrict our attention to monotone access structures. However, it is possible to (inefficiently) realize general access structures using our techniques by having the not of attributes as separate attributes as well.

2.3   Linear Secret Sharing Schemes (LSSS)

To express the access control policy we will make use of LSSS. Here we adopt the definitions from those given in [1].

Definition 2 (Linear Secret Sharing Scheme). *A secret-sharing scheme Π over a set of parties $\mathcal{P}$ is called linear (over $\mathbb{Z}_p$) if*

1. *the shares for each party form a vector over* $\mathbb{Z}_p$,
2. *there exists a matrix A with ℓ rows and n columns called the share-generating matrix for Π. For all $i = 1, \ldots, \ell$, the ith row of A let the function ρ defined the party, labelling row i as $\rho(i)$. When we consider the column vector $v = (s; r_2, \ldots, r_n)$, where $s \in \mathbb{Z}_p$ is the secret to be shared, and $r_2, \ldots, r_n \in \mathbb{Z}_p$ are randomly chosen, then $Av = \lambda$ is the vector of ℓ shares of the secret s according to Π. The share $(Av)_i = \lambda_i$ belongs to party $\rho(i)$.*

In [1] it is shown that every linear secret sharing-scheme according to the above definition also enjoys the *linear reconstruction property*, defined as follows. Suppose that Π is an LSSS for the access structure $\mathbb{A}$. Let $S \in \mathbb{A}$ be any authorized set, and let $I \subset \{1, 2, \ldots, \ell\}$ be defined as $I = \{i | \rho(i) \in S\}$. Then, there exist constants $\{\omega_i \in \mathbb{Z}_p\}_{i \in I}$ such that, if $\{\lambda_i\}$ are valid shares of any secret s according to Π, then $\sum_{i \in I} \omega_i \lambda_i = s$. Furthermore, it is also shown in [1] that these constants $\{\omega_i\}$ can be found in time polynomial in the size of the share-generating matrix A and for unauthorized sets, no such $\{\omega_i\}$ constants exist.

We use the convention that $(1, 0, 0, \ldots, 0)$ is the "target" vector for any linear secret sharing scheme. For any satisfying set of rows I in A, we will have that the target vector is in the span of I, but for any unauthorized set, it is not.

Using standard techniques (see [6] - Appendix G) one can convert any monotonic boolean formula into an LSSS representation. An access tree of ℓ nodes will result in an LSSS matrix of ℓ rows.

2.4 Revocation Scheme for Multi-authority CP-ABE

A multi-authority Ciphertext-Policy Attribute-Based Encryption system with identity-based user revocation is comprised of the following algorithms:

Global Setup$(\lambda) \to GP$. The global setup algorithm takes in the security parameter λ and outputs global parameters GP for the system.

Central Authority Setup$(GP) \to (SK^*, PK^*)$. The central authority runs this algorithm with GP as input to produce its own secret key and public key pair, SK^*, PK^*.

Identity KeyGen$(GP, RL, GID) \to K^*_{GID}$. The central authority runs this algorithm upon a user request for identity secret key. It checks whether the request is valid and if yes, generates K^*_{GID}.

Authority Setup$(GP) \to (PK, SK)$. Each attribute authority runs the authority setup algorithm with GP as input to produce its own secret key and public key pair, SK, PK.

KeyGen$(GP, SK, GID, i) \to K_{i,GID}$. The attribute key generation algorithm takes in an identity GID, the global parameters, an attribute i belonging to some authority, and the secret key SK for this authority. It produces a key $K_{i,GID}$ for this attribute, identity pair.

Encrypt$(GP, \mathcal{M}, (A, \rho), \{PK\}, PK^*, RL) \to CT$. The encryption algorithm takes in a message $\mathcal{M}$, an access matrix (A, ρ), the set of public keys for relevant authorities, the public key of the central authority, the revoked user list and the global parameters. It outputs a ciphertext CT.

Decrypt$(GP, CT, (A, \rho), \{K_{i,GID}\}, K^*_{GID}, RL) \rightarrow \mathcal{M}$. The decryption algorithm takes in the global parameters, the revoked user list, the ciphertext, identity key and a collection of keys corresponding to attribute, identity pairs all with the same fixed identity GID. It outputs either the message $\mathcal{M}$ when the collection of attributes i satisfies the access matrix corresponding to the ciphertext. Otherwise, decryption fails.

2.5   Security Model

We now define (chosen plaintext) security of multi-authority CP-ABE system with identity-based revocation. Security is defined using the following *Security Game* between an attacker algorithm $\mathcal{A}$ and a challenger. We assume that adversaries can corrupt authorities only statically, but key queries are made adaptively. The definition reflects the scenario where all users in the revoked set RL get together and collude (this is because the adversary can get all of the private keys for the revoked set). The game is the following:

Setup. The challenger runs the Global Setup algorithm to obtain the global public parameters GP. $\mathcal{A}$ specifies a set $AA' \subseteq AA$ of corrupt attribute authorities and uses the Authority Setup to obtain public and private keys. For honest authorities in $AA \setminus AA'$ and for the Central Authority, the challenger obtains the corresponding keys by running the Authority Setup and Central Authority Setup algorithms, and gives the public keys to the attacker.

Key Query Phase. $\mathcal{A}$ adaptively issues private key queries for identities GID_k (which denotes the kth GID query). The challenger gives $\mathcal{A}$ the corresponding identity keys $K^*_{GID_k}$ by running the Identity KeyGen algorithm. Let UL denote the set of all queried GID_k. $\mathcal{A}$ also makes attribute key queries by submitting pairs of (i, GID_k) to the challenger, where i is an attribute belonging to a good authority. The challenger responds by giving the attacker the corresponding key, K_{i,GID_k}.

Challenge. The attacker gives the challenger two messages M_0, M_1, a set $RL \subseteq UL$ of revoked identities and an access matrix (A, ρ).

RL and A must satisfy the following constraints. Let V denote the subset of rows of A labelled by attributes controlled by corrupt authorities. For each identity $GID_k \in UL$, let V_{GID_k} denote the subset of rows of A labelled by attributes i for which the attacker has queried (i, GID_k). For each $GID_k \in UL \setminus RL$, we require that the subspace spanned by $V \cup V_{GID_k}$ must not include $(1, 0, \ldots, 0)$ while for $GID_k \in RL$, it is allowed and we only require that the subspace spanned by V must not include $(1, 0, \ldots, 0)$. (In other words, the attacker cannot ask for a set of keys that allow decryption, in combination with any keys that can be obtained from corrupt authorities in case of a non revoked GID_k. For revoked identities we only do not allow corrupted attributes to satisfy the access structure alone.)

The attacker must also give the challenger the public keys for any corrupt authorities whose attributes appear in the labelling ρ.

The challenger flips a random coin $\beta \in (0, 1)$ and sends the attacker an encryption of M_β under access matrix (A, ρ) with the revoked set RL.

Key Query Phase 2. The attacker may submit additional attribute key queries (i, GID_k), as long as they do not violate the constraint on the challenge revocation list RL and matrix (A, ρ).

Guess. $\mathcal{A}$ must submit a guess β' for β. The attacker wins if $\beta' = \beta$. The attacker's advantage in this game is defined to be $\mathbb{P}(\beta' = \beta) - \frac{1}{2}$.

Definition 3. *We say that a multi-authority CP-ABE system with identity-based revocation is (chosen-plaintext) secure (against static corruption of attribute authorities) if, for all revocations sets RL of size polynomial in the security parameter, all polynomial time adversary have at most a negligible advantage in the above defined security game.*

3   Our Results

To build our model we will use the prime order group construction of Lewko and Waters [6], because of its favourable property of having independent attribute authorities. In order to achieve identity-based revocation we supplement the distributed system with a Central Authority. However it seems to contradict with the original aim of distributing the key generation right, this additional authority would generate only secret keys for global identifiers $(GID \in \mathbb{Z}_p)$ of users and the attribute key generation remains distributed. Our Central Authority does not possess any information that alone would give advantage during decryption, in contrast to single authority schemes, where the authority is able to decrypt all ciphertexts. Regarding this, we can say that our system remains distributed, in spite of launching a Central Authority.

Our Approach to the Cloud Storage Scenario. We give a high-level description about a possible application of the algorithms that we proposed in subsection 2.4. Because of efficiency reasons data should be encrypted by a symmetric cipher, always using fresh random number as key, which is also encrypted, but with our scheme and in this form attached to the ciphertext that is stored by the cloud service provider (CSP). Decryption is possible for users, who can obtain the symmetric key, or with other words those, who possess the necessary attributes and were not revoked. Attribute Authorities are run locally on trusted servers of organisations, that are using the system, while the Central Authority is run by the CSP, which also maintains (archives, publishes) the RL revocation list, based on the revocation requests from authorised parties from the organisations. The ABE encryption always uses the fresh RL and ABE decryption is run with the RL at the encryption time of the ciphertext, which are obtained from the CSP. This approach automatically leads to lazy re-encryption of ciphertext, as fresh symmetric key and RL are used whenever data is edited.

Our Technique. We face with the challenges of identity-based revocation. To realize the targeted features, we use some ideas from public key broadcast encryption systems [5]. We use secret sharing in the exponent. Suppose an encryption algorithm needs to create an encryption with a revocation set $RL = GID_1^*, \ldots, GID_r^*$

of r identities. The algorithm will create an exponent $s^* \in \mathbb{Z}_p$ and split it into r random shares $s_1, \ldots, s_r$ such that $\sum_{k=1}^{r} s_k = s^*$. It will then create a ciphertext such that any revoked user with GID_k^* will not be able to incorporate the kth share and thus not decrypt the message.

This approach presents the following challenges. First, we need to make crucial that the decryptor need to do the GID comparisons even if his attributes satisfy the access structure of the ciphertext. Second we need to make sure that a user with revoked identity GID_k^* cannot do anything useful with share k. Third, we need to worry about collusion attacks between multiple revoked users.

To address the first one we are going to take advantage of the technique of [6] that is used to prevent collusion attacks. Here the secret s, used for the encryption, is divided into shares, which are further blinded with shares of zero. This structure allows for the decryption algorithm to both reconstruct the main secret and to "unblind" it in parallel. When we would like to make this algorithm necessary, but not enough for decryption it is straightforward to spoil the "unblinding" of the secret by changing the shares of zero in the exponent to shares of an other random number, $s^* \in \mathbb{Z}_p$. Thus we can require an other computation, namely the comparison of the decryptor's and the revoked users' GIDs. If correspondence is found, the algorithm stops, otherwise reveals the blinding, enabling decryption.

The second challenge is addressed by the following method. A user with $GID \neq GID_k^*$ can obtain two linearly independent equations (in the exponent) involving the share s_k, which he will use to solve for the share s_k. However, if $GID = GID_k^*$, the obtained equations are going to be linearly dependent and the user will not be able to solve the system.

In the third case, the attack we need to worry about is where a user with GID_k^* processes ciphertext share l, while another user with GID_l^* processes share k, and then they combine their results. To prevent collusion, we use $H(GID)$ as the base of the identity secret key, such that in decryption each user recovers shares $s_k \cdot \log_g H(GID)$ in the exponent, disallowing the combination of shares from different users.

3.1   Our Construction

Based on the above principles, the proposed algorithms are the following:

Global Setup$(\lambda) \to GP$
  In the global setup, a bilinear group $\mathbb{G}_0$ of prime order p is chosen. The global public parameters, GP, are p and a generator g of $\mathbb{G}_0$, and a function H mapping global identities $GID \in \mathbb{Z}_p$ to elements of $\mathbb{G}_0$ (this is modelled as a random oracle in the security proof).

Central Authority Setup$(GP) \to (SK^*, PK^*)$
  The algorithm chooses random exponents $a, b \in \mathbb{Z}_p$, keeps them as secret key $SK^* = \{a, b\}$ and publishes $PK^* = \{g^a, g^{1/b}\}$.

Identity KeyGen$(GP, RL, GID, SK^*) \rightarrow K^*_{GID}$

Upon the request of a user it first checks whether the user is on the list of revoked users (RL) or it has been queried before, if yes refuses the request, otherwise computes $H(GID)$ and generates the global identity secret key:

$$K^*_{GID} = H(GID)^{(GID+a)b}.$$

Authority Setup$(GP) \rightarrow (PK, SK)$

For each attribute i belonging to the authority (these indices i are not reused between authorities), the authority chooses two random exponents $\alpha_i, y_i \in \mathbb{Z}_p$ and publishes $PK = \{e(g,g)^{\alpha_i}, g^{y_i} \; \forall i\}$ as its public key. It keeps $SK = \{\alpha_i, y_i \; \forall i\}$ as its secret key.

KeyGen$(GP, SK, GID, i) \rightarrow K_{i,GID}$

To create a key for a GID, for attribute i belonging to an authority, the authority computes:

$$K_{i,GID} = g^{\alpha_i} H(GID)^{y_i}$$

Encrypt$(GP, \mathcal{M}, (A, \rho), \{PK\}, PK^*, RL) \rightarrow CT$

The encryption algorithm takes in a message $\mathcal{M}$, an $n \times \ell$ access matrix A with ρ mapping its rows to attributes, the global parameters, the public keys of the relevant authorities, the user identity public key and the most recent list of revoked users.

It chooses random $s, s^* \in \mathbb{Z}_p$ and a random vector $v \in \mathbb{Z}_p^\ell$ with s as its first entry. Let λ_x denote $A_x \cdot v$, where A_x is row x of A. It also chooses a random vector $w \in \mathbb{Z}_p^\ell$ with s^* as its first entry. Let ω_x denote $A_x \cdot w$.

For each row A_x of A, it chooses a random $r_x \in \mathbb{Z}_p$ and supposed that the number of revoked users is $|RL| = r$ it chooses s_k such that $s^* = \sum_{k=1}^r s_k$. The CT ciphertext is computed as

$$C_0 = \mathcal{M} \cdot e(g,g)^s,$$
$$C_{1,x} = e(g,g)^{\lambda_x} e(g,g)^{\alpha_{\rho(x)} r_x}, C_{2,x} = g^{r_x}, C_{3,x} = g^{y_{\rho(x)} r_x} g^{\omega_x} \quad \forall x = 1, \ldots, n$$
$$C^*_{1,k} = \left(g^a g^{GID^*_k}\right)^{-s_k}, C^*_{2,k} = g^{s_k/b} \quad \forall k = 1, \ldots, r.$$

Decrypt$(GP, CT, (A, \rho), \{K_{i,GID}\}, K^*_{GID}, RL) \rightarrow \mathcal{M}$

We assume the ciphertext is encrypted under an access matrix (A, ρ). If the decryptor is not on the list of revoked users (RL) and has the secret keys K^*_{GID} for his GID and $\{K_{i,GID}\}$ for a subset of rows A_x of A, such that $(1, 0, \ldots, 0)$ is in the span of these rows, then the decryptor proceeds as follows. First chooses constants $c_x \in \mathbb{Z}_p$ such that $\sum_x c_x A_x = (1, 0, \ldots, 0)$ and denoting $r = |RL|$ computes:

$$\frac{\prod_x \left(\frac{C_{1,x} \cdot e(H(GID), C_{3,x})}{e(K_{\rho(x),GID}, C_{2,x})}\right)^{c_x}}{\prod_{k=1}^r \left(e(K^*_{GID}, C^*_{2,k}) e(C^*_{1,k}, H(GID))\right)^{1/(GID-GID^*_k)}} = e(g,g)^s$$

The message then can be obtained as : $\mathcal{M} = C_0/e(g,g)^s$.

To see the soundness of the Decryption algorithm observe the following:

$$\mathscr{A} = \prod_x \left(\frac{C_{1,x} \cdot e(H(GID), C_{3,x})}{e(K_{\rho(x), GID}, C_{2,x})} \right)^{c_x} = \prod_x \left(e(g,g)^{\lambda_x + \omega_x \log_g H(GID)} \right)^{c_x}$$

$$= e(g,g)^{\sum_x \lambda_x c_x} \cdot e(H(GID), g)^{\sum_x \omega_x c_x} = e(g,g)^{s + s^* \log_g H(GID)}$$

$$\mathscr{B} = \prod_{k=1}^r \left(e(K_{GID}^*, C_{2,k}^*) e(C_{1,k}^*, H(GID)) \right)^{-1/(GID - GID_k^*)}$$

$$= \prod_{k=1}^r \left(e(g,g)^{(GID - GID_k^*) s_k \log_g H(GID)} \right)^{-1/(GID - GID_k^*)}$$

$$= e(g,g)^{-\sum_{k=1}^r s_k \log_g H(GID)} = e(g,g)^{-s^* \log_g H(GID)}$$

Remarks. Supposing that we have a honest but curious CSP, which does not collude with the users, it is also possible to achieve indirect revocation (similarly to [9,11]), with simple modifications on our scheme. With other words, the CSP could fully supervise user revocation based on the revocation requests from parties, authorised for this. We only need to modify the Encrypt algorithm to compute $C, C_0, C_{1,x}, C_{2,x}$ as originally and $C'_{3,x} = g^{y_{\rho(x)} r_x} \quad \forall x = 1, \ldots, n$. These values would form CT' that is sent to the CSP, where the collusion resistant CT with the revocation information is computed and published. CT has the same form as earlier, the only difference is that the blinding vector w is chosen by the CSP, so $\omega_x, C_{1,k}^*, C_{2,k}^*$ (as previously) and $C_{3,x} = C'_{3,x} \cdot g^{\omega_x}$ are computed also by the CSP. The main advantage of this approach is that immediate and efficient (partial) re-encryption can be achieved as only $w, s_k, \omega_x, C_{1,k}^*, C_{2,k}^*$ and $C_{3,x}$ need to be recomputed after a revocation event.

Alternatively, it is also possible to give revocation right directly to the encryptor by simply publishing a user list instead of RL. In this case RL would be defined by the user, separately for each ciphertext, and attached to CT.

3.2 Efficiency

Traditional, attribute-based user revocation (e.g. [13,10,15]) affects attributes, thus the revocation of a user may cause the update of all the users' attribute secret keys who had common attribute with the revoked user (a general attribute can affect big proportion of the users) and the re-encryption of all ciphertext the access structure of which contain any of the revoked user's attributes (most of these could not be decrypted by the revoked user).

In our scheme, a revocation event does not have any effect on the attributes as it is based on identity. Although it is a trade-off and in the other hand there is some computational overhead on the encryption and decryption algorithms. In this way the necessary extra computation of authorities is reduced and distributed between the largest set of parties, the users, preventing a possible performance bottleneck of the system. At the same time the extra communication is also reduced to the publication of the revoked user list. Our revocation scheme has the following costs.

The ciphertext has $2r$ additional elements, if the number of revoked users is r. For the computation of these values $3r$ exponentiations and r multiplications are needed in $\mathbb{G}_0$. Alternatively, the revoked user list may contain $g^a g^{GID_i^*}$ instead of the global identifiers. In this case the encryptor need to do only $2r$ additional exponentiations in $\mathbb{G}_0$, compared with the scheme of [6], to compute the ciphertext. The overhead of the decryption algorithm is $2r$ pairing operations, r multiplications and exponentiations in group $\mathbb{G}_1$.

3.3   Security

We point out that from the point of view of a user, whose attributes have never satisfied the access structure defined in the ciphertext, our construction is at least as secure as the one by [6], because the computation of $\mathscr{A}$ is equivalent to the decryption computation given there. However in our case, it is not enough to obtain the message. Changing the first entry of the blinding vector w from zero to a random number (as we did), causes that the blinding will not cancel out from $\mathscr{A}$, but we need to compute $\mathscr{B}$ which can divide it out. $\mathscr{B}$ can be computed with any GID different from any GID_k^* of the revocation list and we ensure that the decryptor must use the same GID both in $\mathscr{A}$ and $\mathscr{B}$ by using $H(GID)$ in the keys.

Theorem 1. *For any adversary $\mathcal{A}$, let q be a bound on the total number of group elements it receives from queries it makes to the group oracles and from its interaction with the security game, described in 2.5. The above described construction is secure according to Definition 3 in the generic bilinear group and random oracle models. The advantage of $\mathcal{A}$ is $\mathcal{O}(q^2/p)$.*

Our construction is proven to be secure in the generic bilinear group model previously used in [2,6], modelling H as a random oracle. Security in this model assures us that an adversary cannot break our scheme with only black-box access to the group operations and H. At an intuitive level, this means that if there are any vulnerabilities in our scheme, then these vulnerabilities must exploit specific mathematical properties of elliptic curve groups or cryptographic hash functions used when instantiating our construction. The formal proof can be found in the full length version of this paper in [4].

4   Future Work

We proposed a scheme for efficient identity-based user revocation in multi-authority CP-ABE. In the future, our work can be continued in several directions.

The method of identity-based user revocation can be the foundation of a future method that allows non monotonic access structures in multi-authority setting. However our scheme cannot be applied directly for this purpose, it may be used to develop ideas in this field.

The security of our construction is proved in the generic bilinear group model, although we believe it would be possible to achieve full security by adapting the dual system encryption methodology, which was also used by Lewko and Waters [6] in their composite order group construction. This type of work would be interesting even if it resulted in a moderate loss of efficiency from our existing system.

Acknowledgements. This work was started as a master thesis at Eötös Loránd University, in the Security&Privacy program of EIT ICT Labs Masterschool. I am thankful to my supervisor, Levente Buttyán from CrySyS Lab for his help and valuable comments.

References

1. Beimel, A.: Secure schemes for secret sharing and key distribution. Ph.D. thesis, Israel Institute of Technology, Technion, Haifa, Israel (1996)
2. Bethencourt, J., Sahai, A., Waters, B.: Ciphertext-policy attribute-based encryption. In: IEEE Symposium on Security and Privacy, pp. 321–334 (2007)
3. Chase, M.: Multi-authority Attribute Based Encryption. In: Vadhan, S.P. (ed.) TCC 2007. LNCS, vol. 4392, pp. 515–534. Springer, Heidelberg (2007)
4. Horváth, M.: Attribute-Based Encryption Optimized for Cloud Computing. Cryptology ePrint Archive, Report 2014/612 (2014), `http://eprint.iacr.org/`
5. Lewko, A., Sahai, A., Waters, B.: Revocation systems with very small private keys. In: IEEE Symposium on Security and Privacy, pp. 273–285 (2010)
6. Lewko, A., Waters, B.: Decentralizing attribute-based encryption. In: Paterson, K.G. (ed.) EUROCRYPT 2011. LNCS, vol. 6632, pp. 568–588. Springer, Heidelberg (2011)
7. Li, Q., Xiong, H., Zhang, F.: Broadcast revocation scheme in composite-order bilinear group and its application to attribute-based encryption. International Journal of Security and Networks 8(1), 1–12 (2013)
8. Li, Y., Zhu, J., Wang, X., Chai, Y., Shao, S.: Optimized Ciphertext-Policy Attribute-Based Encryption with Efficient Revocation. International Journal of Security & Its Applications 7(6) (2013)
9. Liang, X., Lu, R., Lin, X., Shen, X.S.: Ciphertext policy attribute based encryption with efficient revocation. TechnicalReport, University of Waterloo (2010)
10. Ruj, S., Nayak, A., Stojmenovic, I.: Dacc: Distributed access control in clouds. In: IEEE 10th International Conference on Trust, Security and Privacy in Computing and Communications, pp. 91–98 (2011)
11. Sahai, A., Seyalioglu, H., Waters, B.: Dynamic credentials and ciphertext delegation for attribute-based encryption. In: Safavi-Naini, R., Canetti, R. (eds.) CRYPTO 2012. LNCS, vol. 7417, pp. 199–217. Springer, Heidelberg (2012)
12. Sahai, A., Waters, B.: Fuzzy identity-based encryption. In: Cramer, R. (ed.) EUROCRYPT 2005. LNCS, vol. 3494, pp. 457–473. Springer, Heidelberg (2005)
13. Wang, G., Liu, Q., Wu, J., Guo, M.: Hierarchical attribute-based encryption and scalable user revocation for sharing data in cloud servers. Computers & Security 30(5), 320–331 (2011)
14. Waters, B.: Ciphertext-policy attribute-based encryption: An expressive, efficient, and provably secure realization. In: Catalano, D., Fazio, N., Gennaro, R., Nicolosi, A. (eds.) PKC 2011. LNCS, vol. 6571, pp. 53–70. Springer, Heidelberg (2011)
15. Yang, K., Jia, X., Ren, K., Zhang, B.: DAC-MACS: Effective data access control for multi-authority cloud storage systems. In: INFOCOM, 2013 Proceedings IEEE, pp. 2895–2903 (2013)

Trustworthy Virtualization of the ARMv7 Memory Subsystem

Hamed Nemati, Roberto Guanciale, and Mads Dam

KTH Royal Institute of Technology, Stockholm, Sweden
{hnnemati,robertog,mfd}@kth.se

Abstract. In order to host a general purpose operating system, hypervisors need to virtualize the CPU memory subsystem. This entails dynamically changing MMU resources, in particular the page tables, to allow a hosted OS to reconfigure its own memory. In this paper we present the verification of the isolation properties of a hypervisor design that uses direct paging. This virtualization approach allows to host commodity OSs without requiring either shadow data structures or specialized hardware support. Our verification targets a system consisting of a commodity CPU for embedded devices (ARMv7), a hypervisor and an untrusted guest running Linux.The verification involves three steps: (i) Formalization of an ARMv7 CPU that includes the MMU, (ii) Formalization of a system behavior that includes the hypervisor and the untrusted guest (iii) Verification of the isolation properties. Formalization and proof are done in the HOL4 theorem prover, thus allowing to re-use the existing HOL4 ARMv7 model developed in Cambridge.

Keywords: formal verification, hypervisor, memory management.

1   Introduction

Memory isolation is a key requirement of systems executing software at different privilege levels. Inevitable security flaws of COTS OSes make providing trustworthy memory isolation using only OS level security mechanisms infeasible. Alternatively, memory isolation can be provided by leveraging special low-level execution platforms, like hypervisors. The small code-base of hypervisors makes the verification of their memory isolation properties feasible. In that way, a hypervisor can be used to host a commodity OS (which provides the non-critical services) along with several critical components, which are deployed in isolated partitions.

In this paper we present the formal verification of the memory isolation properties guaranteed by a hypervisor for embedded systems. Here, we focus on the main functionality of the hypervisor namely the virtualization of the memory subsystem, which determines the binding of physical memory locations to locations addressable at the application level. In order to properly isolate partitions, the hypervisor takes control of the memory configuration, by configuring the MMU and the corresponding page tables. Moreover, in order for such a hypervisor to host a general purpose OS it is necessary to allow the guest OS to

G.F. Italiano et al. (Eds.): SOFSEM 2015, LNCS 8939, pp. 578–589, 2015.

dynamically reconfigure its internal memory hierarchy and to impose its own access restrictions. The virtualization of the memory subsystem must provide this functionality and play the role of a security monitor for the MMU settings. In fact, since the MMU is the key functionality used by the hypervisor to isolate the security domains, violation of complete mediation of the MMU settings can enable an attacker to bypass the hypervisor policies which could compromise the security of the entire system. This is also what makes a formal analysis of correctness a worthwhile enterprise.

A distinguishing feature of our work is the adoption of direct paging as virtualization mechanism. This mechanism has been previously introduced by Xen [6] and permits to virtualize the memory subsystem without requiring either shadow data structures (e.g., shadow page tables) or specialized hardware (e.g., nested page tables [2,6]). Direct paging allows a virtualization aware (i.e., paravirtualized) guest to directly manipulate the page tables while they are in passive state, i.e., not in active use by the MMU, and then using a dedicated API, verified in this paper, that effectuates and monitors the transition of page tables between passive and active state. Subsequent operations (like mapping specific entries) are done by invoking the corresponding hypercall, until the page tables are freed.

Our verification is done in the HOL4 theorem prover and targets a system consisting of a commodity CPU for embedded devices (ARMv7 [2]), which hosts the hypervisor and an untrusted guest. The verification is done in three steps: (i) We formalize (Section 2) the hosting hardware by extending the existing ARMv7 model developed in Cambridge with the formal model of the ARMv7 MMU, (ii) We formalize (Section 3) the behavior of the complete system by introducing a system model, which interleaves instructions executed by an untrusted guest with a low-level specification of the hypervisor handlers, and (iii) we prove (Sections 4 and 5) the security properties, by decomposing the proof into lemmas that can be reused by other virtualization mechanisms that use direct paging.

The verification is made complex by the level of abstraction. We target a real commodity CPU architecture and the model of the handlers is deliberately low level so that the implementation can be directly derived from the specification.

2   Formal Model of the ARMv7 CPU

In this model a machine state is modeled as a record $\sigma = \langle regs, coregs, mem \rangle \in \Sigma$, where $regs$, $coregs$ and $mem \in 2^{32} \to 2^8$, respectively, represent the registers, co-processors and system memory. In the state σ, the function $mode(\sigma)$ determines the current privilege execution mode, which can be either $PL0$ (user mode, used by the guest) or $PL1$ (privileged mode, used by the hypervisor). Here, the three coprocessor registers $coregs = \langle SCTLR, TTBR0, DACR \rangle \in 2^{32} \times 2^{32} \times 2^{32}$ are the System Control Register, the Translation Table Base Control Register, and the Domain Access Control Register respectively.

The system behavior is modeled by the state transition relation $\xrightarrow{l \in \{PL0, PL1\}}$ $\subseteq \Sigma \times \Sigma$, where a transition realizes the effects of the execution of an ARM instruction. Non-privileged transitions ($\sigma \xrightarrow{PL0} \sigma'$) start and end in $PL0$ states.

$$\frac{\begin{array}{l} \sigma.SCTLR \neq 0 \\ desc = read_{L_1}(\sigma.TTBR0, \sigma, va.l1\_idx) \\ desc.type = SEC \\ (ap\_check_{L1}(desc, PL, accreq)) \end{array}}{mmu(\sigma, PL, va, accreq) = translate_{L1}(desc, va)}$$

$$\frac{\begin{array}{l} \sigma.SCTLR \neq 0 \\ desc_{L1} = read_{L_1}(\sigma.TTBR0, \sigma, va.l1\_idx) \\ desc_{L1}.type = PT \\ desc_{L2} = read_{L_2}(desc_{L1}.pa, \sigma, va.l2\_idx) \\ desc_{L2}.type = SP \\ (ap\_check_{L2}(desc_{L2}, PL, accreq)) \end{array}}{mmu(\sigma, PL, va, accreq) = translate_{L2}(desc_{L2}, va)}$$

Fig. 1. (a) One-step and (b) Two-step address translation

All the other transitions ($\sigma \xrightarrow{PL1} \sigma'$) involve at least one state in the privileged level. A transition from $PL0$ to $PL1$ is done by raising an exception, that can be caused by software interrupts, illegitimate memory accesses and hardware interrupts.

We extended the HOL4 ARMv7 model developed in Cambridge [3] to take into account the behavior of the ARMv7 MMU. The main functionalities of the MMU are virtual-to-physical address mapping and memory access control. The MMU is modeled by the $mmu(\sigma, PL, va, accreq) \rightarrow pa \cup \{\bot\}$ function. The function takes the state σ, a privilege level PL, a virtual address $va \in 2^{32}$ and the requested access right $accreq \in \{rd, wt, ex\}$, for *read*, *write* and *execution* respectively, and returns either the corresponding physical address $pa \in 2^{32}$ (if the access is granted) or an access permission fault ($\bot$).

The SCTLR register controls the MMU. If the MMU is disabled ($SCTLR = 0$), all access permissions are granted and the mmu function yields an identity mapping ($mmu(\sigma, PL, va, accreq) = va$). If the MMU is enabled ($SCTLR \neq 0$), the mmu can execute up to two page table walks to translate a virtual address. The first walk accesses the active first level ($L1$) page table, whose address is identified by the TTBR0 register.

An $L1$ page table contains 4096 entries, each mapping 1MB of contiguous virtual memory. When executing the first table walk, the MMU accesses the proper L1 entry. If the entry is *unmapped*, then the MMU yields a permission fault ($\bot$). If the entry is *Section* (SEC) (Figure 1a) and the requested access is legitimate ($ap\_check_{L1}(desc, PL, accreq)$) then the corresponding 1MB is linearly mapped to 1MB of physical memory ($translate_{L1}(desc, va)$). If the entry is *Page table* (PT), then a second page table walk is needed (Figure 1b). In this case, the L1 entry points to an L2, which contains 256 entries. Each L2 entry, of type *Small page* (SP), linearly maps 4KB of virtual memory.

We introduce some auxiliary definitions to describe the properties guaranteed by an ARMv7 CPU that obeys the access privileges computed by the MMU. We use the predicate $mmu_{phys}(\sigma, PL, pa, accreq)$ to identify the accesses granted to the physical memory. An access to a physical address is allowed if at least one virtual address exists that enables the requested access permission and that maps to pa; $mmu_{phys}(\sigma, PL, pa, accreq) = (\exists va. \ mmu(\sigma, PL, va, accreq) = pa)$.

We say that two states are *MMU-consistent* if their memories differ only for writable physical addresses. Formally, $mmu_c(\sigma, \sigma', PL)$ if $\forall pa. \ \sigma.mem(pa) \neq \sigma'.mem(pa) \Rightarrow mmu_{phys}(\sigma, PL, pa, wt)$.

Two states are *MMU-equivalent* if for any virtual address *va* the MMU yields the same translation and the same access permissions. Formally, $\sigma \stackrel{mmu}{\equiv} \sigma'$ if and only if $\forall va, PL, accreq.\ mmu(\sigma, PL, va, accreq) = mmu(\sigma', PL, va, accreq)$.

The transition relation queries the MMU model to identify when an instruction raises an exception and satisfies the following properties.

Property 1. Let $\sigma \in \Sigma$ such that $mode(\sigma) = PL0$. If $\sigma \xrightarrow{PL0} \sigma'$ then $mmu_c(\sigma, \sigma', PL0)$ and $\sigma.coregs = \sigma'.coregs$.

Property 2. Let $\sigma_1, \sigma_2 \in \Sigma$, A be a set of physical addresses, $mode(\sigma_1) = mode(\sigma_2) = PL0$ and $A \supseteq \{pa \mid \exists accreq.\ mmu_{phys}(\sigma, PL0, pa, accreq)\}$, and $\sigma_1 \stackrel{mmu}{\equiv} \sigma_2$, $\sigma_1.regs = \sigma_2.regs$, $\sigma_1.coregs = \sigma_2.coregs$ and $\forall pa \in A.\ \sigma_1.mem(pa) = \sigma_2.mem(pa)$. If $\sigma_1 \xrightarrow{PL0} \sigma_1'$ and $\sigma_2 \xrightarrow{PL0} \sigma_2'$ then $\sigma_1'.regs = \sigma_2'.regs$, $\sigma_1'.coregs = \sigma_2'.coregs$ and $\forall pa \in A.\ \sigma_1'.mem(pa) = \sigma_2'.mem(pa)$.

In [8] the authors validated the HOL4 ARMv7 model against these properties under the assumption that the address translation is the identity map.

3   Formal Model of the Hypervisor

Our target scenario consists of an ARMv7 CPU hosting the hypervisor and an untrusted guest. The hypervisor uses direct paging to virtualize the memory subsystem: the page tables are allocated by the guest inside its own memory and the guest is allowed to manage its page table while tables are not in active use by the MMU. Subsequent operations on the tables are done by invoking the corresponding hypercall serving the request. In our approach the physical memory is logically fragmented into blocks of 4KB, resulting in 2^{20} possible physical blocks. Since L1 and L2 page tables are 16KB and 1KB respectively, an L1 page table is stored in four contiguous physical blocks and a physical block can contain four L2 page tables. The hypervisor associates a type to each block: (i) D; the block does not contain sensitive data, (ii) L_1: the block contains part of an L1 page table, and (iii) L_2: the block contains four L2 page tables.

To handle guest requests, the hypervisor provides nine hypercalls: *switch* that selects the active L1, *L1create* and *L2create* to change the type of a block to the corresponding table type, *L1free* and *L2free* to free the page tables and changing the type of a block to D, *L1unmap* and *L2unmap* to unmap a page table entry, *L1map* and *L2map* to map a specific entry of a page table.

The hypercalls enforce the page type policy: the guest is allowed to change only blocks of type D. Naively enforcing this policy requires the hypervisor to re-validate the page tables before reactivating them, that is a time consuming task. To make overhead sustainable, the hypervisor maintains a reference counter for each block. The intuition is that the hypervisor changes the type of a physical block (e.g., allocates or frees a page table) only if the corresponding reference counter is zero and that this enables the hypervisor to skip the re-validation tasks.

We model the complete system reusing the formal model of Section 2. A system state is modeled by a tuple $\langle \sigma, h \rangle$, consisting of an ARMv7 state σ

$$bls = \{block(pa) + i \mid i < 4\}$$
$$\forall bl \in bls.\tau \vdash bl : D \land G_{mem} \vdash bl : 0 \land \rho(bl) = 0$$
$$descs = [read_{L_1}(pa, \sigma, j) \mid j < 4096]$$
$$pts = [block(d.pa) \mid d \in descs \land d.type = PT]$$
$$secs_{rd} = [block(d.pa) \mid d \in descs \land d.type = SEC \land (0, rd) \in d.ap]$$
$$secs_{wt} = [block(d.pa) \mid d \in descs \land d.type = SEC \land (0, wt) \in d.ap]$$
$$\forall bl \in pts.\tau \vdash bl : L_2$$
$$\forall bl \in secs_{rd}.\forall idx < 256.G_{mem} \vdash bl + idx : 0$$
$$\forall bl \in secs_{wt}.\forall i < 2^8.bl + i \notin bls \land \tau \vdash bl + i : D \land \rho(bl + i) < MAX - 2^{12}$$
$$\rho' = for\ bl\ \in\ secs_{wt}(for\ i\ < 2^8\ (\lambda \rho_1.\rho_1(bl + i) := \rho_1(bl + i) + 1))\rho$$
$$\rho'' = for\ bl\ \in\ pts(\lambda \rho_2.\rho_2(bl) := \rho_2(bl) + 1)\rho'$$
$$\tau' = (\tau(bl \in bls) := L_1)$$

$$\langle \sigma, \langle \tau, \rho \rangle \rangle \xrightarrow{createL1(pa)} \langle \sigma, \langle \tau', \rho'' \rangle \rangle$$

Fig. 2. Inference rule for hypervisor *createL1* handler

and an abstract hypervisor state h, of the form $\langle \tau, \rho \rangle$. Let $bl \in 2^{20}$ be the index of a physical block and $t \in \{D, L1, L2\}$, $\tau \vdash bl : t$ tracks the type of the block and $\rho(\tau) \in 2^{30}$ tracks the reference counter. We use G_{mem} to statically identify the memory region assigned to the guest: if the block bl is part of the guest memory then $G_{mem} \vdash bl : 0$, otherwise $G_{mem} \vdash bl : 1$

The behavior of the system is defined by a labeled transition relation $\langle \sigma, h \rangle \xrightarrow{\alpha} \langle \sigma', h' \rangle$. The model interleaves standard non-privileged transitions ($\alpha = 0$) with abstract handler invocations (e.g., $\alpha = createL1(pa)$):

- if $\sigma \xrightarrow{PL0} \sigma'$ then $\langle \sigma, h \rangle \xrightarrow{0} \langle \sigma', h \rangle$; instructions executed in non-privileged mode that do not raise exceptions behave equivalently to the standard ARMv7 semantics and do not affect the abstract hypervisor state.
- if $\sigma \xrightarrow{PL1} \sigma'$ then the hypervisor intercepts the exception and its handler atomically transforms the state of the system

The model of the hypervisor handlers is defined by HOL4 functions that describe the hypervisor behavior. These functions are deliberately low level, for example they store the page tables in the system memory instead of using abstract data structures. This enables us to include the page tables in the attack surface taken into account by our verification. The hypervisor handlers check that (i) guest can only change *data* pages (pages of type D), (ii) page table blocks are typed correctly, and (iii) the blocks that are readable/writable by the guest are enclosed in the part of the memory granted to the guest. If a handler fails to validate the guest request, then it terminates without affecting the system state (i.e., $\langle \sigma, \langle \tau, \rho \rangle \rangle \xrightarrow{\alpha} \langle \sigma, \langle \tau, \rho \rangle \rangle$.

In Figure 2 we use an inference-rule to exemplify the behavior of the *L1create* handler. The guest uses the hypercall to request the validation of an L1 pointed to by the address pa. If the validation succeeds the type of the corresponding physical blocks is changed to $L1$. Here, $block(pa)$ returns the block pointed to by the physical address pa and $[f(x) \mid p(x)]$ represents list comprehension. Moreover, $r(\beta) := value$ represents the update of the field β of the record r with $value$. Finally, let d be an entry of a page table, we use $d.pa$, $d.type$ and $d.ap$ to represent the initial physical address pointed to by the entry, its type (either a section SEC or a page table PT) and the set of the granted access rights.

The handler checks that the four blocks containing the new page table have reference zero, are typed D and reside in the guest memory. To accept the request, each PT entry must point to a valid $L2$ in the guest memory. Section descriptors allowing guest read accesses must point to a section encompassing only blocks that are part of the guest memory. Moreover, if the descriptor allows guest write access, then the hypervisor ensures that all the reachable blocks are typed D, that their reference counter is less than the maximum allowed bound and that none of the blocks of the new page table are included in that section.

The hypercall also updates the reference counters of the pointed blocks, by summing the number of references that enable guest write access and the number of references from PT entries.

4   Security Properties

As common, our verification strategy consists in introducing a state invariant ($\mathcal{I}(s)$) that guarantees the desired security properties and in demonstrating that the invariant is preserved by any possible transition. Clearly, the system must start (i.e., the boot must terminate) in a state that satisfies the invariant. We use $\mathcal{Q}_\mathcal{I}$ to identify the set of all possible states that satisfy the invariant.

Theorem 1. *Let* $s \in \mathcal{Q}_\mathcal{I}$, *if* $s \xrightarrow{\alpha} s'$ *then* $s' \in \mathcal{Q}_\mathcal{I}$.

Guaranteeing that the invariant is preserved requires to demonstrate that each handler preserves the invariant ($\alpha \neq 0$) and that the guest is not able to break it. Intuitively, while the hypervisor is inactive ($\alpha = 0$), the only mechanism that can confine the behavior of an arbitrary guest is the MMU. Thus, the hypervisor must play the role of a security monitor of the MMU settings. If *complete mediation* of the MMU settings is violated, then an attacker can bypass the hypervisor policies and compromise the security of the entire system. Enforcing this property is critical in the direct paging mechanism because the page tables are dynamically allocated and released by the untrusted guest and they reside in the guest memory.

Theorem 2. *Let* $s \in \mathcal{Q}_\mathcal{I}$, *if* $s \xrightarrow{0} s'$ *then* $s \overset{mmu}{\equiv} s'$.

Definition 1. *Two states* s *and* s' *do not differ in a physical block of memory (written* $s \overset{bl}{\equiv} s'$*) if for each physical address* pa *if* $block(pa) = bl$ *then* $s.mem(pa) = s'.mem(pa)$.

Definition 2. *Two states* s *and* s' *are t-equivalent (written* $s \overset{G_{mem}:t}{\equiv} s'$*) iff for each physical block* bl *if* $G_{mem} \vdash bl : t$ *then* $s \overset{bl}{\equiv} s'$.

We use the approach of [5] to analyze the data separation properties. The *non-exfiltration* property guarantees that a transition executed by the guest does not modify the secure resources:

Theorem 3. *Let $s \in \mathcal{Q}_\mathcal{I}$, if $s \xrightarrow{0} s'$ then $s \stackrel{G_{mem}:1}{\equiv} s'$.*

The *non-infiltration* property is a noninterference property guaranteeing that a transition executed by the guest depends only on its accessible resources.

Theorem 4. *Let $s_1, s_2 \in \mathcal{Q}_\mathcal{I}$ such that $s_1.regs = s_2.regs$, $s_1.coregs = s_2.coregs$ and $s_1 \stackrel{G_{mem}:0}{\equiv} s_2$. If $s_1 \xrightarrow{\alpha} s_1'$ and $s_2 \xrightarrow{\alpha} s_2'$ then $s_1'.regs = s_2'.regs$, $s_1'.coregs = s_2'.coregs$ and $s_1' \stackrel{G_{mem}:0}{\equiv} s_2'$*

5   Verification Strategy

To describe the verification of the security properties we summarize the structure of the system invariant.

Definition 3. $\mathcal{I}(\langle \sigma, \langle \tau, \rho \rangle \rangle)$ *holds if,*

$$\sigma.SCTLR \neq 0 \wedge \tau \vdash \sigma.TTBR0 : L_1 \wedge \forall bl \in 2^{20}. \ \mathcal{I}_T(\langle \sigma, \langle \tau, \rho \rangle \rangle, bl) \wedge \mathcal{I}_C(\langle \sigma, \langle \tau, \rho \rangle \rangle, bl)$$

$$\text{where } \mathcal{I}_C(\langle \sigma, \langle \tau, \rho \rangle \rangle, bl) \text{ holds if,} \quad \rho(bl) = \sum_{i \in 0...2^{20}} cnt(\sigma, \tau, bl, i)$$

$$\text{and } \mathcal{I}_T(\langle \sigma, \langle \tau, \rho \rangle \rangle, bl) \text{ holds if,} \quad \begin{cases} \tau \vdash bl : L_1 \Rightarrow \mathcal{I}_{T_1}(\sigma, \tau, bl) \\ \tau \vdash bl : L_2 \Rightarrow \mathcal{I}_{T_2}(\sigma, \tau, bl) \end{cases}$$

The invariant requires that the MMU is enabled, the active page table is typed L_1, the reference counter is correctly counting the references to each block and each block is well-typed.

We use Figure 3 to summarize the properties checked by the invariant. The table in the center represents the physical memory and reports the page type (pt), the static type (gm) and the reference counter of each block (rc).

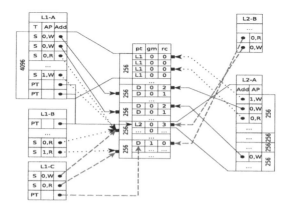

Fig. 3. Invariant

The four top most blocks contain an L1, whose 4096 entries are depicted by the table L1-A. The top entry is a section descriptor $(T = S)$ that grants write permission to the guest $(ap = 0, w)$. The entry points to the second physical section, which consists of 256 blocks. Three other section descriptors are depicted: one grants write accesses to the guest, one grants only read permission to the guest $(0, r)$, and the last one prevents any guest access and enables write permission to the privileged mode $(1, w)$. The last two entries of the L1 are PT-entries. These two entries point to two different L2 page tables that are stored in the same physical block. To satisfy $\mathcal{I}_{T_1}(\sigma, \tau, 0)$, if a section enables guest access then the pointed blocks must all be in the guest memory. Moreover, if a section enables guest write access then each pointed block must be typed D. Finally, PT-entries must point to addresses contained in physical blocks that are typed L_2. The Figure depicts two additional L1 page tables; L1-B satisfies the invariant, L1-C contains three entries that violate $\mathcal{I}_{T_1}$. In fact (i) the first section grants write permission to the guest, but at least one of the pointed blocks is not typed D, (ii) the second section enables guest accesses, but at least one of the pointed blocks is not in the guest memory and (iii) the third entry is a PT-entry, but points to a physical address that is contained by a block typed D.

The table L2-A depicts the content of the 768-th physical block, which contains four L2 page tables. To satisfy $\mathcal{I}_{T_2}(\sigma, \tau, 768)$, if one entry of the four page tables enables guest access then the pointed block must be in the guest memory. Moreover, if the entry also enables guest write access then the pointed block must be typed D.

We use the same figure to illustrate the reference counter. For a physical block bl, if block i is typed L_1 then $cnt(\sigma, \tau, bl, i)$ counts the number of section entries that point to bl and that are writable in user mode plus the number of PT-entries that point to bl. If block i is typed L_2 then $cnt(\sigma, \tau, bl, i)$ counts the number of entries that point to bl and that are writable in user mode. In Figure 3 we use solid arrows to represent the references that are counted and dashed arrows to represent the other references.

Lemma 1. *Let $\langle \sigma, \langle \tau, \rho \rangle \rangle \in \mathcal{Q}_\mathcal{I}$ then $\forall pa.\ mmu_{ph}(\sigma, 0, pa, wt) \Rightarrow (G_{mem} \vdash block(pa) : 0 \wedge \tau \vdash block(pa) : D)$*

Lemma 2. *Let $\langle \sigma, \langle \tau, \rho \rangle \rangle \in \mathcal{Q}_\mathcal{I}$ then $\forall pa.\ mmu_{ph}(\sigma, 0, pa, rd) \Rightarrow (G_{mem} \vdash block(pa) : 0)$*

Lemmas 1 and 2 directly follow from the invariant and show that the MMU setup forbids guest accesses outside the guest memory and guest write accesses to physical blocks that are not typed D. Using Lemma 1 and Property 1 we directly show that the guest can only change physical blocks that are inside its own memory and that are typed D.

Lemma 3. *Let $s = \langle \sigma, \langle \tau, \rho \rangle \rangle \in \mathcal{Q}_\mathcal{I}$ and $s \xrightarrow{0} s'$. For each bl, if $s \overset{bl}{\neq} s'$ then $G_{mem} \vdash bl : 0$ and $\tau \vdash bl : D$.*

Notice that Lemma 3 directly guarantees Theorem 3. Similarly, Lemma 1 and Property 2 guarantee Theorem 4 for guest transitions (i.e., $\alpha = 0$).

Proof of Theorem 2. We must ensure that for each possible address va the MMU translation is equivalent in the states $s = \langle \sigma, \langle \tau, \rho \rangle \rangle$ and $s' = \langle \sigma', \langle \tau, \rho \rangle \rangle$. Property 1 guarantees that the coprocessors registers are not affected by user transitions, thus the address of the active L1 can not be changed by the guest. The first translation walk accesses the L1 entry that maps va. The invariant $\mathcal{I}(s)$ guarantees that the active page L1 is typed L_1 ($\tau \vdash \sigma.c2 : L_1$). Thus, Lemma 3 guarantees that in the two states s and s' the MMU fetches the same L1 descriptor. If the descriptor is either unmapped or a section the proof is completed. If the descriptor is a PT-entry then the translation executes the second walk, accessing an entry of the pointed $L2$. Let bl be the block containing the pointed $L2$. From the invariant, we know that $\tau \vdash bl : L_2$. Again, Lemma 3 guarantees that in the two states s and s' the MMU fetches the same L2 descriptor, thus concluding the proof.

Lemma 4. *Let* $\langle \sigma, \langle \tau, \rho \rangle \rangle \in \mathcal{Q_I}$, *if* $\sigma \overset{bl}{\equiv} \sigma'$ *then* $\mathcal{I}_T(\langle \sigma', \langle \tau, \rho \rangle \rangle, bl)$ *and* $\forall bl'. \; cnt(\sigma, \tau, bl', bl) = cnt(\sigma', \tau, bl', bl)$.

Lemma 4 shows that the well-typeness of a block and its counted references are independent from the content of the other physical blocks.

Lemma 5. *Let* $s \in \mathcal{Q_I}$, *if* $s \overset{0}{\to} s'$ *then* $s' \in \mathcal{Q_I}$.

Lemma 5 shows that the invariant is preserved by the execution of an arbitrary guest instruction, thus guaranteeing that untrusted software can not breach the security of the system. Since the guest can not directly change the abstract hypervisor data-structures and can not directly affect the coprocessor registers (Property 1), in order to reestablish the invariant we need to show that for every block bl both $\mathcal{I}_T(\langle \sigma', \langle \tau, \rho \rangle \rangle, bl)$ and $\mathcal{I}_C(\langle \sigma', \langle \tau, \rho \rangle \rangle, bl)$ hold. If $\tau \vdash bl : D$ then $\mathcal{I}_T$ trivially holds. Otherwise, Lemma 3 guarantees that $s \overset{bl}{\equiv} s'$ and Lemma 4 demonstrates that $\mathcal{I}_T$ holds. For the reference counter we must prove that $\rho(bl) = \sum_{i \in 0...2^{20}} cnt(\sigma', \tau, bl, i)$ knowing that $\rho(bl) = \sum_{i \in 0...2^{20}} cnt(\sigma, \tau, bl, i)$. We directly show that for every block i $cnt(\sigma', \tau, bl, i) = cnt(\sigma, \tau, bl, i)$. If $\tau \vdash i : D$ this equality is trivial, since cnt is zero. Otherwise, Lemma 3 guarantees that $s \overset{i}{\equiv} s'$ and Lemma 4 concludes the proof.

Lemma 6. *Let* $\langle \sigma, \langle \tau, \rho \rangle \rangle \in \mathcal{Q_I}$ *and* $\forall bl'. \; (\tau(bl') \neq \tau'(bl')) \Rightarrow (\rho(bl') = 0)$. *For every* bl *if* $\tau'(bl) = \tau(bl)$ *then* $\mathcal{I}_T(\langle \sigma, \langle \tau', \rho' \rangle \rangle, bl)$ *and* $\forall bl'. \; cnt(\sigma, \tau, bl', bl) = cnt(\sigma, \tau', bl', bl)$.

Lemma 6 expresses that well-typedness and counters are preserved for all hypervisor data changes, as long as the blocks whose types change have reference counter zero. The equality of the reference counter is established by showing that cnt is independent on the type of the block bl'. The strategy used to establish $\mathcal{I}_T$ depends on the type of the block bl. If $\tau \vdash bl : D$ then the proof is trivial. If $\tau \vdash bl : L_2$ we use reductio ad absurdum. Assume that $\mathcal{I}_T$ does not hold, then there must exist an entry i of the page table that grants guest write access to a block bl' that is not typed D in τ'. From the invariant, we know that such an entry does not exist in $\langle \sigma, h \rangle$. Thus, i must point to a block bl' such that $(\tau(bl') \neq \tau'(bl'))$. The hypothesis of the Theorem guarantees that $\rho(bl') = 0$, then for all bl'' (including bl) $cnt(\sigma, \tau, bl', bl'') = 0$. This contradicts the assumption that there is an entry i that points to bl' and that is writable by the guest. For $\tau \vdash bl : L_1$ we use a similar reasoning.

Lemmas 4 and 6 are used to modularize the proof of Theorem 1 for the transitions modeling the hypervisor handlers. For example, to demonstrate that the invariant is preserved by $createL1(pa)$ we first show that the type is changed only for pages having reference zero, then we demonstrate that the content of the blocks containing the page tables is not changed.

For the $\mathcal{I}_T$ predicate, we use the two Lemmas to guarantee that $\mathcal{I}_T$ is preserved for blocks that are not in $bls = \{block(pa) + i \mid i < 4\}$. Then, we demonstrate that the checks performed by the hypercall guarantees $\mathcal{I}_{T_1}(\sigma', \tau', bl)$ for the every block in bls (i.e., the blocks containing the new L1).

For the $\mathcal{I}_C$ predicate, since the invariant guarantees that $\rho(bl) = \sum_{i \in 0...2^{20}} cnt(\sigma, \tau, bl, i)$, then we must prove :

$$\rho'(bl) = \sum_{i \in 0...2^{20}} cnt(\sigma', \tau', bl, i) + \rho(bl) - \sum_{i \in 0...2^{20}} cnt(\sigma, \tau, bl, i).$$

For a block i that is not in bls, the two Lemmas guarantee that $cnt(\sigma', \tau', bl, i) = cnt(\sigma, \tau, bl, i)$, thus our goal is reduced to demonstrate $\rho'(bl) = \sum_{i \in bls} cnt(\sigma', \tau', bl, i) + \rho(bl) - \sum_{i \in bls} cnt(\sigma, \tau, bl, i)$. Since the hypercall checks that the initial type of a block $i \in bls$ is D, then $cnt(\sigma, \tau, bl, i) = 0$. Finally, we must demonstrate that the hypercall correctly updates the counters by adding the references of the new L1 page table: $\rho'(bl) = \rho(bl) + \sum_{i \in bls} cnt(\sigma', \tau', bl, i)$.

6   Evaluation

The verification has been performed using the HOL4 interactive theorem prover. This allows building the formal model of the system on top of the existing ARMv7 model developed in Cambridge [3], by extending the transition relation to take into account the MMU constraints and by substituting the activation of exceptions with the specification of the hypervisor handlers. This specification consists of 500 lines of HOL4 code, intentionally avoids any high level construct and resembles the control flow of the C implementation, with the aim of making this executable specification as close as possible to the C implementation. This increased the difficulty of the proof (e.g., the system invariant consists of 2k lines of HOL4 and the proofs consist of 15k lines of HOL4), which must handle finite arithmetic overflows, page tables not stored into abstract states and forced us to identify the invariants of the loops required to iterate the page tables. However, the low level of abstraction allowed us to identify several bugs of the original design: (i) arithmetic overflow when updating the reference counter, due to the guest being able to create unlimited references to a physical block (ii) bit field and offset mismatch, (iii) missing check that a newly allocated page table prevents the guest from overwriting the page table itself, (iv) usage of the signed shift operator where the unsigned one was necessary, and (v) side channels exploitable by the guest by requesting to validate page tables outside the guest memory.

We implemented a prototype of the hypervisor to experiment the verified design. This prototype consists of 4500 LOC of C (with a minor part of assembly) and is directly derived from the verified specification. Initial benchmarks show promising results; the hypervisor is capable of hosting a paravirtualized Linux and for the LMBench introduces an overhead between 2% (select benchmark) and 495% (fork+bin benchmark), compared to the native Linux. This overhead is much less than what the many other hypervisors for ARM impose [7].

We briefly compare our virtualization mechanism to existing approaches to virtualize the memory subsystem. Functional correctness of mechanisms based on Shadow page tables (SPT) has been verified in [1,4,10]. Using SPT, the hypervisor keeps a shadow copy of the guest page tables. This copy is updated (after validation) by the hypervisor whenever the guest operates on its page tables. The main benefits of direct paging respect to SPT is that the hypervisor does not replicate the guest page tables, thus reducing memory accesses and memory overhead and not requiring dynamic allocation in the hypervisor.

Hardware-assisted virtualization (e.g., nested-paging included in the ARMv7 virtualization extension) frees the hypervisors from implementing a virtualization mechanism of the memory subsystem (e.g., [6,11]). This simplification comes at the cost of enlarging the TCB and moving the verification from software to hardware. Moreover, since hardware virtualization support is still uncommon in embedded systems, then software based virtualization is the only viable option for several platforms (including ARM11, ARM CortexA5 and Intel Quark).

The formal verification of seL4 [9] demonstrated that the verification of a complete microkernel is possible even at the machine code level [12]. A complete commodity OS can be executed on top of a microkernel by mapping the OS threads directly to the microkernel threads, thus delegating completely the process management functionality from the hosted OSes to the microkernel (e.g., L[4]Linux). This generally involves an invasive and error-prone OS adaptation process, however. An alternative approach consists of extending the microkernel with a virtualization mechanism of the memory subsystem, like the one proposed in this paper.

7   Concluding Remarks

We presented a design to virtualize the ARMv7 memory subsystem that requires neither specialized hardware support nor shadow data structures. Together with the machine-assisted proof of its correctness and the spatial isolation provided by the hypervisor, the design represents the first trustworthy virtualization mechanism using *direct paging*, previously introduced by Xen for x86, but not verified.

The design correctness is stated in terms of (i) complete mediation of the MMU settings, (ii) non-exfiltration and (iii) non-infiltration. These properties show that user mode processes (e.g., a possibly malicious guest OS) are incapable of affecting the MMU behaviour and that can not influence (or be influenced by) the resources that are not allocated to the guest.

The low level abstraction of the hypervisor specification increased the complexity of our verification, but allowed us to identify and to correct several bugs. Moreover, since the specification avoids any high level constructs, it has been used to directly drive a prototype.

References

1. Alkassar, E., Hillebrand, M.A., Paul, W.J., Petrova, E.: Automated verification of a small hypervisor. In: Leavens, G.T., O'Hearn, P., Rajamani, S.K. (eds.) VSTTE 2010. LNCS, vol. 6217, pp. 40–54. Springer, Heidelberg (2010)
2. ARMv7-A architecture reference manual,
   http://infocenter.arm.com/help/index.jsp?topic=/com.arm.doc.ddi0406b
3. Fox, A., Myreen, M.O.: A trustworthy monadic formalization of the aRMv7 instruction set architecture. In: Kaufmann, M., Paulson, L.C. (eds.) ITP 2010. LNCS, vol. 6172, pp. 243–258. Springer, Heidelberg (2010)
4. Heiser, G., Leslie, B.: The OKL4 microvisor: convergence point of microkernels and hypervisors. In: Thekkath, C.A., Kotla, R. and Zhou, L. (eds.), ApSys, pp. 19–24. ACM (2010)
5. Heitmeyer, C., Archer, M., Leonard, E., McLean, J.: Applying formal methods to a certifiably secure software system. IEEE Trans. Softw. Eng. 34(1), 82–98 (2008)
6. Hwang, J.-Y., Suh, S.-B., Heo, S.-K., Park, C.-J., Ryu, J.-M., Park, S.-Y., Kim, C.-R.: Xen on ARM: System virtualization using Xen hypervisor for ARM-based secure mobile phones. In: 5th IEEE Consumer Communications and Networking Conference, CCNC 2008, pp. 257–261. IEEE (2008)
7. Iqbal, A., Sadeque, N., Mutia, R.I.: An overview of microkernel, hypervisor and microvisor virtualization approaches for embedded systems. Report, Department of Electrical and Information Technology, Lund University, Sweden, 2110 (2009)
8. Khakpour, N., Schwarz, O., Dam, M.: Machine assisted proof of aRMv7 instruction level isolation properties. In: Gonthier, G., Norrish, M. (eds.) CPP 2013. LNCS, vol. 8307, pp. 276–291. Springer, Heidelberg (2013)
9. Klein, G., Elphinstone, K., Heiser, G., Andronick, J., Cock, D., Derrin, P., Elkaduwe, D., Engelhardt, K., Kolanski, R., Norrish, M., Sewell, T., Tuch, H., Winwood, S.: seL4: formal verification of an OS kernel. In: Matthews, J.N., Anderson, T.E. (eds.), SOSP, pp. 207–220. ACM (2009)
10. Leinenbach, D., Santen, T.: Verifying the Microsoft Hyper-V hypervisor with VCC. In: Cavalcanti, A., Dams, D.R. (eds.) FM 2009. LNCS, vol. 5850, pp. 806–809. Springer, Heidelberg (2009)
11. McCoyd, M., Krug, R.B., Goel, D., Dahlin, M., Young, W.D.: Building a hypervisor on a formally verifiable protection layer. In: HICSS, pp. 5069–5078. IEEE (2013)
12. Sewell, T.A.L., Myreen, M.O., Klein, G.: Translation validation for a verified os kernel. In: Proceedings of the 34th ACM SIGPLAN Conference on Programming Language Design and Implementation, pp. 471–482. ACM (2013)

True Random Number Generators Secure in a Changing Environment: Improved Security Bounds

Maciej Skorski[*]

Cryptology and Data Security Group, University of Warsaw
maciej.skorski@gmail.com

Abstract. Barak, Shaltiel Tromer showed how to construct a True Random Number Generator (TRNG) which is secure against an adversary who has some limited control over the environment.

In this paper we improve the security analysis of this TRNG. Essentially, we significantly reduce the entropy loss and running time needed to obtain a required level of security and robustness.

Our approach is based on replacing the combination of union bounds and tail inequalities for ℓ-wise independent random variables in the original proof, by a more refined of the deviation of the probability that a randomly chosen item is hashed into a particular location.

1 Introduction

1.1 Random Number Generators

Random number generators are used for various purposes, such as simulating, modeling, cryptography and gambling. Below we briefly discuss possible approaches and issues in generating random numbers.

TRUE RANDOM NUMBER GENERATORS. The term "True Random Number Generator" (TRNG) refers to a hardware device that generates random numbers based on some *physical* phenomena (radiation, jitter, ring oscillators, thermal noise etc.). As an example of such an implementation one can mention the HotBits project [13], which is based on timing the decay of Caesuim-137. Such implementations are typically very reliable and hard to tamper. Sometimes as a TRNG one considers also a software application which generates random numbers based on unpredictable human behavior, like mouse movement or typing keyboard keys. Even if they are not completely unpredictable (because knowing an operator's habits helps in predicting the output) the generated results are typically of high quality and have found real-world cryptographic applications (e.g. PGP). True Random Number Generators do not have internal states (hence are not predictable from the sampling history) and produces a high quality output. However, they are usually

---

[*] This work was partly supported by the WELCOME/2010-4/2 grant founded within the framework of the EU Innovative Economy Operational Programme.

slow and not easy to implement (for instance, because of the need of a dedicated hardware).

1.2   Pseudo-Random Number Generators

On the other side we have pseudo-random number generators (PRNG's), algorithms that use mathematical formulas to produce a randomly looking output. They depend on internal states and for this reason needs to be externally "seeded". They are fast and relatively easy to implement. However, seeding them properly is of critical importance. A lot of PRGNs from standard software use predictable seeds. For example, recall the discovered vulnerability in a version of Netscape browser [7].

CLASSICAL TRNG DESIGN. Typically the process of building a TRNG consists of the following stages

(a) Setting an entropy source (physical or non-physical)
(b) Post-processing part
(c) Quality evaluation

The entropy source does not necessarily provide samples of excellent quality and therefore step (b) is needed. Its purpose is to eliminate bias or dependencies. Posprocessing procedure could be very simple as the famous von Nuemann corrector or very complicated. Finally, the whole implementation should be subjected to common statistical tests, for example [9,2].

1.3   TRNGs in Environments under (Partial) Adversarial Control

Imagine a setting where an attacker has some *partial* control over the environment where the sampling device operates. For instance he could influence voltage or temperature. The goal is to build a TRNG which is robust in such a setting.

RESILIENT EXTRACTORS. Slightly simplifying the problem, we can focus on the postprocessing algorithm, that is on how to extract random bits from randomness within a source. Suppose that we have a source that produces samples distributed according to X, where X is unpredictable in the sense that it has high entropy. This assumption is the most general way of capturing "randomness" because we cannot assume that our source, which might be a very complicated physical processes, has any specific nice properties. One can extract almost uniform bits from high entropy source by the use of so called "randomness extractors". However, no deterministic procedure can extract one bit which is close to uniform from *every* high-entropy source [12]. The general purpose randomness extractors, which are guaranteed to work with every source having enough entropy, require additional truly random bits (called the *seed*) as a "catalyst" [10]. While this concept is generally extremely useful, the direct application of a seeded extractor to an entropy source does not provide a way to build a good TRNG:

(a) In real applications, generating even small number of *truly* random bits can be extremely hard.
(b) In some designs there might be correlations between the source and the seed. For instance, when the seed is stored in a TRNG which uses a source of randomness within the computer (like timing events).
(c) If we want a TRNG with some kind of resilience, we should ensure it also for the procedure generating seed.

One can overcome this problem by designing a deterministic extractor which works for a restricted class of sources. Some constructions for specific cases are known; see for instance [8,6] and recall the most famous example - von Neumann sources. However they are not applicable in our case.

RESILIENT EXTRACTORS VIA FIXING SEEDS. Barak, Shaltiel and Tromer [3] came up with the very simple but surprisingly useful idea of *fixing a seed*. Let us discuss it briefly. Suppose for a moment that we have only one source X of entropy k and an arbitrary seeded extractor , that from any X having k-bits of min-entropy extracts m close-to-uniform bits using a random seed. This means that the output is close to uniform *in average* over all possible seeds. Hence running the extractor with a *fixed* seed, for *most of the seeds*, yields an output which is still close to uniform (by the Markov Inequality). Now let us make a realistic assumption that the source X depends on some finite number of boolean[1] environmental variables (corresponding to changes in the voltage, temperature, radiation etc) and suppose that

(a) the adversary controls t of the environmental variables
(b) in every of 2^t possible configurations of the "compromised" states, entropy in the source is big enough (i.e. at least k)

Provided that t is small, by the union bound we conclude that fixing the seed, for most of the seeds, we still obtain a good extractor in every state. Below we summarize this discussion more quantitatively:

Proposition 1 (Resilient extractor from any extractor). *Let $\{X_e\}_{e \in \{0,1\}^t}$ be a collection of n-bit random variables and let $\mathrm{Ext} : \{0,1\}^n \times \mathcal{S} \to \{0,1\}^m$ be a function such that for every e the distribution of $\mathrm{Ext}_S(X_e, S)$, where S if chosen randomly from $\mathcal{S}$, is ϵ-close to uniform. Then for all but a 2^{-u} fraction of $s \in \mathcal{S}$ the distribution $\mathrm{Ext}(X_e, s)$ is $2^{u+t}\epsilon$-close to uniform for every $e \in \{0,1\}^t$.*

Even the best extractors need in worst case at least $k = m + 2\log(1/\epsilon) - \mathcal{O}(1)$ bits of entropy on their input in order to extract m bits which are ϵ-close to uniform [11]. The optimal rate, with $k = m + 2\log(1/\epsilon) - 2$, is achieved for example by 2-universal hashing (the Leftover Hash Lemma).

RESILIENT TRNG FROM THE RESILIENT EXTRACTOR. The assumption that our extractor works only for *small* (of size 2^t) family of distributions in the context of a TRNG is not restrictive. Indeed, imagine a manufacturer who has a device producing samples of a distribution X. The seed s is chosen <u>once and for all</u>

---

[1] Without losing generality, since we can describe more "granulated" properties using more than one boolean variable.

and every single TRNG box is built by composing a copy of the sampling device with the extractor seeded by the same s. Once s is chosen, could be even made public. The confidence level δ ensures that with high probability we can find a good s. After choosing s, the manufacturer tests the implementation against randomness test like NIST [2] and DIEHARD [9]. For more details, we refer the reader to [3]. The above discussion can be summarized by the following result.

Theorem 1 (Simple resilient TRNG, informal). *There exists an efficient seeded extractor* Ext *such that for every source X which in every of 2^t states of the environment has the min-entropy at least*

$$k \geqslant m + 2\log(1/\epsilon) + 2\log(1/\delta) + 2t - 2, \tag{1}$$

for all but at most a δ fraction of the seed s it holds that $\mathrm{Ext}(X, s)$ is ϵ-close to the uniform m-bit string in every state of the environment.

Note that the entropy loss $L = k - m$ must be substantially bigger than $2\log(1/\epsilon)$ if we want non-trivial values of δ and t. Than additional entropy loss is a price we pay for resilience of the extractor.

THE RESILIENT TRNG OF BARAK SHALTIEL AND TROMER Barak et al. showed and implemented a construction of a TRNG which is secure against any adversary who controls t environmental variables. In their proof ℓ-wise independent hash families are used as extractors. Roughly speaking, the assumption on ℓ-wise independence allows estimating higher moments of the statistical distance between output of hashing and the uniform distribution. This way we get significant improvements over the Markov Inequality used in Theorem 1.

Theorem 2 ([3] Resilient TRNG from any ℓ-universal hash family, informal). *Let $\mathcal{H}$ be an ℓ-wise independent family of hash functions from n to m bits. Suppose that an n-bit source X in every of 2^t states of the environment has the min-entropy at least*

$$k \geqslant \frac{\ell + 2}{\ell} \cdot m + 2\log(1/\epsilon) + \frac{2\log(1/\delta)}{\ell} + \frac{2t}{\ell} + \log\ell - 2 + \frac{4}{\ell}. \tag{2}$$

Then for all but δ fraction of $h \in \mathcal{H}$ it holds that $h(X)$ is ϵ-close to the uniform m-bit string in every state of the environment. For $\ell = 2$ we have the better bound $k \geqslant m + 2\log(1/\epsilon) + 2\log(1/\delta) + 2t - 2$.

We remark that the constant -2 in Theorem 2 is slightly better than what is stated in [3]. This is because the authors used a slightly weaker statement of the Leftover Hash Lemma.

OPTIMIZING THE PARAMETERS. The construction of Barak et al. depends on several parameters and gives a lot of freedom to optimize the design for a particular real-world application. For instance, minimizing the entropy loss (i.e. minimizing k) is of the crucial importance when the source produces entropy slowly or expensively (for instance when one uses patterns typing by a user, like mouse clicking, as the source). In such a case, one may prefer the (slightly more complicated) implementation with universal families of a higher degree. In the other hand, when the sampling procedure is more efficient (like thermal noise) one can afford entropy losses and prefer faster running time of the extractor, a higher confidence level for the choice of the seed or to increase the number of the environmental variables under adversarial control.

ADVANTAGES AND DISADVANTAGES. The big advantage of Theorem 2 over trivial Theorem 1 is that one can increase t proportionally to the degree ℓ of hashing family, which is actually a bit surprising. The main disadvantage is the entropy loss $L = k - m$ needs to be bigger than $\frac{2m}{\ell}$ which is $\Omega(m)$ for small ℓ. Theoretically, one can reduce this with ℓ big enough, however this could be inconvenient because of the following two reasons: (a) the running time increases by a factor poly(ℓ) and (b) the description of an ℓ-wise independent hashing family on $\{0,1\}^n$ takes ℓn bits hence there could be a problem with sampling a good function h (note that n could be even much bigger than k, which is the case of low entropy rate).

1.4   Our Results and Techniques

SUMMARY OF OUR CONTRIBUTION. We reduce the entropy loss in Theorem 2 by $2m/\ell$ for any ℓ, saving linear amount of entropy. This matches the RT-bound and hence is tight. Our approach is based on the more refined analysis of the concentration properties of universal hashing.

HASHING INTO A GIVEN SLOT - BOUNDS ON THE DEVIATION. Applying estimates for ℓ-wise independent random variables [1] we prove the following formula

Lemma 1. *Let $\ell \geqslant$ be an even integer, $\mathcal{H}$ be an ℓ-universal family of hash functions from n to m bits and $k - m \gg 2 \log \ell$. Then for any X of min-entropy at least k we have*

$$\mathop{\mathbf{E}}_{h \leftarrow \mathcal{H}} |\Pr\left(h(X) = y\right) - \Pr(U_m = y)|^\ell \lesssim C_\ell \cdot \left(2^{-k-m}\ell\right)^{\ell/2} \tag{3}$$

where $C_\ell = 2\sqrt{\pi\ell} \cdot \mathrm{e}^{1/6\ell} \cdot \left(\frac{5}{2\mathrm{e}}\right)^{\ell/2}$.

The left-hand side of Equation(3) gives the deviation of the probability (over the choice of the hash functions) of hashing a random variable X into a particular slot from its expectation (equal to 2^{-m} by the universal property). Studying such deviations is a natural idea, used essentially in [4].

Remark 1 (Sharp bounds on the deviation). One can get *symptomatically* sharp bounds in Equation(3) with more effort, by expanding the bracket and compute the ℓ-th moment precisely. The improvement is by a factor of c^ℓ and leads to further non-trivial improvements of the results of Barak et al. We find this gain too small and do not optimize the bounds in Equation(3).

IMPROVED BOUNDS ON THE FRACTION OF GOOD SEEDS IN HASHING. We prove the following inequality

Proposition 2. *Let X be an n-bit random variable and $\mathcal{H}$ be an arbitrary family of functions from n bits to m bits. Let $\ell \geqslant 2$ be an even integer and $\epsilon > 0$. Then*

$$\Pr_{h \leftarrow \mathcal{H}}[\mathrm{SD}(h(X); U_m) > \epsilon] \leqslant \frac{\mathbf{E}_{y \leftarrow U_m} \mathbf{E}_{h \leftarrow \mathcal{H}}(\Pr[h(X) = y] - \Pr[U_m = y])^\ell}{2^{-m\ell}(2\epsilon)^\ell}. \quad (4)$$

This estimate allows us to bound the fraction of the seeds (i.e. hash functions) for which the statistical distance is small, in terms of the deviation of the hashing probability. This bound offers a significant improvement over an alternative approach which bounds the deviation $|\Pr[h(X) = y] - \Pr[U_m = y]|$ for every y *separately* and after that uses the union bound to upper-bound the sum (this is essentially the strategy of Barak et al.). Intuitively, the gain could be even of a factor 2^m which should save a linear (in m) amount of entropy. Indeed, unlike Theorem 2 we are able to get meaningful security even for $k < m(1 + 2/\ell)$ (assuming small t and ℓ).

IMPROVED EFFICIENCY AND SECURITY OF THE CONSTRUCTION OF BARAK ET AL. Using the tools discussed above, we prove the following result

Theorem 3 (A resilient TRNG from any ℓ-universal hash family, informal). *Let $\mathcal{H}$ be an ℓ-universal family of hash functions from n to m bits, where ℓ is an even integer. Suppose that a source X which in every of 2^t states of the environment has the min-entropy at least*

$$k \geqslant m + 2\log(1/\epsilon) + \frac{2\log(1/\delta)}{\ell} + \frac{2t}{\ell} + \log\ell - 2. \quad (5)$$

Then for all but at most a δ fraction of $h \in \mathcal{H}$ it holds that $h(X)$ is ϵ-close to the uniform m-bit string in every state of the environment.

The theorem is valid under the assumption $k - m \gg \log\ell$ which we omitted as it is satisfied for interesting values of parameters. Our improvements over [4] can be summarized as follows:

(i) For $\ell = 2$ (the simplest extractor) we save $\log(1/\delta) + t$ bits of entropy. Alternatively, the probability of choosing a bad hash functions in the preprocessing phase gets squared and the number of the states under adversarial control can be doubled. Entropy savings is important for expensive or slow sources. Higher confidence level for the choice of the seed is important if we want to subject the implementation to the statistical tests, like DIEHARD [9]. Finally, the more variables under adversarial control the more robust the PRNG is.

(ii) For $\ell > 2$ (but not too big), in comparison to Theorem 2, we save the *linear amount* of entropy, precisely $\frac{2m}{\ell}$. The case $\ell > 2$ is preferable for slow or expensive entropy sources or when the priority is the high robustness (i.e. big number of states).

(iii) Even for $\ell \gg 2$ our result is still much better in some settings. For example, for $\epsilon = 2^{-\sqrt[10]{m}}$ (reasonable subexponential security) and $\ell \approx \log(1/\epsilon)$ the entropy loss $L = k - m$ becomes close to $L \approx 2\log(\epsilon)$ whereas Theorem 2 gives $L \approx 2\log(\epsilon) + 2m^{0.9}$. In general, the entropy amount $2m/\ell$ we save can be used to increase the number of the adversary's degrees of freedom by m, which is quite a lot.

Remark 2. Even reducing the entropy loss by only *constant* number of bits gives *non-trivial* results! This is because decreasing the minimal k by d is equivalent to increasing t by $d\ell/2$ (keeping ϵ, δ unchanged). In particular, optimizing the bound in Theorem 3 would slightly improve our results (see Remark 1).

2   Preliminaries

STATISTICAL DISTANCE. For any two random variables X, Y taking values in the same space we define the statistical distance of X and Y to be $\Delta(X;Y) = \sum_x |\Pr[X = x] - \Pr[Y = x]|$. When $\Delta(X;Y) \leqslant \epsilon$ we say that X and Y are ϵ-close.

ENTROPY NOTIONS. The min-entropy of a random variable X is defined to be $\mathbf{H}(X) = \log(1/\max_x \Pr[X = x])$.

INDEPENDENT HASH FUNCTIONS. A family $\mathcal{H}$ from n to m bits is called ℓ-wise independent if and only if for every different n-bit strings $x_1, x_2, \ldots, x_\ell$ and h chosen at random from $\mathcal{H}$ the random variables $h(x_1), h(x_2), \ldots, h(x_\ell)$ are independent and uniform.

2.1   Security Definitions

CHANGING ENVIRONMENT - SECURITY GAME. We consider the following ideal setting[3]

(i) An adversary chooses 2^t distributions $X_1, \ldots, X_{2^t}$ over $\{0,1\}^n$, such that $\mathbf{H}_\infty(X) \geqslant k$ for all $i = 1, \ldots, 2^t$.
(ii) A public parameter h is chosen at random and independently of the choices of X_i
(iii) The adversary receives h, and selects $i \in \{1, \ldots, 2^t\}$
(iv) The user computes $\mathrm{Ext}(X)$, where X is sampled from X_i.

Note that in the game defining the security of an extractor, the adversary chooses the distribution and the user chooses (independently) a seed. Here the adversary is in some sense "semi-adaptive", because he can choose an arbitrary distribution but from the class of distributions he had *committed* to before he saw a seed. Of course, the adversary cannot be made fully-adaptive in the sense that he chooses a distribution without any restriction after seeing the seed. Thus, this definition seems to be a reasonable compromise.

RESILIENT EXTRACTOR. We define resilient extractor exactly as in [3] except that we state the confidence level δ explicitly.

Definition 1 (Resilient extractor [3]). *Given* $n, k, m, \epsilon, \delta$ *and* t *an extractor* E *is* t-*resilient with the confidence level* δ *if, in the above setting, with probability*

$1 - \delta$ *over the choice of the public parameter* s *the statistical distance between* $\mathrm{Ext}(X, s)$ *and* U_m *is at most* ϵ. *For shortness, we also call it* (k, ϵ, t, δ)-*resilient extractor.*

This together with the entropy source yields a construction of a TRNG which is robust against some adversarial influences. One possible concern here is how to ensure that the entropy amount, under possible influences, is still sufficient? This is a serious problem but must be solved *independently* on the designing an extractor, because if the adversary had a way to significantly decrease entropy amount then no scheme would be secure anymore, regardless of what an extraction function is applied. We note that, as mentioned in [3], the security definition might be even too strong for real world applications. For example, the assumption that the adversary is computationally unlimited and that all distributions X_i could be completely independent[2]. For long data streams, the extractor can be applied sequentially to consecutive blocks, provided that each block has enough entropy conditioned on all previous blocks.

3   Improved Analysis for Pairwise Independent Hashing

MOTIVATING DISCUSSION. Let $\mathcal{H}$ be a family of 2-universal hash functions from n to m bits and let X be a distribution over $\{0, 1\}^n$ of min-entropy at least k. We will show that if $L = k - m$ is big enough, then the distribution $H(X), H$, where H is a random member of $\mathcal{H}$, is ϵ-close to U_m, H. This result is known as the Leftover Hash Lemma:

Theorem 4. *For* $\mathcal{H}$, H *and* X *as above we have* $\mathrm{SD}((H(X), H); (U_m, H)) \leqslant \sqrt{2^{m-k}}$.

Note that $L = k - m$ needs to be roughly $2\log(1/\epsilon)$ if we want to guarantee that the statistical distance at most ϵ. We will refer to L as the entropy loss, because it equals the difference between the amount of entropy we invest and the length of the extracted output. By the Markov Incquality we trivially obtain the following corollary (see also [5], the remark after Theorem D.5)

Corollary 1. *For all but most a* δ *fraction of the functions* $h \in \mathcal{H}$ *it holds that* $\mathrm{SD}(h(X); U_m) \leqslant \sqrt{2^{m-k}}/\delta$.

This corollary states that for a *fixed* source X, a *fixed* hash function yields a good extractor for all but a small fraction of hash functions. In particular we obtain the existence of an resilient extractor with parameters as in Theorem 1.

IMPROVED ANALYSIS BY THE SECOND MOMENT TECHNIQUE. In Lemma 2 below we will prove a much better result than Corollary 1. We will apply the Markov Inequality for the second moment. Essentially, we bound the deviation of the probability of hashing X into particular value from its expectation which is 2^{-m} (from the universal property).

---

[2] They should be related being a perturbed version of the same distribution.

Lemma 2. *Let $\mathcal{H}$ be a 2-universal family of hahs functions from n to m bits and let X be a distribution of min-entropy at least k. Then for all but an δ fraction of $h \in \mathcal{H}$ we have* $\mathrm{SD}(h(X); U_m) < \sqrt{2^{m-k}/\delta}$.

As an easy corollary we obtain the following theorem, which is much better than Theorem 2.

Theorem 5 (A resilient TRNG from 2-universal family). *Let $\mathcal{H}$ be a 2-universal family of hahs functions from n to m bits and let δ, ϵ be parameters. Then for all but a δ fraction of $h \in \mathcal{H}$, the function h is a (k, ϵ, t, δ)-resilient extractor where*

$$k \geqslant m + 2\log(1/\epsilon) + \log(1/\delta) + t \tag{6}$$

Proof. The proof will follow from the following claims:

Claim 1. For every X we have

$$\Pr_{h \leftarrow \mathcal{H}} [\mathrm{SD}(h(X); U_m) > \epsilon] \leqslant \frac{\mathbf{E}_{y \leftarrow U_m} \mathbf{E}_{h \leftarrow \mathcal{H}} \left(\Pr[h(X) = y] - \Pr[U_m = y]\right)^2}{2^{-2m}\epsilon^2} \tag{7}$$

Claim 2. The expression

$$\mathbf{E}_{h \leftarrow \mathcal{H}} \left(\Pr[h(X) = y] - \Pr[U_m = y]\right)^2$$

over the distributions X of min-entropy at least k is maximized for a flat X, i.e. X uniform over a set of size 2^k.

Claim 3. For every X uniform over a set of size 2^k we have

$$\mathbf{E}_{h \leftarrow \mathcal{H}} \left(\Pr[h(X) = y] - \Pr[U_m = y]\right)^2 \approx 2^{-m-k}. \tag{8}$$

Now we give the proofs.

Proof (Proof of Claim 1). By the definition of the statistical distance and the Markov Inequality we obtain

$$\Pr_{h \leftarrow \mathcal{H}} [\mathrm{SD}(h(X); U_m) > \epsilon] = \Pr_{h \leftarrow \mathcal{H}} \left[\mathbf{E}_{y \leftarrow U_m} |\Pr[h(X) = y] - \Pr[U_m = y]| > 2^{-m}\epsilon\right]$$

$$\leqslant \frac{\mathbf{E}_{h \leftarrow \mathcal{H}} \left(\mathbf{E}_{y \leftarrow U_m} |\Pr[h(X) = y] - \Pr[U_m = y]|\right)^2}{2^{-2m}\epsilon^2} \tag{9}$$

The inequality between the first and the second moment (which follows immediately from the Jensen Inequality) yields

$$\left(\mathbf{E}_{y \leftarrow U_m} |\Pr[h(X) = y] - \Pr[U_m = y]|\right)^2 \leqslant \mathbf{E}_{y \leftarrow U_m} \left(\Pr[h(X) = y] - \Pr[U_m = y]\right)^2. \tag{10}$$

Combining Equation (11) and Equation (10) changing the order of the expectations we obtain

$$\Pr_{h \leftarrow \mathcal{H}} [\text{SD}(h(X); U_m) > \epsilon] \leq \frac{\mathbf{E}_{h \leftarrow \mathcal{H}} \, \mathbf{E}_{y \leftarrow U_m} \, (\Pr[h(X) = y] - \Pr[U_m = y])^2}{2^{-2m} \epsilon^2}$$

$$\leq \frac{\mathbf{E}_{y \leftarrow U_m} \, \mathbf{E}_{h \leftarrow \mathcal{H}} \, (\Pr[h(X) = y] - \Pr[U_m = y])^2}{2^{-2m} \epsilon^2} \quad (11)$$

which finishes the proof. $\qquad\qquad\square$

Proof (Proof of Claim 2). This fact easily follows from the extreme point technique. It is known that every distribution of min-entropy k is a convex combination of flat distributions of min-entropy k. Our expression is a *convex* function of the distribution X. Hence, the maximum is on a flat distribution. $\qquad\square$

Proof (Proof of Claim 3). By expanding the square we get

$$\mathbf{E}_{h \leftarrow \mathcal{H}} \, (\Pr[h(X) = y] - \Pr[U_m = y])^2 = \mathbf{E}_{h \leftarrow \mathcal{H}} \, \Pr[h(X) = y]^2$$
$$- 2 \cdot 2^{-m} \mathbf{E}_{h \leftarrow \mathcal{H}} \, \Pr[h(X) = y] + 2^{-2m} \quad (12)$$

Let X' be an independent copy of X. By the universality of $\mathcal{H}$, we can compute the first term as follows

$$\mathbf{E}_{h \leftarrow \mathcal{H}} \, \Pr[h(X) = y]^2 = \mathbf{E}_{h \leftarrow \mathcal{H}} \, \Pr[h(X) = h(X') = y]$$
$$= \mathbf{E}_{h \leftarrow \mathcal{H}} \, \Pr[h(X) = h(X') = y | X \neq X'] \Pr[X \neq X']$$
$$+ \mathbf{E}_{h \leftarrow \mathcal{H}} \, \Pr[h(X) = h(X') = y | X = X'] \Pr[X = X']$$
$$= 2^{-2m} \Pr[X \neq X'] + 2^{-m} \Pr[X = X']$$
$$\approx 2^{-2m} + 2^{-m-k} \quad (13)$$

where the last approximation follows from $\Pr[X = X'] = 2^{-k} \ll 1$. By the universality we also have

$$\mathbf{E}_{h \leftarrow \mathcal{H}} \, \Pr[h(X) = y] = 2^{-m}. \quad (14)$$

The claim follows by plugging Equation (13) and Equation (14) into Equation (12). $\qquad\square$

The proof is finished. $\qquad\qquad\square$

4   Improved Analysis for ℓ-wise Independent Hashing

It is easy to see that Proposition 2 and Lemma 1, together with the observation that the right hand side Proposition 2 among all X of min-entropy is maximized for flat X (which follows by convexity, see Claim 2), imply the following

Theorem 6 (An resilient from ℓ-universal hashing). *Let $\mathcal{H}$ be an ℓ-universal family of hash functions from n to m bits and let ϵ, δ be parameters. Then for all but a δ fraction of $h \in \mathcal{H}$ the function h is a (k, ϵ, t, δ)-resilient extractor where*

$$k \geqslant m + 2\log(1/\epsilon) + \frac{2\log(1/\delta)}{\ell} + \frac{2t}{\ell} + \log \ell - 2 \tag{15}$$

The proofs of Proposition 2 and Lemma 1 are discussed in the next two subsections. For consistency with some standard notations we denote $\ell = p$.

4.1   Bounds on the Fraction of Good Seeds.

We give the proof of Proposition 2

Proof (Proof of Proposition 2). Let $\delta(y, h) = \Pr[h(X) = y] - \Pr[U_m = y]$. Note that we have $\mathrm{SD}(h(X); U_m) = \frac{1}{2} \cdot 2^m \mathbf{E}_{y \leftarrow U_m} |\delta(y, h)|$. By the Markov Inequality we obtain

$$\Pr_{h \leftarrow \mathcal{H}} [\mathrm{SD}(h(X); U_m) > \epsilon] = \Pr_{h \leftarrow \mathcal{H}} \left[\mathbf{E}_{y \leftarrow U_m} |\delta(y, h)| > 2 \cdot 2^{-m} \epsilon \right]$$

$$\leqslant \frac{\mathbf{E}_{h \leftarrow \mathcal{H}} (\mathbf{E}_{y \leftarrow U_m} |\delta(y, h)|)^\ell}{2^{-mp} (2\epsilon)^\ell}. \tag{16}$$

Since for every h we have $(\mathbf{E}_{y \leftarrow U_m} |\delta(y, h)|)^\ell \leqslant \mathbf{E}_{y \leftarrow U_m} |\delta(y, h)|^\ell$ by the Jensen Inequality, the last equation implies

$$\Pr_{h \leftarrow \mathcal{H}} [\mathrm{SD}(h(X); U_m) > \epsilon] \leqslant \frac{\mathbf{E}_{h \leftarrow \mathcal{H}} (\mathbf{E}_{y \leftarrow U_m} |\delta(y, h)|^p)}{2^{-m\ell} (2\epsilon)^\ell}. \tag{17}$$

The result follows by exchanging the order of the expectations.   □

4.2 L_p-distance between the Output of Hashing and the Uniform Distribution

Proof (Proof of Lemma 1). We can assume that X is flat. We will use the well-known estimate on ℓ-wise independent random variables.

Lemma 3 (ℓ-wise independence moment estimate [1]). *Let $\ell \geqslant 4$ be an even integer. Let $Z_1, \ldots, Z_t$ be ℓ-wise independent random variables taking values in $[0, 1]$. Let $Z = Z_1 + \ldots + Z_n$, $\mu = \mathbf{E} Z$. Then we have*

$$\mathbf{E} |Z - \mu|^\ell \leqslant C_\ell \cdot (\ell\mu + \ell^2)^{\ell/2} \tag{18}$$

where $C_\ell = 2\sqrt{\pi\ell} \cdot e^{1/6\ell} \cdot (5/2e)^{\ell/2} \leqslant 8$.

We will apply Lemma 3 to the random variables $Z_x = \mathbf{1}_{\{h(x)=y\}}$ where $x \in$ supp(X) and y is fixed. Let $\delta(x,y) = \Pr_X[h(X) = y] - \Pr[U_m = y]$. We obtain

$$\underset{h \leftarrow \mathcal{H}}{\mathbf{E}} |\delta(x,y)|^\ell = 2^{-k\ell} \cdot \mathbf{E} \left| \sum_{x \in \mathrm{supp}(X)} Z_x - \mathbf{E}\,Z \right|^\ell$$

$$\leqslant 2^{-k\ell} \cdot C_\ell \cdot \left(\ell \cdot 2^{k-m} + \ell^2 \right)^{\ell/2}$$

$$= 2^{-k\ell} \cdot C_\ell \cdot \left(2^{k-m}\ell \right)^{\ell/2} \cdot \left(1 + 2^{m-k}\ell \right)^{\ell/2}$$

$$\leqslant C_\ell \cdot e^{2^{m-k}\ell^2/2} \cdot \left(2^{-k-m}\ell \right)^{\ell/2}$$

and the result follows.   □

5   Conclusion

We improved the security analysis of the TRNG of Barak et al. by carefully studying the deviation of the probability of hashing into a given location. The loss in the entropy amount seems to be optimal. An interesting problem for the future work is to propose different models for controlling environment.

References.

1. Bellare, M., Rompel, J.: Randomness-efficient oblivious sampling. In: Proceedings of the 35th Annual Symposium on Foundations of Computer Science, SFCS 1994, pp. 276–287. IEEE Computer Society, Washington, DC (1994)
2. Bassham, I.L.E., Rukhin, A.L., Soto, J., Nechvatal, J.R., Smid, M.E., Barker, E.B., Leigh, S.D., Levenson, M., Vangel, M., Banks, D.L., Heckert, N.A., Dray, J.F., Vo, S.: Sp 800-22 rev. 1a. a statistical test suite for random and pseudorandom number generators for cryptographic applications, Tech. report, Gaithersburg, MD, United States (2010)
3. Barak, B., Shaltiel, R., Tromer, E.: True random number generators secure in a changing environment. In: Walter, C.D., Koç, Ç.K., Paar, C. (eds.) CHES 2003. LNCS, vol. 2779, pp. 166–180. Springer, Heidelberg (2003)
4. Barak, B., Shaltiel, R., Wigderson, A.: Computational analogues of entropy. In: Arora, S., Jansen, K., Rolim, J.D.P., Sahai, A. (eds.) RANDOM 2003 and APPROX 2003. LNCS, vol. 2764, pp. 200–215. Springer, Heidelberg (2003)
5. Goldreich, O.: Computational complexity: A conceptual perspective, 1st edn. Cambridge University Press, New York (2008)
6. Gabizon, A., Raz, R.: Deterministic extractors for affine sources over large fields. In: Proceedings of the 46th Annual IEEE Symposium on Foundations of Computer Science, FOCS 2005, pp. 407–418. IEEE Computer Society, Washington, DC (2005)
7. Goldberg, I., Wagner, D.: Randomness and the netscape browser (1996)
8. Kamp, J., Zuckerman, D.: Deterministic extractors for bit-fixing sources and exposure-resilient cryptography. In: Proceedings of the 44th Annual IEEE Symposium on Foundations of Computer Science, FOCS 2003, p. 92. IEEE Computer Society, Washington, DC (2003)

9. Marsaglia, G.: DIEHARD: A battery of tests of randomness, Technical report, Florida State University, Tallahassee, FL, USA (1996)
10. Nisan, N., Zuckerman, D.: Randomness is linear in space. J. Comput. Syst. Sci. 52(1), 43–52 (1996)
11. Radhakrishnan, J., Ta-Shma, A.: Bounds for dispersers, extractors, and depth-two superconcentrators. Siam Journal on Discrete Mathematics 13 (2000)
12. Santha, M., Vazirani, U.V.: Generating quasi-random sequences from semi-random sources. Journal of Computer and System Sciences 33(1), 75–87 (1986)
13. Walker, J.: Hotbits: Genuine random numbers, generated by radioactive decay (2011)

Java Loops Are Mainly Polynomial[*]

Maciej Zielenkiewicz, Jacek Chrząszcz, and Aleksy Schubert

Faculty of Mathematics, Informatics and Mechanics,
University of Warsaw, Poland
{maciej.zielenkiewicz,chrzaszcz,alx}@mimuw.edu.pl

Abstract. Although there exist rare cases where exponential algorithms are used with success, practical software projects mostly consist of polynomial code. We present an automatic analysis tool which divides while-loops in a Java software project into polynomial ones and the rest. The analysis can be useful for example in software quality assurance, maintenance and design of new programming language idioms.

After running our tool on two sets of several medium size Java projects we conclude that almost 80% of while-loops are trivially polynomial.

1 Introduction

Most work on improving existing programming languages, coding habits or program development methodologies are based on problems whose existence comes from common knowledge and does not result from any thorough analysis of existing code. The main goal of this work is to propose an automatic analysis of while-loops to detect ones which, in intent, run in polynomial time (PTIME). Our secondary goal is to obtain a comprehensive picture of how effective are simple criteria for polynomial execution time that do not require complicated analysis of a program and can be applied by a relatively fresh person (e.g. a new developer in a programmers team, a quality assurance person not deeply involved in the project or a person in a software maintenance team).

The polynomial complexity is often identified with the limit of efficient computation, although a number of practical exponential algorithms are in everyday use (e.g. extended regexp matching, see [5]). We establish a number of typical cases, such as counting up to a specified integer, reading an input stream, or iterating over a data structure. In the end it turns out that about 80% of the while-loops are trivially polynomial. Our analysis is rough, it is not backed up by any formal treatment of the loops' complexity. Its goal is to provide an experimental background for a proposal of a new programming language in which polynomial loops would be syntactically or statically distinguished from the other ones. Our aim is an evolutionary change in programmers' habits with a view towards more readable and more maintainable code. In particular we want to be able to statically detect more programming errors such as unintended change of loop control variable. The proposal itself is out of scope of the current paper.

---

[*] This paper is partially supported by Polish government grant No. N206 493138.

G.F. Italiano et al. (Eds.): SOFSEM 2015, LNCS 8939, pp. 603–614, 2015.

It can be reasonably expected that most loops should be polynomial and the others likely have some kind of mistake. In fact it is rare to find correct code which is not polynomial in a regular software project. This work serves as a quantitative base of this assumption; we propose a number of categories of polynomial loops, encode them as XPath expressions over abstract syntax trees and, using a code scanning tool CodeStatistics [4], for a given Java project we can calculate the number of loops that are easily checked as being polynomial.

Our analysis is not a precise one. Unlike the line of research connected with Implicit Computational Complexity [2], we do not aim at giving precise linguistic criteria to select programs in PTIME. Instead, we work on real code and give only an estimate on the number of polynomial loops. Of course there are many loops which are perfectly correct (and polynomial) and do not fit into any of our categories. On the other hand, the number of iterations of some loops in our categories may be bounded by a variable's value, which itself may be exponential with respect to the size of input. Nevertheless we treat them as polynomial.

A similar pragmatic analysis of code is done by many tools such as Find-Bugs [11] or CheckStyle [1], which help limit the quality assurance effort Using these programs as part of regular development toolchain can greatly improve the code quality without inducing any significant cost (a detailed analysis can be found in Section 4 of [10]).

A few static approaches to inference of complexity of loops were described previously. One of them was proposed by Shkaravska, Kersten, and van Eekelen [14] and is based on assumption that running time is given exactly by a polynomial (or combination of polynomials depending on the loop condition), coefficients of which are determined with test runs of the loop. Ermedahl et al. have studied derivation of numerical upper bounds of the execution time of the program ([3]). Hoffman et al. [8] have shown analysis of worst-case polynomial bounds of programs in a simplified functional programming language. One of the other approaches is generation of bounds which are linear combinations of variables, developed by Gulwani et al. [6], which needs some additional user-defined functions for the description of data structures used in the program.

Another approach at computing polynomial time bounds is shown in work of Gulwani and Zuleger [7] who have achieved 76% success rate in analysing .NET binaries. The difference to our works is that (i) we operate on source code with no translation during analysis, so our results can be easily related to source code which would benefit presentation to programmer (ii) we do not limit the iteration control, whereas in the work of Gulwani only arithmetic and boolean control is analysed (iii) we deal explicitly with loops doing input/output operations.

Kasai and Adachi [12] propose a language of essentially for-loops designed in a way that allows syntactic characterization of for polynomial functions.

The paper is organized as follows. Section 2 presents details of our loop categories together with examples of corresponding XPath rules. Section 3 reports the results of the study performed on a set of several open source Java projects together with actual code snippets presenting various loop categories. The numbers resulting from this analysis are compared with the analysis of another set of Java projects. In Section 3.4 we present related work on for-loops that can easily be combined with our analysis and we conclude in Section 4.

2 Analysis

2.1 Cost Model

Analysis of complexity is always relative to a cost model associated with the computation process. The model should describe (1) which data is taken into account as input, (2) which resources are counted (e.g. time, memory, network connections etc.) and (3) which elements of the model cause charging for the resource (e.g. firing of a rule in a Turing machine is charged one time unit).

In this study, we focus on running time analysis of Java programs. As input we take the numerical variables occurring in a loop as well as the graph structure of the objects reachable from the variables in scope. We charge one time unit for a simple operation such as assignment, addition, control-flow split etc.

In general, there are three approaches possible for measuring complexity of programs operating on numerical variables. The first two are strict from the complexity theory point of view, but highly counter-intuitive. The third one is a compromise, which follows the human intuition about program complexity, but has some unexpected properties as far as the complexity theory is concerned.

In the first approach all 32-bit integers are viewed as symbols of a Turing machine alphabet rather than actual input data. In this view, input data is stored in external massive storage devices such as disks since this may make a potentially infinite input size (see e.g. [13]). This model follows our intuition when operations on data-structures are considered, but is rather disconcerting when numeric programs are considered. Indeed, basic operations on integers are here naturally constant-time operations, as they produce one integer as a result. However it follows that all numeric programs run basically in constant time, as they use at most constant number of additional 32-bit integer variables, which all can range over a constant number of values, so the total number of configurations is constant (although this might sometimes be quite a big constant).

In the second model machine integers are bit sequences. In this view basic operations on integers, such as addition, have linear complexity, and any program whose number of operations is linear with respect to the number represented by the input value is in fact exponential, which is again contrary to the intuition. This model is followed quite often when cryptographic algorithms are analysed.

The model that is most often adopted in the analysis of algorithms and programs consists in treating integers as unary values, e.g. an integer of value n has size n, but nevertheless considering basic operations on integers as atomic, i.e., done in constant time. This model agrees with the common intuition, but leads to paradoxes as far as complexity theory is concerned. For example consider the program in Figure 1. It works in linear time with respect to the value (and size) of n, but the final value of k is 2^n, whose size is exponential. So it produces exponential output in linear time. No Turing Machine can do such a thing!

```
k = 1;
i = 0;
while (i < n) {
    k = k + k;
    i++;
}
```

Fig. 1. Linear program calculating an exponential value

We adopt here the view of the second model.[1] It is important to remark that our meaning of polynomial loops amounts only to limiting the number of iterations of the loop. Such polynomial loops do not compose well, in a sense that nested or even subsequent polynomial loops can in fact run in exponential time. Consider the two programs presented in Figure 2 (loops are presented as for-loops for the sake of readability).

```
k = 1;                              k = 1;
for (i = 0; i < n; i++) {           for (i = 0; i < n; i++) {
    k = k + k;                          k = k + k;
    for (j = 0; j < k; j++) {       }
        doSomething();              for (j = 0; j < k; j++) {
    }                                   doSomething();
}                                   }
```

Fig. 2. Two polynomial loops make an exponential program

Although in both programs the number of iterations of both loops is bounded by values of variables that are not modified in their bodies, since the value of k can grow to be exponential, doSomething is executed exponential times as so is bounded the number of iterations of the second loop. At the same time the second loop analysed in isolation is perfectly polynomial since given a bound k it iterates a polynomial number of times with respect to the value.

Having in mind these paradoxes we have consciously decided to follow this inconsistent model in our study. The main reason for this is that we want to obtain a picture based upon the way the programmers view their code. We believe also that the community of programmers is likely to adopt solutions that help in code management in accordance with their point of view and they will disregard methods that claim a program is non-polynomial while they see at first sight it is.

---

[1] A language sensitive to complexity issues should make it possible to distinguish which model is used in a particular piece of code. This is not available in contemporary programming languages.

Table 1. Versions of programs used as sources for experiments. The second set of programs shown below the line was used for verification of rules.

Program	Version
Apache Tomcat	7.0.32
Googla App Engine SDK	1.7.2.1
Apache Hadoop	1.0.4
Hibertate	4.1.7.Final
Oracle Berkeley DB	5.0.58
JEdit	5.0pre1
AspectJ	1.7.2
Spring Framework	3.1.1
Vuze	5.0.0.1

2.2 Scope

Our experiment consists of analysing while-loops in six open-source Java programs: Apache Hadoop, Google App Engine SDK, JEdit, Hibernate, Oracle Berkeley DB, Apache Tomcat (for exact versions see Table 1). This is the same selection as of the previous analysis done on the for-loops with CodeStatistics, which covered the for-loops [4]. The body of the programs outside of the loops is not analysed and recursion is not taken into account.

In the chosen projects almost 35% of loops are while-loops, while the other approximately 65% are for-loops. We do not consider do-while-loops as they constitute less then 1% of the total number of loops. We do not analyse the for-loops because a similar analysis was done previously be Fulara and Jakubczyk [4], but we do relate their categories with our notion of a polynomial loop.

For the sake of simplicity we do not analyse termination of loops by throwing an exception, as it is generally hard to detect in Java given the existence of unchecked exceptions. However an exception can only lead to a decrease in the number of iterations.

The aim of this work is to categorize approximately 80% of the loops. This would give a reasonable ground for introducing to a programming language a syntactic distinction between "safe" (i.e. polynomial) loops and unsafe ones. The unsafe loops, constituting 20% of the total number of loops in the program, could be marked as possibly unsafe and left for more detailed checks by the programmer.

2.3 Tools

The tool used for the analysis of Java code is CodeStatistics [4]. It translates the abstract syntax tree of a Java source file to an XML representation and counts the numbers of matches for each of the specified XPath expressions. For debugging and/or further analysis, CodeStatistics can print each matched fragment together with the matching XPath expression. An example of CodeStatistics XPath rule is presented in Figure 3.

```
<description name="while−read−write" xpath='//WhileStatement[
(
    //MethodInvocation[starts−with(@methodName, "read")]
    or
    //MethodInvocation[starts−with(@methodName, "write")]
    or
    //MethodInvocation/Name/attribute::fullname = "System.in"
)
]'/>
```

Fig. 3. An example of a rule definition. This rule selects all loops which either call method *read* or *write* on any object or call any method of *System.in*

2.4   Methodological Remarks

Real programming languages deal with phenomena that are absent from the focus of complexity theory. We have to fit them somehow into the framework of the theory. Here are the major issues that we take into account.

Polynomial loops. The main goal of the analysis is to identify polynomially bounded loops. We consider a loop as polynomially-bounded if its execution time is bound by a single variable polynomial, for which the variable is a value which either exists in memory or can be polynomially calculated from values in memory.

System interaction. There are loops for which main reason of existence is interaction with the outside world. This includes for example reading or writing data, waiting for other threads and so on. We categorize these loops in a separate category as the complexity of the interaction is generally not defined in the usual computation models such as Turing machines.

We assume that loops that interact with system by the methods outlined above are polynomial, which is backed by a simple argument that a loop which reads some data in every iteration does need at least as much amount of data as the number of iterations. The same applies, in a slightly more convoluted way, to writes — the usual notion of polynomial execution time refers to decision problems on Turing machines, so to express a problem of producing output we must include the output of the program in the input of the decision problem. We see the inter-process communication the same way — a communication can be seen as transfer of some information between threads, which a Turing machine would have to include in the input tape.

Aliasing. In all places where this may be a problem potential aliasing of variables is not taken into account, i.e. we consider all objects to be different instances. For more detailed view of the problem see for example [9].

3 Results

3.1 Rules of Categorization

System interaction: sleep-wait-notify. This rule selects all loops in which a call to *Thread.sleep* or *wait, notify, notifyAll* on any object is done. An example of a loop which is categorized in this category is shown in Figure 4(a).

System interaction: read-write. This rule catches all loops where method *read* or *write* of any object is called or there is a call to any method of *System.in*; see Figure 4(b). Only calls to method *write* on *System.out* are caught as it turned out that other methods are mostly used for debugging or logging of messages, which would categorize otherwise innocent loops as system interactions.

```
while (!stopAwait) {
  try { Thread.sleep(10000); }
  catch (InterruptedException ex) { } }
```

(a) An example of loop in the category System interaction: sleep-wait-notify.

```
while ((line=br.readLine()) != null) {
  if (version == null && line.contains(key)) {
    version=line.substring(line.indexOf(key) + key.length(),
        line.length() - 1); } }
```

(b) An example of loop in the category System interaction: read-write.

```
while (st.hasMoreTokens()) {
  String pane=st.nextToken();
  addOptionPane(pane);
}
```

(c) An example of loop in the category Linear iteration: iterators.

Fig. 4. Categories of loops

The following categories include loops that satisfy a particular condition but do not belong to any of the "system interaction" categories.

Linear iteration: iterators. This category looks for calls of methods which are either advancing an iterator or checking if a collection is empty, in which case it can be reasonably expected that elements are removed from the collection as they are processed. More specifically, the rule catches all loops where the condition either is just an invocation of a method with one of the names *hasNext, hasPrevious, hasMoreElements, hasMoreTokens, isEmpty* on an object or an expression which includes invocation of method *size* or *length*, for example like in Figure 4(c).

This should catch most of iterations with standard collections' iterators, the *Enumerable* interface and usage of *StringTokenizer*. As in general it is not possible to tell if a collection is not altered in loop body, 100 loops matching this rule were selected at random and hand-checked for additions to the iterated collection, and it was found out that there were none. It was found out as well that 99% of checked loops had a *next* (or corresponding) call on each if-branch.

Linear iteration: local counter. The existence of this rule is motivated by the following example loop:

```
while (end > start) {
    byte temp=buf[start];
    buf[start]=buf[end];
    buf[end]=temp;
    start++;
    end--;
}
```

Assuming non-negative starting values and no overflows the condition eventually becomes false, as the direction of increasing/decreasing of variables is towards the check failing. The example was generalised to select all loops in which at least one of the check operands is a local variable that is increased or decreased in the desirable way (only operators -- and ++ are taken into account), is not assigned to, and if the other operand is a local variable as well it is not assigned to and does not increase/decrease in the undesirable way.

3.2   Categorization of Loops

The results of categorization are given in Figure 5. It should be noted that the intersections between categories are not always empty (but the "linear" categories explicitly exclude the "system interaction" categories); the total number of 80.87% of loops which are categorized is the difference between 100% and the number of uncategorised loops, so that it is correct with non-zero intersections as well.

To make the intersections more clear additional calculation of the number of loops which are categorized simultaneously in two categories was performed. Results of that analysis are shown in Table 2. The rules which excluded system interactions from other categories were turned off for that analysis, therefore numbers for some categories are bigger than in Figure 5.

After developing the set of rules a second evaluation was done on additional set of programs in order to check if the rules were not overtrained to specific examples used during the test. The results are shown in Figure 6; in total 72% of loops were categorized.

3.3   Sights

During the experiments a number of interesting sights could be noticed. First, some of the loops have a termination condition which depends on some non-trivial invariant property of certain objects, for example on acyclity of the graph

program	all loops	uncategorized	linear iteration		system interaction	
			iterators	local counter	read write	sleep-wait notify
appeng	6	2	1			3
hadoop	1274	137	82	47	890	472
hibernate	739	94	464	8	170	5
berkdb	830	241	123	42	319	175
jedit	394	143	65	36	138	12
tomcat	1017	198	219	65	461	187
total	4260	815	954	198	1978	854
percent of all		19.13%	22.39%	4.65%	46.43%	20.05%
categorized loops			**80.87%**			

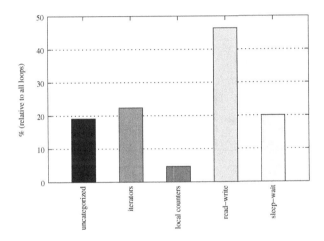

Fig. 5. Results of categorization of while-loops with rules given in Section 3.1

of relevant objects in the heap. We feel that those loops should be at least commented in some way by the programmer. We have also found that 0.28% of loops are not terminated by neither loop condition nor usual flow control (break, return etc.), and are probably using exceptions for terminating the loops. We think that those loops should be clearly commented as well.

3.4   For-Loops

After categorizing over 80% of while-loops it is a natural extension to check if a similar result can be obtained for the for-loops. Building onto work of Fulara and Jakubczyk [4] let's recall their result of generating *decreases* formula for 74.4% of for-loops. Considering their categories and how the expression which shall decrease is built we can see what is its relationship to other variables in the program, and therefore infer an upper bound on the complexity. All of the rules used in the referenced results excluded any additional modifications of control variables;

program	all loops	uncategorized	linear iteration		system interaction	
			iterators	local counter	read write	sleep-wait notify
aspectj	1453	626	93	176	532	74
spring	187	98	55	8	9	8
vuze	1485	151	426	51	704	318
total	3125	875	574	235	1245	400
percent of all		28.00%	18.37%	7.52%	39.84%	12.80%
categorized loops			**72.00%**			

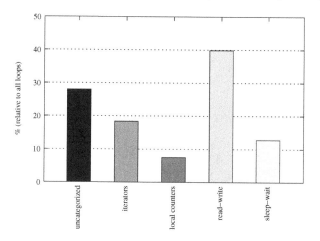

Fig. 6. Results of categorization of while-loops for second set of test programs

Table 2. Intersections of categories as absolute numbers and percentages of their categories

	iterators	local counter	read-write	sleep-wait-notify
iterators	0	12	1050	166
local counter		0	230	117
read-write			0	532
sleep-wait-notify				0

		iterators	local counter	read-write	sleep-wait-notify
		1537	478	1978	854
iterators	1537	0	3% / 1%	53% / 68%	19% / 11%
local counter	478		0	12% / 48%	14% / 24%
read-write	1978			0	62% / 27%
sleep-wait-notify	854				0

that means that it is only necessary check the other operand of the comparison in loop guard. It turns out that in every case this is known upon entering the loop:

Literal. A literal cannot be changed in any way, so literal-bounded loop has complexity $\mathcal{O}(1)$.

Constant. Using a constant is practically equivalent to using a literal (with a type), so the complexity is still $\mathcal{O}(1)$.

Final field. A final field in a program is effectively a constant known upon entering the loop.

Local expression. This category consists of loops for which the number of loop body executions is bounded by an expression which uses only local variables not modified in the loop body. The complexity of such loop depends on the expression which is used: it is polynomial if the expression is polynomial. As we do not check functions which may be called, the only way to write an expression which is not polynomial in our case is to use bit shift or unary negation operators, which can produce numbers exponentially big with respect to their operands. Additional check was ran for that category to check if the expressions contain aforementioned operators, and it was found out that there are no uses of those operators in the studied projects.

3.5   Recapitulation of Results

Taking the above analysis of for-loops into account it turns out that all of the previously categorized for-loops have polynomial complexity. That means than in total we have

$$74.4\% \cdot 65\% + 80.8\% \cdot 35\% \approx 76.6\%$$

of loops categorized.

Our rules and instructions to reproduce the results, with links to the tools used, are available on the webpage `http://codestatistics.mimuw.edu.pl/`. The webpage also contains an extended version of this paper with additional examples.

4   Conclusions and Future Work

We have presented an automatic analysis tool which detects clearly polynomial while-loops in a Java software project. The polynomial loops are divided into several categories, such as iterating to a specified integer value, reading an input stream, or visiting a data structure. We ran our tool on two sets of several medium size open source Java projects and, as expected, it turned out that almost 80% of while-loops are trivially polynomial. We believe that this is also the amount of polynomial time loops that can be judged as such by programmers or software engineers that are not deeply involved in the project.

The analysis we presented can be extended in various ways. One of the extensions would be to transform the categories of while-loops into the corresponding ones for the for-loops, which would improve the analysis described in [4].

The presented analysis constitutes a necessary starting point for further investigations on how to apply implicit computational complexity methods to real languages such as Java. For example a programming language could introduce separate keyword for loops which are possibly non-polynomial and reject or warn about loops which do not use the keyword but do not seem to be polynomial. Another useful conclusion that can be drawn from the analysis is that a useful category that could have its own separate syntax is the category of loops that deal with input and output.

References

1. Checkstyle, http://checkstyle.sourceforge.net/
2. Bellantoni, S., Cook, S.: A new recursion-theoretic characterization of the polytime functions. Computational Complexity 2(2), 97–110 (1992)
3. Ermedahl, A., Sandberg, C., Gustafsson, J., Bygde, S., Lisper, B.: Loop bound analysis based on a combination of program slicing, abstract interpretation, and invariant analysis. In: Rochange, C. (ed.) 7th Intl. Workshop on Worst-Case Execution Time Analysis (WCET). Internationales Begegnungs- und Forschungszentrum für Informatik (IBFI), Schloss Dagstuhl, Germany, Dagstuhl, Germany (2007)
4. Fulara, J., Jakubczyk, K.: Practically applicable formal methods. In: van Leeuwen, J., Muscholl, A., Peleg, D., Pokorný, J., Rumpe, B. (eds.) SOFSEM 2010. LNCS, vol. 5901, pp. 407–418. Springer, Heidelberg (2010)
5. Garey, M.R., Johnson, D.S.: Computers and Intractability: A Guide to the Theory of NP-Completeness. W. H. Freeman & Co., New York (1979)
6. Gulwani, S., Mehra, K.K., Chilimbi, T.M.: Speed: precise and efficient static estimation of program computational complexity. In: Shao, Z., Pierce, B.C. (eds.) POPL, pp. 127–139. Press (2009)
7. Gulwani, S., Zuleger, F.: The reachability-bound problem. In: Zorn, B.G., Aiken, A. (eds.) PLDI, pp. 292–304. ACM (2010)
8. Hoffmann, J., Aehlig, K., Hofmann, M.: Multivariate amortized resource analysis. ACM Trans. Program. Lang. Syst. 34(3), 14 (2012)
9. Hogg, J., Lea, D., Wills, A., deChampeaux, D., Holt, R.: The Geneva convention on the treatment of object aliasing. SIGPLAN OOPS Mess. 3(2), 11–16 (1992)
10. Hovemeyer, D.: Simple and Effective Static Analysis to Find Bugs. Ph.D. thesis, University of Maryland (College Park, Md.), College Park, Maryland (2005)
11. Hovemeyer, D., Pugh, W.: Finding bugs is easy. SIGPLAN Notices 39(12), 92–106 (2004)
12. Kasai, T., Adachi, A.: A characterization of time complexity by simple loop programs. Journal of Computer and System Sciences 20(1), 1–17 (1980)
13. Kosovskiy, N.K.: Polynomial-time program conditions for three programming languages. In: Proceedings of Fourth Workshop on Developments in Implicit Computational Complexity, DICE 2013 (2013)
14. Shkaravska, O., Kersten, R., van Eekelen, M.: Test-based inference of polynomial loop-bound functions. In: Krall, A., Mössenböck, H. (eds.) PPPJ 2010: Proceedings of the 8th International Conference on the Principles and Practice of Programming in Java. ACM Digital Proceedings Series, pp. 99–108 (2010)

Author Index